HEAT PUMPS

Eugene Silberstein

Australia • Brazil • Japan • Korea • Mexico • Singapore • Spain • United Kingdom • United States

Heat Pumps
Eugene Silberstein

Business Unit Director: Alar Elken

Executive Editor: Sandy Clark

Acquisitions Editor: James DeVoe

Development Editor: John Fisher

Executive Marketing Manager: Maura Theriault

Channel Manager: Fair Huntoon

Marketing Coordinator: Brian McGrath

Executive Production Manager: Mary Ellen Black

Production Editor: Ruth Fisher

Assistant: Mary Ellen Martino

Library of Congress Control Number: 2002018070

ISBN-13: 978-0-7668-1959-7

ISBN-10: 0-7668-1959-0

Delmar
Executive Woods
5 Maxwell Drive
Clifton Park, NY 12065
USA

Cengage Learning is a leading provider of customized learning solutions with office locations around the globe, including Singapore, the United Kingdom, Australia, Mexico, Brazil, and Japan. Locate your local office at **international.cengage.com/region**

Cengage Learning products are represented in Canada by Nelson Education, Ltd.

For your lifelong learning solutions, visit **www.cengage.com/delmar**

Visit our corporate website at **www.cengage.com**

Printed in the United States of America
9 10 11 12 13 14 13 12 11 10
ED188

CONTENTS

PREFACE

As the concept of vapor-compression refrigeration has grown in both popularity and complexity since its inception, so did the technologies that were utilized in heat pump applications. Since its initial introduction to our industry in the 1960s, the heat pump system has taken on a major role in providing users of HVAC equipment with an economical and efficient alternative to meeting their heating and cooling needs.

Generally speaking, heat pump systems are designed to provide comfort cooling in the warm summer months as well as heating in the colder months. Based on the basic vapor-compression refrigeration system, the heat pump, with the aid of check valves and reversing valves, mechanically reverses the direction of refrigerant flow through major system components. This, in effect, causes heat to be rejected into the conditioned space during periods when heating is desired and heat to be absorbed from the occupied space when a cooling effect is desired. This technology has also been adapted to facilitate the heating of swimming pools, spas, domestic water, and other fluids.

As heat pump technologies advanced, so did the need for accurate sources of information regarding the installing, servicing, and troubleshooting of these systems. This text will serve to provide this necessary information. Intended for use by those just beginning the study of HVAC, those who already have a working knowledge of this equipment, those who are well acquainted and experienced with this aspect of our ever-changing industry, this text will provide a comprehensive source of information. Through its sample service calls, over fifty discussion topics, more than 300 review questions, and printed text material, the reader will be exposed to all aspects of this ever-changing industry.

ORGANIZATION AND FEATURES OF THIS TEXT

This text is divided into 19 chapters, each covering a specific topic that will prove useful to even the most experienced field technician. Each chapter begins with a list of the learning objectives within the chapter, followed by a brief introduction and overview of the material that follows. Each chapter ends with a series of review questions which are intended to test the reader's knowledge, and discussion topics, which can be utilized on both the professional level and in the classroom setting. Where applicable, lists of key terms are provided. The key terms and other terms used in our industry are defined in the glossary at the end of the text.

Unlike other heat pump textbooks, this work begins with chapters on the vapor-compression refrigeration system, upon which the heat pump concept is based. Information regarding compressors, condensers, metering devices, and evaporators is included to provide a comprehensive piece of reference material for personal as well as field use. This text also includes over thirty sample service calls, most of which follow the service technician step-by-step as the system problem is located and remedied. The remainder of the service calls rely on the reader to become the service technician and to provide possible solutions to the service problems presented.

The reader should be aware that it is impossible for a single text to discuss the equipment produced by all manufacturers. An attempt has been made to provide information that will be specific yet general enough to provide the maximum benefit to the reader. For this reason, it is strongly recommended that any service or installation literature from a specific piece of equipment be obtained and consulted to ensure that the manufacturer's recommendations are followed.

SUPPLEMENTS

An Instructor's Guide (ISBN 0-7668-1960-4) is available to accompany this text. This supplement contains discussion questions for each chapter as well as answers to the end-of-chapter review questions. Explanation and clarification as to why an answer is correct are provided.

ACKNOWLEGMENTS

I would like to thank the many companies who have provided artwork and or printed material for use in this text. A special thanks goes to the following individuals:

Reed Wilson, Calorex USA
David Foster, Uniweld Products
Mindy Phelps, Copeland Corporation
Vito Costanza, Industrial Cooling
Tina Klobe, Alco Controls
Cynthia Houtz, Honeywell, Inc.

I would also like to acknowledge the following individuals for serving as proofreaders and reviewers for the text:

Nephtaly Rodriguez
Barry Burkan
Linda Barrow, Lanier Technical Institute, Oakwood, GA
Harold Benton, Advanced Technology Institute, Virginia Beach, VA
Allen Clay, Vatterott College, St. Louis, MO
Richard McDonald, Sante Fe Community College, Gainsville, FL
Joseph Moravek, Lee College, Baytown, TX

In addition to these people who provided support with the printed matter in the text, I would also like to thank the following people for assistance provided with some of the artwork and photography that appears in this book:

Bill Johnson
Chester Marcin
Nephtaly Rodriguez
Barry Burkan
Rosanna Luna
Maria Reyes
Alex John

Finally, a very special thank you goes out to two very special individuals. Without their continued support, I would not have been able to bring this project to fruition: Selma Silberstein for giving the love that only a mother can, Nephtaly Rodriguez for support beyond words, 831. It is to you both that this book is dedicated.

Eugene Silberstein

CHAPTER 1

Vapor-Compression Refrigeration System Components

OBJECTIVES After studying this chapter, the student should be able to:

- Describe the basic vapor-compression refrigeration cycle.
- Explain the difference between a repeating cycle and a nonrepeating cycle.
- List the basic components that make up a vapor-compression refrigeration system.
- Describe the function of a compressor.
- List various types of compressors.
- Describe the function of the condenser.
- Explain the difference between water-cooled and air-cooled condensers.
- Describe the function of the metering device.
- List three commonly used metering devices.
- Describe the function of the evaporator.
- Name the various evaporator styles based on construction and operation.

INTRODUCTION

In today's society, it is the knowledgeable, well-trained technician who helps to ensure the continued satisfactory operation of not only air-conditioning and refrigeration equipment for comfort and food preservation but medical testing equipment and computer systems as well. To do this, the heating, ventilating, and air-conditioning/refrigeration (HVAC/R) technician needs to fully understand the fundamental concepts of the vapor-compression refrigeration system and the four key components that, together, allow the system to operate effectively and efficiently. In this chapter, you will learn about the closed vapor-compression system. The term *closed* indicates that this system will continually repeat itself without depleting the refrigerant contained within it. You will also learn about the four key components of a vapor-compression refrigeration system—the compressor, the condenser, the metering device, and the evaporator—and the role that each plays in the closed, repeating cycle. Once you have a solid understanding of this basic system, you will be ready to move on and learn about additional components that allow the system to operate effectively under a wide range of conditions and applications. The understanding of these additional components, together with the knowledge gained from this chapter, will enable you to better understand the concepts of more complicated refrigeration systems, including heat pumps.

OVERVIEW

The term **vapor compression** refers to the method by which the refrigerant is circulated through the system. A high-pressure fluid will always travel toward a lower-pressure fluid, provided that an open path is available. Vapor is compressed by the **compressor** in order to increase its pressure, thereby creating a pressure difference that allows flow to occur. The term *vapor compression* is somewhat redundant because a liquid can never be compressed. A given amount of liquid will always have the same volume and occupy the same amount of space. A vapor, on the other hand, will expand or contract in order to completely fill the vessel or container that it occupies. A roomful of air could theoretically be compressed and stored in a small box, but the pressure of the air in the smaller vessel would be higher than the pressure of the same amount of air in a larger container. For this reason, when a fluid enters the compressor, it is understood that, under normal operating conditions, the refrigerant is in the vapor state. The refrigerant enters the compressor via the **suction line** as a low-pressure, low-temperature vapor and leaves as a high-pressure, high-temperature vapor through the **discharge line**. Note that, as the refrigerant is compressed, the heat contained in the refrigerant is concentrated and the refrigerant leaves the compressor at temperatures that can easily exceed 200°F.

The compressed refrigerant vapor then travels to the **condenser**. The condenser is where heat is rejected from the refrigeration system. The high-pressure, high-temperature vapor leaving the compressor undergoes three changes in the condenser:

1. The refrigerant gives up **sensible heat** in preparation for condensing.
2. The refrigerant condenses from a vapor into a liquid, rejecting **latent heat**.
3. After condensing to 100 percent liquid, the refrigerant subcools and continues to reject sensible heat.

Sensible heat is defined as a temperature change that can be measured with a thermometer. You can observe changes in outside temperature by looking at a thermometer or by opening the window and sticking your head out. Latent heat, on the other hand, is a temperature change that cannot be measured with a thermometer. For example, as ice changes to water at 32°F, there is a heat transfer even though the temperature remains constant. Latent heat is therefore also referred to as "hidden heat." **Subcooling** refers to the amount of sensible heat that is removed from a liquid after it has condensed. The temperature at which the refrigerant condenses is called the **condenser saturation temperature**. If the condenser saturation temperature is 110°F and the temperature of the refrigerant falls to 100°F, the condenser is operating with 10 degrees of subcooling. When the refrigerant is **saturated** in the process of changing from a vapor to a liquid, it follows a pressure/temperature relationship, that is accurate only for latent heat transfers within a refrigerant. This relationship states that at a specific pressure the refrigerant will be at a specific temperature. The pressure/temperature relationship for commonly used refrigerants is shown in Figure 1-1. From this chart it can be seen that, in the above example, the pressure of refrigerant 22 will be 226.4 **psig** (pounds per square inch gauge) since it condenses at 110°F. Superheated and subcooled refrigerants exhibit sensible heat changes and, therefore, *do not* follow the pressure/temperature relationship.

The high-pressure, high-temperature, subcooled liquid from the condenser then travels to the **metering device**. The piping or tubing that connects the condenser to the metering device always carries liquid and is, therefore, called the **liquid line**. The metering device is responsible for controlling the flow of refrigerant to the evaporator and creating a pressure drop between the high and low sides of the system. Too much or too little refrigerant entering the evaporator can result in one or more of the following:

- Reduced system efficiency (not enough cooling)
- Evaporator coil freeze-up
- Compressor damage
- Excessive power consumption

A properly sized metering device must be used to ensure that the system operates within its design range to maximize efficiency. When operating correctly, the metering device and the pressure drop that it creates change the high-pressure, high-temperature, subcooled liquid from the condenser to a low-pressure, low-temperature, saturated liquid. A saturated liquid is a mixture of mostly liquid and some vapor but can, on occasion, be 100 percent liquid. The liquid in the evaporator is at a temperature below the temperature of the conditioned space and is now able to absorb heat from it. This saturated liquid also follows the pressure/temperature relationship.

The low-pressure, low-temperature, saturated liquid then flows to the **evaporator**. The refrigerant in the evaporator is responsible for absorbing heat, both latent and sensible, from the conditioned space. When absorbing latent heat, the liquid is removing moisture, or humidity, from the air passing over or through the evaporator. When absorbing sensible heat, the liquid is lowering the measurable temperature of the air passing over the evaporator coil. As the liquid flows through the evaporator and absorbs heat, it begins to vaporize, or boil. During this boiling process, the temperature of the refrigerant remains constant indicating that the refrigerant is experiencing a latent heat transfer. Once the refrigerant completely boils into a vapor, it continues to absorb a small amount of heat from the conditioned space. The vapor begins to pick up additional sensible heat, which increases the temperature of the vapor leaving the evaporator coil. A vapor that is heated above its boiling point is considered to be superheated.

Temperature	Refrigerant				Temperature	Refrigerant				Temperature	Refrigerant			
°F	12	22	134a	502	°F	12	22	134a	502	°F	12	22	134a	502
−60	*19.0*	*12.0*		7.2	12	15.8	34.7	13.2	43.2	42	38.8	71.4	37.0	83.8
−55	*17.3*	*9.2*		3.8	13	16.4	35.7	13.8	44.3	43	39.8	73.0	38.0	85.4
−50	*15.4*	*6.2*		*0.2*	14	17.1	36.7	14.4	45.4	44	40.7	74.5	39.0	87.0
−45	*13.3*	*2.7*		1.9	15	17.7	37.7	15.1	46.5	45	41.7	76.0	40.1	88.7
−40	*11.0*	0.5	*14.7*	4.1	16	18.4	38.7	15.7	47.7	46	42.6	77.6	41.1	90.4
−35	*8.4*	2.6	*12.4*	6.5	17	19.0	39.8	16.4	48.8	47	43.6	79.2	42.2	92.1
−30	*5.5*	4.9	*9.7*	9.2	18	19.7	40.8	17.1	50.0	48	44.6	80.8	43.3	93.9
−25	*2.3*	7.4	*6.8*	12.1	19	20.4	41.9	17.7	51.2	49	45.7	82.4	44.4	95.6
−20	0.6	10.1	*3.6*	15.3	20	21.0	43.0	18.4	52.4	50	46.7	84.0	45.5	97.4
−18	1.3	11.3	*2.2*	16.7	21	21.7	44.1	19.2	53.7	55	52.0	92.6	51.3	106.6
−16	2.0	12.5	*0.7*	18.1	22	22.4	45.3	19.9	54.9	60	57.7	101.6	57.3	116.4
−14	2.8	13.8	.3	19.5	23	23.2	46.4	20.6	56.2	65	63.8	111.2	64.1	126.7
−12	3.6	15.1	1.2	21.0	24	23.9	47.6	21.4	57.5	70	70.2	121.4	71.2	137.6
−10	4.5	16.5	2.0	22.6	25	24.6	48.8	22.0	58.8	75	77.0	132.2	78.7	149.1
−8	5.4	17.9	2.8	24.2	26	25.4	49.9	22.9	60.1	80	84.2	143.6	86.8	161.2
−6	6.3	19.3	3.7	25.8	27	26.1	51.2	23.7	61.5	85	91.8	155.7	95.3	174.0
−4	7.2	20.8	4.6	27.5	28	26.9	52.4	24.5	62.8	90	99.8	168.4	104.4	187.4
−2	8.2	22.4	5.5	29.3	29	27.7	53.6	25.3	64.2	95	108.2	181.8	114.0	201.4
0	9.2	24.0	6.5	31.1	30	28.4	54.9	26.1	65.6	100	117.2	195.9	124.3	216.2
1	9.7	24.8	7.0	32.0	31	29.2	56.2	26.9	67.0	105	126.6	210.8	135.0	231.7
2	10.2	25.6	7.5	32.9	32	30.1	57.5	27.8	68.4	110	136.4	226.4	146.4	247.9
3	10.7	26.4	8.0	33.9	33	30.9	58.8	28.7	69.9	115	146.8	242.7	158.5	264.9
4	11.2	27.3	8.6	34.9	34	31.7	60.1	29.5	71.3	120	157.6	259.9	171.2	282.7
5	11.8	28.2	9.1	35.8	35	32.6	61.5	30.4	72.8	125	169.1	277.9	184.6	301.4
6	12.3	29.1	9.7	36.8	36	33.4	62.8	31.3	74.3	130	181.0	296.8	198.7	320.8
7	12.9	30.0	10.2	37.9	37	34.3	64.2	32.2	75.8	135	193.5	316.6	213.5	341.2
8	13.5	30.9	10.8	38.9	38	35.2	65.6	33.2	77.4	140	206.6	337.3	229.1	362.6
9	14.0	31.8	11.4	39.9	39	36.1	67.1	34.1	79.0	145	220.3	358.9	245.5	385.0
10	14.6	32.8	11.9	41.0	40	37.0	68.5	35.1	80.5	150	234.6	381.5	262.7	408.4
11	15.2	33.7	12.5	42.1	41	37.9	70.0	36.0	82.1	155	249.5	405.1	280.7	432.9

Vacuum (in Hg) – Italic Figures
Gage Pressure (psig) – Bold Figures

FIGURE 1-1 Pressure/temperature relationship chart for some common refrigerants

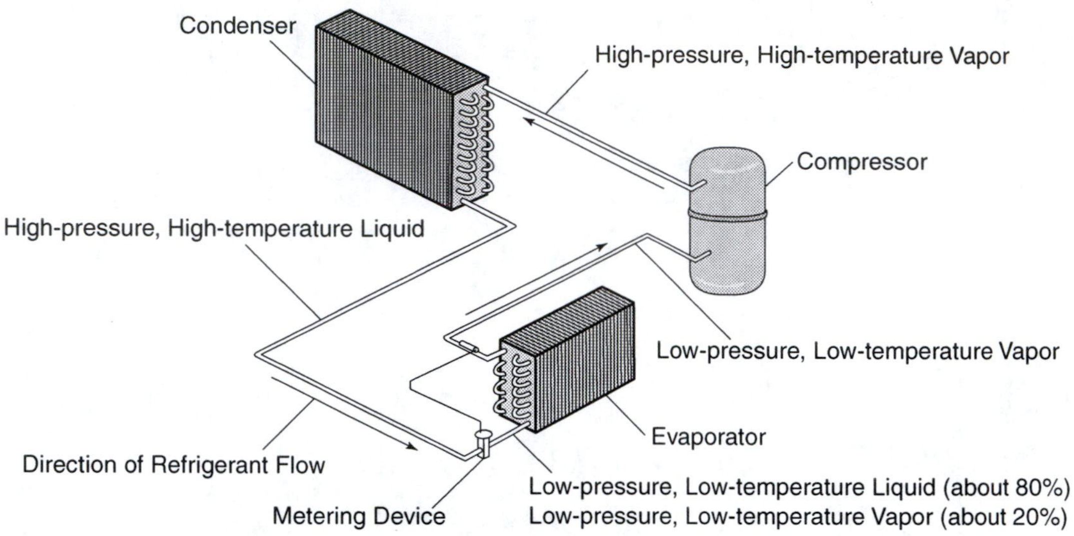

FIGURE 1-2 Pictorial diagram of the basic refrigeration cycle

Superheat can be defined as the amount of sensible heat that is added to a vapor after it has completely boiled. The refrigerant leaves the evaporator as a low-pressure, low-temperature, superheated vapor and enters the suction line.

The low-pressure, low-temperature vapor from the evaporator now returns to the compressor. These four basic system components (compressor, condenser, metering device, evaporator), along with the interconnecting piping, form a **repeating cycle** in the **closed system**. A repeating cycle is one that has the ability to perform a series of tasks over and over without depleting any of its resources. For example, an automobile engine does not represent a repeating cycle because it violates one of the requirements. The engine's crankshaft rotates, so it can easily duplicate its cycle, but the fuel supply needs to be replenished when it is exhausted. If it were possible to turn the exhaust fumes back into gasoline, the process, or cycle, could continue indefinitely. A vapor-compression, air-conditioning/refrigeration system is a repeating cycle for the following reasons:

- The fluid, or refrigerant, supply that flows through the system is not depleted, and
- When one cycle is completed, the refrigerant is in the correct state to begin another cycle.

A pictorial diagram of a basic vapor-compression refrigeration system is shown in Figure 1-2.

COMPRESSORS

The compressor is the component in a refrigeration system that is responsible for raising the pressure and temperature of the refrigerant that leaves the evaporator. This is necessary for two reasons. First, increasing the pressure of the vapor creates the pressure difference that is needed for refrigerant flow to occur. As stated earlier, a high-pressure fluid will always flow toward a low-pressure fluid. In conjunction with a metering device, the compressor increases the pressure of the refrigerant on one side of the system and decreases the pressure on the other side in order to create this difference in pressure. Second, increasing the temperature of the refrigerant will enable it to easily reject system heat to another medium. By increasing the temperature of the refrigerant, the rate of this heat transfer is increased. The direction of heat transfer is from a higher-temperature substance to a lower-temperature substance. If the heat from the system is to be rejected to the condensing medium, it must be at a higher temperature than the medium. The condensing medium, usually air or water, is the substance that absorbs the heat that the system must reject. Condensing medium and heat rejection are covered in more detail in the condenser section of this text.

The compressor is classified as a vapor pump since liquid entering the component is not desirable.

FIGURE 1-3 The welded, hermetic compressor is used in the smaller compressor sizes, from 1 ton to 24 tons. The suction line is usually piped directly into the shell and is open to the crankcase. The discharge line normally is piped from the compressor inside the shell to the outside of the shell. The compressor shell is typically thought of as a low-side component (*Courtesy Bristol Compressors, Inc.*)

Liquid entering the compressor will result in reduced efficiency and capacity and, in many cases, can result in mechanical damage to the compressor's internal components. To avoid the possibility of damage to the compressor, the refrigerant entering the compressor should be 100 percent vapor. To ensure that only vapor enters the compressor, the refrigerant leaving the evaporator must be superheated.

The most popular types of compressors are rotary, reciprocating, and scroll. Compressors can be hermetically sealed, semihermetic, or open type. Hermetically sealed compressors cannot be serviced internally and are normally replaced when internal failure occurs (Figure 1-3). Large hermetic compressors can be rebuilt by companies specializing in this type of work as long as doing so is economically feasible. Semihermetic compressors are bolted together and can be serviced in the field (Figure 1-4). Open

FIGURE 1-4 Serviceable hermetic (semihermetic) compressor designed in such a manner that it can be serviced in the field (*Courtesy Copeland Corporation*)

compressors have an external drive mechanism, such as a motor, which can be serviced or replaced without disturbing the refrigerant circuit. Although the methods used for compression are different, the function of the compressor is the same. Open compressors are also equipped with shaft seals that separate the refrigerant circuit from the surrounding atmosphere, thereby containing the refrigerant within the system.

Rotary Compressors

A rotary compressor can be identified by its cylindrical shape and the fact that the discharge line is located in the center at the very top of the compressor (Figure 1-5). This type of compressor operates with an eccentric roller connected directly to the shaft of the motor. This roller rotates at the same speed as the motor within the shell of the compressor, rubs against the inside of the compression chamber (Figure 1-6), and forms a compartment in which refrigerant vapor is trapped. This movable vane creates a seal with the roller due to a spring pressure that constantly pushes the vane up against the roller as it rotates. The compression process in a rotary compressor can be divided into four intermediate steps:

1. The introduction of suction gas into the compression chamber
2. The sealing off of the suction chamber, trapping refrigerant inside

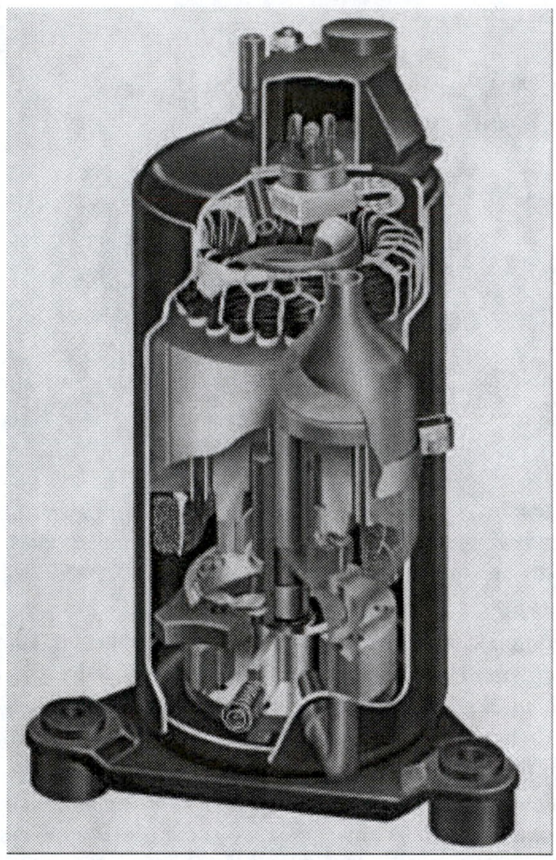

FIGURE 1-5 Rotary compressor (*Reprinted with permission of Motors and Armatures, Inc.*)

3. The compression of the refrigerant
4. The discharge of the high-pressure refrigerant from the compressor

To illustrate the compression process in a rotary compressor, we start at the point where the roller is blocking off the discharge port and the refrigerant from the suction port is permitted to enter the cylinder (Figure 1-7a). As the roller rotates in a counterclockwise direction, the suction or intake port is now sealed off from the refrigerant in the cylinder. The refrigerant is now trapped in the cylinder by the vane, the discharge valve, and the roller's contact with the cylinder wall (Figure 1-7b). As the roller continues to turn, the space that the refrigerant occupies decreases and the pressure and temperature of the refrigerant increase (Figure 1-7c). Once the pressure of the refrigerant in the cylinder rises above the pressure of the refrigerant in the discharge line, the discharge valve will open and the high-pressure, high-temperature refrigerant vapor will be pushed out of the compressor (Figure 1-7d). While the refrigerant is being compressed, the other side of the cylinder is open to the suction side, so new suction gas is entering the chamber in preparation for the next compression cycle. Once all of the high-temperature, high-pressure vapor has been discharged from the cylinder, the suction gas that has entered will be compressed and the cycle will repeat itself. In this type of rotary compressor, the vane is stationary and

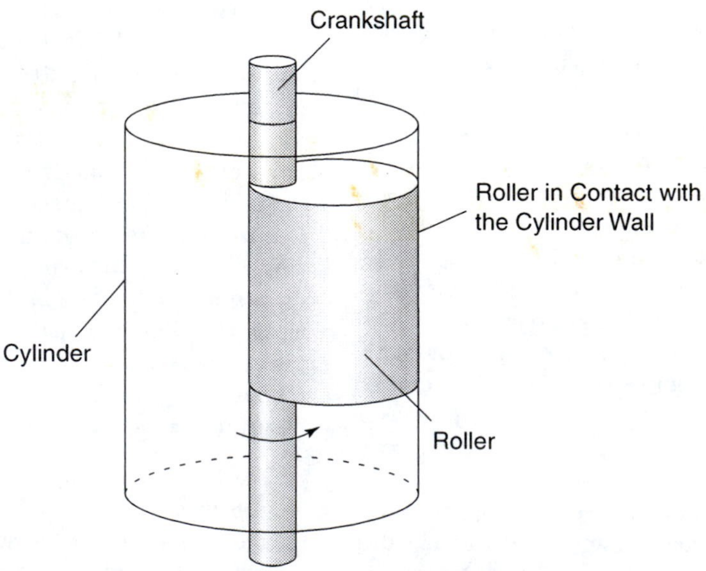

FIGURE 1-6 Roller scraping the interior surface of the cylinder

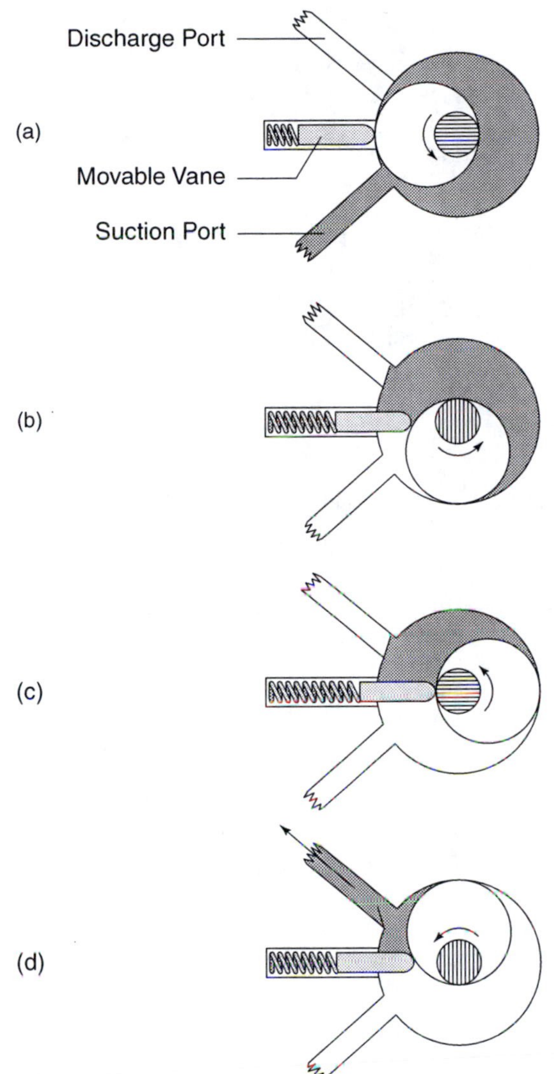

FIGURE 1-7 Compression process in a rotary compressor: (a) Vapor refrigerant enters the chamber through the suction port (b) The refrigerant is trapped in the chamber (c) The refrigerant is being compressed (d) The high-pressure refrigerant is discharged

simply moves in and out to provide a continuous seal for the vapor refrigerant.

Another type of rotary compressor has multiple vanes that rotate with the roller. The roller in this type of compressor is offset from the center but is not eccentric. As the roller spins, the vanes move in and out of the roller to provide barriers, creating the sealed space for compression to occur. As the roller continues to turn, this space gets smaller and smaller, causing the pressure and temperature of the vapor refrigerant to increase. The refrigerant is pushed toward the discharge port where it is exhausted from the compressor. The operation of a rotary compressor with rotating vanes is shown in Figure 1-8. Rotary compressors may have as many as eight rotating vanes and operate on the same basic principle as the one just described. The more vanes that a rotary compressor has, the more efficient the compressor will be.

Reciprocating Compressors

A reciprocating compressor utilizes pistons, cylinders, and valves to accomplish the compression of the refrigerant. The **piston** (Figure 1-9a), moves in a back-and-forth, or reciprocating, path inside the **cylinder**. Reciprocating compressors vary in size and are equipped with anywhere from one to sixteen pistons and cylinders depending on the design requirements of the compressor. The pistons on larger compressors, typically over 3 horsepower, are equipped with O-rings to prevent refrigerant leakage into the crankcase. Defective rings in a compressor result in reduced operating efficiency. The cylinder is the location where the refrigerant is actually compressed and is located under the **compressor head** (Figure 1-9b). The compressor head is a dividing point between the high- and low-pressure sides of the system and houses other integral components of the compressor. These components include:

- Head gasket(s)
- Valve-plate gasket(s)
- Suction valve(s)
- Discharge valve(s)
- Valve plate(s)

Head gaskets and valve-plate gaskets are used to make certain that a leak-tight seal is maintained between the head and the valve plate and the valve plate and the compressor body, respectively. The gaskets compensate for any imperfections in the machining processes used to manufacture these components. They are usually made of cardboard, rubber, or neoprine. The suction and discharge valves open and close depending on the pressure difference across them and allow the vapor refrigerant to enter and leave the compression chamber at the proper point in

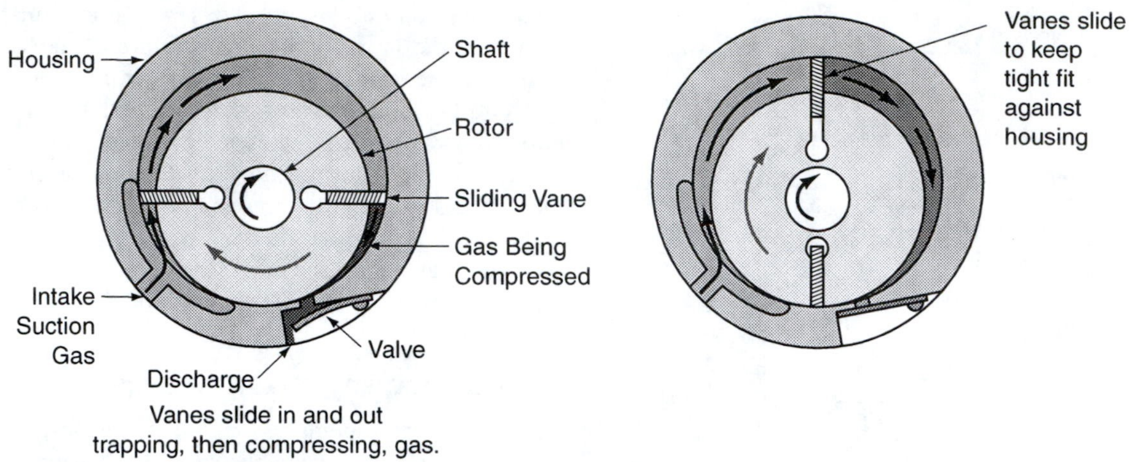

FIGURE 1-8 Operation of a rotary compressor with rotating vanes

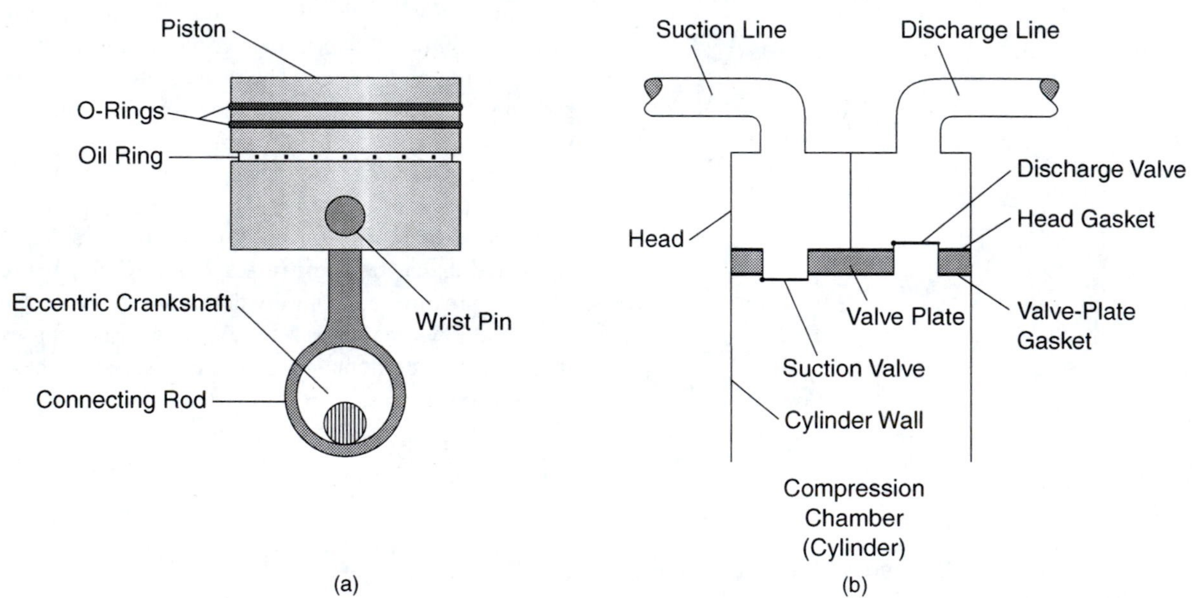

FIGURE 1-9 (a) Piston. (b) Compression chamber and integral components. The piston moves up and down in the cylinder to compress vapor refrigerant.

time. The valve plate holds the suction and discharge valves in place. The suction valves are located on the underside of the plate, while the discharge valves are located on the top of the plate.

A crankshaft is used to convert the rotating motion of the motor to the *reciprocating* motion of the piston. The crankshaft is usually a *crank-throw type* or an *eccentric type*. The crank-throw type (Figure 1-10a) has offset arms, or throws, to facilitate a larger piston stroke in the cylinder. The eccentric type (Figure 1-10b) is manufactured as a straight shaft with off-center eccentrics onto which the piston rods are positioned. The pistons are connected to the crankshaft with connecting rods. For ease in explanation, the compression process can be divided into four distinct "miniprocesses":

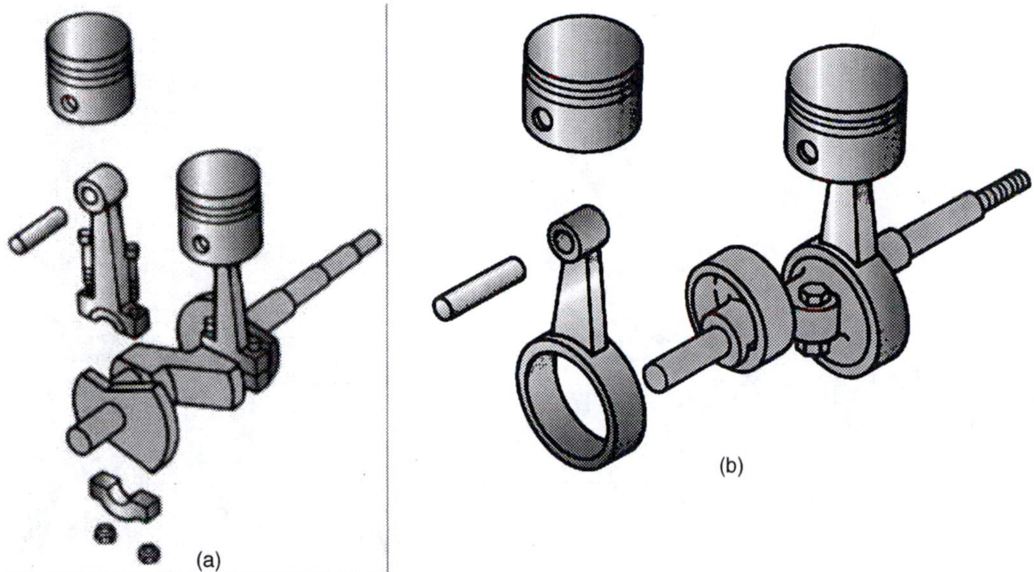

FIGURE 1-10 (a) Crank-throw type—Relationship of the pistons, rods, and crankshaft. (b) Eccentric type—This crankshaft obtains the off-center action with a straight shaft and an eccentric. To remove the connecting rods, the crankshaft must be taken out of the compressor. (*Reproduced courtesy of Carrier Corporation*)

- Reexpansion
- Suction or intake
- Compression
- Discharge

REEXPANSION. As a reference point, we will start our discussion about the compression process at the point where the piston is at the highest possible position within the cylinder. This position is referred to as *top dead center.* At top dead center, both the suction and the discharge valves are in the closed position and the refrigerant in the compression chamber is equal to the discharge pressure (Figure 1-11a). Since both valves are closed, no refrigerant can enter or leave the compression chamber. As the crankshaft continues to turn, the piston starts to move down in the cylinder. This increases the volume of the cylinder, and the pressure of the refrigerant in the cylinder begins to decrease. The refrigerant in the cylinder goes through a process of expansion. This part of the cycle is actually referred to as **reexpansion**, because the refrigerant that occupies the space above the piston at top dead center gets expanded during the first part of every downward stroke. The refrigerant that remains in the cylinder when the piston reaches top dead center is called **clearance vapor**, and the space it occupies is called **clearance volume**. The smaller the clearance volume, the more efficiently the compressor will operate. Zero clearance volume is impossible to achieve in reciprocating compressors because the pistons would have to come in contact with the valve plate, causing damage to the valve plate, pistons, and valves.

SUCTION (INTAKE). The pressure of the refrigerant in the cylinder will continue to drop until it reaches a point just below the suction pressure of the system. At this pressure, the suction pressure will now be greater than the compression chamber pressure and the suction valve will open (Figure 1-11b). At this point, the pressure in the suction line is equal to the pressure in the compression chamber. As the piston continues to move downward, suction gas is drawn into the compression chamber. This portion of the cycle is referred to as **suction** or **intake**. Suction will continue until the piston stops moving in the downward direction. When the piston reaches its lowest position in the cylinder, bottom dead center, the suction portion of the cycle ends.

COMPRESSION. As the crankshaft continues to turn, the piston starts to move upward in the cylinder. The refrigerant in the cylinder pushes the suction valve closed, and the refrigerant is trapped in the compression

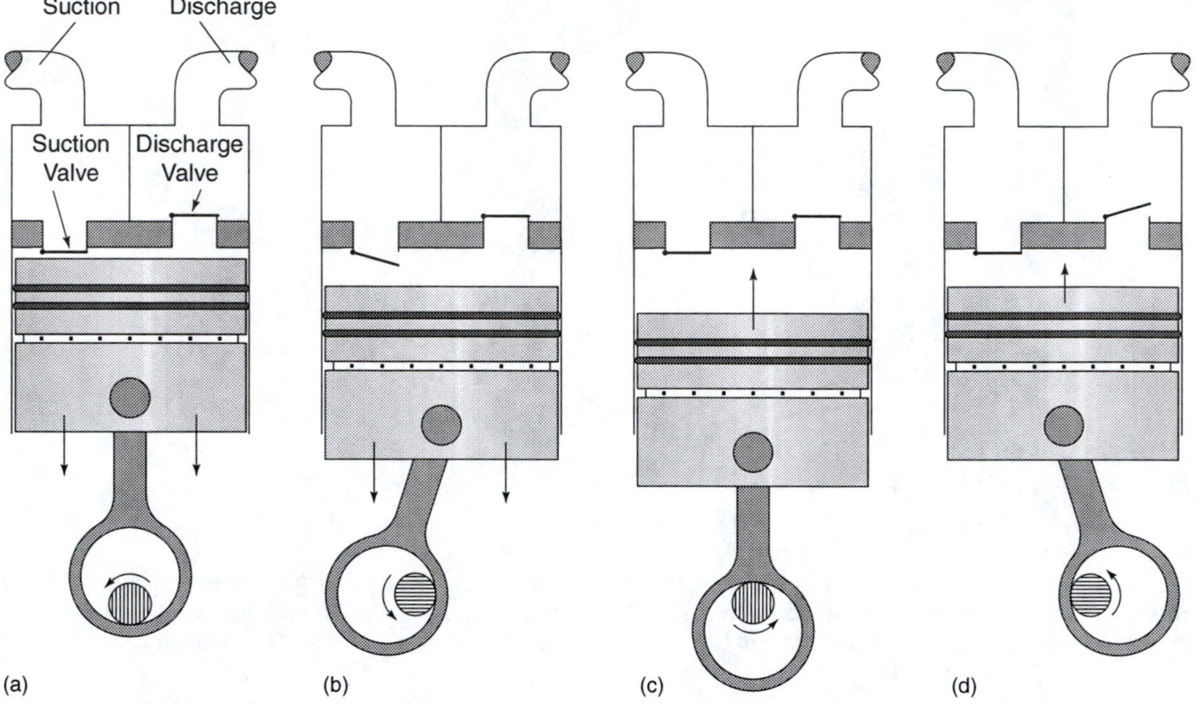

FIGURE 1-11 (a) The piston is at top dead center and is beginning to move down in the cylinder. Both the suction and the discharge valves are closed. (b) The suction valve opens when the cylinder pressure drops below the suction pressure. Refrigerant from the suction line now enters the compression cylinder. (c) The piston is at bottom dead center and is beginning to move up in the cylinder. The suction valve is pushed closed, and compression begins. (d) When the pressure in the cylinder is greater than the discharge pressure, the discharge valve is pushed open. The high-pressure, high-temperature vapor is now discharged to the condenser.

cylinder. The piston continues to move upward, reducing the volume of the cylinder and increasing the pressure of the refrigerant. This part of the cycle is referred to as **compression** (Figure 1-11c). Compression will continue until the pressure of the refrigerant inside the cylinder is slightly greater than the pressure of the refrigerant in the discharge line.

DISCHARGE. When the cylinder pressure is greater than the discharge pressure, the discharge valve will be pushed open, allowing the high-pressure refrigerant to be pushed out of the cylinder into the discharge line as the piston continues to move upward. This part of the cycle is referred to as **discharge** (Figure 1-11d). Discharge will continue until the piston reaches top dead center, where the discharge refrigerant will push the discharge valve closed as the piston again starts to move downward. The cycle then repeats itself as long as the compressor is energized.

Scroll Compressors

Another type of compressor that is rapidly growing in popularity is the scroll compressor (Figure 1-12). The concept used in this type of compressor has been known for quite some time but was ahead of the technology needed to manufacture it. The successful and efficient operation of the scroll is reliant on the production of two perfectly machined spirals or scrolls (Figure 1-13). One of the scrolls is stationary, while the other scroll vibrates or wobbles. As this scroll vibrates, refrigerant vapor is pushed and compressed toward the center of the compressor where it is discharged from the device. The compression process is illustrated in Figure 1-14. The configuration of the scrolls forms multiple chambers, each of which is at a different stage of compression, allowing the compressor to operate smoothly and continuously.

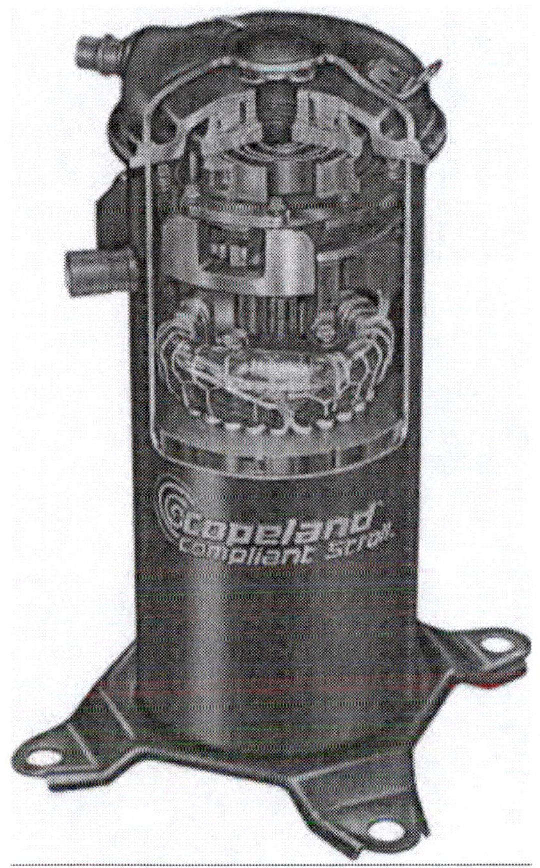

FIGURE 1-12 Scroll compressor. (*Courtesy Copeland Corporation*)

FIGURE 1-13 Two identical scrolls that form crescent-shaped pockets when nested together. (*Courtesy Copeland Corporation*)

Scroll compressors are becoming a popular choice for replacement because they are more forgiving when it comes to liquid that may enter the device. Because one of the scrolls is stationary and the other floats, the scrolls have some *play*. This play will allow the floating scroll to move if liquid, which is not compressible, should enter the compressor, preventing damage to the device. Damage to the compressor is adverted since the scrolls movement increases the volume of the chamber, thereby accommodating the volume of the liquid. Since liquid cannot be compressed, the scroll compressor is much more forgiving than the reciprocating compressor with regard to liquid refrigerant. The pistons on a reciprocating compressor move in a well-defined path and the introduction of liquid could result in major component failure. Liquid entering the scroll compressor will, however, reduce the capacity of the system.

Another benefit of scroll compressors is that they do not utilize suction and discharge valves. The scrolls, as they age, tend to "wear in," making them more efficient, as opposed to valves, which wear out with time. Since there are no suction or discharge valves, scroll compressors are equipped with a low-mass, disc-type check valve at the discharge port that prevents the high-pressure refrigerant from traveling back through the compressor during the off cycle. The valve prevents the compressor from running backward for more than one second.

CONDENSERS

The condenser is a heat-exchange surface that permits the transfer of heat from the system refrigerant to the condensing medium. This heat comes from the load that is being cooled and from the heat generated during the compression process. The condensing medium, which is usually air or water or a combination of both,

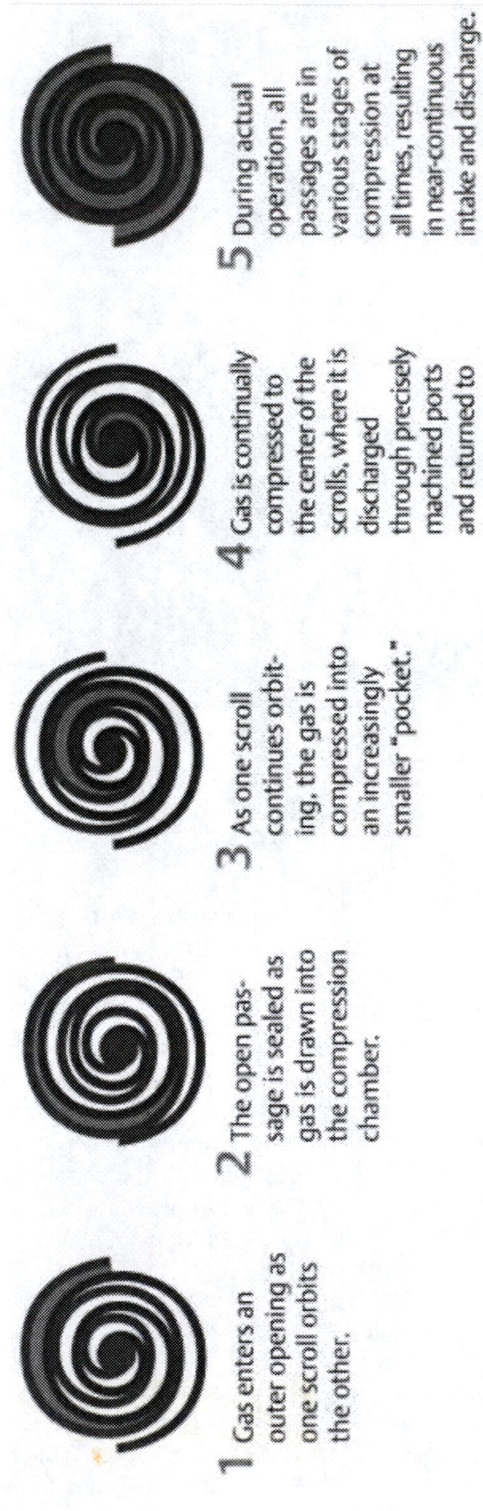

1 Gas enters an outer opening as one scroll orbits the other.
2 The open passage is sealed as gas is drawn into the compression chamber.
3 As one scroll continues orbiting, the gas is compressed into an increasingly smaller "pocket."
4 Gas is continually compressed to the center of the scrolls, where it is discharged through precisely machined ports and returned to the system.
5 During actual operation, all passages are in various stages of compression at all times, resulting in near-continuous intake and discharge.

FIGURE 1-14 Compression process in a scroll compressor. Compression is created by the interaction of an orbiting spiral and a stationary spiral. Gas enters the outer openings as one of the spirals orbits. As the spiral orbits, gas is compressed into an increasingly smaller pocket, reaching discharge pressure at the center port. Actually, during operation, all six gas passages are in various stages of compression at all times, resulting in nearly continuous suction and discharge (*Courtesy Copeland Corporation*)

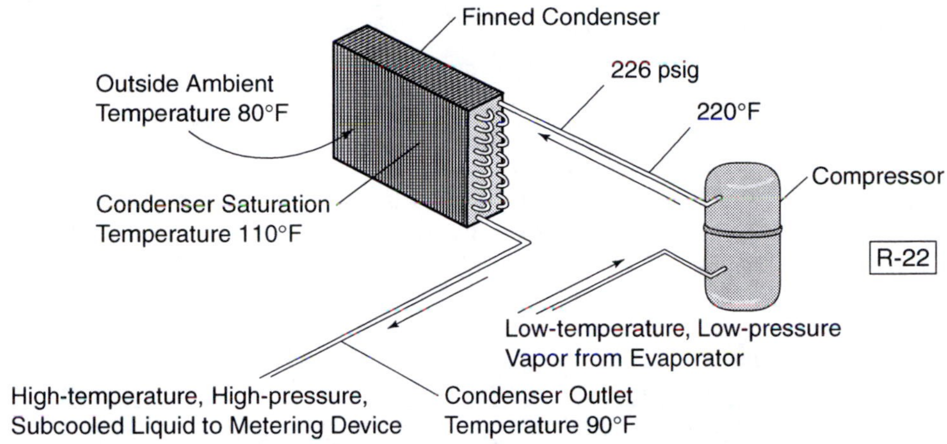

FIGURE 1-15 The superheated discharge gas from the compressor must desuperheat from 220°F to 110°F before it can begin to condense

must be at a lower temperature than the refrigerant to allow heat transfer to occur, causing the refrigerant to condense. When you go into a diner and are served a piping-hot cup of coffee that is just too hot to drink, how do you cool it off? Chances are you blow on it. This action causes a lower-temperature medium, air, to move across the surface of a higher-temperature medium, coffee. This causes the heat from the coffee to transfer to the air, thus cooling off the coffee. Even though the air blown across the surface is approximately 98°F, it is still cooler than the surface of the coffee.

The refrigerant leaving the compressor is superheated and is well above the saturation temperature, so it must first be cooled down to the refrigerant's saturation temperature before it can condense. The process of removing the superheat from the discharge refrigerant is referred to as **desuperheating**. The desuperheating process is the removal of superheat and is a sensible heat exchange since this change can be measured with a thermometer. This is the first function of the condenser.

For example, if an R-22 air-conditioning system is operating as in Figure 1-15, the superheated refrigerant is leaving the compressor at 220°F and must be cooled down to 110°F before it will begin to condense. The condenser saturation temperature of 110°F corresponds to the high-side pressure of 226 psig on a pressure/temperature chart (Figure 1-1). Once the refrigerant's temperature falls to 110°F, it begins to change state. This is a latent heat transfer and continues until all of the refrigerant has become a liquid. This transfer of heat causes the 110°F vapor to change into a 110°F liquid. Since the 110°F liquid is still warmer than the condensing medium, it will continue to give up heat to the medium and begin to subcool. One degree of subcooling is equivalent to one degree below the condenser saturation temperature. In this example, the refrigerant is condensing at 110°F and leaving the condenser at 90°F, so the condenser is operating with 20 degrees of subcooling. The overall purpose of the condenser can be summed up in the chart shown in Figure 1-16.

Air-Cooled Condensers

In a condenser, heat is transferred from one location to another. Air-cooled condensers transfer heat into the air that passes over the condenser coil (Figure 1-17). Many air-cooled condensers are located outdoors because the heat rejected from indoor condensers can raise the space temperature to an uncomfortable level. Air-cooled condensers that are located indoors must have ductwork to bring outside air to the coil as well as ductwork to remove the heat-laden discharge air from the space. Smaller, air-cooled condensers used for refrigeration applications are very often discharged into surrounding areas, such as basements and storage rooms in restaurants. Air-cooled condensers can be equipped with one or more fans to increase the quantity of air moving over the coil and to increase the operating efficiency of the condenser. The condenser in Figure 1-15 shows outside, or ambient, air passing

Condenser Process	Type of Heat Transfer	Purpose	Location in Condenser
Desuperheating	Sensible	Reduces refrigerant to condenser saturation temperature	Discharge line and top portion of the condenser
Condensing (change of state)	Latent	Allows refrigerant to change from a vapor to a liquid	Middle portion of the condenser
Subcooling	Sensible	Cools refrigerant to a temperature below the condenser saturation temperature	Usually the last 10% of the condenser

FIGURE 1-16 Description of the basic condenser functions

FIGURE 1-17 Larger, air-cooled condenser. This condenser draws air in from the bottom and discharges it at the top (*Courtesy Heatcraft, Inc., Refrigeration Products Division*)

over the coil at 80°F. For a standard-efficiency condenser, the condenser saturation temperature is roughly 30 to 35 degrees higher than the outside ambient temperature. In this example, the condenser saturation temperature is 110°F, indicating a 30-degree difference. Air must be allowed to flow freely through the coil to ensure maximum efficiency. If the coil becomes dirty or a fan motor burns out, the system's efficiency will be greatly reduced and it may stop operating altogether.

Higher-efficiency condensers, which are typically larger than standard-efficiency models, can operate with saturation temperatures as low as 20 degrees higher than outside ambient temperature. The larger coils increase the effective surface area and increase the rate of heat transfer. They also allow for more subcooling, which further reduces the temperature of the refrigerant in the liquid line, leading to more efficient system operation. Another method that manufacturers can use to increase the condenser's efficiency is to increase the number of **fins per inch**, which increases the effective heat-transfer surface of the coil. Fins are the metallic heat-transfer surfaces, often in the form of aluminum sheets, that are connected to the refrigerant piping material to make up the coil.

Whether the condenser is standard or high efficiency, the condenser saturation temperature is still determined by the outside ambient temperature. When it is hot, the condenser saturation temperature will be higher and the system efficiency lower. On cooler days, the condenser saturation temperature will be lower and the system efficiency and effectiveness higher. This leads to an interesting problem: All refrigeration equipment becomes less efficient in hot weather, when it is needed most. One alternative to utilizing air-cooled equipment is to utilize water-cooled equipment that addresses this problem.

Water-Cooled Condensers

Since increased outside ambient conditions can reduce the cooling capacity of air-conditioning and refrigeration systems that use air-cooled condensers, water-cooled condensers are an attractive alternative.

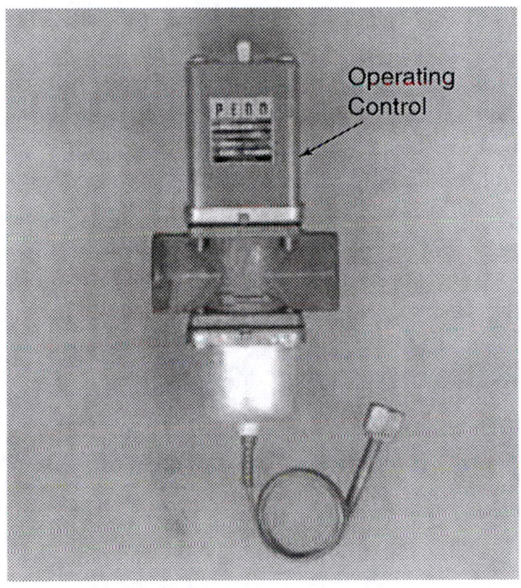

FIGURE 1-18 Water-regulating valve used to maintain the head pressure on water-cooled systems (*Photo by Bill Johnson*)

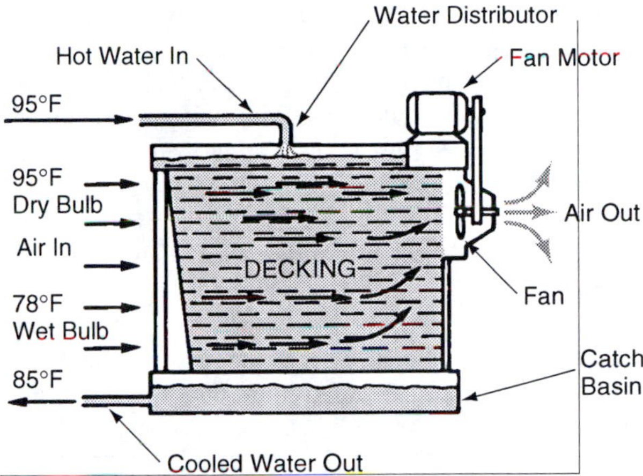

FIGURE 1-19 Typical cooling-tower operating temperatures

They differ from air-cooled condensers in that they do not employ fans or blowers and they use water instead of air as the condensing medium. They are, therefore, less sensitive to changes in the outside ambient temperature. Since the water temperature can be maintained at a more or less constant temperature, water-cooled systems are less likely to experience large fluctuations in high-side pressure and condenser saturation temperature. Hence, they tend to be much more efficient than air-cooled systems. Two distinct types of water systems can be used in conjunction with water-cooled condensers. They are the **waste-water** and the **recirculating systems**.

WASTEWATER SYSTEM. In a wastewater system, the water, also known as tap water, is supplied by the local municipality. This water passes through the condenser one time and is then expelled down a drain. The water is therefore "wasted" and cannot be reused. To limit the amount of water consumed, a water-regulating valve is used (Figure 1-18). This valve senses the high-side or **head pressure** of the system and adjusts the water flow accordingly. If the head pressure is too high, the valve opens and allows more water to flow through the coil. If the head pressure is too low, the valve closes, allowing the head pressure to rise. The water-regulating valve is considered to be a **modulating** valve since it can start, stop, or adjust the flow of water. A **snap-action** valve, on the other hand, can be only open or closed, with no intermediate positions. Under normal operating conditions water will enter the wastewater condenser at about 75°F and leave the condenser at about 95°F. This results in a temperature differential across the condenser of 20 degrees. The required water flow through a condenser utilizing a wastewater system is roughly 1.5 gallons per minute per ton. One advantage of a wastewater system is that the water tends to be relatively cold, so less of it would need to be used. One major disadvantage of this type of system is that water can be very expensive and operating this equipment would be cost-ineffective. Some localities set requirements and guidelines that strictly limit the usage of wastewater systems.

RECIRCULATING SYSTEM. In a recirculating system, water is passed through the condenser, cooled down, and is then recirculated through the condenser again and again. Recirculating-water systems utilize cooling towers (Figure 1-19), which cool the water that leaves the condenser. Under normal operating conditions, the tower can cool the water to within 7 degrees of the wet-bulb temperature of the air moving through the tower. The water cooled by the tower is the same water that returns to the condenser for another cycle. Under normal operating conditions, water will enter the recirculating condenser at about 85°F and leave the condenser at about 95°F. This results in a temperature differential across the condenser of 10 degrees. Since the temperature of the supply water in a recirculating-water system is warmer than the water in a wastewater system, more water flow is needed. The required water flow through a condenser utilizing a recirculating system is roughly 3.0 gallons per minute per ton. This type of system obviously uses much less "new" water than the wastewater system, even though more water must be circulated through the system. One problem with recirculating-water systems is that mineral deposits, also referred to as **scale**, can form on the inside surfaces of the water pipes. This is because heat in the system allows the minerals to separate from the water and adhere to the inner surfaces of the water-carrying piping. Scale acts as an insulator between the water and refrigerant and must therefore be removed to ensure proper condenser, and system, operation. Proper cooling-tower maintenance and proper chemical treatment of the water can minimize the formation of scale. Figure 1-20 compares a wastewater and a recirculating-water system.

	Wastewater System	**Recirculating System**
Approximate required water flow per ton	1.5 gallons per minute (gpm)	3.0 gpm
Approximate temperature of water supplied to the condenser	75°F	85°F
Approximate temperature of water leaving the condenser	95°F	95°F
Temperature differential (return water temperature–supply water temperature)	20°F	10°F
Source of supply water	Local municipal water supply (tap water)	Cooling tower
Advantages	• Inexpensive to operate if water is free. • Greater heat transfer. • Less water is circulated through the condenser. • Lower installation costs.	• Uses less "new" water. • Water costs are less. • Widely acceptable by local governments. • Many systems have constant flow, so water-regulating valves are not needed.
Disadvantages	• Uses large amounts of "new" water. • Expensive to operate in urban areas where water is expensive. • Strict government regulations regarding water usage and draining.	• Scale issue must be addressed. • Requires more maintenance. • Proper equipment operation relies on the proper operation of the cooling tower. • Periodic chemical treatment of tower is required. • Higher installation costs. • Water must be tested periodically.

FIGURE 1-20 Comparison of wastewater and recirculating-water systems

There are three common types of water-cooled condensers, each of which is classified by its method of construction, cleaning the water tubes, and heat transfer. They are the **tube-in-tube**, the **shell and coil** and the **shell and tube**.

TUBE-IN-TUBE CONDENSER. The tube-in-tube, or pipe-in-pipe condenser, is constructed with one pipe located inside the other and can be configured as either a coiled spiral (Figure 1-21a) or straight pipes (Figure 1-21b). This type of condenser is the least expensive to manufacture of the three types discussed and is commonly found on air-conditioning and refrigeration systems ranging from 1 ton to about 10 tons. The refrigerant flows in the outer pipe while the water flows in the inner pipe in the opposite direction (Figure 1-22). This is called **counterflow** and is designed to provide maximum heat transfer. The refrigerant flows in the outer tube so that it can give up its heat to both the water in the inner tube and the air that is surrounding the coil. If the refrigerant were flowing in the inner tube, the heat from the refrigerant *and* the heat from the surrounding air would be absorbed by the water, making the condenser less effective. The counterflow design ensures that the coolest water comes in contact with the refrigerant as it leaves the condenser to maximize subcooling. The more subcooling with which a condenser operates, the more efficient the condenser. If the refrigerant and water were traveling in the same direction, called **parallel flow**, the rate of heat transfer would reduce as the refrigerant and water flowed through the coil, and the efficiency of the condenser and the system would drop.

The water circuit in the straight-pipe version of the tube-in-tube condenser can be cleaned mechanically by removing the end plates of the condenser. The purpose for cleaning the interior walls of the condenser is to remove any scale or minerals that may accumulate in the water circuit. As stated earlier, the formation of scale reduces the rate of heat transfer and reduces the condenser's ability to reject heat. The removal of the end plates will not open the refrigeration circuit. When both end plates are removed, special brushes can be used to remove any scale.

The water circuit in the spiral tube-in-tube condenser must be cleaned chemically since brushes cannot completely access the interior surfaces of the tubes. Care must be taken when using chemical cleaners. Many cleaning solutions contain acid and must be handled with extreme care. The manufacturer's directions should always be followed exactly as indicated. Many air-conditioning and refrigeration firms choose to have professional coil-cleaning companies perform this task for them. Chemical cleaners containing acid must be completely flushed from the coil to prevent damage to the condenser walls.

SHELL-AND-COIL CONDENSERS. The shell-and-coil condenser is constructed of a steel shell with one or more coils of tubing running inside

(a)

(b)

FIGURE 1-21 Two types of tube within a tube condenser (a) A pipe within a pipe (b) A flanged type of condenser. The flanged condenser can be cleaned by removing the flanges. Removal of the flanges opens only the water circuit, not the refrigerant circuit ([a] *Courtesy Noranda Metal Industries, Inc.* [b] *Photo by Bill Johnson*)

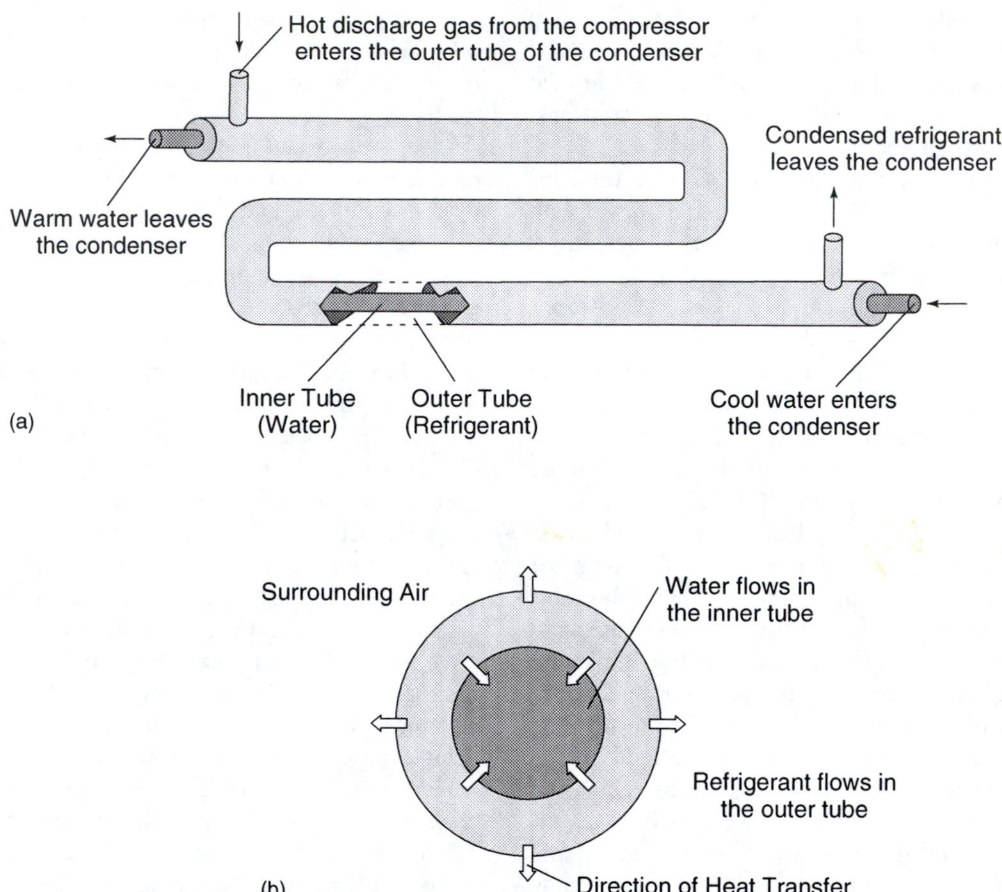

FIGURE 1-22 Cutaway of a tube in tube condenser (a) The water and refrigerant flow in opposite directions to maximize heat transfer (b) The refrigerant flows in the outer tube so it may give up heat to the air as well as to the water in the inner tube

(Figure 1-23). This type of condenser is commonly used in applications ranging from 3 tons up to approximately 15 tons. The hot discharge gas from the compressor is piped into the shell while water is circulated through the internal tubing coils. As the hot gas comes in contact with the cool surface of the coils, it begins to desuperheat and condense to form small droplets of liquid on the coil's surface. As these droplets increase in size, they fall from the coil and accumulate at the bottom of the shell. A portion of the water coil is submerged below the level of the liquid, and here the liquid refrigerant is subcooled before leaving the shell.

The subcooled liquid refrigerant leaves the condenser either from the bottom or from the top, depending on the construction of the device. If the condenser is designed for the refrigerant to enter the liquid line from the top, a dip tube is required to ensure that 100 percent liquid leaves the device (Figure 1-24). Condensers that have the refrigerant leave from the bottom of the device are not equipped with dip tubes.

The shell-and-coil condenser also serves as a storage vessel for refrigerant. Once the refrigerant has condensed and subcooled, it remains in the bottom of the shell until the metering device, discussed in the next section, calls for it. The condenser is therefore acting as a **receiver**, which is a component commonly found in air-conditioning and refrigeration systems and will be discussed in Chapter 2. The purpose of the receiver is to store refrigerant until needed by the evaporator and metering device.

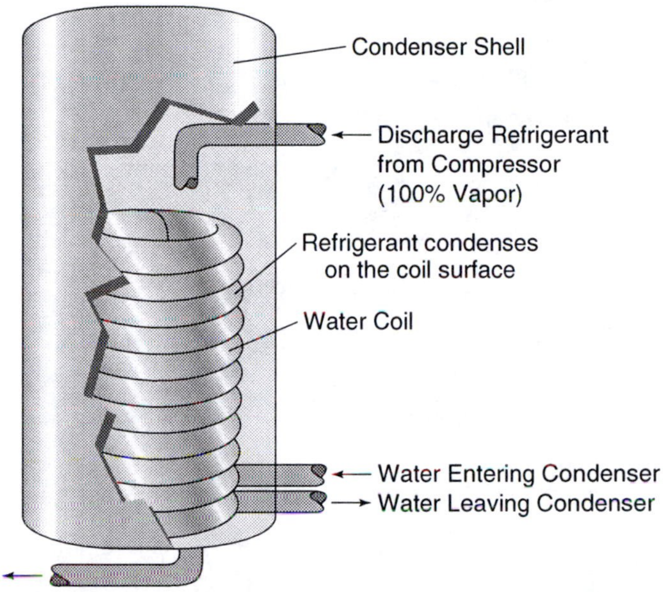

FIGURE 1-23 Shell-and-coil condenser. The hot gas is piped into the shell, and the water circulates through the coil

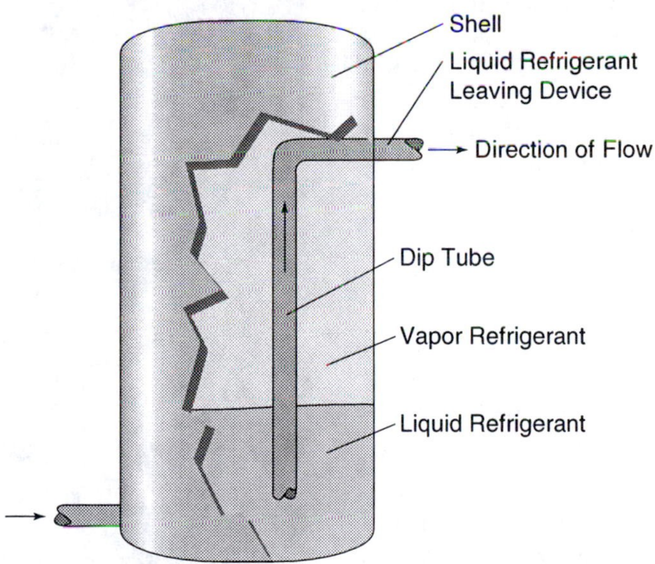

FIGURE 1-24 Function of a dip tube

Because of the construction of the shell-and-coil condenser, it is not possible to clean the water circuit mechanically. Therefore, this type of condenser must be cleaned chemically and all safety precautions must be taken.

SHELL-AND-TUBE CONDENSERS. The third type of water-cooled condenser is the shell and tube. This type of condenser is the most expensive to manufacture and can be described as a cross between the mechanically cleanable tube-in-tube condenser and

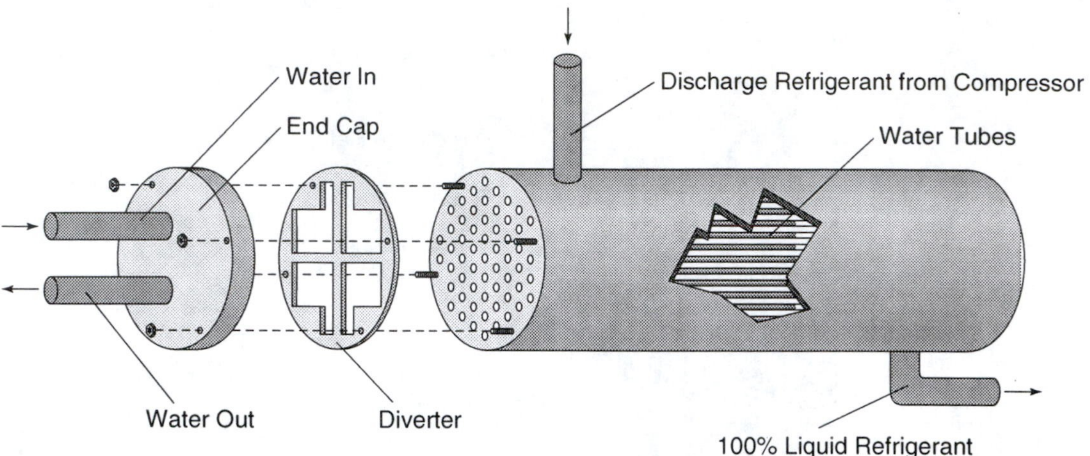

FIGURE 1-25 Shell-and-tube condenser. The device is equipped with end caps that can be removed to clean the water tubes

the shell-and-coil condenser. It is found in larger systems, is usually over 20 tons, and is constructed of a steel shell that has a series of straight tubes passing through from one end to the other (Figure 1-25). The shell is equipped with end plates, sometimes called *water boxes* that can be removed from both ends so that the interior of the water tubes can be cleaned mechanically (Figure 1-26). As in the tube-in-tube configuration, the removal of these end plates does not affect the refrigerant circuit. The shell-and-tube and the shell-and-coil condensers have a number of things in common:

- They both use the bottom of the shell to subcool the liquid refrigerant.
- They can both act as a receiver to store the refrigerant until needed.
- They are both equipped with dip tubes when the refrigerant leaves the device from the top.

Although the three types of water-cooled condensers that have just been discussed have some similarities and differences with regard to construction, configuration, and method of heat transfer, the basic function of the condenser is the same. They all desuperheat, condense, and subcool the refrigerant so that it may pass on to the metering device and continue the refrigeration cycle. A summary of water-cooled condensers is given in Figure 1-27.

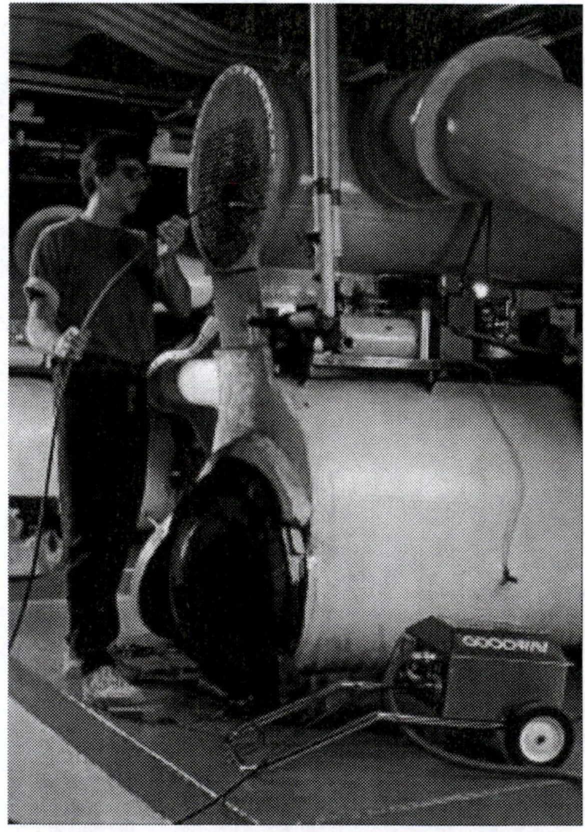

FIGURE 1-26 Shell-and-tube condenser being cleaned with specially designed brushes (*Courtesy Goodway Tools Corporation*)

	Tube in Tube	Shell and Coil	Shell and Tube
Discharge refrigerant from compressor	Piped into the outer tube of the condenser	Piped into the shell of the condenser	Piped into the shell of the condenser
Water flow	• Piped into the inner tube of the condenser • Flows in the opposite direction of the refrigerant	Piped through the coil or coils located in the shell of the condenser	Piped through the tubes located within the shell of the condenser
Method of cleaning	• Spiral-type condensers cleaned chemically • Straight-pipe condensers cleaned mechanically	Cleaned chemically	Cleaned mechanically

FIGURE 1-27 Overview of styles of water-cooled condensers

EXPANSION OR METERING DEVICES

As previously discussed, the refrigeration system uses a closed cycle that repeats itself. For this to occur, whatever is done must later be "undone" at another point in the cycle. If pressure is increased, it must be decreased. If heat is absorbed, it must be rejected. Just as the compressor increased the temperature and pressure of the refrigerant entering it, another component is responsible for reducing the pressure and temperature of the refrigerant in the system. This component is the **expansion device** or metering device. The expansion device is responsible for feeding the proper amount of refrigerant to the evaporator. The refrigerant that enters the expansion device is a high-temperature, high-pressure, subcooled liquid and leaves the device as a low-temperature, low-pressure, saturated liquid. The saturated liquid is roughly 80 percent liquid and 20 percent vapor; and, since the refrigerant is saturated, there is a pressure/temperature relationship for that refrigerant. For an air-conditioning system operating with R-22 as its refrigerant, the pressure of the refrigerant entering the expansion device is approximately the same as the high-side pressure, which for an ideal air-conditioning system at design conditions is about 226 psig, assuming an air-cooled condenser is used. The temperature of the refrigerant entering the expansion device will be roughly 95°F, assuming that the condenser is operating with 15 degrees of subcooling (Figure 1-28, *top*). The pressure of the refrigerant as it leaves the expansion device, at design conditions, should be 68.5 psig and its boiling temperature should be 40°F (refer to the pressure/temperature chart for R-22).

NOTE: *Longer refrigerant lines will result in a larger pressure drop through the liquid line, resulting in lower pressure at the inlet to the expansion device.*

The reduction in pressure and temperature of the refrigerant as it passes through the expansion device is accomplished by a restriction in the liquid line in front of the inlet to the evaporator. This restriction prevents the compressor discharge gas from flowing through to the evaporator. The pressure on the low side of the system is also further reduced by the suction created by the compressor. A simplistic example follows. Consider a four-lane highway on which traffic is moving freely. The traffic volume is constant at all points on the roadway. Now imagine the same four-lane highway with three of the four lanes blocked due to a car accident. Obviously, there will

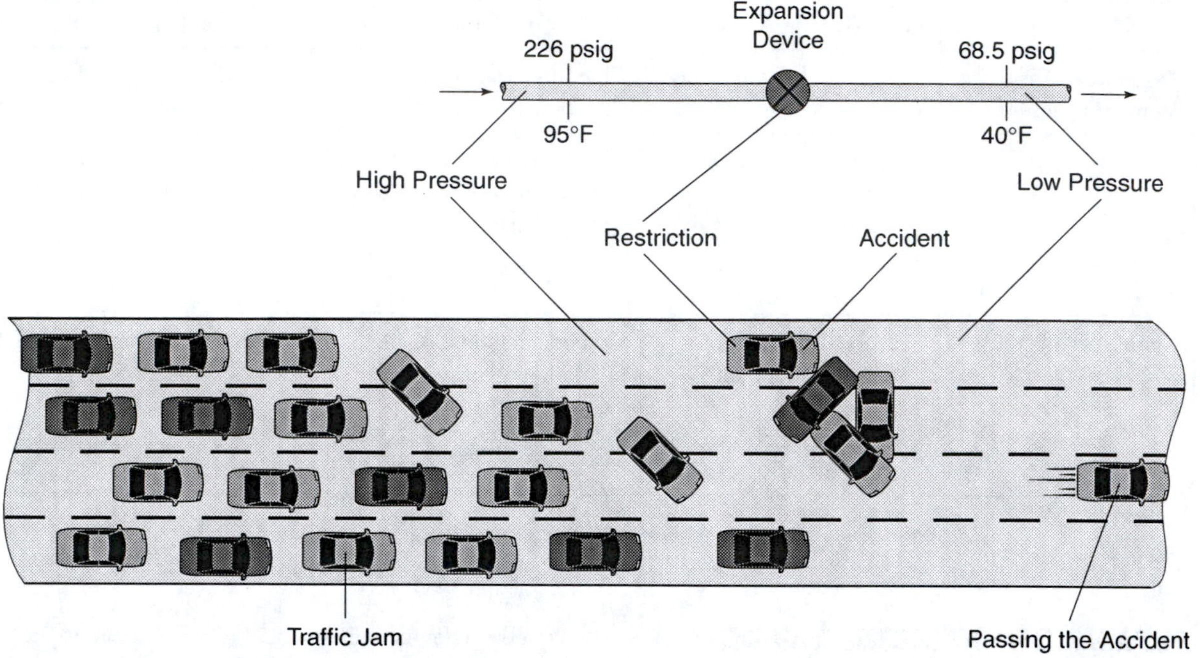

FIGURE 1-28 The expansion device is similar to a car accident in that it restricts normal flow and separates high pressure from low pressure

be a major traffic jam at the accident location as four lanes of traffic attempt to squeeze together into one single lane in an effort to pass. The number of cars drastically increases, and the space between cars reduces as well. This is very similar to the pressure increasing as the refrigerant waits to pass through the expansion device. The number of cars approaching the accident is large and the pressure is high. Passing the accident location, the number of cars is much lower. As soon as a car passes the accident, away it goes! What happened to the traffic? Now the road is free and clear and the distance between cars has increased drastically. The "pressure" is now reduced, and the car is free to continue its trip. If this concept is clear, the concept of the expansion device will also be clear (Figure 1-28, *bottom*). The reduction in pressure after refrigerant passes the expansion device is created by the combined effects of the metering device and the compressor's suction.

Three types of expansion devices are mentioned in this text, and, although the basic purpose of each is similar, their operation is very different. These devices are the **capillary tube**, the **automatic expansion valve**, and the **thermostatic expansion valve**. The capillary tube is the least complicated.

Capillary Tube

The capillary tube is a **fixed-bore** device, meaning that the opening that the refrigerant flows through does not change in size (Figure 1-29). Capillary tubes are found on **critically charged** systems, which, by definition, have the **total-system charge** moving through the system whenever it is operating. Total-system charge is the total amount of refrigerant that the system holds. Capillary-tube systems operate differently from other systems that store refrigerant until it is needed by the evaporator, so the amount of refrigerant in the system must correspond precisely to the manufacturer's specifications.

In operation, as the high-temperature, high-pressure, subcooled liquid refrigerant from the liquid line enters

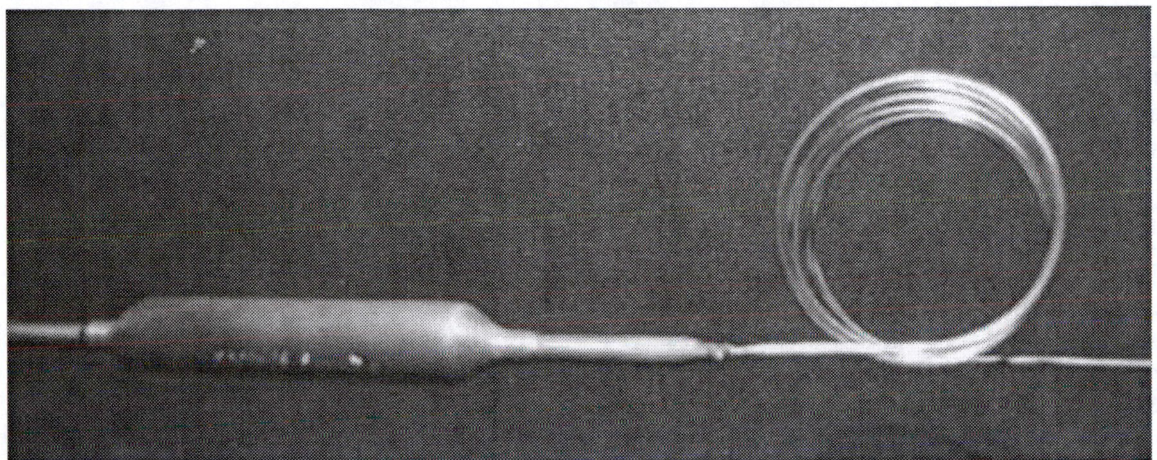

FIGURE 1-29 Drier/strainer, *left,* prevents moisture and particulate matter from entering the capillary tube, *right* (*Photo by Eugene Silberstein*)

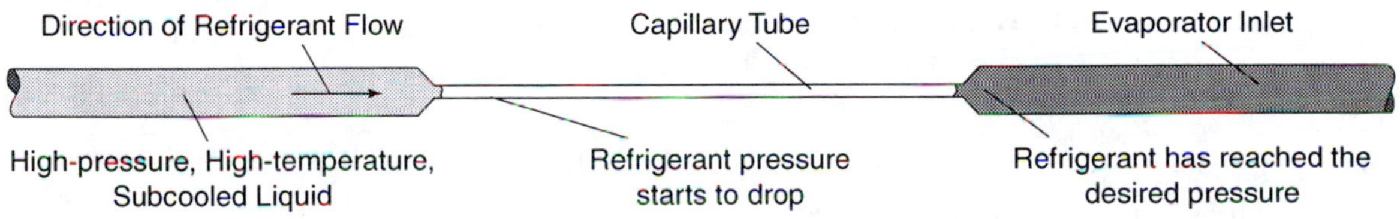

FIGURE 1-30 When the refrigerant exits the capillary tube, its pressure has dropped to the desired level

the capillary tube, its flow is restricted through the device. As the refrigerant flows through the capillary tube, its pressure begins to drop and some of the liquid boils into a vapor. When the refrigerant reaches the end of the capillary tube, the pressure of the refrigerant has dropped significantly. When the refrigerant leaves the capillary tube as a saturated liquid, it expands, since the tubing size of the evaporator is larger, and the pressure drops down to the desired evaporator saturation pressure (Figure 1-30). More liquid refrigerant boils into a vapor upon leaving the capillary tube and helps cool the remaining vapor by absorbing heat from it. The refrigerant that boils off is referred to as **flash gas** because it immediately flashes to a vapor. This flashing of the liquid to a vapor occurs within the capillary tube as well as at its outlet.

The length of the capillary tube is critical. The longer the capillary tube, the greater the pressure drop from inlet to outlet will be; and, conversely, the shorter the capillary tube, the smaller the pressure drop will be. The length of the capillary tube is carefully determined when the system is initially designed. For this reason, if the capillary tube is damaged, simply cutting out the damaged section is not good practice. The entire tube should be replaced. A capillary tube with the same bore size should be used, and it must be cut to the exact length of the old tube. Adjustment charts are also readily available that provide equivalent lengths if a capillary tube with the same bore size cannot be obtained. If the replacement capillary tube has a larger bore, it will be longer than the original. Similarly, if the new capillary tube has a smaller bore, the new capillary tube will be shorter.

When making repairs to a capillary tube, care must also be taken so that the brazing or filler material used during repair does not restrict the bore. The capillary tube does not have any moving parts and, therefore, does not require regular maintenance. It can, however, get clogged or partially blocked if any particulate matter finds its way into the system. For this reason, a filter drier and/or strainer is usually located immediately before the capillary tube to prevent any moisture or foreign matter from entering the device (Figure 1-29). Moisture can also cause a blockage of the capillary tube because water droplets can freeze when undergoing a pressure drop inside the tubing. At this point in the system, the refrigerant is saturated, part liquid and part vapor, so, as the pressure of the refrigerant drops, the temperature also drops.

Automatic Expansion Valves

The automatic expansion valve (AEV) modulates the flow of refrigerant into the evaporator to keep the evaporator pressure constant (Figure 1-31). Unlike the capillary tube, which is not capable of adjusting the flow of refrigerant, the automatic expansion valve opens and closes to either increase or decrease the amount of refrigerant feeding into the evaporator in response to the coil's pressure.

The valve operates on a "needle-and-seat" mechanism that modulates the amount of refrigerant that is permitted to pass through the valve. The position of the needle is determined by the difference between two pressures:

- The spring pressure
- The evaporator pressure

The **spring pressure** is the pressure that opens the valve. This pressure is adjustable and set at the desired evaporator pressure. The higher the spring pressure, the higher the evaporator pressure will be. The **evaporator pressure** is the pressure that closes the valve.

The spring and evaporator pressures push in opposite directions, and the valve position changes accordingly (Figure 1-32a). For example, if the spring pressure is 70 psig and the evaporator pressure is only 60 psig, the spring pressure, which is larger than the evaporator pressure, will cause the valve to open in order to feed more refrigerant to the evaporator (Figure 1-32b). This is done to increase the evaporator pressure. When the evaporator pressure reaches the desired level, the valve will maintain the proper flow in order to maintain that pressure. When the spring pressure is equal to the evaporator pressure, the valve is said to be in equilibrium (Figure 1-32a). If, on the other hand, the spring pressure is set at 70 psig and the evaporator pressure is 80 psig, the evaporator pressure, which is greater than the spring pressure, will close the valve. This will limit the amount of refrigerant flowing into the evaporator (Figure 1-32c), reducing the evaporator pressure.

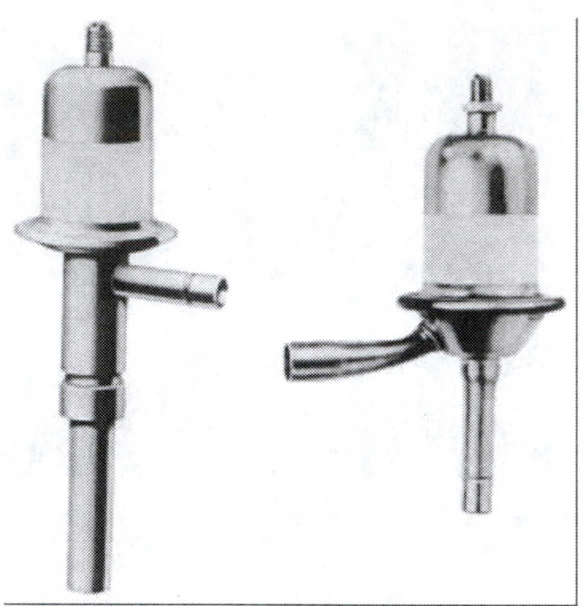

FIGURE 1-31 Automatic expansion valves are designed to maintain a constant evaporator pressure (*Courtesy Singer Controls Division*)

Thermostatic Expansion Valves

The thermostatic expansion valve (TEV) is a modulating valve that opens and closes in order to feed the proper amount of refrigerant to the evaporator. Unlike the capillary tube, which does not modulate fluid flow, and the automatic expansion valve, which is designed to maintain a constant pressure in the evaporator, the thermostatic expansion valve is designed to maintain a constant evaporator superheat

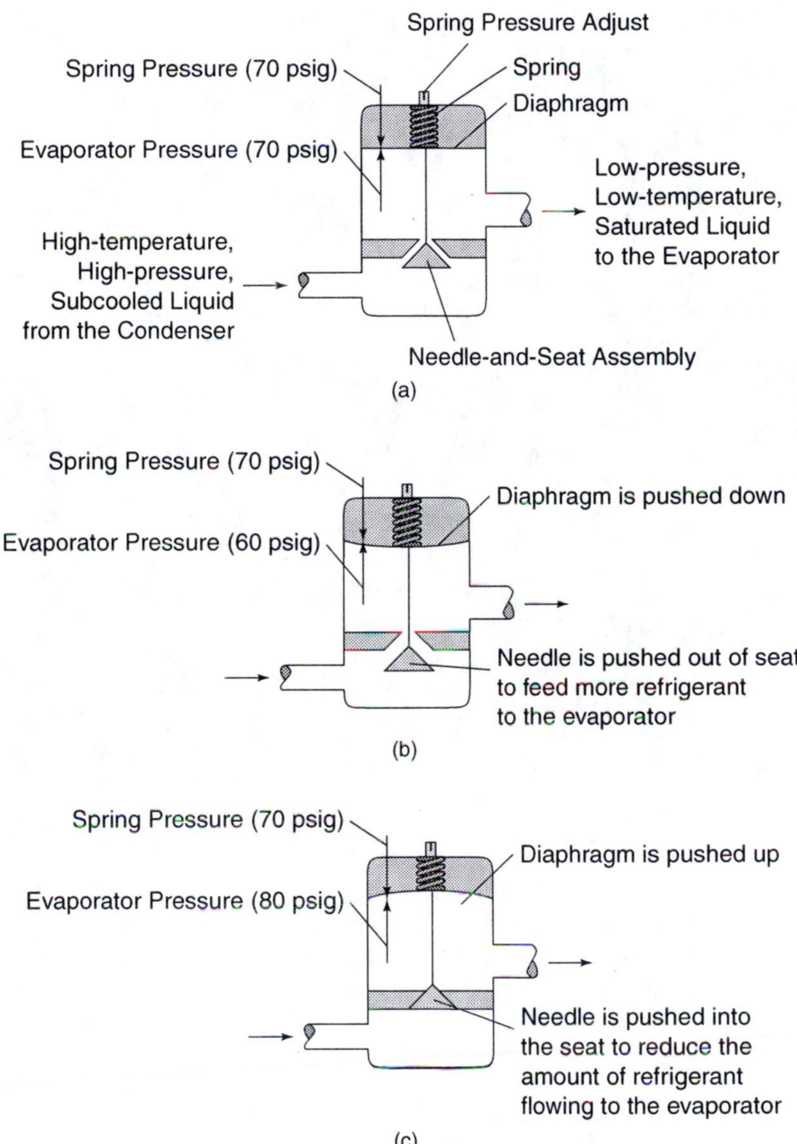

FIGURE 1-32 (a) The spring and evaporator pressures push in opposite directions. This valve is in equilibrium. (b) The evaporator pressure is lower than desired, so the valve is opening to allow more refrigerant flow to the evaporator. (c) The evaporator pressure is higher than desired. This valve is closing to reduce refrigerant flow to the evaporator

(Figure 1-33). Modulating devices, such as the automatic expansion valve and the thermostatic expansion valve, can open and close in response to external conditions to regulate refrigerant flow through the devices. The thermostatic expansion valve operates on a needle-and-seat concept that is very similar to the automatic expansion valve. The main difference between the TEV and the AEV is that the TEV will close as the system load is reduced, while the AEV will close as the system load is increased.

To briefly recap, superheat is the amount of sensible heat that is added to the refrigerant after it has

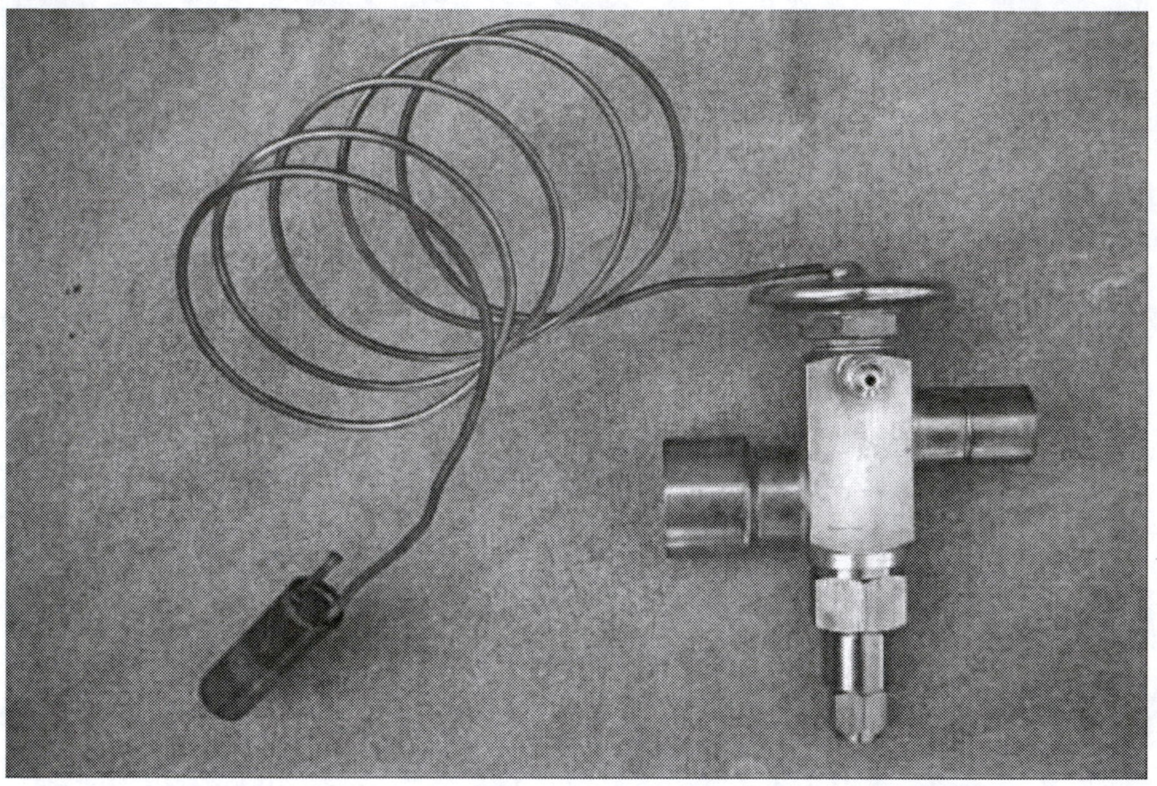

FIGURE 1-33 An externally equalized TEV (*Photo by Eugene Silberstein*)

all boiled off into a vapor. To give a conceptual example, consider the following. At sea level and atmospheric pressure, water boils at 212°F. As the water boils, it turns from a liquid to a vapor. This vapor is the same temperature as the boiling water, namely, 212°F. As the steam rises, it cools down and condenses back into a liquid. For visual clarification, you can hold a lid over a pot of boiling water. You will notice that as the steam comes in contact with the lid it turns back into water. If, instead of allowing the steam to cool, you add additional heat to the steam, it will simply get hotter. It is therefore possible to have steam at a temperature higher than 212°F. To turn this steam back into water, you need to remove superheat from the steam until it cools down to 212°F, at which point it begins to condense.

As stated in the section on compressors, the refrigerant entering the compressor needs to be 100 percent vapor to prevent component damage. The thermostatic expansion valve helps ensure that this is indeed the case. The amount of superheat that will be maintained by the valve depends on the setting of the superheat spring and the size of the valve itself. A cutaway view of a thermostatic expansion valve is shown in Figure 1-34.

To understand how the thermostatic expansion valve operates, it is important to understand how evaporator superheat is measured. Superheat is the difference between the evaporator saturation temperature and the evaporator outlet temperature. For example, if the low-side pressure in an air-conditioning system operating with R-22 as its refrigerant is 70 psig, the refrigerant is boiling at approximately 40°F (from the pressure/temperature relationship). If the temperature of the refrigerant at the outlet of the evaporator is 50°F, then the superheat can be calculated by subtracting the evaporator saturation temperature from the evaporator outlet temperature. In this case, 50 degrees – 40 degrees = 10 degrees of superheat (Figure 1-35). An evaporator superheat between 8 and 15 degrees is acceptable for most high-temperature, air-conditioning applications. A superheat of less than 8 degrees will

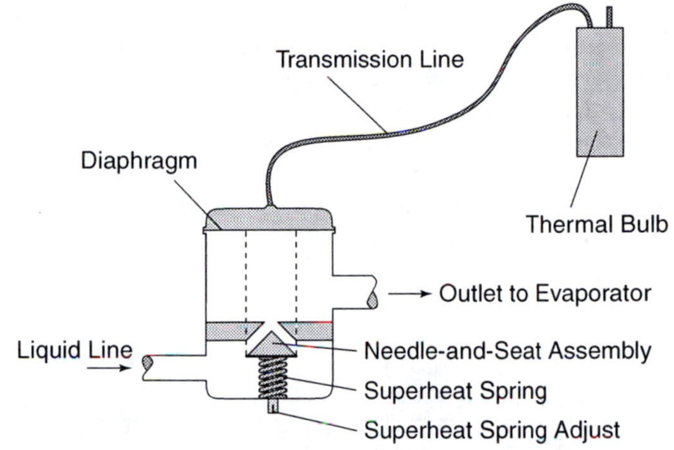

FIGURE 1-34 Cutaway illustration of a TEV

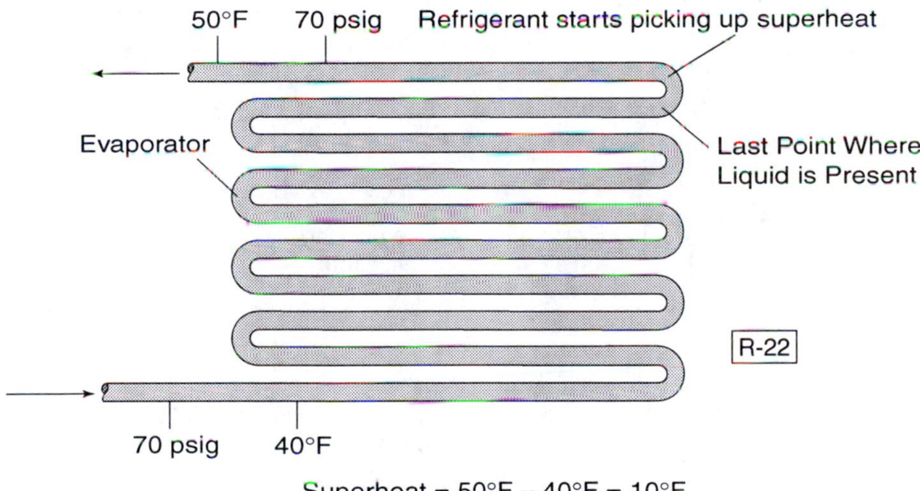

FIGURE 1-35 Evaporator superheat calculation

increase the possibility of refrigerant condensing back into a liquid before entering the compressor. A superheat above 15 degrees will cause the efficiency of the evaporator to drop.

The thermostatic expansion valve's needle-and-seat assembly is controlled by three pressures that properly position the needle in order to feed the correct amount of refrigerant to the evaporator. The pressures push on a diaphragm, which is a thin and very flexible piece of steel, whose position determines the position of the needle in the seat. The three pressures are:

- Evaporator pressure
- Spring pressure
- Bulb pressure

EVAPORATOR PRESSURE. The evaporator pressure is one of the pressures that pushes to help close the valve. This pressure attempts to push the needle into the seat in order to reduce the flow of refrigerant into the evaporator. The evaporator pressure can be taken from either the inlet or the outlet of the coil. If the pressure drop (difference between the inlet and outlet pressures) across the evaporator coil is small, the inlet and outlet pressures are relatively close to each other and the pressure reading can be taken from the inlet of the coil. A thermostatic expansion valve that senses the inlet pressure is called an **internally equalized** valve (Figure 1-36a), while a thermostatic expansion valve that senses the evaporator outlet pressure is called an **externally equalized** valve (Figure 1-36b). An example illustrating the difference in operation of internally and externally equalized valves can be found at the end of the section.

SPRING PRESSURE. The spring pressure, also known as the superheat spring pressure, determines how much superheat with which the evaporator will operate. The higher the spring pressure, the higher the amount of superheat. The spring comes factory

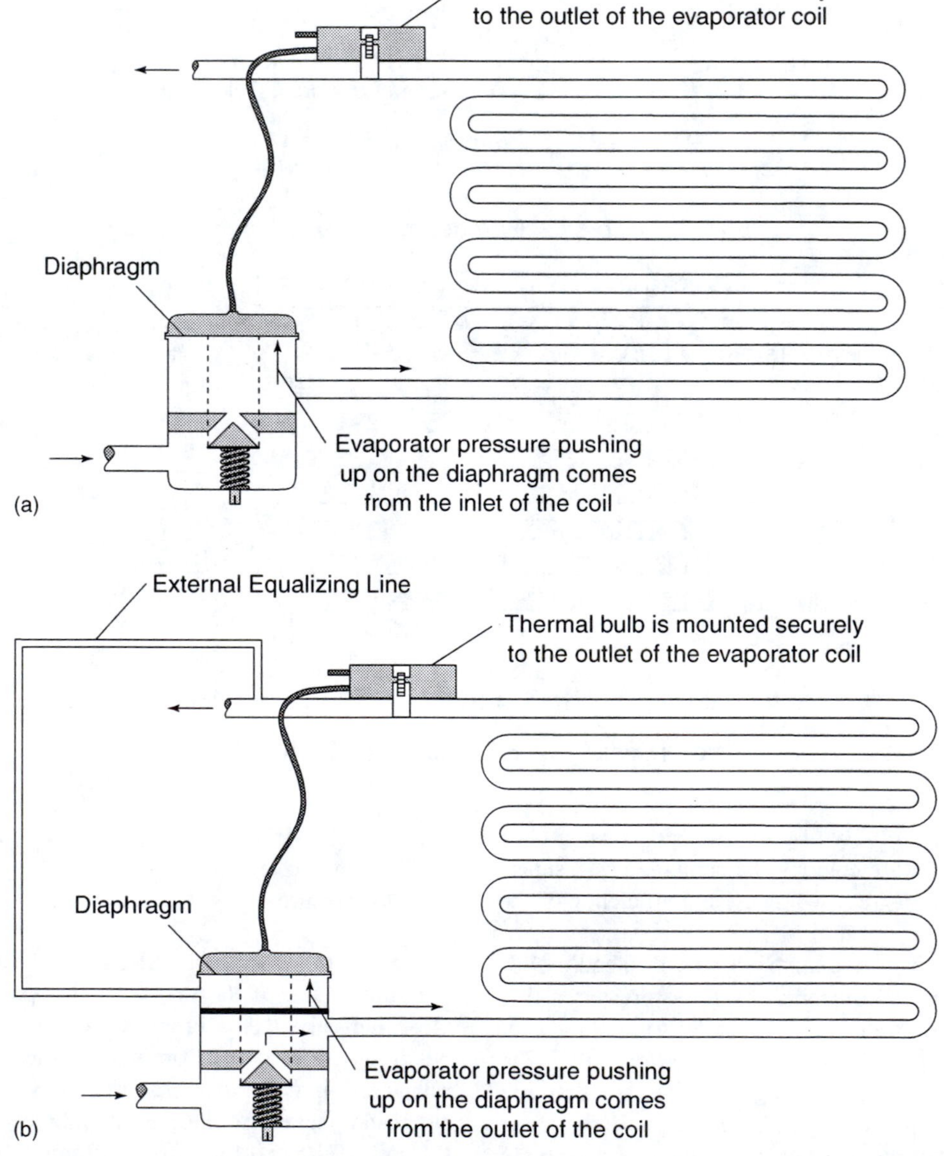

FIGURE 1-36 (a) An internally equalized TEV. The valve senses the evaporator pressure from the inlet of the coil. (b) An externally equalized TEV. The valve senses the evaporator pressure from the outlet of the coil

set and should only be adjusted by trained professionals since improperly adjusted superheat springs can cause major system damage, including compressor failure. The spring pressure is the other pressure that closes the valve, reducing the amount of refrigerant flowing into the evaporator. The evaporator pressure added to the spring pressure provides the total *closing pressure* for the valve.

BULB PRESSURE. The **bulb pressure** is the only pressure that opens the valve. This pressure is generated inside a thermal bulb that is mounted at the outlet of the evaporator. It is filled with a fluid, or refrigerant, that follows a pressure/temperature relationship and is isolated to prevent mixing with the system's refrigerant. The temperature that this bulb senses is the evaporator outlet temperature. The refrigerant at this temperature exerts a specific amount of pressure (pressure/temperature relationship) that pushes down on the diaphragm, opposing the evaporator and spring pressures (Figure 1-37).

Proper operation of the thermostatic expansion valve relies on the proper mounting of the thermal bulb. The following guidelines should be followed:

1. Make certain the bulb is strapped securely to the suction line, as close to the outlet of the evaporator as possible.
2. Make certain that the bulb is secured to a clean, straight section of hard-drawn pipe.
3. Do not mount the thermal bulb on a fitting or other component that will prevent good thermal contact.
4. Whenever possible, use the strapping material that is supplied with the valve.
5. Never solder the bulb to the suction line.

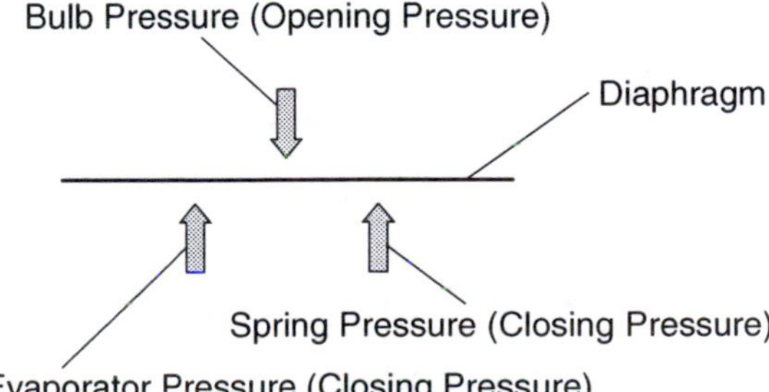

FIGURE 1-37 The bulb pressure pushes to open the thermostatic expansion valve. The spring and evaporator pressures push to close the valve

6. Do not use glues, tapes, string, or other adhesive materials to secure the bulb.
7. Always wrap the bulb with a good-quality insulation tape to ensure accurate readings.
8. On evaporators with small suction lines (up to about three-fourths inch), the thermal bulb is usually strapped to the top of the suction line.
9. Always follow manufacturer's instructions to ensure proper installation of the specific valve you are using.

TEV EXAMPLE 1 (VALVE BEING PUSHED CLOSED). In the system illustrated in Figure 1-38, you are given the following:

- Evaporator pressure is 70 psig.
- Spring pressure is 10 psig.
- Bulb pressure is 73 psig.

You can conclude the following:

- Opening pressure = 73 psig (bulb pressure)
- Total closing pressure = 80 psig (evaporator pressure + spring pressure)

The valve in this example is being pushed closed since the closing pressure is larger than the opening pressure. If the refrigerant in the system and the bulb is R-22, you can determine the amount of superheat in the evaporator by converting the pressures into temperatures using the pressure/temperature relationship for R-22. If the evaporator pressure is 70 psig, the evaporator saturation temperature is 41 degrees. If the bulb pressure is 73 psig, this means that the temperature the bulb is sensing is 43 degrees at the outlet of the evaporator. Therefore, the superheat in the evaporator is 2 degrees, which is the result of subtracting the evaporator saturation temperature from the evaporator outlet temperature (43°F – 41°F). You can logically conclude that the TEV should indeed be closing since the evaporator superheat is only 2 degrees.

TEV EXAMPLE 2 (VALVE BEING PUSHED OPEN). In the system illustrated in Figure 1-39, you are given the following:

- Evaporator pressure is 70 psig.
- Spring pressure is 10 psig.
- Bulb pressure is 84 psig.

This example is used as a class discussion exercise at the end of the chapter.

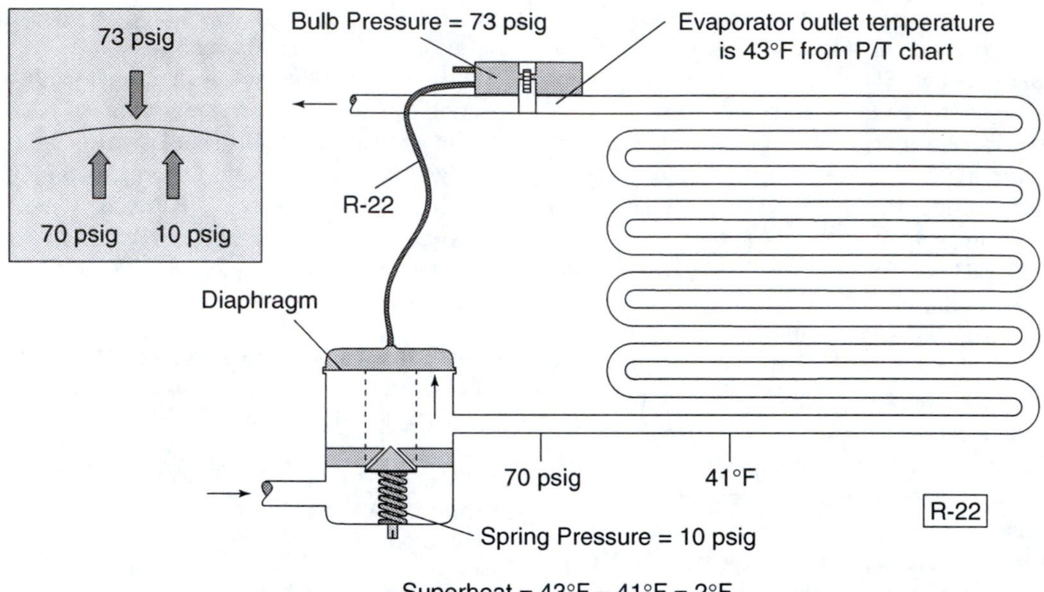

FIGURE 1-38 This valve is closing to increase the superheat in the evaporator

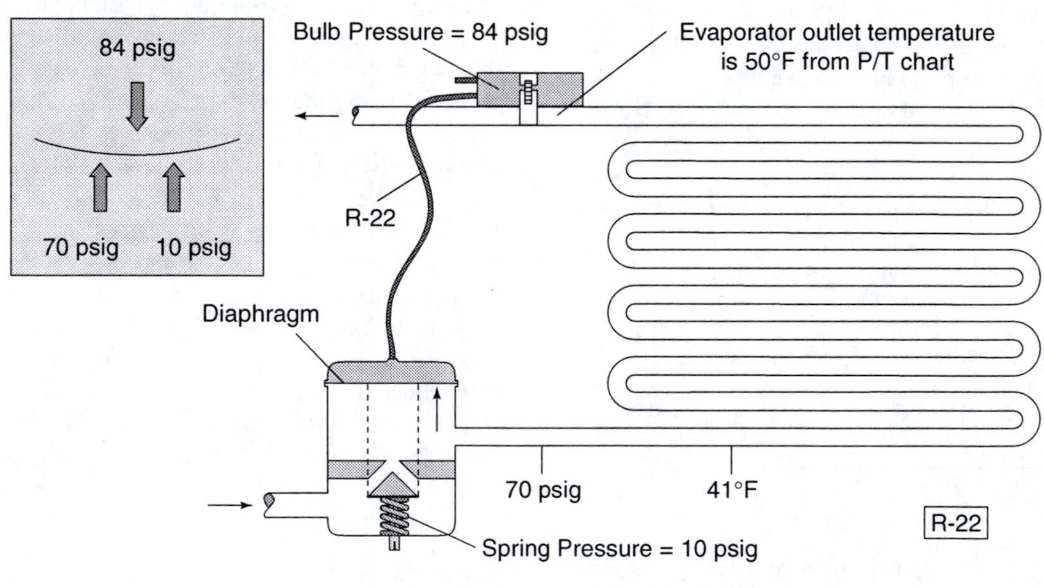

FIGURE 1-39 This valve is opening to reduce the superheat in the coil

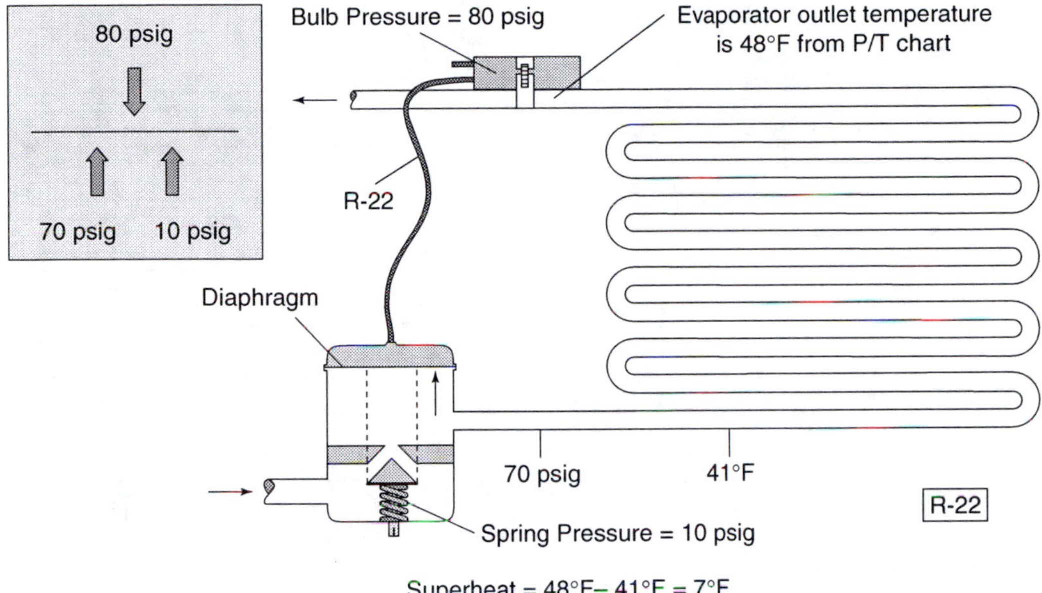

FIGURE 1-40 This valve is in equilibrium. It is feeding the proper amount of refrigerant to maintain 7°F of superheat in the coil

TEV EXAMPLE 3 (VALVE IN EQUILIBRIUM). In the system illustrated in Figure 1-40, you are given the following:

- Evaporator pressure is 70 psig.
- Spring pressure is 10 psig.
- Bulb pressure is 80 psig.

This example is used as a class discussion exercise at the end of the chapter.

EVAPORATOR PRESSURE DROPS AND THE TEV. Now examine a situation in which the evaporator pressure drop is small, less than 2 psig, and compare it to an evaporator that has a large, more than 2 psig, pressure drop. You can then see when an internally equalized valve is desirable and when an externally equalized valve is desirable.

Consider an ideal air-conditioning evaporator that has no pressure drop. This is an ideal case, but a pressure drop of less than 2 psig through the coil is considered negligible. The evaporator pressure remains constant at 68.5 psig through the entire coil and is operating with R-22 as its refrigerant. The thermostatic expansion valve operates with R-22 in its bulb and has a spring pressure of 15.5 psig and a bulb pressure of 84 psig. The evaporator saturation temperature is 40°F, and the evaporator outlet temperature is 50°F (from the pressure/temperature chart). The evaporator superheat is 10°F and the valve is in equilibrium since the opening pressure (84 psig) is equal to the closing pressure (15.5 psig + 68.5 psig = 84 psig), as shown in Figure 1-41.

Now consider the same evaporator, but this time assume that the evaporator has a pressure drop across it of 11 psig. This indicates that the pressure at the inlet of the evaporator is 68.5 psig and the pressure at the outlet of the evaporator is 57.5 psig. Assume that an internally equalized valve is being used and you are, therefore, reading the evaporator pressure at the inlet of the coil. Since the refrigerant is saturated and follows a pressure/temperature relationship, as the pressure of the refrigerant drops, the boiling temperature of the refrigerant will also drop. Assume that the last of the refrigerant boils to a vapor at 32°F (57.5 psig). At this point, the refrigerant will start picking up superheat. If the valve is to reach equilibrium, assuming the same spring pressure of 15.5 psig, the bulb pressure will have to reach 84 psig (68.5 psig + 15.5 psig), which corresponds to a 50°F evaporator outlet temperature (Figure 1-42). This evaporator is now operating with 18 degrees of superheat, compared to the

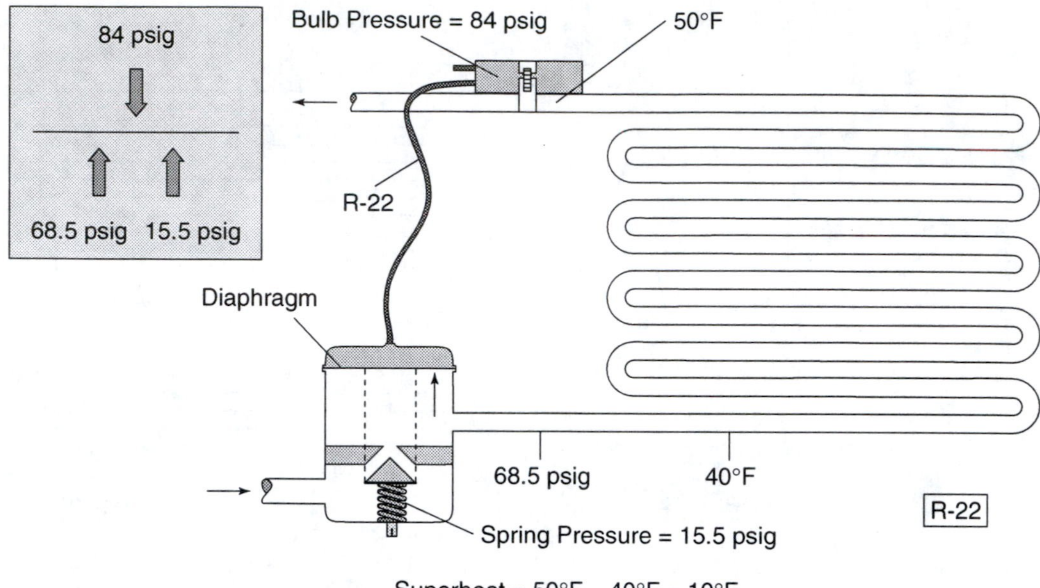

FIGURE 1-41 An ideal evaporator with no pressure drop across it. The inlet pressure and outlet pressure are both 68.5 psig

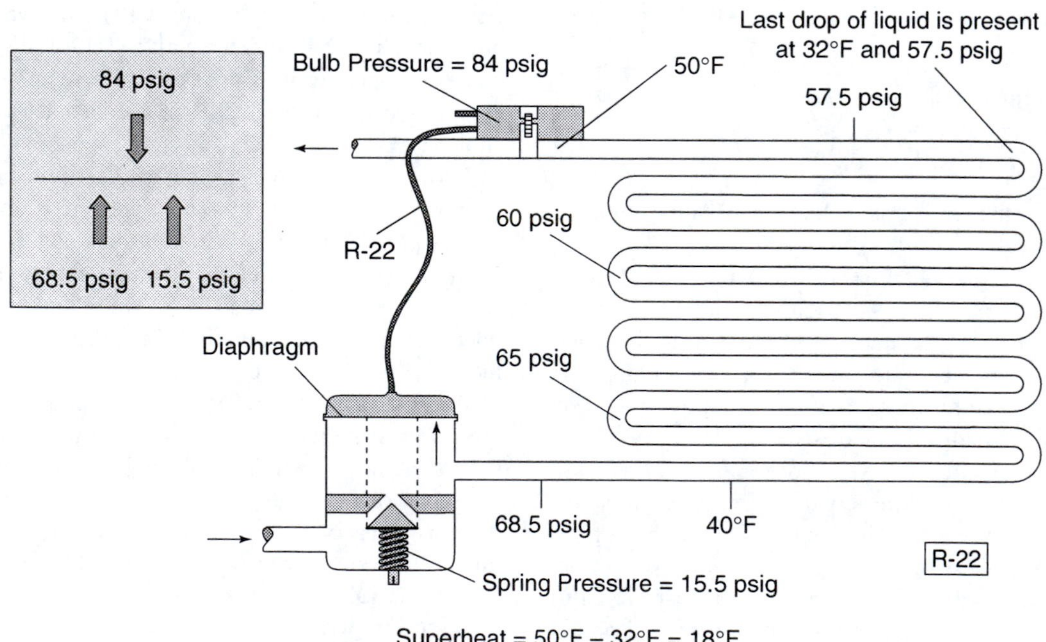

FIGURE 1-42 This evaporator has a large pressure drop across it. The evaporator must operate with excess superheat in order to reach equilibrium

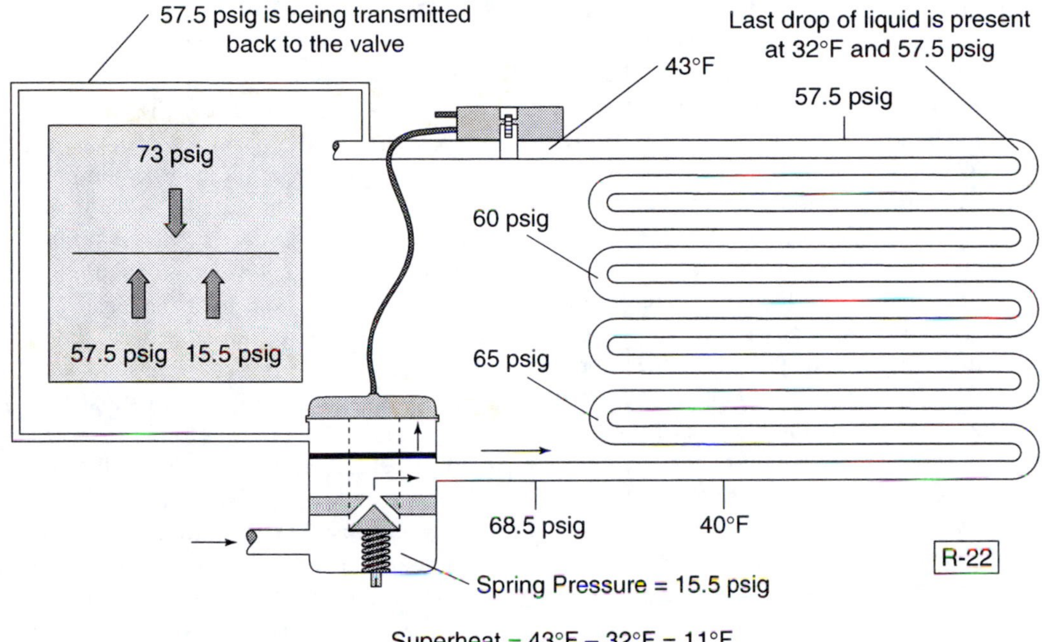

FIGURE 1-43 Using an externally equalized TEV on evaporators with large pressure drops helps to reduce the amount of superheat

10 degrees of superheat in the previous evaporator with no pressure drop. Now look at the same coil again, but this time assume that an externally equalized valve is being used.

Instead of measuring the pressure at the inlet of the evaporator coil, now measure the evaporator pressure at the outlet of the coil. This evaporator pressure is 57.5 psig. The spring pressure is still 15.5 psig, so the total closing pressure is 73 psig (57.5 psig + 15.5 psig). For the valve to reach equilibrium, the opening pressure will therefore have to be 73 psig. This pressure corresponds to about 43°F, so the temperature at the outlet of the evaporator coil is 43°F. The superheat in the evaporator will therefore be 11°F (43°F – 32°F), which will allow the evaporator to operate more efficiently (Figure 1-43).

You can see from the preceding examples that if the pressure drop across the evaporator coil is large, an externally equalized thermostatic expansion valve should be used, since it compensates for the pressure drop through the coil. When the pressure drop across the coil is small, either an internally equalized or an externally equalized valve can be used. The internally equalized valve is usually preferred when the pressure drop is small, less than 2 psig, because there are two less piping connections. This reduces the possibility of future refrigerant leaks.

THERMAL BULB CHARGE. So far, some specific examples regarding the basic operation of the thermostatic expansion valve have been illustrated and discussed. Next, take a closer look at the thermal bulb and the fluid contained inside it, referred to as the *charge.* There are four different types of charge that can be contained within the bulb; each is addressed separately. The four charges are:

- Liquid charge
- Vapor charge
- Crossed-liquid charge
- Crossed-vapor charge

The Liquid-Charged Thermal Bulb. The liquid-charged thermal bulb most closely resembles the bulb that has been referred to in the previous examples. As the name implies, liquid is always present inside the bulb and the refrigerant inside it always follows the corresponding pressure/temperature relationship. As the temperature of the bulb increases, the liquid inside the bulb begins to boil to a vapor but some liquid is always present. The refrigerant in the bulb will always be saturated, even though the *amount* of liquid inside the bulb may change. If the outlet of the evaporator becomes very hot, the pressure in the bulb will rise according to the pressure/temperature relationship. This could potentially damage the valve, so the desired application should be evaluated carefully before a valve is selected. The refrigerant in a liquid-charged bulb will also match the refrigerant that is in the refrigeration system. In other words, if the refrigeration system is operating on R-22, then the refrigerant inside the thermal bulb assembly will also be R-22.

The Vapor-Charged Thermal Bulb. The vapor-charged thermal bulb, also referred to as a critically charged bulb, contains the same refrigerant as the system, just like the liquid-charged bulb. The vapor-charged bulb has only a small amount of liquid refrigerant contained in it. As the temperature of the bulb increases, the liquid boils off into a vapor. At some predetermined temperature, all of the liquid turns into a vapor. When the bulb contains only vapor refrigerant, it will no longer follow the pressure/temperature relationship since only saturated refrigerants follow the relationship. Once all of the liquid has vaporized, any additional increases in bulb temperature will have very little effect on the bulb pressure. This will limit the amount of opening pressure and help protect the valve in the event that the evaporator outlet gets too hot. In the case of the heat-pump system, this type of bulb charge is very popular since the bulb will get quite hot during the heating mode.

The Crossed-Liquid- and Crossed-Vapor-Charged Thermal Bulbs. The crossed-liquid- and crossed-vapor-charged bulb have properties similar to the liquid- and vapor-charged bulbs with one major exception. *The refrigerant contained in the thermal bulb assembly is different from the refrigerant contained in the refrigeration system.* These bulbs therefore, do not follow the pressure/temperature relationship of the system. Thermostatic expansion valves with crossed-charge thermal bulbs are used for special applications and are designed and specified accordingly.

EVAPORATORS

The **evaporator** is the system component that is responsible for performing the actual *cooling* or *refrigerating* of the occupied area or product. The evaporator, while reducing the temperature of a given space, does not actually make the space cold. *Cold* can be defined as the absence of heat. Therefore, to cool down a room or some other space, you must not add cold to it but, instead, you must remove heat from it. The purpose of the evaporator, therefore, is to absorb heat from the space that is to be cooled (Figure 1-44) and to satisfy the load on the system.

Evaporators can be used to cool down an occupied space and provide air conditioning or "comfort cooling" by absorbing heat from the air in that space. Evaporators can also be used to absorb heat from liquids such as water. Water fountains and water coolers utilize refrigeration systems to cool the water. The water comes in contact with the cool evaporator surface in order to reduce the water temperature. Cooled water can also be used as a secondary refrigerant to provide comfort cooling by circulating the water through coils located in remote locations. Blowers or fans are then used to move air through these coils. This type of system is called a chilled-water system.

Consider the following. You have just returned from the supermarket with your packages. Among other things, you purchased a bottle of juice. The juice is at room temperature, and you are very thirsty. Your next action would be to get a glass, fill it with ice, and then pour the juice into the glass. You now have a glass of cold juice. But how did the juice get cold? You may answer that the ice made the juice cold, but, actually, this is not true! The juice really makes the ice warm! Heat normally travels from a warmer substance to a cooler substance. In this instance, the heat travels from the juice to the ice, removing heat from the juice. This is why the ice melts so fast. The operation of the evaporator is basically the same. One major difference is that, once the ice melts, there is no more "cooling." An evaporator cools continuously because there is a constant flow of refrigerant (from the expansion device) through the coil that does not melt or deteriorate.

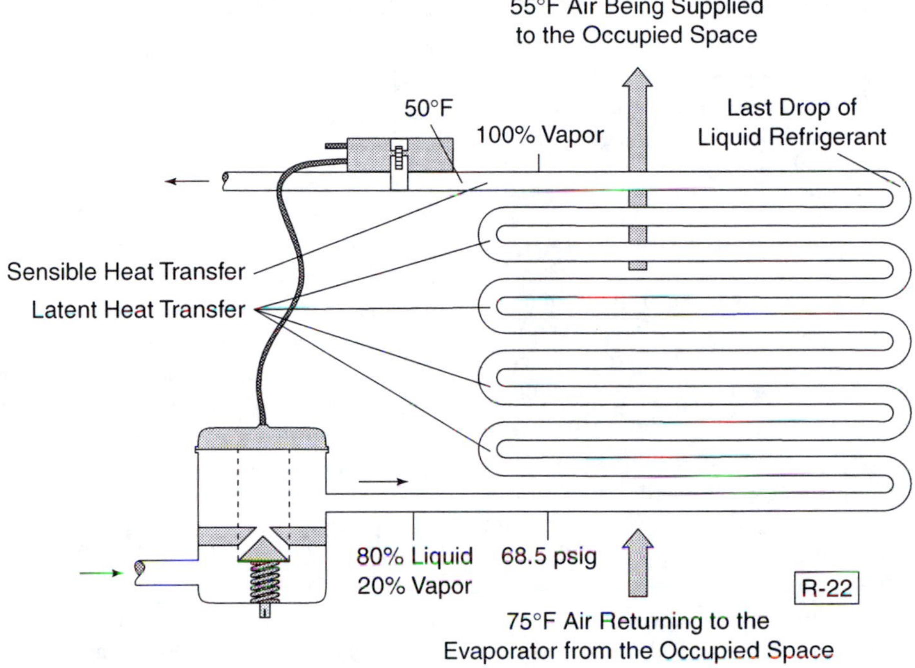

FIGURE 1-44 Evaporator at design conditions for an R-22 air-conditioning application

Since heat normally travels from a warmer substance to a cooler substance, the evaporator must be cooler than the medium to be cooled, usually air or water, in order to absorb heat from it. The evaporator is also responsible for **dehumidifying**. The evaporator operates at a temperature below the dew-point temperature, so it causes moisture in the air—humidity—to condense out of the air. This principle is the same as when you take a glass of ice water and set it on the table. After a period of time, a water ring appears on the table. This water did not come from inside the glass. It comes from the air surrounding the glass. When the air comes in contact with the cool glass, the moisture in the air condenses and accumulates on the table.

The refrigerant in the evaporator is able to maintain a low temperature because its pressure is greatly reduced by the combined effects of the compressor and the expansion device. Remember, the refrigerant follows a pressure/temperature relationship that determines a refrigerant's temperature at a given pressure as long as the refrigerant is saturated. When the refrigerant enters the evaporator, it is saturated and some of the liquid refrigerant leaving the expansion device flashes into a vapor upon leaving the device. The refrigerant entering the evaporator is approximately 80 percent liquid and 20 percent vapor.

As the refrigerant flows through the evaporator, heat is transferred from the air or water passing over the coil to the refrigerant flowing inside the coil. As the heat is absorbed by the refrigerant, more and more of the liquid refrigerant boils, or evaporates, into a vapor. This heat transfer is latent since the temperature of the refrigerant is not changing, but the state of the refrigerant is changing from a liquid to a vapor. This process will continue until all of the liquid refrigerant boils into a vapor. After all of the refrigerant boils into a vapor, it begins to superheat. Nevertheless, it will continue to absorb heat. The refrigerant superheats because it is now 100 percent vapor and is still much cooler than the medium passing over the coil so it continues to absorb heat. Figure 1-44 illustrates a typical evaporator on an air-conditioning system operating with R-22 as its refrigerant. Air-conditioning systems operating with different refrigerants can easily be adapted by converting the temperatures to the correct pressures using a pressure/temperature chart.

NOTE: *If there is no load on the system, it is possible that the refrigerant will remain saturated. In this case, the refrigerant will not be superheated.*

Flooded and Dry-type Evaporators

There are many different styles and configurations of evaporators, depending on the specific application. Evaporators can be either **flooded** or **dry-type**. Flooded evaporators operate with no superheat, which means that the refrigerant that leaves the evaporator coil is saturated. Some provision must exist to ensure that no liquid refrigerant enters the compressor. These evaporators utilize special metering devices known as float controls and are used to maximize the efficiency of the evaporator.

Dry-type evaporators, also known as direct-expansion evaporators, are designed so that, under proper operating conditions, all of the refrigerant will boil off into a vapor before leaving the coil. This prevents liquid refrigerant from entering the compressor, protecting it from damage. Dry-type evaporators, therefore, operate with a certain amount of superheat, which is determined by the heat load on the coil as well as by the type of expansion device used. Dry-type evaporators are also configured differently based on their application. Some configurations of dry-type evaporators are stamped plate, bare tube, finned tube, natural draft, and mechanical draft.

Stamped-Plate Evaporators

Stamped-plate evaporators are constructed of two pieces of metal that are "stamped" with channels through which the refrigerant flows. These two plates are mirror images of each other and, when they are secured together, form the refrigerant path (Figure 1-45). This type of evaporator is commonly found on domestic refrigerators. The stamped-plate evaporator is desired for domestic applications because its surface is easily cleaned, provides a smooth shelf for food storage, and is relatively safe compared to other evaporator coils with respect to sharp edges that can cause injuries. Stamped-plate evaporators also have large heat-transfer surfaces to help increase operating efficiency.

Bare-Tube Evaporators

Bare-tube evaporators were one of the first type of evaporators used. These coils were constructed of long runs of pipe that had refrigerant flowing through them. This type of evaporator was commonly used in large coolers, and the systems required large amounts of refrigerant. For this reason, they are not used as often on newer systems. Bare-tube evaporators are **natural draft** and rely on the fact that colder air falls and warmer air rises. In operation, the air that comes in contact with the coil is cooled and therefore falls. This pushes the warm air upward so that it can come in contact with the coil. These evaporators are large and cumbersome, so they are normally mounted high off the floor so that valuable storage space is not lost.

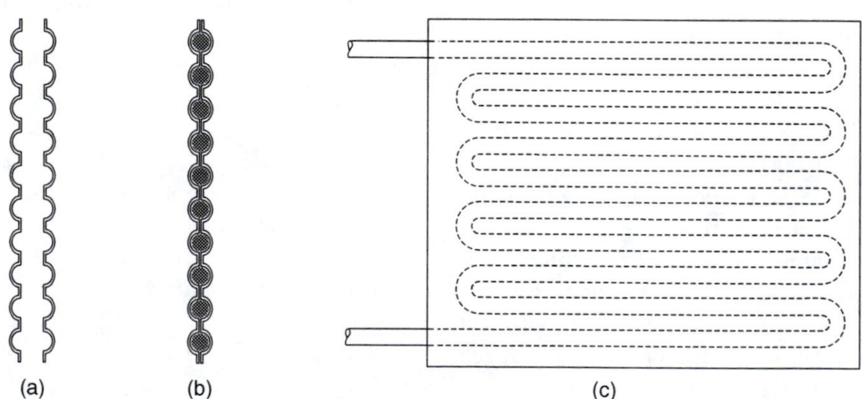

FIGURE 1-45 A stamped-plate evaporator. (a) Two mirror-imaged, impressioned plates. (b) Plates secured to each other, creating a refrigerant path. (c) Pattern of flow

Natural-Draft Evaporators

As just mentioned, natural-draft evaporators rely on the fact that colder air falls and warmer air rises. Natural-draft evaporators are most often located high in the conditioned space, and natural air currents move air across its surface. As air comes in contact with the evaporator, the air is cooled and falls toward the floor. This pushes the warmer air up toward the evaporator where it is cooled as well. Natural-draft evaporators are generally much larger than fan-assisted coils because the velocity of the air moving across the coil is very low. The larger physical coil size helps compensate for the reduction in velocity. Natural-draft evaporators are typically maintenance free, because there are no electrical components—namely, fans—to service.

Mechanical-Draft Evaporators

Mechanical-draft evaporators can be either **induced draft** or **forced draft**. Both induced-draft and forced-draft evaporators utilize fans or blowers to move the air through and across the coil (Figure 1-46). Induced-draft evaporators have the air pulled through the coil while forced-draft coils have the air pushed through the coil. The use of blowers and fans increases the amount of air moving over the coil, so the physical size of a mechanical-draft evaporator can be smaller than a natural-draft evaporator of the same capacity.

Evaporator Pressure Drop

As evaporators get larger, the amount of piping and the number of tubing bends within the coil increase. This causes the pressure drop through the evaporator to increase. If the pressure of the refrigerant entering the evaporator coil is, for example, 68 psig and the coil has many bends, the pressure of the refrigerant may be 58 psig at the outlet due to the friction in the coil. This condition is not desirable and is avoided by the use of a **multicircuit evaporator** (Figure 1-47). A multicircuit evaporator is made up of two or more parallel circuits instead of one large circuit. This type of evaporator uses a **refrigerant distributor** to ensure that the refrigerant is distributed evenly between all of the individual evaporator circuits. (Refer to the section on refrigerant distributors in Chapter 2 for more details.) Consider the following example which illustrates the concept of a multicircuit coil. An evaporator is manufactured as a single circuit. The refrigerant must pass through 200 tubes in the evaporator, and the pressure drop across the evaporator is large. Now consider the alternative. The same evaporator is divided into two, 100-tube coils. The refrigerant entering the coils splits, and one half of the refrigerant flows through each of the coils. The refrigerant must now only flow through 100 tubes. The amount of refrigerant flowing through each coil is cut in half, and the velocity of the refrigerant is reduced. The pressure drop through this evaporator has been greatly reduced.

FIGURE 1-46 Forced-draft evaporator. This coil has air pushed through it (*Courtesy Bally Case and Cooler, Inc.*)

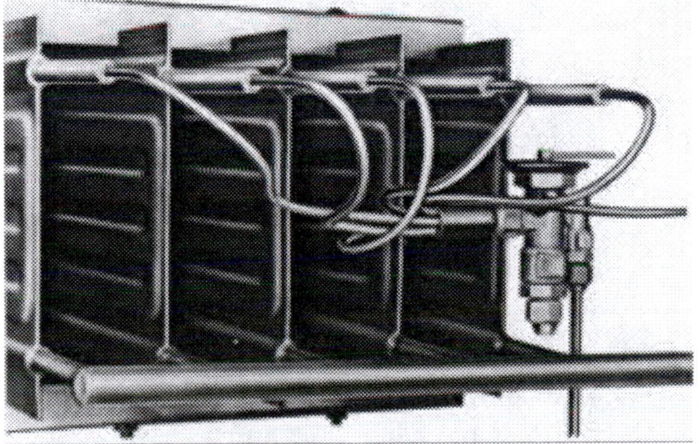

FIGURE 1-47 A multicircuit, stamped-plate evaporator (*Courtesy Sporlan Valve Company*)

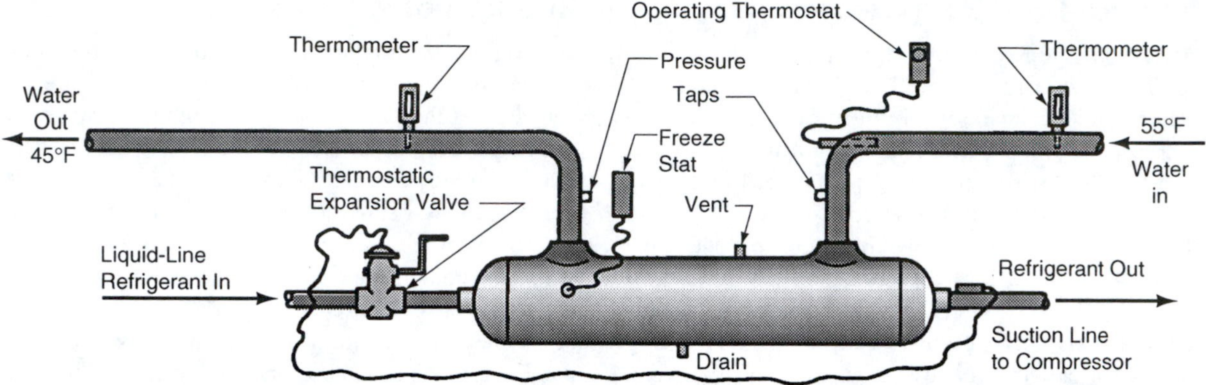

FIGURE 1-48 Evaporator used to cool water. The 45°F water is circulated through remote coils to provide cooling. Note that the water temperature is nearly the same as a 40°F air-conditioning coil operating with a refrigerant such as R-22

One other method used by design engineers to reduce the effect of large pressure drops in the evaporator is to use externally equalized thermostatic expansion valves. Externally equalized valves compensate for the pressure drop by sensing the pressure at the outlet of the coil, allowing the evaporator to operate with more saturated refrigerant. This increases the efficiency of the coil. Externally equalized valves sense the actual evaporator pressure. Refer to the section on thermostatic expansion valves for more on this topic.

EVAPORATORS USED FOR LIQUID COOLING. One special type of evaporator is designed to cool liquids. Small evaporators used to cool the water in drinking fountains can be designed as a metal storage tank that has the refrigerant piping running around the outside. This creates a heat-exchange surface that resembles a stamped-plate evaporator. Some liquid-cooling evaporators are used to provide comfort cooling. These special evaporators are referred to as **chiller barrels**. The chiller barrels can be configured the same as a tube-in-tube or shell-and-coil condenser. Water used for comfort cooling is normally cooled down to 45°F before it is circulated. This circulated water is referred to as a secondary refrigerant since the primary refrigerant is used to cool the water, which, in turn, is circulated to the area to be cooled. This 45°F water functions in a manner similar to the 40°F coil that is desired on standard air-conditioning systems (Figure 1-48).

Once the refrigerant leaves the evaporator, it may pass directly to the compressor or it may pass to other system components depending on the specific system application. These components are discussed later on in the text. For the time being, assume that the superheated refrigerant that leaves the evaporator flows directly back to the compressor. This refrigerant is 100 percent vapor (low temperature, low pressure) and is now ready to start the compression process again. The devices that are discussed in the next chapter are designed to control the operation of the system's major components. These devices determine the start and stop time of the components to ensure that the system's proper operating parameters are maintained.

SUMMARY

- The vapor-compression refrigeration cycle is a repeating cycle, consisting of a compressor, a condenser, a metering device, and an evaporator.
- The compressor changes the refrigerant from a low-temperature, low-pressure, superheated vapor to a high-temperature, high-pressure, superheated vapor.
- Three common types of compressors are the rotary, the reciprocating, and the scroll.

- The condenser rejects heat from the system and changes the refrigerant from a high-temperature, high-pressure vapor into a high-temperature, high-pressure liquid.
- Condensers can be air-cooled or water-cooled.
- Water-cooled condensers are more efficient than air-cooled condensers.
- Latent-heat transfers cannot be measured with a thermometer.
- When the temperature of a substance changes, a sensible-heat transfer takes place.
- The metering device controls the flow of refrigerant to the evaporator. It changes high-temperature, high-pressure liquid into a low-temperature, low-pressure liquid.
- Three common metering devices are the capillary tube, the automatic expansion valve, and the thermostatic expansion valve.
- The refrigerant in the evaporator absorbs heat into the refrigeration system.
- Evaporators can be dry-type, flooded, mechanical, or natural draft.

KEY TERMS

Automatic expansion valve
Bulb pressure
Capillary tube
Chiller barrels
Clearance vapor
Clearance volume
Closed system
Compression
Compressor
Compressor head
Condenser
Condenser saturation temperature
Condensing medium
Counterflow
Critical charge
Cylinder
Dehumidifying
Desuperheating
Discharge
Discharge line
Dry-type evaporator
Evaporator
Evaporator pressure
Expansion device
Externally equalized
Fins per inch
Fixed bore
Flash gas
Flooded evaporator
Forced draft
Head pressure
Induced draft
Intake
Internally equalized
Latent heat
Liquid line
Metering device
Modulating
Multicircuit evaporators
Natural draft
Parallel flow
Piston
psig
Receiver
Recirculating system
Reexpansion
Refrigerant distributor
Repeating cycle
Saturated
Scale
Sensible heat
Shell and coil
Shell and tube
Snap action
Spring pressure
Subcooling
Suction
Suction line
Superheat
Thermostatic expansion valve
Total-system charge
Tube-in-tube
Vapor compression
Wastewater system

FOR DISCUSSION

1. Referring to Figure 1-39, determine the total opening and closing pressures within the TEV. Explain how the valve reacts to these conditions and how it ultimately achieves equilibrium, if applicable.
2. Referring to Figure 1-40, determine the total opening and closing pressures within the TEV. Explain how the valve reacts to these conditions and how it ultimately achieves equilibrium, if applicable.
3. Discuss why and how clearance volume affects the operating efficiency of a reciprocating compressor.
4. Discuss why rotary compressors do not have a clearance volume.

REVIEW QUESTIONS

1. Describe a repeating cycle.
2. Ideally, refrigerant enters the compressor as a:
   a. low-temperature vapor.
   b. low-temperature liquid.
   c. high-temperature vapor.
   d. low-pressure liquid.
3. The piping between the compressor and the condenser is called the
   a. suction line.
   b. discharge line.
   c. expansion line.
   d. liquid line.
4. What causes the suction and discharge valves to open and close in a reciprocating compressor?
5. Water-cooled condensers are more efficient than air-cooled condensers. Why?
6. What three processes take place in the condenser?
7. Under normal operating conditions, the pipe leaving the condenser is always a:
   a. subcooled liquid.
   b. saturated liquid.
   c. superheated vapor.
   d. saturated vapor.
8. Explain the concept of counterflow.
9. How do the water and refrigerant flow through a tube-in-tube condenser?
   a. Water in the shell, hot gas in the tubes
   b. Water in the inner tube, hot gas in the outer tube
   c. Hot gas in the inner tube, water in the outer tube
   d. Low-pressure refrigerant in the outer tube, water in the inner tube
10. Refrigerant enters the expansion device as a:
    a. low-pressure liquid.
    b. high-pressure liquid.
    c. low-pressure vapor.
    d. high-pressure vapor.
11. List the three pressures that control the operation of a thermostatic expansion valve.
12. What does the automatic expansion valve maintain in an evaporator?
13. What does the thermostatic expansion valve maintain in an evaporator?
14. What are the two most important factors with respect to the size of a capillary tube?
15. Under what conditions will a thermostatic expansion valve be in the open position?

16. The refrigerant in the evaporator:
    a. absorbs heat from the air passing over it.
    b. removes moisture from the air passing over the coil.
    c. removes both latent heat and sensible heat from the air passing over the coil.
    d. All of the above are correct.
17. Multicircuit evaporators:
    a. usually have a low pressure drop across them.
    b. use a refrigerant distributor.
    c. Both *a* and *b* are correct.
    d. None of the above are correct.
18. Mechanical-draft evaporators:
    a. can be forced or induced draft
    b. use at least one fan to move air across the coil
    c. move larger amounts of air than a natural-draft evaporator
    d. All of the above are correct.
19. The evaporator saturation temperature for an R-22 evaporator operating at 68.5 psig is approximately
    a. 68.5°F.
    b. 40°F.
    c. 22°F.
    d. 40°C.
20. If the thermal bulb on a TEV comes loose from its mounting at the outlet of the evaporator, which of the following will occur?
    a. The valve will open, feeding more refrigerant to the evaporator
    b. The valve position will remain unchanged, and the pressure in the bulb will remain the same
    c. The valve will close, starving the evaporator
    d. The suction pressure will drop

CHAPTER 2

Vapor-Compression Refrigeration System Accessories and Controls

OBJECTIVES After studying this chapter, the reader should be able to:

- List four common positions of a service valve and explain when each is used.
- Explain why filter driers are used.
- Explain why refrigerant receivers are used.
- Describe the operation of a solenoid valve.
- Explain under what circumstances crankcase heat is needed.
- Describe the function and operation of a cooling tower.
- Describe the function of a thermostat.
- List several different types of thermostat.
- Explain how low voltage is obtained for use in control circuits.
- List several different types of pressure control.
- Explain why low ambient controls are needed for systems that operate year-round.
- Explain the differences between a relay and a contactor.

INTRODUCTION

The previous chapter discussed, in detail, the four integral components that make up a vapor-compression refrigeration system. These components, together with the interconnecting piping, can produce a refrigeration effect without the introduction of any other devices to the refrigeration circuit. This type of system, however, would have to operate within a very narrow range of parameters and would, without a doubt, operate very inefficiently. To provide air conditioning or refrigeration in the most economic and reliable manner possible, **accessories** and **controls** are added to the basic system.

The term *accessories* refers to components that are not necessary to provide refrigeration but enhance the operation of the basic system and allow it to operate more efficiently under a wider range of conditions. *Controls* are devices that "tell" the system and/or individual components when to operate and when not to, depending on system requirements, safety considerations, and the like. They are, simply stated, switches that either allow or do not allow electrical current to flow to the device. The light switch on the wall can be considered a control since the occupant of the room must flip the switch to turn the light on. The individual is "telling" the light to turn on. Controls tell which components to turn on and off at the correct time. After completing this chapter, the reader will have the knowledge needed to begin the study of more complicated refrigeration systems.

OVERVIEW

Remember that the vapor-compression refrigeration system is a closed-loop system that has the ability to continually repeat its cycle without depleting the refrigerant contained within it. To recap, this is when, at the completion of one cycle, the system is in exactly the same configuration it was at the start and is now ready to repeat itself. This process will continue as long as the system is operating. As illustrated earlier, the key to a continuous repeating cycle is that whatever is "done" in the system must later be "undone," eventually finishing in the same place it started. If pressure is increased, it must be decreased. If heat is absorbed, it must be rejected. The following statements summarize the basic refrigeration cycle:

- The compressor increases the pressure and temperature of the refrigerant while the expansion device causes the pressure and temperature of the refrigerant to drop.

- The condenser changes the refrigerant from a vapor to a liquid while the evaporator changes the refrigerant from a liquid into a vapor.
- The condenser rejects heat while the evaporator absorbs heat.

The reason for adding accessories and controls to the system is not to alter but, instead, to ensure that the system components operate as expected, even when exposed to changing external factors and conditions. If these external conditions (heat load, outside air temperature, humidity, etc.) remained relatively constant, there would be less of a need for these devices.

An individual who enjoys ice cream will purchase the product throughout the year. The quality of the product is expected to be the same no matter when the product is bought. The refrigeration equipment must also then produce the same result year-round. How is it then that air-cooled refrigeration equipment is able to operate properly in the dead of winter as well as in the heat of summer when such drastic changes occur in the outside ambient temperature? Systems that are required to operate year-round are equipped with devices called **low ambient controls** that help to simulate "ideal" design conditions. In geographic locations where extreme temperature swings are not an issue, low ambient controls are generally not needed. Accessories that permit air-conditioning and refrigeration systems to operate safely and efficiently under a wide range of conditions include:

- Low ambient controls
- Refrigerant receivers
- Crankcase heaters

Other types of accessories are designed to help make the servicing of the equipment quicker and easier. Any device that may help speed the troubleshooting and service time required on any given job will benefit both the service company and the customer. Although the initial installation of these devices may be expensive, they tend to pay for themselves in a very short time. These devices include:

- Service valves
- Sight glasses
- Permanently mounted gauges
- Permanently mounted thermometers

It must be determined if it is economically feasible to install these devices on a system. For example, small domestic refrigerators and air conditioners are hermetically sealed systems, designed to provide years of uninterrupted service. The installation of any of the aforementioned devices would therefore be unnecessary. On larger systems, however, where system pressures and temperatures are constantly monitored by building maintenance staff, the installation of these devices becomes necessary.

Still other components are designed to help protect the refrigerant and the technician. Some devices automatically de-energize a system if the operating pressures reach unsafe levels. Some devices help to remove moisture, acid, and particulate matter from the refrigerant as it passes through the system. Still other components tell the system when to turn on and off to maintain proper space temperature. Other devices play an integral role in the overall operation of a piece of equipment. A cooling tower, for example, is not a component part of the air-conditioning or refrigeration system, but it supplies the much needed condensing medium for water-cooled condensers.

Just as understanding the basic refrigeration cycle and its components is important, equally important is having a good grasp of the information contained in this chapter. Accessories and controls will be encountered on every piece of air-conditioning or refrigeration equipment that is worked on. All systems, no matter how small, will be equipped with at least one accessory or control. As the systems get larger, they tend to employ more accessories. The primary goal is to make the overall service, maintenance, and operation of equipment easier, more economical, and more reliable.

SERVICE VALVES

For the field technician to effectively troubleshoot and evaluate a particular system, obtaining the system's operating pressures may be necessary. If the system is hermetically sealed, this task becomes an extremely difficult one since there is no access to the refrigerant circuit. On larger systems, obtaining the needed pressure readings is very easy since they are normally equipped with **service valves** (Figure 2-1).

Service valves are located at strategic points throughout the refrigeration system and enable the technician to perform a number of tasks:

- Taking accurate system pressure readings
- Adding or removing refrigerant from the system
- Properly evacuating the system
- Pumping down the system in preparation for repair

Accurate pressure readings, taken from the service valves, are necessary to determine if the system is operating correctly. The system pressures relate directly to the saturation temperatures, both evaporator and condenser, of the refrigerant and provide useful troubleshooting data. It would be impossible, for example, to determine the evaporator superheat or the condenser subcooling without obtaining the system's operating pressures.

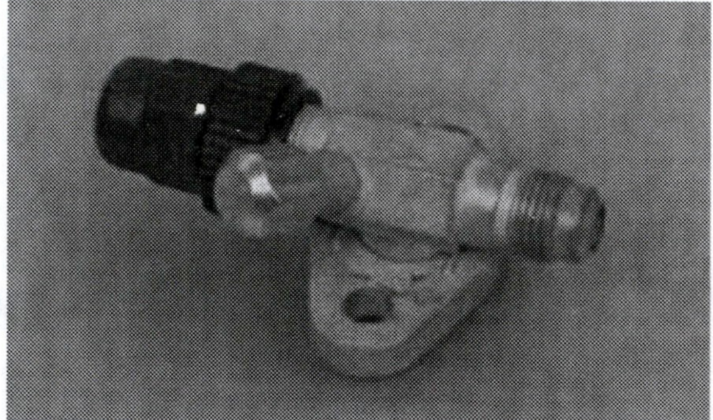

FIGURE 2-1 Typical service valve. The cap on the left of the valve covers the stem that is used to adjust the valve's position *(Photo by Bill Johnson)*

Since the service valves provide access to the actual refrigerant circuit, the technician is able to add or remove refrigerant from the system to correct the refrigerant charge. **System evacuation**, which is the removal of air and moisture from the system after a repair or prior to charging, is also accomplished through the service valves. **System pump-down**, the process of storing the refrigerant in the high side of the system while a repair is made on the low side, would not be possible without these valves.

Service valves are usually located on the suction and discharge ports of the compressor—called **suction service valves** and **discharge service valves**, respectively—and the receiver outlet. Service valves located on the receiver are often called **king valves**. The service valve has three ports and can be adjusted with a **refrigeration service wrench**, which is a ratchet-type wrench that is used to turn square valve stems (Figure 2-2). The three valve ports are the:

FIGURE 2-2 A service technician uses a refrigeration service wrench on a service valve *(Photo by Eugene Silberstein)*

- Service port
- Device port
- Line port

A cutaway view of a typical service valve is shown in Figure 2-3. The **service port** is a ¼-inch male flare fitting that is used to connect hoses from the gauge manifold. The **device port** is the connection to either the compressor or the receiver, which are referred to as the system "devices." The **line port** is connected to the refrigerant line, which either feeds into the device or carries refrigerant from the device. If the service valve is connected to the suction side of the compressor, the line port is connected to the suction line, which feeds refrigerant into the compressor. If the service valve is connected to the discharge of the compressor, the line port is connected to the discharge line, which carries the refrigerant from the

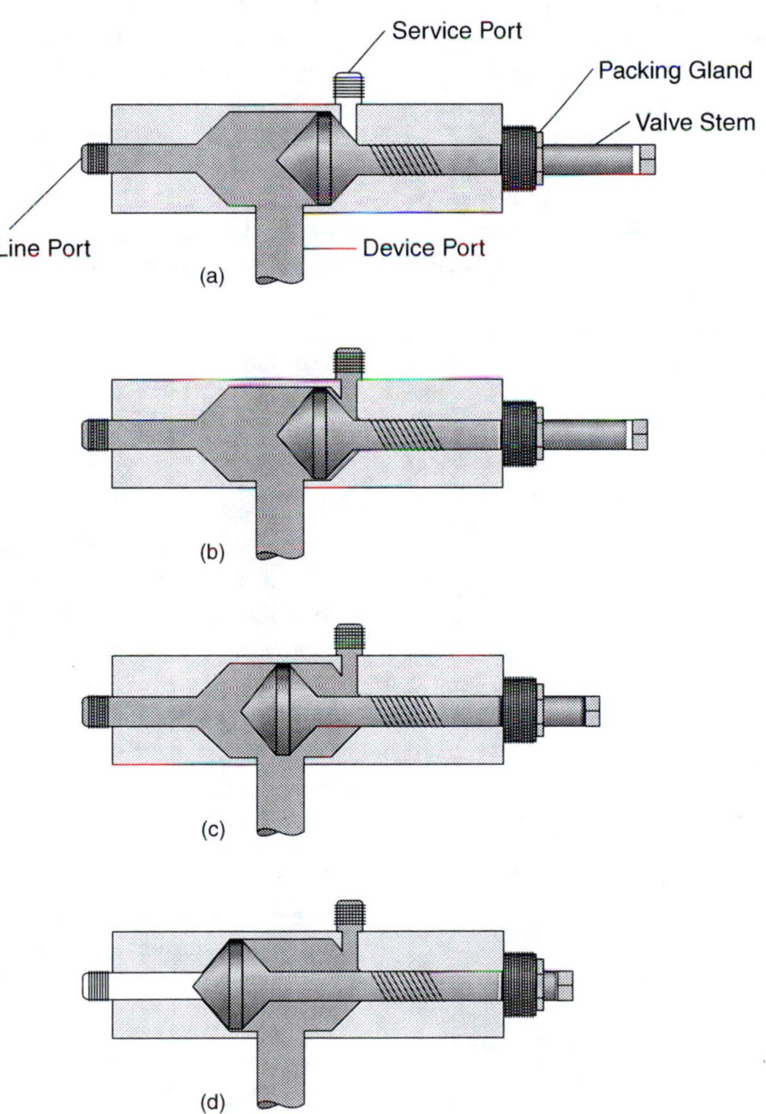

FIGURE 2-3 Cross-sectional view of a service valve. (a) Backseated position. (b) Cracked-off-the-backseat position. (c) Midseated position. (d) Frontseated position

compressor. The valve can be manually set to one of four positions:

- Backseated
- Cracked off the backseat
- Midseated
- Frontseated

Backseated is the normal operating position on the service valve. When looking directly at the valve stem, it should be turned all the way counterclockwise to be in the backseated position. All service valves should be in this position unless they are being serviced or have an external pressure control connected to the service port. The valve in Figure 2-3a is backseated. Notice that the service port is sealed off and the device port is open to the line port. This allows refrigerant to flow freely between the line and the device, without having refrigerant leak from the system through the service port. This is the correct position for the valve to be in when removing a set of gauges from the system.

The **cracked-off-the-backseat** position is shown in Figure 2-3b. This position is achieved by turning the valve stem approximately one-half to one turn clockwise from the backseated position. This position is used when the technician needs to take pressure readings from the system. Notice that refrigerant can flow freely between the device and the line but there is now pressure at the service port. Refrigerant can also be added to the system when the valve is in this position. *Note:* The gauge manifold, when connected to a valve that is cracked off the backseat, cannot be removed without allowing the system refrigerant to escape to the atmosphere.

The **midseated** position is shown in Figure 2-3c. A service valve is midseated when the stem is midway between the backseated and frontseated positions. This position is used primarily for system evacuation. This position has all ports open to each other to help ensure that the best possible vacuum is being pulled on the system.

SERVICE NOTE: *The gauge manifold, when connected to a valve that is midseated, cannot be removed without allowing the system refrigerant to escape to the atmosphere.*

The **frontseated** position is shown in Figure 2-3d. The valve is frontseated when the stem is turned completely clockwise when looking directly at the stem. In this position, the line port is sealed off and the device port is open to the service port. This position is used to pump a system down prior to system repair, to valve off a compressor, and to check the compressor valves. System pump-down is the method by which the system refrigerant is stored in the condenser and/or receiver during a repair on the low-pressure side of the system. Frontseating the king valve on the receiver will prevent the refrigerant from leaving the receiver, trapping it inside. Once all of the refrigerant is stored in the receiver, the service valve at the inlet of the compressor would also be frontseated, trapping the refrigerant within the compressor, condenser, and receiver.

SERVICE NOTE: *The gauge manifold, when connected to a valve that is frontseated, cannot be removed without allowing the system refrigerant to escape to the atmosphere, unless the system is completely pumped down.*

Safety Tip

Compressor discharge service valves should never be frontseated while the compressor is operating. Severe system damage, personal injury, or even death could result.

FILTER DRIERS

When refrigeration systems are initially installed and serviced, foreign matter and moisture have a tendency to enter the system. **Filter driers** (Figure 2-4) are installed in the system to help keep the refrigeration circuit and the refrigerant itself free from these undesirables, which can lead to acid formation and component damage. The driers are designed to:

- Remove moisture from the system
- Absorb system acid
- Filter small particulate matter

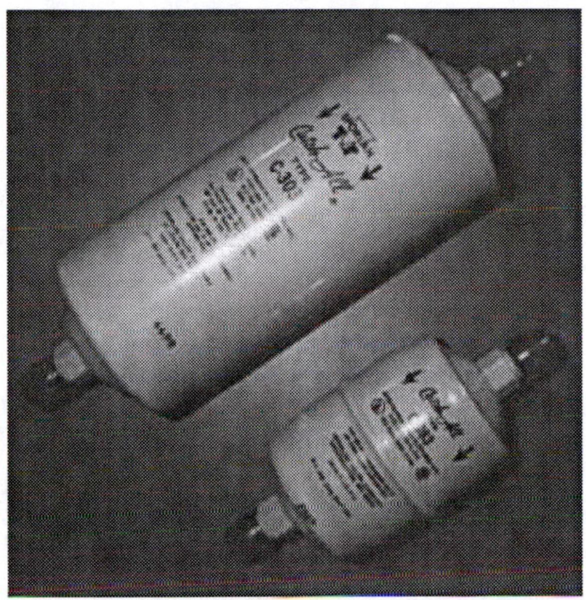

FIGURE 2-4 Typical liquid-line filter driers. This type of drier must be discarded and replaced if the desiccant becomes saturated. The arrows on the drier must point in the direction of refrigerant flow *(Photo by Eugene Silberstein)*

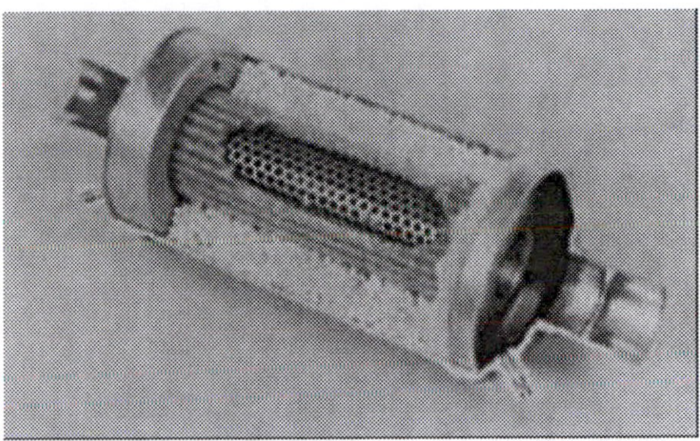

FIGURE 2-5 Desiccants are designed to remove moisture and acid from the refrigeration system. They can be in the form of solid blocks or pellets *(Courtesy Alco Controls)*

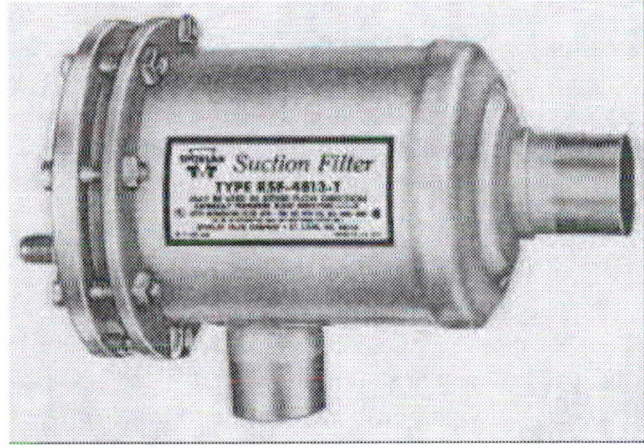

FIGURE 2-6 This suction-line filter drier has a removable core. The canister is permanently installed in the system *(Courtesy Sporlan Valve Company)*

The substance in the drier that is responsible for the actual removal of moisture and/or acid from the system is called a **desiccant** (Figure 2-5). Commonly used desiccants include activated alumina, calcium sulfate, and silica gel. Silica gel, for example, is the same substance that is packaged in small sugar-like envelopes found packaged with new clothing and other items to reduce the accumulation of moisture. Some filter driers use a combination of two different desiccants to increase their moisture-holding capacity.

Mesh screens within the drier accomplish the removal of small particulate matter. Particulate matter may include residual solder, flux, dust, or dirt. Filter driers that have reached their moisture- and particle-holding capacity are no longer effective and must be tended to. Smaller driers must be removed from the system and replaced, while larger, permanently installed, canister-type driers are manufactured in a manner that permits the desiccant, or core, to be replaced (Figure 2-6). These driers have bolted covers that can be removed to gain access to the desiccant. The system must be pumped down before attempting to remove the cover of the drier canister. Two types of filter driers, named for their location in the system, are the liquid-line drier and the suction-line drier.

Liquid-Line Filter Driers

The liquid-line filter drier is normally located in the liquid line after the receiver, if the system has one (Figure 2-4). As well as removing moisture and acid, this drier helps prevent foreign matter from entering and, potentially, clogging the expansion device. As

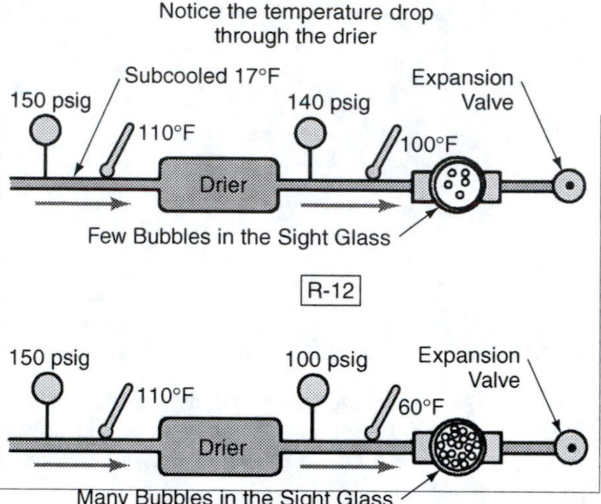

This drier has a noticeable temperature drop that can be felt. It may even sweat or freeze at the drier outlet.

FIGURE 2-7 The driers each have a restriction. One of them is very slight. Where there is pressure drop in a liquid line, there is temperature drop. If the temperature drop is very slight, it can be detected with a thermometer. Sometimes gauges are not easy to install on each side of a drier to check for pressure drop

the drier picks up more and more debris, it begins to reach its saturation point, at which it becomes ineffective. A clogged drier can impede normal refrigerant flow and affect system operation. A noticeable temperature drop across the drier is an indication that this is indeed the case (Figure 2-7). The drier should then be replaced. In an extreme case of clogging, the outlet of the drier may begin to frost, indicating a large restriction in the drier. In effect, the drier is acting as a metering device. One indication that the drier is becoming clogged is the appearance of vapor bubbles in the sight glass. Sight glasses are explained in further detail a little later on in this chapter.

Suction-Line Filter Driers

Located in the suction line, this type of drier is designed to remove moisture, acid, and particulate matter from the refrigerant before it enters the compressor, reducing the possibility of component damage. Suction-line driers are usually installed after a compressor motor burnout but can also be a part of a new system installation (Figure 2-6). Suction-line driers are physically larger than liquid-line driers and are manufactured with service ports at both the inlet and the outlet of the device. The purpose of these ports is to enable the service technician to read the pressure drop across the drier. An excessive pressure drop, usually over 3 psig, across the drier indicates that the drier is clogged and should either be replaced or have its desiccant replaced, depending on the type of drier used. A clogged suction-line filter drier will reduce system capacity. Check the manufacturer's specifications for actual maximum pressure drop values since they can vary from one manufacturer to the next.

SIGHT GLASSES

As discussed earlier, the service valve provides the field technician with access to the refrigeration circuit to monitor system pressures, adjust the refrigerant charge, or evacuate the system. Another component that helps the technician to properly evaluate an air-conditioning or refrigeration system is the *sight glass* (Figure 2-8). As the name implies, the sight glass enables the technician to look into the refrigeration piping and actually see the refrigerant flowing in the system. The sight glass is normally installed in the liquid line between the liquid-line drier and the expansion device. The expansion device operates more efficiently when 100 percent liquid enters it, and the sight glass enables the technician to determine if a full column of liquid is reaching the device.

The sight glass also helps the technician determine if the liquid-line drier is beginning to clog. Some liquid may flash into a vapor as it passes through the drier if there is a large enough pressure drop across it. If this is the case, vapor bubbles may appear in the sight glass.

NOTE: *The appearance of bubbles in the sight glass does not always indicate that the drier is clogged. Vapor bubbles may appear on systems that are operating properly depending on the heat load that they are experiencing.*

FIGURE 2-8 Sight glass used only to view the liquid refrigerant to be sure no vapor bubbles are in the liquid line *(Courtesy Henry Valve Company)*

SERVICE NOTE: *Checking system operating pressures and operating conditions will help the technician determine the cause for sight glass bubbles.*

Some sight glasses are also equipped with moisture indicators, which alert the technician if unsafe moisture levels exist. The indicator contains a chemical salt that changes color depending on the amount of water it senses. Moisture affects system operation and can also lead to the formation of acid and sludge, so the amount present should always be as low as possible. In addition, moisture in the system can result in system freeze-ups and copper plating of the compressor cylinder walls. Copper plating is a result of acid formation, which wears away copper from the refrigerant piping and deposits this copper on the warm interior surfaces of the compressor cylinders.

REFRIGERANT RECEIVERS

Refrigerant receivers are located in the liquid line at the outlet of the condenser and are simply storage tanks that hold refrigerant until needed by the system (Figure 2-9). Receivers are used on systems that have modulating-type metering devices such as automatic or thermostatic expansion valves. As the expansion valve closes, the refrigerant backs up into the receiver.

Receivers are not found on critically charged systems with capillary-tube or fixed-bore metering devices. The installation of a receiver on this type of system would reduce the force needed to push the refrigerant through the capillary tube, thereby reducing flow. The system would operate with a slightly higher than normal head pressure and a very low suction pressure.

The liquid refrigerant from the condenser feeds into the receiver and leaves through a service valve, usually located near the top of the device. This valve resembles the service valve discussed at the beginning of this chapter and is often referred to as a king valve. If this valve is frontseated while the compressor is operating, the system will pump down, storing all of the system refrigerant in the condenser and receiver. This is useful when a repair must be made on the low-pressure side of the system. For this reason, refrigerant receivers are designed to hold the entire system charge plus about 10 percent. Once the repair is complete, the part of the system that was opened to the atmosphere can be evacuated and the refrigerant can then be released back into the system. Systems that are not equipped with receivers normally have oversized condensers that can hold the system's refrigerant charge.

Receivers are usually manufactured with a dip tube that extends down toward the bottom of the canister to ensure that 100 percent liquid refrigerant is present in the liquid line (Figure 1-24).

SOLENOID VALVES

The previous section noted that an air-conditioning or refrigeration system can be pumped down by frontseating the king valve on the receiver. A refrigeration service wrench must be used to place the valve in the frontseated position. Some equipment is designed to pump itself down automatically when the cooling cycle is ending. To accomplish this, a valve must close to prevent refrigerant from leaving the receiver. It would be an economic disaster to keep a technician at the ready to frontseat and backseat the valve whenever pump-down is to begin and end. For this reason, an automatic valve, called a **solenoid valve**, is used (Figure 2-10).

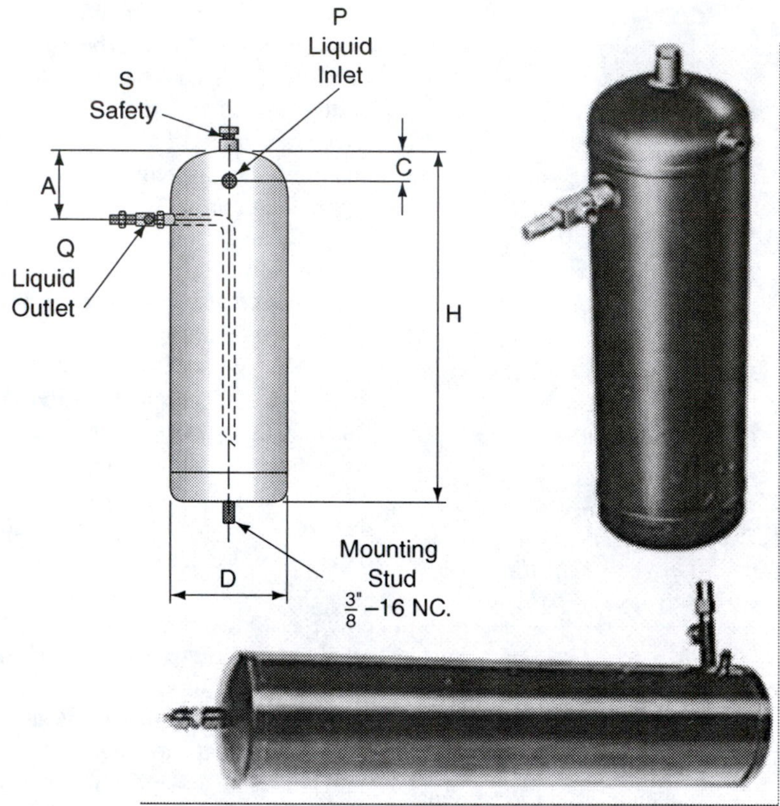

FIGURE 2-9 Vertical and horizontal receivers *(Courtesy Refrigeration Research)*

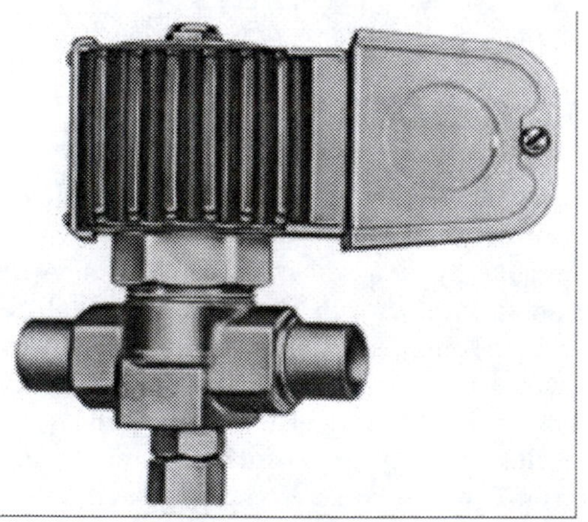

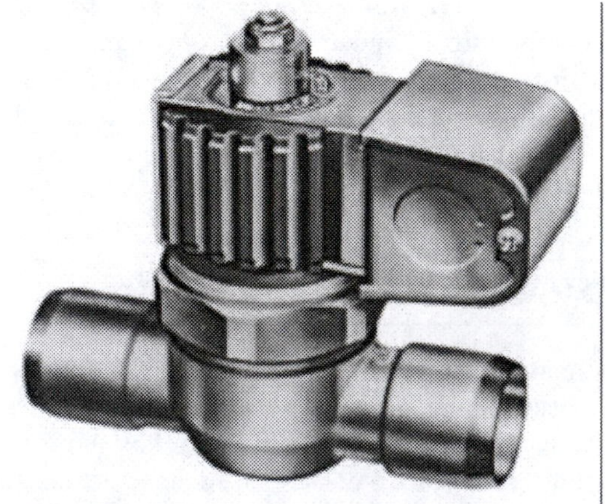

FIGURE 2-10 Electrically operated solenoid valves. The valve is moved by the plunger attached to the seat. The plunger moves into the magnetic coil when the coil is energized to either open or close the valve *(Courtesy Sporlan Valve Company)*

A **solenoid** is a coil of wire that generates a magnetic field when an electric current is passed through it. This magnetic field is what causes the valve to open and close in order to start or stop the flow of a fluid. The solenoid is referred to as an **electromagnetic device**, the operation of which is covered in greater detail in the next chapter.

Solenoid valves can be either normally open or normally closed. A normally closed valve is in the closed position whenever the coil is de-energized. When the coil is energized, the valve will open. On the other hand, the normally open valve is always in the open position when the coil is de-energized. It closes when the coil is energized.

The solenoid valve can be used to ensure that a full column of liquid enters the expansion device on systems in which the compressor is located above the evaporator. The solenoid in this case is installed in the liquid line near the expansion device. When the system is ready to shut down, the solenoid will close and the system will pump itself down. Liquid refrigerant will be trapped in the line upstream from the solenoid valve. When the system cycles back on, the solenoid valve will open and a full column of liquid will flow through the expansion device.

The solenoid valve can also be used to prevent refrigerant from backing up into the system evaporator during the off cycle. When a system is off, the refrigerant tends to migrate to the coolest portion of the system. During periods of warm ambient temperatures, the coolest location is the evaporator. The solenoid valve prevents this migration from occurring. If the refrigerant were able to migrate and accumulate in the evaporator, a large amount of refrigerant in both the liquid and vapor state would be rushing back to the compressor upon start-up, leading to potential compressor failure.

PRESSURE-RELIEF VALVES

It must be reiterated at this point that in this text the term *accessory* is being used to describe a component that is not required to provide a refrigeration effect. The term does not imply that the component can be removed from a system without sacrificing safety or system integrity. Many air-conditioning and refrigeration systems are equipped with safety devices that prevent them from operating if system pressures, temperatures, or other governing factors are not within design ranges. The **pressure-relief valve** is one such device (Figure 2-11). It is normally mounted on the refrigerant receiver and is a required component on larger systems. Local codes and authorities must be consulted within specific geographic locations to obtain proper guidelines regarding these devices.

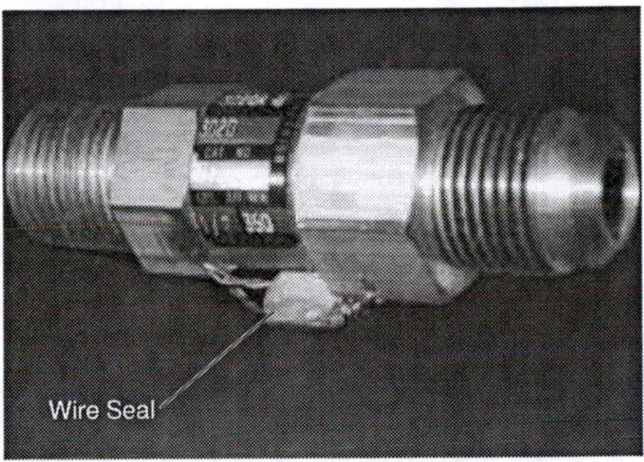

FIGURE 2-11 Spring-loaded relief valve. The flare fitting on the right side of the valve can be piped to an evacuated storage vessel. Notice the wire seal to prevent tampering *(Photo by Eugene Silberstein)*

The relief valve is designed to open and release refrigerant if dangerous and unsafe system pressure levels exist. These devices are intended to protect the equipment and, more importantly, the technician and those in close proximity to the system. Relief valves are designed to open at different pressures, so the proper valve must be chosen to ensure that the valve does not open prematurely. The three common types of relief valve are the:

- Rupture disc
- Fusible plug
- Spring-loaded valve

The rupture disc and the fusible plug are referred to as **one-time valves**. When these valve-sense unsafe pressure levels, they open and release the system refrigerant. Since the disc shatters or the plug is pushed out of the valve, these devices must be replaced after one use. Before replacing the valve, the cause for the excessive pressure must be found and corrected. If permitted to vent to the atmosphere, the release of a one-time relief valve will result in a total loss of refrigerant.

SERVICE NOTE: *New government guidelines regarding refrigerant releases, along with the increasing cost of refrigerant, are making the spring-loaded type of relief valve a popular alternative.*

The **spring-loaded valve** is a device that will reseat itself once the system pressure has dropped into a safe range. This valve allows the system to operate within a safe range until the problem is discovered and corrected. This valve is a permanent part of the system and does not need to be replaced after it releases pressure. Since this valve reseats itself when a lower pressure is sensed, less refrigerant is lost from the system.

Pressure-relief valves are manufactured with flare connections at the outlet so that the released refrigerant can be piped to a storage vessel. This prevents unnecessary and excessive refrigerant loss and also reduces the possibility of venting refrigerant to the atmosphere. If the storage vessel is properly evacuated, the refrigerant that is released can be reused.

Safety Tips

- The flare connection at the outlet of the valve should NEVER be capped. The ability of the valve to relieve pressure would be eliminated, and the system could explode!
- Relief valves have a wire seal (Figure 2-11) to prevent tampering. DO NOT remove or tamper with this seal.
- If you notice that the wire seal has been cut or removed, replace the relief valve immediately.

REFRIGERANT DISTRIBUTORS

The majority of the components discussed so far in this chapter have been **high-side** components. These components are intended to ensure the safe and efficient operation of the refrigeration system, as well as ensuring that a full column of liquid reaches the expansion device. Once the liquid reaches and passes through the expansion device, system efficiency can be increased further by using a multicircuit evaporator, discussed in Chapter 1. For this type of evaporator to operate effectively, the amount and *quality* of the refrigerant that flows through each circuit should be exactly the same. The refrigerant leaving the expansion device is a saturated liquid, and the term *quality* is used to describe the percentages of liquid and vapor that enter each circuit of the evaporator. The device that ensures that refrigerant is distributed equally, with respect to both quantity and quality, is called the **refrigerant distributor** (Figure 2-12). If the evaporator has two or more parallel circuits, a distributor will be located at the outlet of the expansion device.

Distributors are factory installed and are a component part of the evaporator. Refrigerant distributors are usually made out of brass and are connected to the outlet of the expansion device. Various types of distributors include the:

- Venturi type
- Centrifugal type
- Pressure-drop type
- Manifold type

The **venturi-type distributor** operates on the **venturi effect** and is popular since it can be installed in any direction and the pressure drop across it is small. A venturi is a constriction in a pipe or tube that is designed to increase the velocity of the refrigerant passing through it. This helps to reduce the pressure drop across the device. The **centrifugal-type distributor** operates best when the velocity of the refrigerant passing through it is high. A swirling effect is created to evenly divide the refrigerant to each evaporator circuit. The **pressure-drop-type distributor** utilizes interchangeable nozzles. The nozzle is designed to increase the velocity of the refrigerant in order to properly mix the liquid and vapor refrigerant to reduce the effects of gravity. The **manifold-type distributor** is commonly used, but it must be installed level to ensure proper refrigerant distribution. The velocity of the refrigerant in this type of distributor should be low.

CRANKCASE HEATERS

One of the major enemies of compressors, especially reciprocating compressors, is the introduction of liquid to the device. As was discussed earlier in the text, compressors are vapor pumps, since liquids cannot be compressed. This poses problems for systems that

FIGURE 2-12 Venturi-type refrigerant distributor mounted to the outlet of the TEV. The refrigerant flow to each evaporator circuit should be the same *(Photo by Eugene Silberstein)*

have the compressor located outdoors when the outside ambient temperature drops:

- Refrigerant mixes with the compressor oil very well at low temperatures.
- Compressor oil may foam when the compressor is started.
- Liquid refrigerant migrates to the coldest part of the system during the off cycle.

First, as the temperature drops, the ability of the compressor oil to mix with the refrigerant increases. As suction gas is pulled into the compression chamber, the chance of compressor damage is higher since the amount of liquid entering the compressor is greater.

Second, since the refrigerant and oil tend to mix, the reduction in pressure will cause the oil to foam. This foaming is a result of the rapidly boiling refrigerant at a lower pressure. The foam can then travel through the refrigeration system, reducing the amount of oil in the compressor, possibly causing component damage.

Third, liquid refrigerant tends to migrate to the coldest portion of the system during the off cycle. For systems that are designed to operate during periods of low temperatures, lower than the occupied-space temperature, the liquid refrigerant will migrate outdoors during the off cycle. If the compressor is energized, there will be an inrush of liquid refrigerant to the compression chamber and compressor damage is likely to occur. To reduce the possibility of introducing liquid refrigerant to the compressor, **crankcase heaters** are used (Figure 2-13). The purpose of the device is to heat the crankcase in order to boil the refrigerant from the oil.

These heaters are usually designed to be energized whenever the compressor is off to ensure that the refrigerant and oil are not mixed. Some systems are designed to have the crankcase heater energized all the time, even when the compressor is running. Having as few electrical controls as possible in the crankcase heater circuit is a good idea to ensure its continued operation. Two types of crankcase heaters are commonly used; they are the:

- Strap-on heater
- Insertion-type heater

Strap-on heaters are usually found on hermetic compressors and are externally mounted. This type of

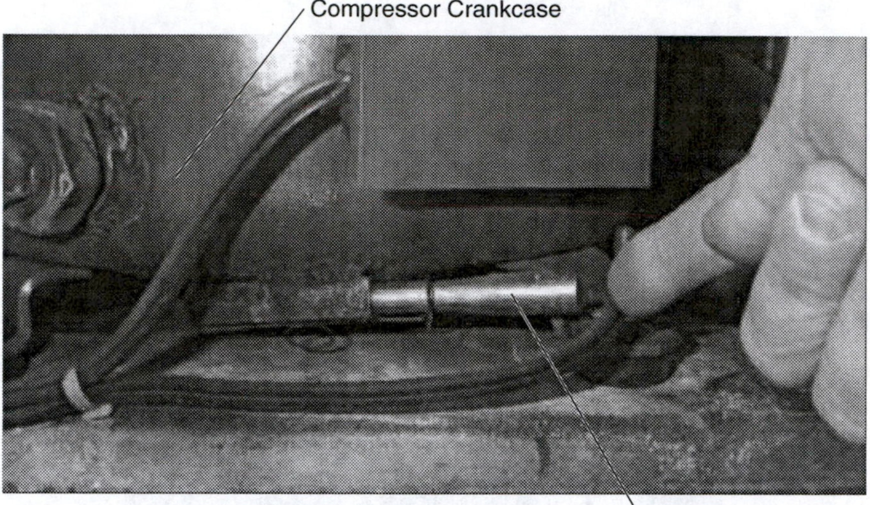

FIGURE 2-13 Insertion-type crankcase heater *(Photo by Eugene Silberstein)*

heater is often added to compressors when a low ambient problem surfaces and must be remedied. Replacement of defective strap-on heaters is very easy since the device is simply clamped around the lower portion of the compressor's shell.

Insertion-type heaters are found in larger compressors, usually semihermetic, and are manufactured with the compressor. This type of heater is typically more complicated to replace, since the refrigerant circuit may need to be accessed to do so.

COOLING TOWERS

Cooling towers are an integral part of a recirculating-water system. The function of the cooling tower is to cool down the water coming from the condenser so that it can be used again. The water is cooled mainly by evaporation. Water leaving the condenser is pumped to the tower where it flows over a series of decking, or fill, material that is designed to increase the surface area of the water (Figure 2-14). As air passes over the water, some of the water evaporates and helps to cool the remaining water. This is similar to the flash gas in the evaporator that helps to cool down the liquid refrigerant entering the coil.

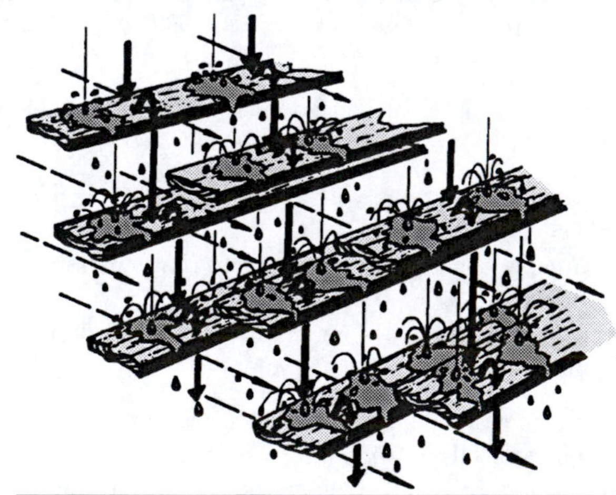

FIGURE 2-14 The water trickles down through the fill material *(Courtesy Marley Cooling Tower Company)*

Once the water is cooled, it accumulates in a reservoir or sump where it can be circulated back to the condenser. The reservoir holds a great deal of water, which also helps to dissipate and distribute any residual heat in the water. Generally, cooling towers can lower the temperature of returning water down to within 7 degrees of the **wet-bulb temperature** of the

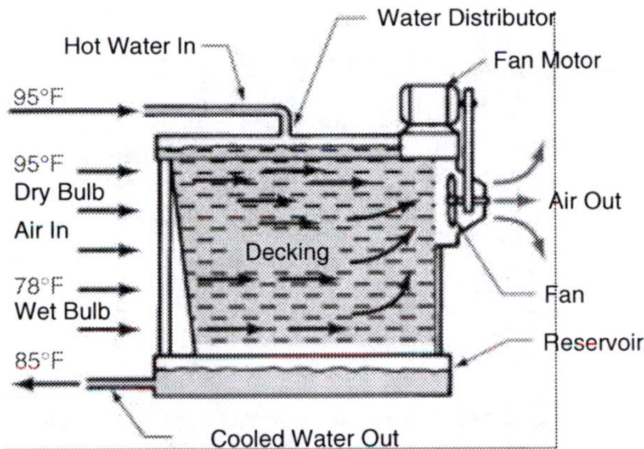

FIGURE 2-15 Relationship of a forced-draft cooling tower to the ambient air. Cooling tower performance depends on the wet-bulb temperature of the air. This relates to the humidity and the ability of the air to absorb moisture

outside air. This temperature is called the **approach temperature**. The wet-bulb temperature relates to the moisture content in the air, not necessarily the temperature that is read with a thermometer. It then follows that the lower the wet-bulb temperature, the more efficiently the cooling tower will operate. For this reason, cooling towers are more efficient in geographic areas with dry climates.

For example, if the outside temperature is 95°F (dry-bulb temperature) and the wet-bulb temperature is 78°F, the tower would be able to reduce the temperature of the water returning to it to about 85°F (Figure 2-15).

Several different types of cooling towers can be used. The most common are the:

- Natural-draft tower
- Mechanical, induced-draft tower
- Mechanical, forced-draft tower

The **natural-draft tower** relies on breezes and normal air movement to move air across the water flowing through it. The location for this type of tower must be chosen carefully. If it is placed where there is not enough natural airflow, the equipment that relies on it will not function properly. If the prevailing winds are strong, water could possibly be blown out of the tower. To reduce the chance of this occurrence, the tower is manufactured with inwardly pitched slats to direct the water back into the reservoir (Figure 2-16).

Because the amount of air flowing through the tower at any given point can vary considerably, the approach temperature tends to be higher than that of a mechanical-draft tower. Natural-draft towers can normally cool the return water to a temperature within 10 to 12 degrees of the wet-bulb temperature of the ambient air, compared to 7 degrees for a fan-assisted tower. This type of tower does not require the use of fans, belts, and, in most cases, electricity, so it tends to require less maintenance than a mechanical-draft tower.

The **mechanical-draft tower** utilizes fans or blowers to move the air through it. This tower allows the air-conditioning or refrigeration system to operate more efficiently because it moves more air across the water and can, therefore, operate with a lower approach temperature. The approach temperature for mechanical-draft towers, as stated earlier, is about 7 degrees. Mechanical-draft towers can be either forced or induced draft. **Forced-draft towers** are designed to have the air pushed over the water, while **induced-draft towers** have the air pulled over the water surface. A typical cooling-tower setup used for air conditioning is shown in Figure 2-17.

Cooling-tower water evaporates at a rapid rate, so **makeup water** must be provided to maintain proper fluid levels in the reservoir. *Makeup water* is supplied through a fresh-water line connected to the tower. The water level is controlled by a float valve, similar in operation to a toilet bowl tank, that feeds water into the reservoir when the level drops (Figure 2-16).

One problem with recirculating-water systems is the occurrence of scale buildup and mineral deposits. As the water evaporates, it leaves behind minerals. These mineral deposits form on the interior of the water circuit piping and reduce heat transfer between the water and the condensing refrigerant. Some of the problems that can result from excessive mineral deposits are:

- High head pressure
- Reduced system and cooling capacity
- Increased compressor discharge temperatures
- Increased power consumption
- Reduced condenser efficiency

To prevent this, the water should be chemically treated to reduce the rate of scale buildup. Another way to reduce the accumulation of solid matter is to

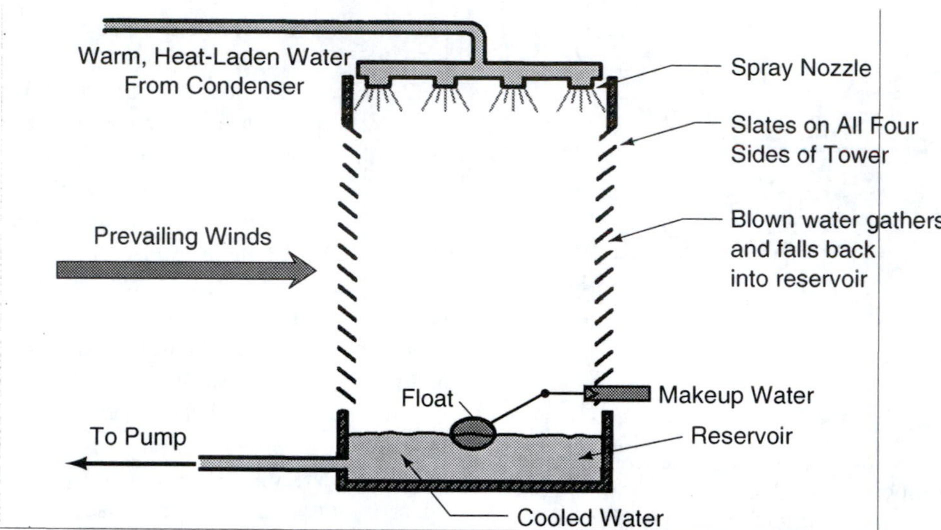

FIGURE 2-16 Slats on the sides of the natural-draft cooling tower keep the water inside the tower when the wind is blowing

utilize **blowdown**. Blowdown is the process of bleeding a small amount of water from the system on a continuous basis. This lowers the concentration of solid matter in the system and reduces the rate of buildup. The water that is removed from the system carries with it a portion of the minerals that would have otherwise ended up in the piping circuit. Blowdown must be implemented properly since both water and the expensive treatment chemicals are being disposed of. Too much blowdown will increase the amount of money spent on chemical treatment of the system. These chemicals must also be disposed of in accordance with all local regulations and in an environmentally safe manner.

SERVICE NOTE: *When performing cooling-tower maintenance, be sure to check the chemical treatment process. Properly treated water will reduce scale buildup and undesirable bacteria growth in the tower.*

SERVICE NOTE: *Checking the strainer on the cooling tower is probably the single most important service task when it comes to preventive maintenance. A clean strainer helps ensure proper water flow through the condensers. At a minimum, the strainer should be cleaned during the seasonal system start-up.*

SYSTEM CONTROLS

Thus far, a variety of devices that are utilized in various pieces of equipment to enhance the operation, efficiency, and reliability of the system have been covered. Now take a look at a number of different types of system controls whose purpose is to start and stop either a single component or the entire system, depending on the temperature, time, and/or safety requirements set forth by the equipment designers and operators.

Thermostats

Thermostats are temperature-sensing devices that are responsible for maintaining the space temperature at the desired level. They open and close electric circuits at the correct time to energize and de-energize air-conditioning and refrigeration system components. Commonly controlled system components are the:

- Compressor
- Condenser fan motor
- Evaporator fan motor
- Heating circuit

Thermostats can also be used as safety devices that can energize or de-energize system components if temperatures reach unsafe or undesirable levels. For example, if you do not want ice to form on an evaporator coil, you can place a thermostat on the

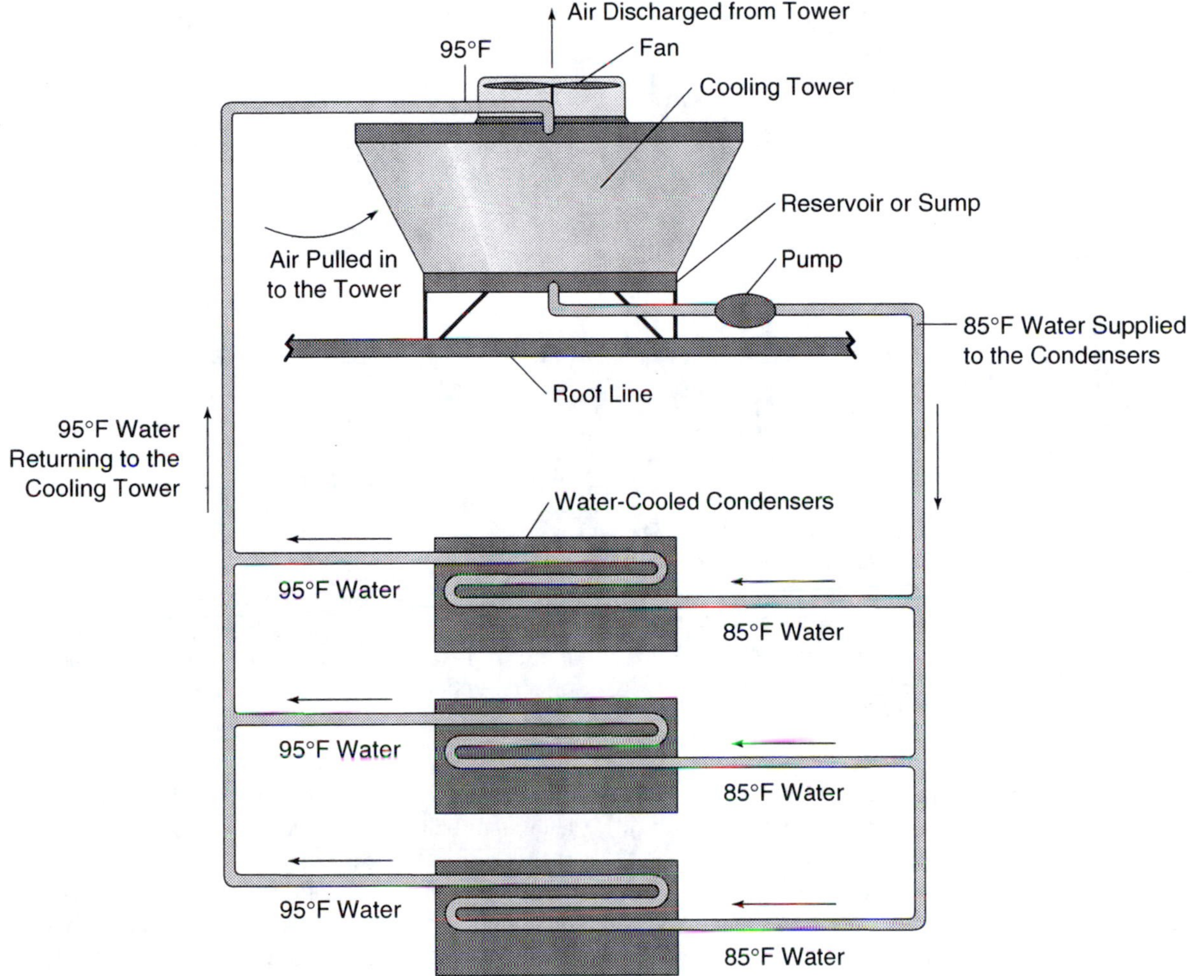

FIGURE 2-17 The cooling tower can supply water to multiple condensers. These condensers are piped in parallel with each other

coil that will open its contacts and de-energize the compressor if the coil temperature falls below 32°F. This type of thermostat is referred to as a device that closes on temperature rise, which is the same as one that opens on a drop in temperature.

Several methods are employed to convert the temperature that the device senses into mechanical motion that causes the thermostat to open and close its contacts. The various types of thermostat mechanisms are the:

- Bimetal type
- Sensing bulb
- Solid-state or electronic

Bimetal thermostats operate on the principle that different metals expand and contract at different rates when exposed to temperature changes. When two dissimilar metals are connected to each other side-by-side and are either heated or cooled, bending will occur (Figure 2-18). The longer the metal strips are, the more exaggerated the bending will be. Unfortunately, longer bimetal strips occupy more space. For this reason, it is quite common to see bimetal strips bent into various shapes including coils and spirals (Figure 2-19a).

By connecting a mercury-filled glass bulb to the bimetal coil, an electric circuit can be closed or opened depending on the temperature that the device

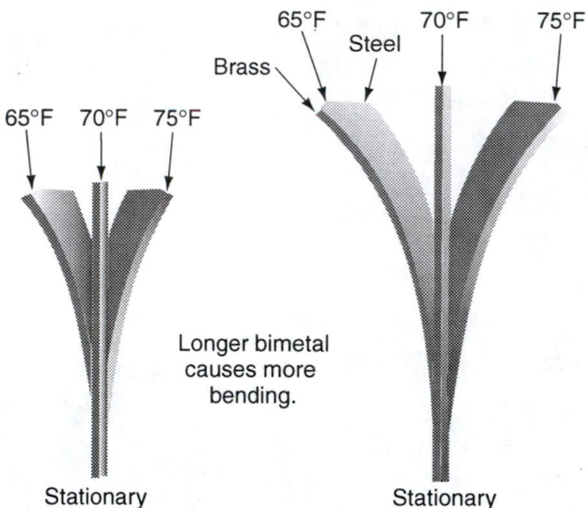

FIGURE 2-18 This bimetal is straight at 70°F. The brass side contracts more than the steel side when cooled, causing a bend to the left. The brass side expands faster than the steel side on a temperature rise, causing a bend to the right. The bend is a predictable amount per degree of temperature change. The longer the strip, the more the bend

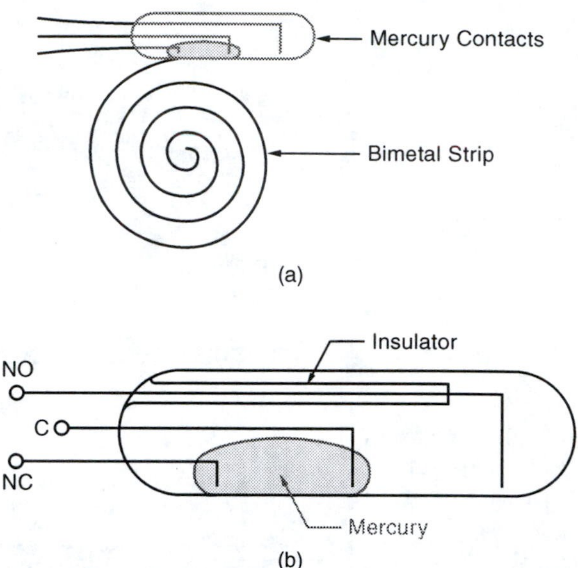

FIGURE 2-19 (a) Bimetal strip formed into spiral to save space. This strip has a mercury bulb attached to it. (b) The mercury shifts in the glass tube depending on the temperature sensed by the bimetal strip

is sensing (Figure 2-19b). Mercury bulbs consist of a "ball" of mercury, which is an excellent conductor of electricity, inside a sealed glass tube that has had all of the air removed to prevent oxidation. Electric conductors pass through the glass into the chamber, and contacts are opened or closed depending on the position of the mercury within the tube. The bulb in Figure 2-19b can either complete a circuit between terminals *C* and *NO* or between *C* and *NC,* but not both. The notations on the contact terminals are defined as follows:

- *C* terminal—Common (This terminal is common to both contacts)
- *NC*—Normally closed
- *NO*—Normally open

NOTE: *To ensure proper thermostat operation, devices containing mercury bulbs must be installed perfectly level. Improperly mounted thermostats will cause the actual space temperature to differ from the desired set point.*

The **sensing bulb** thermostat uses a copper or aluminum bulb that is filled with a **volatile liquid** that follows a pressure/temperature relationship and expands when heated and contracts when cooled. The bulb is connected to a bellows, which can open or close a set of contacts, depending on the amount of pressure exerted on it (Figure 2-20). The sensing bulb is located within the space for which the air-conditioning or refrigeration system is maintaining the temperature. On window air conditioners, for example, the sensing bulb is located in the return air stream.

The **solid-state**, or electronic, thermostat uses temperature-sensitive resistors called **thermistors**. The resistance of the thermistor changes as the surrounding air temperature changes. If the resistance of the thermistor increases as the temperature it senses increases, the device is said to have a **positive temperature coefficient** (PTC). If, on the other hand, the resistance of the thermistor decreases as the temperature it senses increases, it is said to have a **negative temperature coefficient** (NTC). PTCs are commonly used as safety controls and starting components on electric motors and other devices.

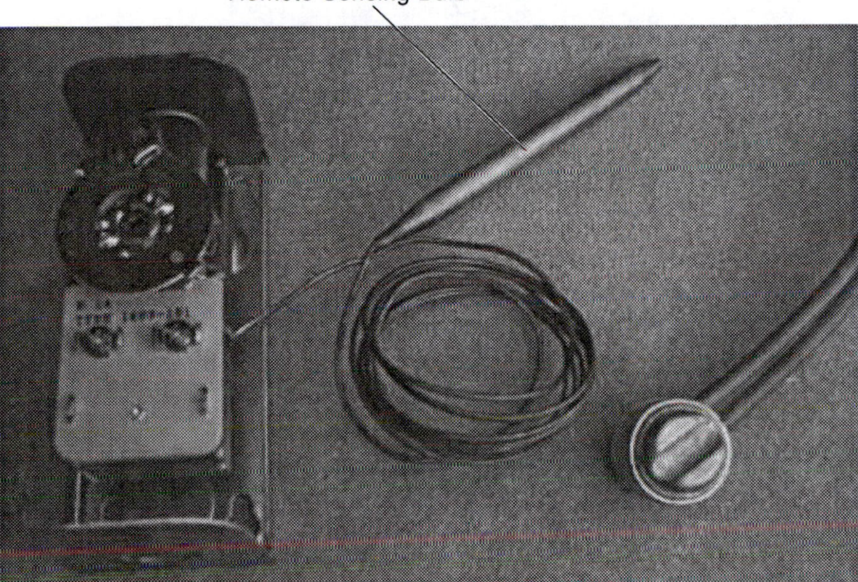

FIGURE 2-20 Thermostat with a sensing bulb. The temperature switch on the right is a nonadjustable, bimetal-type thermostat *(Photo by Eugene Silberstein)*

The electronic circuitry senses the change in resistance and adjusts the system accordingly. Changes in resistance cause the amount of current flowing in the circuit to change. Lower resistance translates to higher current flow and vice versa. Electronic thermostats are desired by many because they tend to be more reliable and last longer than mercury bulb devices. Part of this is due to the fact that there are no moving parts to wear out.

Line-Voltage Thermostats

Line-voltage thermostats typically control 115-volt or 208/220-volt circuits, while low-voltage thermostats typically control 24-volt circuits. The thermostats in Figure 2-20 are typical line-voltage devices. When intended to be located within the occupied space, they are manufactured with tightly secured covers to reduce the risk of electric shock. A licensed electrician may be needed to run electrical lines that connect line-voltage thermostats in accordance with local codes.

Depending on the individual system, the current flowing through line-voltage thermostats can be relatively high when wired in series with the electric load, so there is a chance of personal injury both to the service technician and to the equipment owner if the device is not properly installed. A **series circuit** is one in which the current has one and only one possible path to take, so the current in a series circuit is the same at all points throughout the circuit. For this reason, thermostats wired in series with the electric load (Figure 2-21a) must be able to withstand the current that the load draws.

Thermostats can be also be used to energize and de-energize **holding coils**, which are located in relays and contacts and are used to control electric contacts in series with the major loads (Figure 2-21b). This greatly reduces the amount of current flowing through the thermostat. In operation, when the thermostat closes its contacts, current flows through the holding coil of the relay or contactor, creating a magnetic field. This circuit is referred to as a **control circuit**. The current in this control circuit is very low (Ammeter 2), allowing thermostats with lower amperage ratings to be used. Once energized, the holding coil generates a magnetic field, pulling the two sets of contacts closed, completing the electric path to the main load. The

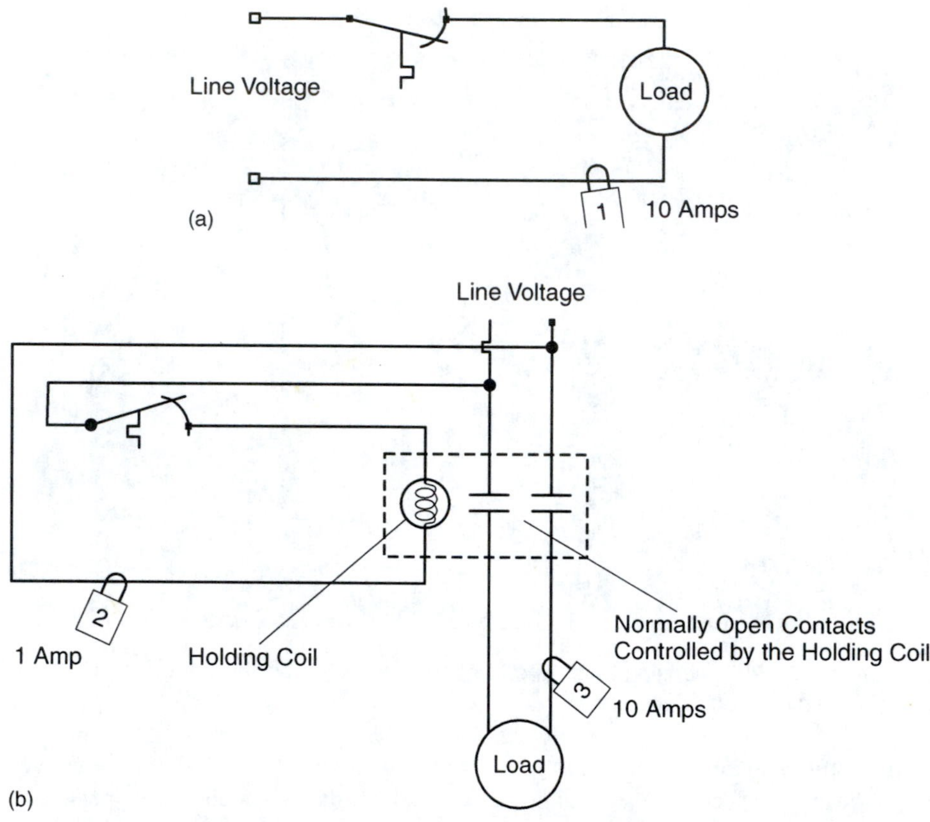

FIGURE 2-21 (a) This line-voltage thermostat is connected in series with the load. When the thermostat closes, the amperage (current) that flows through the load will also flow through the thermostat. The amperage read by Ammeter 1 is the amperage of the entire circuit. (b) Line-voltage thermostat in series with the holding coil. The amperage read by Ammeter 2 may be as low as 0.5 amps. The holding coil, when energized, causes the contacts to close, energizing the load. The amperage read by Ammeter 2 remains low, while the amperage read by Ammeter 3 is high

amperage read by Ammeter 3 represents the amperage drawn by the main load and is much higher than the amperage in the control circuit. The circuit containing the main load is referred to as the **power circuit** and is parallel to the control circuit. A **parallel circuit** is one in which the electric current has more than one possible path to take. As can be seen in Figure 2-21b, the current can either flow through the circuit containing the thermostat and the holding coil or the circuit containing the main load.

Low-Voltage Thermostats

Another type of thermostat that is commonly used is the low-voltage variety (Figure 2-22). It is used to properly route current through low-voltage control

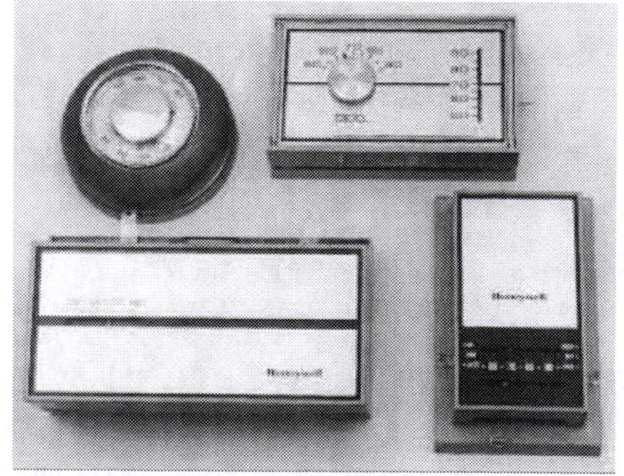

FIGURE 2-22 Low-voltage thermostats

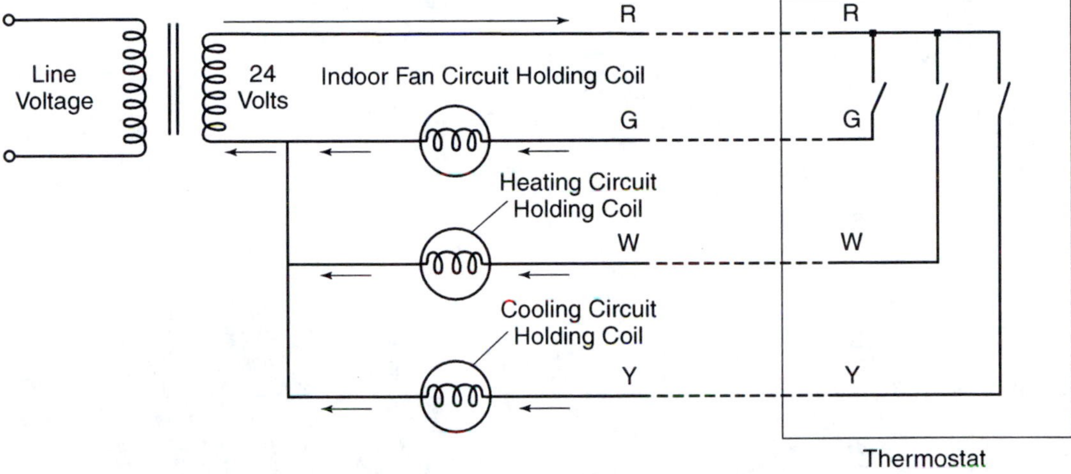

FIGURE 2-23 Simplified low-voltage control circuit. The indoor fan operates when the thermostat closes the contact between *R* and *G*. The system is in the cooling or heating mode when contacts *R* and *Y* or *R* and *W*, respectively, are closed

circuits that, in turn, energize and deenergize the major system loads (Figure 2-23). Low-voltage wires are colored for easy distinction between them, and most manufacturers in the air-conditioning industry have accepted the following color-coding system:

R terminal	Red	Hot (power coming from the transformer)
R_C terminal	Red	Hot (power feeding the cooling circuit)
R_H terminal	Red	Hot (power feeding the heating circuit)
W terminal	White	Heating circuit
Y terminal	Yellow	Cooling circuit
G terminal	Green	Indoor fan circuit

NOTE: *The R_C and R_H terminals are found on* **isolating subbases**, *which separate the heating and cooling contacts when different control-circuit power supplies are used for each mode of operation. If only one power supply is used, a jumper wire can be placed between the R_C and R_H terminals.*

Low-voltage thermostats are connected in a circuit parallel to that of the main loads and, therefore, are not subjected to the same current draws or voltage requirements that the main loads are. These thermostats are often preferred over the line-voltage version because:

- They reduce the risk of severe electric shock because of the low amperage draw.
- They tend to be less expensive than line-voltage thermostats.
- In most geographical areas, the associated wiring can be installed without an electrician's license.

Low-voltage thermostats reduce the risk of electric shock because both the supply voltage and the current flow through the circuit are very low. Since the current flow is low, the electric contacts in the device can be manufactured for lighter duty, reducing the overall cost. In addition, the installation of low-voltage thermostats does not generally require the services of a licensed electrician because most cities do not have electrical codes regarding low-voltage wiring. For these reasons, low-voltage thermostats are usually preferred even though they require a transformer to provide the desired voltage.

Low-voltage thermostats are usually mounted on a **subbase**, (Figure 2-24). The subbase is the mounting plate on which the electric connections are made. The subbase is attached directly to the wall or electric junction box, and the wires enter the plate from behind through an access hole provided for this purpose.

Subbases are equipped with manual switches that allow the user to switch between the heating and cooling modes of operation, as well as to select either continuous or automatic fan operation (Figure 2-25).

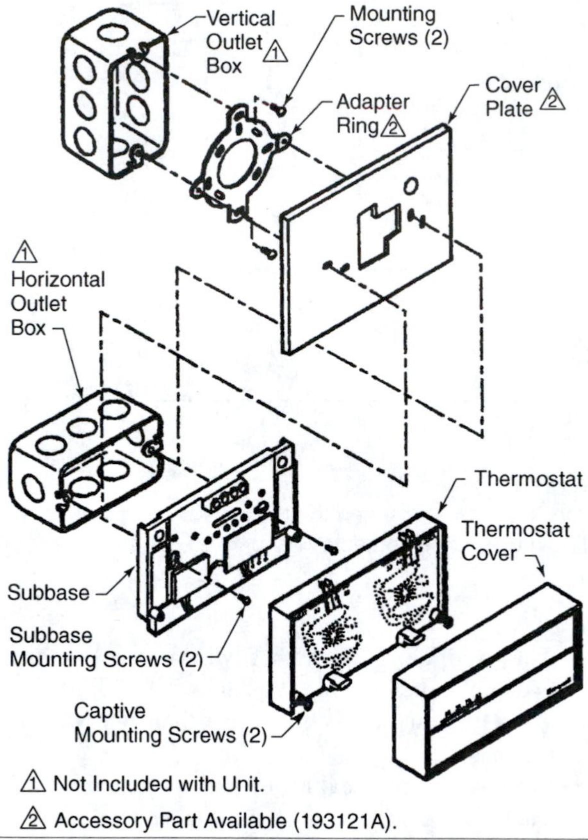

FIGURE 2-24 The subbase mounts to either the wall or an electric outlet box. The electric connections are made on the base, and the thermostat body is secured to the subbase *(Courtesy Honeywell, Inc.)*

Letter Designation	
Voltage to Transformer	R-V
Heating	W-H
Cooling	Y-C
Fan	G-F
Heating Damper (seldom used)	B
Cooling Damper (seldom used)	O

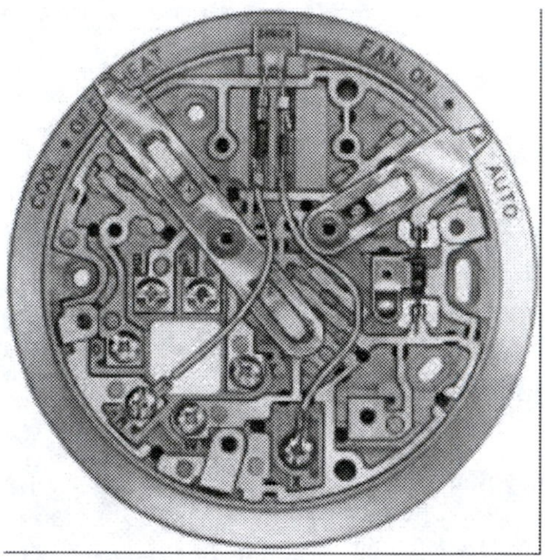

FIGURE 2-25 This subbase has manual switches that allow the user to set the *HEAT-OFF-COOL* switch to the desired mode as well as select automatic or constant fan operation *(Courtesy Honeywell, Inc.)*

Various types of thermostats are categorized by the type of equipment on which they are used. They are the:

- Heating only
- Cooling only
- Heating and cooling combination

Heating-only thermostats are used on independent heating systems, such as baseboard. *Cooling only* thermostats are used on air-conditioning systems that are not equipped with a heating coil. *Heating-and-cooling* thermostats control both the heating and cooling cycles of a system equipped with both heating and cooling components. Figure 2-25 shows a heating/cooling subbase.

Cooling Thermostats

Cooling thermostats are designed to close their contacts as the space temperature rises above the desired set point. The cooling thermostat is shown in electrical schematics as in Figure 2-21a. The electrical symbol is drawn in a manner that is representative of the way in which the device operates. The line segment representing the thermostat "moves" up and down as the temperature it senses rises and falls. The line segment on the cooling thermostat's electric symbol is below the main circuit line, and, as the temperature rises, the line moves up until, at some predetermined temperature, the thermostat closes the circuit. Similarly, as the temperature falls, the circuit opens. For example, if a room thermostat is set to maintain 74°F, referred to as the **set point**, the cooling thermostat closes its contacts if the room temperature rises above 74°F, energizing the cooling portion of the control circuit.

This type of thermostat can be either line voltage or low voltage. If a line-voltage thermostat is used, it may be wired directly in series with the compressor (Figure 2-21a); or, if the compressor is large and draws a large amount of current, the thermostat will most likely energize the holding coil of the cooling circuit. This holding coil will be rated at the supply voltage. If the thermostat is a low-voltage device, the closing of its contacts will energize the holding coil of the cooling circuit. In this case, the holding coil will be rated at 24 volts.

If the thermostat closed its contacts at a temperature just above the set point and opened its contacts just below the set point, the system would be turning on and off constantly. To alleviate this situation, most thermostats have a **temperature differential** built into them to provide a temperature swing of about 1 degree above and below the set-point temperature. This means that if the thermostat is set to maintain 74°F, the system will turn on when the space reaches 75°F and turn off when the temperature drops to 73°F.

Heating Thermostats

Heating thermostats are designed to close their contacts as the space temperature falls below the desired set point. The heating thermostat is shown in electric schematics as in Figure 2-21b. This electrical symbol is also drawn in a manner that is representative of the way in which the device operates. The line segment representing the heating thermostat "moves" up and down as the temperature it senses rises and falls. The line segment on the heating thermostat's electrical symbol is drawn above the main circuit line. As the temperature drops, the line moves down until, at some predetermined temperature, the thermostat closes the circuit. Similarly, as the temperature rises, the circuit opens. For example, if a room thermostat is set to maintain 68°F, the heating thermostat will close its contacts if the room temperature falls below 68°F, energizing the heating system.

Heating thermostats can also be line-voltage or low-voltage devices. They can be wired in series with the heating equipment or used to energize the holding coil of the heating circuit. Like the cooling thermostats, heating thermostats have a temperature differential built into them to prevent rapid system cycling.

Heating/Cooling Combination Thermostats

Combination thermostats are used on equipment that has both heating and cooling components. The occupant determines if the equipment is to run in the cooling mode or the heating mode by means of a manual switch. More sophisticated thermostats have what is referred to as an **automatic changeover**, which switches the system automatically between the heating and cooling modes. Many thermostats are equipped with fan switches that give the user the opportunity to run the indoor fan continuously, when switched to the *ON* position, even when the system is off or the room thermostat is satisfied (Figure 2-26a,b). The fan switch can also be set to the *AUTO* position, which is used to cycle the indoor fan on and off with the system compressor (Figure 2-26c,d). Since moving air is somewhat cooler than stagnant air, it is recommended that the fan be in the *ON* position during the warmer summer months and in the *AUTO* position during the cooler winter months.

The thermostat can be programmable or nonprogrammable, digital or analog. Nonprogrammable thermostats are less expensive than programmable models and require more attention from the occupant of the space. Digital thermostats utilize solid-state circuitry, while analog devices rely on mechanical means to operate. Digital programmable thermostats are generally more expensive than their analog counterpart but can often save the user money in the long run (Figure 2-27). These thermostats can be programmed to maintain different temperatures at different times of the day. During summer operation, the system can be programmed to maintain a higher temperature when the space is not occupied and to turn on shortly before the occupants return. Newer models of programmable thermostats can be set to maintain up to five different temperatures throughout the day and can even be programmed to turn off on the weekends. This helps benefit companies that are only open for business Monday through Friday. Since these devices turn the system on and off automatically, they can be mounted with locking covers, which helps reduce tampering by company employees. This also helps save energy dollars.

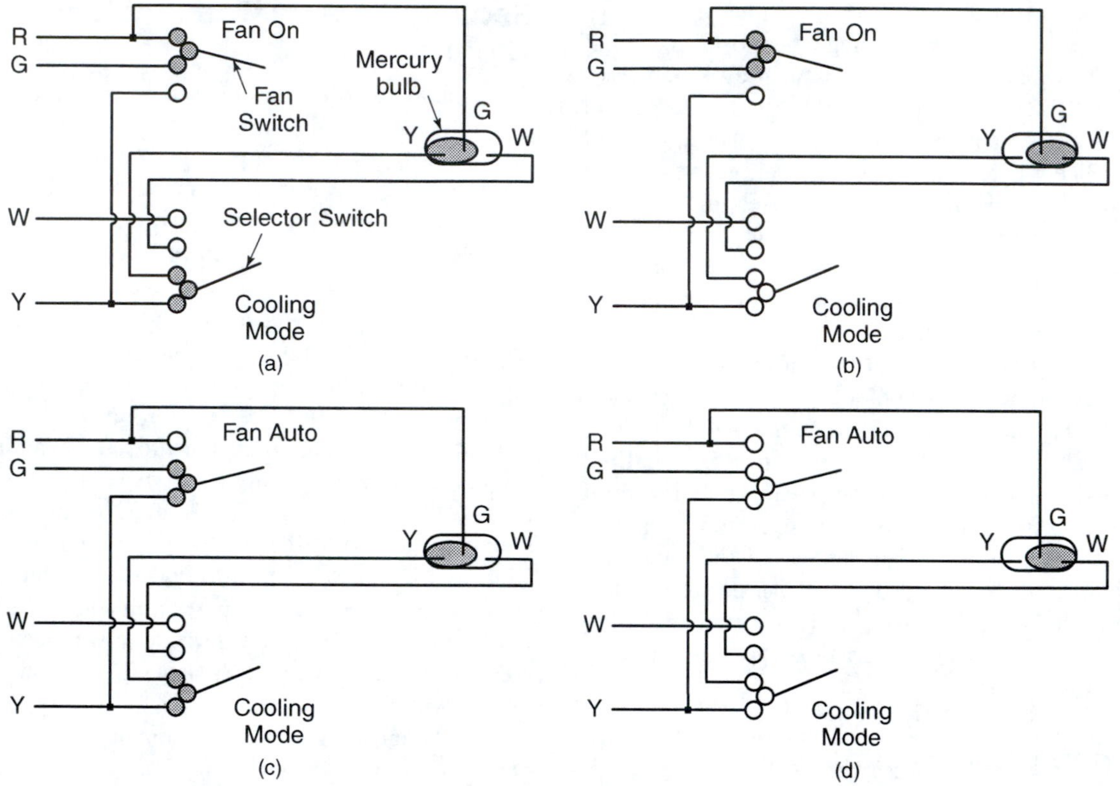

FIGURE 2-26 (a) Cooling mode with continuous fan operation. (b) Continuous fan operation even though the mercury bulb has shifted to open the cooling-circuit contacts. (c) Fan operates only when the cooling circuit is energized. (d) Mercury bulb senses that the room has reached the desired temperature and opens the cooling circuit and de-energizes the fan circuit

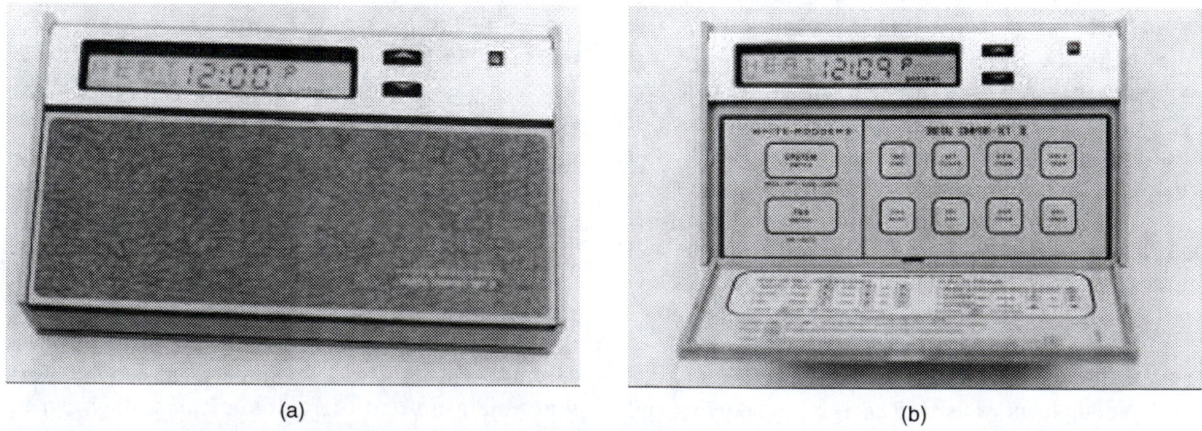

(a) (b)

FIGURE 2-27 (a) Digital programmable thermostat. (b) The device can be easily programmed with the keypad to allow for multiple temperature settings throughout the day

Multistage Thermostats

For smaller applications, thermostats with one heating and/or one cooling set of contacts are sufficient to energize the heating and cooling modes, respectively. As the applications get larger and heat loads fluctuate more and more, however, operating only portions of the system at different times may be economically feasible. For example, a 20-ton air-conditioning system made up of two, 10-ton circuits may not need to run at full capacity during the early morning hours when the heat load is low. Ten tons may be enough to cool the space while the sun is not at its peak and the building is not crowded with occupants. The **multistage thermostat** allows each circuit to be controlled independently, energizing the second stage only when needed.

Multistage thermostats can have two or more stages of heating and/or two or more stages of cooling. Common configurations of multistage thermostats include the following:

- Two cooling stages, one heating stage
- One cooling stage, two heating stages
- Two cooling stages, two heating stages

A thermostat with two stages of cooling and one stage of heating would have two *Y* terminals, denoted Y_1 and Y_2, but only one *W* terminal. Similarly, a thermostat with two stages of heating would have two *W* terminals, namely W_1 and W_2. A multistage thermostat with two heating and two cooling stages would have the following wiring connections:

R	Hot (power supplied from the transformer)
G	Indoor fan circuit
W_1	Heating—Stage 1
W_2	Heating—Stage 2
Y_1	Cooling—Stage 1
Y_2	Cooling—Stage 2

The *Y* contacts would open and close at different temperatures, allowing independent operation. The contact for Stage 1 may be set to close at 74°F, while the contact for Stage 2 may be set to close at 76°F. If Stage 1 is capable of handling the heat load, Stage 2 will not be needed and will remain off. If, however, the temperature of the space begins to rise, the Stage 2 contact will close at 76°F, energizing the second half of the system. When the space temperature begins to drop, Stage 2 will de-energize, while Stage 1 continues to operate.

Two-stage heating thermostats are commonly found on heat-pump systems. The first stage of heating controls the vapor-compression cycle (compressor-operated heat cycle) and the second heating stage controls supplementary heating strips. Stage 1 may be set to close its contacts at 70°F, while Stage 2 is set to close its contacts at 68°F. If single-stage heating is not sufficient to satisfy the heating requirements of the occupied space, the temperature will continue to drop and Stage 2 will energize the supplementary heating strips. *When the space temperature rises to the desired set point, the second stage will drop out until needed at a later time.*

Anticipators

In an attempt to minimize the swing in temperature of an occupied space, thermostats are often equipped with **cold anticipators** and/or **heat anticipators**. They allow for a more even conditioning of the space by reducing the extreme temperatures at the beginning and end of the operating cycle.

Cold anticipators are fixed resistances wired in parallel with the cooling contact in the thermostat. When the contacts are open, a small amount of current flows through the anticipator, generating a small amount of heat. This causes the thermostat to close its contacts and energize the cooling cycle shortly before it normally would (Figure 2-28), preventing the space from getting any warmer.

Heat anticipators are usually variable resistances wired in parallel with the heating contacts of the thermostat. Adjustable anticipators must be properly set upon initial installation. The setting on the device must be equal to the current flowing through the heating control circuit (Figure 2-29). The best way to measure the control-circuit current is to use a coil of wire with ten turns, wrapped around the jaw of a clamp-on ammeter. This wire is then connected across the heating contacts of the thermostat. The current in the circuit will be the reading on the meter divided by 10 (Figure 2-30). A clamp-on ammeter that is capable of accurately measuring very low amperages can also be used. The purpose of the heat anticipator is to prevent the occupied space from getting too warm during the heating cycle. This excess

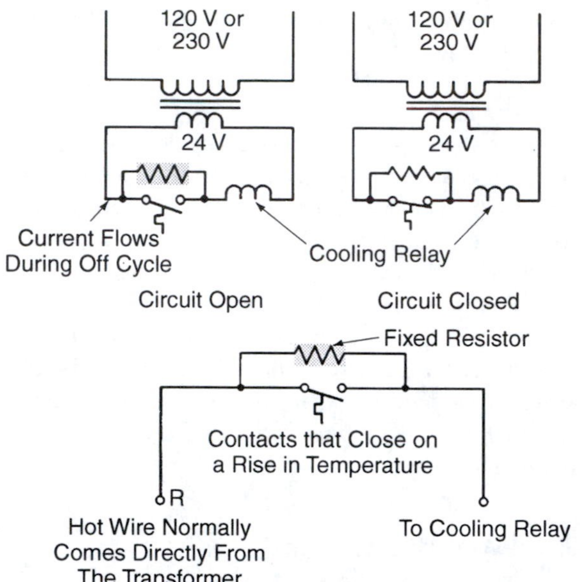

FIGURE 2-28 The cold anticipator is wired in parallel with the cooling contacts of the thermostat. This allows a small amount of current to flow through the device in the off cycle, generating heat. This causes the cooling circuit to be energized early, eliminating the extreme temperatures at the beginning of the cycle

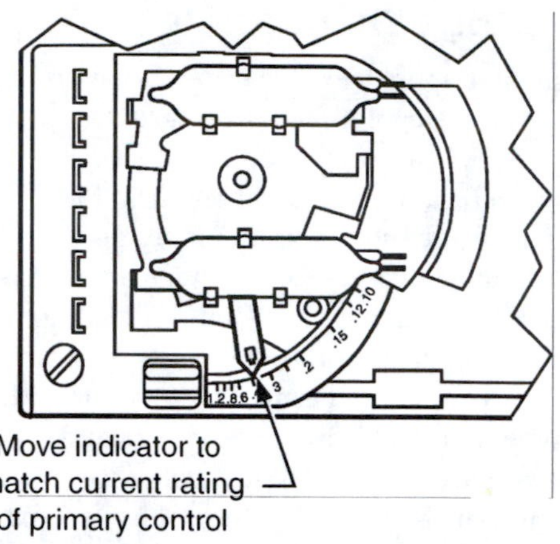

FIGURE 2-29 Adjustable heat anticipator. The indicator must match the current rating of the heating control circuit *(Courtesy Honeywell, Inc.)*

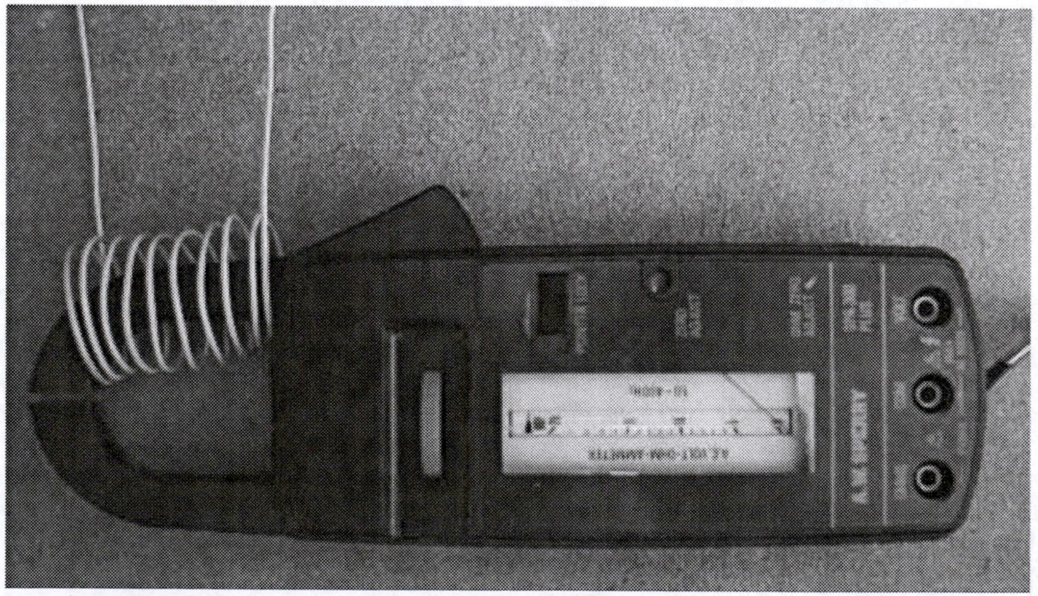

FIGURE 2-30 The amperage that is read using the 10-wrap method must be divided by 10 to get the proper circuit current used in setting the heat anticipator *(Photo by Eugene Silberstein)*

heat, or temperature above the desired space temperature, is called *overshoot.* The anticipator will cause the thermostat's contacts to open shortly before it normally would, allowing residual system heat to dissipate. If the anticipator was not there, the thermostat would open at the set point and the residual heat would create a large overshoot.

TRANSFORMERS

As described earlier, some thermostats are designed as low-voltage devices, operating at 24 volts. Since a 24-volt power supply is not readily available from a wall outlet, it must be produced from within the system. The component used to produce this voltage is called a **transformer** (Figure 2-31).

The transformer typically has two windings called the **primary winding** and the **secondary winding**. The windings are made up of coils of wire, with a specific number of turns in each, wrapped around a **laminated core** (Figure 2-32). The laminated core is made up of stacked iron plates that are insulated from each other with shellac.

Line voltage, usually 115 volts or 220 volts, is supplied to the primary winding of the device; a magnetic field is then generated in the winding; and the desired low voltage is produced at the secondary winding. The strength of the magnetic field depends on the number of turns in the primary winding.

For example, an insulated copper wire is wrapped around an iron nail 25 times. The ends of the conductor are connected to the terminals of a lantern battery. The nail is then lowered into a pile of paper clips and then lifted. A number of paper clips are attracted to the nail, which has become an electromagnet. The experiment is repeated, but this time the number of turns of wire around the nail is increased to 50 turns. The number of paper clips that are attracted to the nail has increased. This is because the strength of the electromagnet has increased, due to the increased magnetic

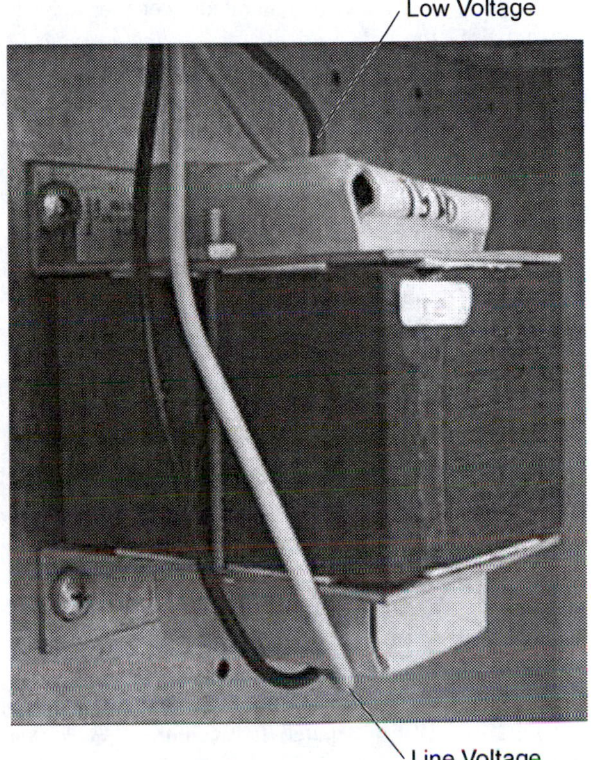

FIGURE 2-31 Typical control transformer *(Photo by Eugene Silberstein)*

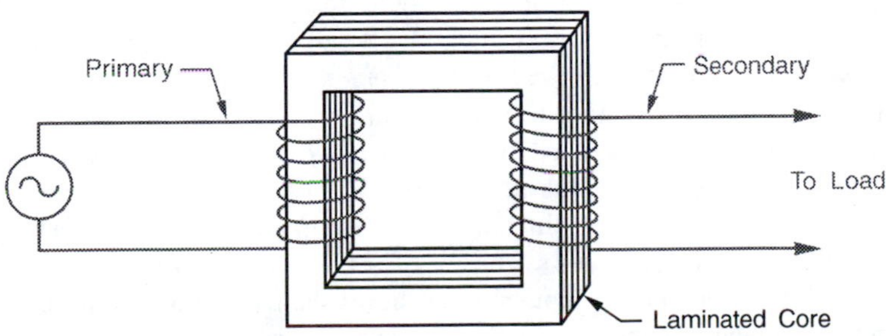

FIGURE 2-32 Primary and secondary windings are wrapped around the laminated core. Note that the two windings are not electrically connected to each other

field. A magnetic field is generated whenever there is current flow through a conductor, but the strength of this field is magnified by coils with many turns.

The secondary winding, which is located close to the primary winding, experiences the effects of the magnetic field. The magnetic field created by the primary winding causes a voltage, or potential, to be generated in the secondary winding. Note that at this point the primary and secondary windings are not electrically connected to each other. The voltage that is produced in the secondary winding is said to be an **induced voltage**.

Step-Up and Step-Down Transformers

Just as the strength of the electromagnet in the earlier example was determined by the number of turns in the coil, the voltage that is produced at the secondary winding is determined by the number of turns in the coil. The lower the number of turns, the lower the induced voltage. If the primary has 10 times the number of turns as the secondary, the secondary voltage will be one-tenth that of the primary. This transformer is said to have a 10 to 1 ratio, denoted 10:1 (Figure 2-33a). This type of transformer is called a **step-down transformer**. The voltage is stepped down from a higher voltage to a lower voltage. Transformers may also be designed with more turns in the secondary winding than the primary winding. These transformers are called **step-up transformers** since the voltage is stepped up to a higher potential (Figure 2-33b).

Each transformer is designed to accept a specific primary voltage, which must be within design parameters—usually plus or minus 10 percent of nameplate voltage—to function properly. For example, a transformer with a primary voltage rating of 115 volts will operate effectively within the range from 103 to 127 volts. Damage to the device could occur if incorrect voltage were applied to the primary winding. Because of this, being aware of the specific job application is extremely important before selecting a transformer. If a defective transformer is to be replaced, use the exact replacement whenever possible. This may prove troublesome to the service technician since a number of different transformers would need to be stocked to properly service the customer's equipment. If the exact replacement is not available, the new transformer must match the specifications of the old one.

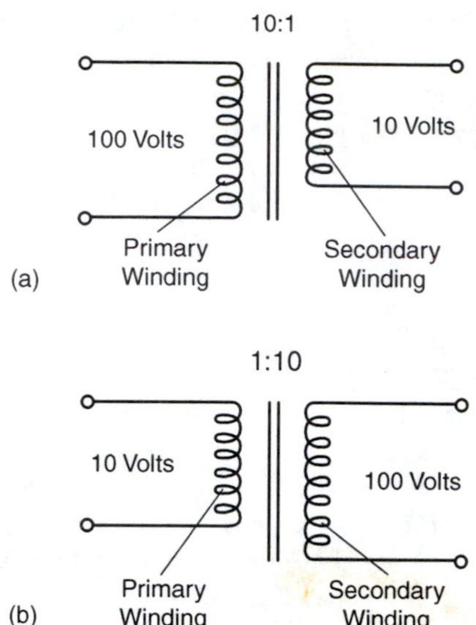

FIGURE 2-33 (a) Step-down transformer. This transformer has more turns in the primary winding, so the secondary voltage is lower than the primary voltage. (b) Step-up transformer. This transformer has more turns in the secondary winding, so the secondary voltage is higher than the primary voltage

Multitap Transformers

The **multitap transformer** is designed with multiple wires, or taps, connected to the primary winding. This enables the device to be used with different supply voltages (Figure 2-34). Care must be taken to use the correct wires or damage to the device could occur. These primary wires are color-coded and a legend is provided to reduce the possibility of incorrect installation. All unused taps must be isolated from each other and capped to prevent contact with other taps, grounds, or other conductors.

When installing transformers and other electric components, all safety rules should be observed to avoid the possibility of personal injury and damage to the components. The low-voltage circuit connected to the secondary winding must be free from short circuits to prevent winding damage. If a short circuit exists, the circuit resistance becomes very low and causes the current to rise to levels beyond the normal operating range. This can cause the device to overheat and break down. If the secondary winding opens, the

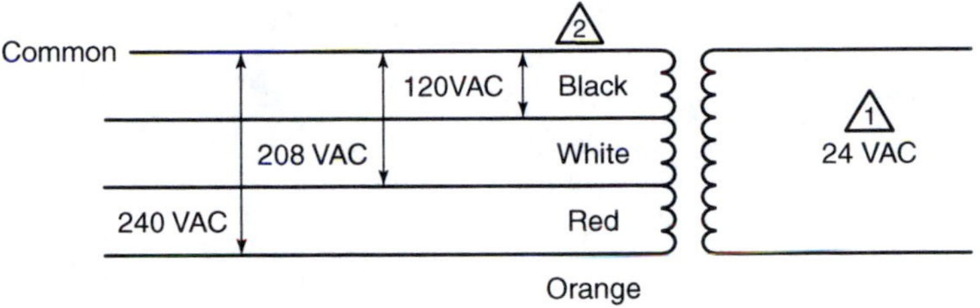

FIGURE 2-34 Multitap transformer schematic *(Courtesy Honeywell, Inc.)*

device must be replaced. To prevent this, an **in-line fuse** should be installed in the circuit. Then, if the current in the low-voltage circuit is too high, the fuse will blow and the device will not be damaged. This fuse must be sized properly so that it does not blow during normal system operation. The cause of the failure should be located and corrected before replacing the fuse.

Safety Tip

Always exercise caution when working on electric circuits. Disconnect power whenever possible to avoid component damage as well as serious personal injury.

Transformer Rating

Transformers are rated in units called *volt-amperes* (VA). This rating represents the power of or work that can be done by the transformer. From basic electricity theory and Ohm's law:

$$\text{Power} = \text{Voltage} \times \text{Current}$$

where current is measured in amperes. It directly follows that the VA rating of a transformer is its power rating. The VA rating refers to the secondary winding of the device and is used to determine its output capability. For example, a 40-VA transformer supplying 24 volts can safely operate with a low-voltage-circuit current of 1.67 amps (Figure 2-35). If too many loads are connected in parallel, the control-circuit resistance will be low and the current flowing in the

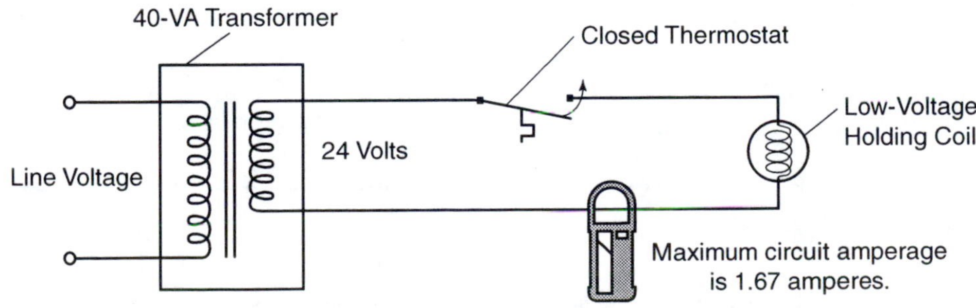

FIGURE 2-35 Control-circuit amperage multiplied by the voltage must not exceed the VA rating of the transformer

circuit will rise. If this current exceeds the rating of the transformer, it can be damaged. If multiple devices need to be connected, a transformer with a larger VA rating must be used. Properly sizing the transformer is important to ensure proper system operation.

PRESSURE CONTROLS

Pressure controls are switches that open and close electric contacts depending on the pressure within a specific part of the system. These devices can be used to control the normal operation of a system or as safety elements. These controls help to ensure that the system operates within its designed pressure ranges. Three distinct classifications of pressure controls based on their location in the system and the pressures they sense are the:

- High-pressure control
- Low-pressure control
- Dual-pressure control

High-pressure Controls

High-pressure controls, also known as *high-pressure switches,* are located on the high-pressure side of the system and are designed to open their contacts if the pressure that is sensed is higher than the desired level (Figure 2-36). These controls are said to open on pressure rise and are drawn in schematic diagrams as shown in Figure 2-37c,d. The schematic symbol is drawn above the main circuit line, indicating that the circuit will open when the system pressure rises.

FIGURE 2-36 Commonly used pressure control *(Photo by Bill Johnson)*

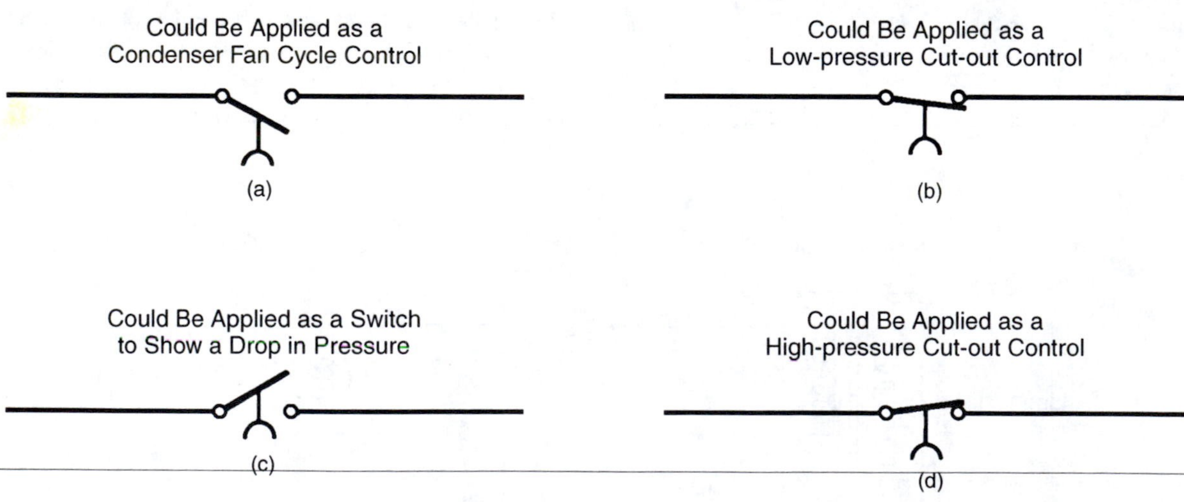

FIGURE 2-37 These symbols show how pressure controls connected to switches appear on control diagrams. The two symbols in (a) and (b), indicate that the circuit will close on a rise in pressure. (a) Indicates the switch is normally open (NO) when the machine does not have power to the electric circuit. (b) Shows that the switch is normally closed (NC) without power. The two symbols in (c) and (d) indicate that the circuit will open on a rise in pressure. (c) NO, (d) NC

Consider this example. An air-conditioning system is equipped with a mechanical-draft, air-cooled condenser, and the fan motor burns out. What will happen to the system? The compressor, not knowing that the motor is no longer running, will continue to operate, causing the head pressure to rise. At this point, the excess pressure in the system can cause refrigerant lines to burst, possibly causing serious personal injury. The high-pressure control prevents this from happening. The device senses that the system pressure is higher than desired and disables the system. High-pressure controls can be **manually reset** or **automatically reset** (Figure 2-38).

Manual-reset controls prevent the system from operating until the device is reset by hand. If the high-pressure switch trips, the system will stop running and the space temperature will rise. The customer will place a no-cooling call to the service company. Upon arrival, the technician should notice the tripped control, signaling that a pressure problem exists. The problem should be identified and corrected before resetting the control.

SERVICE NOTE: *Some field technicians are under the misconception that they are doing the equipment owner a service by pointing out the manual reset on the pressure control to them. The equipment owner then, in an effort to save money on future service calls, may choose to continually reset the control with the hope that the problem will go away. Remember that the high-pressure control is a safety device and the continual resetting of the device puts the equipment and those around it in a potentially dangerous situation.*

High-pressure switches that reset automatically will close their contacts once the pressure drops below a predetermined set point. The pressure at which the contacts open is called the **cut-out pressure**, and the pressure at which the contacts close is called the **cut-in pressure**. The difference between the cut-out and cut-in pressures is called the **differential**. Most high-pressure switches come factory set with a 100 psig differential. For example, if the pressure control is set to cut out at 400 psig, it will automatically reset once the pressure drops down below 300 psig. If the pressure control is manually reset, it can only be reset once the pressure drops below 300 psig (Figure 2-39). The cut out pressure must be set high enough so that it will not open during normal system operation and low enough to protect the equipment, service technician and occupants.

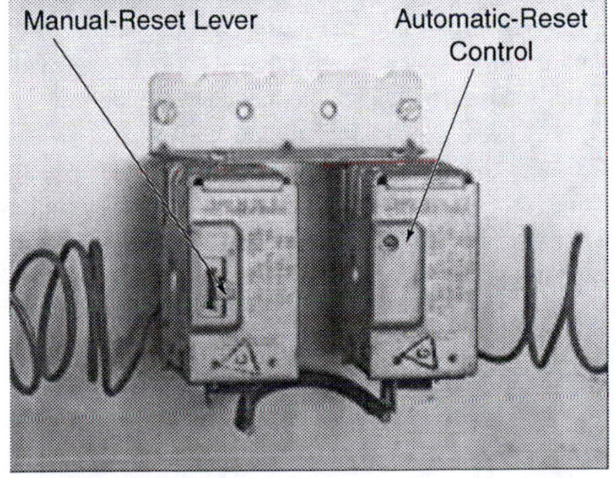

FIGURE 2-38 The control on the *left* is a manual-reset control, and the one on the *right* is an automatic-reset control. Note the push lever on the left-hand control *(Photo by Bill Johnson)*

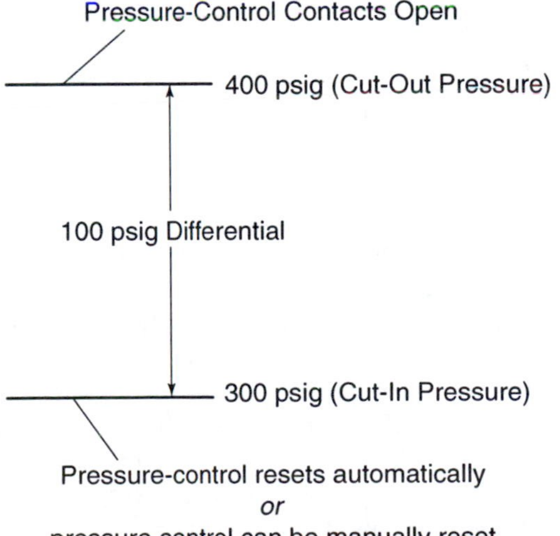

FIGURE 2-39 High-pressure controls normally have a 100 psig differential built into the device: Cut-out – cut-in = differential

Low-pressure Controls

Law-pressure controls are located on the low-pressure side of the system and are designed to open their contacts if the pressure drops below a predetermined point. These devices can be used in many applications, including:

- Low-charge protection
- Temperature control
- Freeze control
- Automatic pump-down cycles

The schematic symbol for low-pressure controls, also known as *low-pressure switches,* is shown in Figure 2-37a,b. These controls are drawn below the circuit line, indicating that they will open when the pressure they sense drops.

When used as a low-charge protector, the cut-out pressure must be well below the normal operating pressure of the system but still above atmospheric pressure to make certain that no atmosphere or moisture is permitted to enter the system. For example, for an R-22 air-conditioning system, the low-side pressure is normally in the range of 60 to 70 psig. If the low-pressure, cut-out set point was 30 psig, the system would cut out only if the system lost all or part of its refrigerant charge, since the pressure would not normally drop that low. When used for low-charge protection, the device is usually manually reset, which would prevent the compressor from running and reduce the possibility of contaminating the system. Some refrigeration applications, however, utilize an automatic-reset control so that the system will continue to run, however sporadically, to provide at least some refrigeration to protect the product.

When used as a freeze control, the low-pressure switch is set to cut out at a pressure corresponding to a temperature just below 32°F. For example, if an R-22 air-conditioning system normally operates with a 40°F coil, the corresponding suction pressure is 68.5 psig. The coil will not freeze since its temperature is above 32°F. If the air flow through the coil were restricted for any reason, the temperature and pressure of the refrigerant inside the evaporator coil would drop, causing the temperature of the coil surface to drop as well. The coil will start to frost at a refrigerant pressure of 57 psig. To prevent the coil from freezing, the cut-out set point on the freeze control would therefore be set just below 57 psig. This type of control is normally automatically reset.

Another common use for low-pressure switches is on a system equipped with an automatic pump-down cycle. This is very popular in refrigeration systems and is used every so often in air-conditioning applications. The concept behind the automatic pump-down cycle is to remove most of the refrigerant from the evaporator before the compressor is able to cycle off. The reason for this is that in the off cycle, refrigerant tends to migrate to the coldest part of the system. Depending on the system application, location of the condenser, and the outside ambient temperature, this location can, on occasion, be the evaporator. If the evaporator is filled with refrigerant, usually saturated liquid, and the compressor starts up with the expansion valve wide open, a very large amount of refrigerant will be pulled into the compressor and can cause an overload condition and possible component damage. The pump-down cycle helps prevent this and is accomplished as follows.

When the space reaches the desired temperature, the thermostat deenergizes an electrically controlled valve called a solenoid valve. This solenoid valve is located in the liquid line and only allows refrigerant to pass to the metering device and evaporator when it is energized. When the valve is de-energized, it closes and stops the flow of refrigerant to the metering device. The compressor is still operating, and the refrigerant is pumped out of the evaporator and into the condenser and/or receiver, where it is stored. The low-side pressure begins to drop; and, when it reaches the desired pressure, the low-pressure switch's contacts open, de-energizing the compressor (Figure 2-40).

When the space temperature rises above the desired temperature, the thermostat closes its contacts, energizing the solenoid valve, and causing it to open. The high-pressure refrigerant now flows through the solenoid, through the metering device, and into the evaporator. The compressor is still not energized, and the pressure in the suction line starts to increase. When the pressure reaches the preset cut-in set point, the pressure control closes its contacts, energizing the cooling circuit (Figure 2-41). Since the low-pressure control constantly opens and closes, you can readily see that this device, when used in conjunction with an automatic pump-down cycle, resets automatically.

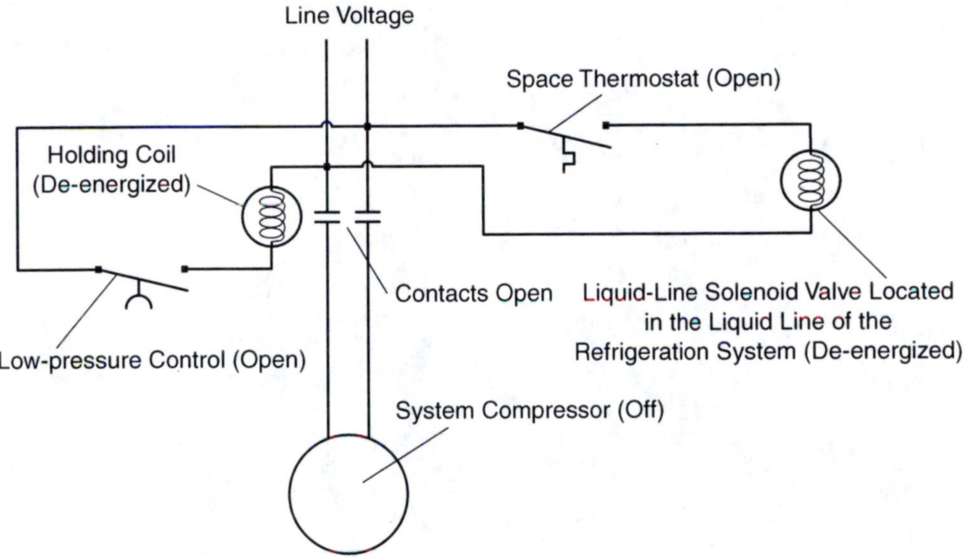

FIGURE 2-40 Basic circuit for an automatic pump-down cycle on shutdown

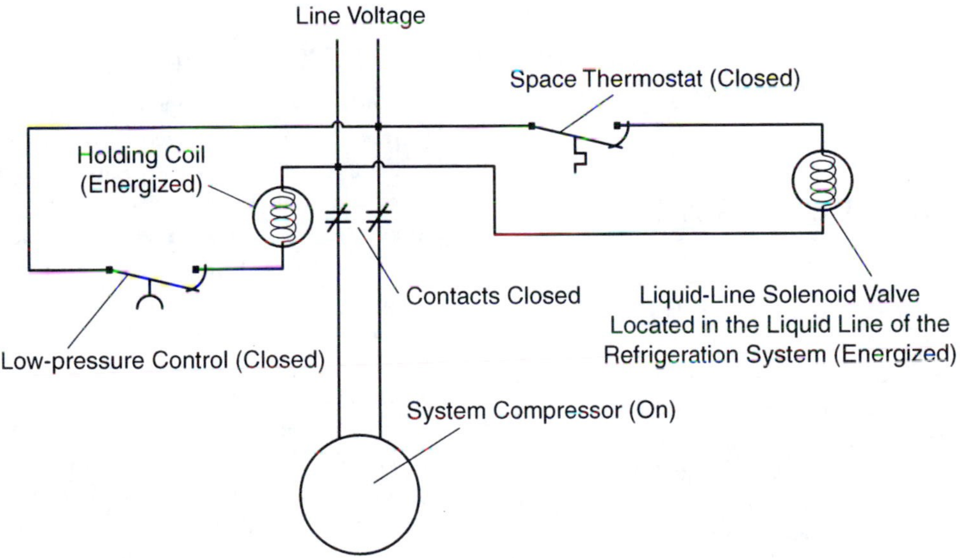

FIGURE 2-41 Basic circuit for an automatic pump-down cycle on start-up

Dual-pressure Controls

Quite often, refrigeration systems are equipped with low-pressure controls as well as high-pressure controls (Figure 2-42). To facilitate the installation and centralize the controls, dual-pressure controls are often used. The dual-pressure control is a combination of a low-pressure control and a high-pressure control in a single unit. For the contacts of the dual-pressure control to be closed, both the high and low-pressures must be within the specified range. If either the low pressure or the high pressure is out of range, the contacts will be open. It is possible to have the low-pressure component of the control reset automatically, while the high-pressure

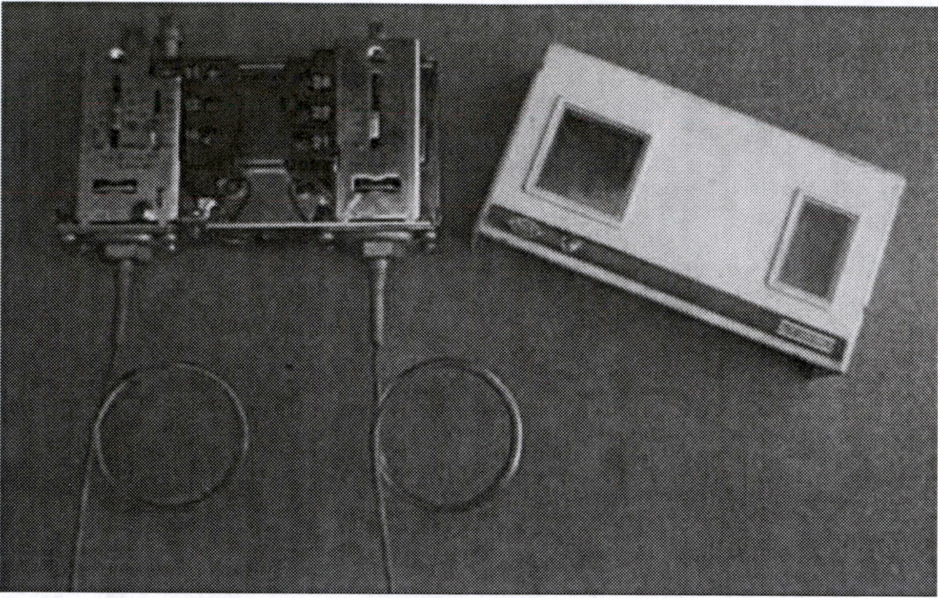

FIGURE 2-42 Dual-pressure controls are made up of a low-pressure control and a high-pressure control *(Photo by Eugene Silberstein)*

component resets manually. This allows the system to operate as though the high- and low-pressure controls are separate devices.

LOW AMBIENT CONTROLS

In many applications, air-conditioning and refrigerant equipment must operate year-round. Two examples of this are supermarket refrigeration equipment and large office buildings. If the system's air-cooled condenser is located outdoors, the unit will fail to operate properly if the outside ambient temperature is low. Supermarket customers expect the quality of products to be consistent, no matter what time of year the purchase is made. Low ambient controls are designed to eliminate this problem by helping the system simulate warmer outdoor conditions when they do not exist. Low ambient controls can be controlled by either temperature or pressure, depending on the specific system application. Commonly used methods for maintaining head pressure are:

- Condenser fan cycling
- Condenser flooding
- Variable-condenser fan speed
- Shutters

Fan-Cycling Control

Air-conditioning and refrigeration systems that are equipped with mechanical-draft, air-cooled condensers use fans to help move air through the coil. If one or more of these fans fail, the system's head pressure will rise. This can lead to a reduction in system efficiency and possible system damage. Designers use this concept to increase the head pressure during periods of low outside ambient temperatures. The fan-cycling control is designed to start and stop the condenser fan motor, or motors, in order to maintain the head pressure within the desired range (Figure 2-43). The fan runs to lower the head pressure and stops to raise the head pressure. The device is usually connected to the high-pressure side of the system and turns the fan on and off.

To raise the head pressure, the low ambient control does not permit the fan to come on. Once the head pressure reaches the desired level, the control closes its contacts and energizes the fan, lowering the head pressure. The fan continuously cycles on and off to maintain the head pressure within acceptable range limits. If a refrigeration-gauge manifold is connected to the system, a technician can observe the fluctuating head pressure. The farther apart the cut-in and cut-out pressures are from each other, the more the

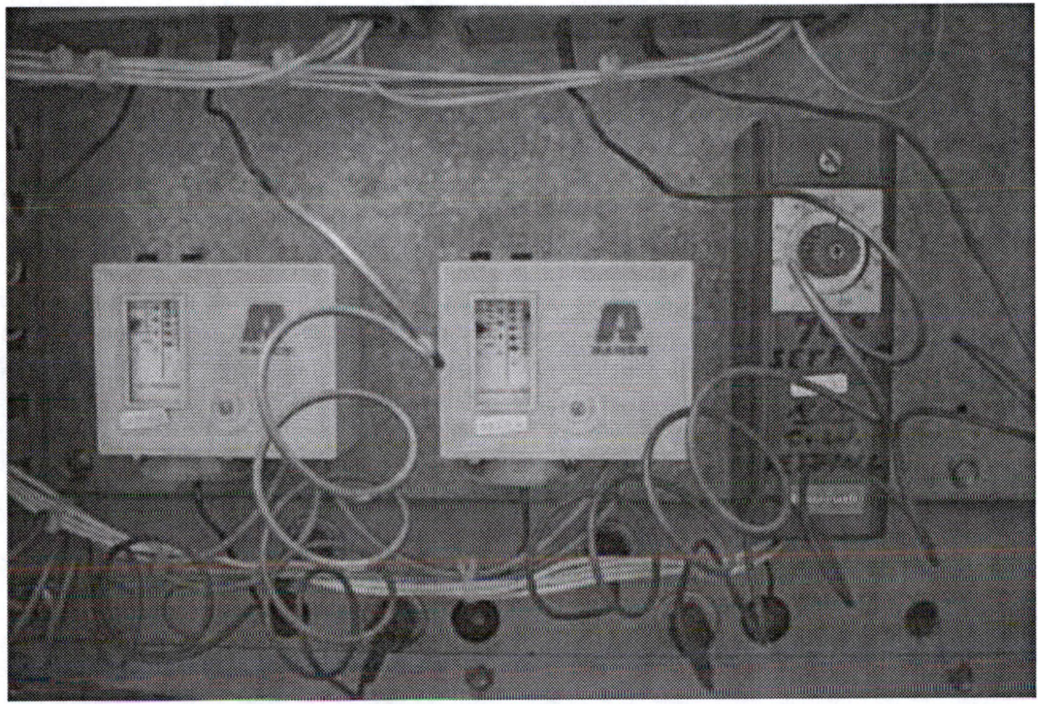

FIGURE 2-43 These devices are used to cycle condenser fans, depending on outside ambient temperature *(Photo by Eugene Silberstein)*

head pressure will fluctuate. During periods of warm outside ambient temperatures, the low ambient control will have its contacts closed, allowing continuous fan operation. On larger systems with multiple condenser fans, only some of the fans are connected to low ambient devices to provide a more constant head pressure.

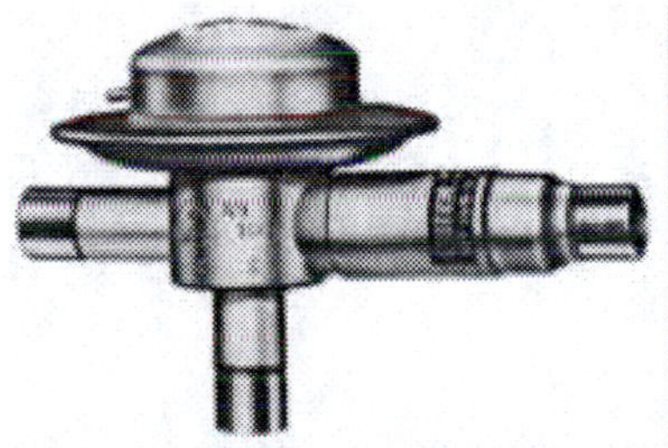

FIGURE 2-44 Head-pressure control for condenser flooding. This valve allows the refrigerant to flood the condenser during both mild and cold weather. This method requires enough refrigerant to flood the condenser and has a large receiver to hold the refrigerant during the warm season when it is not needed to keep the head pressure up *(Courtesy Sporlan Valve Company)*

Condenser-Flooding Valve

Another type of low ambient control is the condenser-flooding valve, also known as a modulating head-pressure control (Figure 2-44). It allows a portion of the refrigerant from the compressor to bypass the condenser in order to increase the head pressure during low ambient operation. Refrigerant bypassing the condenser coil accomplishes two things. First, if the refrigerant does not pass through the condenser, it cannot give up as much heat. If this heat stays within the system, the head pressure will go up. Second, when the refrigerant bypasses the coil, a backup is created at the outlet of the coil. The liquid refrigerant in the condenser then backs up in the coil. This excess liquid in the coil reduces the effective surface area of the condenser and causes the head pressure to increase.

When the outside ambient temperature is high, the condenser-flooding valve is closed and all of the discharge refrigerant is directed into the condenser. As the ambient temperature drops, the head pressure drops and the valve opens, in a modulating fashion, to allow some of the refrigerant to bypass the coil. As the head pressure drops more and more, the condenser-flooding valve opens more and more as well. For example, when the outside temperature is 50°F, the condenser may be 50 percent filled with liquid and, when the outside temperature is 0°F, the condenser may be 90 percent filled with liquid. In the first part of the example, the bypass valve is only partially open; while, in the second part, the bypass valve is open nearly all the way.

In the winter, the condenser is nearly filled with liquid; while, in the summer, the condenser is filled with mostly vapor. So what happens to all of the liquid refrigerant in the warmer months? Does the technician have to come to the job and remove the excess refrigerant? Does the technician have to return in the fall to add more refrigerant to the system? The answers to these questions are quite simple. The refrigerant stays in the system year-round and simply shifts from one location to the other. Systems equipped with condenser-flooding valves are also equipped with oversized receivers. In the summer months, the receiver stores the excess refrigerant until needed. In the colder months, the refrigerant shifts from the receiver to the condenser, in order to reduce its effective surface area, thereby increasing the head pressure.

FIGURE 2-45 Electronic device used to sense high-side pressure and modulate the speed of the condenser fan motor *(Courtesy Alco Controls)*

Variable-Condenser Motor Fan Speed

Another method commonly used to maintain adequate head pressure during low ambient conditions is by controlling the speed of the condenser fan motor. This is usually accomplished with the aid of electronic devices (Figure 2-45). The device is equipped with a pressure sensor mounted in the high-pressure side of the system that, in turn, sends a signal to the condenser motor, modulating its speed.

Shutters

One final method that can be used to maintain the head pressure in air-cooled systems is to use shutters. Just as the water-regulating valve controlled the flow of water through a water-cooled condenser, shutters control the amount of air that is permitted to flow through the condenser coil (Figure 2-46). During periods of low ambient, the shutter will move toward the closed position, reducing the air flow through the coil. When the outside ambient is warm, the shutters are completely open, allowing maximum air circulation.

FIGURE 2-46 Condenser with air shutter. With one fan, it is the only control. With multiple fans, the other fans can be cycled by temperature with the shutter controlling the final fan *(Courtesy of Trane Company)*

RELAYS AND CONTACTORS

In previous sections, the concept of a magnetic field has been discussed. Figure 2-47 shows how a magnetic field is created. Electromagnets accomplish many tasks, including turning motors and opening and closing electric contacts (Figure 2-48). The device in Figure 2-48 and those shown in Figure 2-49 are called **relays** and are used primarily in control circuits to energize and de-energize system components.

On the device shown in Figure 2-48, it can be seen that there is a spring that pulls the **armature**, or movable arm, and causes contact to be made between the movable contact and the top stationary contact. When voltage is applied to the coil, a magnetic field will be generated, creating an electromagnet. This magnet overcomes the spring pressure and pulls the armature over to close the set of contacts between the movable contact and the bottom stationary contact. If the coil is de-energized, the spring pulls the strip back to its original position. This position is called the de-energized position, which is the normal position of the contacts, and the top stationary contact is called the normally closed contact, denoted *NC*. The bottom stationary contact is called the normally open contact, denoted *NO*.

The coil of the relay is designed to operate at a specific voltage, and this rating is marked on the device itself. If the incorrect voltage is applied, damage to the component could occur.

The circuit that energizes the coil and the circuit that is controlled by the contacts can be at different voltages. This enables a 24-volt control circuit to control a line-voltage circuit (Figure 2-50). There are many different types and styles of relays, so the individual requirements of the system must be evaluated before selecting the device. The ratings printed on the relay provide the coil voltage as well as the current capacity of the contacts. These ratings should be adhered to.

Relays are used primarily to control circuits carrying up to approximately 15 amperes. Relays designed for applications above 15 amperes are called **contactors**. Contactors are designed for heavier-duty applications but operate on the principle already described. Figure 2-51 shows a bank of contactors that are used to open and close three sets of contacts. The sets of contacts are referred to as *poles*. So these devices are 3-pole contactors. These contacts are normally open, so they close only when the correct voltage is applied to the coil.

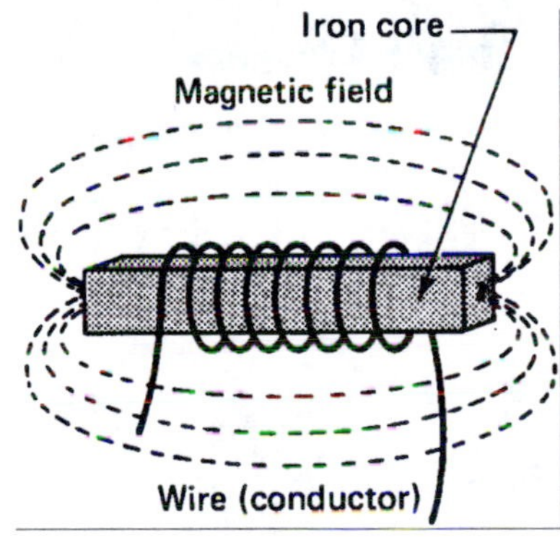

FIGURE 2-47 Magnetic field of an iron core when a current-carrying conductor is wound around the core

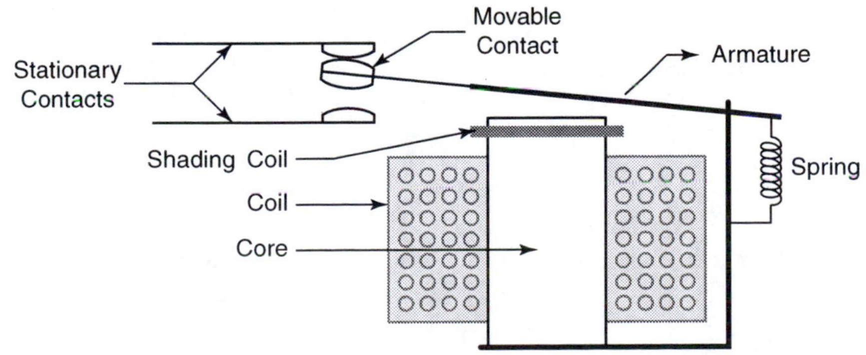

FIGURE 2-48 Simple relay

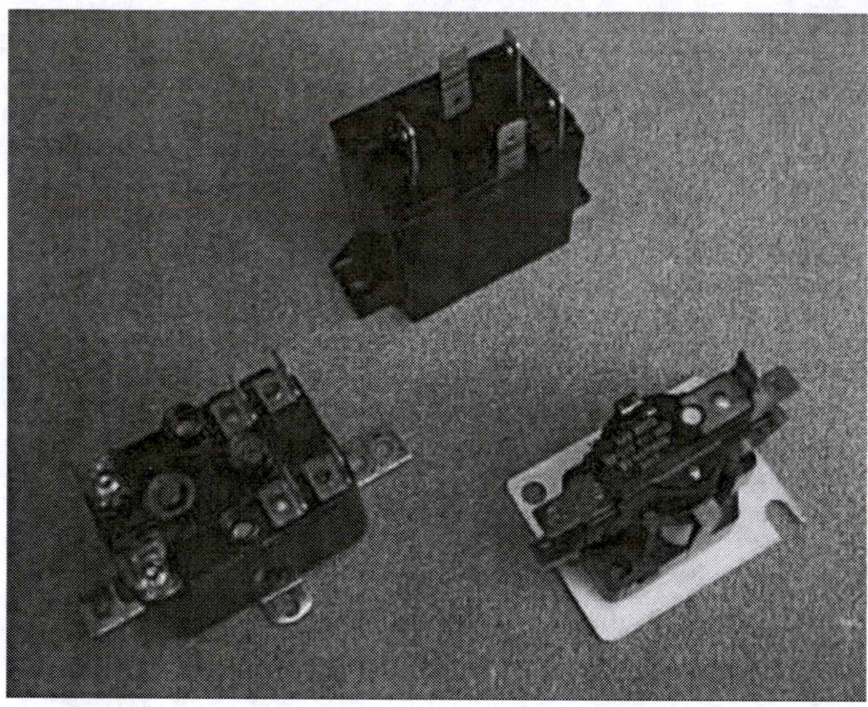

FIGURE 2-49 Common relays used in control circuits *(Photo by Eugene Silberstein)*

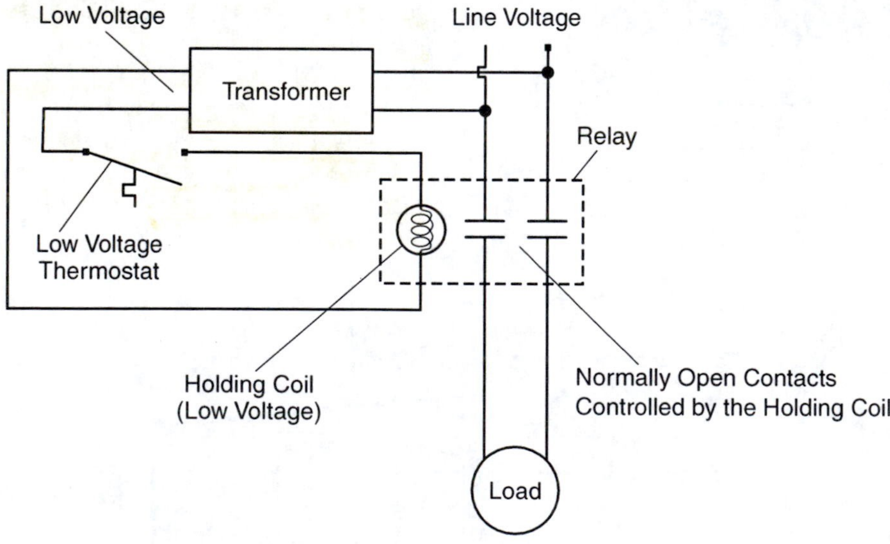

FIGURE 2-50 A low-voltage control circuit controls a line-voltage circuit with the help of a control relay

FIGURE 2-51 A bank of contactors used to control a series of condenser fan motors *(Photo by Eugene Silberstein)*

SUMMARY

- Refrigeration accessories are designed to enhance system operation.
- Service valves and sight glasses help the technician evaluate a refrigeration system.
- Filter driers remove moisture, acid, and dirt from the system.
- Refrigerant receivers store liquid refrigerant until needed by the evaporator.
- Refrigerant flow can be stopped and started with solenoid valves.
- Crankcase heaters should be energized during the compressor off cycle.
- Cooling towers cool the water returning from the water-cooled condenser.
- Thermostats can use bimetal strips, sensing bulbs, or electronic circuits.
- Heat and cold anticipators help reduce large temperature swings in the space.
- Transformers produce an induced voltage in the secondary winding.
- Pressure controls can be used as safety or operational devices, which can be manually or automatically reset.
- Low ambient controls are used to simulate design conditions.
- Common low ambient controls are fan cycling, condenser flooding, variable fan speed, and shutters.
- Relays and contactors open and close electric contacts by generating a magnetic field.

KEY TERMS

Accessories
Approach temperature
Automatic changeover
Automatic reset
Backseated
Bimetal thermostat
Blowdown
Centrifugal-type distributor
Cold anticipator
Contactor
Control
Control circuit
Cracked off the backseat
Crankcase heater
Cut-in pressure
Cut-out pressure
Desiccant
Device port
Differential
Discharge service valve
Evacuation
Electromagnetic device
Filter drier
Forced-draft tower
Frontseated
Heat anticipator
High side
Holding coil
Induced-draft tower
Induced voltage
In-line fuse
Isolating subbase
King valve
Laminated core
Line port
Low ambient controls
Makeup water
Manifold-type distributor
Manual reset
Mechanical-draft tower
Midseated
Multistage thermostat
Multitap transformers
Natural-draft tower
Negative temperature coefficient
One-time valve
Parallel circuit
Positive temperature coefficient
Power circuit
Pressure-relief valve
Pressure-drop-type distributor
Primary winding
Refrigerant distributor
Refrigeration service wrench
Relay
Secondary winding
Sensing bulb
Series circuit
Service port
Service valve
Set point
Solenoid
Solenoid valve
Solid-state
Spring-loaded valve
Step-down transformer
Step-up transformer
Subbase
Suction service valve
System evacuation
System pump-down
Temperature differential
Thermistor
Thermostat
Transformer
Venturi effect
Venturi-type distributor
Volatile liquid
Wet-bulb temperature

FOR DISCUSSION

1. Discuss why safety controls and devices should never be jumped out and removed from the effective control circuit.
2. Explain why pumping a system down prior to a repair can save the customer both time and money.
3. Can systems without refrigerant receivers be pumped down? Explain.
4. Explain the importance of energizing the crankcase heater before initial system start-up.

REVIEW QUESTIONS

1. Service valves help the service technician accomplish which task?
   a. Reading system operating pressures
   b. Adjusting the refrigerant charge in the system
   c. Evacuating a refrigeration system
   d. All of the above are correct

2. Under normal operating conditions, the service valves are in the:
   a. frontseated position.
   b. backseated position.
   c. midseated position.
   d. cracked-off-the-backseat position.
3. What part of a filter drier is responsible for removing moisture from a refrigeration system?
   a. Strainer
   b. Desiccant
   c. Chemical salt
   d. Activated sieve
4. A large temperature drop across a liquid-line filter drier is an indication that:
   a. the drier is too large for the system.
   b. the drier is clogged with debris.
   c. the liquid-line insulation has come loose.
   d. the sight glass is defective.
5. Refrigerant receivers are:
   a. located on the low-pressure side of the system.
   b. designed to hold one half of the system's total refrigerant charge.
   c. refrigerant storage tanks.
   d. used to store excess refrigeration oil.
6. True or False: Solenoid valves can be normally open or normally closed devices.
7. What system component is energized in the compressor off cycle to remove refrigerant from the compressor oil?
   a. Solenoid valve
   b. Drain plug
   c. Crankcase heater
   d. Expansion valve
8. True or False: Pressure-relief valves should be set to release at a pressure that is slightly higher than the suction pressure of the system.
9. What is the function of a refrigerant distributor?
10. When are cooling towers most efficient?
    a. When they are located in areas where there is high humidity
    b. When they are located in areas where there is low humidity
    c. When the condenser subcooling is high
    d. Both *b* and *c* are correct
11. True or False: Forced-draft cooling towers rely on fans to move air across the water.
12. What are three common mechanisms used in thermostats?
    a. Remote bulb, sensing bulb, pressure bulb
    b. Solid-state, remote bulb, sensing bulb
    c. Electronic, solid-state, bimetal
    d. Electronic, sensing bulb, bimetal

13. Low-voltage thermostats are preferred over line-voltage thermostats because:
    a. they are typically less expensive than line-voltage thermostats.
    b. they can be used outdoors.
    c. they are sealed to prevent electric shock.
    d. all of the above are correct.
14. The purpose of a transformer is to:
    a. transform high pressure to low pressure.
    b. transform low voltage to high voltage.
    c. transform high voltage to low voltage.
    d. both *b* and *c* are possible.
15. A transformer that can be used at more than one primary voltage is called a:
    a. multicircuit transformer.
    b. multimeter.
    c. multitap transformer.
    d. bimetal transformer.
16. The differential setting on a pressure control is:
    a. the difference between cut-out and cut-in pressures.
    b. cut-out pressure added to the cut-in pressure.
    c. equal to the cut-out pressure.
    d. higher than the head pressure of the system.
17. High pressure controls:
    a. are used for low-charge protection.
    b. are used primarily as safety devices.
    c. open on a rise in pressure.
    d. both *b* and *c* are correct.
18. Three common low ambient control devices used on air-cooled equipment are:
    a. evaporator flooding, condenser flooding, dampers.
    b. condenser flooding, dampers, fan cycling.
    c. water-regulating valves, condenser flooding, fan cycling.
    d. condenser flooding, dampers, water-regulating valves.
19. Contactors
    a. can have holding coils and sets of normally open contacts.
    b. can have holding coils and sets of normally closed contacts.
    c. can have holding coils and normally open and normally closed sets of contacts.
    d. all of the above are correct.
20. True or False: The current in a series circuit is the same at all points in the circuit.

CHAPTER 3

Mechanical Troubleshooting of Vapor-Compression Refrigeration Systems

OBJECTIVES After studying this chapter, the reader should be able to:

- List several common reasons why an air-conditioning/refrigeration system may fail.
- List some reasons why refrigerant escapes from a refrigeration system.
- Identify the effects of a defective condenser fan motor on system operation.
- Identify the effects of a defective evaporator fan motor on system operation.
- Explain how a refrigerant overcharge affects system operation.
- Explain how a refrigerant undercharge affects system operation.
- Explain the importance of a correctly mounted thermostatic expansion valve thermal bulb.
- Explain the effects of reduced water flow through a water-cooled condenser.

INTRODUCTION

Having obtained a general understanding of the basic vapor-compression refrigeration system, the technician now needs to begin acquiring skills to keep these systems up and running in the most efficient manner possible. A system that is operating efficiently will provide years of service to the equipment owner. It is the responsibility of the field technician to identify and repair minor system problems in an effort to prevent them from becoming major system failures. This chapter deals with mechanical problems within the refrigeration circuit that will prevent the system from operating properly. Identifying these problems is a major task that needs to be performed in the field. Later on in the text, electrical failures that often occur within a system are addressed.

OVERVIEW

Two important issues that separate the excellent service technician from the average technician are the speed and the accuracy with which the technician is able to properly evaluate a system and determine the cause for system failure. Although the technician may not be able to immediately repair the problem, an important first step is to identify the problem and decide what path to take to resolve it. Good troubleshooting techniques enable the technician to accomplish this goal. The accurate diagnosis of a system is a key element in the prompt repair of the customer's unit. This diagnosis allows other company personnel to order needed parts, schedule the repair, get customer approval for the repair, and see it to its completion. On the other hand, an inaccurate diagnosis leads to the ordering of incorrect parts, increased repair time, lost company revenue, and, more important, a dissatisfied customer.

System troubles can usually be categorized as either electric or mechanical. Mechanical problems include both the refrigerant and the system in which it is contained, as well as physical component damage. It is important that service technicians be able to distinguish between the two, because an electrical problem may give the appearance of being a mechanical problem and vice versa. For example, a motor that is not operating when it is supposed to may be experiencing internal mechanical failure or there may be a problem with the electric circuit that feeds power to the device. A lightbulb may not light when the wall switch is turned on because the switch or associated circuit is defective or, more likely, the bulb has burned out. This represents a failure of the device, not the circuit supplying power to it. Mechanical problems can refer to:

- Physical damage to a component resulting from age or mishandling
- Foreign matter that enters a device and prevents proper operation
- Improper installation of a component that leads to improper operation
- Improper refrigerant charge (either overcharge or undercharge)

System problems stemming from electrical circuit problems are covered later in the text.

EVAPORATOR AND CONDENSER FAN MOTOR PROBLEMS

When it is established that the correct voltage is supplied to a motor and the windings and starting components are functioning properly, there is most likely a mechanical problem with the motor if it fails to operate. The most common problems with motors arise from improper airflow, improper lubrication, improper pulley alignment, or improper belt tension. Proper **preventive maintenance** on an air-conditioning or refrigeration system should help eliminate mechanical motor failure. It is also important that the proper type of motor be used for the specific application. For example, if a motor is to be used in a wet environment it should be sealed to prevent water from entering it. If an open motor is used instead of a sealed motor, its expected service life will be greatly reduced. A motor that is used within its design and application range will provide years of satisfactory operation; and one that is properly maintained will last even longer.

Improper Airflow

Motors may exhibit symptoms that lead the technician to conclude that the device is defective when, in fact, another system problem is at fault. A common factor that leads to the unnecessary replacement of motors is insufficient airflow. Many motors rely on air passing over them to help keep them cool and within safe operating ranges. If the airflow is restricted, the motor may overheat and cut off on an internal **overload protector.** The overload protector is a device that will open its contacts and de-energize a motor when temperatures exceed the design ratings. The technician may conclude that the motor was not operating properly, causing the reduction in airflow, when the opposite was actually the case.

Before concluding that the motor is defective, the technician must first establish that the air path is not restricted. To accomplish this, the following must be checked. For an *evaporator fan motor,* make certain that:

- *All air filters are clean.* If the system has no air filters, the return side of the evaporator coil must be visually inspected.
- *The evaporator coil is clean.* Dust and the condensation accumulating on the coil will form a glue-like substance on the coil surface and restrict airflow. Since the evaporator coil is usually inaccessible from both sides, shining a light through the coil to visually inspect it may be possible. If the light shines through the coil, it should be clean.
- *All duct lining is intact.* If not properly installed, acoustical lining has a tendency to come loose and block the airstream. Duct lining can be found in both the supply and return ducts.
- *The return-air grills are not blocked with furniture, boxes, or other obstructions.*
- *The supply registers are open and not blocked.* Closed or blocked supply registers will restrict airflow.
- *The fan blades are clean and not caked with dirt.* If the blades are caked with dirt, the ability of the blade to scoop and move air is greatly reduced. This usually occurs on the blades of forward-curved centrifugal fans, or squirrel cages (Figure 3-1).

For a *condenser fan motor,* make certain that:

- *The condenser coil is not dirty and/or blocked.* If the condensing unit is located outside, leaves, dirt, bushes, and other debris must be cleared from the coil's surface. It is good field practice to mount the condensing unit on a pad or frame that will lift it above ground level (Figure 3-2).
- *The condensing unit has at least as much clearance as indicated by the manufacturer.* If too little space is provided, the hot discharge air may recirculate back through the coil (Figure 3-3).

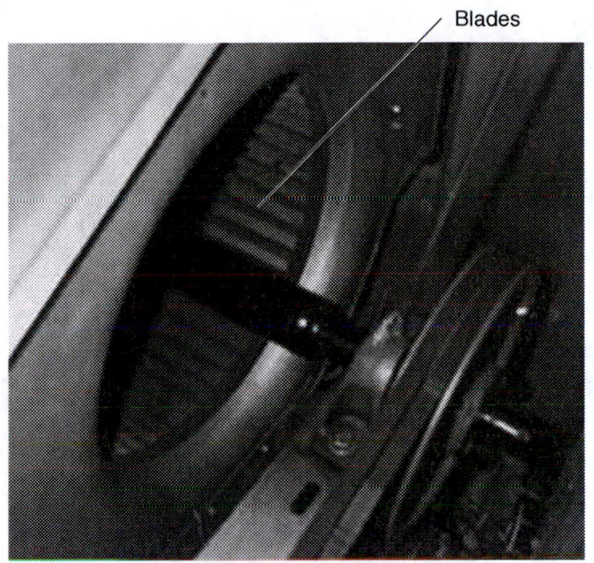

FIGURE 3-1 The blades of a squirrel cage can get clogged with dirt, reducing its ability to move air. *(Photo by Eugene Silberstein)*

- *If multiple condensing units are located next to each other, they do not discharge into the other units.* This can cause high-pressure-related problems and inefficient cooling.
- *The fan blades are clean and not caked with dirt.* If the blades are caked with dirt, the ability of the blade to scoop and move air is greatly reduced. This usually occurs on the blades of forward-curved centrifugal fans, or squirrel cages.
- *The condensing unit is not located under any low overhangs or other overhead obstructions.* They will also cause the condenser's discharge air to recirculate through the coil (Figure 3-4).

Once the air path has been cleared, the motor should operate properly. If the air path is unobstructed to begin with, other mechanical problems may be the cause of a motor's premature failure.

Improper Motor Lubrication

Depending on the type and style, motors may require periodic lubrication. Motors that are installed in equipment that is not easily accessible are usually "permanently lubricated" and do not need periodic oiling. These motors are not equipped with **oil ports,**

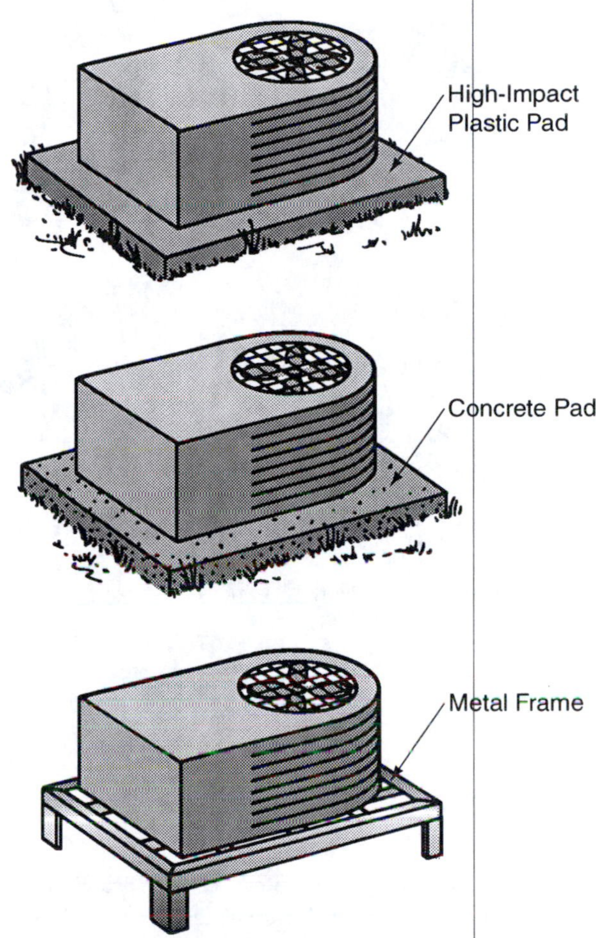

FIGURE 3-2 Various types of pads used to support condensing units

FIGURE 3-3 Condensing unit for a split system located so that it has adequate airflow and room for service. *(Courtesy Carrier Corporation)*

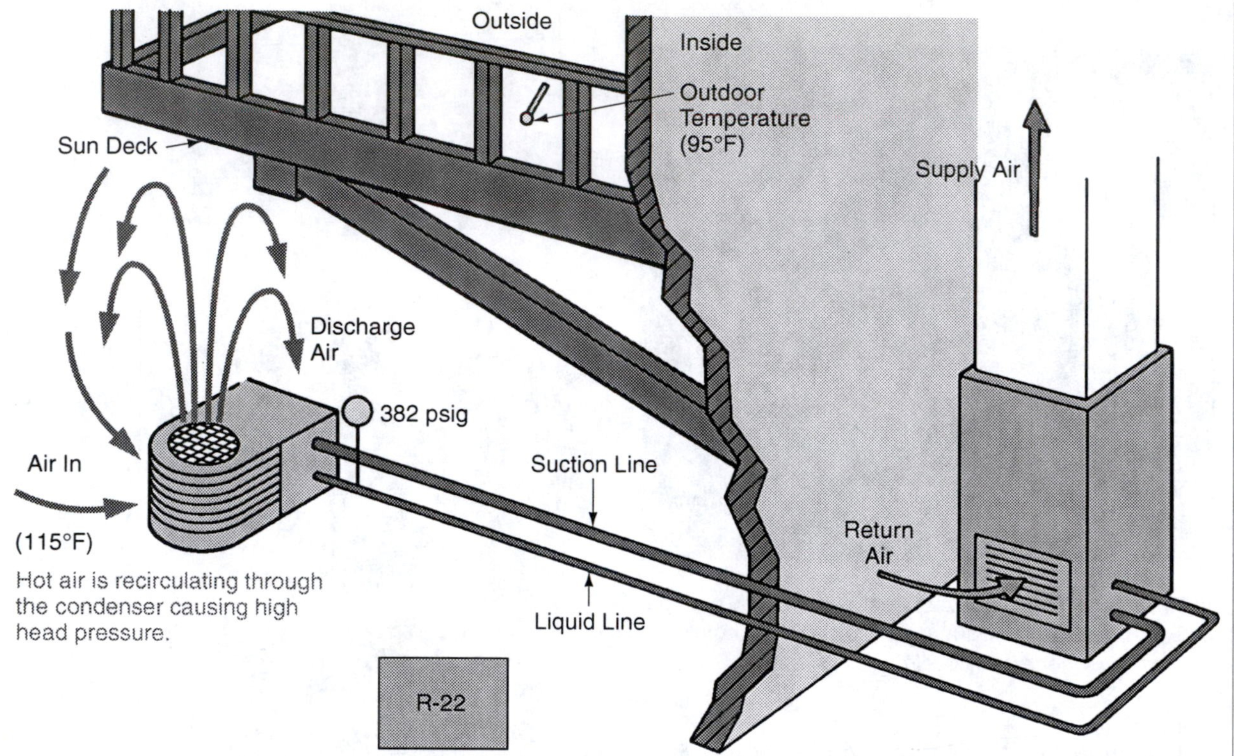

FIGURE 3-4 Condensing unit installed so that the discharge air is hitting a barrier in front of it

which are openings in the motor casing or frame that permit the oil to reach the bearings (Figure 3-5). Some other motors are equipped with **sleeve bearings,** which tend to be very quiet in operation, and should be oiled as part of annual or semiannual preventive maintenance. The type and the amount of oil to be used are usually specified by the manufacturer. If no lubrication data are available, four or five drops of a medium-grade motor-lubricating oil should be used in each oil port. Motors with sleeve bearings are usually located near the occupied space because of the low noise level and easy access.

Heavier-duty motors are often equipped with **ball bearings.** These bearings are noisier than sleeve bearings and are normally lubricated with grease instead of oil. Motors with ball bearings are equipped with **grease fittings** instead of oil ports. A grease gun charged with the proper grease must be used, and the relief plug (Figure 3-6) must be removed to prevent the grease seal from being pushed out of place if too much pressure is built up. After lubricating, the plug must be replaced.

Unfortunately, many equipment owners do not call for service until a system problem arises. Upon the technician's arrival on a service call, improper motor lubrication will become evident if the motor does not turn freely when all pulleys have been removed. Saving the motor may be possible by properly oiling and lubricating the bearings, provided the motor has not operated for an extended period of time without proper lubrication. If the motor has been overheating and bearing damage has occurred, the motor will need to be either repaired or replaced, depending on the size of the motor. Small motors, fractional to about 5 horsepower, are usually replaced, while larger motors are often rebuilt. Rebuilding smaller motors is not economically feasible. To determine if excessive bearing damage has occurred, the shaft of the motor should be inspected. The shaft should have some play in and out of the motor but almost no play from side to side. This side-to-side play is referred to as **endplay.** If there is endplay, the bearings of the motor are no longer

FIGURE 3-5 Oil ports are covered with plastic or rubber plugs to prevent dirt and dust from entering the motor. *(Photo by Bill Johnson)*

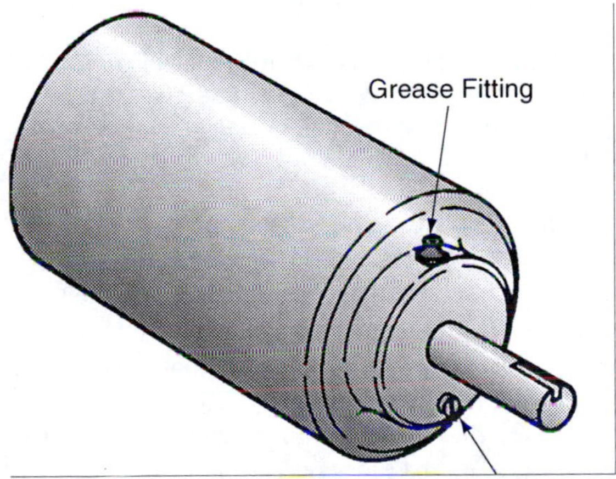

FIGURE 3-6 Motor bearing using grease for lubrication. Notice the relief plug.

functional and the motor must be repaired or replaced. Oiling or lubricating a motor that has defective bearings will not fix the problem.

Improper Pulley Alignment

For a motor to operate properly throughout its expected useful life, the pulleys and any other devices connected to the shafts of the motor should provide evenly distributed force on the shaft and bearings. One way to ensure that the force that is exerted on the shaft is even is to make certain that the pulleys, both the **drive pulley** and the **driven pulley,** are aligned properly. The drive pulley is the pulley connected directly to the shaft of the motor, while the driven pulley is connected to the shaft that turns the fan or blower (Figure 3-7). If these pulleys are not aligned properly, excessive pulley wear and bearing damage can occur.

Excessive pulley wear can be identified when an actual groove has been worn into the interior wall of the pulley. If inspecting the pulley is visually difficult, feeling the groove of the pulley will quickly indicate any imperfections in its surface, which should normally be flat with no waves, gaps, or notches. When pulley wear is observed, the pulley must be replaced. Upon replacement of the pulley, the alignment between the drive and driven pulley must be checked with a straightedge (Figure 3-8). Bearing damage, as stated earlier, can be identified by side-to-side play in the motor shaft. Again, after the repair or the replacement of the motor has been performed, the pulley alignment must be carefully checked to reduce the possibility of another failure. Key indicators that pulleys are not aligned properly include:

- Uneven belt wear
- Excessive belt breakage
- Belts coming loose from the pulleys without breaking

FIGURE 3-7 The drive pulley is located on the motor shaft *(left)*, while the driven pulley is connected to the blower *(right)*. *(Photo by Eugene Silberstein)*

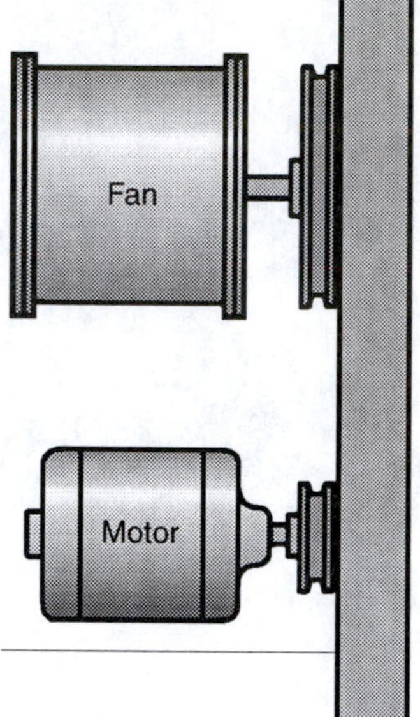

FIGURE 3-8 Pulleys must be in proper alignment, or belt and bearing wear will occur. The pulleys may be aligned using a straightedge.

Normal preventive maintenance should always include belt inspection. Any imperfections or uneven wear patterns in the belts should alert the technician to a potential problem. Changing the defective belt is only part of what should be done at this point. The cause for the uneven wear must be identified and corrected as well. The term *preventive maintenance* implies maintenance that is geared toward the prevention of future system problems. System downtime is always shorter during a preventive maintenance service call than during an emergency repair service call. For this reason, system downtime is greatly reduced when proper preventive maintenance is performed.

Excessive belt breakage should also alert the technician that an adjustment problem exists. If the broken belt shows uneven wear or imperfections, chances are that an alignment problem exists. If the belt shows even wear, the belt may have been adjusted too tightly. Improper belt tension is discussed in the next section. In either case, the cause for the breakage must be identified and corrected to prevent future failure.

On occasion, a technician will arrive on a service call to find that the belt has come loose from the pulleys. This is usually the result of a pulley misalignment and a belt installed too loosely. This problem is more common than one may expect for a very simple reason. When pulleys are misaligned, the belt tends to squeal or make a loud whining sound. To alleviate this sound, some technicians loosen the belt tension to reduce the noise level. This is not good field practice, since it remedies the symptom but not the problem. The belt may now come loose from the pulleys very easily, leading to a "no-cooling" call sometime in the future.

Improper Belt Tension

Once the pulleys are aligned correctly, the belt tension must then be checked. The rule of thumb that most field technicians use to approximate correct belt tension is as follows:

> There should be approximately one inch of play in the belt when the belt is installed correctly. For larger belts with pulleys located far apart from each other, there should be a little more play. For smaller belts with pulleys located close together, there should be a little less play in the belt.

Incorrect belt tension can lead to a number of system problems including the following:

Belt Tension Too Tight

- Bearing damage
- Excessive belt breakage
- Excessive noise levels
- Motor overheating, premature motor failure

Belt Tension Too Loose

- Belt slippage
- Reduced cooling
- Reduced airflow (possibly leading to low suction pressure and a frosted evaporator coil)
- Belt coming loose from the pulleys
- Excessive pulley wear
- Motor overheating

The service technician can easily determine if the belt tension is too loose by inspecting the grooves of the pulley. If the grooves have been polished to a near-mirror finish, the belt is slipping and too loose. The belt tension must be increased, and the pulleys should be replaced. The reason for replacing the pulleys is simple. If the next technician arrives at the location and sees the mirror-like surface, the tension may be increased when, in fact, the tension was correct to begin with. In addition, pulleys tend to grip belts better when the groove surfaces are not perfectly smooth. One tool that can be used to ensure proper belt tension is the **belt tension gauge,** which is more precise than the rule of thumb stated earlier. If the problem is caught early enough, a simple adjustment of the belts and pulleys can eliminate a very costly repair and excessive system downtime.

Safety Tips:

When working on or around belts and pulleys, always keep the following in mind:

- Never try to stop a motor or blower by hand.
- Never grab a belt that is turning.
- Always keep fingers away from the area between the belt and the pulley.
- Do not wear loose-fitting clothing, especially neckties, when working around moving machinery.
- Make certain all belt guards and other safety devices are replaced securely after performing service.

REFRIGERANT-CHARGE-RELATED PROBLEMS

For an air-conditioning or refrigeration system to operate properly, the amount of refrigerant in the system—referred to as the **system charge**—must be correct. The field technician must be familiar with the system and all of its components to ensure that the amount of refrigerant added to the system is well within the acceptable range. Improper refrigerant charge can lead to many system problems, including the following:

Excessive Refrigerant Charge (System Overcharge)

- Reduced condenser efficiency
- Reduced evaporator efficiency (reduced cooling)
- Reduced system efficiency (increased power consumption)
- Reduced evaporator superheat
- Increased possibility of **floodback** to compressor (liquid refrigerant getting back to the compressor)
- Higher operating pressures
- Higher compressor operating temperatures

Low Refrigerant Charge (System Undercharge)

- Reduced evaporator efficiency (reduced cooling)
- Increased evaporator superheat
- Low operating pressures
- Reduced system efficiency (increased power consumption)
- Higher compressor operating temperatures

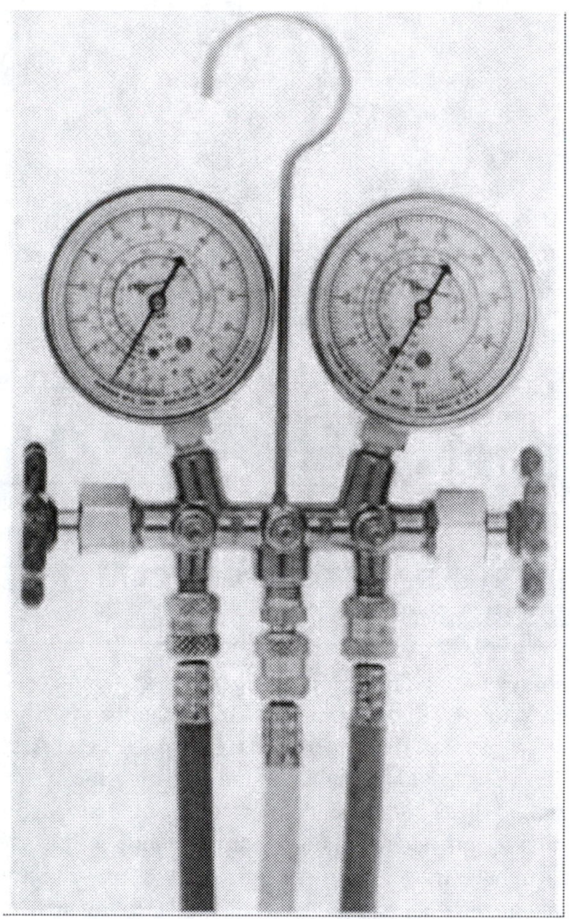

FIGURE 3-9 Standard gauge manifold. *(Courtesy Robinaire Division, Sealed Power Corporation)*

System Pressure Readings

The installation of a set of refrigeration gauges (Figure 3-9) on the system will provide the technician with information about what is taking place inside the system. A technician taking system pressures can be equated to a medical doctor taking a patient's blood pressure. The **gauge manifold** consists of two gauges—the high-pressure gauge, which is color-coded red, and the low-pressure gauge, which is color-coded blue. The outer, black-numbered scales on the face of the gauges provide the pressure in **pounds per square inch gauge (psig).** The inner scales provide the corresponding saturation temperatures for various refrigerants.

The data on the gauge face are exactly the same as the data found in the pressure/temperature chart

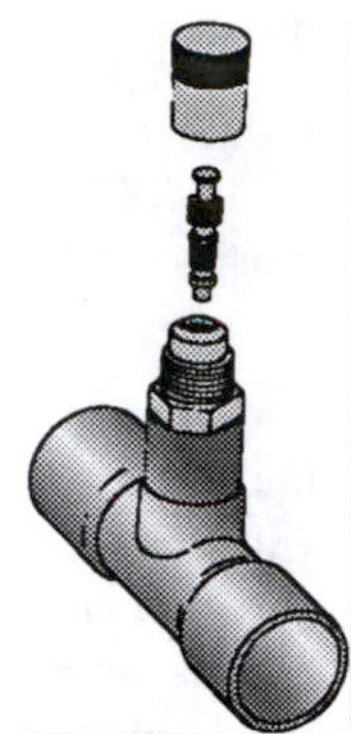

FIGURE 3-10 Schrader valve assembly. *(Courtesy J/B Industries)*

(Figure 1-1). If the gauges are being used to evaluate a system that contains a refrigerant not shown on the gauge, a separate pressure/temperature chart must be used to obtain the desired information. This valuable data, if interpreted correctly, will help lead the technician to a rapid and accurate system diagnosis. The gauge manifold enables the technician to:

- Read the system's high-side pressure (also referred to as head pressure)
- Read the system's low-side pressure (also referred to as suction pressure or back pressure)
- Read the condenser saturation temperature
- Read the evaporator saturation temperature
- Determine amount of condenser subcooling
- Determine evaporator superheat

If the system is equipped with **Schrader valves** (Figure 3-10), the high- and low-side pressure readings of a system can be obtained by connecting the appropriate gauge hose to any available port in the system. The Schrader valve operates in a manner similar to that of the valve on car or bicycle tire. When the valve stem is pushed in, the valve opens. When pressure on the stem is released, the valve pushes closed (Figure 3-11). High-side service ports are most commonly located on the discharge valve of the compressor or the liquid line leaving the condensing unit. Low-side service ports are commonly located in the suction line near the compressor or at the evaporator outlet.

The next sections of this text provide detailed procedures for the installation and removal of a gauge manifold for an air-conditioning system equipped

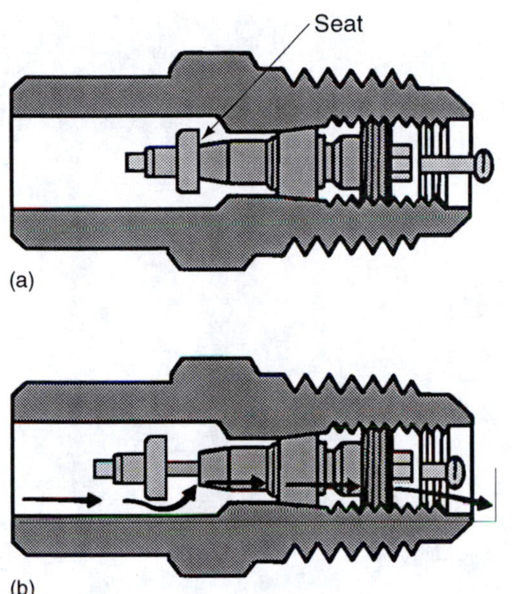

FIGURE 3-11 The Schrader valve operates in a manner similar to the valve on a car tire. The valve is a NC device (a). When the center pin is pushed in, the valve opens (b). Gauge-manifold hoses are equipped with depressors that push the pins in. *(Courtesy J/B Industries)*

with service valves. If the system is **hermetically sealed,** with no service valves or ports, a line-tap valve must be installed to take pressure readings.

Line-tap valves come in a variety of styles. Some are designed to be used as a tool and must be removed after use (Figure 3-12a), while others are designed to be permanently installed on a system (Figure 3-12b). Line-tap valves that are to be removed after use should be installed on **process tubes** (Figure 3-13) to make removing the valve possible. Process tubes are stems in the refrigerant piping that are designed to allow system access. These process tubes must be properly pinched off and resealed after use to ensure that the refrigerant charge of the system is not lost. A **pinch-off tool** is shown in Figure 3-14.

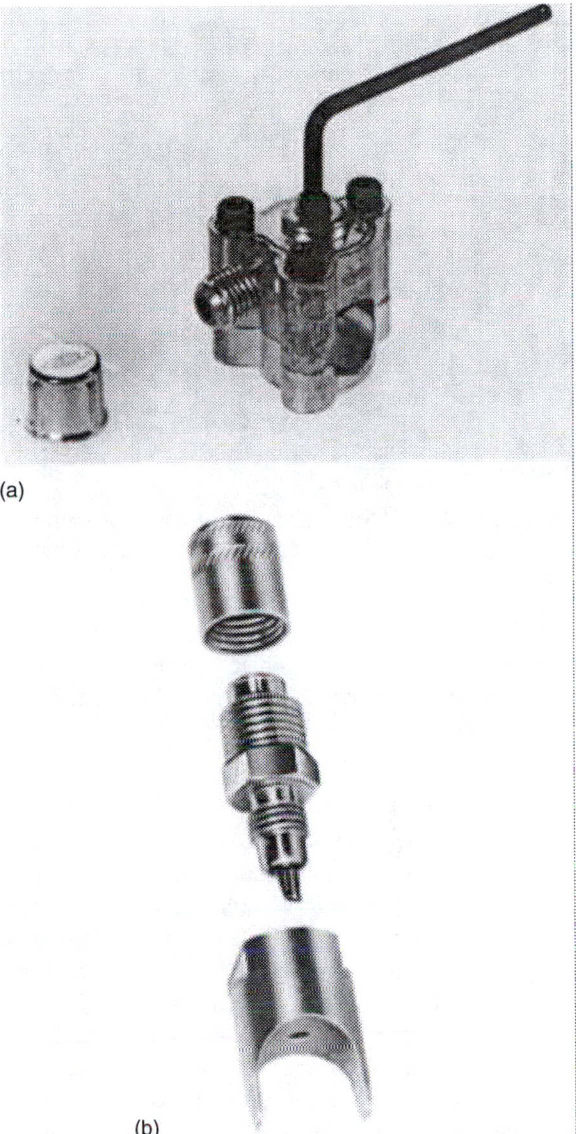

FIGURE 3-12 (a) Line-tap valve that must be removed after use. This type of valve must be installed on a process tube. *(Photo by Bill Johnson)* (b) This type of line-tap valve can be permanently installed on the system. *(Courtesy J/B Industries)*

Gauge-Manifold Installation Procedure

If the air-conditioning system being worked on contains Schrader-type access ports, a gauge manifold can be connected to the system very easily. All that is entailed is the tightening of the hose on the valve. The pin depressor located inside the hose will push the valve pin in, providing immediate system access to the technician.

However, if the air-conditioning system is equipped with service valves, standardized procedures for installing a gauge manifold will reduce the possibility

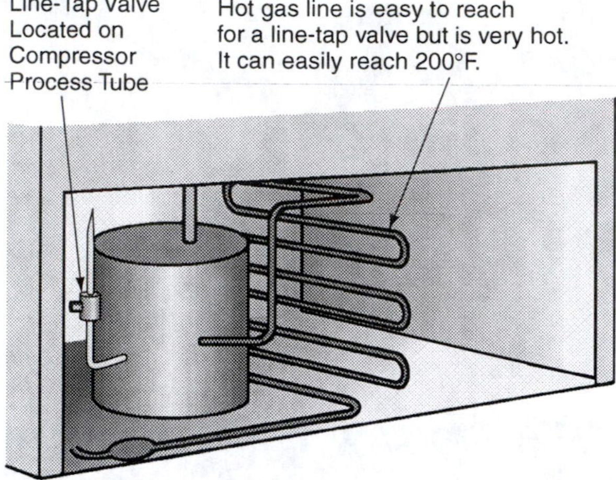

FIGURE 3-13 Process tubes are commonly found on hermetic compressors. This is where line-tap valves must be installed.

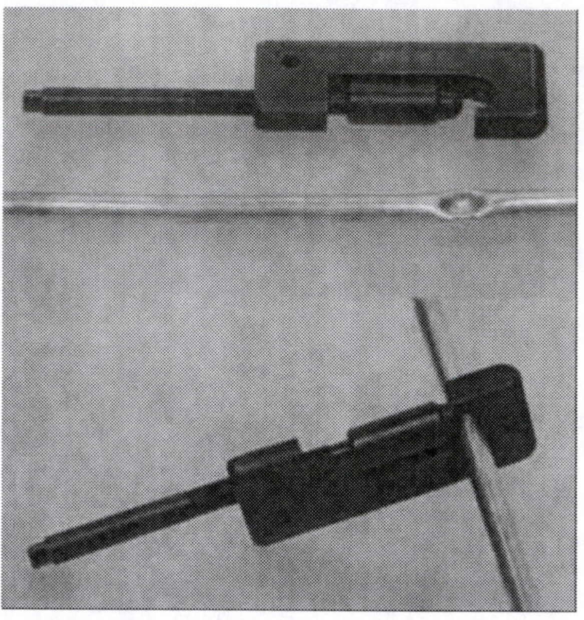

FIGURE 3-14 Special pinch-off tool used to seal off process tubes. *(Photo by Bill Johnson)*

of allowing atmosphere to enter the refrigerant circuit. The following is a step-by-step procedure that, if followed correctly, ensures the proper installation of a gauge manifold on the system.

Gauge-Installation Procedures for Systems Equipped with Service Valves

1. Make certain that both gauges on the manifold are properly **calibrated.** Calibrate if necessary (Figure 3-15a); both the high- and low-side indicating needles should point directly to zero. To properly check the calibration of the gauges, the high- and low-side hoses must be removed from the **blank ports** of the manifold. The blank ports are those that have no internal connections to the center hose of the manifold. The calibration screw on the face of the gauge can be turned slowly in the direction opposite to that of desired movement.
2. Remove the stem cap from the suction service valve and make certain that the valve is in the backseated position (Figure 3-15b). Remember that backseated is the position when the stem is turned completely counterclockwise.
3. Remove the port cap from the suction service valve and place it on the blank port of the gauge manifold (Figure 3-15c).
4. Connect the low-side hose from the gauge manifold to the service port of the suction service valve.
5. Remove the stem cap from the discharge, or liquid-line, service valve, and make certain that the valve is backseated.
6. Remove the port cap from the high-side service valve, and place it on a blank port of the gauge manifold.
7. Connect the high-side hose from the gauge manifold to the high-side service port (Figure 3-15d).
8. Make certain that the middle hose on the gauge manifold is secured to the blank port.
9. Make certain that both valves on the gauge manifold are in the closed position.
10. Place the high-side service valve in the cracked-off-the-backseated position.
11. Open both valves on the gauge manifold.
12. Loosen the hose connection on the suction service valve for approximately 2 seconds. This will purge any air from the gauge manifold. (The hissing sound of escaping vapor at this point is completely normal and expected.)
13. Tighten the hose connection on the suction service valve.
14. Close both valves on the gauge manifold.
15. Place the suction service valve in the cracked-off-the-backseated position.

FIGURE 3-15(a) Calibrating screw on the gauge manifold. The screw is turned clockwise to move the needle counterclockwise, and vice versa. *(Photo by Eugene Silberstein)*

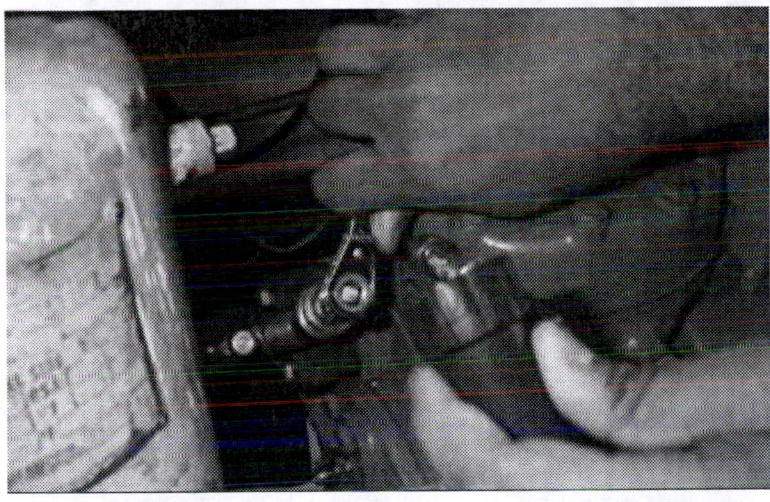

FIGURE 3-15(b) The service valve must be backseated before removing the port cap. *(Photo by Eugene Silberstein)*

FIGURE 3-15(c) The port caps should always be placed on the blank ports of the manifold so they are not lost. *(Photo by Eugene Silberstein)*

FIGURE 3-15(d) Connecting the low-side hose to the suction service valve. *(Photo by Eugene Silberstein)*

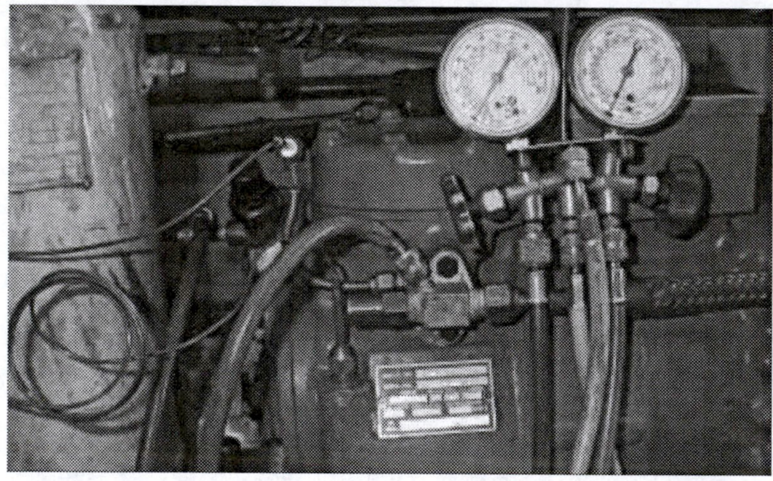

FIGURE 3-15(e) The gauge manifold is installed on the system. *(Photo by Eugene Silberstein)*

16. Replace both stem caps on the low-side and high-side service valves.
17. The gauge manifold is now properly installed on the system (Figure 3-15e).

This procedure can be used whether the system is operating or not. If no refrigerant flows through the manifold in Step 12, the system is **flat**, or void of refrigerant, and should be checked to locate and repair the leak. Locating system leaks is covered later in this chapter.

Gauge-Removal Procedures for Systems with Service Valves (System Operating with Low Side in a Vacuum)

1. Remove the stem cap from the high-side service valve, and backseat the valve.
2. Open the high-side and low-side valves on the gauge manifold until the pressure drops to about 10 psig (do not allow the high side of the manifold to pull into a vacuum).
3. Close the low-side valve on the gauge manifold.
4. Remove the stem cap from the suction service valve, and backseat the valve.
5. Briefly open and close the low-side valve on the gauge manifold.
6. Close the high-side valve on the manifold.
7. Remove the high-side hose from the high-side service valve.
8. Remove the low-side hose from the suction service valve.
9. Replace the service-port caps on the high- and low-side valves.

10. Replace the hoses on their proper blank ports.
11. Replace the stem caps on the high- and low-side service valves.
12. The gauge manifold has now been removed from the system.

Gauge-Removal Procedures for Systems with Service Valves (System Operating with Low-side Pressure between 0 and 40 psig)

1. Remove the stem cap from the high-side service valve, and backseat the valve.
2. Open the high-side and low-side valves on the gauge manifold until the manifold pressures equalize.
3. Remove the stem cap from the suction service valve, and backseat the valve.
4. Close the high- and low-side valves on the manifold.
5. Remove the high-side hose from the high-side service valve.
6. Remove the low-side hose from the suction service valve.
7. Replace the service-port caps on the high- and low-side valves.
8. Replace the hoses on their proper blank ports.
9. Replace the stem caps on the high- and low-side service valves.
10. The gauge manifold has now been removed from the system.

Gauge-Removal Procedures for Systems with Service Valves (System Operating with Low-side Pressure above 40 psig)

1. Remove the stem cap from the high-side service valve, and backseat the valve.
2. Open the high-side and low-side valves on the gauge manifold, and allow pressures to equalize.
3. Remove the stem cap from the suction service valve, and frontseat the valve.
4. Allow the manifold pressure to pull into as low a vacuum as the compressor can pull.
3. Close the low-side valve on the gauge manifold.
4. Backseat the suction service valve.
5. Open the low-side valve on the gauge manifold, allowing the manifold pressures to equalize.
6. Close the high- and low-side valves on the manifold.
7. Remove the high-side hose from the high-side service valve.
8. Remove the low-side hose from the suction service valve.
9. Replace the service-port caps on the high- and low-side valves.
10. Replace the hoses on their proper blank ports.
11. Replace the stem caps on the high- and low-side service valves.
12. The gauge manifold has now been removed from the system.

Gauge-Removal Procedures for Systems with Service Valves (System Compressor Is Inoperative)

1. Remove the stem cap from the high-side service valve, and backseat the valve.
2. Remove the stem cap from the low-side service valve, and backseat the valve.
3. Make certain the high- and low-side valves on the manifold are closed.
4. Remove the high-side hose from the high-side service valve.
5. Remove the low-side hose from the suction service valve.
6. Replace the service-port caps on the high- and low-side valves.
7. Replace the hoses on their proper blank ports.
8. Replace the stem caps on the high-and low-side service valves.
9. The gauge manifold has now been removed from the system.

Evaluating and Interpreting System Pressures

Once the gauge manifold has been properly installed on the system and pressure readings have been obtained, these data must be properly interpreted to reach the correct conclusion regarding system status. Pressure readings can be taken whether or not the system is operating. Both situations provide the technician with useful troubleshooting information. Taking pressure readings on an inoperative system will help to:

- Determine if the refrigerant charge has been completely lost.
- Determine if noncondensable gases are in the system.

Taking system pressure while the system is operating will enable the technician to:

- Determine evaporator and condenser saturation temperatures and pressures.
- Help evaluate evaporator and condenser effectiveness and efficiency.
- Help evaluate the pumping effectiveness of the compressor.
- Determine if the refrigerant charge needs to be adjusted.

COMPLETE LOSS OF REFRIGERANT CHARGE. Installing a gauge manifold on an inoperative system will, as stated earlier, tell the service technician if the system is holding pressure. If, upon installation of the manifold, the gauges register 0 psig, the system has lost its charge and is said to be, as stated earlier, flat. The leak must be located and repaired, before evacuating and recharging, in order to resume normal system operation. Systems develop leaks for a number of reasons. Some of the most common follow:

- Refrigerant piping surfaces that rub together and wear holes in the lines
- System vibration that causes stress fractures in fittings and lines
- Poor-quality soldered or brazed joints
- Refrigerant lines inadvertently getting damaged or cut
- Service valves not fully backseated
- Stem caps and port caps left off service valves

A number of different methods can be utilized to locate the leak, including:

- Listening for an audible hiss from the leak
- Using a liquid leak detector, or bubble solution
- Using a halide leak detector
- Using an electronic leak detector
- Using an ultraviolet (UV) light

If the system has a total loss of charge, chances are that the leak is very large. Quite often in this case, a fitting has cracked, a solder joint has come loose, a compressor terminal has blown off, or some other system fracture has developed. For this reason, it may be very easy to locate the leak by pressurizing the system with dry nitrogen and listening for the escaping gas (Figure 3-16a).

SERVICE NOTE: *Do not pressurize the system with refrigerant for leak-detection purposes. This will violate laws set forth by the Environmental Protection Agency (EPA).*

One very popular method for leak detection is the use of a soap-bubble solution. By pressurizing the system with dry nitrogen and applying the solution to all solder joints and seams in the refrigerant piping circuit, the leak will be identified by the formation of bubbles (Figure 3-16b). This method is very popular mainly because smaller leaks, which are generally harder to locate, tend to produce the largest and most visible bubbles.

Halide leak detectors (Figure 3-16c) operate on the concept that burning refrigerant will cause a flame to change color. The device usually utilizes propane, or mapp gas; an open flame; and a hose that is guided over the piping in search of the leak. The flame gets a portion of its needed oxygen through the rubber hose. Refrigerant is pulled through the tubing with the surrounding air, and when a leak is reached, the flame on the device will change color. This type of leak detector is more effective when a larger leak exists. It is a very slow method but still relatively popular, due in part to the fact that the equipment is relatively inexpensive and the propane refills cost only a few dollars.

Electronic leak detectors have become the method of choice by many field technicians, (Figure 3-16d). Technological advances over the years have made these devices more reliable and accurate. To utilize the electronic leak detector, the system must contain at least some refrigerant. If the system is operating with a small leak, plenty of pressurized refrigerant is left in the system. If, however, the system is flat, it must first be pressurized with nitrogen, to which a trace of R-22 has been added. The trace gas must be added since the device will not detect the nitrogen by itself. Most electronic leak detectors have the capability of detecting several different refrigerants, so the literature accompanying the detector should be read.

One final method for leak detection that has grown in popularity over the last few years is the UV leak-detection system. This system (Figure 3-16e) utilizes a UV light, which causes a refrigerant additive to glow when it is present on the surface of a refrigerant

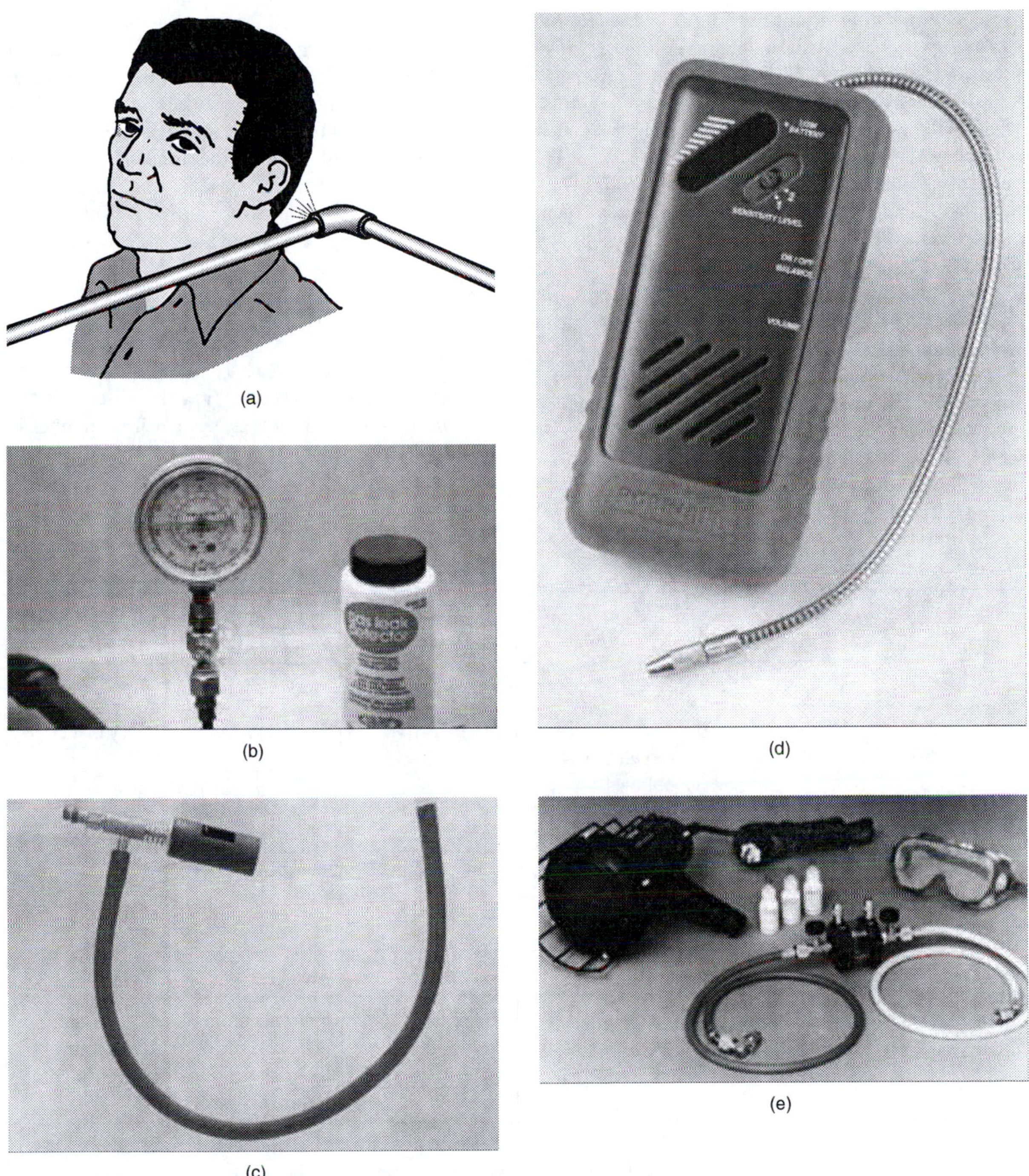

FIGURE 3-16 Various methods and equipment used for locating refrigerant leaks. ([a], [b] *and* [c] *Photos by Bill Johnson;* [d] *Courtesy White Industries,* [e] *Spectronics Corporation*)

pipe. This additive is normally introduced into the system upon initial installation or at some point in time after that. If a leak has developed, the technician simply shines the light on the refrigerant piping, which will glow when the leak has been located. This method is very quick and can even locate leaks in the center of evaporator and condenser coils. This method of leak detection tends to be the most expensive from the "first cost standpoint" but tends to pay for itself in a very short time by locating leaks faster. One major drawback with the UV leak-detection method is that its effectiveness is greatly reduced when used outdoors in bright sunlight.

The aforementioned methods can be used on systems to located small and/or large leaks, and the circumstances of each particular job should determine which is the best method to employ. A summary of the leak-detection methods described is given in Figure 3-17.

NONCONDENSABLE GAS. Technicians define a **noncondensable gas** as one that cannot condense into a liquid at all, or at least within the normal operating ranges of the equipment they work on. Since these gases cannot condense into a liquid, they simply take up useful space in the condenser, reducing the condenser's effective surface area. This causes the system operating pressures to rise. (On systems with an automatic expansion valve, the low-side pressure will remain the same while the head pressure increases.) Two common noncondensables are:

- Air
- Nitrogen

Air in the Refrigeration System. Air can find its way into an air-conditioning system in a number of different ways. The most common are:

- Improper system evaluation before evacuation
- Insufficient evacuation time

	Audible	Soap Bubbles	Halide Detector	Electronic Detector	UV
Large Leaks	Very effective	Effective	Very effective	Very effective	Very effective
Very Small Leaks	Not effective	Very effective (creates larger bubbles)	Not effective	Very effective	Very effective
Ease of Use	Very easy	Easy	Easy	Easy	Very easy
Portability of Equipment	No equipment required	Extremely portable (small bottle)	Very portable	Very portable	Bulky, & UV lamps are fragile
Speed of Use	Very fast	Fast	Very slow	Slow	Very fast
Advantages	No equipment necessary	Thick liquid adheres well to pipe surfaces	Reliable	Very accurate & effective on leaks as small as ½ ounce a year	Can locate leaks in the center of coils
Drawbacks	Only good for large leaks	Not very effective on very large leaks	Very slow & not very effective on small leaks	Cannot locate leaks in the center of a coil	Bulky equipment, & additive must be added before a leak develops

FIGURE 3-17 Comparing the various options available for leak detection

- Improper gauge installation/removal
- Low-pressure-side leak

Most of these situations can be easily avoided by using common sense and, maybe, an inexpensive pressure control.

Before a system is evacuated prior to charging, the entire piping circuit must be properly evaluated to ensure that all sections of piping are open to each other. This avoids the possibility of trapping air in a completely isolated portion of the system. Solenoid valves, manual flow valves, and other flow controls should all be in the open position during evacuation. Service valves should all be placed in the midseated position to ensure that the best possible evacuation is performed. Figure 3-18 shows a system equipped with a liquid-line solenoid that is closed during evacuation. Note the path that must be taken and all of the restrictions that exist. Figure 3-19 shows the same system, but this time the solenoid valve is open.

NOTE: *If the solenoid valve is to be energized during evacuation, the compressor circuit must be deenergized to ensure that the compressor does not cycle on during the evacuation process.*

One common mistake that is often made in the field relates to insufficient evacuation time. Many field technicians utilize their gauge manifolds for system evaluation as well as evacuation. During evacuation, they read the vacuum scale on the low-side gauge to determine if an adequate vacuum has been reached. The commonly used gauge manifold displays the vacuum region as a measurement from 0 psig down to a "perfect vacuum" of 30 inches of mercury (actually 29.921 inches), which corresponds to a range of **absolute pressure** from 14.7 pounds per square inch absolute (psia) down to 0 psia. (psia = psig + 14.7). The absolute pressure, denoted in psia, takes into account the pressure exerted by the some 50-mile-high column of air over a one-square-inch area on the surface of the earth (Figure 3-20). The gauge pressure, denoted in psig, does not take this element into account and therefore displays 0 psig at atmospheric pressure (Figure 3-21).

This method of measuring the system vacuum is no longer the acceptable field standard! In addition to the gauge manifold, the field technician must connect a **micron gauge,** which reads in **microns,** to the system to determine if the vacuum is sufficient. The micron scale divides the vacuum range, from 0 psig down to 29.921 inches Hg (inches of mercury), into approximately 760,000 subdivisions, each referred to as a *micron.* On the micron scale, therefore, atmospheric pressure will be displayed as 760,000 microns, while a perfect vacuum will be displayed as 0 microns. The widely acceptable level of vacuum that should be reached and maintained on a system before charging is about 1000 microns, which corresponds to a vacuum pressure of 29.88 inches Hg. This ensures that the moisture from the system has boiled off and has been removed from the system.

Improper gauge installation and removal can also allow air to enter the air-conditioning system. All hoses on the gauge manifold should be properly purged to prevent the introduction of air into the refrigerant lines. Proper purging entails pushing a small amount of refrigerant from the high side of the system or the refrigerant tank through the gauges as they are installed. It is also good field practice to keep the gauge manifold charged with refrigerant when not being used to minimize the chances of having foreign matter, including air, enter the hoses. If air is present in the hoses, it will be pushed into the lines as refrigerant is added to the system.

Systems that are not equipped with low-pressure switches are at risk of having the atmosphere enter the refrigerant circuit in the event of a low-side pressure leak. The refrigerant will be released from the system until the pressure in the system reaches 0 psig. When the compressor cycles on, the low-side pressure begins to drop, pulling a vacuum. This pulls atmosphere into the system. The low-pressure switch, when used as a low-charge protector, prevents the system from operating when the refrigerant charge has been lost.

Nitrogen in the Refrigeration System. Nitrogen is not normally found in a refrigeration system after initial start-up. It is, however, found in the system throughout the installation process. Nitrogen is used:

- By equipment manufacturers to pressurize various pieces of air-conditioning equipment prior to shipment. The release of the nitrogen by the system installer ensures that the equipment arrived at the site leak-free. Self-contained, or package,

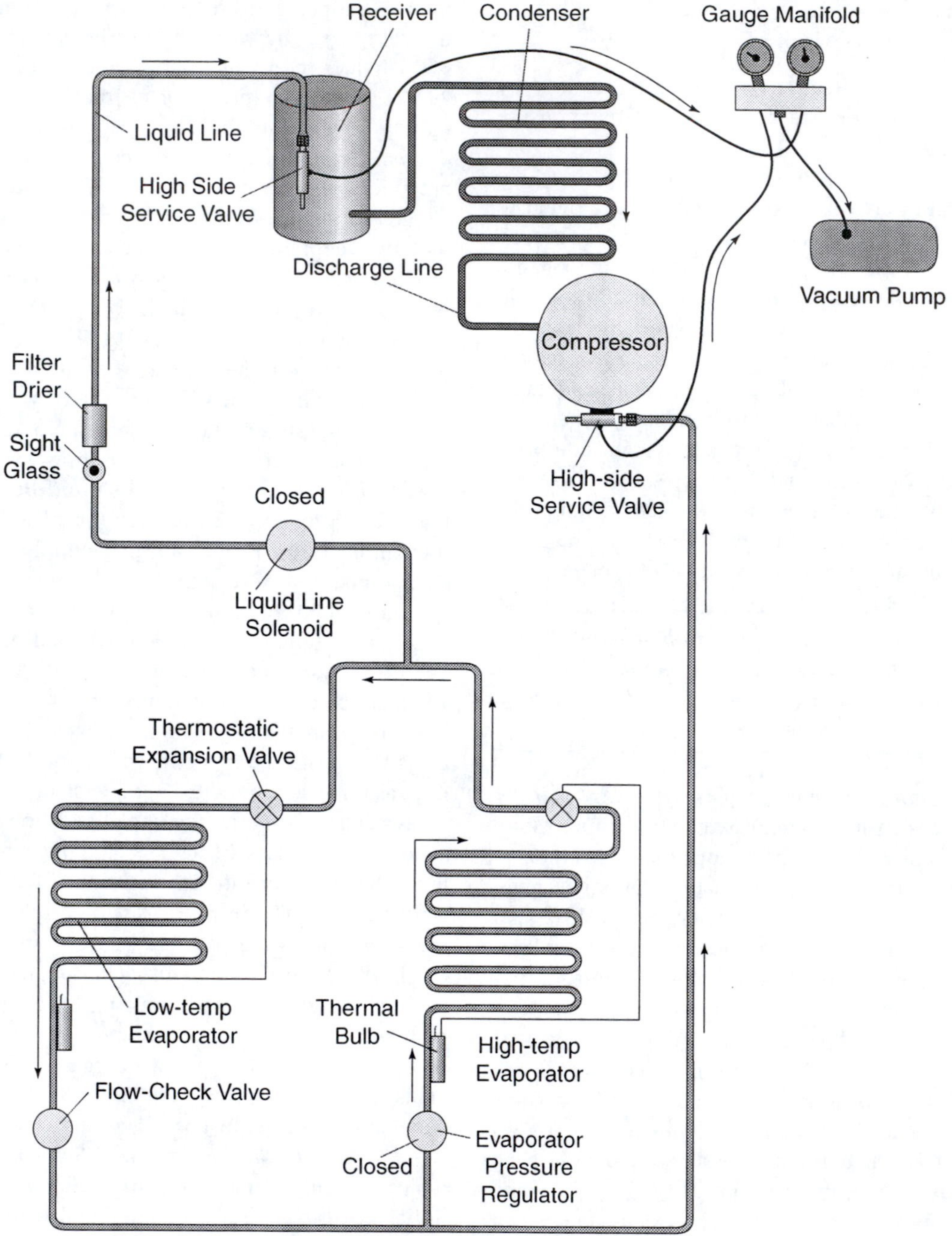

FIGURE 3-18 The evacuation process will take much longer if the liquid-line solenoid is closed. The *arrows* indicate direction of flow with the valve closed. The evaporators are being evacuated in series with each other, which tends to take much longer.

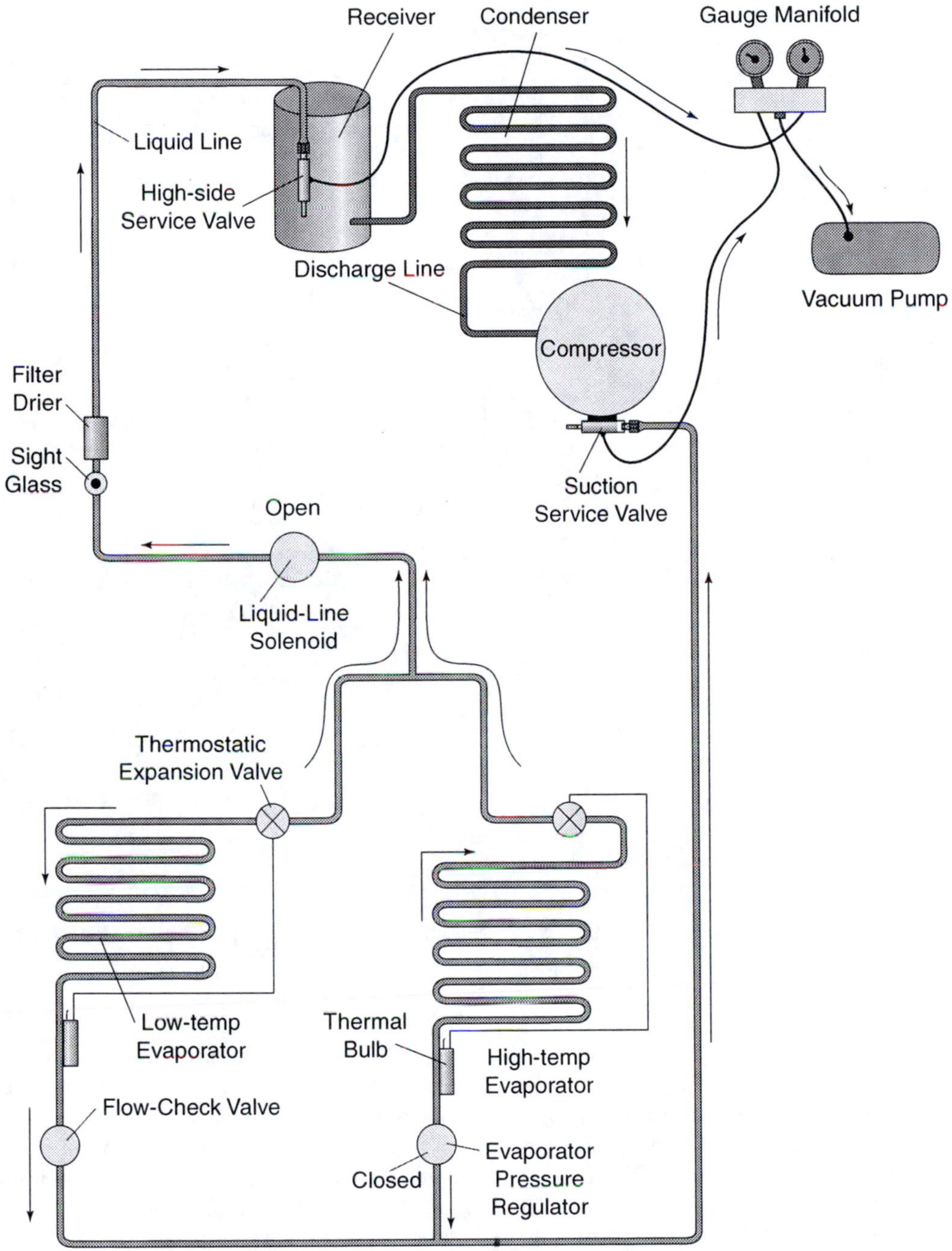

FIGURE 3-19 The evacuation process will go much faster if the liquid-line solenoid is open during evacuation. Notice how the evaporators are evacuated from both sides as opposed to only one in Figure 3-18.

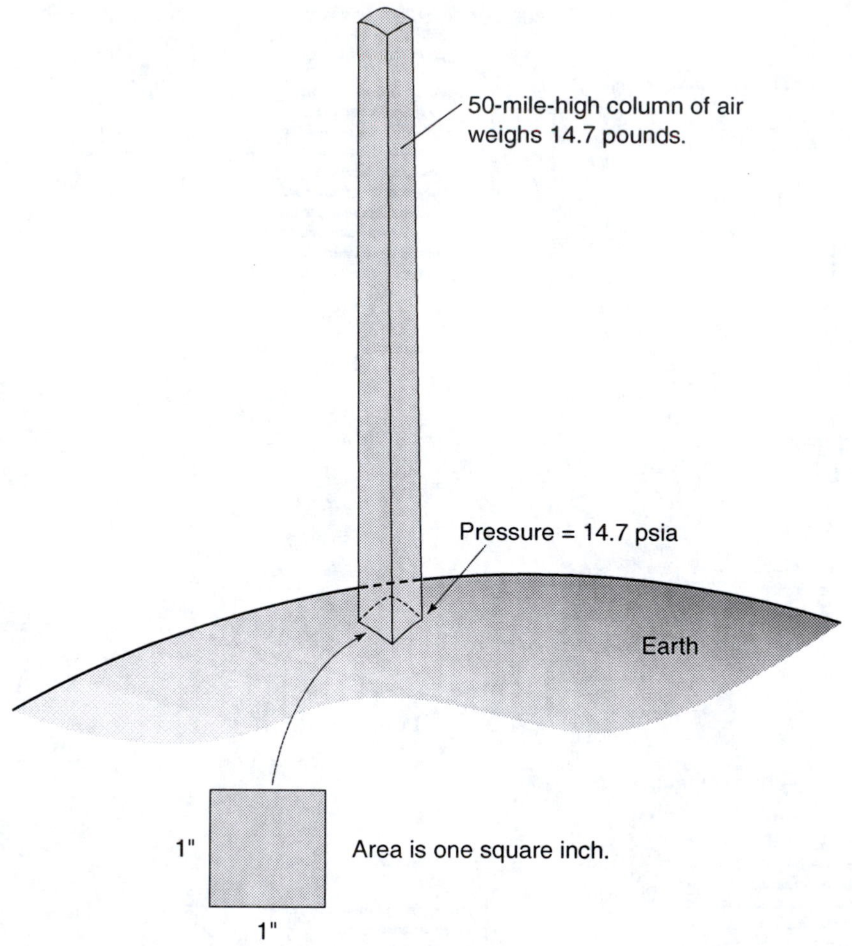

FIGURE 3-20 Absolute pressure readings take into account the 50-mile column of air that exists over every square inch of surface area on the earth.

units are not shipped with nitrogen as they come charged with refrigerant directly from the factory. Carefully read all literature accompanying the piece of equipment since some components come with a holding charge of refrigerant, not nitrogen.

- By field technicians and installation crews to pressure-test refrigerant lines before system evacuation is performed. Pressurizing with dry nitrogen is a popular and preferred method of leak-checking a system. Small leaks can be identified with the aid of liquid solutions that blow bubbles when leaks are present.

In either case, the nitrogen must be completely removed prior to evacuation and the addition of valuable refrigerant to the system.

How do you know if the system you are working on contains noncondensable gas? If, upon installing the gauge manifold on an inoperative system, the pressures are above 0 psig, the system is holding at least some pressure. While off, the high- and low-side pressures should correspond to each other unless the system is in pump-down or is equipped with a liquid-line solenoid. These pressures should also correspond to a saturation temperature that matches the actual ambient temperature surrounding the unit. If these

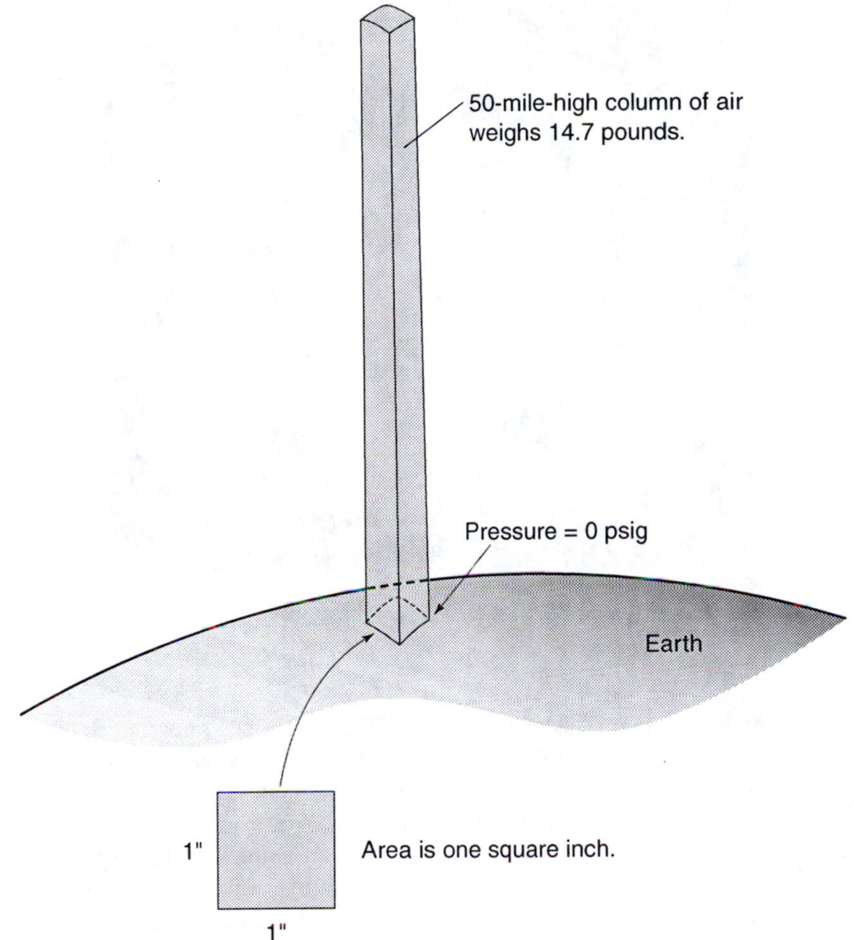

FIGURE 3-21 Gauge pressure readings do not take into account the 14.7 pounds of air that are present. 0 psig = 14.7 psia

two temperatures are the same, the system does not contain noncondensables.

If the system is operating upon arrival at the service call, turn the unit off and allow it to sit for approximately 30 minutes, longer if at all possible, before comparing the ambient and saturation temperatures. This allows the system pressures to equalize, giving more accurate readings.

If the corresponding saturation temperature is higher than the ambient, chances are that the system contains noncondensable gas. If this is indeed the case, the refrigerant charge must be properly recovered from the system. The system must then be evacuated and recharged with new refrigerant. **Recovery** involves the removal of the refrigerant from the system, which can then be turned in to an approved facility for reprocessing.

NOTE: *Only technicians possessing proper certification granted by the Environmental Protection Agency (EPA) can handle and work on systems containing refrigerant.*

DETERMINING THE EVAPORATOR SATURATION PRESSURE AND TEMPERATURE. The evaporator saturation pressure is read directly from

FIGURE 3-22 This low-side gauge reads 68.5 psig, which corresponds to 40°F for R-22. *(Photo by Eugene Silberstein)*

the face of the low-side, or blue, gauge. This gauge is typically located on the left-hand side of the manifold; and the outer, black dial provides the technician with the suction, or back, pressure, depending on the position of the needle on the gauge. The low-side gauge in Figure 3-22 indicates a pressure of 68.5 psig on the low side of the system.

The evaporator saturation temperature is also read directly from the face of the gauge. If an air conditioning system is operating with R-22 as its refrigerant and the low-side pressure is 68.5 psig, the evaporator saturation temperature will be 40°F (Figure 3-22). In this case, the refrigerant is boiling in the evaporator at 40°F. The data contained on the face of the gauge are the same as those provided on the pressure/temperature chart (Figure 1-1).

DETERMINING THE CONDENSER SATURATION PRESSURE AND TEMPERATURE. The condenser saturation pressure and temperature are obtained from the face of the high-side, or red, gauge, in a manner similar to that of the evaporator saturation pressure and temperature. This gauge is located on the right-hand side of the manifold, and the system head pressure is indicated on the outer, black scale. The high-side gauge in Figure 3-23 indicates a pressure of 226 psig on the high side of the system.

The condenser saturation temperature is also read directly from the face of the gauge. If an air-conditioning system is operating with R–22 as its refrigerant and the high-side pressure is 226 psig, the condenser saturation temperature will be 110°F. In this case, the refrigerant is condensing at 110°F. Just as on the low side of the system, this information on the gauge is the same as that provided on the pressure/temperature chart (Figure 1-1).

EVALUATING EVAPORATOR EFFECTIVENESS AND EFFICIENCY. For air-conditioning applications, the ideal, or design, evaporator saturation temperature is roughly 40°F. This temperature is desired because it is cool enough to provide proper air conditioning and is above the freezing point, which eliminates the need for a defrost cycle. Under normal system operation, frost should never accumulate on the coil surface.

As with the rest of the refrigeration system, it is impossible to visually look into the evaporator coil and determine at which point the liquid refrigerant completely vaporizes. For this reason, we rely on the evaporator superheat to give us this information. The normal range of evaporator superheat is between 8 and 12 degrees. Two temperature readings are needed to determine the evaporator superheat. They are:

FIGURE 3-23 This high-side gauge reads 226 psig, which corresponds to 110°F for R-22. *(Photo by Eugene Silberstein)*

1. The evaporator outlet temperature, measured near the thermal bulb
2. The evaporator saturation temperature, measured near the thermal bulb

As stated earlier in the text, the evaporator superheat is determined by subtracting the evaporator saturation temperature from the temperature of the suction line at the outlet of the evaporator. Under many system conditions and configurations, taking a suction pressure reading at the outlet of the evaporator may not be possible because a service port may not be located there. If the suction service valve is not located too far from the evaporator, the evaporator saturation temperature can be read from it but the superheat calculation may be off by 1 or 2 degrees.

Excessive Superheat. If the evaporator superheat is excessive, over 15 degrees, the evaporator is not operating at its maximum potential. It is said to be a **starved,** or **underfed evaporator.** Common reasons for high superheat are:

- Refrigerant undercharge
- Improper superheat spring setting on the thermostatic expansion valve
- Blocked or clogged strainer on the thermostatic expansion valve
- Blocked or clogged capillary tube
- Blocked or clogged strainer at the capillary-tube inlet
- Indoor fan motor speed too high

Low Superheat. If the evaporator superheat is too low, less than 5 degrees, the evaporator is operating very effectively but there is an increased possibility of liquid refrigerant flowing back to the compressor. This is called **floodback.** The evaporator is called an **overfed evaporator.** Common reasons for low superheat are:

- Refrigerant overcharge
- Overfeeding metering device
- Improper superheat spring setting on the thermostatic expansion valve
- Improperly mounted thermostatic expansion valve thermal bulb
- Restricted airflow across the evaporator (blocked coil or inoperative fan motor)
- Dirty air filters
- Reduced water flow through the evaporator (chiller application)

EVALUATING CONDENSER EFFECTIVENESS AND EFFICIENCY. As with the evaporator evaluation, we must rely on some external method to determine how effectively the condenser is operating. By calculating the subcooling in the condenser, we are able to determine how effectively system heat is being rejected. The larger the amount of subcooling, the more efficient the condenser and vice versa. As with the evaporator, two measurements are needed to calculate subcooling. They are:

1. The condenser saturation temperature
2. The condenser outlet temperature

To briefly recap, condenser subcooling is calculated by subtracting the condenser outlet temperature from the condenser saturation temperature. The normal range for condenser subcooling is from 10 to 20 degrees. Higher-efficiency condensers can operate with somewhat higher subcooling.

Low Subcooling. If the amount of condenser subcooling is low, below 8 degrees, the condenser is not rejecting its heat effectively. A number of factors can cause the condenser subcooling to drop. A few of them are:

- Dirty or blocked condenser coil
- Defective condenser fan motor
- Service panels removed from condensing unit
- Reduced water flow through the condenser (water-cooled applications)
- Defective water-regulating valve (water-cooled applications)

Excessive Subcooling. As mentioned earlier, a condenser's effectiveness is measured by the amount of subcooling; however, too much subcooling could be an indication of a system deficiency. Too much subcooling can actually reduce the effectiveness of the condenser because the heat-transfer surface has been reduced. High subcooling could be a result of:

- System overcharge
- System operating in low ambient conditions
- Refrigerant-flow restriction
- Noncondensable gas in the system
- Overfeeding water-regulating valve (water-cooled applications)

EVALUATING THE PUMPING EFFECTIVENESS OF THE COMPRESSOR. The compressor is the single most expensive component in an air-conditioning system, both from the initial cost and from installation standpoints. It is also the component that is the easiest to misdiagnose. When the compressor motor experiences a mechanical or electric failure and the compressor fails to operate completely, the diagnosis is relatively straightforward. However, when the compressor is operating, how effectively it is pumping the system refrigerant is a more difficult question to answer. A very effective way to evaluate the compressor is to remove it from the system and perform a bench test. This is, however, very time-consuming and, especially for large compressors, very impractical. A number of things can be done by technician in the field to determine if the compressor is the cause for system failure. These include:

- Determining the compression ratio of the system
- Evaluating the volumetric efficiency of the compressor
- Performing a vacuum test on the compressor

Compression Ratio and Volumetric Efficiency. Just as evaporator superheat and condenser subcooling are used to evaluate those system components, we the **compression ratio** helps to evaluate the compressor. The compression ratio is obtained by dividing the system's high-side pressure by the system's low-side pressure:

$$\text{Compression ratio} = \frac{\text{system high-side pressure (absolute)}}{\text{system low-side pressure (absolute)}}$$

The pressures used to calculate the compression ratio are the absolute pressures, not the gauge pressures that are obtained during normal system evaluation. The reason for this is to avoid obtaining a negative value, which would be completely ambiguous and useless.

For the compressor in Figure 3-24, the suction pressure and the discharge pressure are 70 psig and 225 psig, respectively. The pressures can be easily converted to absolute pressures by adding 15 to each pressure, resulting in a low-side pressure of 85 psia and a high-side pressure of 240 psia. To obtain the compression ratio, divide 240 by 85. The compression ratio of 2.82 to 1 is denoted 2.82:1. This ratio indicates that, for every pound of pressure on the low side, 2.82 pounds of pressure are on the high side.

Air-conditioning systems operating with R-22 as their refrigerant will, under design conditions, have a

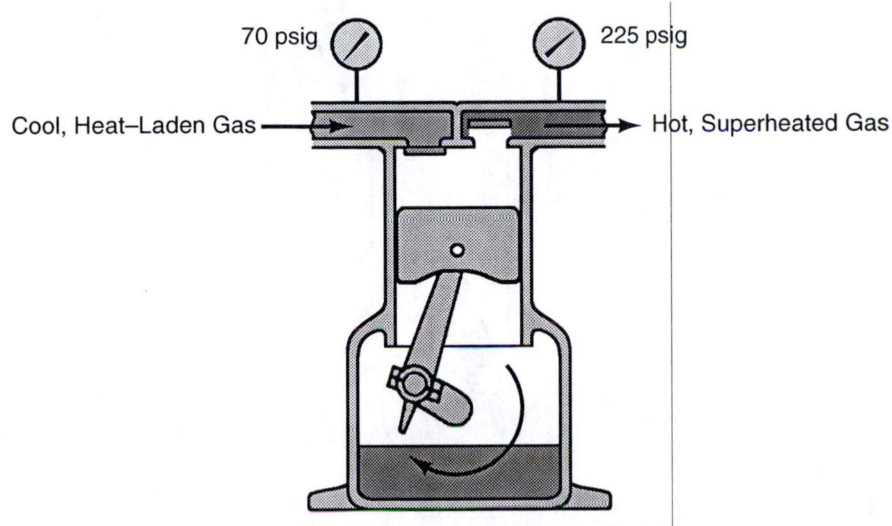

FIGURE 3-24 The compression ratio for this system is 2.8:1.

compression ratio of around 3:1. Since different refrigerants operate at different pressures, the compression ratios will also be different. Systems operating in different temperature ranges—medium- and low-temperature refrigeration, for example—will also have different compression ratios.

The **volumetric efficiency** is inversely proportional to the compression ratio and relates to the amount of refrigerant that the compressor pumps. Since the job of the compressor is to increase the vapor refrigerant from suction pressure to discharge pressure and vice versa, more work must be done if the differential between them is high.

If, in the preceding example, the suction pressure dropped from 70 psig to 60 psig, it would take the compressor longer to reduce the pressure from 225 psig to 60 psig and to increase the pressure from 60 psig to 225 psig. The compressor would, therefore, pump less refrigerant through the system because more of the piston's stroke would be used to increase and decrease the pressure. In this case, the compression ratio has increased and the volumetric efficiency of the compressor has decreased. An efficient compressor has a low compression ratio and a high volumetric efficiency.

High Compression Ratio. As mentioned earlier in this section, the average compression ratio for an R-22 air-conditioning system is approximately 3:1. Several factors within the system can cause the compression ratio to rise:

- Defective evaporator or condenser fan motor
- Dirty air filter
- Underfeeding metering device
- Liquid-line filter/strainer blockage
- Dirty evaporator or condenser coil
- System restriction
- System overcharge (automatic expansion valve applications)
- Insufficient water flow through the condenser (water-cooled applications)
- Defective water-regulating valve (water-cooled applications)

In short, the compression ratio will increase when the suction pressure drops, the head pressure rises, or when a combination of both is present. Consider the following example. The head pressure in Figure 3-25 is 225 psig, and the suction pressure is 8″ Hg. To compute the compression ratio, we must first convert these pressures to psia values. The head pressure becomes 240 psia, by simply adding 15 to the 225 psig. To convert the 8″ Hg, we must use the following formula:

$$\text{psia} = \frac{29.92''\text{Hg} - \text{gauge reading}}{2}$$

In this case, we get:

$$\text{psia} = \frac{29.92''\text{Hg} - 8''\text{Hg}}{2} = \frac{21.92}{2} = 10.96$$

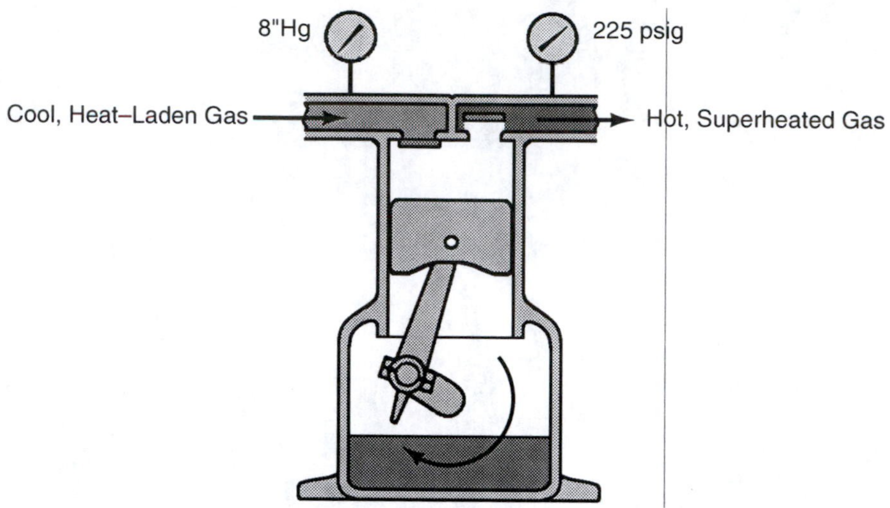

FIGURE 3-25 The compression ratio for this system is 22:1.

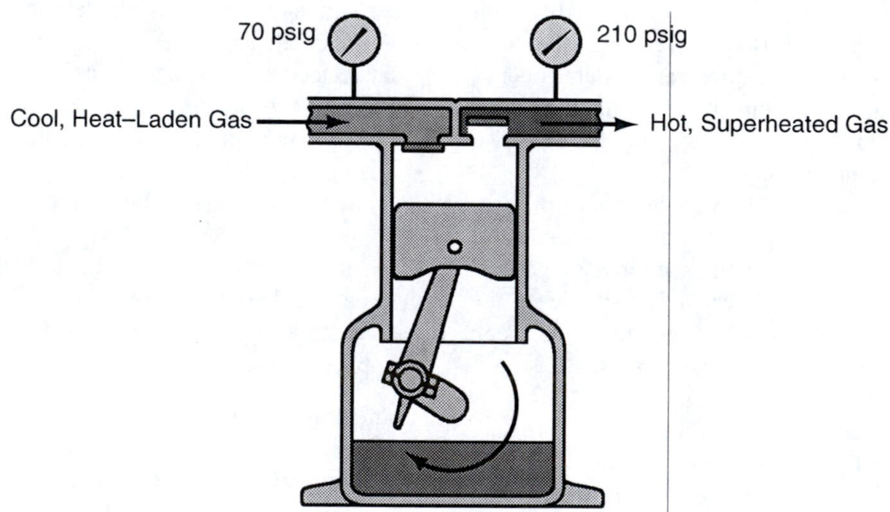

FIGURE 3-26 The compression ratio for this system is 2.6:1.

The compression ratio is then 240 divided by 10.96, which is approximately 22:1. This value is much too high and should send up a red flag immediately. Compression ratios over 11:1 are considered very high and could result in total system failure.

Low Compression Ratio. Water-cooled systems tend to operate with lower head pressures and lower compression ratios than air-cooled systems. Consider the compressor in Figure 3-26. The suction pressure is 70 psig, and the discharge pressure is 210 psig. As an exercise, verify that the compression ratio for this system is 2.6:1. This is somewhat lower than the system in Figure 3-24, which represented an identical system equipped with an air-cooled condenser.

This discussion of the concepts of compression ratio and volumetric efficiency has established that the lower the compression ratio, the higher the volumetric efficiency of the compressor. Unfortunately, this is not *always* the case. If, for example, the compression ratio was 1:1, which is as low as it can get,

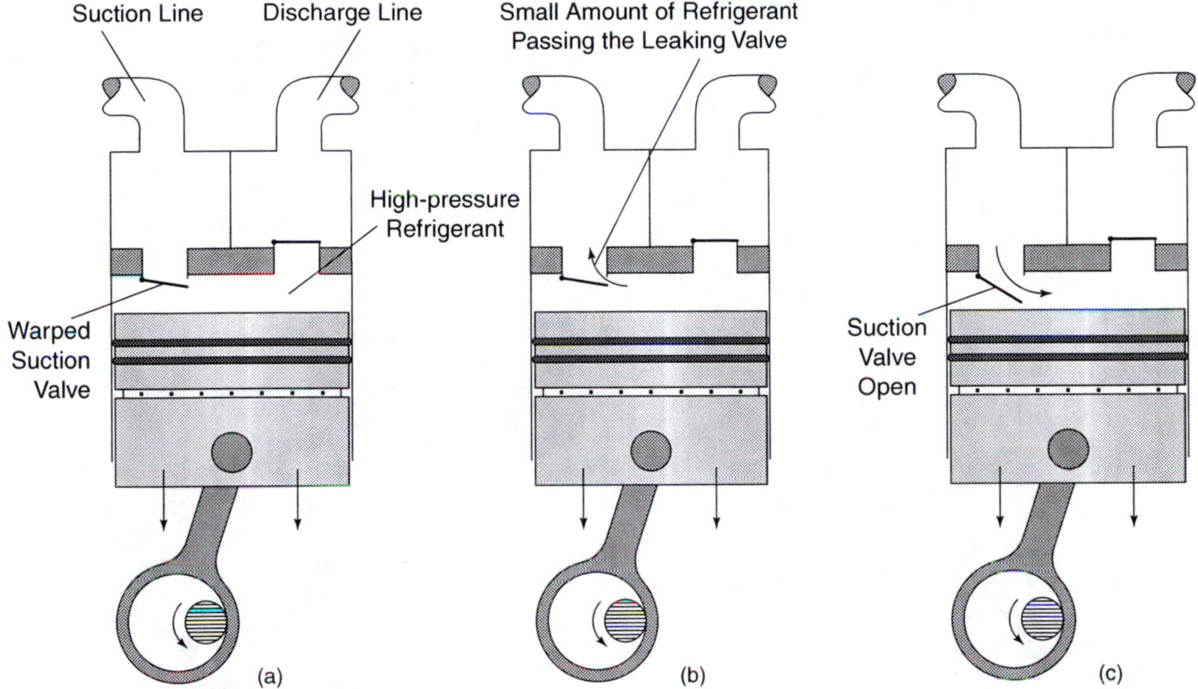

FIGURE 3-27 (a) Even though the suction and discharge valves are supposed to be closed, the suction valve is leaking and cannot close all the way. (b) The high-pressure vapor in the compression cylinder is escaping into the suction line. (c) When the suction valve opens, the same refrigerant that escaped is sucked back into the cylinder, reducing the amount of "new" refrigerant entering the cylinder.

the compressor would be completely inefficient. This is because the discharge pressure and the suction pressure would be exactly the same. In this instance, no refrigerant is being pumped through the system and volumetric efficiency would be nonexistent.

In reciprocating compressors, leaking suction and/or discharge valves can cause the compression ratio to drop. In this case, the volumetric efficiency will decrease instead of increase. Consider the case of a leaking compressor suction valve (Figure 3-27a). As the piston begins to move downward in the cylinder from top dead center, the pressure in the cylinder is equal to the discharge pressure of the system. Because the suction valve is leaking, a small amount of refrigerant escapes from the cylinder and enters the suction line (Figure 3-27b). When the suction valve finally opens and refrigerant from the suction line enters the cylinder, a portion of this refrigerant is the same refrigerant that escaped from the cylinder earlier (Figure 3-27c). This reduces the amount of "new" refrigerant entering the compression cylinder. If less new refrigerant enters the cylinder, less new refrigerant can leave the cylinder via the discharge line, thereby reducing the amount of refrigerant being pumped by the compressor. This reduces the volumetric efficiency of the compressor. A similar scenario can be set up illustrating a leaking discharge valve and is left as an exercise for the reader.

DETERMINING IF THE REFRIGERANT CHARGE NEEDS TO BE ADJUSTED. Upon initial inspection of the system, pressure and temperature readings will play an important role in the proper evaluation of an air-conditioning system. Superheat and subcooling calculations, along with other information provided in the text, enable the field technician to determine if the refrigerant charge is in need of adjustment. Note that multiple readings are often necessary to properly evaluate the system. Reaching intelligent conclusions about a system is difficult by taking, for example, a low-side pressure reading and nothing else.

EVALUATING THE METERING DEVICE

The metering device is responsible for controlling the flow of refrigerant to the evaporator. If this device does not feed properly, the evaporator can become either **flooded** or starved. A flooded evaporator is one that has saturated refrigerant at its outlet. This is not desirable because system damage could result if the liquid refrigerant should reach the compressor. A metering device that overfeeds could cause the following:

- Liquid floodback to the compressor
- Low compression ratio
- High suction pressure
- Low head pressure
- Reduced cooling capacity

A metering device that underfeeds the evaporator can cause the following:

- Excessive superheat
- Reduced cooling capacity
- Low suction pressure
- High head pressure
- Increased compression ratio
- Ice formation on the evaporator coil

Evaluating the Capillary Tube

Since the capillary tube is a fixed-bore metering device, it has no moving parts to go bad, no adjustments that need to be made, and no general maintenance to be performed. The number-one enemies of the capillary tube, though, are moisture and dirt.

If moisture is present in the system, it commonly freezes as it enters the capillary tube. This is because there is a pressure drop in the tube. Ice crystals will cause a restriction in the tube, affecting system operation. If the system is turned off, the ice will melt and, when energized again, the system will operate until the ice forms again. The moisture must be removed from the system by recovering the refrigerant charge by evacuating and recharging the system, or by replacing the filter driers.

Dirt or debris in the capillary tube (Figure 3-28) will also cause a restriction in the line. Wax, found in system contaminants, has a tendency to solidify in the capillary tube. For this reason, strainers at the inlet to the device are needed (Figure 3-29). System problems that can result from a restricted capillary tube include those listed for underfeeding the evaporator. Capillary-tube systems do not experience problems with overfeeding evaporators unless the capillary tube was replaced with a shorter tube or one with a larger bore.

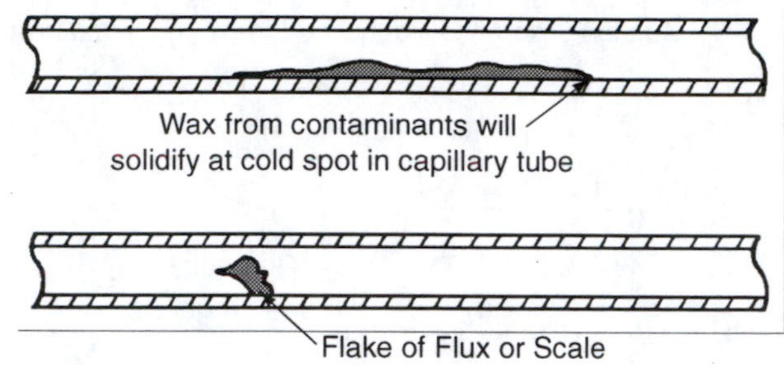
Wax from contaminants will solidify at cold spot in capillary tube
Flake of Flux or Scale

FIGURE 3-28 Wax, flux, and scale can restrict the capillary tube, causing an underfed condition.

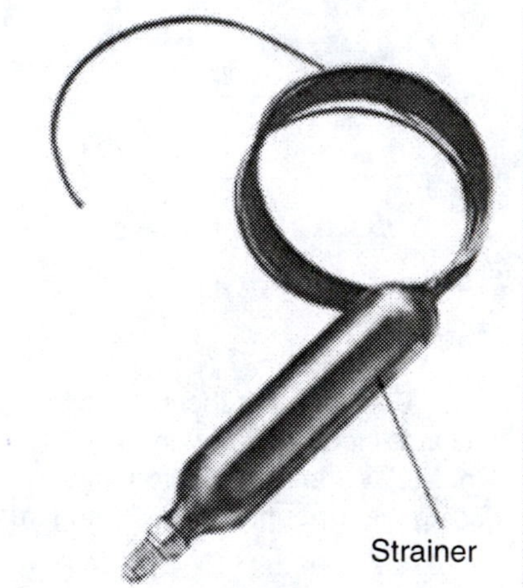
Strainer

FIGURE 3-29 Strainers are located at the inlet to the capillary tube to help reduce the possibility of foreign matter entering the tube. *(Courtesy Parker Hannifin Corporation)*

Evaluating the Automatic Expansion Valve

The automatic expansion valve is designed to maintain a constant evaporator pressure and is not generally serviced in the field. If the pressure the device is maintaining is different from the desired pressure, turning the adjustment screw (Figure 3-30) can change the valve setting. The screw is turned clockwise to increase the evaporator pressure and counterclockwise to reduce the pressure. Adjustments should be made slowly to ensure an accurate and precise setting, as well as reducing the possibility of compressor damage. If the device fails to maintain the proper pressure after adjustment and the system charge is correct, the valve may need to be replaced.

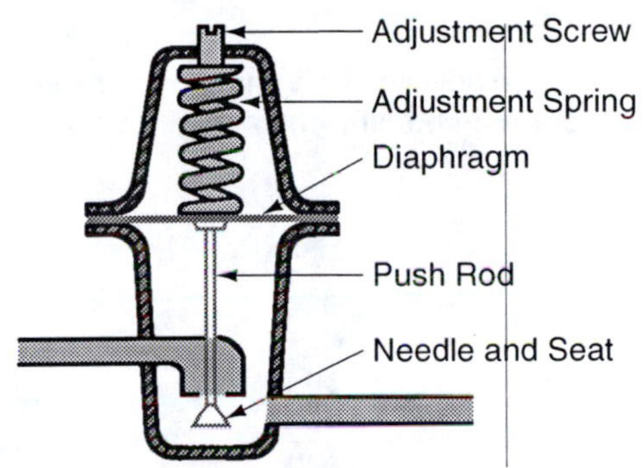

FIGURE 3-30 The adjusting screw on an automatic expansion valve

Evaluating the Thermostatic Expansion Valve

Unlike the capillary tube and the automatic expansion valve, the thermostatic expansion valve can be serviced in the field, depending on the size and configuration of the valve. The thermostatic expansion valve is designed to maintain constant evaporator superheat and, therefore, must be able to effectively measure the evaporator outlet temperature. The following should be checked when evaluating this control:

- Make certain the thermal bulb is mounted securely to the suction line at the outlet of the evaporator, according to manufacturer's instructions (Figure 3-31).
- Make certain the thermal bulb is wrapped with insulation tape to ensure accurate readings.
- Make certain the strainer at the inlet of the valve is clean and free from debris.
- If the valve fails to maintain superheat, the thermal bulb should be checked for possible loss of charge. Gently warming the bulb should cause the suction pressure to rise, since the bulb pressure pushes the valve open. Heat the bulb with your hand, or place the bulb briefly in a glass of warm water. If the

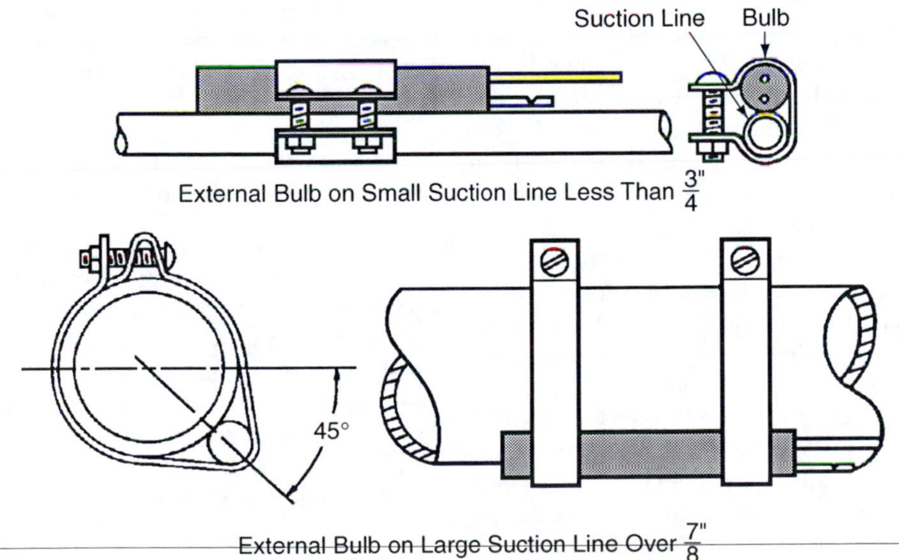

FIGURE 3-31 Best positions for mounting the thermostatic expansion valve's thermal bulb. *(Courtesy ALCO Controls Division, Emerson Electric Company)*

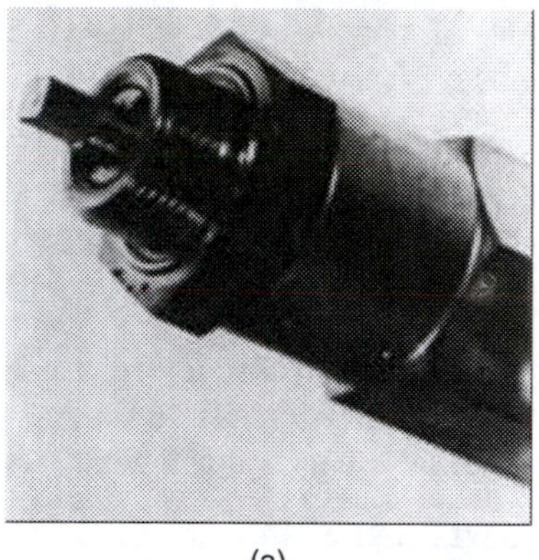

(a)

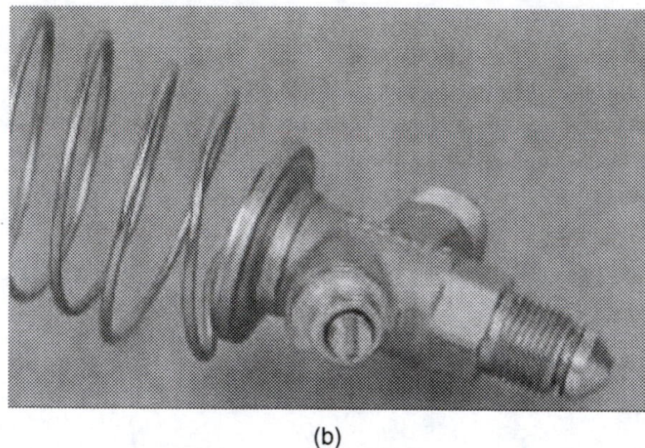

(b)

FIGURE 3-32 Adjustment stems on expansion valves. Some of them are adjusted with (a) a refrigeration service wrench and some with (b) a screwdriver. One full turn of the stem will normally change the superheat ½ to 1 degree. *(Photos by Bill Johnson)*

evaporator pressure does not change, the power head has lost its charge and must be replaced. If the valve does not have interchangeable parts, the entire valve must be replaced.

THERMOSTATIC EXPANSION VALVES THAT OVERFEED THE EVAPORATOR. Thermostatic expansion valves that overfeed the evaporator tend to operate with a superheat that is lower than desired. To alleviate this problem, the superheat spring adjustment (Figure 3-32) should be turned clockwise to increase the spring pressure. This will increase the amount of evaporator superheat. Before adjusting the valve, make certain that the four items listed for evaluating this control were checked first.

Superheat adjustments on thermostatic expansion valves should be done with care. They should be made slowly, and ample time must be allowed for the system to stabilize itself after each adjustment. Improper valve adjustment can result in major component—including compressor—damage. One full turn of the superheat spring adjusting screw generally changes the superheat approximately ½ to 1 degree.

THERMOSTATIC EXPANSION VALVES THAT UNDERFEED THE EVAPORATOR. Underfeeding of the evaporator generally causes a reduction in cooling efficiency. An evaporator that is underfed has a higher-than-desired superheat, which can be caused by:

- A system undercharge
- A clogged inlet strainer at the thermostatic expansion valve
- Moisture in the system
- Improper superheat spring adjustment

FIGURE 3-33 Most valves have some type of inlet screen to strain any small particles out of the liquid refrigerant before it reaches the small opening in the expansion valve. This helps to keep the needle-and-seat assemblies clear of any obstructions.

A refrigerant undercharge will result in high superheat, low suction pressure, and low discharge pressure. Adjusting the charge should alleviate the problem. Obviously, the reason for the loss of refrigerant must be located and corrected before proceeding. A low refrigerant charge also reduces the pressure drop across the thermostatic expansion valve and affects its performance.

A clogged inlet strainer (Figure 3-33) reduces the amount of refrigerant that flows through the thermostatic expansion valve. The evaporator will be starved and the superheat will be high; but, unlike the low-charge scenario, the head pressure will be high.

Moisture in the system can freeze in the valve, causing the valve to freeze closed. This will create the same symptoms as the clogged strainer, but the evaporator pressure will rise if the valve body is gently warmed, allowing the ice to melt. As in the example with the capillary tube, the filter driers should be replaced to absorb the moisture in the system.

If all else fails, the superheat spring may be in need of adjustment. As stated earlier, care must be taken when making adjustments to the valve. Adjusting the superheat spring counterclockwise will reduce the spring pressure, which reduces the superheat, and allows more refrigerant to flow to the evaporator.

SUMMARY

- Motor problems can result from improper airflow, lubrication, pulley alignment, or belt tension.
- A refrigerant overcharge can cause reduced system efficiency and cooling capacity.
- A refrigerant undercharge can cause reduced system efficiency and cooling capacity.
- Gauge manifolds must be installed and removed properly to reduce the possibility of allowing atmosphere to enter the system.
- Refrigerant leaks can be caused by rubbing or vibrating surfaces or poor-quality soldering or brazing.
- Refrigerant leaks can be located by audible noise, bubble solutions, electronic leak detectors, halide torches, or UV light.
- Noncondensable gases, such as air and nitrogen, increase operating pressures, reduce system performance, and reduce cooling capacity.
- A vacuum gauge should be used during system evacuation to ensure that proper vacuum levels are reached.
- Excessive evaporator superheat reduces cooling capacity.
- Compressors are evaluated by the compression ratio and volumetric efficiency.
- Moisture in the air-conditioning system can cause the expansion device to malfunction.

KEY TERMS

Absolute pressure
Ball bearings
Belt tension gauge
Blank ports
Calibrating
Compression ratio
Drive pulley
Driven pulley
Endplay
Flat system
Floodback
Flooded evaporator
Gauge manifold
Grease fittings
Hermetically sealed
Micron
Micron gauge
Noncondensable gas
Oil ports
Overfed evaporator
Overload protector
Pinch-off tool
Pounds per square inch gauge (psig)
Preventive maintenance
Process tube
Recovery
Schrader valve
Sleeve bearings
Starved evaporator
System charge
Underfed evaporator
Volumetric efficiency

FOR DISCUSSION

1. Calculate the compression ratio for a system operating with a high-side pressure of 200 psig and a low-side pressure of 0 psig using gauge pressures only. Why doesn't this result produce any useful information for the technician?
2. Repeat the preceding example using absolute pressures. Discuss why this result is more realistic and useful to the field technician.

3. Why is it important for technicians to verify that the airflow through both the evaporator and the condenser is sufficient before altering a system's refrigerant charge? How can doing this save valuable time and money on a service call?

4. Explain how an improperly installed set of gauges can result in excessive refrigerant loss and possible compressor damage.

REVIEW QUESTIONS

1. Which of the following protects a motor from overheating?
   a. Crankcase heater
   b. Internal overload protector
   c. Low-pressure control
   d. Low ambient control
2. Reduced airflow through an evaporator could be caused by:
   a. Dirty air filters.
   b. Dirty evaporator coil.
   c. Dirty condenser coil.
   d. Both *a* and *b* are correct.
3. Permanently lubricated motors are equipped with:
   a. Oil ports.
   b. Grease fittings.
   c. Relief plugs.
   d. None of the above.
4. Endplay in a motor shaft is an indication that:
   a. The motor needs to be lubricated.
   b. The bearings have been damaged.
   c. The grease fitting has come loose.
   d. All of the above.
5. The drive pulley is connected to the:
   a. Motor bearings.
   b. Motor shaft.
   c. Blower shaft.
   d. Blower wheel.
6. Belt tension that is too tight can result in:
   a. Belt slippage.
   b. Reduced airflow.
   c. Excessive noise.
   d. The "polishing" of the pulley groove.
7. A belt that is too loose can lead to
   a. Belt slippage.
   b. Bearing damage.
   c. Reduced airflow.
   d. Both *a* and *c* are correct.

8. True or False: A dirty return air filter can cause reduced airflow in the supply duct.

9. Which of the following can cause reduced airflow through an outdoor condensing unit's coil?
   a. Blocked or closed supply registers
   b. Excessive amounts of fallen leaves or shrubs
   c. Dirty evaporator coil
   d. Both *a* and *b* are correct

10. True or False: The relief plug must never be removed when adding grease to a motor equipped with grease fittings.

11. Excessive belt breakage is an indication that:
   a. The belt tension is too tight.
   b. The belt tension is too loose.
   c. The belt may be too small.
   d. Both *a* and *c* are possible.

12. Which of the following information can be obtained directly from the dial of a low-side pressure gauge?
   a. Evaporator saturation temperature
   b. Evaporator superheat
   c. Evaporator outlet temperature
   d. All of the above are correct

13. A low refrigerant charge will cause:
   a. The discharge pressure to rise.
   b. The suction pressure to rise.
   c. The evaporator superheat to increase.
   d. Excessive cooling capacity.

14. Liquid floodback occurs when:
   a. The evaporator produces a large quantity of condensation.
   b. Liquid refrigerant travels back to the compressor.
   c. Liquid refrigerant floods the floor.
   d. Liquid refrigerant travels from the compressor to the evaporator.

15. Refrigerant leaks can be detected by which of the following methods?
   a. UV light
   b. Electronic leak detectors
   c. Halide torch
   d. All of the above are correct

16. True or False: Large leaks can be detected by pressurizing the system with refrigerant and listening for the leak.

17. True or False: A noncondensable gas is one that cannot condense into a liquid.

18. If the thermal bulb of a thermostatic expansion valve comes loose from the suction line:
   a. The evaporator superheat will go up.
   b. The evaporator superheat will go down.
   c. The evaporator superheat will remain the same.
   d. The system refrigerant will leak out.

19. One turn of the superheat spring adjust will generally change the superheat how many degrees?
    a. 0.5 to 1.0 degrees
    b. 1.0 to 2.0 degrees
    c. 2.0 to 2.5 degrees
    d. 2.5 to 3.5 degrees
20. Reduced water flow through a water-cooled condenser will cause:
    a. The head pressure to rise.
    b. The condenser saturation temperature to rise.
    c. The condenser saturation pressure to drop.
    d. Both *a* and *b* are correct.

CHAPTER 4

Electric Motors and Starting Components

OBJECTIVES After studying this chapter, the reader should be able to:

- Name five types of electric motors commonly found in air-conditioning equipment.
- Explain the operation of a shaded-pole motor.
- Explain the characteristics of the start and run windings in an electric motor.
- Describe the operation of a split-phase or induction-start-induction-run (ISIR) motor.
- Describe the function of start and run capacitors.
- Explain the operation of a current magnetic relay (CMR).
- Explain the operation of a potential magnetic relay (PMR) or potential relay.
- Describe the operation of a capacitor-start-induction-run (CSIR) motor.
- Describe the operation of a capacitor-start-capacitor-run (CSCR) motor.
- Describe the operation of a permanent-split-capacitor (PSC) motor.
- Describe the basic configuration of a three-phase motor.
- Name and describe three commonly used methods to start three-phase motors.
- Explain the difference between a wye and a delta configuration.
- Draw electrical schematics for various types of electric motors.

INTRODUCTION

Thus far in the text, when reference was made to electric motors, it was done so in a strictly mechanical manner. Belt tension, pulley alignment, motor lubrication, and other important factors integral to proper motor operation were discussed. However, many different types of electric motors are classified based on a number of factors, including motor winding configuration, design application, starting devices used, horsepower, and the number of phases. This chapter focuses on various types of electric motors, as well as the components that help the devices operate as effectively and efficiently as possible, both on initial start-up and after reaching a steady-state condition.

OVERVIEW

The function of electric motors is to change electrical energy into mechanical energy. Just as contactors, relays, solenoids, and transformers utilize a magnetic field to accomplish their tasks, electric motors also rely on magnetic fields to create a rotating motion. This motion is in turn used to turn blowers, pumps, gears, and other rotating devices. To obtain this desired rotating motion, an imbalance in magnetic fields and forces must be present. Consider the following example.

Two individuals are competing in an arm-wrestling match. When the match begins, both competitors exert a great deal of force but neither one, in many cases, is able to move the arm of the other. This is because the force exerted by one arm is canceled out by the force exerted by the other. As one competitor tires, the force exerted is reduced and the other competitor is declared the victor. An imbalance in forces existed that allowed one competitor to push the arm of the other down. If this imbalance did not exist, arm-wrestling matches would go on indefinitely. Electric motors create this needed imbalance in a number of different ways, depending on the type and construction of the motor. Each will be described in due time.

No matter what specific type of air-conditioning equipment a field technician is servicing, electric motors will be encountered on a continual basis.

Numerous motors are found on even the simplest of systems, so a good understanding of the various types of motors, as well as their operation, is of vital importance. Motors are commonly utilized:

- In compressors to pump refrigerant through the system
- To move air through an evaporator coil
- To move air through a condenser coil
- To pump water or other liquid through a piping system
- To pass air through a cooling tower
- To turn gears in timers and other devices

To ensure continued satisfactory motor operation, the type of motor chosen must have been constructed in a manner that matches the specific conditions under which it is intended to function. Various types of motor construction include the following:

- Open motors
- Enclosed motors
- Drip-proof motors

For example, *open motors* (Figure 4-1) are designed to have their windings cooled by air moving across them and have openings in the motor shell for that purpose. This type of motor would *not* be desirable in a wet location. *Enclosed motors* (Figure 4-2) are designed to be used in dirty locations. Since air cannot pass over the windings directly, alternative methods must be utilized to cool them. *Drip-proof motors* are intended for use in wet locations.

In addition to the construction and the appearance of the device, motors are also classified by their ability to do usable work. This directly relates to the device that the motor is being used to operate. For example, a motor that is used to turn a large blower must be larger and able to do more work than one used to turn a small fan. The term that describes the motor's power is called **horsepower**. It is noted from Ohm's law that power, equal to voltage times current, is measured in **watts**. The conversion factor between horsepower and wattage is 746. This means that 746 watts = 1 horsepower. The larger the horsepower rating, the more work the motor can do and, typically, the more **starting torque** that is exerted on initial motor start-up.

Starting torque refers to the force that allows the motor to reach its desired speed effectively and efficiently from the stopped position. Consider this example: If a soccer ball were kicked by an individual who was permitted to swing her leg back as far as she needed, it would travel much farther than if she were not permitted to swing her leg back at all. Starting torque is the extra *push* that is given to the motor that helps it overcome any resistance offered by belts, pulleys, bearings, and internal motor resistance to allow the motor to start properly.

Starting torque is generated in the motor by an imbalance in the magnetic field that is generated by current flowing through the motor's electric windings. These windings, called the start and run, are constructed differently, have different resistances,

FIGURE 4-1 Open motor. *(Courtesy Magnatek, Inc., St. Louis, Mo.)*

FIGURE 4-2 Totally enclosed motor. *(Courtesy Grainger.)*

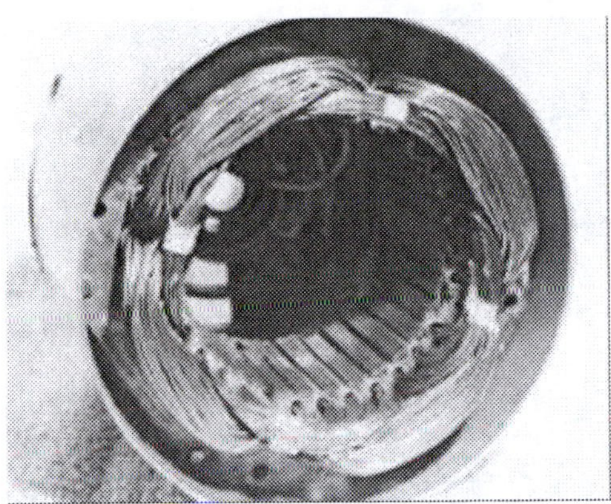

FIGURE 4-3 Stator made up of start and run windings that do not rotate

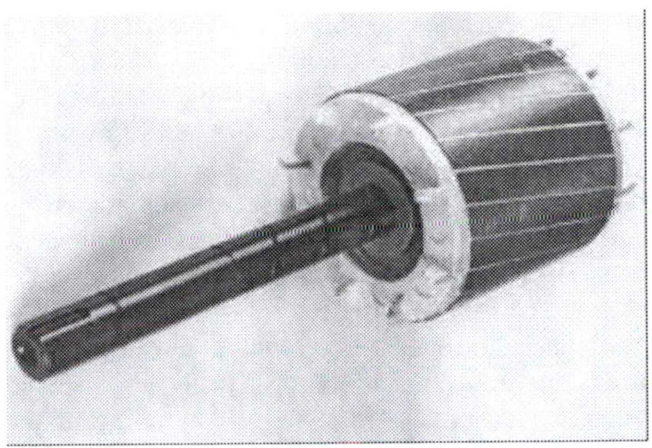

FIGURE 4-4 Squirrel-cage rotor. *(Courtesy Magnatek, Inc., St. Louis, Mo.)*

and, therefore, generate magnetic fields with different strengths. The larger the difference in magnetic field, the larger the starting torque. The **start winding** has relatively high resistance and is made up of thin wire formed into coil with many turns. The **run winding**, on the other hand, has lower resistance and is made up of thicker wire formed into a coil with fewer turns than the start winding. The start and run windings are wired in parallel with each other, permitting different amounts of electrical current to flow through each.

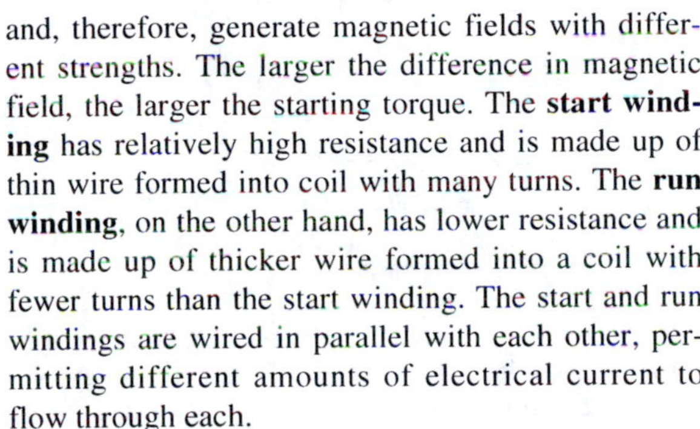

The start and run windings are located in the section of the motor referred to as the **stator** (Figure 4-3). The stator is the section of that motor that does not turn and is located directly inside the shell of the motor. The portion of the motor that actually turns is called the **rotor** (Figure 4-4). The motor's shaft is connected to the rotor, which rests inside the stator. The rotor, also called a *squirrel-cage rotor* because of its appearance, is not electrically connected to the stator and turns as a result of being exposed to magnetic fields produced in the stator. This is similar to a transformer in which the secondary winding is able to produce voltage without being electrically connected to the primary winding.

Motors used to turn small and lightweight devices with low mechanical resistance, therefore, can be smaller since they require less torque. Using a motor that is smaller than required by the application is not a very good idea since the motor will burn out prematurely. This chapter discusses various types of motors based on their operating range of service as well as the amount of starting torque they exert, relative to each other. They include:

- Shaded-pole motors
- Capacitor-start-induction-run (CSIR) motors
- Capacitor-start-capacitor-run (CSCR) motors
- Permanent-split-capacitor (PSC) motors
- Three-phase motors

Shaded-pole motors are designed for very light duty and are normally used to turn small fan blades connected directly to the shaft of the motor. They have very low starting torque. **Permanent-split-capacitor (PSC) motors** have more starting torque than the shaded-pole motor and can, therefore, be used in somewhat larger applications. **Capacitor-start-induction-run (CSIR)** and **capacitor-start-capacitor-run (CSCR) motors** are used for even larger applications and employ starting components that allow the motor to start, run, and function efficiently. These starting components include:

- Capacitors
- Current magnetic relays
- Potential relays

Capacitors are devices that store electrical charge and help give a motor the extra starting torque it needs during start-up. Two commonly encountered types of capacitor, named for their function in the circuit, are the **start capacitor** and the **run capacitor**. The *start capacitor* is intended to be in the electrical circuit only to assist the motor during start-up. After the motor is up and running to speed, or at least close to running speed, it is removed from the circuit. *Run* *capacitors,* on the other hand, are designed to remain in the circuit as long as the motor is energized.

Capacitors and even motor windings can be electrically removed from the circuit as needed by utilizing specially designed relays. These relays include the **centrifugal switch**, the **current magnetic relay**, and the **potential relay**. The centrifugal switch opens and closes its contacts depending on the speed of the motor. The current magnetic relay opens and closes its contacts in response to the current flowing through the run winding of the motor. Finally, the induced voltage measured across the start winding of the motor controls the potential relay's contacts. The term *induced* was discussed earlier in the text during the discussion on transformers, and the concept applied here is exactly the same.

Of all the motor types discussed in this text, the **three-phase motor** has the highest starting torque. Unlike the PSC, the CSIR, and the CSCR motors, which have a start-and-run winding, the three-phase motor has three constantly energized run windings. These motors have three "hot" power lines feeding into them, and they do not utilize capacitors or starting relays.

Three-phase motors are used primarily on larger applications for which a great deal of starting torque is required. Large blowers that are turned by multibelt pulleys, for example, are prime applications for three-phase motors. These motors utilize **starters**, instead of relays and contactors, to control their operation. Starters contain built-in overload protection for the motor, as well as the electrical contacts that open and close to energize and de-energize the motor. This built-in overload protection is needed because electrical current can still flow through the motor if one of the power-line circuits opens. This can cause the motor's amperage draw to rise, in turn causing the motor to overheat. If excessive heat is sensed by the overload, the motor will be shut down. These devices are also manually reset, preventing the motor from running until the problem is corrected.

SHADED-POLE MOTORS

The shaded-pole motor (Figure 4-5) has the lowest starting torque of all the motor types discussed in this text. This type of motor is relatively inexpensive and is used to turn very small fan blades connected directly to the shaft of the motor. The basic construction of this motor is very simple since no start winding is present. The imbalance in magnetic field necessary to produce rotation is obtained by *shading* a portion of the run winding with a heavy copper wire or band (Figure 4-6). When the motor is energized, the strength of the magnetic field will be different in the area of the main poles and in the area of the shaded poles, allowing the rotor to turn. The direction of rotation of a shaded-pole motor is determined by the orientation of the main poles and the shaded poles. The rotor will turn in the direction indicated by the *arrows* in Figure 4-7.

FIGURE 4-5 Shaded-pole motor. *(Courtesy Magnatek, Inc., St. Louis, Mo.)*

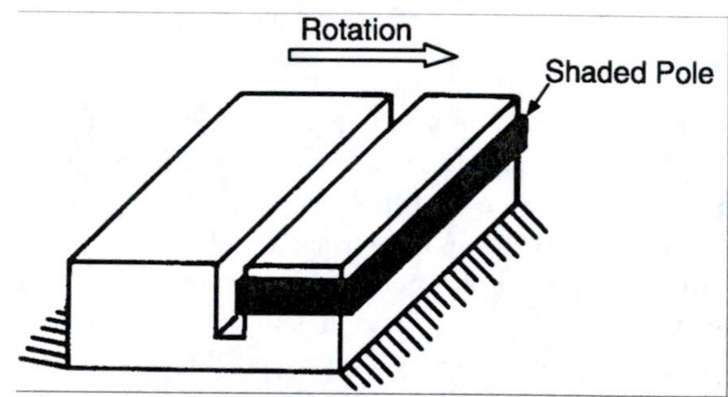

FIGURE 4-6 A portion of the pole is shaded to create an imbalance in magnetic field. The direction of rotation is from the unshaded portion of the pole to the shaded portion of the pole.

Shaded-pole motors, since they are used for such small applications, are rated in watts instead of horsepower. They normally range in size from 6 watts to roughly 35 watts. This is the equivalent of 1/125 horsepower to 1/20 horsepower.

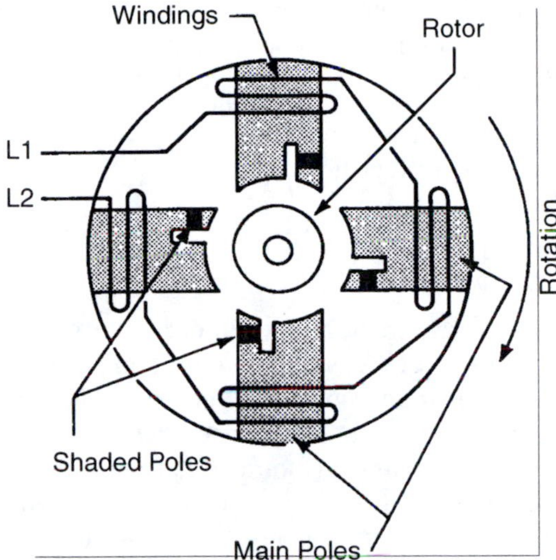

FIGURE 4-7 Layout of a single-speed, shaded-pole motor. Note that there is actually one large winding in this type of configuration.

SPLIT-PHASE MOTORS

Split-phase motors, also referred to as **induction-start-induction-run (ISIR) motors**, have a relatively low starting torque compared to the motors that are discussed later on but more torque than the shaded-pole motor. They range in size from 1/20 horsepower to about 1/3 horsepower. Split-phase motors get their name from the fact that a single power supply is *split* between two individual windings, the run and start, to produce the necessary torque to start the motor (Figure 4-8). The windings and motor wires are labeled as follows:

- COMMON *(C)*: The wire or terminal that is common to both windings
- START *(S)*: The wire or terminal that is unique to the start winding
- RUN *(R)*: The wire or terminal that is unique to the run winding

The **run winding** is energized whenever the motor is energized. This winding has a lower resistance than the **start winding**, which is only in the circuit long enough to help the motor start. For the motor to start, both the start and run windings must be energized. When both windings are energized, the current flows

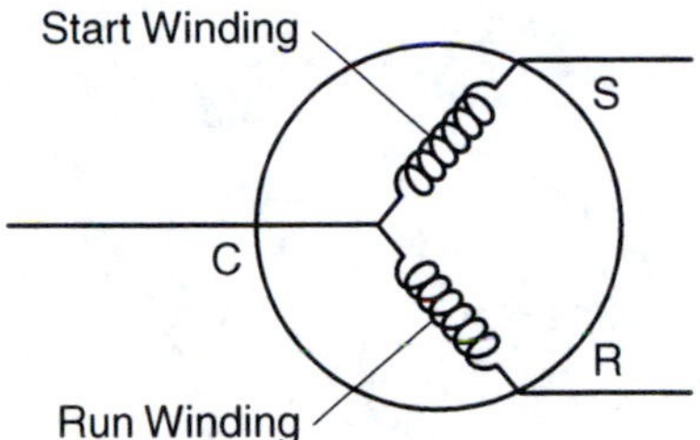

FIGURE 4-8 Schematic diagram of a split-phase motor

through each of the windings at a different rate, creating a *phase shift.* Briefly stated, this phase shift is what creates the imbalance needed to start the motor. Phase shift is measured in electrical degrees and is often referred to as the phase angle. The larger the phase angle, the more starting torque a motor is said to have. For a point of future reference, a split-phase motor, as just described, has a phase angle of about 30 degrees. If the run and start winding were constructed and configured exactly the same, with the same size wire and the same number of turns, the phase angle would be zero, the magnetic field would have no imbalance, and the motor would not start.

Once the split-phase motor has started, the start winding must be removed from the electric circuit. The start winding is designed to be energized for only a short time and can become damaged if it is not de-energized. One commonly used device to remove the start winding from the circuit is called the **centrifugal switch**.

The centrifugal switch (Figure 4-9) opens and closes its contacts depending on the speed of the motor. The electrical contacts on the switch are connected in series with the start winding and are normally closed. Figure 4-10 shows the wiring configuration. When voltage is initially applied to the motor, both the start and run windings are energized and the motor begins to turn. Once the motor has reached a speed equal to about 70 percent of its rated speed, the contacts of the centrifugal switch open, de-energizing the start winding. The run winding, or main winding, is now the only winding energized, and the motor continues to run in this fashion, since the turning rotor now creates the imbalance needed to keep the motor running. When the motor is de-energized, it begins to slow down and the centrifugal switch closes in preparation for the next start-up.

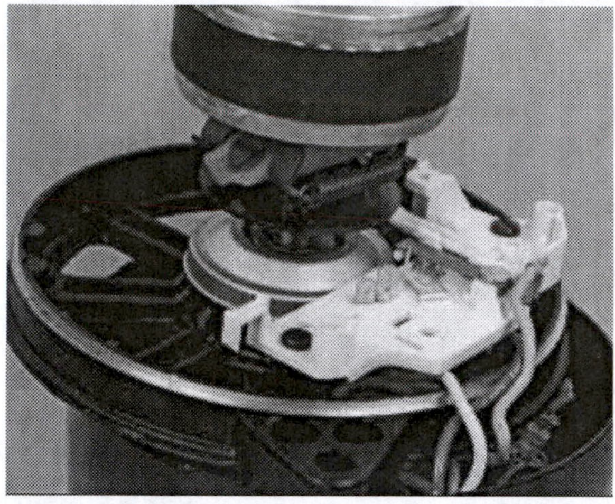

FIGURE 4-9 Centrifugal switch located at the end of the motor. *(Photo by Bill Johnson)*

Other components can be added to the split-phase motor to increase its torque, as well as its range of applications. These components include:

- The capacitor
- The current magnetic relay
- The potential relay

Capacitors

As mentioned earlier, starting torque is the push that is given to a motor enabling it to start up effectively. Up to this point, torque was obtained by the difference in magnetic fields generated between the run and start windings. An additional way that this torque can be increased is by the use of the **capacitor** (Figure 4-11). As already noted, the capacitor is a device that stores electrical charge and helps

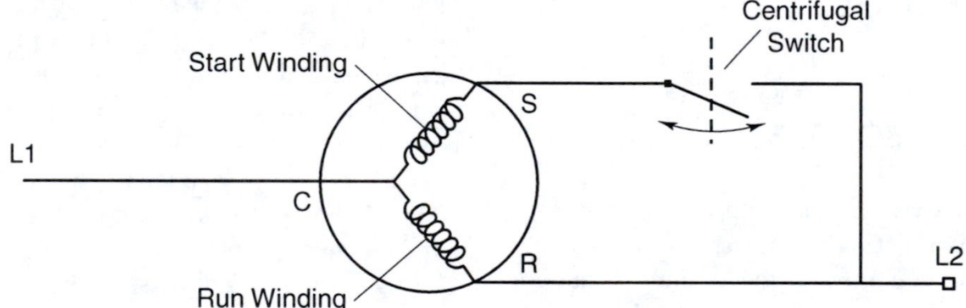

FIGURE 4-10 Split-phase motor wired with a centrifugal switch

(a)

(b)

FIGURE 4-11 (a) Start capacitors. (b) Run capacitors. *(Courtesy Aerovox, Inc.)*

increase the phase angle in the motor, thereby increasing the starting torque and, in some cases, the running efficiency of the motor. Capacitors are rated in *microfarads,* denoted μ, indicating how much electrical charge they can store. The larger this rating, the more charge a capacitor can store. The capacitor also has a voltage rating, which represents the capacitor's ability to withstand voltage without breaking down.

Capacitors are constructed of two parallel, metal—usually aluminum—plates separated by an insulator called a *dielectric* (Figure 4-12). The electric symbol for a capacitor (Figure 4-19) indicates these two plates. The dielectric can be any one of a number of different substances, including air, paper, and aluminum oxide. The ability of the capacitor to store electrical charge is dependent on the size of the plates, the dielectric used, and the thickness of the dielectric. Two commonly used types of capacitors that are encountered by the field technician based on their construction and their application are the:

- Start capacitor
- Run capacitor

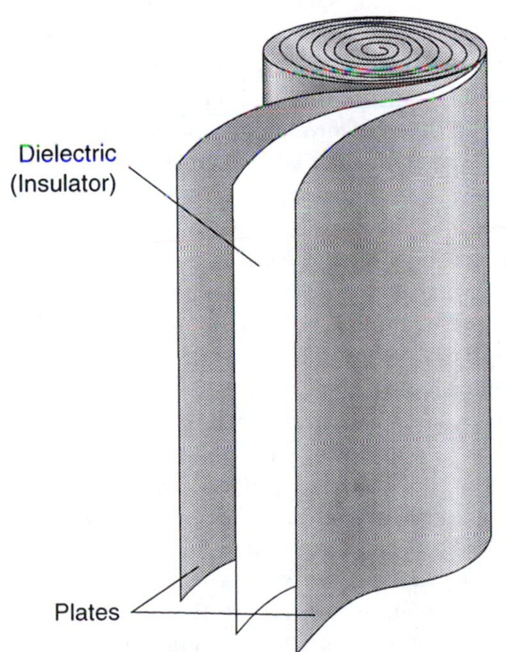

FIGURE 4-12 An insulator called a dielectric separates the metal plates in the capacitor.

START CAPACITORS. Start capacitors are used primarily to add starting torque on motor start-up. They are commonly found in ranges from about 75 to 600 microfarads and are easily identified by their *Bakelite* casing, Figure 4-11(a), which is a dark plasticlike substance. Start capacitors are considered *dry* because the plates and the dielectric are placed in the casing with no surrounding fluid. (Obviously, another type of capacitor is referred to as being *wet,* which is discussed shortly.) The start capacitor is wired in series with the start winding and is removed from the circuit along with the start winding when the centrifugal switch, or other starting device, opens.

RUN CAPACITORS. Run capacitors, Figure 4-11(b), are intended to remain in the circuit whenever the motor is energized and are found in the range of 2 to about 60 microfarads. Although the primary function of the capacitor is to increase starting torque, the run capacitor is used primarily to increase the running efficiency of the motor. These capacitors are used on motors that are specially designed to have the start winding remain in the circuit whenever the motor is energized. The run capacitor helps correct and improve the *power factor* of the motor, increasing its efficiency. The power factor is a number, ranging from 0 to 1, that relates the power used, or paid for, to the output power obtained. For example, if 100 watts of power were paid for and 100 watts were actually obtained and used in a useful manner, the power factor would be 1. If, on the other hand, only 50 watts of useful power were obtained, the power factor would be 0.5, or one-half.

The run capacitor, being in the active electrical circuit whenever the motor is energized, has the tendency to get hot. For this reason, it is enclosed in a metal case, which is filled with oil. The oil helps absorb the heat from the plates and transfer it to the metal shell of the device. The heat is then transferred to the air surrounding the device. For this reason, the run capacitor is referred to as a *wet* capacitor. The distinctions between start and run capacitors are summarized in Figure 4-13.

Current Magnetic Relays

Another device that can be used in conjunction with split-phase motors is the **current magnetic relay (CMR)** (Figure 4-14). The CMR is commonly found on fractional-horsepower motors that require low

	Run Capacitors	Start Capacitors
Range	2–60 microfarads	75–600 microfarads
Construction	• Wet type • Oil-filled • Metal casing	• Dry type • Bakelite casing
Function	• In the active circuit whenever the motor is energized • Designed to increase running efficiency of the motor	• Only in the circuit for a few seconds • Designed to increase motor's starting torque

FIGURE 4-13 Summary of start and run capacitors

FIGURE 4-14 The current relay is identified by the size of the wire in the holding coil. *(Photo by Bill Johnson)*

starting torque. This relay can be used to remove the start winding and the start capacitor from the circuit or, in the cases in which the start winding remains energized during normal operation, just the start capacitor. Unlike the centrifugal switch, which opens and closes its contacts depending on the speed of the motor, the CMR's contacts are opened and closed depending on the current flowing through the run winding of the motor. When a motor is initially energized, the amperage draw is very high, roughly 5 times the amount of current flowing through the motor when it is up and running up to speed. This amperage is called the **locked-rotor amperage (LRA)**. The amperage that the motor draws after start-up is only a fraction of the LRA and is called the *running-load amperage (RLA)*. When the RLA is close to the full-load condition of the motor, it is referred to as the full-load amperage (FLA). The CMR is made up of two parts: a coil of wire and a normally open set of contacts.

The wire in the coil is rather thick, and the number of turns in the coil is very low, making the resistance of this coil very low, usually less than 1 ohm. The coil is wired in series with the run winding and, therefore, has the same current flowing through it as the run winding. The coil, with its low resistance, has a small voltage drop across it, making its effect on the voltage across the run winding negligible. The current flowing through the run winding is either zero, when the motor is de-energized; RLA, when the motor is up and running to speed; or LRA, when the motor is in the process of starting up. When the current drawn by the run winding is equal to the LRA, a strong magnetic field is generated, compared to the field generated when the motor is drawing RLA. Obviously, when the motor draws no current, no magnetic field is present.

The set of electrical contacts in the relay is held open by either spring pressure or, more commonly, by gravity. This set of contacts is called the bridging contact and is connected to an **armature**, which moves back and forth within the coil (Figure 4-15). These contacts are wired in series with the start winding and, when open, remove this winding from the circuit. The contacts of the relay should be open when the motor is de-energized, as well as when the motor is up and running to speed.

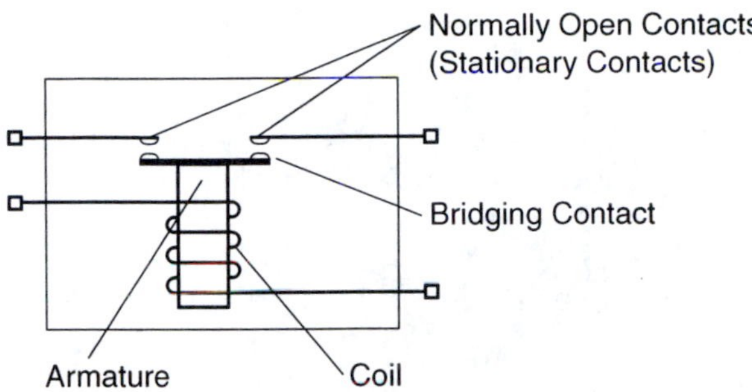

FIGURE 4-15 Coil and contacts in a current magnetic relay. The bridging contact is connected to the armature, which moves within the relay's coil.

In operation, when the motor is initially energized, the run winding is energized but the start winding is not. This will prevent the motor from turning, because no imbalance exists in the magnetic field. The amount of current flowing through the run winding at this point will be LRA. The increased current draw creates a large magnetic field in the coil of the CMR. This force pushes the normally open contacts of the relay closed, completing the electric circuit through the start winding. Having energized the start winding, the necessary imbalance is created, allowing the motor to start. Once started, the current draw of the motor quickly drops from LRA to RLA. This reduced current reduces the strength of the magnetic field holding the contacts closed, and the contacts fall open. The motor then continues to run with only the main winding energized. When the motor is de-energized, the contacts remain open until the next start-up.

Potential Relays

Another device that is commonly used in various types of split-phase motors is the **potential relay** (Figure 4-16), sometimes referred to as a potential magnetic relay (PMR). The function of the PMR is similar to that of the CMR in that it can de-energize the start capacitor and/or the start winding; but, instead of sensing the current in the run winding, the potential relay controls its contacts by sensing the induced voltage, or potential, across the start winding. The PMR is commonly found on CSCR motors requiring more starting torque than those that utilize the CMR.

The potential relay consists of a normally closed set of contacts and a coil. The coil is wired in parallel with the start winding and is the part of the device that is used to sense the induced voltage across the start winding. *Terminal 2* to *Terminal 5* on the device (Figure 4-17) identifies the coil. Since the installation of the relay does not reduce the current flow through the start winding, the resistance of the coil is obviously very high, since current takes the path of least resistance. The coil resistance is typically in the range of 15,000 to 50,000 ohms. This concept is similar to that of a **voltmeter**, where the resistance of the test instrument is very high to allow only a very small amount of current to flow through the meter. The normally closed set of contacts on the PMR is identified by Terminal 1 to Terminal 2 (Figure 4-17).

The potential relay's coil acts as a voltmeter and "measures" the induced voltage across the start winding. The coil then causes the contacts to open and close, depending on this reading. The voltage that is sensed across the start winding can easily be greater than the voltage supplied to the motor. This is because the total voltage that the coil senses is the sum of the actual voltage across the coil and the induced voltage generated by the magnetic field of the run winding. Just as the primary winding of a transformer generates an induced voltage in the secondary winding, the run winding generates an induced voltage in the start winding. The relay is sensitive to two specific voltages called the **pick-up voltage** and the **drop-out voltage**. The pick-up voltage is the voltage at which the relay's contacts open, and the drop-out voltage is the voltage at which the relay returns its contacts to their normally closed position.

In operation, when the motor is initially energized, both the start and run windings are energized and the motor starts. As the motor picks up speed, the induced voltage across the start winding increases, since the voltage-generating effect increases. Once this voltage increases to the pick-up voltage, the PMR's contacts open. Generally, this removes only the start capacitor from the circuit. If the motor experiences difficulty running and begins to slow down, the induced voltage will drop and the relay's contacts will close when the voltage falls to the drop-out voltage. This will put the start capacitor back into the circuit and allow the motor to regain its speed. Once

FIGURE 4-16 Potential relays

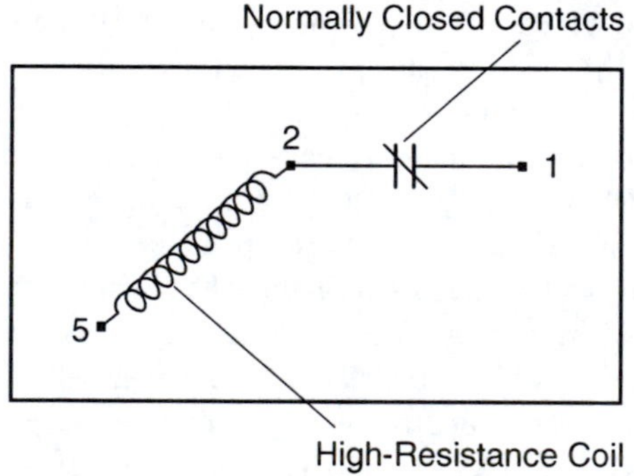

FIGURE 4-17 Wiring diagram of a potential relay

again, when the induced voltage rises to the pick-up voltage, the contacts will open. A summary of the centrifugal switch, the CMR, and the potential relay is given in Figure 4-18.

CAPACITOR-START-INDUCTION-RUN MOTORS

The **capacitor-start-induction-run (CSIR) motor** is similar in operation to the ISIR motor. The start winding is energized long enough to assist the motor in starting and then removed from the circuit using either a centrifugal switch or a CMR. Potential relays are not commonly used in conjunction with a CSIR motor, because they are designed for motors requiring more starting torque.

The main difference between an ISIR motor and a CSIR motor is the addition of a start capacitor. The start capacitor, the start winding, and the contacts of the switching device are all wired in series with each other so, when the contacts open, the start winding and the start capacitor are both removed from the circuit. Figure 4-19 shows a wiring diagram of a CSIR motor that is wired with a current magnetic relay.

The CSIR motor has a phase angle that approaches 90 degrees, compared to the 30-degree phase angle obtained by an ISIR motor without the start capacitor. The CSIR motor is often used for small compressors that must be started under full load, since they provide more starting torque and range in size from about ⅙ horsepower to ¾ horsepower.

PERMANENT-SPLIT-CAPACITOR MOTORS

Of all the motors that utilize capacitors, the **permanent-split-capacitor (PSC) motor** has the lowest starting torque. This type of motor has very good running efficiency and is a popular choice for low-torque applications since no starting relays or switches are needed. The PSC motor uses a small run capacitor that, along with the start winding, is in the active circuit whenever

	Centrifugal Switch	Current Relay	Potential Relay
Controlled by	Speed of the motor	Current flow through the run winding	Induced voltage across the start winding
Coil Characteristics	N/A	• Low resistance • Thick wire • Few turns	• High resistance • Thin wire • Many turns
Coil Location	N/A	Wired in series with the run winding	Wired in parallel with the start winding
Normally Open or Normally Closed Contacts	Normally closed	Normally open	Normally closed
Operation	• Contacts open when 70% of total motor speed is reached • Contacts close when motor is de-energized	• Contacts close when locked-rotor amperage (LRA) is drawn • Contacts open when the motor amperage falls to running-load amperage (RLA)	• Contacts open when the induced voltage reaches the pick-up voltage • Contacts close when the voltage drops to the drop-out voltage
Application	Used primarily on open motors	• Used primarily on low-torque and low-horsepower motors • Used primarily on CSIR motors	• Used primarily on motors requiring high starting torque • Used on larger-horsepower motors • Used on CSCR motors
Installation Requirements	• Not position sensitive • Must be located on the motor	• Must be mounted right-side up • Must be connected directly to the motor	• Normally not position sensitive • Does not need to be located near the motor

FIGURE 4-18 Summary of starting switches and relays

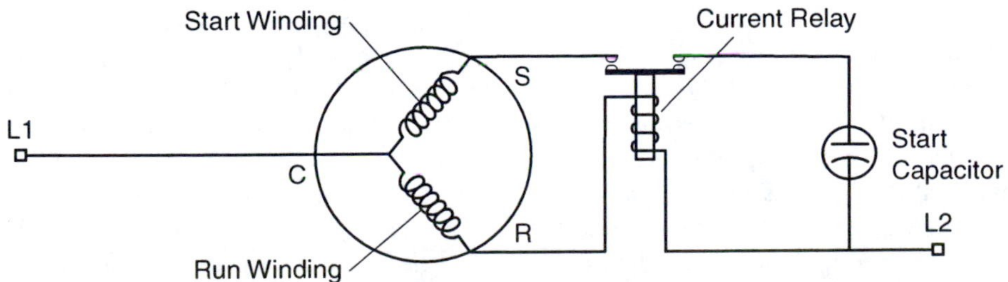

FIGURE 4-19 Capacitor-start-induction-run (CSIR) motor using a current magnetic relay. When the relay's contacts open, the start winding and the start capacitor are removed from the circuit.

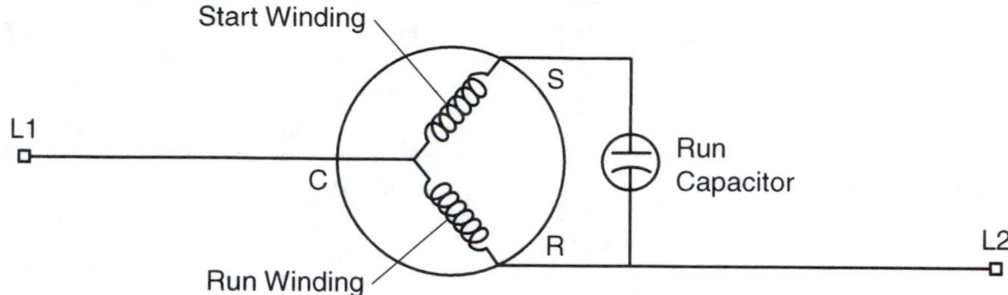

FIGURE 4-20 Schematic of a permanent-split-capacitor (PSC) motor. Note that the start and run windings are energized whenever the motor is running.

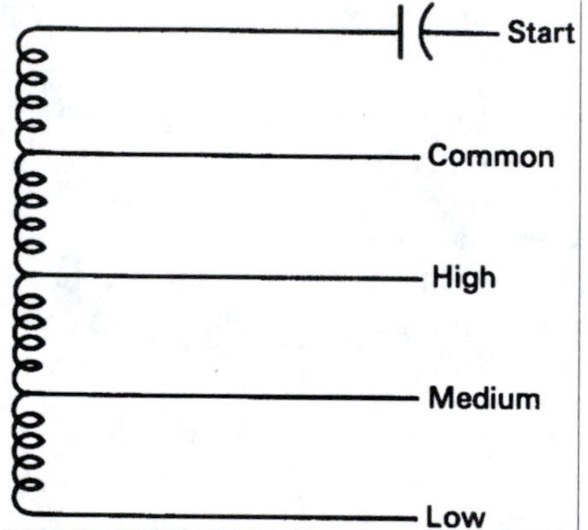

FIGURE 4-21 Schematic of a three-speed PSC motor

the motor is energized. The run capacitor helps to increase the power factor. A wiring diagram of a single-speed PSC motor is shown in Figure 4-20. The start winding in this type of motor is specially designed to remain in the circuit at all times. Permanent-split capacitor motors are widely used in applications in which the blower is located close to the occupied space, since the low-starting-torque characteristic of the motor causes it to start slowly and quietly.

Permanent-split-capacitor motors can be one-speed or multispeed components. If the motor operates at only one speed, typically it has three wires protruding from the motor's shell, labeled Common, Start, and Run. If the motor has multiple speeds, it has a common wire, a start wire, and one wire for each speed. For example, if the motor has three speeds, the wiring diagram will look similar to that in Figure 4-21. Note that the start winding, between the common and the start wire, remains unchanged no matter which speed is selected. The run winding, however, gets larger and larger as the motor speed is reduced. This is because, as the winding resistance increases, the current draw drops and the strength of the magnetic field drops, thereby reducing the motor speed. It can be seen from the figure that the resistance between the common and the low wires is greater than the resistance between the common and the high wires.

CAPACITOR-START-CAPACITOR-RUN MOTORS

Of all the single-phase motors discussed in this text, the **capacitor-start-capacitor-run (CSCR) motor** has the highest starting torque; hence, the largest phase angle; and excellent running efficiency. It is found in the range from one-half to 10 horsepower. This motor commonly uses a potential relay as a starting device and uses both a large start capacitor and a small run capacitor. The start winding is designed to remain energized all the time. A wiring diagram for a CSCR motor is shown in Figure 4-22. Note that when the potential relay's contacts open, the motor resembles a PSC motor. This makes sense since, as already noted, the PSC motor has excellent running efficiency but very low starting torque. Adding starting torque to the PSC motor creates in effect, the CSCR motor. The figure also shows that, on initial start-up, the run and the start capacitors are connected in parallel with each other. When wired in this fashion, the effective capacitance obtained is the sum of the values of the individual capacitors. If, for example, the run capacitor is rated at 5 microfarads and the start capacitor is rated at 100 microfarads,

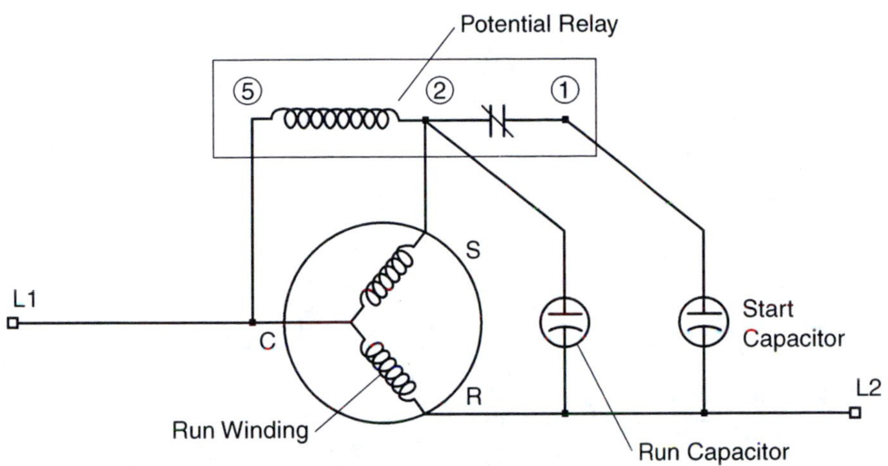

FIGURE 4-22 Wiring diagram of a CSCR motor using a potential relay. The relay removes the start capacitor from the circuit.

then the total capacitance is 105 microfarads on start-up but drops to only 5 microfarads once the relay's contacts open.

THREE-PHASE MOTORS

The motors discussed so far in the text are all single-phase motors, meaning that the voltage supplied to the device is all in the same phase. The difference in resistance between the start and the run windings, as well as the start capacitors, was used to increase the phase angle, creating the effect of more than one phase, to generate starting torque. At best, under those conditions, two-phase power could be achieved when a phase angle of 90 degrees was established. Now examine a type of motor that utilizes three phases of power, generating a phase angle of 120 degrees without the use of capacitors or starting relays. This motor is called the **three-phase motor** and generates the highest amount of starting torque. It has extremely high running efficiency and is the motor of choice whenever three-phase power is available for use. Since three-phase power is not available—by law—in residential locations, it is found exclusively in commercial and industrial applications.

Three-phase motors are made up of three run windings, as opposed to the start and run windings found on single-phase motors. Three hot lines, identified L1, L2, and L3, feeding into the device power the motor. For a 240-volt, three-phase power supply (Figure 4-23), the voltage measured between any two of the hot "legs" is 240 volts. Similarly, a 220-volt, three-phase power supply would provide 220 volts:

- Between L1 and L2
- Between L2 and L3
- Between L1 and L3

Also note that each of the three hot legs is responsible for generating a portion of two of the phases. For example, L1 and L2 generate one phase, while L1 and L3 generate a second phase. The third phase, generated by L2 and L3, does not involve L1 at all. This poses a potential problem.

If one power leg of a single-phase motor opened for whatever reason, the entire motor would shut down because there would be no path for the current to take. In a three-phase motor, this would not be the case. If one leg opened, only two of the three phases would be disabled, leaving the third phase energized. In this case, the motor may or may not continue to run, depending on the specific application. The motor will, however, overheat due to an increase in the current draw of the motor. The motor will attempt to continue to operate on one phase and will need excessive current to do so. This can cause severe motor damage, and, since these motors are typically very expensive, every attempt should be made to prevent this from occurring. This phenomenon is referred to as single phasing and cannot be

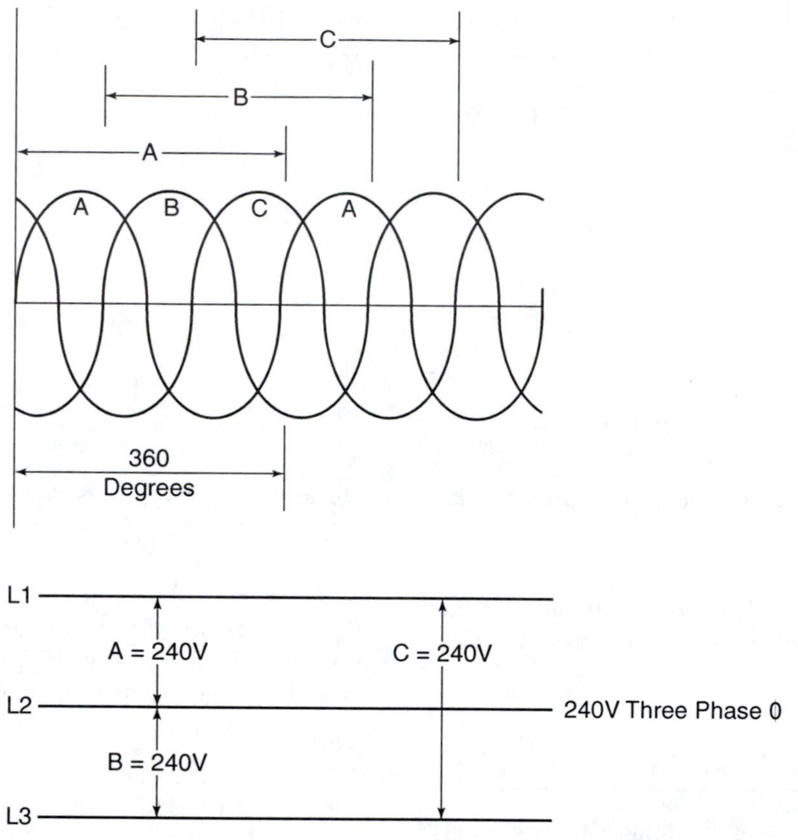

FIGURE 4-23 Three-phase power is made up of three single-phase supplies, each of which is 120 degrees out of phase with the next.

avoided since something as simple as a blown fuse can cause it. Although a single-phase condition is difficult at best to avoid, motor damage is not. By utilizing a device called a motor starter, three-phase motors are protected from this condition.

Motor Starters

Motor starters are, simply stated, contactors with built-in overload protection (Figure 4-24). The built-in overload consists of **heaters** and **switches**. The heaters are located in the power circuit, in series with the motor, while the switches are in the control circuit, in series with the holding coil of the starter. If the heaters detect a temperature that is higher than normal, they will cause the switches to open, de-energizing the holding coil of the starter. This in turn will open the contacts feeding power to the motor, de-energizing it. The heat that is sensed is a result of

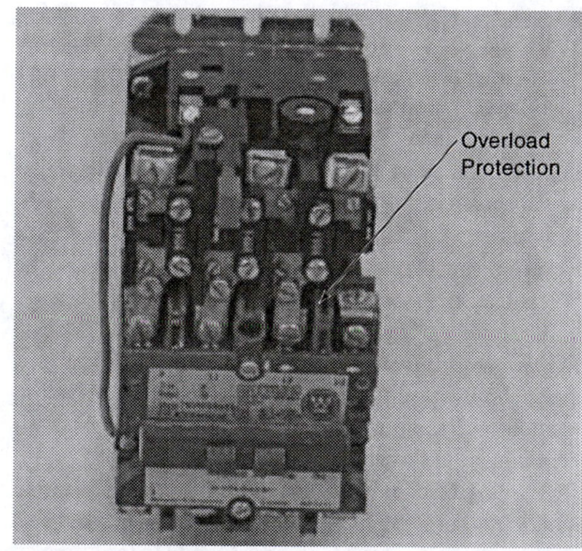

FIGURE 4-24 Typical motor starter. *(Photo by Bill Johnson)*

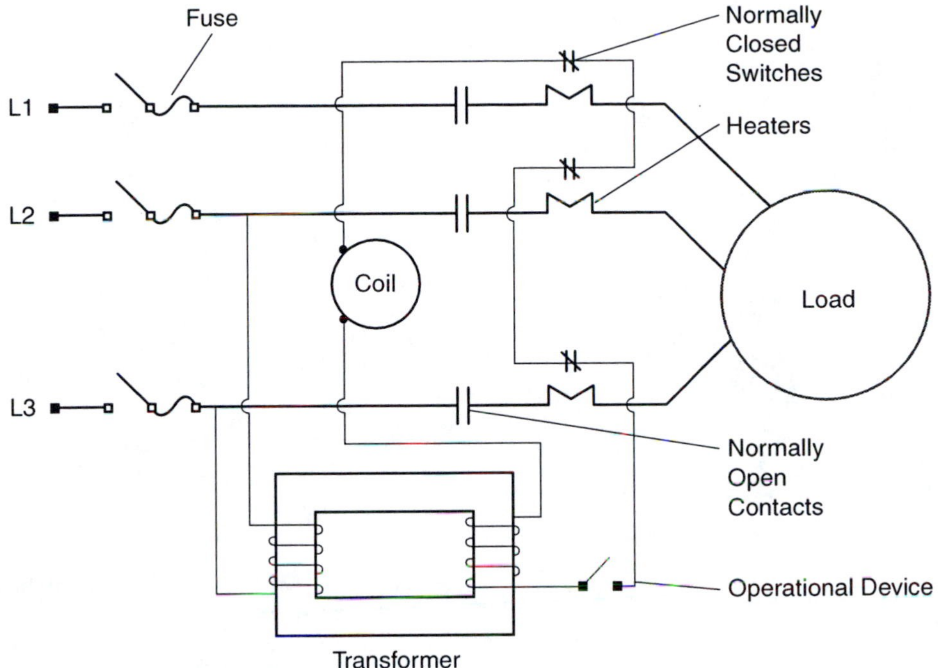

FIGURE 4-25 Wiring diagram for a low-voltage starter

the excessive amperage drawn by the motor attempting to operate on one phase, since heat is a by-product of current. A wiring schematic for a starter is shown in Figure 4-25.

The holding coil for the starter cannot be energized unless the operational devices (thermostat or other control) and all three switches in the overload are closed. If, after start-up, any one of the fuses blows or one power leg fails to supply voltage, the current draw in that leg will be zero but the current in the other two legs will increase. The heaters in those two lines will sense the excess heat and cause the normally closed switches in those lines to open, deenergizing the holding coil. The starter is a manually reset device, so the motor will be unable to run until the problem is located and corrected. The starter can then be reset to put the motor back into operation. Once the starter has been reset, current is permitted to flow to the motor windings, and operation resumes. The three-phase motor windings can be configured in one of two ways:

- Wye configuration
- Delta configuration

Stator Windings: Wye Configuration

As mentioned earlier, three-phase motors have three run windings, but no mention was made of how these windings were connected to each other. In reality, a number of different configurations exist in which the windings can be arranged, and each one is addressed separately. The first configuration discussed is the **wye configuration**. The wye configuration gets its name from the *Y* shape in the wiring diagram. Figure 4-26 shows a simple, three-phase motor with a wye-configured stator. Each of the three power lines supplying the motor is connected to wires T1, T2, and T3. This type of three-phase motor is designed to operate at a specific voltage and can be easily identified by the fact that only three wires or terminals are available for external wiring connections. Other types of three-phase motors can be used at different supply voltages. These motors are called dual-voltage motors.

For example, three-phase motors that can operate at 220 volts as well as 440 volts are very common. In the case of dual-voltage motors, nine wires protrude from the motor shell instead of the usual three. Each of these wires is numbered, and care must be taken

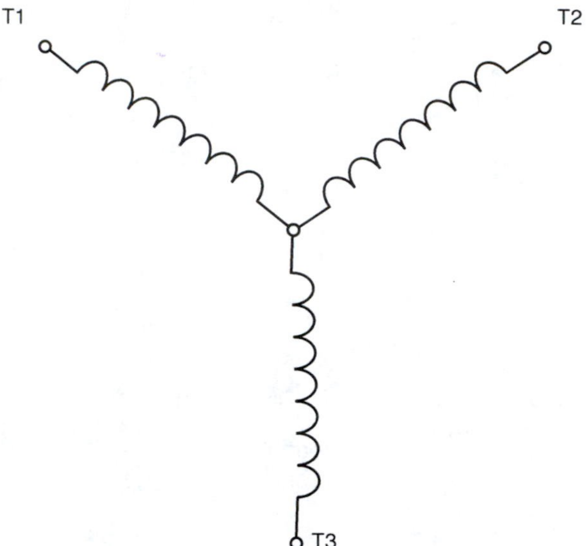

FIGURE 4-26 Three windings connected in a wye configuration

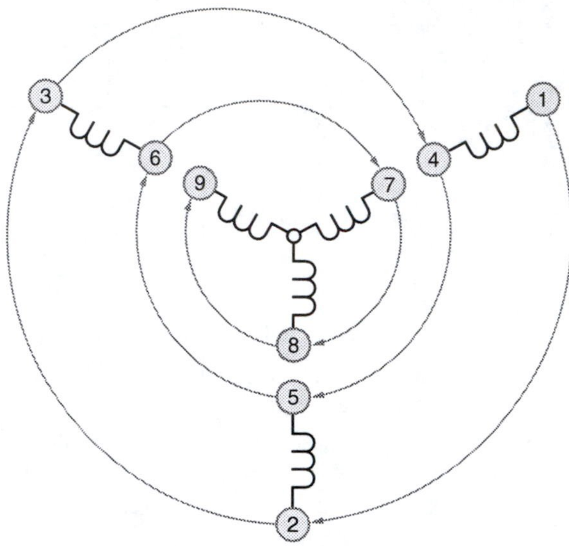

FIGURE 4-27 Method used to number the wires in a dual-voltage stator

when wiring the motor or serious component damage could occur. The three windings in a dual-voltage motor are each divided into two smaller windings, making a total of six. Figure 4-27 shows how the windings appear on a wiring diagram and how the wires are numbered. Dual-voltage motors normally have wiring diagrams on the nameplate, providing the field technician with the proper wiring connections that must be made for the desired voltage application. The diagrams are normally labeled high voltage and low voltage. In the case of a 220-volt/440-volt motor, high voltage refers to 440 volts and the low voltage refers to 220 volts.

HIGH-VOLTAGE CONNECTION. When operating at the higher of the two voltages specified on the motor nameplate, the windings are connected in series with each other, forming a large *Y*. Figure 4-28 shows the wiring diagram as it may appear on the motor nameplate. Figure 4-29 shows how the windings look after the connections are made. In this configuration, the windings resemble those in the simple three-phase motor in Figure 4-26. When operating at the higher voltage, the current draw of the motor is one-half that of the same motor operating at one-half the voltage.

LOW-VOLTAGE CONNECTION. When wired to operate at the lower of the two voltages, the windings are connected in parallel with each other, forming two smaller *Y* configurations. Figure 4-30 shows the wiring diagram used for the low-voltage wiring connections. Figure 4-31 shows how the connections may look. Notice in the figure that *L3* is connected to *Wire 3* and *Wire 9,* putting those two windings in parallel with each other. Similarly, the other windings are also in parallel with each other. Since *Wire 4, Wire 5,* and *Wire 6* are all connected together,

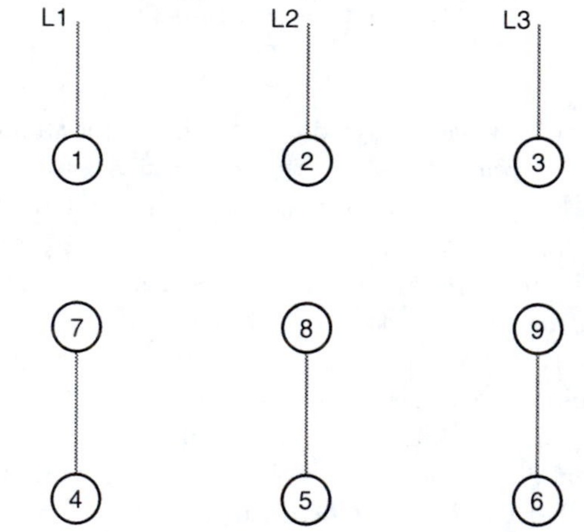

FIGURE 4-28 Wiring diagram for the high-voltage connections on a dual-voltage motor

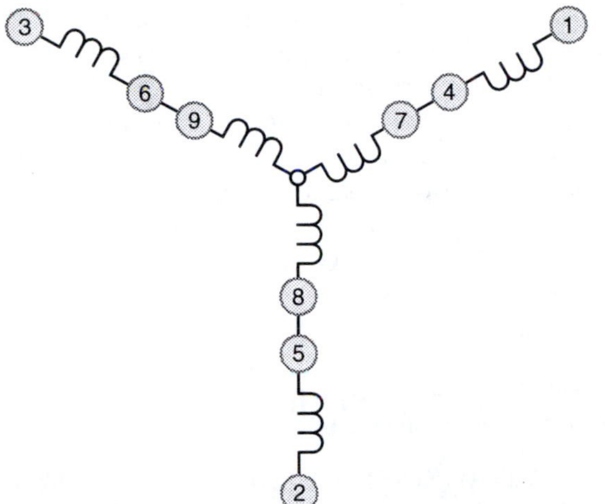

FIGURE 4-29 Windings connected in series

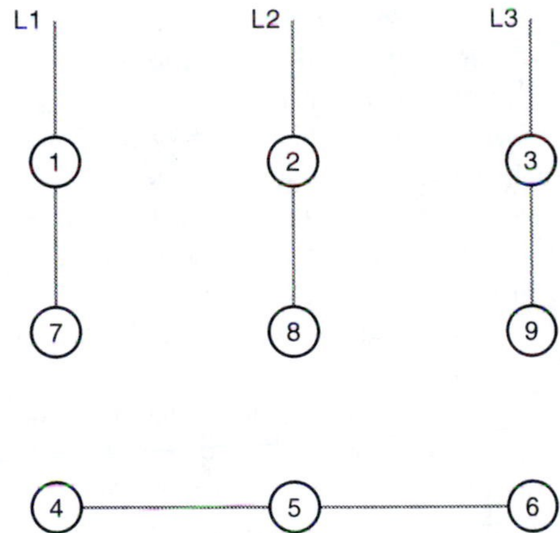

FIGURE 4-30 Wiring diagram for the low-voltage connections on a dual-voltage motor

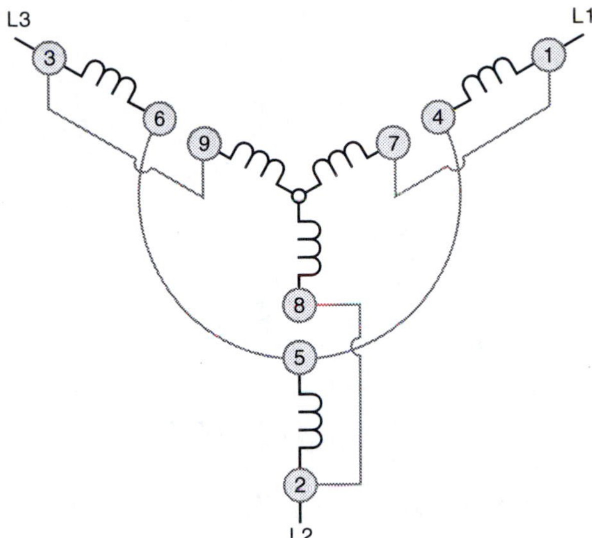

FIGURE 4-31 Windings connected in parallel

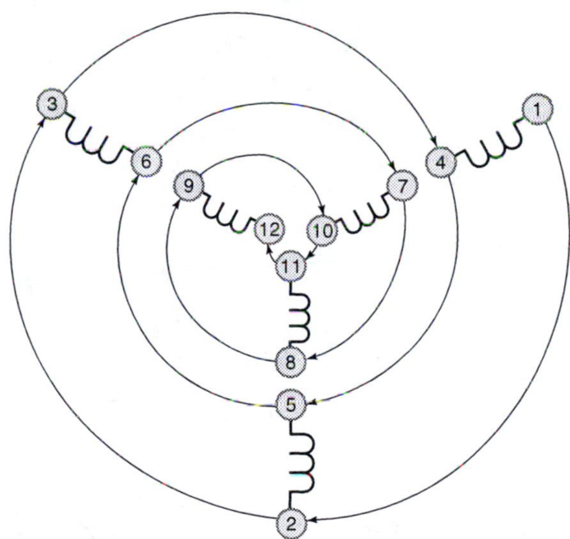

FIGURE 4-32 Numbering system used for a dual-voltage motor with a delta-configured stator

another *Y* connection is formed. The amperage draw of a dual-voltage motor that is wired to operate at the lower voltage will be twice that of an identical motor wired to operate at the higher voltage. This is because the resistance of the windings is lower, since the windings are connected in parallel.

Stator Windings: Delta Configuration

Another common layout for the windings in three-phase motors is the **delta configuration**. A three-phase motor that has the stator wired in the delta configuration will have 12 wires protruding from the motor's shell, instead of the 9 wires in the case of the wye configuration. The numbering system used in the delta configuration is similar to that used in the wye configuration and is shown in Figure 4-32. Just as in the wye configuration, the high-voltage application places the windings in series, while the low-voltage application arranges the windings in parallel.

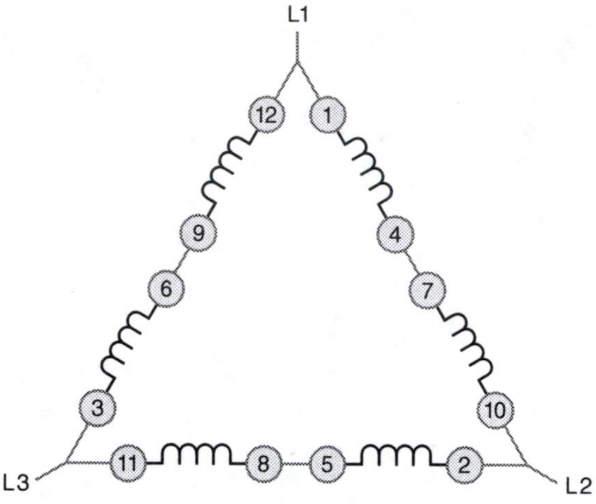

FIGURE 4-33 High-voltage delta connection

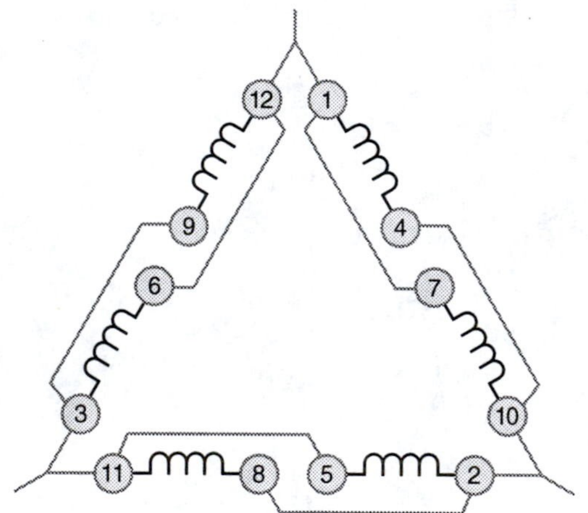

FIGURE 4-34 Low-voltage delta connection

Figure 4-33 shows the high-voltage connections, while Figure 4-34 shows the low-voltage connections.

Three-Phase Motor Starting

Since three-phase motors are commonly used for large commercial and industrial applications, the motors themselves tend to be very large. They draw large amounts of current during normal operation and even larger currents during initial start-up. For example, a motor that draws 60 amperes during normal operation may draw over 300 amperes during start-up. This excessively high start-up amperage can cause the contacts on the starters to become pitted and fail prematurely. For this reason, a number of methods have been implemented to reduce the amount of current that a motor draws on start-up. Smaller, three-phase motors can be effectively started using a method similar to that shown in Figure 4-25. The larger the motor, the more important the effects of the start-up amperage become. The following applications are employed primarily on large motors, usually exceeding 25 horsepower. Two of the most common methods used to reduce this amperage are the:

- Part-wind start
- Reduced-voltage start

PART-WIND START. To reduce the amperage drawn by the motor on initial start-up using a **part-wind start**, only half of the motor's windings are energized. Figure 4-35 shows a wiring diagram of a part-wind start. The part-wind start is configured as a dual-voltage motor, with a wye-type stator, wired for the low-voltage application. This creates two small wyes, each of which can be energized at a different time. If the two wyes, which are connected in parallel with each other, are energized at the same time, the resistance will be low and the current will be high. (Remember from Ohm's law that two resistors connected in parallel with each other will have a combined total resistance that is less than the resistance of each individual resistor.) The LRA of this motor will therefore be very high. By energizing only half of the windings, the resistance is higher and, therefore, the current is reduced. The LRA drawn by energizing only half of the motor is therefore reduced. Once the motor begins to rotate, the second half of the windings are introduced into the circuit. The sequence of operations of this type of motor starter is as follows:

1. The operational device (thermostat or other control) closes.
2. As long as the six normally closed switches in the overloads are closed, the holding coil of Starter 1 will be energized.
3. When Coil 1 is energized, the four sets of normally open contacts will close. (Three sets of contacts feed current to the first half of the windings, and one set of contacts completes the circuit through the time-delay relay.)
4. The first half of the windings are energized, as well as the coil on the time-delay relay.
5. The motor starts turning with only one half of the windings energized.

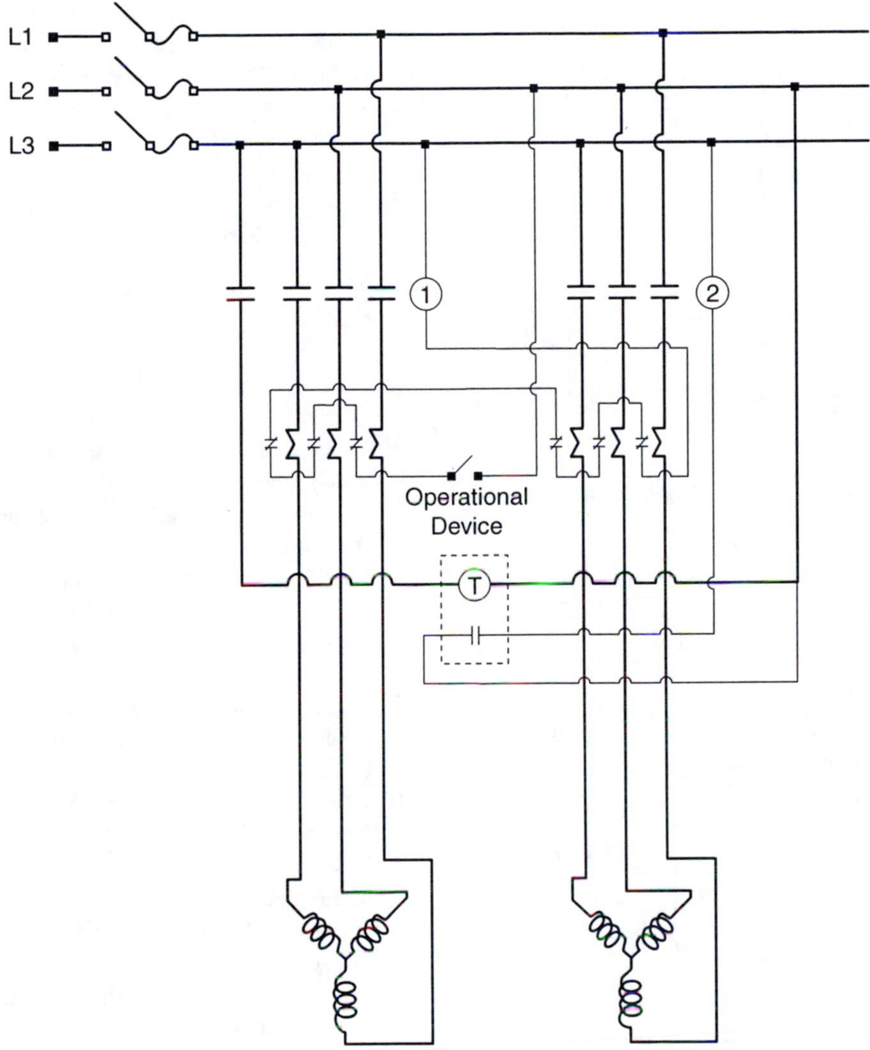

FIGURE 4-35 Part-wind start

6. After the time-delay period, usually one second, the contacts on the time-delay relay are closed.
7. This completes the circuit through the holding coil of Starter 2.
8. The three normally open sets of contacts on Starter 2 are closed.
9. The second half of the windings are energized.

If any one of the six power leads feeding the motor becomes defective and stops feeding current to the motor, the other heaters will cause the entire motor to shut down. Again, the overloads on the starters are manually reset, so the motor will not operate until the problem is found and corrected.

REDUCED-VOLTAGE STARTER. In the part-wind start, all of the voltage was applied to part of the windings to reduce the current draw on initial start-up. The **reduced-voltage starter** accomplishes the same end result but by different means. This type of motor starter supplies a lower voltage to all of the windings until the motor starts, then the voltage is increased to the normal operating voltage. Figure 4-36 shows a schematic of a reduced-voltage starter. The underlying concept of the reduced-voltage starter is to place a large resistor in series with each winding of the motor. Since a voltage drop will occur across the resistors, the voltage that is ultimately applied to the motor will be lower than the supplied line voltage. In addition,

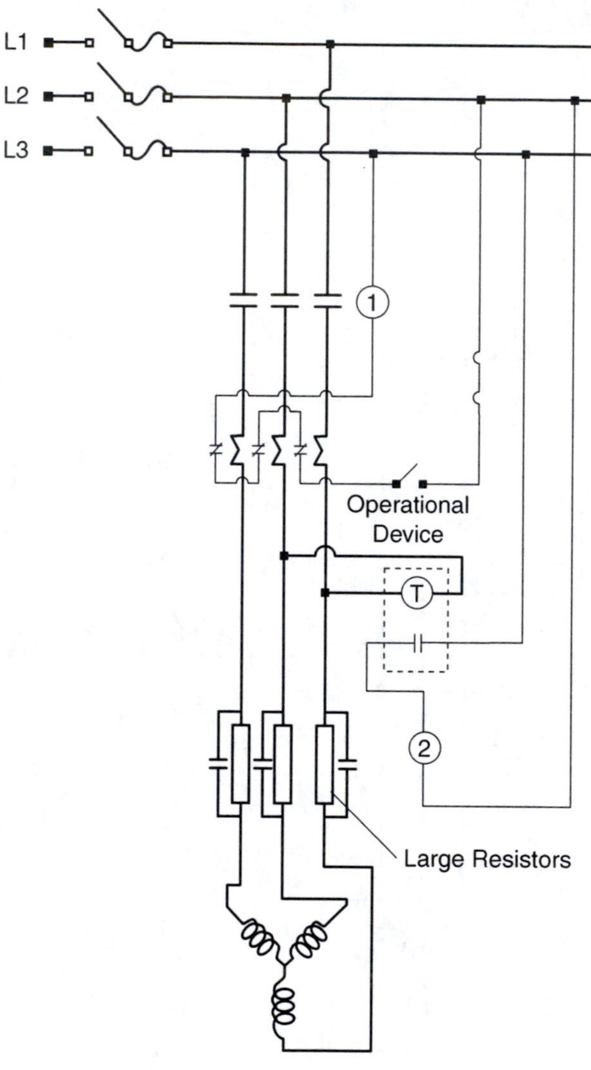

FIGURE 4-36 Reduced-voltage start

since the resistors are added to the circuit, the current in the circuit will be reduced. The motor is started in this fashion. Once the motor begins to turn, the resistors are removed from the circuit. This causes the circuit resistance to drop, thereby increasing the current flowing through—and the voltage supplied to—the motor windings. The sequence of operations of this type of motor starter is as follows:

1. The operational device (thermostat or other control) closes.
2. As long as the three normally closed switches in the overloads are closed, the holding coil of Starter 1 will be energized.
3. When Coil 1 is energized, the three sets of normally open contacts will close. These contacts feed current through the three large resistors to the motor windings.
4. Once the three normally open sets of contacts close, the motor starts at a reduced voltage since a substantial voltage drop occurs across the large resistors.
5. The coil of the time-delay relay is also energized when the normally open contacts on the starter close.
6. After the time-delay period has elapsed, usually 2 seconds, the relay's contacts close.
7. The circuit through Holding Coil 2 is now completed.
8. The normally open contacts on Contactor 2 are now closed.
9. Since current takes the path of least resistance, the current now bypasses the large resistors.
10. The circuit resistance is reduced. The circuit current increases. The voltage applied to the motor is now equal to the supply voltage.

SUMMARY

- Commonly found motors include the shaded-pole motor, the split-phase motor, and the three-phase motor.
- Starting torque is the electric push given to motors to help them start.
- Motors used in lightweight applications typically have low starting torque.
- Motors are made up of a stator, which does not turn, and the rotor, which rotates and is connected directly to the shaft of the motor.
- Split-phase motors have a start winding and a run winding.
- The start winding is a high-resistance coil made up of thin wire with a large number of turns.
- The run winding is a low-resistance coil made up of thick wire with a small number of turns.
- Common types of split-phase motors include the ISIR, the CSIR, the CSCR, and the PSC.

- A split-phase motor can use centrifugal switches, current magnetic relays, or potential relays to remove the start winding and/or start capacitor from the circuit.
- Centrifugal switches operate depending on the speed of the motor.
- Current magnetic relays operate based on the current in the run winding.
- Potential relays operate depending on the induced voltage across the start winding.
- Three-phase motors utilize motor starters, not starting relays or capacitors.
- Motor starters have built-in overload protection.
- Three-phase motors often use part-wind or reduced-voltage starts.

KEY TERMS

Armature
Capacitor
Capacitor-start-capacitor-run (CSCR) motor
Capacitor-start-induction-run (CSIR) motor
Centrifugal switch
Current magnetic relay (CMR)
Delta configuration
Drop-out voltage
Heaters
Horsepower
Induction-start-induction-run (ISIR) motor
Locked-rotor amperage (LRA)
Motor starter
Part-wind start
Permanent-split-capacitor (PSC) motor
Pick-up voltage
Potential relay
Reduced-voltage start
Rotor
Run capacitor
Run winding
Split-phase motors
Start capacitor
Start winding
Starters
Starting torque
Stator
Switches
Three-phase motor
Voltmeter
Watt
Wye configuration

FOR DISCUSSION

1. Explain the importance of properly wiring a dual-voltage motor. Discuss possible results if a dual-voltage motor is incorrectly wired.
2. Review the basics of Ohm's law and discuss how the amperage draw of a dual-voltage motor changes depending on the wiring configuration.
3. Discuss the importance of properly sized heaters in a motor starter.
4. Discuss the differences between a wet and a dry capacitor.

REVIEW QUESTIONS

1. The component part of a motor that is directly connected to the shaft is called the:
   a. Stator.
   b. Pulley.
   c. Rotor.
   d. All of the above are correct.
2. The start and run windings of a motor can be found in the:
   a. Rotor.
   b. Current magnetic relay.
   c. Potential relay.
   d. Stator.

3. The centrifugal switch opens and closes its contacts depending on the:
   a. Amperage flowing through the run winding.
   b. Speed of the motor.
   c. Induced voltage across the start winding.
   d. Induced voltage across the run winding.
4. The coil of a current magnetic relay is wired:
   a. In series with the start winding.
   b. In series with the run winding.
   c. In parallel with the common winding.
   d. In parallel with the run winding.
5. The potential relay:
   a. Has a low-resistance coil.
   b. Has a normally closed set of contacts.
   c. Is used primarily on larger motors requiring high starting torque.
   d. Both *b* and *c* are correct.
6. Which of the following motors has the highest starting torque?
   a. The CSCR motor
   b. The CSIR motor
   c. The Shaded-pole motor
   d. The PSC motor
7. What is the function of a capacitor?
   a. Stores electric charge
   b. Provides starting torque for a motor
   c. Helps to correct and maximize the power factor of a motor
   d. All of the above are correct.
8. Which of the following motors utilizes both a start and a run capacitor?
   a. The three-phase motor
   b. The CSIR motor
   c. The CSCR motor
   d. The ISIR motor
9. Why are run capacitors considered to be "wet"?
10. True or False: Permanent-split-capacitor motors use a CMR to remove the start winding from the circuit.
11. A split-phase motor has which of the following windings?
    a. Common and run
    b. Start and run
    c. Three run windings
    d. Common and start
12. Match the motor characteristics in the left column with the correct motor type in the right column:

a. Heavy copper bands	1. Three-phase
b. Highest starting torque	2. CSCR
c. Run capacitor, no starting relay or switch	3. Shaded pole
d. Start capacitor, CMR	4. PSC
e. Start capacitor, run capacitor	5. CSIR

13. Three-phase motors operate with a phase angle of:
    a. 0 degrees.
    b. 30 degrees.
    c. 90 degrees.
    d. 120 degrees.
14. Explain the condition called *single phasing.*
15. True or False: The purpose of the motor starter is to protect a three-phase motor from single phasing.
16. A three-phase, dual-voltage motor with a wye-configured stator:
    a. Will have nine wires that must be field connected.
    b. Will have 12 wires that must be field connected.
    c. Can only operate at one voltage.
    d. Both *a* and *c* are correct.
17. The overload portion of a motor starter has the heaters installed in the ___________ circuit and the normally closed switches in the __________ circuit.
18. True or False: The heaters in the starter are wired in series with the normally closed switches.
19. A part-wind start utilizes:
    a. Two contactors.
    b. Two starters.
    c. One starter and one contactor.
    d. One starter.
20. The part-wind start and the reduced-voltage start are designed to:
    a. Reduce the amperage draw of a three-phase motor on initial start-up.
    b. Reduce the effects of wear and tear on the starting components often caused by excessive heat.
    c. Protect expensive pieces of equipment from premature failure.
    d. All of the above are correct.

CHAPTER 5

Electrical Wiring Diagrams

OBJECTIVES After studying this chapter, the reader should be able to:

- Explain the importance of a solid understanding of the operation of electrical circuits.
- Explain how contacts in switches and thermostats route electrical current to the major system components.
- Describe the modes of operation of a cooling-only system with line-voltage controls.
- Follow an electrical schematic of a cooling-only system with line-voltage controls.
- Explain the difference between a schematic diagram and a ladder diagram.
- List components that are generally found in the power circuit.
- List components that are generally found in the control circuit.
- Describe the modes of operation of a cooling-only system with low-voltage controls.
- Follow an electrical schematic of a cooling-only system with low-voltage controls.

INTRODUCTION

As can be readily seen, no air-conditioning or refrigeration system can function properly unless the system's controls, or "brains," effectively coordinate the starting, stopping, and throttling of its components. To ensure this, the electrical circuits that allow current to flow to these controls and the major system components must, therefore, be trouble free. It is the job of the successful service technician to keep these circuits in tip-top shape. This chapter takes a close look at various control and power circuits that are commonly encountered in the field. By the end of the chapter, a common thread will be woven between the circuits described, giving the reader the basics needed to evaluate and navigate through any wiring schematic, regardless of its complexity, by taking a simple, commonsense approach.

OVERVIEW

The functions of electrical circuits are to direct and control current flow, at the correct time, to the various components in the air-conditioning or refrigeration system. A broken wire, open switch, or wiring error can stop the current from flowing, thereby preventing the system, or a part of the system, from functioning.

As shown earlier in the text, safety devices exist that are designed to open if the system conditions reach unsafe levels. Also, operational controls exist that open and close their contacts to energize and de-energize the system components at the proper time. These types of controls, the holding coils and other circuit loads, the power source, and the interconnecting wiring make up electrical circuits. A thorough and complete understanding of basic circuit theory is essential to be able to properly evaluate the circuits that, in effect, "give the orders" that the system components must follow. If, for example, an army general gives incorrect orders, the soldiers will carry out tasks that they may *think* are correct but do not reflect the wishes of their commanding officer. The soldiers are simply doing what they think is correct. The same holds true in electrical circuits. Electrical current will follow the path that is provided, whether or not it is correct.

Probably the easiest way to understand an electrical circuit and, hence, basic circuit theory is to visualize the wiring diagram as an automobile road map. Consider the following situation. Joan is in her automobile and wishes to drive from her house to the store. Only one road connects her house to the store. This road passes over a river and across a drawbridge. If Joan attempts to drive to the store and the drawbridge is open, she cannot, under any circumstances, cross the river since the path that she wishes to take is not complete. If another path were available to her that was passable, she would take that path in order to complete her trip. If there is no alternative road to take, she must wait until the drawbridge closes. The same holds true for electrical circuits. If a switch is open, current cannot flow through that path. If, however, an alternative, complete path is available for the current to take, it will do so. If not, current cannot flow and the load in that circuit cannot be energized. Consider the following situation as well.

Chester is driving home from work and wishes to get there in a hurry. He would like to travel at 55 miles per hour, but, because there are too many cars on the road, he cannot travel any faster than 25 miles per hour. The extra cars on the road have provided Chester with resistance, and his speed has therefore been reduced. Similarly, in electrical circuits, an increase in circuit resistance results in a reduction in current flow.

Although these two brief examples may seem trivial at first, the key is that electrical current cannot be seen, smelled, heard, touched, or tasted to be studied. (Although, in reality, electrical current *can* be touched, but severe personal injury, severe discomfort, and death are possible results.) Therefore, relating the unknown to something tangible and easily understood by all is important. By following this simplistic approach, we will be able to analyze and evaluate electrical circuits, determine the sequence of operations of various systems, and also determine where a system problem is originating. The underlying concepts that must be completely understood to be successful in the overall evaluation of electrical circuits include the following:

- When multiple switches are connected in series with each other, they must all be in the closed position for the circuit to be energized and have current flowing in it, thereby energizing the load.
- The current in a series circuit is the same at all points in the circuit.
- Parallel circuits operating with the same power supply have the same voltage applied to them.
- Parallel circuits operating with the same power supply can, and normally do, have different currents flowing through each branch.
- Loads connected in parallel can be, and often are, controlled by different switches and controls.
- A complete electrical circuit consists of a power supply, a closed switch or switches, a load, and an unbroken path for the current to take.

Electrical circuits are represented by wiring diagrams that are located in service literature provided with the equipment or on the specific piece of equipment being worked on. These diagrams can take on different appearances depending on their intention and purpose. The **schematic diagram**, for example, is designed to give the technician the following information:

- Component placement and location within the control panel (on some diagrams)
- A representation of each and every wire in the circuit
- The color of each wire in the circuit (when necessary)
- A detail of all **factory wiring** and **field wiring**

Providing the color-coding of the wires in the circuit helps to give the field technician a visual advantage. For a technician to try to locate a red wire in a bundle of multicolored wires is much easier than tracing out a black wire in a sea of other black wires. Since every wire is accounted for in the schematic diagram, a technician can find a wiring mistake more easily. The schematic also differentiates between factory and field wiring. Factory wiring is that which is installed in the factory before the unit is shipped out. Window and wall air conditioners have all of their wiring installed at the factory. Other systems come with only some of the wiring factory installed. The most common is the split system, which has the condensing unit in one location and the air handler in another. In these cases, the installation crew must install some of the wiring, referred to as field wiring. Factory and field wiring are easily distinguished from one another in the wiring diagrams.

The schematic diagram, although a very accurate and complete representation of the actual circuits

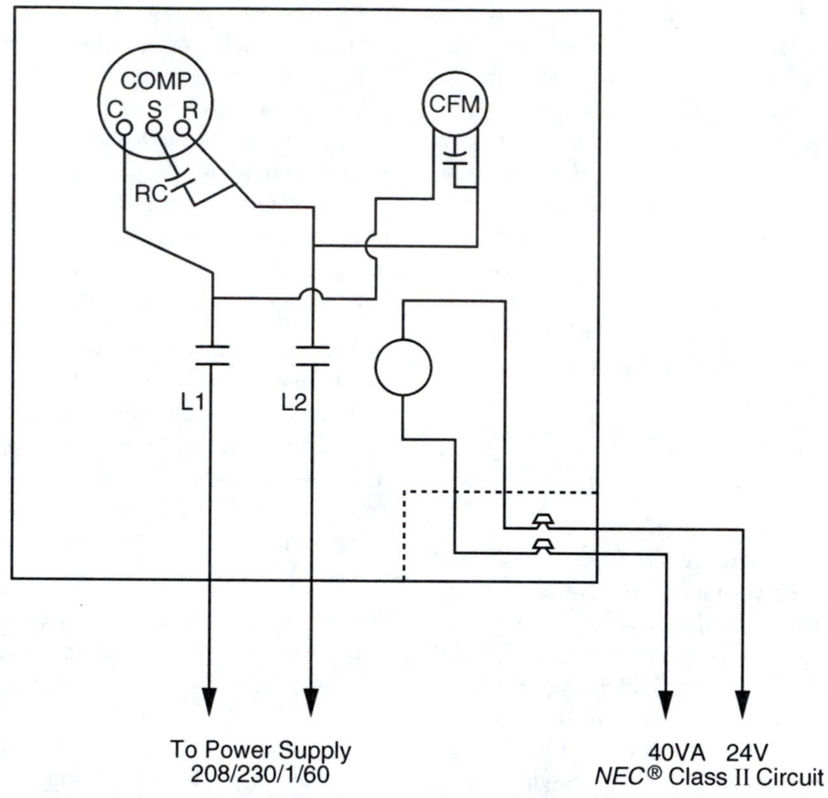

FIGURE 5-1 Schematic diagram of a 208/230-volt condensing unit controlled by a 24-volt control circuit

encountered on the specific system, tends to become confusing and more difficult to navigate as the system becomes more and more complex. Figure 5-1 is an example of a schematic diagram. A simpler wiring diagram that is often encountered is the **ladder diagram,** also referred to as the **line diagram.** This type of diagram does not often provide all of the information found in the schematic diagram but is very useful in the troubleshooting of the system. This type of diagram depicts each individual circuit separately on its own line, so the diagram tends to look like a ladder. By following each individual circuit from left to right, or right to left, the technician can successfully locate the load that is not operating properly, as well as all switches, contacts, and controls that are in series with the device, or the holding coil that ultimately energizes it. When used together, the schematic and ladder diagrams become a powerful tool that, when effectively used, gives the service technician the ability to provide an accurate and quick diagnosis of the system problem.

SCHEMATIC DIAGRAMS

As mentioned in the chapter overview, the **schematic diagram** is intended to provide the field technician with a detailed replica of the actual wiring that is present in the specified system. Figure 5-1 shows a portion of a schematic diagram of a split-type air-conditioning system. A **split system** has the condensing unit (compressor and condenser) in one location, while the air handler (evaporator, blower, and expansion device) is in another location. Schematic diagrams usually have a **legend** that describes any abbreviations used in the diagram. The legend for this diagram might look something like this:

COMP	Compressor
C	Compressor—Common terminal
S	Compressor—Start terminal
R	Compressor—Run terminal
CFM	Condenser fan motor
RC	Run capacitor

From this schematic, the following observations can be made:

- Since this compressor is equipped with only a run capacitor permanently connected between the start and run terminals, we can properly conclude that this compressor operates with a permanent-split-capacitor (PSC) motor.
- The condenser fan motor is also a PSC motor.
- The compressor and the condenser fan motor are wired in parallel with each other.
- The condenser and the condenser fan motor are controlled by a low-voltage control circuit.
- This condensing unit utilizes a two-pole contactor with a 24-volt holding coil.

Using this diagram, you can trace out the circuit and determine how this portion of the system is supposed to operate. Upon initial inspection of the circuit, you can see that the power supply feeds the circuit at Terminal L1 and Terminal L2. You can also see that a normally open set of contacts is in series with each of the lines entering the unit. This means that, unless both of those sets of contacts are closed, no current can flow through the circuit. The next question you may ask yourself is, "What will make those contacts close?" The answer is that the contacts will close when the holding coil of the contactor is energized. This holding coil operates on 24 volts and gets its voltage from a location that cannot be seen on this diagram. You do know, however, that when 24 volts is supplied to the holding coil, both sets of contacts will close. Once these contacts close, you can continue to trace out the circuit. Following L1, you can see that after the contacts the line splits up to feed both the compressor, through Terminal C, and the condenser fan motor. The current flows through each of these devices, and then the two lines leaving them come together again. The current then flows through the other set of contacts, back to the power supply at L2.

This explanation shows that both the compressor and the condenser fan motor are designed to operate at the same time. If one of the components fails, however, the other will continue to run since they are connected in parallel and the current will have an alternative path to take. If the holding coil is not energized, neither the compressor nor the condenser fan motor can operate. As systems get more and more complicated, so do the schematic diagrams. For this reason, the ladder diagram proves to be extremely useful.

LADDER DIAGRAMS

As an air-conditioning or refrigeration system utilizes more and more components, the interconnecting wiring between the components becomes more complicated as well. The number of wires increases, and the schematic diagram becomes more difficult to read. The lines representing the individual wires seem to all come together as one and, to the inexperienced technician, may resemble a plate of spaghetti. The **ladder diagram** is designed to allow the technician to isolate the desired circuit and to determine all of the switches, controls, and wiring connections in that circuit. The ladder diagram in Figure 5-2 is a representation of the same schematic that was shown in Figure 5-1. As can be readily seen, L1 and L2, representing the power source, are located at the left and right sides of the diagram while the individual circuits make up the *rungs* of the ladder. As more components are added to the system, more rungs are added to the ladder diagram. The basic construction and appearance of the ladder diagram will remain the same.

Tracing the ladder diagram is relatively simple, since you can predetermine the direction of current flow by simply looking at the diagram. This is not as easy with the schematic diagram. In Figure 5-2, start tracing the circuit from L1. Following the line from L1, you first reach the normally open set of contacts. The holding coil that is located in the control circuit controls these contacts. Once the current flows through this set of contacts, assuming the contacts are closed, the current then splits and flows to the condenser fan motor and to the compressor. Once the current leaves the devices and flows through the run capacitors, the lines are connected. The current then flows through the other set of contacts and back to L2. Note that the description of this circuit is the same as the description for the equivalent circuit in Figure 5-1. Most manufacturers provide both a schematic and a ladder diagram on their equipment to ease the job of the servicing technician. In this case, one legend is usually provided that encompasses both diagrams.

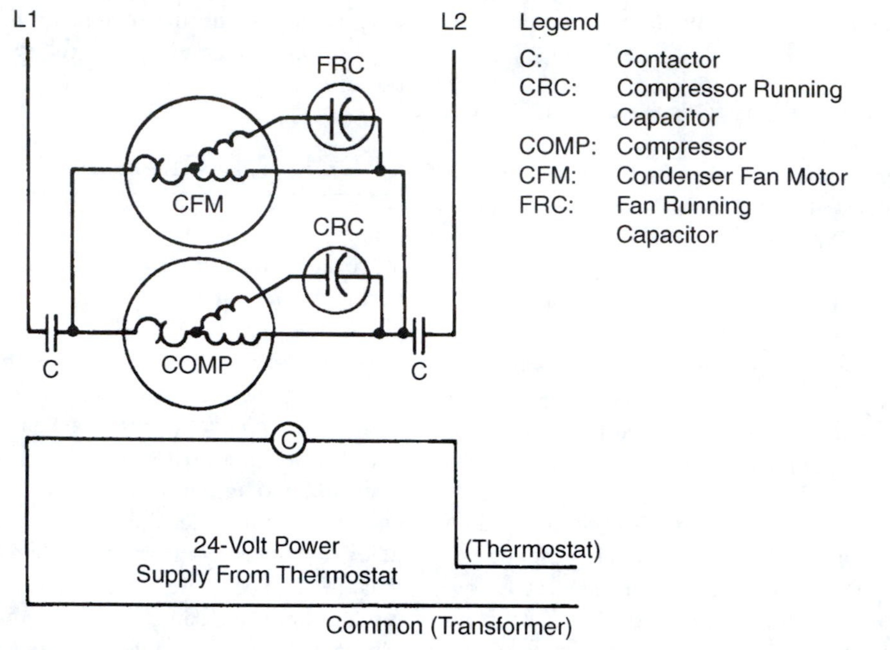

FIGURE 5-2 Ladder diagram of a condensing unit

COOLING-ONLY SYSTEM WITH DIRECT-ACTING, LINE-VOLTAGE CONTROLS (SCHEMATIC)

Figure 5-3 shows a basic air-conditioning wiring diagram that is commonly found on window-type and wall units. The controls for this type of unit are typically line-voltage devices and are connected in series with the system's main loads. The term direct acting indicates that these controls are located in series with the main load and must, therefore, be designed to withstand the current draw of the load. Although many window and wall units have only one fan motor, this schematic diagram has two motors.

The selector switch determines in what mode the system will operate and can be configured in a number of different ways, depending on the position of the switch. The terminals on the switch can be identified as follows:

- Terminal 2—Cooling terminal (since it feeds current to the compressor)
- Terminal 3—High fan terminal (connected to the high-speed wire on the evaporator fan motor)
- Terminal 4—Low fan terminal (connected to the low-speed wire on the evaporator fan motor)

Note that, whenever the compressor and condenser fan motor are operating, the evaporator fan motor should also be operating. If the evaporator fan motor is not operating when the compressor is running, the evaporator coil will freeze up, causing system problems. The various modes of operation for a system with this type of switch can be described as follows:

- OFF — L1 is not connected to any of the other switch terminals. (No components are operating at this point).
- HIGH FAN — L1 is connected only to Terminal 3. (Only the high fan speed is energized.)
- LOW FAN — L1 is connected only to Terminal 4. (Only the low fan speed is energized.)
- LOW COOL — L1 is connected to both Terminal 2 and Terminal 4. (The low fan speed and the cooling terminals are both energized.)
- HIGH COOL — L1 is connected to both Terminal 2 and Terminal 3. (The high fan speed and the cooling terminals are both energized.)

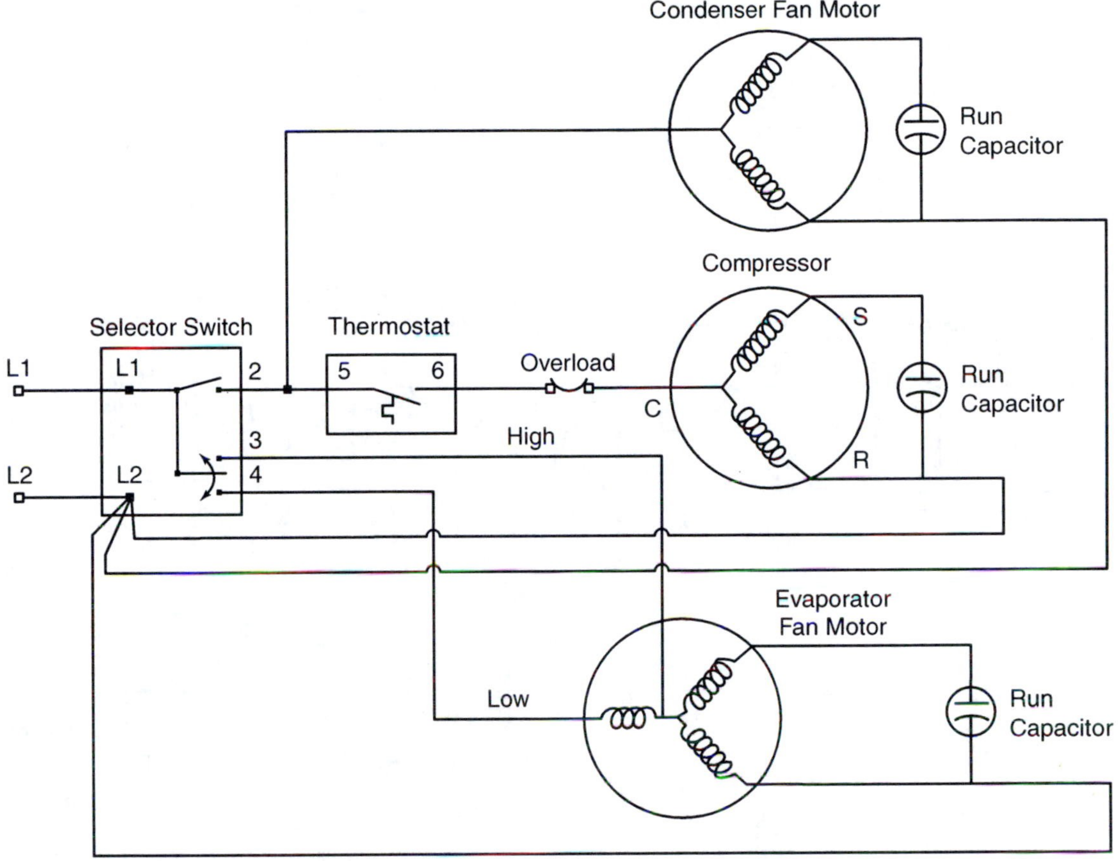

FIGURE 5-3 Schematic diagram of a window-type air-conditioning unit with line-voltage, direct-acting controls. This unit is in the OFF position.

The thermostat, shown connected between Terminal 5 and Terminal 6, can be adjusted by the occupant of the conditioned space. Turning the thermostat dial clockwise will lower the space temperature, while turning the dial counterclockwise will raise the space temperature. Once set, the thermostat will open and close its contacts automatically depending on the temperature of the space. We will now take a closer look at each of the modes of operation, depending on the position of the selector switch and the thermostat. It is important for the reader to understand that the schematics accompanying these next sections have been redrawn to reflect the positions of the switch and the thermostat but, in the field, the technician must determine the position of a specific switch or control.

Off

When the switch is in the OFF position, as determined earlier, no electrical connections exist between the L1 terminal and any other terminal on the switch (Figure 5-3). If we begin to trace the circuit from L1, we can immediately see that L1 is not connected to Terminal 2, Terminal 3, or Terminal 4. We can then conclude that the condenser fan motor, the compressor, and the evaporator fan motor are not operating. The system is, therefore, off.

High Fan

When the selector switch is turned to the HIGH FAN position, the contacts inside the switch now take on the configuration shown in Figure 5-4. By tracing out the circuit as before, we now see that the L1 terminal

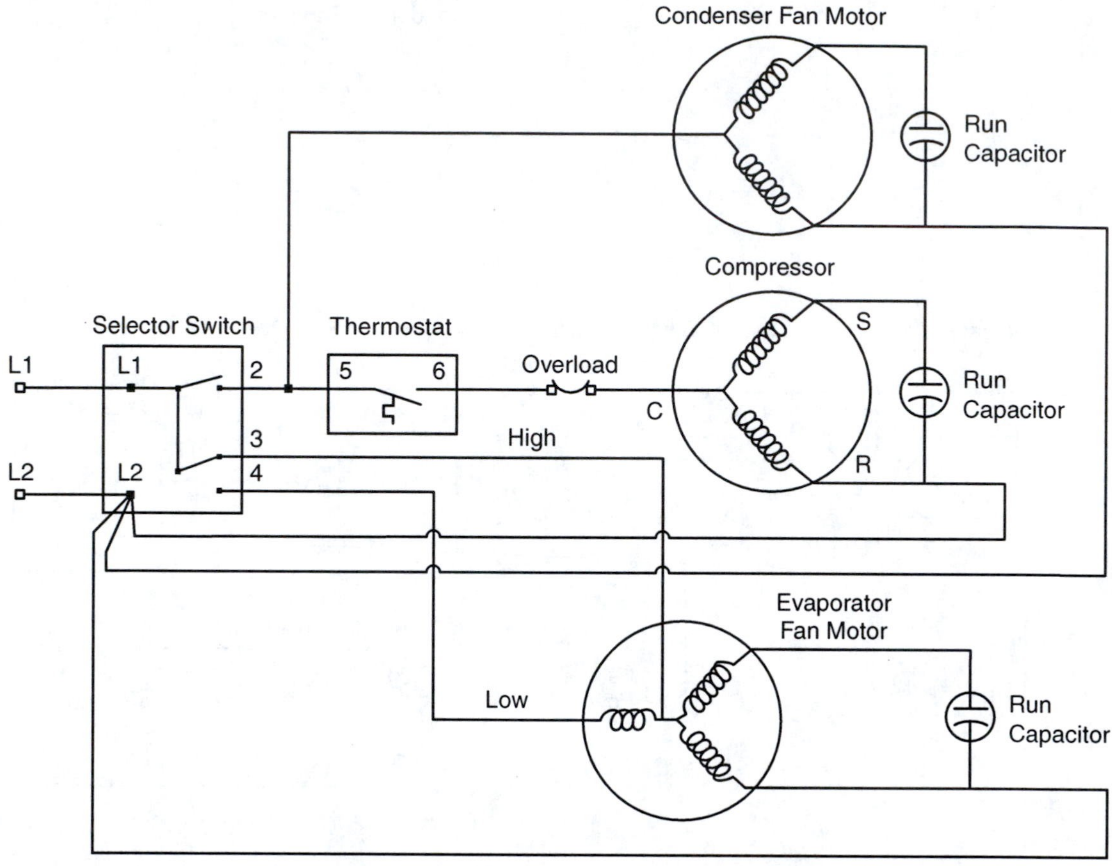

FIGURE 5-4 This unit has the evaporator fan motor operating at high speed.

is connected electrically to Terminal 3. By following the wire connected to Terminal 3, we see that this wire leads to the evaporator fan motor. This is the high-speed wire and is normally color-coded black. Continuing through the motor, the current flows through the start and run windings, through the run capacitor, and then back to the selector switch, L2. The wire connected to L2 then leads back to the power source, completing the circuit through the fan motor.

Low Fan

When the LOW FAN operation is selected, the contacts in the switch appear as in Figure 5-5. We now see that Terminal 4 is electrically connected to the L1 terminal. Tracing out the wire connected to Terminal 4 leads us again to the evaporator fan motor. This time, however, when the current enters the motor, it encounters an additional resistance. This added resistance allows the motor to operate but at a reduced speed. This added resistance also reduces the amount of current that the motor draws.

Low Cool

When the occupant of the conditioned space selects the LOW COOL setting on the selector switch, Terminal 2 and Terminal 4 are now electrically connected to the L1 terminal as shown in Figure 5-6. The low speed of the evaporator fan motor is energized just as in Figure 5-5, but Terminal 2 is now energized

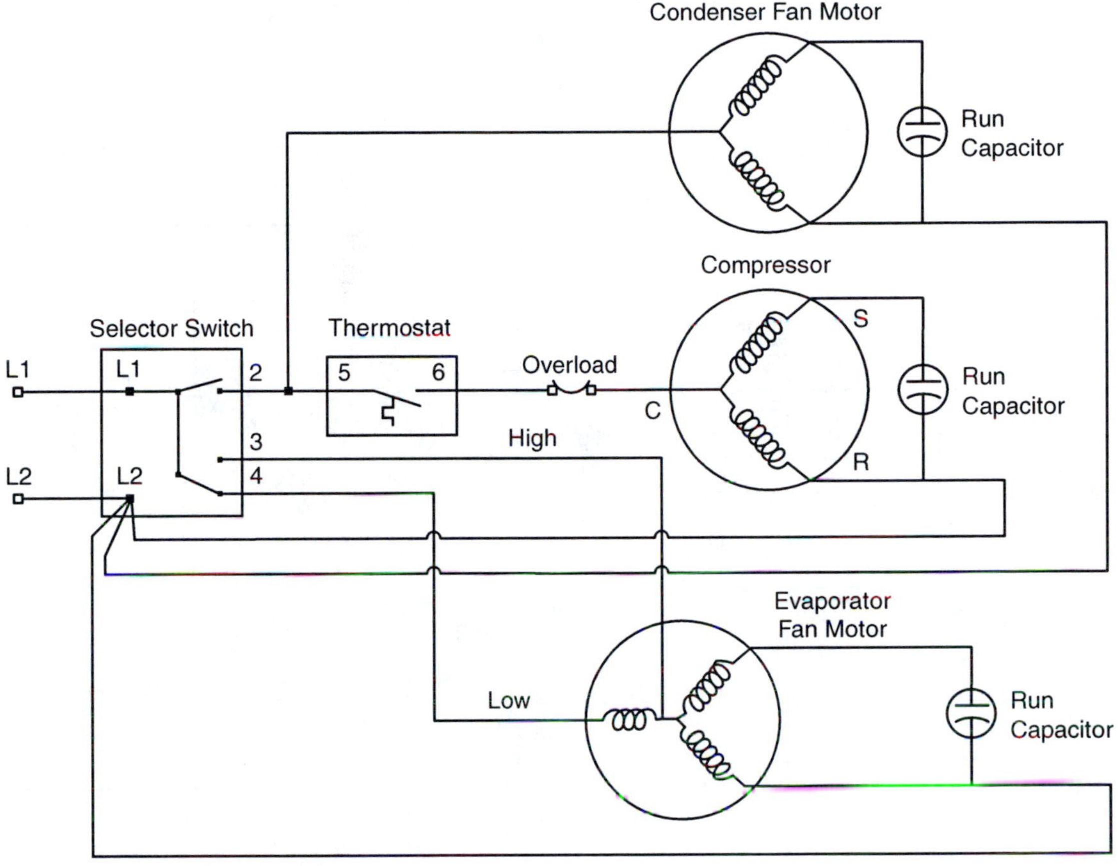

FIGURE 5-5 This unit has the evaporator fan motor operating at low speed.

as well. By tracing the wire connected to Terminal 2, we can see that this wire leads to the condenser fan motor as well as Terminal 5 on the room thermostat. The room thermostat and the condenser fan motor are connected in parallel with each other. We will look first at the condenser fan motor circuit. The path to the condenser fan motor is complete, so the current flows through the start and run windings of the motor, through the run capacitor, and back to Terminal L2. The condenser fan motor is therefore operating.

Looking at the other path through the thermostat, we can see that the thermostat's contacts are open and current cannot pass through the device onto the compressor. Since there is no complete path for the current to take, the compressor is not operating. This means the space is at a temperature that is at or below that which is desired by the occupant.

As the room temperature rises and reaches a level that is higher than the desired temperature, the thermostat's contacts will close, as in Figure 5-7. From this figure, we can see that the condenser fan motor and the evaporator fan motor are energized as in Figure 5-6, but there is now a complete current path through the compressor. We can easily follow the circuit from Terminal L1 to Terminal 2, then on to Terminal 5, through the thermostat, and then to the compressor. Once the current flows through the compressor, it then flows back to Terminal L2, thereby completing the circuit.

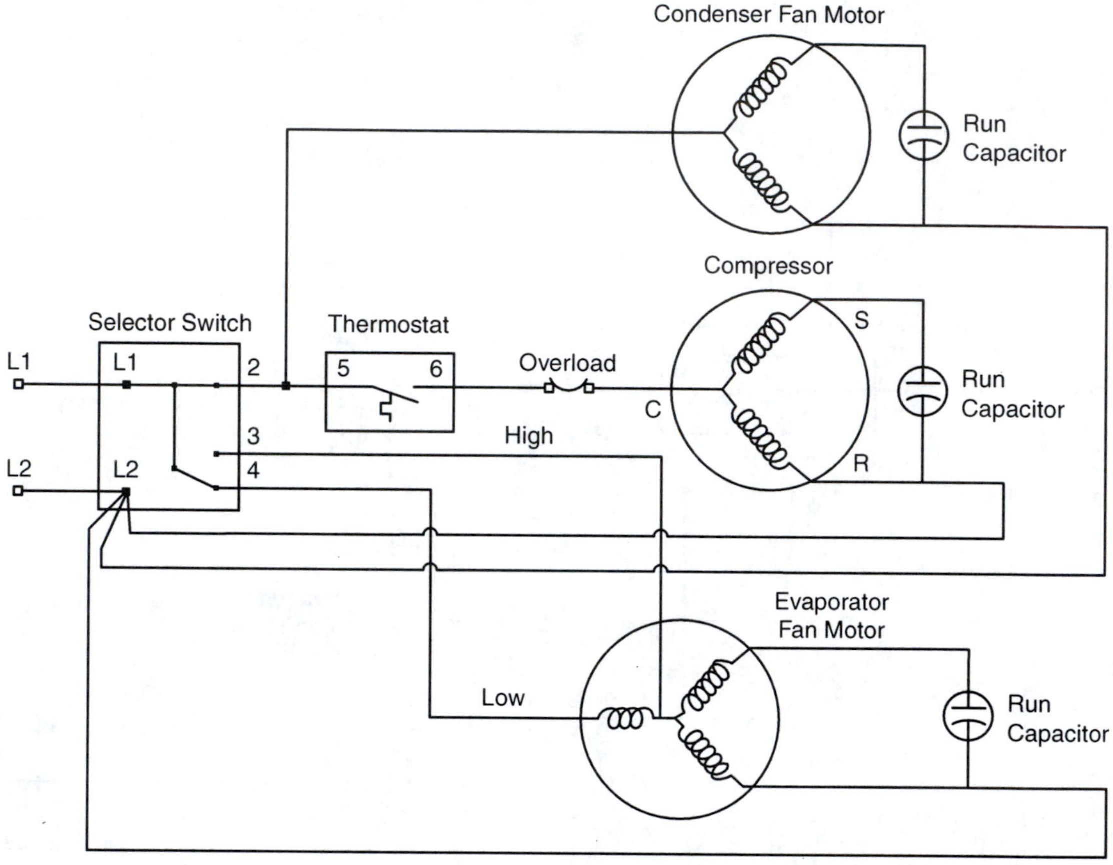

FIGURE 5-6 This unit has its selector switch set to LOW COOL. The evaporator fan motor is operating at low speed, and the condenser fan motor is operating. The compressor is not running because the thermostat's contacts are open.

High Cool

The HIGH COOL operation is exactly the same as that for LOW COOL, except for the speed of the evaporator fan motor. Instead of energizing Terminal 4, Terminal 3 is energized. The thermostat operation is the same as just described.

COOLING-ONLY SYSTEM WITH DIRECT-ACTING, LINE-VOLTAGE CONTROLS (LADDER)

By following the schematic diagrams in the previous sections, the system's sequence of operations can be determined by evaluating when switches and contacts open and close. To properly evaluate the system being worked on, the knowledge of the proper sequence of operations is vital. The ladder diagram makes this process easier to comprehend. The schematic diagram in Figure 5-3 has been redrawn in the form of a ladder diagram and is shown in Figure 5-8. Note the power supply labels L1 and L2 that are located on the left and right sides of the diagram respectively. Each system component—the condenser fan motor, the evaporator fan motor, and the compressor—has its own rung on the ladder. This type of diagram reduces the number of wires crossing over each other and makes each individual branch circuit stand out from the rest.

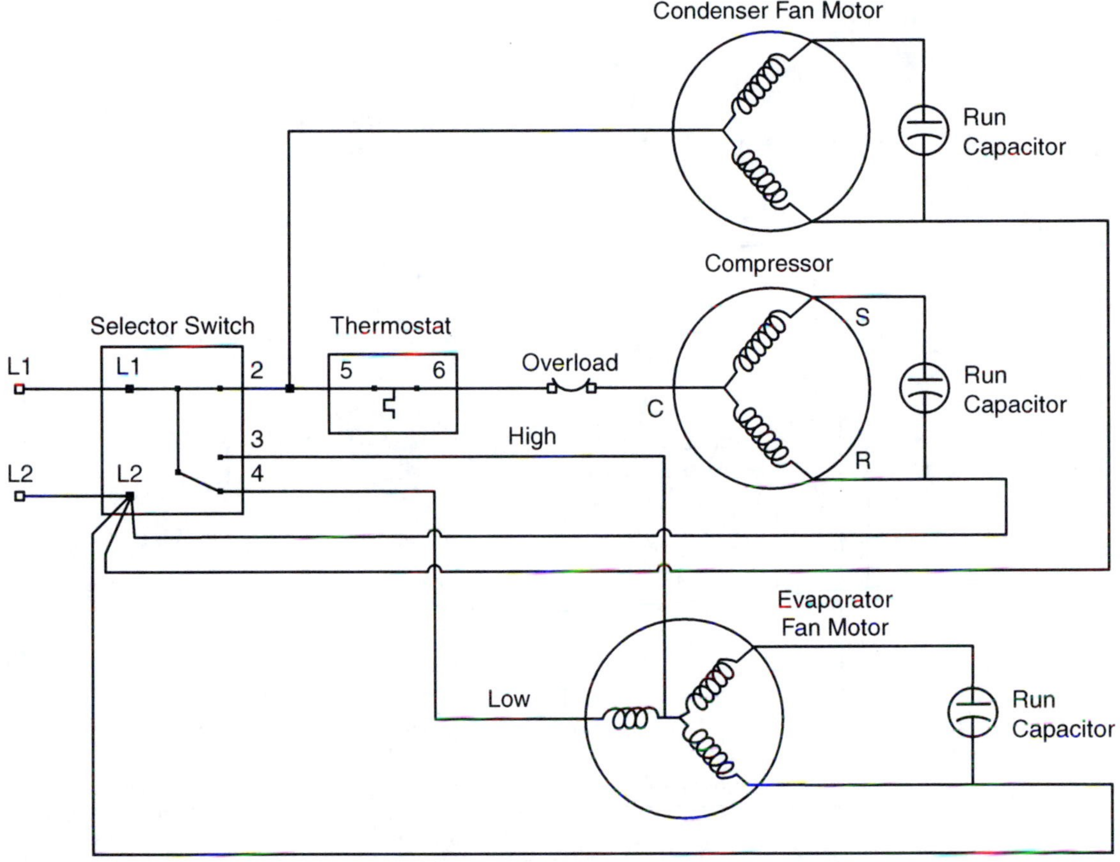

FIGURE 5-7 This unit is operating in the LOW COOL mode. The compressor and the condenser fan motor are both operating as is the low speed of the evaporator fan motor.

Referring to Figure 5-8, if we wanted to determine why, for example, the compressor was not running, we would examine all of the switches and controls that are wired in series with the compressor. For the compressor to operate, the following conditions must be met:

- There must be proper voltage at the power supply, between L1 and L2
- The contacts between Terminal L1 and Terminal 2 must be closed.
- The thermostat Contact 5 and Contact 6 must be closed.
- The compressor overload must be closed.
- The compressor windings must be intact.
- All interconnecting wiring must be intact.
- The run capacitor must be good.

This list simply notes all switches, controls, and devices in series with and including the compressor. Obviously, making these observations from the ladder diagram is much easier than obtaining the same information from the schematic diagram.

COOLING-ONLY SYSTEM WITH LINE-VOLTAGE CONTROL CIRCUIT (SCHEMATIC)

In the previous examples and diagrams, the selector switch and the thermostat are wired in series with the main system components. Since the current in a series circuit is the same at all points throughout the

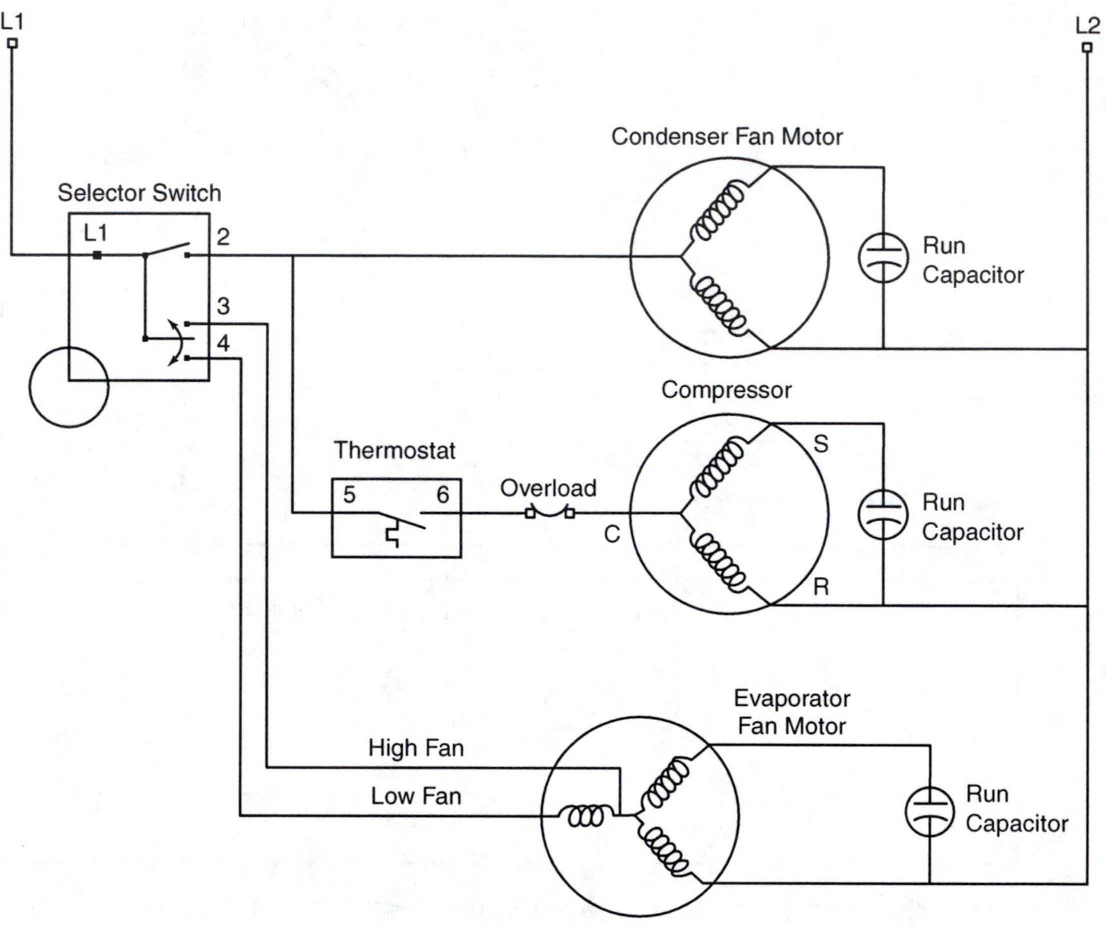

FIGURE 5-8 Ladder diagram of a window-type air-conditioning unit. The circuits are easier to trace out and follow.

circuit, the switch and the thermostat must be rated at a current that exceeds the current draw of the components being controlled. This does not create a problem when the amperage draws are relatively low, but the situation does need to be addressed when the components and, in turn, their amperage draws get larger. Refer to the section in the text on relays and contactors for more on this topic.

We will now duplicate the same system that was discussed earlier in this chapter and represented in Figure 5-3 using contactors and relays to control the compressor, the condenser fan motor, and the evaporator fan motor. To simplify the wiring diagram, we will now have the condenser fan motor and the compressor cycle on and off together. By using contactors and relays, we now have distinct control and power circuits. Figure 5-9 shows the schematic diagram for the control circuit of the system. The legend for this diagram may look like this:

CC	Cooling-contactor holding coil
CC1	Cooling-contactor contacts—Set 1
CC2	Cooling-contactor contacts—Set 2
FRH	Evaporator fan motor holding coil—High speed
FRL	Evaporator fan motor holding coil—Low speed
H	Evaporator fan motor—High-speed contacts
L	Evaporator fan motor—Low speed contacts

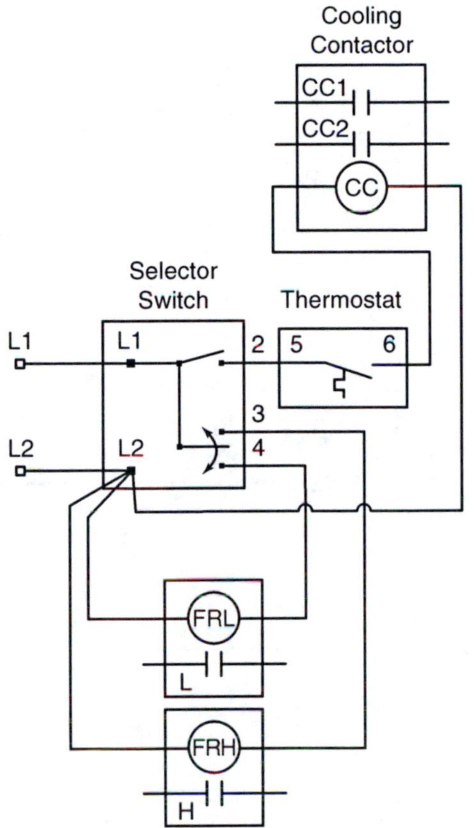

FIGURE 5-9 Line-voltage control circuit for a system using contactors and relays

As can be seen from the wiring diagram, the selector switch and the thermostat are identical to those depicted earlier. However, instead of energizing the main system components, the holding coils of the contactor and relays are energized. Take a look at the HIGH FAN switch position as an example. It can be seen from this diagram that when the HIGH FAN position is selected, Terminal 3 is energized on the selector switch. The coil labeled FRH is then energized, which in turn closes the contacts labeled H, energizing the high speed on the evaporator fan motor.

When the system calls for cooling, the holding coil CC is energized, which in turn closes contacts CC1 and CC2. These two sets of contacts energize the compressor and the condenser fan motor, which are wired in parallel with each other. The complete schematic wiring diagram is shown in Figure 5-10. The schematic very quickly becomes quite complicated and intricate.

COOLING-ONLY SYSTEM WITH LINE-VOLTAGE CONTROL CIRCUIT (LADDER)

The circuit in Figure 5-10 has been redrawn as a ladder diagram and is shown in Figure 5-11. When drawn in this manner, the control circuit—consisting of the selector switch, the thermostat, and the holding coils—is easily isolated from the power circuit. Again, this type of wiring diagram makes evaluating the system and each individual circuit much easier.

An important characteristic of the ladder diagram is that the holding coil and the contacts that it controls do not have to be located together. It is important, though, that the contacts be properly labeled so that the technician or anyone else reading the diagram can easily determine which coil controls which contacts. Referring to Figure 5-11, note that holding coil CC is located in the very top line, or rung, of the diagram while the contacts that it controls, CC1 and CC2, are located at the bottom of the diagram.

COOLING-ONLY SYSTEM WITH LOW-VOLTAGE CONTROL CIRCUIT (SCHEMATIC)

When system components get larger and contactors and relays are utilized to control them, low-voltage devices are often used. The low-voltage control circuit is easily distinguishable from the power circuit mainly because of the size of wire used. Low-voltage wire is much thinner—usually 18 to 22 gauge—and is multicolored for easy identification. A schematic diagram that shows how this can be accomplished is shown in Figure 5-12. The basic system is the same as the one that has been used so far in this chapter. The difference is that here a transformer is used to provide the 24-volt power source used to power the control circuit. Again, there are distinct control and power circuits but, in this case, the control circuit operates on 24 volts.

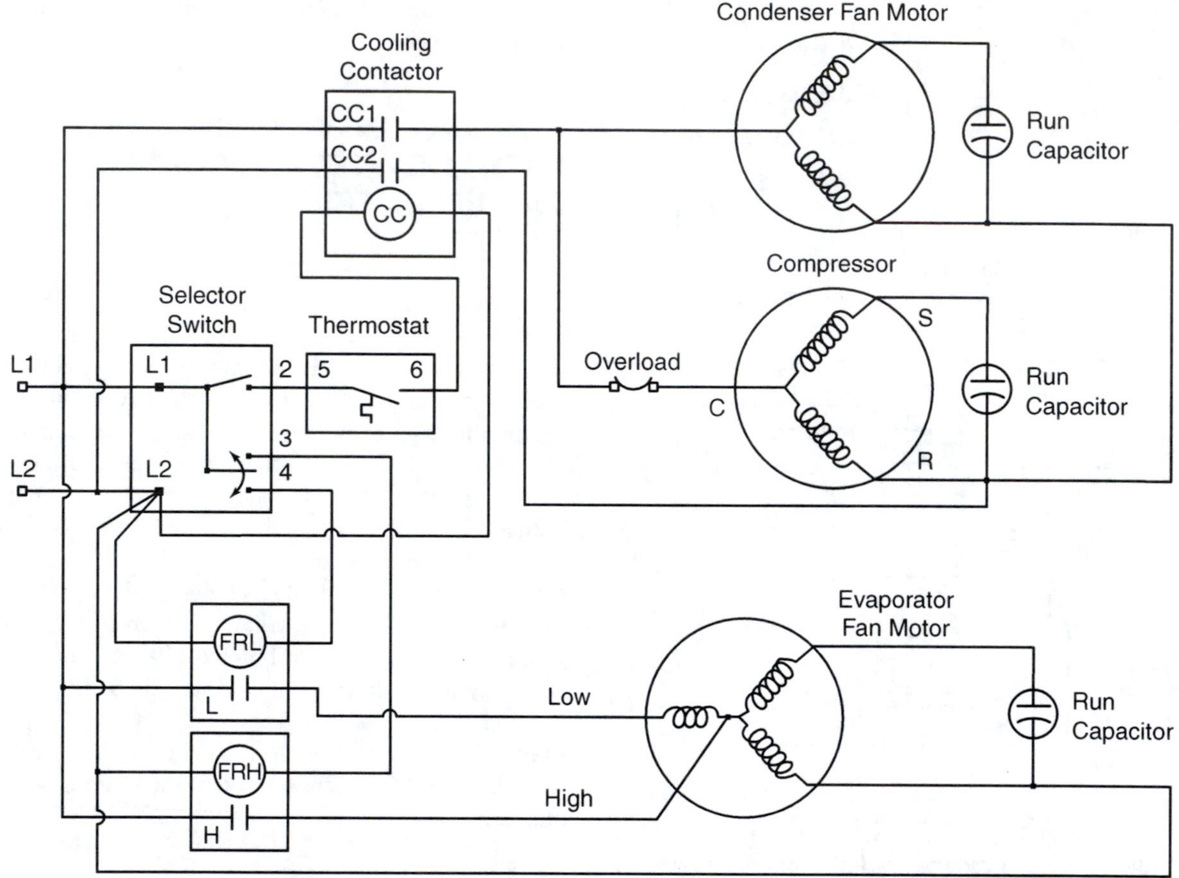

FIGURE 5-10 Complete wiring diagram for a system using a line-voltage control circuit with contactors and relays

COOLING-ONLY SYSTEM WITH LOW-VOLTAGE CONTROL CIRCUIT (LADDER)

To simplify the evaluation process once again, a ladder diagram (Figure 5-13) has been created. Note that the control circuit at the bottom of the diagram is separate from the power circuit. The control circuit is also separate from the line-voltage power supply that feeds the rest of the circuit. Note how the addition of the transformer in the power circuit adds an additional rung to the ladder diagram. As was mentioned earlier, the more components that are added, the taller the ladder will be.

HEATING-COOLING SYSTEM WITH LOW-VOLTAGE CONTROL CIRCUIT

Up to this point, the wiring schematics and the ladder diagrams were of the generic form and were intended primarily to set the groundwork and provide the basic understanding needed to fully comprehend and appreciate real-life systems. A commonly encountered type of system employs both heating and cooling sections and is controlled by a low-voltage thermostat. Common low-voltage thermostats were shown in Figure 2-22. These devices controlled both the heating and cooling functions and were mounted

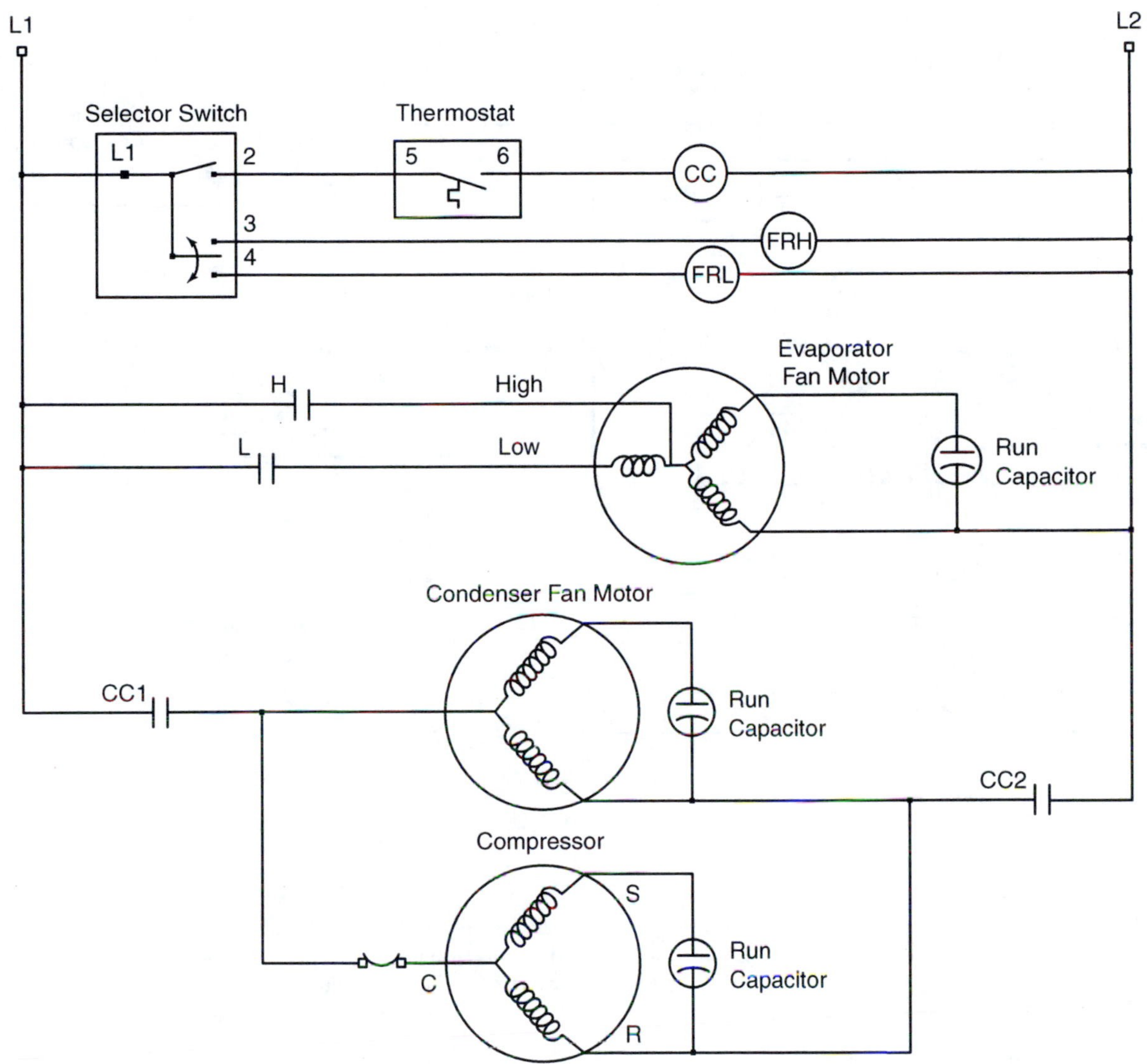

FIGURE 5-11 Ladder diagram for a system with a line-voltage control circuit

on a subbase (Figure 2-25). The subbase is the plate used to electrically and physically connect the thermostat to the system's electrical circuits and to the wall on which it is to be located.

As we have seen in the last few sections, the power circuit remained the same in each. The control circuit was reconfigured as we changed the way the system was controlled. For this reason, we will now examine the control circuit only, by looking at an actual wiring diagram of a low-voltage control circuit. Figure 5-14 shows two common subbases that are found in the field. These devices have one heating stage and one cooling stage. In the previous diagrams, we kept the thermostat and selector switches separate from each other to show the basic operation of the controls. In reality, as seen in the diagram, the selector switch and the actual thermostat are together and operate as one. These subbases

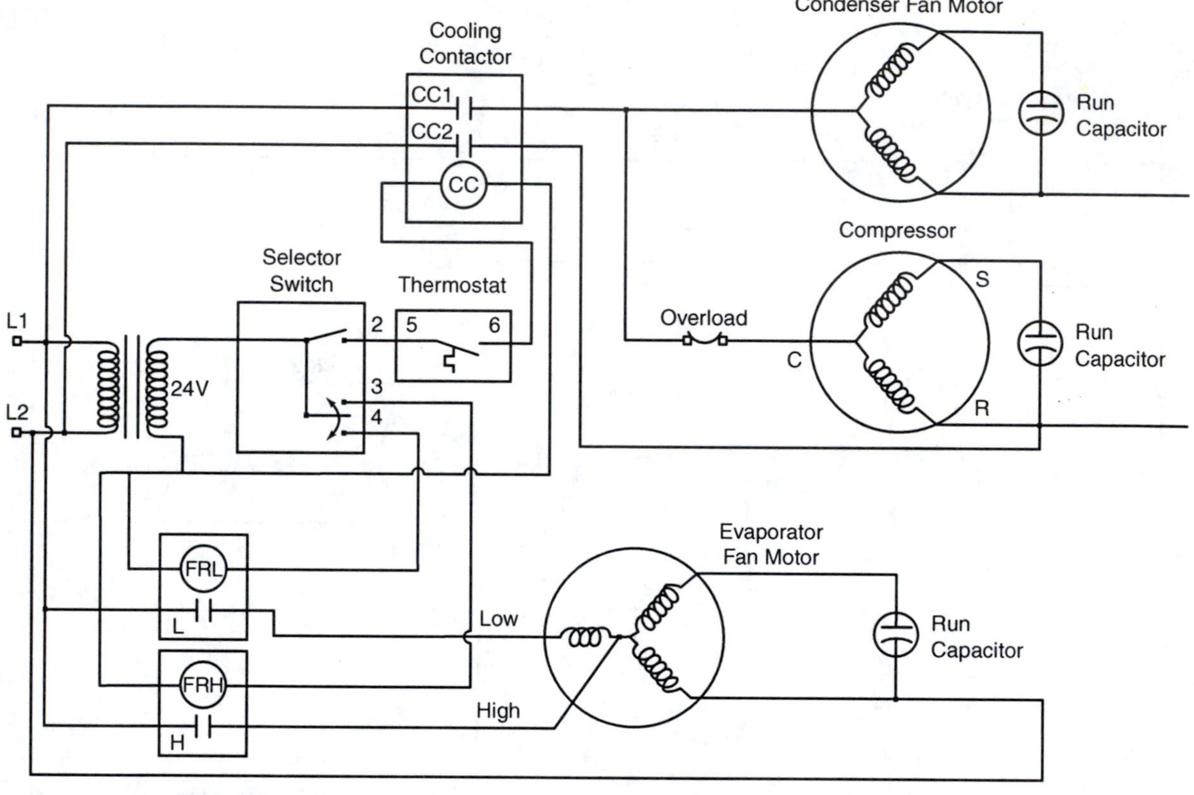

FIGURE 5-12 Schematic of a system with a low-voltage control circuit

have both a heating and a cooling terminal, which are energized depending on the desired mode of operation. The terminal designations used on these thermostats are as follows:

G	Indoor fan
Y	Cooling
W	Heating
B	Heating damper motor (if used)
R	Power from transformer
O	Cooling damper (if used)
X	Switch terminal

It is important for the field technician to remember that each manufacturer may employ a different terminal designation code, so manufacturer's literature should always be consulted when in doubt.

A low-voltage, heating-cooling thermostat that can be used in conjunction with a multispeed evaporator fan is shown in Figure 5-15. This thermostat has two G terminals, G1 and G2, which allow the motor to operate at three different speeds depending on which G terminals are energized. The three-speed fan schematic at the right-hand side of the figure shows how this is accomplished. The speeds are determined as follows:

K1 relay energized	Low speed
K2 relay energized	High speed
K1 and K2 relays energized	Medium speed

MULTISTAGE HEATING-COOLING SYSTEM WITH LOW-VOLTAGE CONTROL CIRCUIT

Systems also exist that utilize more than one stage of heating and/or more than one stage of cooling. These thermostats and subbases are equipped with additional heating and cooling terminals denoted Y2, Y3, W2, W3, and so on. Multistage thermostats were discussed earlier in the text.

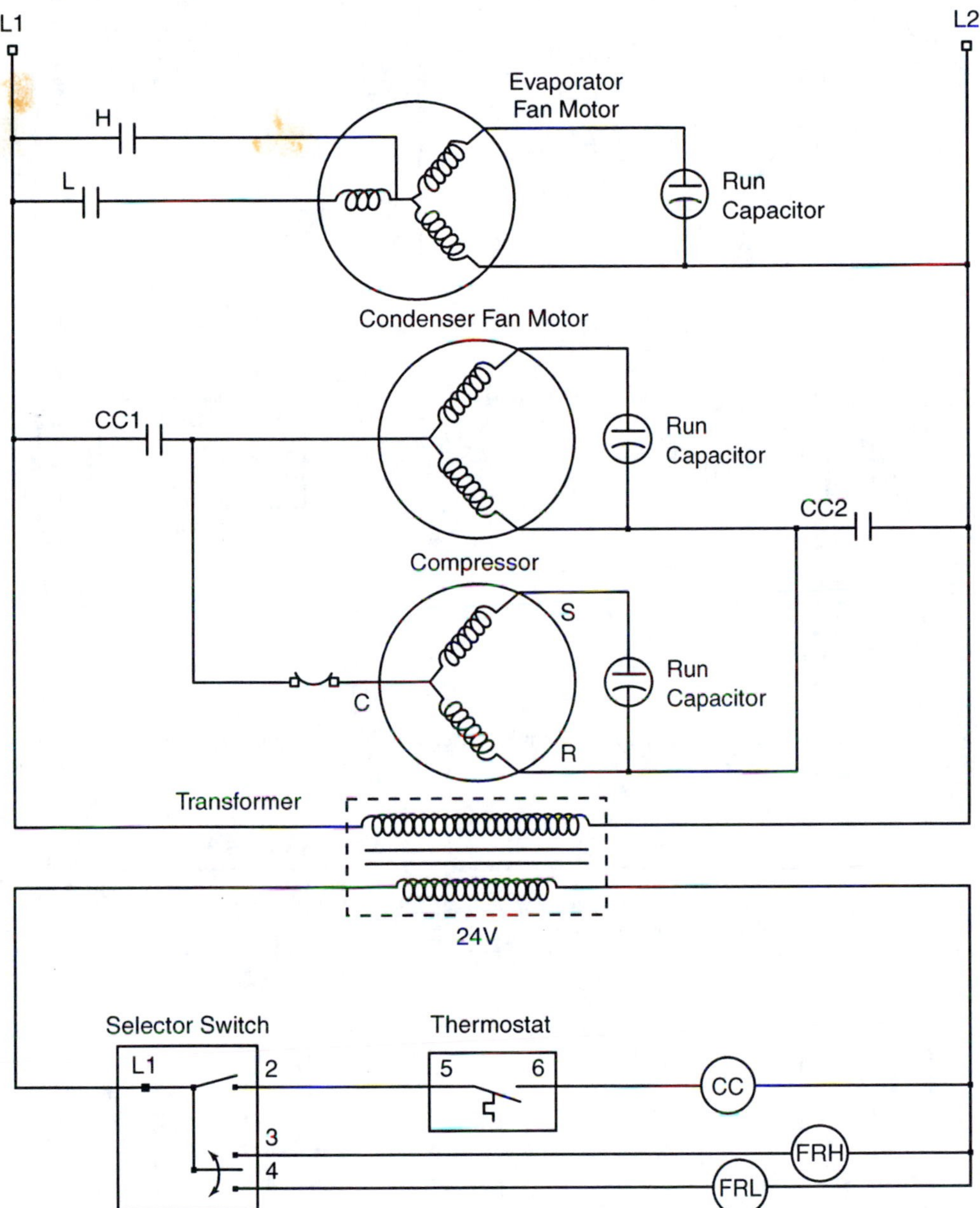

FIGURE 5-13 Ladder diagram of a system with a low-voltage control circuit

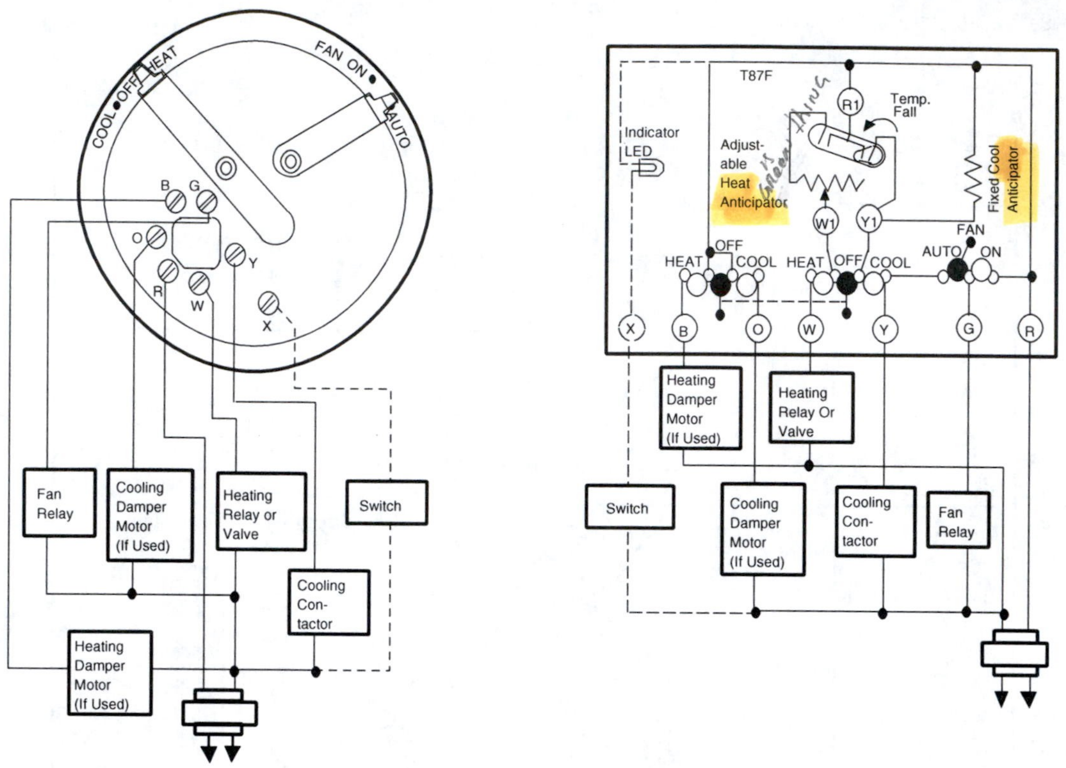

FIGURE 5-14 Low-voltage subbases and wiring connections. These heating-cooling subbases have a common power supply. *(Courtesy Honeywell)*

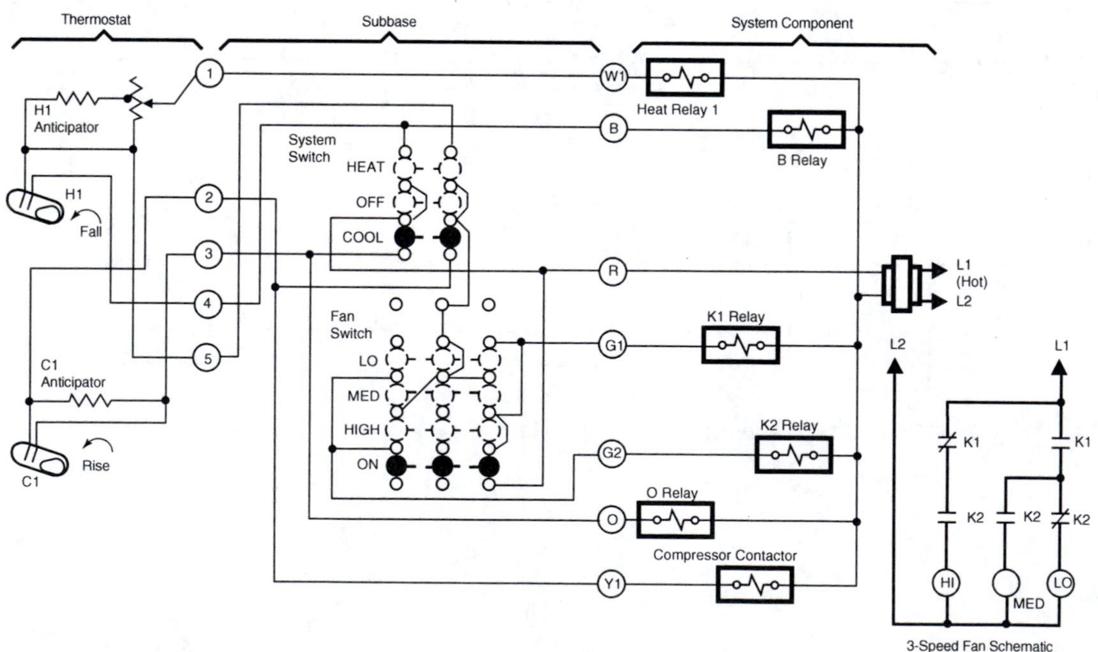

FIGURE 5-15 Internal schematic and typical hookup of a low-voltage thermostat with multispeed capability. *(Courtesy Honeywell)*

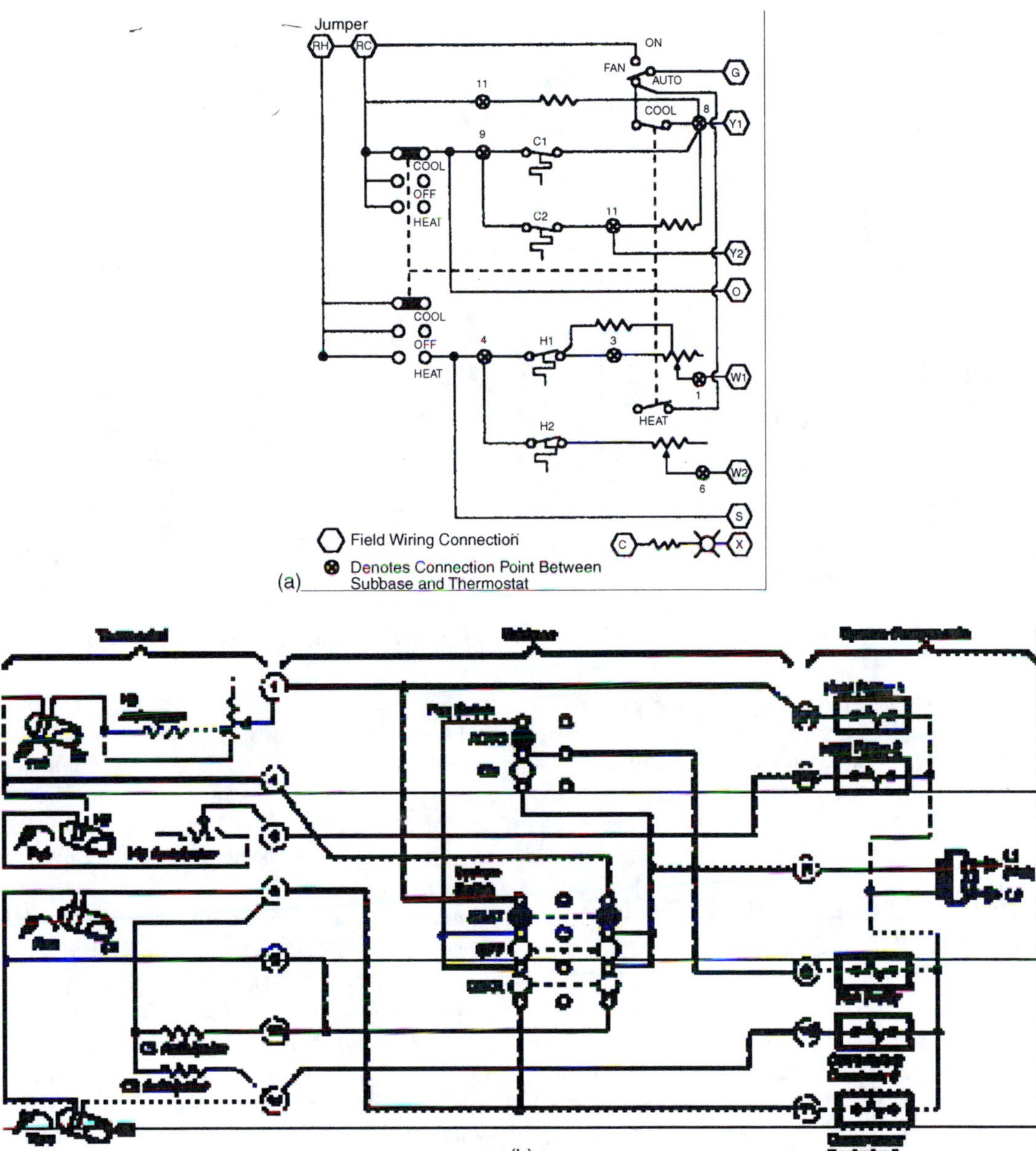

FIGURE 5-16 Internal schematic (a) and typical hookup (b) of a two-stage-heat/two-stage-cool, low-voltage thermostat

For a system that operates with two stages of heating and two stages of cooling, additional wiring is required to accommodate the extra components and contacts that must be controlled. A typical diagram for a two-stage-heat/two-stage-cool system is shown in Figure 5-16. The top portion of the figure shows an internal schematic of the device, while the typical field hookup is shown on the bottom of the figure. Although heating-cooling systems are often wired with only one transformer, seeing them operate with two transformers is quite common—one for the heating mode and one for the cooling mode. These types of thermostats have what is called an **isolating subbase,** which provides a separate R terminal for the heating and cooling cycles. These terminals are denoted RH and RC respectively. A typical wiring

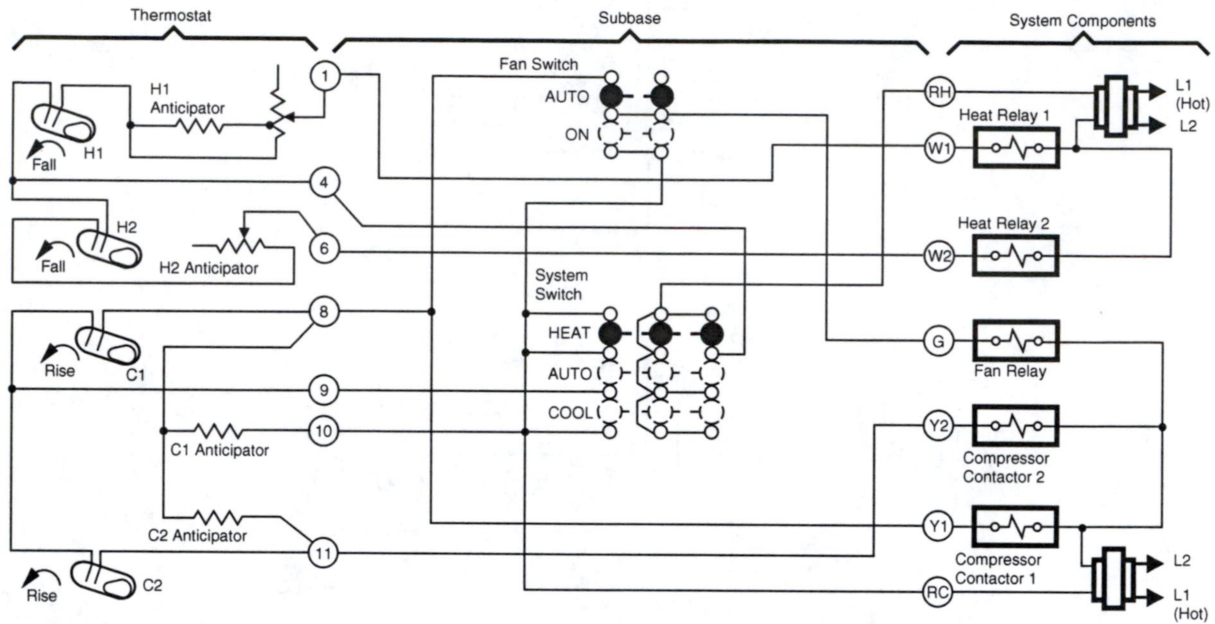

FIGURE 5-17 Schematic and hookup for a low-voltage thermostat with an isolating subbase and two transformers.

layout for this type of system is shown in Figure 5-17. When mercury bulbs are used, a bulb is used for each stage of heating or cooling. In the case of a two-stage-heat/two-stage-cool thermostat, four mercury bulbs are used.

OTHER CONTROLS AND SAFETIES ADDED TO THE CIRCUITS

This chapter has discussed the basics of schematic and ladder diagrams and how they are used to understand the sequence of operations of various system configurations. Understanding the sequence of operations of these systems is one of the first steps in being able to effectively troubleshoot air-conditioning systems, regardless of their differences in control and/or power circuit wiring and the components used in each. To bring these basic concepts to the forefront, the discussion did not include some very basic and necessary controls that should not be disregarded. These are the safety devices, such as high-pressure controls, low-pressure controls, high-temperature cutout switches and the like. These controls are normally wired in the control circuit and should not have any effect on the operation of the system unless there is a problem with the system. The control circuit, located at the top of Figure 5-18, includes both high- and low-pressure controls. Notice that these devices are wired in series with the cooling contactor. If a pressure problem arises in the system, the holding coil of the contactor will be de-energized and the compressor will be disabled. Note that, even though both the high- and low-pressure switches are intended to disable the compressor, they do so in different manners. The high-pressure control, for example, is tied in to a lockout relay that, once tripped, will not reset until the power to the system has been interrupted and then restored. This prevents a high-pressure condition from causing serious personal injury or damage to the system. The low-pressure switch, on the other hand, most often resets itself automatically. This device is often used as a means to control temperature and to provide protection from damage caused by a loss of system refrigerant.

Other safety devices can be utilized in a system, depending on the system requirements. A unit located in a ceiling over an occupied office, for example, may have a float switch in series with the cooling contactor that would disable the compressor if the condensate was not draining from the air handler properly. This not only could prevent damage to the ceiling but could prevent the office workers from taking an unexpected midday shower.

Legend	
CMT	Condenser Motor Thermostat
CAP	Capacitor
CC	Compressor Contactor
CCH	Crankcase Heater
CFM	Condenser Fan Motor
COMP	Compressor
EFC	Evaporator Fan Contactor
F	Fuse
RR	Reset Relay
T	Transformer
TB	Terminal Block
TOP	Thermal Overload Protector
TS	Terminal Strip
EFM	Evaporator Fan Motor
LPC	Low-Pressure Control
HPC	High-Pressure Control
OL	Overload Protector

Warning

Disconnect Electrical Power Source to Prevent Injury or Death from Electrical Shock

Caution

Use Copper Conductors Only to Prevent Equipment Damage

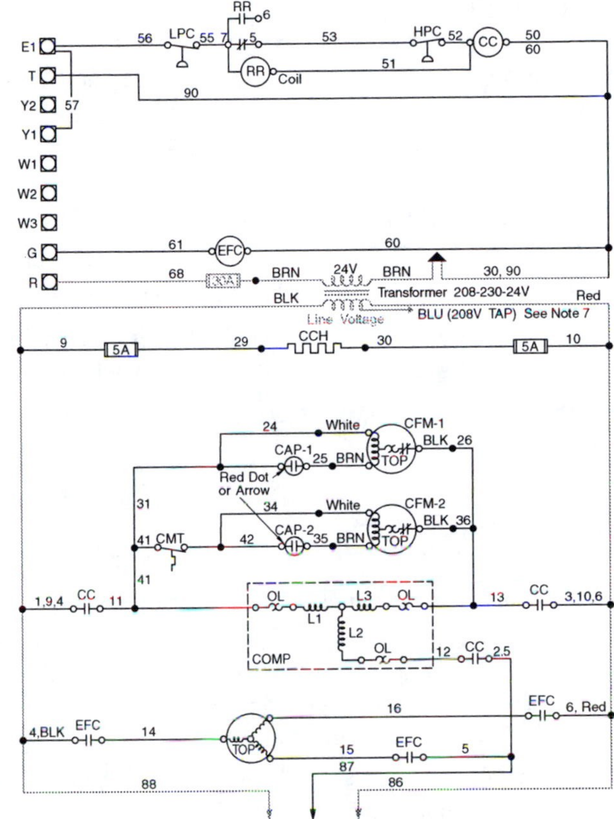

FIGURE 5-18 Power and control circuits. The low-voltage control circuit is wired with both low-pressure and high-pressure controls. *(Courtesy Trane Corp.)*

SUMMARY

- A wiring diagram can be described as an electrical road map for current to follow.
- Schematic diagrams provide a detailed picture of the system's wiring circuits, showing all of the wires and connections.
- Ladder, or line, diagrams are very useful troubleshooting tools.
- A system's sequence of operation can be determined and evaluated using a schematic and a ladder diagram.
- When switches and controls are wired in series with a load, they must have a current rating higher than the current draw of the device.
- Relays and contactors are used when the amperage draw of the main system components is high or when the control and power circuits operate at different voltages.
- Systems with low-voltage control circuits use a transformer to provide the low-voltage power source.
- Systems that have both heating and cooling components can use either one transformer or two. When two transformers are used, an isolating subbase is needed.
- Multistage thermostats have more than one set of heating and cooling terminals.
- Always follow the manufacturer's literature and directions when wiring—or evaluating the wiring—of a system.

KEY TERMS

Factory wiring
Field wiring
Isolating subbase
Ladder diagram
Legend
Line diagram
Schematic diagram
Split system

FOR DISCUSSION

1. Discuss the importance of safety controls as they relate particularly to high pressure.
2. Under what circumstances would a ladder diagram be more beneficial than a schematic to a field technician? When would a schematic be more beneficial?
3. Why is the ability to read wiring diagrams an essential skill that all successful field technicians should possess?

REVIEW QUESTIONS

1. When an electrical switch is in the open position:
   a. Current can pass through the contacts freely.
   b. Current will flow if an alternative, complete path is available.
   c. The device in series with the switch will be always be energized.
   d. None of the above are true.
2. The type of wiring diagram that shows the location of the system components is the:
   a. Pictorial diagram.
   b. Line diagram.
   c. Schematic diagram.
   d. Ladder diagram.
3. The type of diagram that is most useful in troubleshooting a specific circuit is the:
   a. Line diagram.
   b. Schematic diagram.
   c. Ladder diagram.
   d. Both *a* and *c* are correct.
4. A complete circuit consists of the following:
   a. A load, a thermostat, a subbase, and a power supply
   b. A load, an uninterrupted current path, a power supply, and a lightbulb
   c. A load, an uninterrupted current path, a power supply, and a switch
   d. A switch, a load, a thermostat, and a power supply
5. If two switches are wired in series with each other:
   a. Both switches must be in the closed position for current to flow through them.
   b. Current will flow through them if at least one of them is in the closed position.
   c. Current will always flow through both of them.
   d. The first switch will control the operation of the second switch.

6. When referring to schematic diagrams, the legend is a(n):
   a. Old story that is handed down from generation to generation.
   b. List that describes the abbreviations used in the diagram.
   c. Picture that shows how long the wires in the circuit are.
   d. Both *b* and *c* are correct.
7. A direct-acting control is:
   a. Wired in series with the load.
   b. Wired in parallel with the load.
   c. Typically used when the current draw of the load is relatively low.
   d. Both *a* and *c* are correct.
8. What makes the contacts of a relay close and/or open?
   a. Voltage applied to the holding coil of the device
   b. The current flowing through the contacts
   c. The voltage flowing through the power circuit
   d. Both *a* and *c*
9. True or False: Holding coils are most often found in the system's power circuit.
10. True or False: A relay's contacts are most often found in the system's power circuit.
11. True or False: If an electrical circuit with only one possible current path has a broken wire, the current flowing in this circuit will be zero.
12. Which of the following is an example of a multistage thermostat?
    a. Two heating stages and one cooling stage
    b. Two heating stages and two cooling stages
    c. One heating stage and two cooling stages
    d. All of the above
13. An isolating subbase must have which of the following terminals?
    a. *W2* and *RH*
    b. *RC* and *Y2*
    c. *RC* and *RH*
    d. *Y1* and *Y2*
14. True or False: The speed of a fan motor can be reduced by adding resistance in series with the motor.
15. True or False: When drawing a ladder diagram, the holding coil and the contacts it controls must be drawn together to ensure proper circuit evaluation.

CHAPTER 6

Troubleshooting Vapor-Compression Refrigeration System Electrical Components

OBJECTIVES After studying this chapter, the reader should be able to:

- Explain how to determine if a contactor or relay's holding coil is defective.
- Describe how to evaluate a thermostat and determine if it is functioning properly.
- Explain how to determine if a transformer is operating correctly.
- Explain how to establish the cause of a blown control-circuit fuse.
- Describe how to determine the proper operation of pressure and other safety devices.
- Explain how to determine the reason for a blown power-circuit fuse or tripped circuit breaker.
- Explain how to establish if a motor has a grounded, shorted, or open winding.
- Describe how to field-test a capacitor.

INTRODUCTION

Being able to recognize and understand the different types of wiring diagrams, as presented in the preceding chapter, is a major step in becoming an effective and efficient troubleshooting technician. This information, together with a good working knowledge of the sequence of operations of the system and circuit construction, helps ensure that the diagnosis made by the technician will be an accurate one.

The inexperienced service person can easily fall into the trap of becoming a parts changer or a wrench turner. Visually determining that a motor is not operating is not difficult. An inexperienced technician may replace the motor, thinking that it is defective because it isn't turning. Obviously, this quite often is not the case. Associated wiring, components, and controls may be to blame. A newly installed motor may not operate either. At this point, the untrained service person starts grasping at straws, replacing parts one by one until the problem is finally resolved. Imagine all the wasted time and money, as well as the customer's and, equally important, the boss's dissatisfaction with the performance of the technician and lost revenues. This chapter is geared toward the prevention of this scenario. A number of different situations presented here will provide a firm base on which to start building a solid troubleshooting mentality. As mentioned at the beginning of the text, troubleshooting is an acquired skill. Proficiency comes with extensive and continued practice as well as continued education and training.

OVERVIEW

Knowing what controls and loads are in any given electric circuit is the first step in effective troubleshooting. Knowing the system's sequence of operations is the next step. Another important step, the application of basic circuit theory, is being able to electrically examine a given circuit and determine if the individual components in this circuit are performing as they should. To be able to accomplish this, the technician must know the characteristics of these components. For example, what resistance reading is too high or too low when checking the coil of a contactor? How does a field technician electrically check a run capacitor to determine whether or not it is defective?

These questions and many others may need to be answered by the servicing technician in order to reach a correct diagnosis of the system problem. The intention of system troubleshooting is to evaluate the system and its components *prior* to making a physical repair. By first evaluating the system, the technician avoids the situation presented in the introduction. Imagine a patient in a doctor's office. The patient complains of a pain in her arm, so the doctor cuts the patient's arm open looking for a problem. Not seeing anything, the doctor then proceeds to make another incision an inch below the first. Still finding nothing wrong, the doctor makes yet another incision one inch below the second. Does this sound crazy? I hope so! Medical doctors spend a large portion of their schooling learning how the human body works. They learn about *what is right* with the human body. Then, when they notice something unusual, they act quickly, professionally, and, more importantly, accurately.

Although not quite as critical, air-conditioning and service technicians should be well aware of what is right in an air-conditioning system. Then, when something *not quite right* surfaces, it is obvious. Simply changing parts with the attitude "I'll get it right eventually" mirrors the attitude of the doctor in the previous paragraph that continually slices open the patient in search of a problem. Changing parts at random is just as ridiculous.

Electrical troubleshooting is often performed with the aid of a voltmeter, an ammeter, and an ohmmeter (Figure 6-1). These three pieces of test equipment are just as important as a doctor's stethoscope or a fisherman's hook. Electrical troubleshooting cannot be performed without these instruments. By examining a system's schematic or ladder diagram, the servicing technician can determine which switches and controls are in series with each individual component. If a component, such as a motor, is not operating correctly, the technician will be able to determine *why* this is the case. Is the circuit open? Is a winding in the motor defective? Is power being supplied to the circuit?

Probably the easiest way to determine the condition of a circuit is to begin with the obvious. Is power being supplied to the circuit? To answer this question, the technician would use a **voltmeter** to check for a **potential difference** at the power source. For example, if James has a toaster that does not work, he may plug something else into the wall outlet to see if there is power. Whether James knows it or not, he is in fact troubleshooting the situation. He is able to determine if the problem lies within the toaster or in the wall outlet that supplies the power to the toaster. With the aid of a voltmeter, plugging other appliances into the outlet would not be needed. The test leads of the meter would be placed into the socket, and the meter would read the potential difference between the line and the neutral connections in the

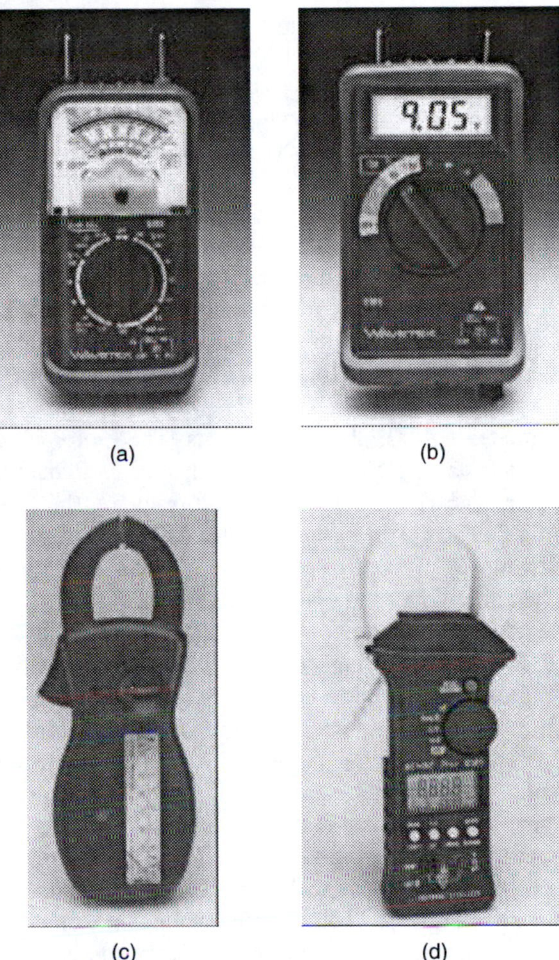

(a) (b) (c) (d)

FIGURE 6-1 Instruments used to troubleshoot the electrical part of an air-conditioning system. (a) Analog volt-ohm milliammeter (VOM). (b) Digital volt-ohm milliammeter (VOM). (c) Clamp-on ammeter. *([a] and [c] Courtesy Amprobe Instrument Division of Core Industries Inc., [b] Courtesy Wave Tek)*

outlet. Once this question is answered, the technician knows in which direction to move. If power was not being supplied, the next logical place to check would be the fuse box or the circuit-breaker panel. If power was being supplied, further investigation would take place within the unit itself.

Using an **ammeter** is another important method that the technician can use to help evaluate a circuit. The ammeter can be used to read the current that is being drawn by a circuit. This can lead the technician to the problem very quickly. For example, if a motor has difficulty starting, it will draw locked-rotor amperage (LRA) for a longer period of time. The ammeter will be able to read this. The motor itself and the starting components can then be evaluated to determine if they are functioning correctly. The ammeter can also be used to determine if current is flowing in a given circuit. This is commonly done, especially when the components being evaluated are not close by and cannot be physically inspected to see if they are operating or not. Electric heating strips (Figure 6-2) provide another example of when an ammeter is used to check for current flow. For obvious reasons, a technician would not touch an electric heater to see if it was energized or not. Checking the current flow indicates to the technician whether or not the heater is energized.

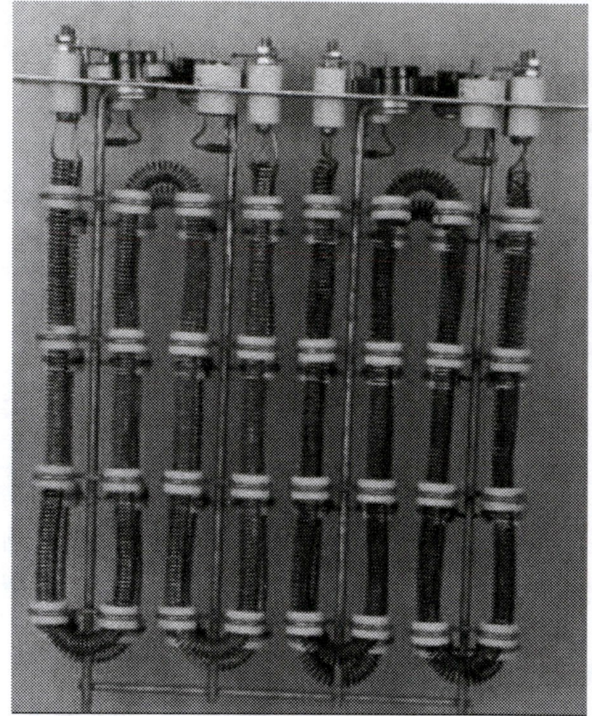

FIGURE 6-2 Electric duct, or strip, heaters. *(Photo by Bill Johnson)*

The **ohmmeter** is widely used to test circuits that are de-energized. With the ohmmeter, the technician can determine if a switch or set of contacts is in the open or closed position. The ohmmeter can also provide accurate resistance readings, which in turn help the technician evaluate individual components. If a split-phase motor was being checked and the technician suspected that the start winding was partially shorted, an ohmmeter would be used. Assume that the following readings were taken on a motor:

Common to start	12 Ohms	(Start winding only)
Common to run	12 Ohms	(Run winding only)
Start to run	24 Ohms	(Both windings in series with each other)

From this information, it can be determined that the technician's initial assumption was correct. Going back to basic motor theory, the start winding is designed to have a higher resistance than the run winding in order to create a difference in magnetic field, producing starting torque, that allows the motor to start, as shown in Figure 6-3(a). The resistance of both windings, however, is 12 ohms, as shown in Figure 6-3(b), indicating that the resistance of the start winding is lower than it should be. The winding is partially shorted.

A number of different situations will be looked at in this chapter, providing the reader with information that will prove useful when applied to specific field problems.

CONTROL-CIRCUIT PROBLEMS

As shown earlier in the text, most electrical schematics are divided into two circuits: the control circuit and the power circuit. This section examines components commonly found in the control circuit and discusses how to properly evaluate them. As in any circuit, the control circuit consists of the following:

- Power supply
- Switches and/or controls
- One or more loads
- Path for the current to take

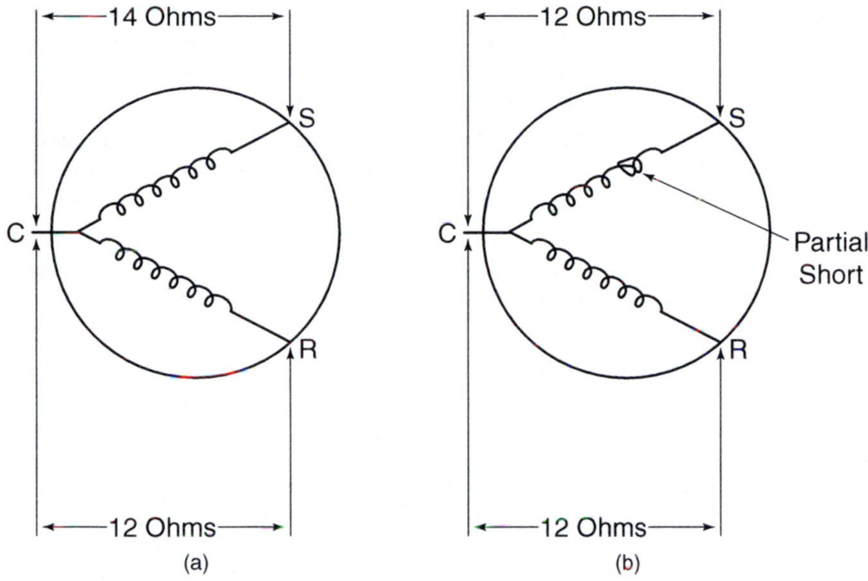

FIGURE 6-3 (a) Motor with good windings, (b) This start winding is partially shorted.

In low-voltage control circuits, the *power supply* is the transformer. Reading a voltage, usually 24 volts, at the transformer's secondary winding verifies that the control circuit is being powered. *Switches and controls* determine where the current in the circuit flows. The *loads* in control circuits are normally the holding coils of contactors and relays. These holding coils, in turn, control the operation of sets of contacts in the power circuit. The *path* is the interconnecting wiring between the power supply, the switches, and the loads. In this text, the focus is on low-voltage control circuits, although the control circuit can just as easily be 115, 208, 220, or 240 volts.

Low-voltage control circuits are easily distinguishable from the power circuit. The wire used in these circuits is much thinner, since the voltage is low; and, more importantly, the current flowing in the circuit is low, normally in the range of 1.7 amperes for a 40VA transformer. This comes from the power rating of the transformer, VA, which is equal to the voltage times the current. For the 24-volt control circuit, the amperage would then be 40VA/24V, which is 1.7 amperes. The wire used in low-voltage control circuits is commonly brightly color-coded and comes encased in either white or brown vinyl (Figure 6-4). This makes the running of the wire from one point to another much easier, while keeping all of the individual conductors together.

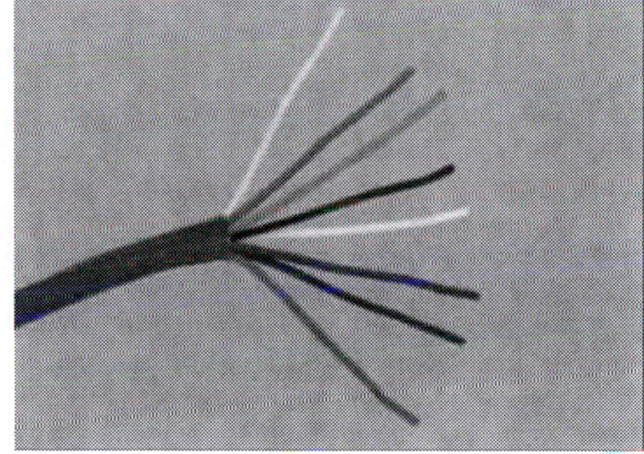

FIGURE 6-4 Thermostat wire. *(Photo by Bill Johnson)*

Most localities do not require technicians to have an electrician's license to work on or install low-voltage wiring.

Holding Coils

Just as a motor has a voltage rating, the holding coil of a contactor or relay also has a voltage rating. It is important that the voltage supplied to the holding coil matches this rating. If the voltage and the rating do not match, the device will not work properly and, in many cases, can become damaged beyond repair. The relay, for example, is considered to be a throwaway device since it is not economically feasible to repair it if it becomes damaged. If the coil burns out, the entire relay must be replaced. The contactor, on the other hand, may, in some instances, be repaired if the coil becomes defective. Smaller contactors are usually treated like the relay and discarded. Larger contactors, though, tend to be very expensive. Within a very short period of time, the field technician can replace the holding coil of a contactor (Figure 6-5), provided the proper replacement coil is available. But how can a technician determine if the holding coil is defective or not?

The holding coil can best be evaluated when the power is disconnected and the wires leading to the coil are removed. Removing the wires ensures that the readings taken are accurate and reliable by eliminating the possibility of reading through a parallel circuit. When evaluating the coil, the technician can come to one of three possible conclusions:

- The coil is good.
- The coil is shorted.
- The coil is open.

To properly check the holding coil, an ohmmeter should be used. The meter should be set to the lowest scale that will register actual resistance. Using the continuity scale, for example, will only rule out if the coil is open or not but will not distinguish between a good coil and a coil that is shorted.

The resistance reading of the coil will determine whether or not the coil is good. If the resistance reading across the coil is 0 ohms, the coil is shorted and should be replaced. Again, if the coil is in a relay or a small contactor, the entire device should be discarded. If the resistance reading of the coil is infinity, the coil is open and should be replaced as well. If the reading is infinity, double-check the coil by turning the dial on the meter to the next lowest resistance range just in case the resistance of the coil is higher than the top limit of the lowest range. If the reading is still infinity, discard the coil. If the ohmmeter registers a resistance value, chances are that the coil is good. Exact resistance values cannot be provided here, since they vary depending on the manufacturer and the voltage rating of the coil. The manufacturer of a specific relay or coil can be consulted if the actual resistance value is needed.

FIGURE 6-5 Replaceable contactor coil

Thermostats

Many inexperienced technicians find that the thermostat is a very confusing system component—both in being able to understand its operation and being able to troubleshoot the device. Part of this confusion arises from the fact that many different types and configurations of thermostats are produced by a number of different companies. The easiest way to understand the thermostat is to realize that it acts as a series of switches that open and close in response to some external and internal conditions. Most of these conditions are manually determined by the end user of the equipment. For example, the owner decides if the system is going to operate in the cooling or heating mode and whether the indoor fan is going to operate continuously or just when the compressor, for example, is operating.

The system thermostat is normally comprised of two separate components: the thermostat proper and the subbase. Most troubleshooting of the thermostat takes place at the subbase, which is where the electrical connections to the device are made.

A simple, low-voltage control circuit operates with the following wires and terminals:

- Red R terminal Hot wire from transformer
- Yellow Y terminal Cooling circuit
- White W terminal Heating circuit
- Green G terminal Indoor fan circuit

Most low-voltage thermostats adhere to the preceding color-coding and terminal designation, which makes the evaluation of the circuits much easier. When the thermostat is removed from the subbase, these terminals can be seen (Figure 6-6). These terminals are labeled screw terminals to which the color-coded, low-voltage wires are connected. With the thermostat removed, the majority of the system can be evaluated from this point. It is often helpful to realize that the R terminal feeds power to the thermostat and the G, W, and Y terminals carry power from the thermostat to the fan, heating, and cooling circuits. A small jumper wire connected to the R terminal (Figure 6-6) is all that is needed to determine if the thermostat and subbase are operating correctly.

The first thing that is normally checked on a thermostat is the indoor fan operation. This is done for several reasons. First, the indoor fan can usually be heard when it is energized and de-energized because the thermostat is often located near the return-air grill—hence, often near the fan itself. Another reason for checking the fan first is to verify the presence of low-voltage power being supplied to the thermostat. Remember that the indoor fan motor cannot operate if there is no low voltage. The following represents a typical but by no means the only procedure to evaluate the operation of the indoor fan motor:

- With the thermostat removed, turn the fan switch to the ON position.
- The indoor fan should begin to operate.
- If the fan starts, the following conclusions can be made:
  - The transformer is producing the desired low voltage.
  - The R terminal on the thermostat has a potential of 24 volts.
- By turning the fan switch to the AUTO position, the fan should turn off and there should be a 24-volt reading between the R and the G terminals on the subbase (Figure 6-7).

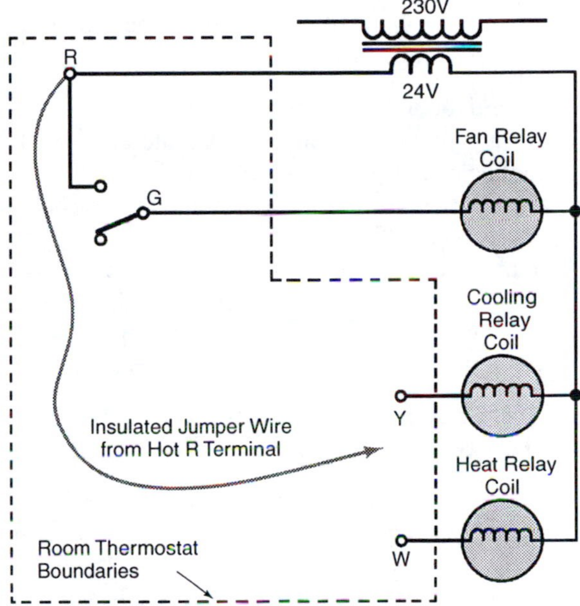

FIGURE 6-6 Terminal designation in a heat-cool thermostat subbase. A jumper from the Terminal R to Terminal G should start the fan. A jumper from R to Y should start the compressor. A jumper from R to W should start the heating cycle.

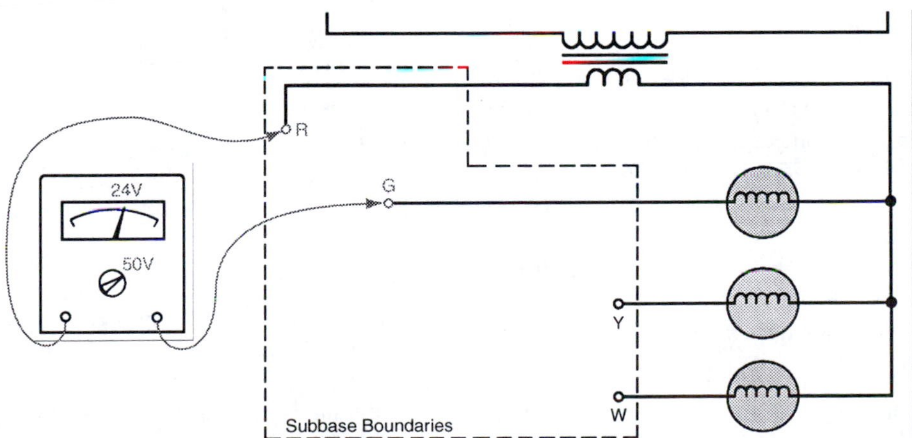

FIGURE 6-7 A multimeter or VOM can be used to test the system at the subbase with the thermostat removed.

- If the fan does not start, place a jumper from the R terminal to the G terminal. If the fan now starts, the subbase is defective and needs to be replaced. If the fan does not start, there is most likely a problem with the system transformer or the associated wiring. In most cases, this indicates that the problem does *not* lie within the thermostat or subbase.

Evaluating the cooling portion of the thermostat can be accomplished as follows:

- If the indoor fan operated when checked before, placing a jumper from R to Y should energize the cooling circuit (Figure 6-8).
- Install the jumper to make certain that the cooling circuit is energized.
- If the cooling circuit is energized (compressor runs) when the jumper is in place but is not energized when the thermostat is replaced and set to the cooling mode, the thermostat is defective.

Evaluating the heating portion of the thermostat can be accomplished as follows:

- If the indoor fan operated when checked before, placing a jumper from R to W should energize the heating circuit.
- Install the jumper to make certain the heating circuit is energized.
- If the heating circuit is energized when the jumper is in place but is not energized when the thermostat is replaced and set to the heating mode, the thermostat is defective.

Transformers

Transformers are among the easiest components in the control circuit to evaluate. They have no moving parts and nothing to oil, tighten, adjust, or calibrate (Figure 6-9). When voltage is applied to the primary winding, an induced voltage is produced at the secondary. The transformer can be easily damaged if care is not taken when the device is being installed or when the system is being serviced. If the transformer is a multitap-type transformer, the following guidelines should be observed:

- Make certain that the device is wired according to the wiring diagram on the device.

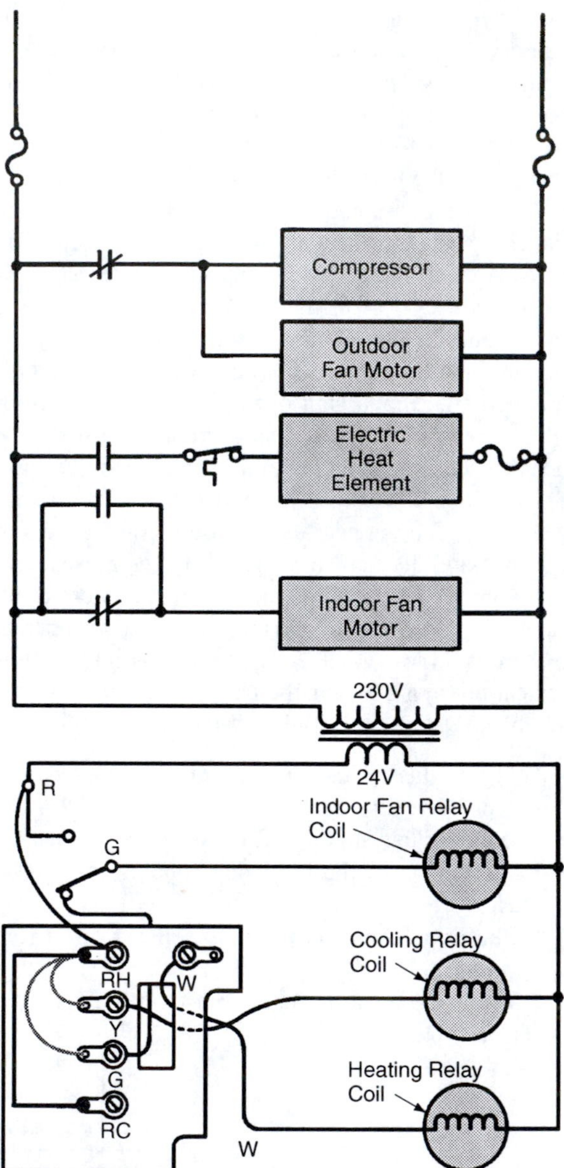

FIGURE 6-8 The complete cooling cycle should be energized when jumpers are placed from the R terminal to the Y and G terminals.

- Make certain that the proper wire leads are used for the voltage supplied to the device.
- Make certain that all unused wire leads are capped individually and taped to prevent the possibility of a short circuit.

FIGURE 6-9 Transformer used in residential and small commercial systems. *(Courtesy of Honeywell, Inc.)*

When troubleshooting the transformer, both the primary and the secondary windings can be evaluated. The transformer can be checked both while installed in the system with voltage applied to it and when removed entirely from the system. The most obvious method of checking the transformer is to apply the correct voltage to the primary winding and take a voltage reading at the secondary. If a 24-volt reading is obtained at the secondary, then the transformer is functioning correctly. If there is no secondary voltage, further investigation is necessary and the device should be disconnected from the system and checked as follows:

- Using a multimeter set to read resistance, take a reading of the primary winding. If the resistance is infinity, the winding has burned. Applying the incorrect voltage to the primary winding of the device commonly causes this. If there is a measurable resistance across the primary winding, the primary winding is okay.
- Using a multimeter set to read resistance, take a reading across the secondary winding. If the reading is infinity, the winding is open and the transformer needs to be replaced. An open secondary winding is often the result of a short in the control circuit. Before replacing the transformer, the short circuit must be located and repaired. A measurable resistance across the secondary winding is an indication that the secondary winding is okay.

Control fuses

Good field practice includes installing a small fuse, in the range of 3 amperes, in series with the secondary winding of the transformer. This **in-line fuse** will blow if a short develops in the control circuit. This fuse acts as a safety and will protect the transformer from future damage. When the circuit is operating correctly, this in-line fuse will simply act as a closed switch.

Control Fuses, just like power-circuit fuses, are designed to protect the components in the circuit. The primary device that is being protected in the control circuit is the transformer. These control fuses are also referred to as in-line fuses. They are usually encased in a plastic holder that provides the means by which they are connected in the control circuit. Just as with any other type of fuse, the reason that the fuse blew must be discovered and corrected before replacing it. The most common cause of a blown control-circuit fuse is a shorted holding coil, either on a contactor or on a relay.

Three ways that a control circuit fuse can be checked are:

- Visually
- With an ohmmeter
- With a voltmeter

Since many in-line fuses are made of glass, the element can be clearly seen. If the element is intact, the fuse is good. Even though this is sufficient in most cases, checking the fuse by other means may be necessary, especially when the fuse is not made of glass.

Checking the fuse with an ohmmeter is a very simple matter. When checking a fuse using an ohmmeter, the fuse must be removed from the circuit. The ohmmeter should be set to read continuity, and a test lead should be placed on each end of the fuse. If there is continuity through the fuse, it is good. If there is no continuity, the fuse needs to be replaced.

Checking fuses with a voltmeter is also a very effective method. When checking a fuse in this manner, the fuse is left in the circuit and the power to the circuit is left on. Since a fuse acts as a switch, there should be 0 volts across a good fuse (closed switch) and 24 volts across a blown fuse (open switch).

Pressure Controls and Safety Devices

Pressure switches and other pressure-actuated devices can act as safety controls as well as operational controls (Figure 6-10). Safety controls are designed to disable the system if unsafe conditions arise. Just as with any other switch, these devices can be checked with a voltmeter to determine if they are in the open or closed position. A closed pressure switch will have a potential difference across it of 0 volts. Evaluating these controls electrically is not enough to ensure proper operation of the device.

The mechanical action of the contacts and the bellows in the control must also be checked to ensure that the contacts of the device open and close at the correct time. To check that the contacts open and close in a timely manner, a gauge manifold should be installed on the system. By placing a voltmeter across the control, the exact pressure at which the contacts of the control open and close can be determined by observing the display on the meter. For low-pressure controls, the evaluation process would be conducted as follows:

- Install a gauge manifold on the system with the power off.
- Connect a voltmeter across the low-pressure control.
- Make note of the pressure at which the control is set to cut out.
- Turn the system on.
- The voltmeter should be reading 0 volts since the contacts of the control are closed.
- Frontseat the liquid-line service valve, causing the system to begin pumping down.
- Observe the needle on the low-pressure gauge as the pressure drops.

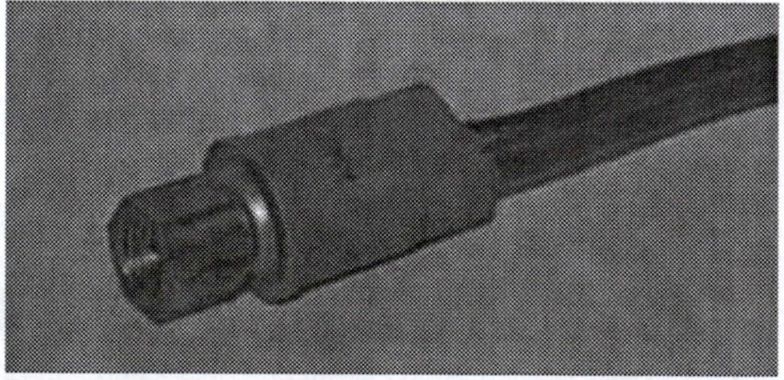

FIGURE 6-10 High-pressure control commonly found in air-conditioning applications. This device is factory set and cannot be adjusted in the field.

- When the display on the voltmeter reads 24 volts, the contacts of the low-pressure control have opened. If the low-pressure control is in series with the compressor contactor's holding coil, it should be obvious that the contacts open at the same instant that the compressor cycles off.
- Compare the pressure at which the contacts opened to the pressure that was set on the control.
- Adjust as needed.
- Continue to make adjustments until the contacts open and close at the desired pressures.

For high-pressure controls, the evaluation process is more difficult. Since high-pressure controls are primarily used to disable the system if excessively high pressure exists, simulating that pressure is difficult. These controls can, however, be set with at least some degree of accuracy and certainty with the following guidelines:

If the high-pressure control is already installed on the system:

- Set the control to cut out at a pressure about 20 psig higher than the normal operating head pressure of the system.
- Turn the system on.
- The contacts on the high-pressure control should be closed, and the system should operate normally.
- Slowly adjust the pressure control setting to a lower pressure until the system shuts down.
- Compare the pressure at which the system shut down to the pressure setting on the control.
- If these pressures are the same, the setting of the control can be raised to the desired pressure.
- Since the control is properly calibrated, the cutout pressure should be very close to the reading on the control's face.

If the high-pressure control is not installed on the system:

- Connect the transmission line from the pressure control to the high-side port of the gauge manifold.
- Connect the center hose of the gauge manifold to the regulator on a nitrogen tank.
- Open the nitrogen tank and set the regulator to a pressure just higher than the desired cutout pressure of the high-pressure control.
- Slowly open the high-side valve on the gauge manifold to allow the nitrogen to enter the high-pressure control.

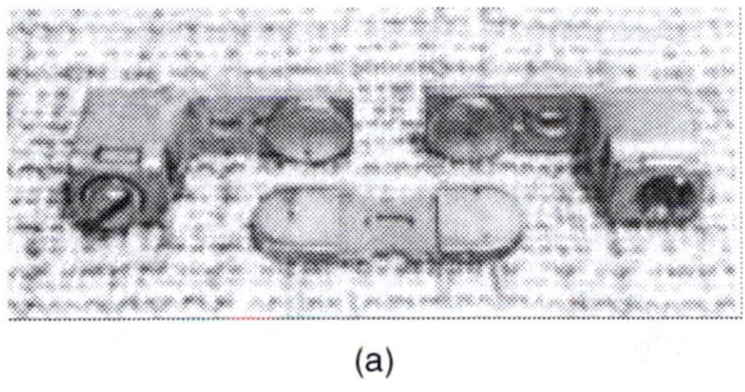

(a)

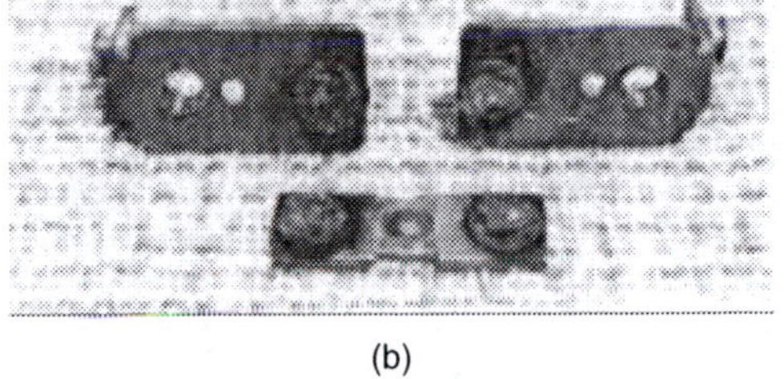

(b)

FIGURE 6-11 (a) Good set of contactor contacts. (b) Pitted contacts.

- An audible click will be heard when the contacts of the control open. An ohmmeter can also be connected across the device. While the contacts are closed, there should be continuity. When the contacts open, there should be no continuity.
- Compare this pressure to the desired cutout pressure.
- Adjust the control as needed.

POWER-CIRCUIT PROBLEMS

This section examines components commonly found in the power circuit and discusses how to properly evaluate them. The power circuit is generally easier to evaluate and troubleshoot than the control circuit. This is because the loads, be they compressors or fan motors or pumps, are connected directly to the line-voltage power supply. The devices that cause these loads to be energized and de-energized are the sets of contacts that are controlled by the holding coils of the control circuit. Generally speaking, all power circuits are basically configured the same for similar types of equipment.

Contactor and Relay Contacts

When a set of electrical contacts closes, current flows to the load that is ultimately being controlled. These contacts are made up of smooth, metallic surfaces that, when closed, offer very little resistance to current flow. Because of this, the voltage reading across a set of closed contacts should be 0 volts. Since the terminals on the line side of a contactor are labeled L1 and L2 and the terminals on the load side of a contactor are T1 and T2, the potential difference between Terminal L1 and Terminal T1 would therefore be 0 volts. Similarly, the potential difference between Terminal L2 and Terminal T2 would also be 0 volts. If a resistance reading was taken between Terminal L1 and Terminal T2, it would be 0 ohms, or very close to it.

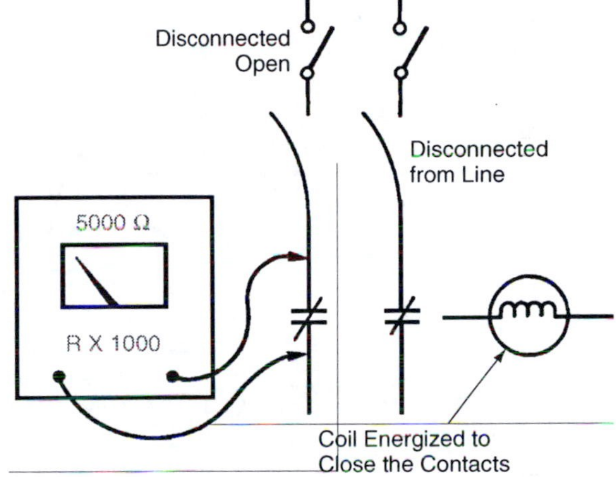

FIGURE 6-12 Checking contacts with an ohmmeter shows resistance in the contacts caused by pitting.

As relays and contactors age, the contacts undergo a process called **pitting**. This is when the smooth surfaces of the contacts become rough. Pitting (Figure 6-11) results from the electrical arcing that takes place whenever electrical contacts open and close. This arcing generates high temperatures, which slowly cause the contact surfaces to deteriorate. As the pitting gets worse, the operation of the main load being controlled is affected. The resistance across a set of pitted contacts is measurable. If the resistance is larger than 2 or 3 ohms (Figure 6-12), the contacts should be replaced. If these contacts are enclosed in a relay, the relay should be discarded and replaced.

Resistance across a set of contacts will result in a potential difference greater than zero across the contacts. This can cause problems with a motor or other load, because these contacts are in series with the load and a voltage drop across the contacts will reduce the voltage being supplied to the load. Consider the following simple circuit.

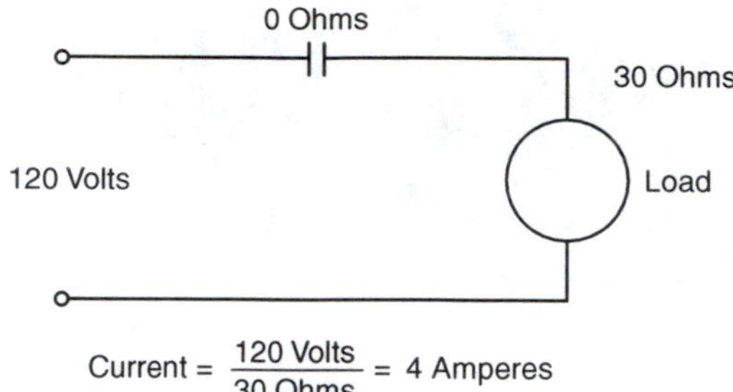

Current = $\frac{120 \text{ Volts}}{30 \text{ Ohms}}$ = 4 Amperes

Voltage across the Load = 4 Amperes x 30 Ohms = 120 Volts

FIGURE 6-13 Good contacts add very little or no resistance to the circuit.

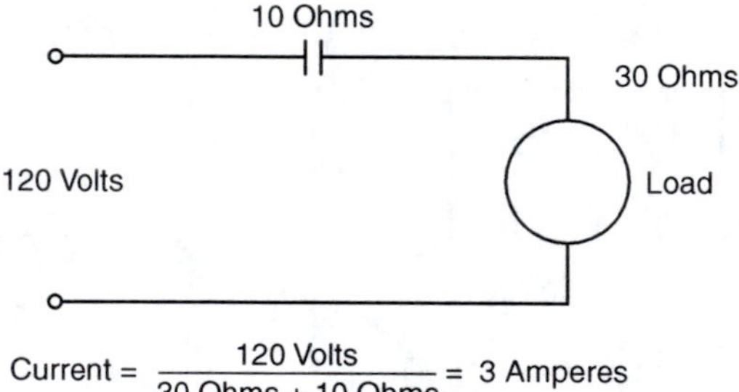

Current = $\frac{120 \text{ Volts}}{30 \text{ Ohms} + 10 \text{ Ohms}}$ = 3 Amperes

Voltage across the Load = 3 Amperes x 30 Ohms = 90 Volts

FIGURE 6-14 Pitted contacts add resistance to the circuit, reducing the voltage supplied to the desired load.

A 120-volt power supply is feeding a 30-ohm load (Figure 6-13). The load is controlled by a normally open set of contacts. When closed, the contacts offer no resistance to current flow. The current in the circuit is calculated, using Ohm's law, as:

$$I = 120/30 = 4 \text{ amperes}$$

The load has 4 amperes flowing through it and is supplied with the full 120 volts as provided. Now consider the following. The contacts that control the load are very old and pitted (Figure 6-14). The resistance across them is 10 ohms (very exaggerated). The total circuit resistance is now:

$$30 + 10 = 40 \text{ ohms}$$

The circuit current is now given by:

$$I = 120/40 = 3 \text{ amperes}$$

The voltage that is now supplied to the load can be calculated, again using Ohm's law, by:

$$V = 30 \text{ ohms} \times 3 \text{ amperes} = 90 \text{ volts}$$

If the load is rated at 120 volts, it will safely operate at 120 volts plus or minus 10 percent. 120 volts minus 10 percent is 108 volts. This load is operating well out of its designed range and will experience operational problems. In this case, the contacts should be replaced.

Circuit Breakers and Fuses

Just as control-circuit fuses protect the components in the control circuit, circuit breakers and fuses can also be used to protect the components in the power circuit (Figure 6-15, Figure 6-16). Since the amperage

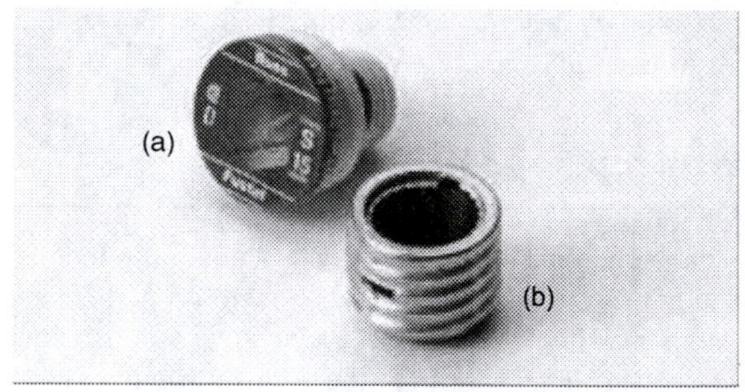

FIGURE 6-15 (a) Type S base plug fuse. (b) Type S fuse adaptor. *(Reprinted with permission by Bussman Division, McGraw-Edison Company)*

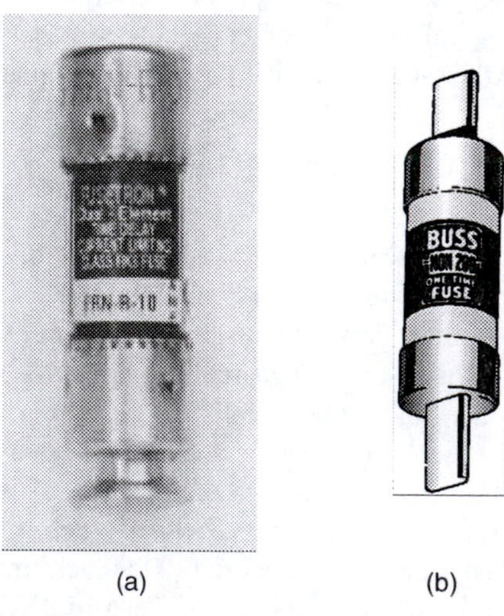

FIGURE 6-16 (a) Ferrule-type cartridge fuses. (b) Knife-blade cartridge fuse. *(Courtesy Cooper Bussman Division)*

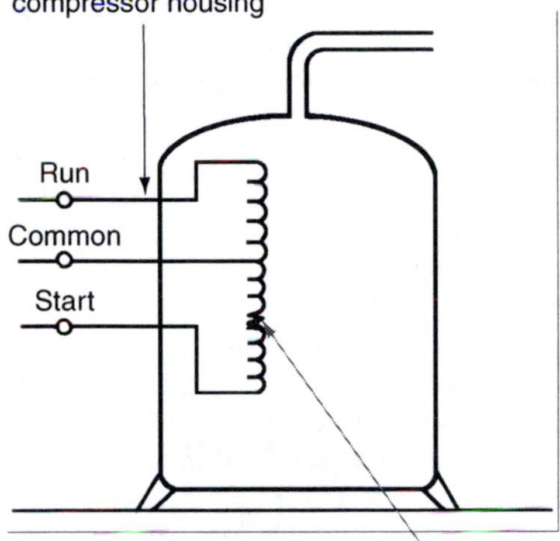
The motor terminals are insulated where they pass through the compressor housing
Run
Common
Start
Some of the windings are touching each other and reducing the resistance in the start winding

FIGURE 6-17 Compressor with a shorted winding

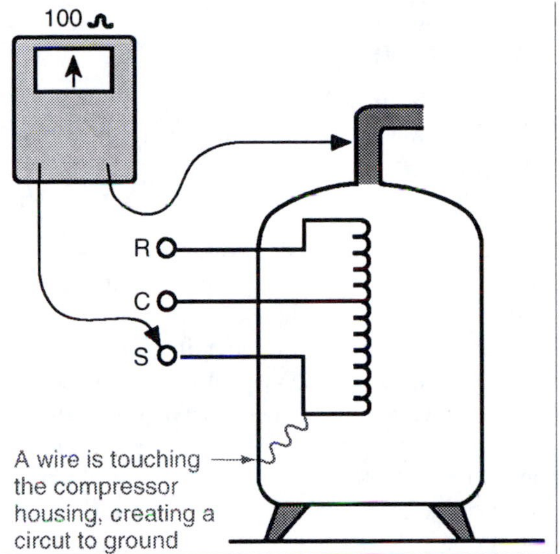
100 Ω
R
C
S
A wire is touching the compressor housing, creating a circut to ground

FIGURE 6-18 Compressor with a winding shorted to the shell

draw in the power circuit is generally much greater than the current draw of the control circuit, the power-circuit fuses and circuit breakers are much larger than those in the control circuit. A blown fuse can be identified relatively easily, as described earlier. A good fuse will allow current to flow freely through it, while a voltage reading taken across it will register 0 volts. Again, a good fuse acts as a closed switch. A reading of line voltage across the fuse indicates that the fuse is bad and needs to be replaced.

Just as with any other component, the reason for the failure needs to be determined and corrected before the device is replaced. Fuses blow and circuit breakers trip because of excessive current flowing in the circuit. Excessive current can be a result of a number of situations, including:

- A shorted device (load) such as a fan motor or a pump
- A short in the associated component wiring
- A short circuit to ground

A load or device that is shorted will offer little or no resistance to current flow (Figure 6-17). This increases the current draw of the device and will cause the power-circuit fuse to blow or the circuit breaker to trip. In this case, the device will need to be either repaired or replaced, depending on the economics of the situation.

A short circuit in the associated wiring is often less costly to the customer but can be very difficult to isolate and correct. A short circuit in the wiring can be the result of:

- Improper field wiring by a technician
- Improper installation of system components or accessories
- Worn wire insulation (conductors making contact with other conductors)

In any case, each circuit should be evaluated with an ohmmeter to determine where the short circuit exists. Each circuit should have at least some degree of measurable resistance. If not, chances are that a dead short must be isolated and repaired. As equipment ages, the wire insulation commonly becomes worn or damaged. When this damage occurs, short circuits are likely to surface.

A **short circuit to ground (short to ground)** is a special type of short circuit. This occurs when a current-carrying conductor makes contact with a ground or zero potential metallic surface (Figure 6-18). This is quite common in our industry, since there are a number of such surfaces that make a perfect ground. They include:

- The compressor shell
- The refrigerant-carrying copper tubing and piping
- The evaporator and condenser coils
- The casing of the condensing unit and air handler

To help prevent possible shorts to ground, all wiring should be inspected regularly and all electrical connections should be tight and sealed properly.

A short to ground can be identified by a measurable resistance between a current-carrying conductor and a grounded surface. When checking for a short to ground, the power should be disconnected and the ohmmeter should be set to a high range. If the meter registers resistance between a conductor and ground, the range can always be lowered to get a more accurate reading. If the range selected is too low, a high resistance may not register on the display, leading the technician to conclude that there is no short circuit when in fact there is. Short circuits to ground are easily remedied when they are in the associated wiring but not when they exist within the motor, pump, or compressor. Shorted wires can be either replaced or insulated. Shorted components often need to be replaced. The next section deals with some specifics on the evaluation of fan motors and compressor motors.

Fan Motors and Compressor Motors

Fan motors and compressors are the most expensive components in an air-conditioning or refrigeration system to replace. Therefore, these components must be properly evaluated. An improper determination can lead to artificially high repair costs to the customer.

Motors can fail for a number of reasons that can be broken down into two categories: mechanical and electrical. The mechanical problems were addressed earlier in the text. The electrical problems that are often encountered are, just as with other devices made up of wire coils, the open winding and the shorted winding. Both of these situations often result in costly repairs; and, in either case, the device often needs to be replaced or rebuilt.

OPEN MOTOR WINDINGS. The windings of an electrical motor can best be checked with an ohmmeter. The power supply should be off, and the wires supplying power to the motor should be disconnected. All capacitors, if there are any, should be properly discharged and removed from the circuit. Before checking the actual windings, though, it may be suspected that a motor has an open winding when voltage is supplied to the device and it does not operate.

Single-phase, split-phase motors, as we already know, have both a start and a run winding. The resistance of the start winding can be determined by taking a reading between the common wire and the start wire of the motor. These terminals are easily identified on a compressor motor, while the wiring diagram on the body of a fan motor may need to be checked to identify the proper wires. The resistance of the run winding can also be determined in a similar manner by taking a reading between the common and the run terminals of the motor. The resistance of the start winding should be greater than the resistance of the run winding. If either or both of the readings is infinity, the motor has an open winding (Figure 6-19) and the device needs to be replaced or repaired.

Simple, three-phase motors have three windings instead of two. The terminals on these three-phase motors are labeled T1, T2, and T3. The resistances of these windings should all be the same. If any resistance reading between the terminals is infinity, the motor has an open winding and must be replaced or repaired as well.

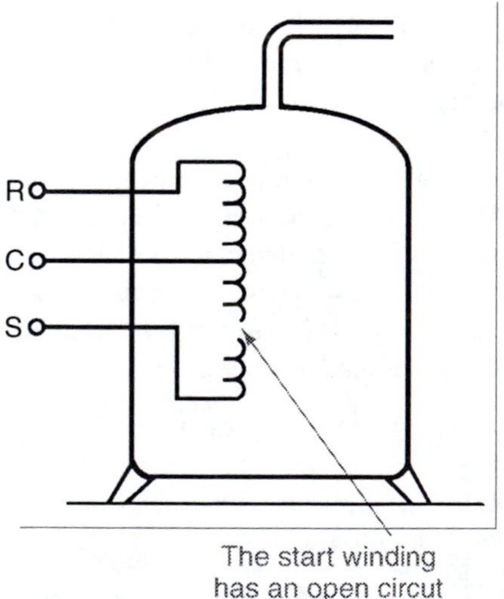

FIGURE 6-19 Compressor with an open start winding

Open windings can be caused by a number of factors, including motor age, overheating, improper voltage supply, and improper application. If the technician finds that a motor does, indeed, have an open winding, the associated wiring, power supply, and intended application need to be checked as well to make certain that the new motor does not fail prematurely.

SHORTED MOTOR WINDINGS. When the resistances of the windings are lower than normal, there is a good possibility that the windings are shorted. For example, if the resistance of the start winding is 8 ohms and the resistance of the run winding is 12 ohms, there is most likely a partially shorted start winding. Part of the coil has been removed from the circuit by a short circuit.

A resistance reading of 0 ohms across a winding indicates a complete short. This often becomes obvious when power is initially fed to the device. The circuit breaker will trip, or the fuse will blow, indicating the presence of a short circuit.

GROUNDED MOTOR WINDINGS. Another problem that occurs within motors is the grounded winding (Figure 6-18). This is often the result of internal motor overheating. The insulation on the winding breaks down and comes in contact with the shell of the device or other metallic surfaces or conductors, causing a short circuit. In essence, this is the same as a shorted winding; but, in this case, the connection is made to the shell of the device. Again, using an ohmmeter can identify a short to ground.

The meter should be set to read high resistance, and readings should be taken between the conductors and the shell of the device. Any measurable resistance reading is an indication that the motor is grounded and needs to be replaced or repaired.

Capacitors

Often, technicians new to the field determine that a motor is defective when, in reality, the capacitor is at fault (Figure 6-20). A defective capacitor can prevent a motor from operating and can also lead to more serious motor problems resulting from overheating. A successful technician must be able to field-test a capacitor to determine if the device is functioning properly. Several test instruments are available that field-test capacitors, but they can be quickly and effectively tested and evaluated with the use of an ohmmeter. An appropriate field check for a capacitor is as follows:

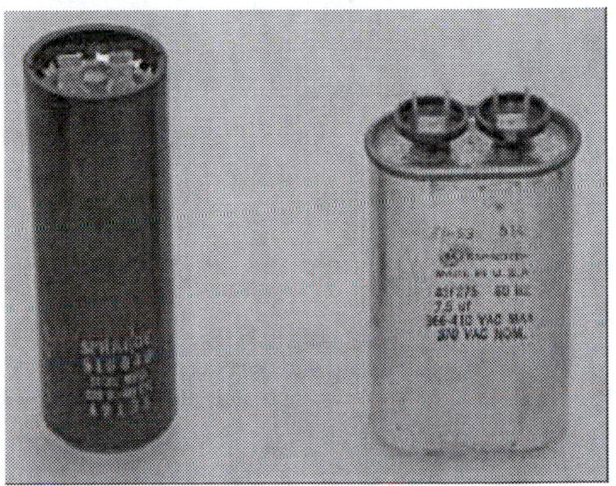

FIGURE 6-20 Start and run capacitors. *(Photo by Bill Johnson)*

- Make certain that the capacitor is properly discharged.
- Remove wiring from the capacitor, including bleed resistors, making certain to note the location of the wires to ensure proper reinstallation of the device.
- Set the resistance range to the R × 1000 or R × 100 range for start capacitors or the R × 1000 or R × 10,000 range for run capacitors.
- Touch the meter's leads to the capacitor terminals.
- If the capacitor is good, the display should fall quickly toward 0 ohms and then slowly go back toward infinity.
- If the display reads 0 ohms and stays there, the capacitor is shorted and needs to be replaced.
- If the display falls toward 0 ohms and then registers a measurable resistance, the capacitor has a partial short and should be replaced as well.
- If the capacitor is to be checked again, the test leads should be reversed.
- There should be infinite resistance between the capacitor terminals and the casing of the device. If there is continuity, the device should be replaced.

This process is outlined in Figure 6-21.

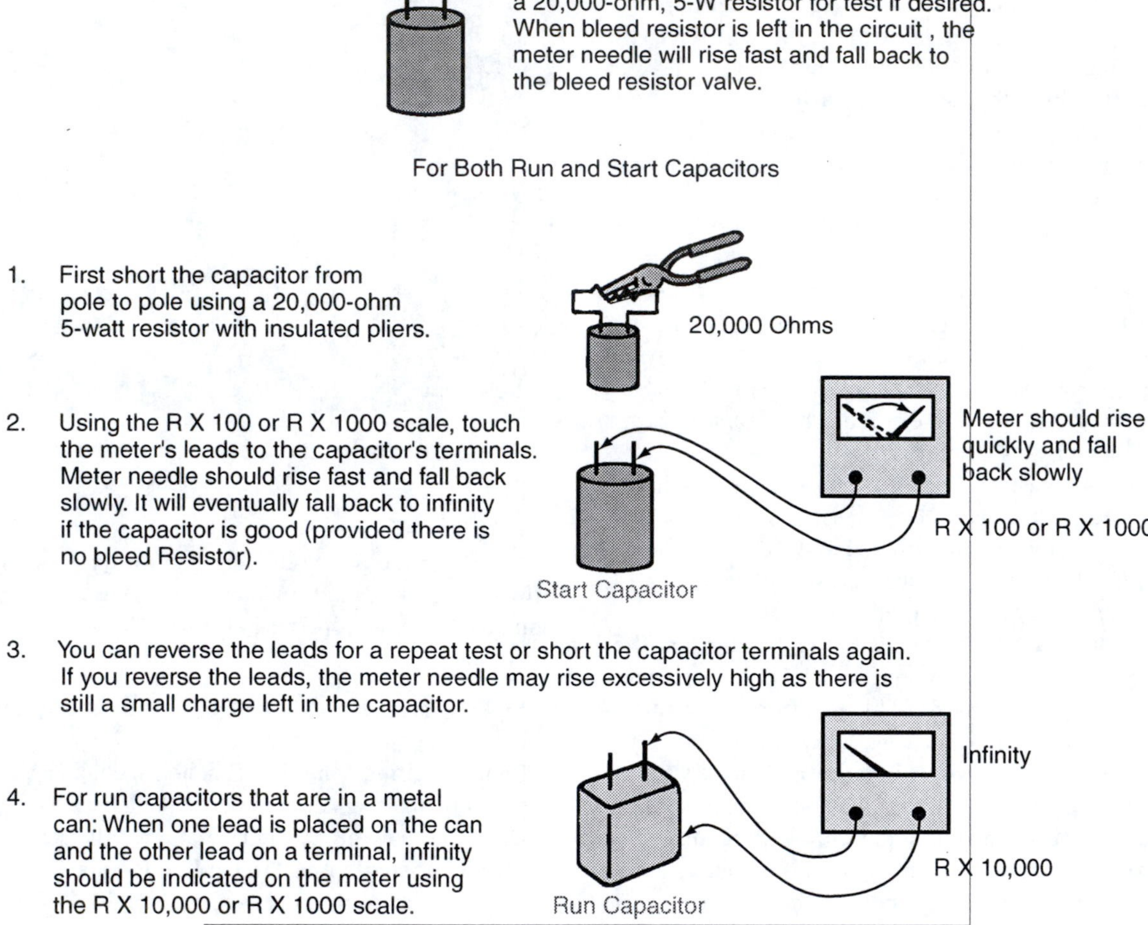

Disconnect bleed resistor after bleeding a 20,000-ohm, 5-W resistor for test if desired. When bleed resistor is left in the circuit , the meter needle will rise fast and fall back to the bleed resistor valve.
For Both Run and Start Capacitors
1. First short the capacitor from pole to pole using a 20,000-ohm 5-watt resistor with insulated pliers.
20,000 Ohms
2. Using the R X 100 or R X 1000 scale, touch the meter's leads to the capacitor's terminals. Meter needle should rise fast and fall back slowly. It will eventually fall back to infinity if the capacitor is good (provided there is no bleed Resistor).
Meter should rise quickly and fall back slowly
R X 100 or R X 1000
Start Capacitor
3. You can reverse the leads for a repeat test or short the capacitor terminals again. If you reverse the leads, the meter needle may rise excessively high as there is still a small charge left in the capacitor.
Infinity
4. For run capacitors that are in a metal can: When one lead is placed on the can and the other lead on a terminal, infinity should be indicated on the meter using the R X 10,000 or R X 1000 scale.
R X 10,000
Run Capacitor

FIGURE 6-21 Procedure to field test a capacitor

SUMMARY

- Electrical troubleshooting is accomplished with ammeters, voltmeters, and ohmmeters.
- Contactor and relay holding coils can be checked by reading the coils' resistance.
- The voltage rating of the holding coil must match the supplied voltage.
- A defective holding coil can be either shorted or open.
- Thermostats can be easily evaluated by placing a jumper across their contacts.
- In-line fuses are often installed in the control circuit to protect the transformer.
- Pressure controls must be checked to ensure that they open and close their contacts at the correct pressures.
- The resistance across a set of closed contacts should be 0 ohms.
- The potential difference across a closed set of contacts should be 0 volts.
- When contacts become pitted, both the resistance and the potential difference across them increase.
- Fuses blow and circuit breakers trip when a short circuit or ground is present.
- Short circuits can result from improper field wiring, improper component installation, or worn or damaged wire insulation.

- Defective motor windings can be open, shorted, or grounded.
- Capacitors can be field-tested with an ohmmeter.

KEY TERMS

Ammeter
Control fuse
In-line fuse
Ohmmeter
Pitting
Potential difference
Short circuit to ground (short to ground)
Voltmeter

FOR DISCUSSION

1. List as many scenarios as possible that could cause a power-circuit fuse to blow.
2. List as many scenarios as possible that could cause a control-circuit fuse to blow.
3. How could a field technician be fooled into thinking that a motor is defective when, in reality, it is not?
4. How would an ohmmeter's readings differ when comparing those obtained from a motor with an open winding with one with a shorted winding?

REVIEW QUESTIONS

1. If the technician reads low voltage at the holding coil of a 24-volt cooling contactor and the normally open contacts are not closed, which of the following is most likely?
   a. The transformer is defective.
   b. The holding coil is open.
   c. The thermostat is not calling for cooling.
   d. The compressor is grounded.
2. A technician notices that the normally open contacts of a cooling contactor are not closed. After taking a voltage reading at the coil, the technician determines that no voltage is being supplied to the holding coil. Which of the following is a possible cause for this situation?
   a. The transformer is defective.
   b. The thermostat is not calling for cooling.
   c. The holding coil is open.
   d. Both *a* and *b* are possible causes for this situation.
3. A technician takes a resistance reading of a contactor's holding coil and reads 0 ohms. A resistance reading of infinity is also measured between the coil and ground. These two readings indicate that:
   a. The coil is good.
   b. The coil is shorted.
   c. The coil is open.
   d. The coil is grounded.
4. A reading of 24 volts between the R terminal and the G terminal of a low-voltage thermostat is an indication that:
   a. The evaporator fan motor is not operating.
   b. The transformer is defective.
   c. The compressor is operating.
   d. The thermostat's indoor fan motor contacts are closed.

5. True or False: Under normal operating conditions, when the compressor is operating the indoor fan motor can be in the OFF position.

6. True or False: Under normal operating conditions, when the indoor fan motor is operating the compressor must be operating as well.

7. If the __________ is operating and the ___________ is not, there is most likely a problem with the evaporator fan motor circuit.

   a. Evaporator fan motor, condenser fan motor
   b. Evaporator fan motor, compressor
   c. Compressor, indoor fan motor
   d. Compressor, condenser fan motor

8. A reading of 0 volts between the R and W terminals on a low-voltage thermostat is an indication that:

   a. The system is operating in the cooling mode.
   b. The system is operating in the heating mode.
   c. The control transformer is defective.
   d. Both *b* and *c* are possible.

9. Why are in-line fuses commonly found in control circuits?

10. A blown control fuse can be caused by:

    a. A shorted holding coil.
    b. An open holding coil.
    c. Both *a* and *b* can cause the fuse to blow.
    d. Neither *a* nor *b* can cause the fuse to blow.

11. Which of the following is a possible indication of a defective control fuse?

    a. The compressor does not run, but the condenser fan motor does.
    b. The evaporator fan runs, but the compressor does not.
    c. The evaporator fan does not operate.
    d. The compressor and the condenser fan are drawing excessive amperage.

12. A line-voltage reading across a power-circuit fuse:

    a. Is impossible.
    b. Indicates that the fuse is defective.
    c. Indicates that the fuse is good.
    d. Indicates that the compressor should be operating.

13. The potential difference between Terminal L1 and Terminal T1 on a contactor that has its holding coil energized should be:

    a. 0 volts.
    b. 24 volts.
    c. 115 volts.
    d. 220 volts.

14. True or False: The pitting of contacts is desirable because it increases the resistance of the power circuit.

15. True or False: There should always be continuity between a run capacitor's terminals and the shell of the device.

CHAPTER 7

Servicing Vapor-Compression Refrigeration Systems

OBJECTIVES After studying this chapter, the reader should be able to:

- List the seven steps in performing a successful service call.
- Explain how to determine if an air-conditioning system has an overcharge of refrigerant.
- Explain how to determine if an air-conditioning system has a refrigerant undercharge.
- Describe how a technician can determine the cause of a blown fuse or tripped circuit breaker in the control or power circuit.
- Explain how to determine if a control transformer is operating properly.
- Describe how to determine if a contactor or relay's holding coil is defective.
- Explain how to determine if a motor winding is burned or open.
- Explain how to determine if a compressor motor or other system component is grounded.

INTRODUCTION

As the air-conditioning and refrigeration student acquires general knowledge about the components, circuits, and sequence of operations of various systems, a broad mental database is being created. This pool of information, in effect, becomes a large "toolbox" that can be accessed as needed to help the technician solve various field problems. Just as a technician carries a complete set of hand tools, including wrenches, gauges, screwdrivers, and the like, a mental toolbox is just as important in order to troubleshoot an air-conditioning or refrigeration system effectively. The commonsense aspect of troubleshooting helps the technician determine what to do with the available tools. Therefore, the more complete this toolbox is, the more effective the service technician will be.

This chapter takes some of the basic concepts covered throughout the text and presents them in the context of actual service calls. Each section begins with the customer complaint, giving details from the point of view of the customer. This section is then followed by the evaluation process as performed by the service technician. Finally, a summary of the service call is presented, making reference to the important concepts that have been addressed in each.

OVERVIEW

As mentioned on more than one occasion in this text, troubleshooting is a skill that one becomes more proficient at with practice and continued technical education. As technicians encounter problems time and time again, they are able to recognize and solve particular problems faster. This is referred to as the **learning curve**. In basic terms, the concept of the learning curve states that as people perform a task over and over, they are able to perform this task better and in less time. The same holds true for servicing air-conditioning and refrigeration systems. As technicians become familiar with common system problems, system configurations, reading schematic diagrams, and component evaluation, problems can be isolated more quickly and more efficiently. New technicians often get frustrated because they experience difficulty in identifying components, troubleshooting problems, or performing a repair correctly. Rest assured, this is quite common. It takes time, patience, and desire to become a *seasoned professional.* There is, however, something that the new technician can do to help speed the process. This is to take a commonsense and logical approach to troubleshooting.

TROUBLESHOOTING STEPS

The troubleshooting process can be broken down into seven simple steps. These seven steps should be performed in order, and each should be completed before moving on to the next. If followed correctly, even the newest field technician will be able to experience success in the field very quickly. The steps are:

1. Verify the complaint.
2. Gather information.
3. Perform a visual inspection.
4. Isolate the problem.
5. Correct the problem.
6. Test the system operation.
7. Complete the service call.

Verify the Complaint

This step may seem obvious, but it is essential. The technician must realize that the customer is not the service expert. If that were the case, the service company would not have been called in the first place. What the customer sees or thinks may or may not be what is actually taking place within the system. Therefore, the service technician must verify the information that the customer is providing. For example, the customer may tell the technician, "The unit doesn't blow cold air." If this is accepted as the truth, the technician may make the following assumptions:

- The evaporator fan is operating.
- The condensing unit is not running.
- The system is calling for cooling.

The technician may reach the conclusion that the condensing unit is not working properly. This may or may not be correct. The following may, in reality, be the case:

- The thermostat is set to the cooling mode.
- The fan switch is in the ON position.
- The room thermostat is satisfied, and the condensing unit is de-energized.
- Since the fan is operating all the time, even when the compressor is not operating, the air coming from the registers will be warm.

It can be seen from this example that the system is operating correctly, but the customer is under the impression that the air coming from the registers must *always* be cold. He felt the air and, since it was warm, came to the conclusion that something must be wrong with the system. A service call was then placed to the HVAC contractor.

Gather Information

Once the service technician has verified the complaint, other information that may help get to the root of the problem quickly must be obtained from the customer. Pertinent information may be obtained by asking questions similar to the following:

- "When did you first notice this problem with the system?"
- "Was anybody working on or around your equipment at the time you first noticed the problem?"
- "When was the last time this unit was serviced?"
- "Prior to this problem, had the system been working reliably?"
- "Have there been any recent power failures?"

Depending on the customer complaint, additional questions may shed more light on the situation at hand. For example, if the customer says that the system just isn't cooling enough, asking when the problem was first noticed may help solve the problem, depending on the answer received. If the customer responds with something like, "It stopped cooling well toward the end of last summer, but since the weather started getting cool, I figured I would wait until this year to have it checked out. Now it's not cooling much at all." This would lead the astute technician to believe that there is most likely a refrigerant leak in the system because the system cooled, but not well, last summer but cools even less now.

Other workers that may have been around the equipment may have had something to do with the improper system operation. For example, if the customer was having the house painted, supply registers may have been covered with plastic and taped closed to prevent paint drips from damaging the grills. If the outside of the house was being painted, the condensing unit may have been covered with a drop cloth for the same reason. Another situation may arise if, for example, the customer's alarm company was at the house checking out the system and the technician, while crawling around the attic, inadvertently pulled on a low-voltage control wire, causing damage to it.

Knowing when the system was last serviced can also provide useful information. Motor lubrication, filter changes, belt adjustment, and other tasks performed during a preventive-maintenance service call help keep the system in tip-top working order. Neglecting these items can cause system problems that would normally be avoided with proper maintenance.

Asking the homeowner if the system was working properly prior to this problem may or may not provide useful information. It is not possible for a layperson to know whether or not the system was working correctly. The answer to this question will give you the answer to the question, "Was the house comfortable prior to the system failure?" If the answer to the question reflects that the system performed "perfectly" prior to this, a mechanical failure of some type or a total loss of refrigerant can *usually* be suspected but, as with any other probable cause, would need to be confirmed through a thorough system check.

Neighborhood power failures can also result in a system that fails to operate as expected. Programmable thermostats may lose their program, and other digital devices may need to be reset. Knowing if the power was interrupted for any length of time may help reduce service time.

Obviously, these questions are only a small sampling of the possible questions that will provide the technician with the needed extra information to properly evaluate the system. It is up to the technician to decide what questions would best help him to help the customer.

Perform a Visual Inspection

Once the technician has spoken to the customer and has gathered information regarding the system, the technician must gather additional information independently. This information is obtained by performing a visual inspection of the equipment and includes:

- Checking the thermostat settings
- Checking the air being discharged from the supply grills
- Inspecting air filters at the return grill
- Inspecting the condensing unit

By checking the thermostat settings, the technician can determine if the system is actually calling for the desired mode of operation. If the customer is not getting any cooling, for example, and the thermostat is set to the heating mode, it is quite obvious that there will be no cooling. If the thermostat is set to provide cooling and the fan is not operating, a low-voltage problem may be suspected. The fan blowing warm air may lead the technician to believe the problem lies within the condensing unit. Dirty or blocked filters may also give the technician useful information. If the filters are completely blocked, this means that not only is there an airflow problem with the system but there may also be a motor-lubrication or belt-tension problem, since the system has obviously not been serviced for some time. Inspecting the condensing unit will tell the technician if the condensing unit is operating when the thermostat is set to the cooling mode. If the thermostat is set to the cooling mode and the condensing unit is not operating, an electrical problem should be suspected. This would include the opening of a pressure switch caused by a loss of refrigerant charge.

Isolate the Problem

Once the visual inspection has been completed, the technician can then concentrate on the particular problem. If, for example, on a no-cooling call, it was determined that the indoor fan motor was operating when the thermostat's fan switch was turn to the FAN ON position but turned off when the fan switch was switched to the AUTO setting, the transformer would not need to be checked, since it has already been established that low voltage is being provided by the device. If, on the other hand, the fan failed to operate, this would be the first thing that should be checked. If a problem exists with a condensing unit and it was observed that the condenser fan motor is operating but the compressor is not, checking whether or not there is voltage being supplied to the condensing unit is not necessary. The technician already knows that voltage is being supplied because the condenser fan motor is operating. The technician also knows that the cooling contactor is operating properly. The problem, therefore, lies within the compressor or its associated wiring.

Correct the Problem

After the specific problem has been identified, the repair should be performed in a manner that observes all safety guidelines and government laws and regulations. Safety guidelines include:

- Power to the circuits should be disconnected when repairing or replacing defective components.
- Safety glasses and work boots should always be worn.
- Loose-fitting clothing and jewelry should not be worn.
- Long hair should be tied back.

Government laws and regulations as outlined by the Clean Air Act and the Montreal Protocol Act should be followed, and all service technicians working on air-conditioning and refrigeration equipment must be certified by the Environmental Protection Agency and possess EPA Section 608 certification. The type of equipment that is worked on determines the type of certification required. For example, low-pressure systems require Type III certification under Section 608, while technicians who work on automotive air-conditioning systems are required to possess Section 609 certification.

In correcting the problem, the service technician must always be aware of and respect the customer's property. The technician should work quickly, efficiently, and neatly and should consciously strive to provide the best possible repair in the shortest possible time, thereby providing the best possible value for the customer.

Test the System Operation

Once the repair has been completed, the system must be tested. Good field practice includes checking all modes of operation and allowing the system to operate under supervision for at least 15 minutes. Waiting until the unit cycles automatically to ensure proper operation is a good idea. System pressures and temperatures should be checked, including evaporator superheat, condenser subcooling, and evaporator temperature differential. The amperage draw of the compressor and motors should also be checked after the system repair to ensure that they are within design parameters. While the system is being monitored, final cleanup and paperwork preparation can be done in order to make good use of this time. This helps ensure that, upon completion of the job, the technician will be ready to proceed to the next call.

Complete the Service Call

No service call is complete until the paperwork is filled out and the customer signs off on the job. Paperwork must be complete and include all required information. A sample *work order,* or service ticket, is shown in Figure 7-1. Descriptions of the work performed should be detailed and written legibly. The technician should not simply write "Replaced Compressor" on the work ticket but should make certain to list everything that was done including refrigerant recovery, system evacuation, and so on. A complete list of all materials and parts used must be provided. Remember, this is the only written record of the work performed. The billing department will rely on this work ticket to prepare a detailed bill for the customer. Time spent on the job and materials used are vital pieces of information that *cannot* be left out.

In addition to the paperwork, the technician must also explain to the customer exactly what was done to the system. After all, the service technician is the initial—and often the only—contact between the customer and the service company. Making the company look good by acting in a professional and courteous manner is the technician's duty.

Before leaving the job site, the technician should perform an inspection of the work area and remove any garbage and debris from the area. Dirt should be cleaned up. The customer cannot see the work performed inside a sealed unit, but fingerprints on or around a thermostat indicate that the technician is a sloppy worker, no matter how well a job was performed. This final inspection also helps reduce the possibility of leaving tools and materials behind. Precious time will be lost if a tool is left behind and is needed on the next job.

SERVICE CALL 1: SYSTEM OVERCHARGE

Customer Complaint

A customer places a call to her air-conditioning service company and tells the service manager that her split-type, air-conditioning system is not cooling her house properly. She explains that the air coming from the supply registers is warm, and that her thermostat is displaying a space temperature of 88°F.

ICACS
Industrial Cooling Air Conditioning Service Agency

Industrial Cooling Inc.
30 South Ocean Avenue, Suite 304
Freeport, NY 11520
(516) 546-0202

№ 0737

Job # ____________

Date ____________ 20 ____

Job Name ____________ Phone ____________

Address ____________ City ____________

See Mr. ____________ Of ____________ At ____________

Bill To ____________ Phone ____________ Person ____________

Product ____________ Tag No. ____________ Model # ____________ Serial # ____________

Action Taken ____________

SERVICE TOOLS, KITTING	MATERIAL USED	MATERIAL REQUIRED
Precision Instrument Dial Indicators		
Portoblast Chemical Cleaner		
6 CFM Vaccuum Pump Kitting		
Megohm Kitting		
Micron Gauge Kitting		
Centrifugal Recovery Kitting		
Unitary Recovery Kitting		
125 lb. Recovery Tanks		
50 lb. Recovery Tanks		
800 lb. Recovery Tanks		
½ Ton Rigging Assembly		
1 Ton Rigging Assembly		
2 Ton Rigging Assembly		
Field Assembled Gantry		
Halide Map Detector		
Acetylene Kitting/Rig		
VFC Charge		
Electric Lube Pump		
Manual Lube Pump		
Centrifugal Bulldog Recovery		
Tube Cleaner, Electric		
Hyperwatt Chiller Heater		
Electronic Detector		

REFRIGERANT MANAGEMENT		
Refrigerant Type		
Refrigerant Removed		Lb.
Removed Refrigerant Recharged		Lb.
New Refrigerant Added		Lb.
Pressure Tested At		Psi
Micron Evacuation		Microns
Time of Micron Evacuation		Minutes
Leaks Found:		
Leaks Repaired:		

LABOR SUMMARY	
# TECHS	
Hours ST	
Hours OT	
Total Hours	

Tech signature

Tech print

Date

Customer signature

Customer print

FIGURE 7-1 Sample service ticket. *(Courtesy Industrial Cooling Inc.)*

Two hours later, Pedro, the service technician, arrives at the house. The home owner explains the situation to him. She also adds that at the end of last year's cooling season the system developed a refrigerant leak. The system was repaired but, shortly after the repair was made, the weather turned cool and the system was not used any more that year. She adds that she has used the unit 2 or 3 times this season, during relatively warm weather, but this is the first time it has been used on a very hot day. The outside temperature is 95°F.

Service Technician Evaluation

After hearing the customer's complaint, Pedro makes his way over to the thermostat and checks

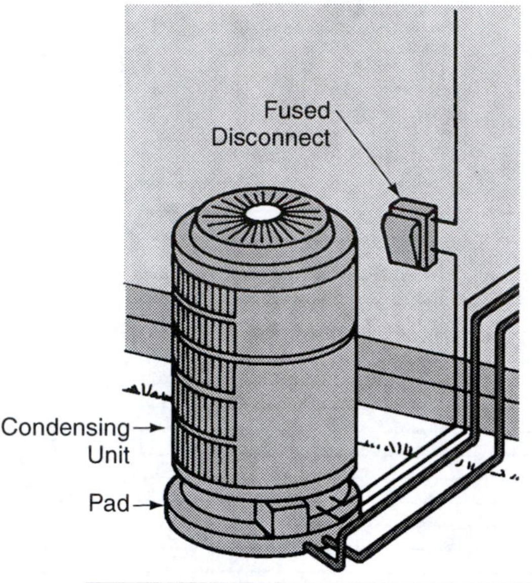

FIGURE 7-2 The fused disconnect is located at the condensing unit location.

FIGURE 7-3 The fused disconnect is equipped with a manual disconnect switch as well as replaceable fuses.

to make certain that the thermostat is set to the cooling mode and that the thermostat is set to maintain a temperature below the present room temperature. The room temperature is 88°F, and the thermostat is set to maintain a temperature of 74°F. He then goes over to a ceiling supply register and feels the air coming from it. Just as the customer stated, the air is warm. From this, he is able to establish the following:

- The indoor, evaporator fan motor is operating.
- The low-voltage control circuit is being powered. (If the transformer were defective, the holding coil of the indoor fan relay could not be energized and the indoor fan would not be operating.)
- The problem most likely exists in the outdoor unit or the wiring leading to it.

Pedro then asks the customer to show him the location of the outdoor condensing unit. The customer leads Pedro outside to the side of the house. He immediately notices that neither the compressor nor the condenser fan motor unit is operating. He puts on his safety glasses, turns off the disconnect switch (Figure 7-2), and carefully opens the cover (Figure 7-3). Turning the disconnect switch to the ON position and using a voltmeter set to the proper scale, Pedro checks for voltage being supplied to the unit. He reads 221 volts on the meter between the two hot legs feeding the unit (denoted by locations *C* and *D* in Figure 7-4), which is fine since the condensing unit is rated at 220 volts. He then removes the service panel from the unit and locates the low-voltage control wiring that feeds power to the condensing unit. After loosening the wire nuts, Pedro checks for low voltage. His meter reads 23 volts, which is within the acceptable range. From this, he determines that:

- The condensing unit is being supplied with the proper line voltage.
- The thermostat is calling for cooling and is sending 24 volts to the condensing unit.
- The problem *definitely* lies within the condensing unit.

Pedro then uses his voltmeter to check the voltage at the holding coil of the cooling contactor (Figure 7-5). The meter reads 0 volts. He is now able to establish that one of the controls is open or the interconnecting low-voltage wiring is damaged.

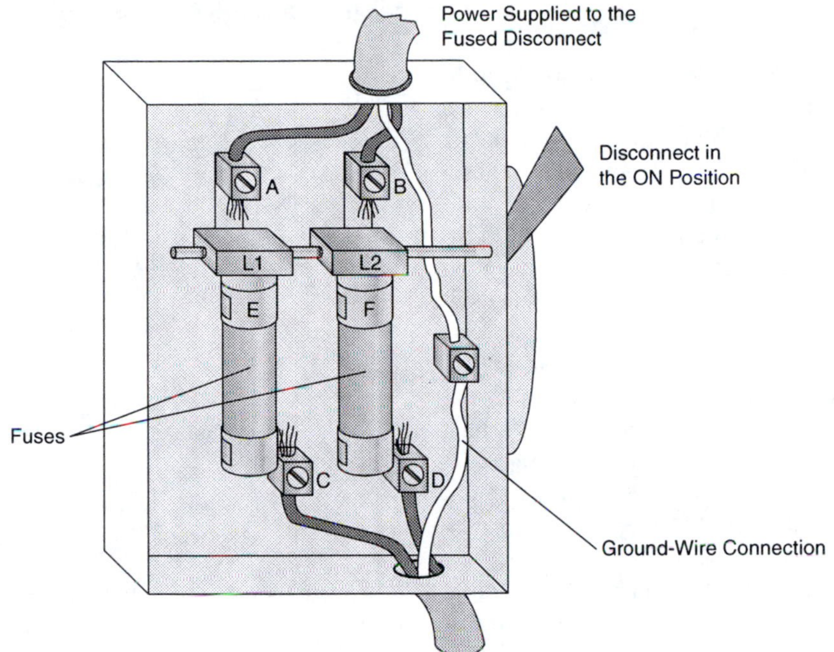

FIGURE 7-4 Power is supplied to the disconnect switch at points *A* and *B*. Power leaves the disconnect at points *C* and *D*, provided that both fuses are good.

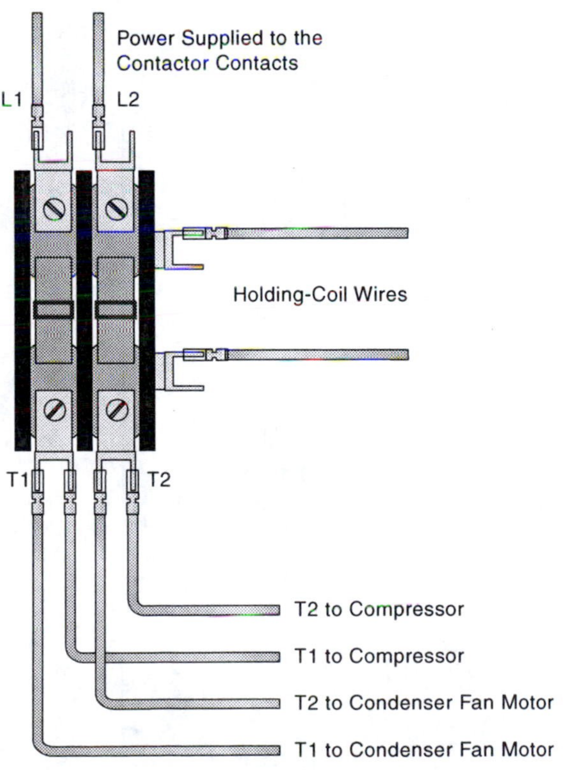

FIGURE 7-5 Taking a voltage reading at the holding-coil terminals will determine if voltage is being supplied to the coil.

The reason he is able to establish this should become clear after referring to Figure 7-6. Since *Meter A* reads 24 volts, Pedro establishes that the transformer is producing 24 volts and that the thermostat is calling for cooling. Since *Meter B* reads 0 volts, he knows that a control, either the high-pressure switch or the low-pressure switch, must be open or the interconnecting wiring is damaged. He begins by checking the voltage across the low-pressure control. He reads 0 volts across this device, which indicates to him that the switch is in the closed position (see *Meter C* in Figure 7-7). He then proceeds to check the voltage across the high-pressure control. This time he reads 24 volts across the switch. This indicates that the switch is in the open position (see *Meter D* in Figure 7-7).

Using the information obtained from the customer as well as on his own, Pedro is able to establish that the system pressure reached unsafe levels and the manually reset high-pressure control opened in order to disable the condensing unit. His next task is to determine what caused the high-pressure condition.

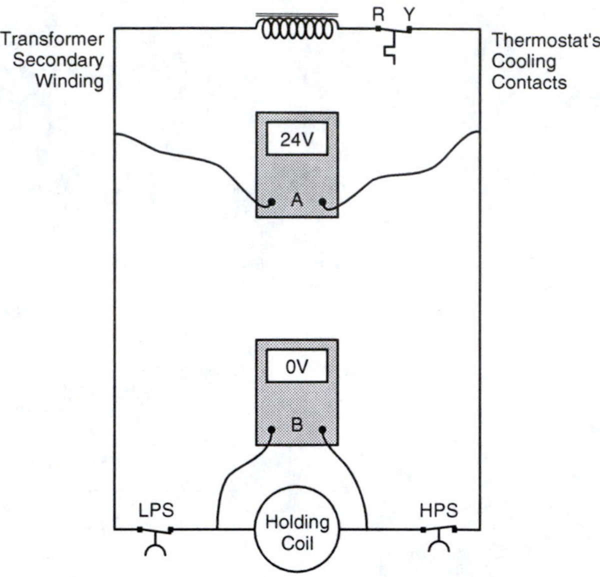

FIGURE 7-6 Twenty-four volts are being supplied to the condensing unit, but 0 volts are being supplied to the holding coil. The problem lies within the condensing unit itself.

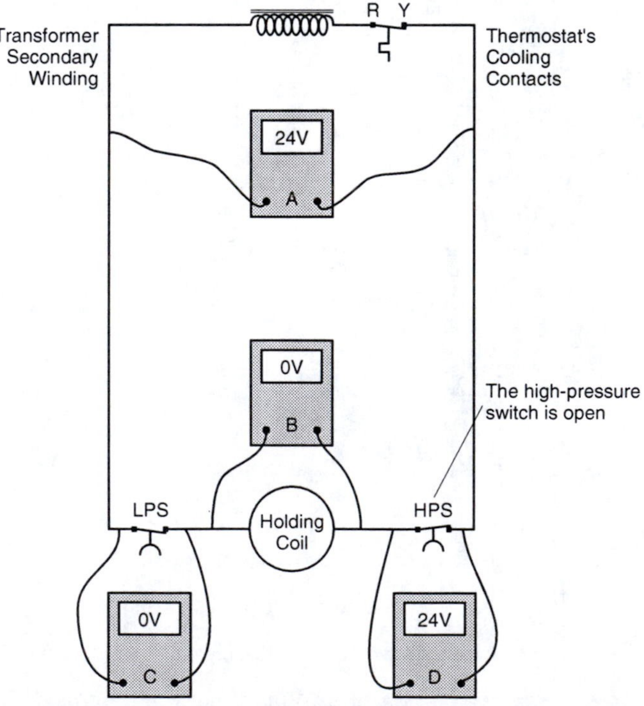

FIGURE 7-7 The high-pressure switch is in the open position.

FIGURE 7-8 Digital thermometers are the tool of choice when taking field-temperature measurements. *(Photo by Eugene Silberstein)*

He knows that one or more of the following may be the cause for a system to operate with high head pressure, ultimately causing the high-pressure switch to open:

- Dirty or blocked condenser coil
- Noncondensable gases in the system
- Defective condenser fan motor
- Refrigerant overcharge

Pedro now needs to determine which of the foregoing possible causes is to blame for the system failure. He wants to proceed cautiously, since high system pressure can result in personal injury as well as system component damage. His actions are as follows.

First, he removes his pocket flashlight and visually inspects the condenser coil. He is able to see clear through the coil and notices no dirt deposits between the coil's fins. He also notices that no leaves or other debris is blocking the coil, and he concludes that the condenser coil is not dirty or blocked. The reason for the failure has now been narrowed down to the following:

- Noncondensable gases in the system
- Defective condenser fan motor
- Refrigerant overcharge

To address the issue of noncondensable gases in the system, the system must have been off for at least 30 minutes, which it obviously has been, and a set of service gauges must be installed on the system. Using proper gauge-installation procedures, Pedro installs the manifold on the system. He then takes a reading of the outside ambient temperature, which in this case is 96°F (Figure 7-8) and compares this temperature to the condenser saturation temperature on the high-side gauge of his manifold, which is 95°F. This corresponds to the actual ambient, so Pedro rules out the possibility that noncondensable gases are in the system.

Up to this point, Pedro has been able to successfully eliminate two of the possible causes for the system failure without starting up the system. Because he knows that the high-pressure situation is potentially dangerous, he has done as much as he can without having to energize the system—but now he has to. He needs to determine if the system has a defective condenser fan motor or has an overcharge of refrigerant. Before starting the unit, he places his clamp-on ammeter around one of the wire leads leading to the condenser fan motor (Figure 7-9) and notes from the nameplate that the running load amperage on the motor is 2.1 amperes. Pedro then turns off the service-disconnect switch that feeds power to the condensing unit and resets the high-pressure control. After making certain his safety glasses are secure, he turns the service-disconnect switch to the ON position.

The condenser fan motor and the compressor both begin to operate. The amperage draw on the condenser fan motor is 1.9 amperes, which is well within the desired range. The system pressures, however, are not. The suction pressure is in the range of 85 psig, and the head pressure begins to rise steadily until the high-pressure control opens its contacts when the pressure reaches approximately 375 psig (Figure 7-10). Pedro has now eliminated the possibility of the defective condenser fan motor and correctly concludes that the system has been overcharged. He then turns off the disconnect switch.

FIGURE 7-9 Securing a clamp-on ammeter around a condenser fan motor wire. *(Photo by Bill Johnson)*

Pedro places a call to his service manager to discuss his findings. The customer's file is consulted, and it is found that the system was repaired on a cool day at the end of last year's cooling season. The technician that performed the original work had overcharged systems several times before, primarily due to a lack of proper training, and has since found employment with another service company.

Using a properly evacuated, DOT-approved recovery tank (Figure 7-11), Pedro begins to remove refrigerant from the high side of the system and store it in the tank (Figure 7-12). Once he has removed enough refrigerant from the system, he then restarts the system and adjusts the charge. Since the outside ambient temperature is 95°F, he knows that the condenser saturation temperature for this system should be approximately

FIGURE 7-10 This manifold registers a low-side pressure of 85 psig and a high-side pressure of 375 psig. This system has a high-pressure problem. *(Photo by Eugene Silberstein)*

FIGURE 7-11 Color-coded, refrigerant-recovery cylinders. The bodies of the tanks are gray, and the tops are yellow. *(Courtesy White Industries)*

125°F (Figure 7-13). He takes temperature readings of the return and supply air within the occupied space and finds the **temperature differential** (TD) to be 19 degrees (Figure 7-14). (Return air = 88°F; Supply air = 69°F; 88 – 69 = 19 degrees). Pedro knows that the normally acceptable temperature differential across the evaporator is between 17 and 20 degrees.

SERVICE NOTE *Systems with higher SEER ratings will generally operate with lower condenser saturation temperatures, resulting in a lower temperature differential across the condenser coil. If systems with higher SEER ratings are charged to a 30-degree differential, they will most likely be overcharged. A 20-degree differential is more the order when dealing with more efficient systems.*

Pedro monitors the system for about 15 minutes while he completes his paperwork for this job. He calls his manager and inquires about the service charges for this call and is told that, since an improper repair last season caused the failure, the customer has no financial obligation. Pedro puts the customer on the phone with his manager so that the manager can explain the situation to the customer. While he waits, Pedro opens the return-air grill and inspects the air filters.

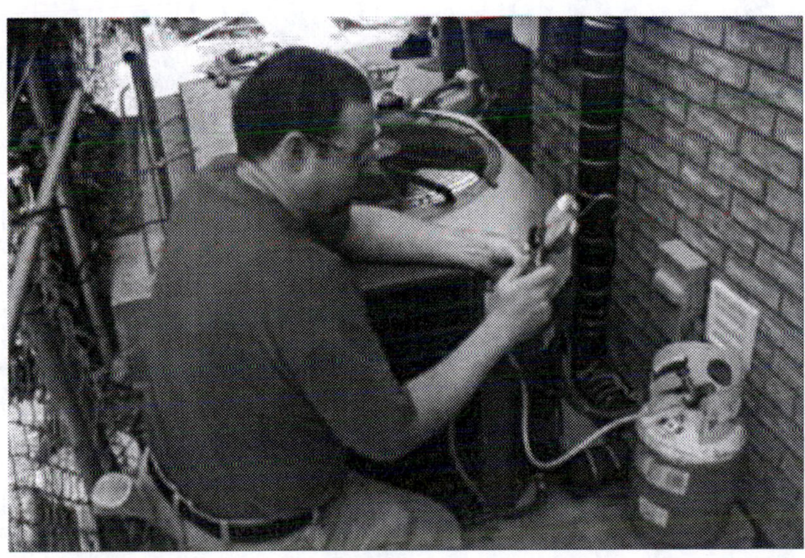

FIGURE 7-12 Removal of refrigerant from an overcharged system is best accomplished from the high side of the system. *(Photo by Maria Reyes)*

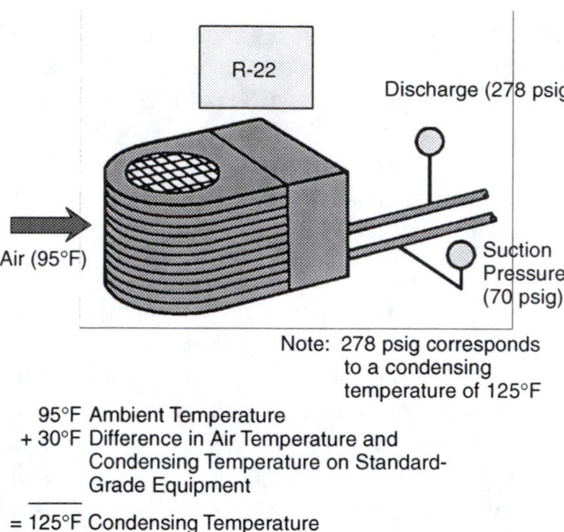

FIGURE 7-13 For standard-efficiency condensers, the condenser saturation temperature is roughly 30 degrees higher than the outside ambient.

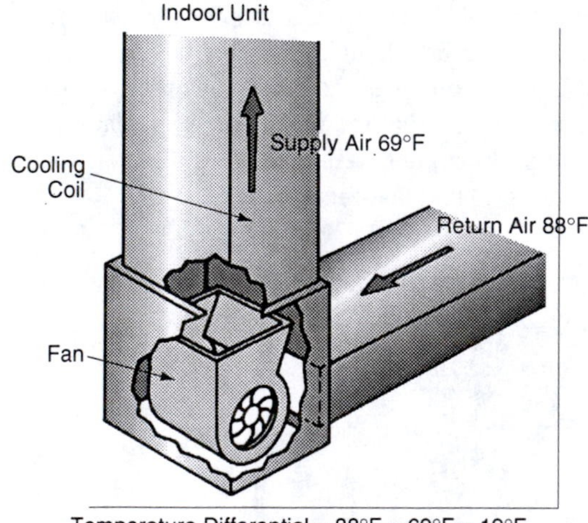

FIGURE 7-14 Measuring the temperature differential across the cooling coil

Service Call Discussion

This service call provides a perfect example of how to follow the seven-step procedure for successfully completing a service call. Pedro first listened to the customer complaint and then verified what he heard. By doing this, he was quickly able to eliminate a number of possible reasons for the system failure. When Pedro went outside, he first checked the power lines that fed the condensing unit to determine if the problem existed within the condensing unit or in the wiring leading to it. By establishing that both line and low voltage were supplied to the unit, he was quickly able to isolate the problem to the condensing unit. In reality, this initial process does not take very long at all, since the processes of gathering information and performing a visual inspection become second nature to the experienced field technician.

Pedro then used his voltmeter to determine the exact *problem* and was then able to determine the *cause* of the problem. Remember that the problem and the cause of the problem are not the same. This is extremely important because, if the cause of the problem is not remedied, it will surface again and again. Notice how Pedro did not simply reset the high-pressure control and leave the job. Instead, he determined *why* the switch opened. Using basic refrigeration theory, common sense, and the process of elimination, he was able to establish that the system was overcharged. Pedro performed his job in a professional manner and used proper techniques to remove the excess refrigerant from the system. If he had been unable to remove enough refrigerant from the system by simply bleeding it from the high side of the system, he would have had to use a recovery unit similar to those in Figure 7-15.

SERVICE NOTE: *Many acceptable methods exist for determining proper refrigerant charge, including weighing in the charge, using the manufacturer's charging chart, and using a charging calculator, which interprets system superheat and subcooling measurements. Weighing in the proper refrigerant charge is most reliable and effective when the system is of the package type with no field-installed refrigerant piping. Using the manufacturer's charging table and/or a charging calculator will help guarantee that the proper refrigerant charge is obtained.*

Another important aspect of this service call is the handling of the customer. Although everything seemed very calm, the situation could have become difficult. If Pedro, for example, had told the customer that one of his coworkers had overcharged the system, the customer possibly would not be very quick to accept the credibility of the company and its technicians in the future. More important, she would

(a)

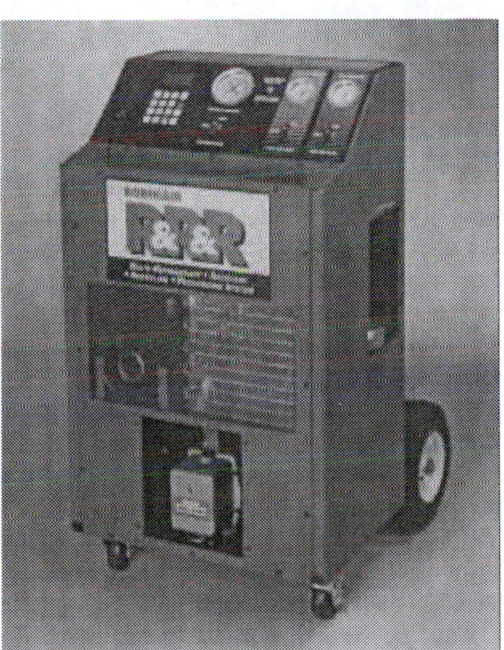

(b)

FIGURE 7-15 Examples of refrigerant-recovery systems. *(Courtesy Robinair Division, SPX Corporation)*

likely have demanded an explanation from Pedro of how this could have happened. Although the overcharging of the system might have been caused by unavoidable circumstances, Pedro could have found himself caught in an awkward situation trying to explain events unknown to him. Instead of placing himself in this situation, he correctly decided to put the customer on the phone with his service manager, who had much more experience in dealing with customers in situations like this and had access to the customer's entire service history file.

This sample service call illustrates that the service technician needs to be somewhat of a public relations person as well as technically knowledgeable. Remember that the service technician actually represents the company and must always display a professional, courteous, and intelligent attitude toward the work *and* customers.

SERVICE CALL 2: SYSTEM UNDERCHARGE

Customer Complaint

A residential customer calls his air-conditioning contractor and tells the dispatcher that, while out in his garden that day, he noticed that the condensing unit ran continuously. The unit did not shut down at all the entire time he was outside. He tells the dispatcher that the system is keeping the house at the desired temperature, but he is concerned that the system did not cycle off. He is also concerned that the system may not be able to keep the house cool as the weather gets warmer and is concerned about a possible increase in power consumption that can lead to higher utility bills.

Service Technician Evaluation

Linda is given the service call by her dispatcher, and she goes over to check out the system as soon as she finishes writing up the paperwork and explaining the charges to the customer at her previous service call. She arrives at the customer's house and finds the home owner in the garden. She approaches him and introduces herself, and he explains the problem to her. He tells her that, prior to placing the service call, he thought that the air filters might be dirty so he replaced them. The system still has not cycled off. Since the system seems to be working and Linda's shoes are already somewhat muddy from the damp grass, she decides to check the outdoor condensing unit first to reduce the possibility of tracking dirt into the customer's house. Linda goes to her truck to retrieve her tools and returns to the side of the house to begin to her evaluation of the system.

Linda can tell immediately from a visual inspection that both the compressor and the condenser fan motor are operating. She decides to install her gauge manifold on the system so that she can monitor the system pressures. After following the correct gauge-installation procedure, she is able to determine that the head pressure is 180 psig and the low-side pressure is 45 psig. Using a good-quality thermometer, Linda takes a reading of the outside temperature and determines that, at the present 80°F temperature, the head pressure for this R-22 system should be approximately 225 psig. She also recalls that the back pressure on an R-22 air-conditioning system ideally should be in the range of 70 psig. Since both the low-side and the high-side pressures are lower than she expected, she is able to determine that the system:

- Has restricted airflow through the evaporator coil or
- Is short of refrigerant

Reduced airflow through the evaporator coil can be the result of a number of situations, including:

- Dirty air filters
- Blocked return-air grill or ductwork
- Defective evaporator fan motor
- Broken or loose belt
- Blocked or closed-off supply registers or ductwork
- Frozen evaporator coil
- Loose or missing air-handler access panels
- Dirty or blocked evaporator coil

Since the customer just replaced the air filters prior to Linda's arrival at the job, she is able to eliminate that possibility relatively quickly. She now needs to enter the house and inspect the air handler for any of the aforementioned conditions. She carefully cleans her shoes before entering. Making her way to the air handler's location in the basement, she is able to hear the evaporator fan running and the movement of air through the ductwork. She turns off the power switch to the unit and removes the **service panel** (Figure 7-16) at the blower. These panels are provided to give the servicing technician access to the various system components. Noticing that the belt is intact and its tension is correct, these possibilities

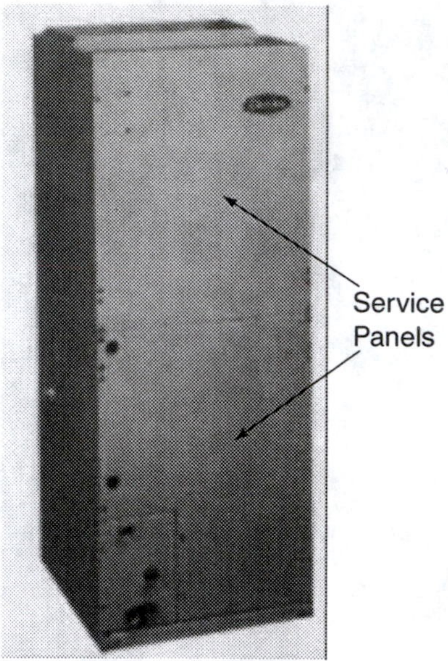

FIGURE 7-16 Service panels give technicians access to the system components. *(Courtesy Carrier Corporation)*

are eliminated as well. Confirming that all service panels are properly installed and that the **return plenum** (Figure 7-17), the section of ductwork connected directly to the return portion of the furnace or air handler, is clear, she is able to eliminate that possibility as well. She then removes the service-access panel at the evaporator coil and inspects it and the **supply plenum** (Figure 7-17). The supply plenum is the section of supply ductwork connected directly to the furnace or air handler. The coil and plenum are clean. She notices, however, that a small portion of the evaporator coil is frosted. The rest of the coil is warm. After replacing the service panels, she makes her way up the living area of the house and inspects the supply registers and finds them to be open. Upon turning the power switch to the furnace back on, she can feel air being discharged from the registers.

Having eliminated the causes for a reduced airflow through the evaporator coil and having noticed that a portion of the evaporator coil is frosted, Linda concludes that the system is short of refrigerant. Before adding refrigerant to the system, she uses her electronic leak detector (Figure 7-18) to look for a refrigerant leak. She turns the detector on and hears the slow beeping of the device. She begins moving the probe slowly around the soldered copper fittings

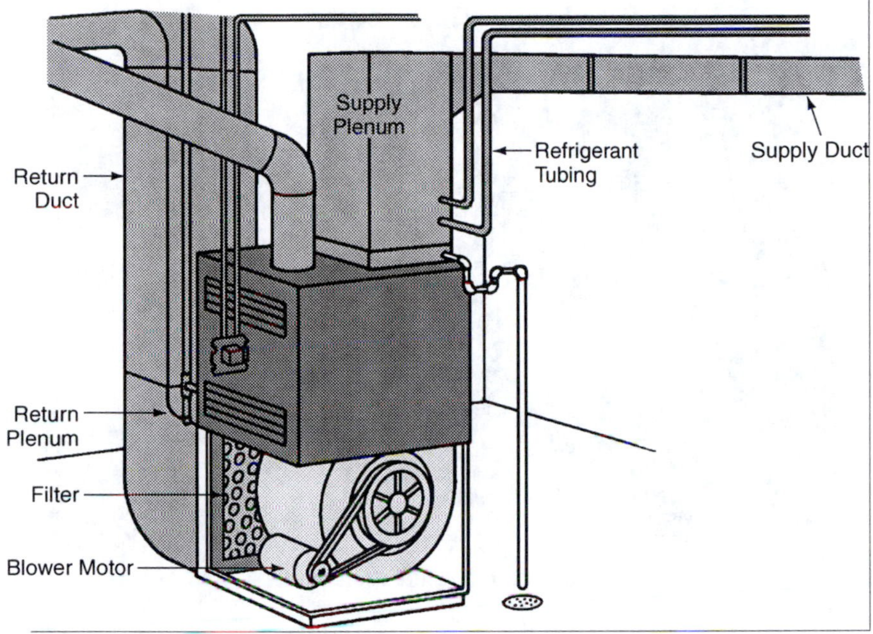

FIGURE 7-17 Plenums connected directly to the air handler

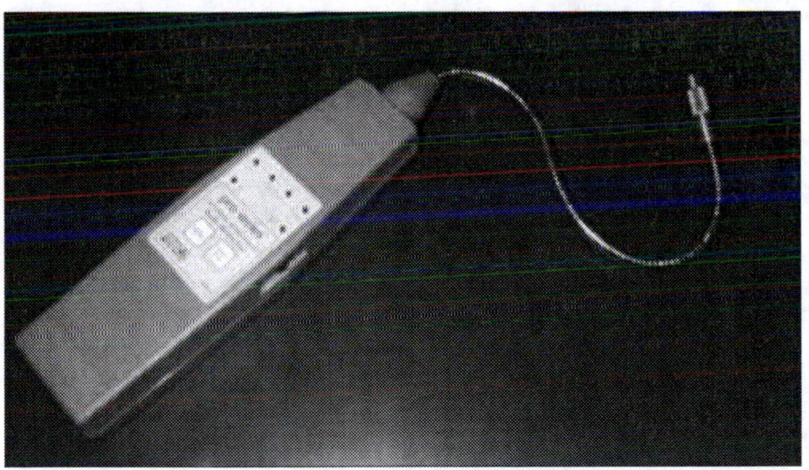

FIGURE 7-18 Electronic leak detector. *(Photo by Eugene Silberstein)*

and across the suction and liquid lines that carry refrigerant to and from the air handler (Figure 7-19). After checking the condensing unit and finding no leaks, she needs to go to the air handler to continue leak checking. After wiping her feet thoroughly, she enters the house and continues checking for leaks. She does not find any.

Knowing at this point that if a leak does exist, it is very small and will be difficult to find, Linda speaks with the home owner and explains that the system is short of refrigerant and that her attempt to locate the leak has failed. Linda tells the home owner that she will add refrigerant as needed but will also introduce an **ultraviolet** (UV) **solution** (Figure 7-20) to the

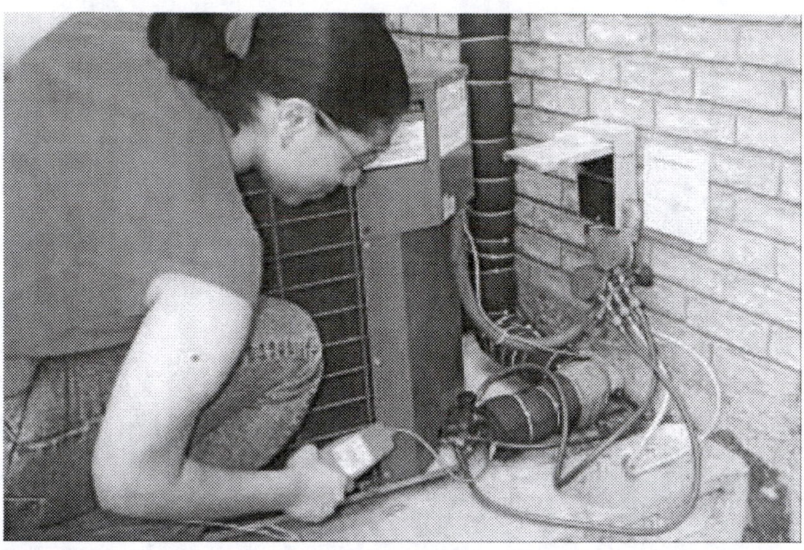

FIGURE 7-19 Using the electronic leak detector to locate refrigerant leaks. *(Photo by Eugene Silberstein)*

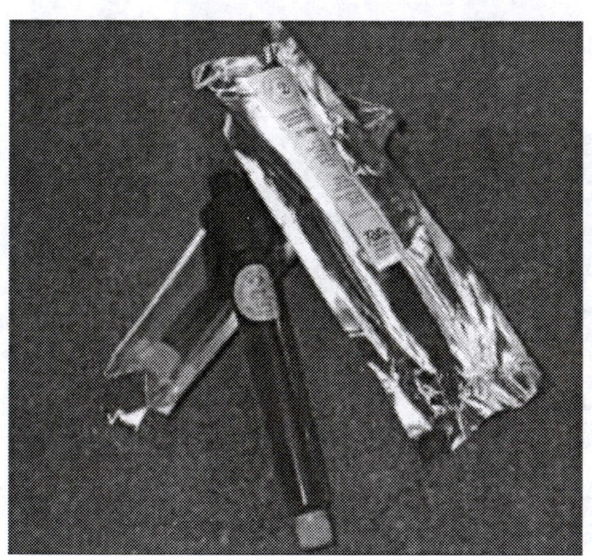

FIGURE 7-20 Ultraviolet solution. *(Photo by Eugene Silberstein)*

system. She explains that this solution will circulate through the system with the refrigerant and, if it escapes from the system, will stain the piping at the point of the leak. This stain will glow when seen under UV light (Figure 7-21). This will make the leak very easy to find, if, indeed, one exists. The customer agrees, and Linda prepares to add the solution and charge the system.

After determining the total system charge, Linda adds the correct amount of UV solution to the system. She makes certain that all of the ice has been removed from the evaporator coil and then begins to add refrigerant to the system. To ensure that no liquid refrigerant enters the compressor, she introduces vapor refrigerant into the suction service valve of the unit (Figure 7-22).

SERVICE NOTE: *When working with blended refrigerants, the refrigerant must leave the tank as a liquid in order to maintain the proper proportions of the various refrigerant components of the blend. Therefore, when charging or adding refrigerant to a system, care must be taken to ensure that the compressor is not flooded with liquid.*

Both the head pressure and the suction pressure begin to rise. Linda brings the head pressure up to approximately 220 psig and the suction pressure to about 68 psig. She tapes a thermometer probe to the suction line at the inlet of the condensing unit and reads a temperature of 55°F. Since the evaporator saturation temperature for the system is 40°F (pressure-temperature relationship), she calculates that the system superheat is 55 – 40 = 15 degrees (Figure 7-23), which is acceptable. She enters the house, again cleaning her feet, and checks the air

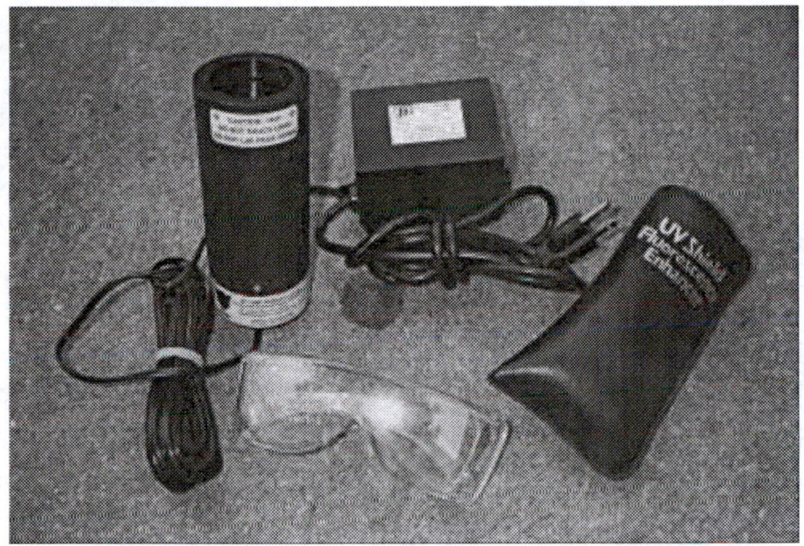

FIGURE 7-21 A UV light and special goggles make finding the leak much easier. *(Photo by Eugene Silberstein)*

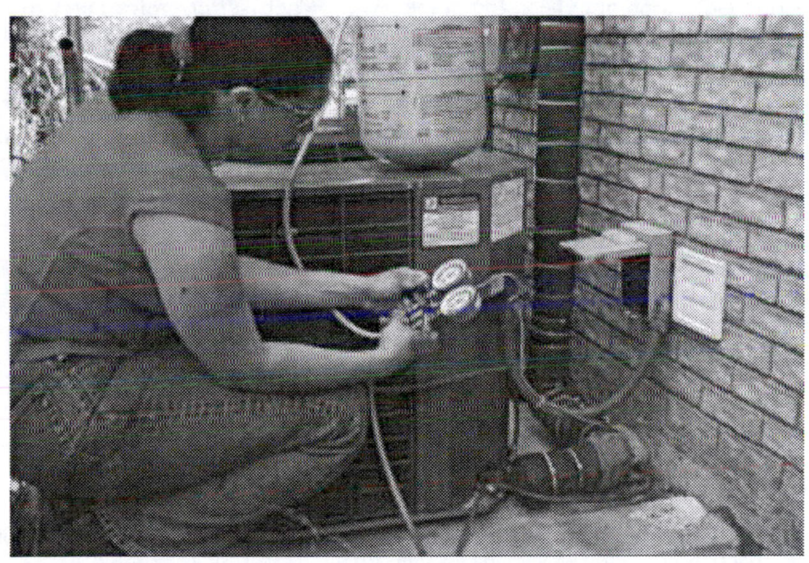

FIGURE 7-22 Adding R-22 to the air-conditioning system through the low-side service valve. *(Photo by Eugene Silberstein)*

handler. She has a temperature differential across the coil of 20 degrees. Again, this value is acceptable. After monitoring the system for approximately 15 minutes, she removes her gauge manifold from the system, packs up her tools, writes up her work order, gets her paperwork signed, and proceeds to her next service call.

Service Call Discussion

In this service call, Linda was very concerned and respectful of the customer's property. She was conscious of the fact that she would most likely track dirt into the house if she went inside first. She tried to minimize the number of times she needed to enter the

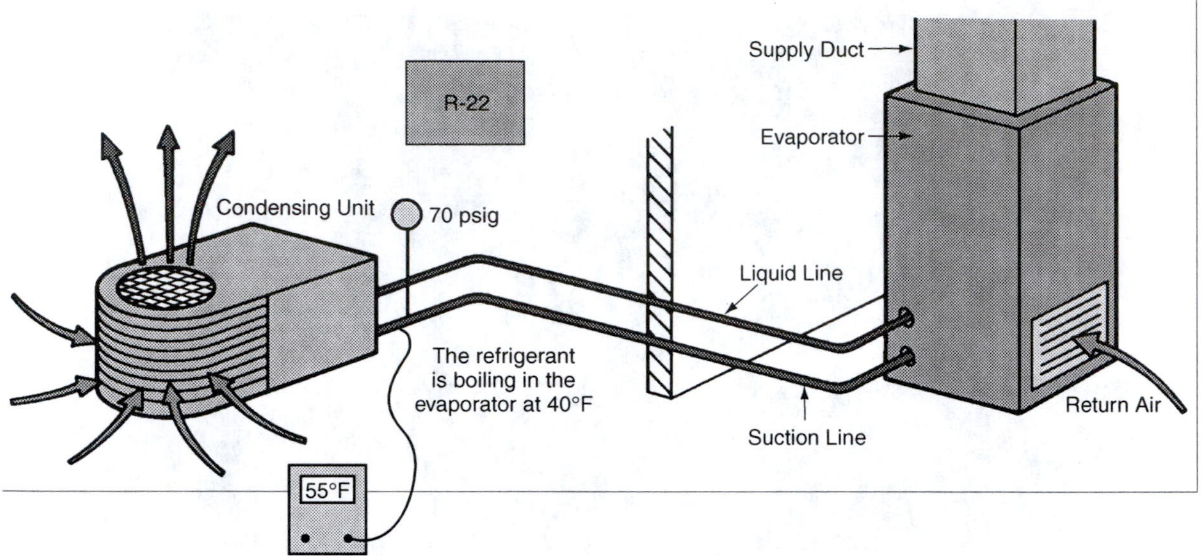

FIGURE 7-23 This refrigerant is picking up 15 degrees of superheat before returning to the compressor.

residence. She also made it a point to keep the home owner informed as to what was being done and what must be done in the future.

SERVICE NOTE: *Some customers may not be interested in the play-by-play explanation of the repair but, instead, may say something to the effect of, "Just let me know when it's fixed." The technician much evaluate just how much involvement the customer desires, and adjust accordingly.*

As far as the technical aspect of this service call is concerned, Linda was extremely conscious of the fact that refrigerant does not simply disappear. If the system is short of refrigerant, it *must* have leaked out. Government regulations require that system leaks be found and repaired. Although she could not locate and repair the leak, she took the proper steps by adding the UV solution. This way, if the system loses refrigerant again, the leak will be found relatively easily. This shows compliance with government policies as well as respect for the customer's wallet. Linda could have easily spent hours and hours, which are billable, looking for a leak that may not even exist. What if, when the unit was last serviced, a service cap was loose, allowing a portion of the refrigerant charge to escape from the system? Linda spent a reasonable amount of time looking for the leak but did not overdo it.

When adding refrigerant to the system, she used outside ambient temperature, head pressure, back pressure, evaporator temperature differential, and system superheat to help evaluate the system charge. Using only one of these may speed the process of system charging but can easily lead to an improper amount of refrigerant being added. Remember that ideal system and design conditions are just that—ideal—and do not necessarily apply to all systems and installations.

SERVICE CALL 3: BLOWN LINE-VOLTAGE FUSE

Customer Complaint

A customer calls with a no-cooling complaint. It is the middle of the summer, and the customer tells the service manager that the system has worked fine all season with no problems. He had a preventive maintenance service call at the beginning of the spring, at which time the refrigerant charge, air filters, and overall system operation were checked.

Service Technician Evaluation

John arrives at the customer's home and asks the home owner about the problem that he is experiencing with his air-conditioning system. After listening

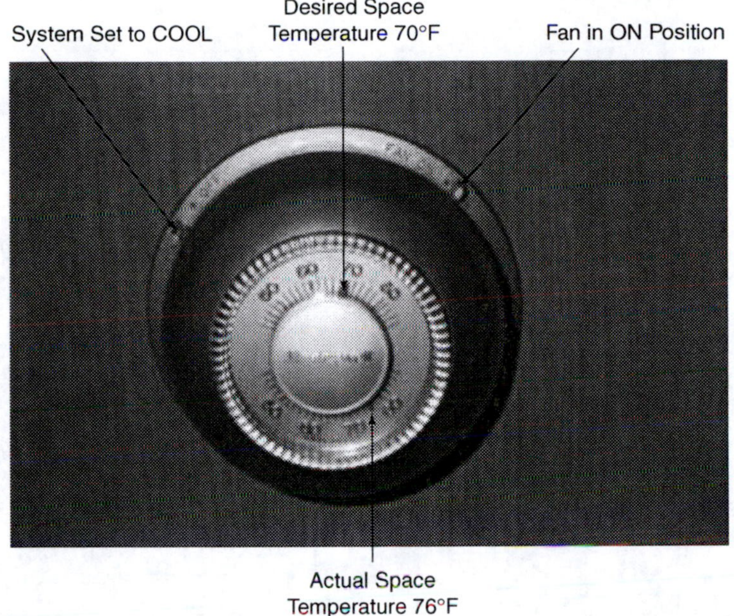

FIGURE 7-24 Low-voltage thermostat calling for cooling. *(Photo by Eugene Silberstein)*

attentively, John goes over to the room thermostat and checks the setting. It is set to the cooling mode, and the thermostat is set to maintain a temperature that is lower than the present space temperature (Figure 7-24). The top arrow on the face of the thermostat, which represents the desired space temperature, is set at 70°F, while the bottom arrow shows the actual space temperature. The fan switch is in the ON position. He then goes over to a wall supply grill and feels the air coming from it. The air is warm, but he can tell that the fan is operating. From the information he was given and his visual inspection, John has already made some conclusions:

- The transformer is functioning.
- The low-voltage control circuit is being energized.
- The holding coil on the indoor fan relay is energized.
- The indoor fan is operating.
- The problem most likely is in the condensing unit or the wiring supplying power to it.

John leaves the thermostat set to the cooling mode and sets the desired temperature to the lowest setting. He then goes outside to the condensing unit and finds that it is not operating. After putting on his safety glasses, he turns off the switch on the fused disconnect at the condensing unit and opens the cover. He then carefully turns the disconnect switch back on. Using his voltmeter set to the proper range, he checks for voltage at the load side of the fuses in the box. He reads 0 volts (Figure 7-25). Since this system is rated at 220 volts, John has established that power is not being supplied to the condensing unit.

To check further, he knows that each hot leg of a 220-volt power supply will give a reading of approximately 115 volts to ground. He then takes one of the leads from his voltmeter and holds it to the ground terminal in the box. He takes his other test lead and touches it to the load side of the fuse on L1. He reads 113 volts (Figure 7-26). He then does the same for L2. This time, he reads 0 volts (*Meter A* in Figure 7-27). He now takes his test lead and touches the line side of the same fuse. He reads 113 volts (*Meter B* in Figure 7-27). In a schematic diagram (Figure 7-28), John's findings can be easily visualized.

John has now established that the fuse on L2 is bad. He turns the disconnect switch off and removes the defective fuse. To make himself perfectly sure, he takes his multimeter and sets it to read resistance. He sets it to the lowest resistance range and checks across the fuse. There is no **continuity** through the fuse; so, again, it is proven that the fuse is defective

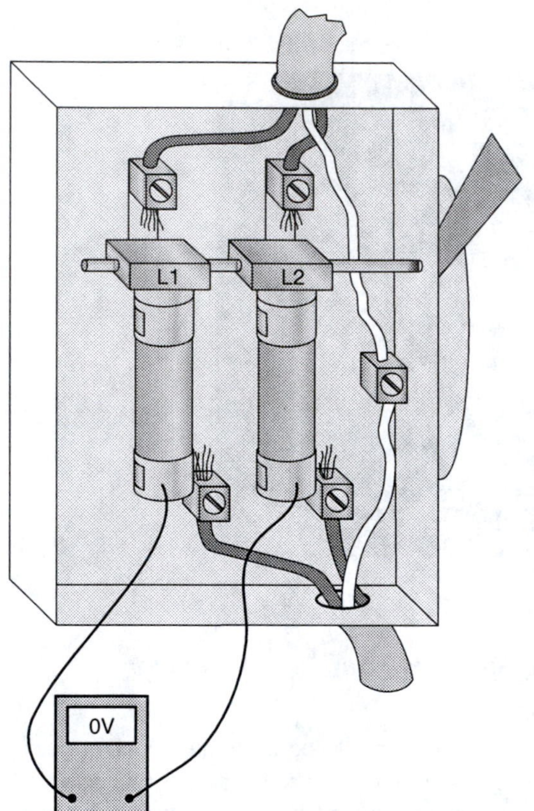

FIGURE 7-25 Power is not being supplied to the condensing unit.

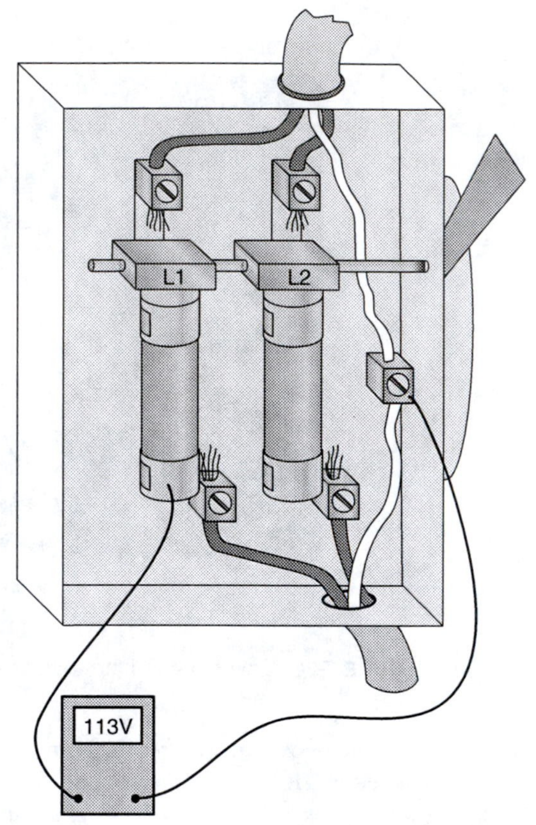

FIGURE 7-26 The fuse on the L1 side is good since power is able to pass through it.

(Figure 7-29). *Continuity* means that there is an electrical connection between two points, so *no continuity* simply means that there is no connection.

John must now ask himself the all-important question, Why did the fuse blow? John knows that the fuse did not blow for no reason, so he *must* determine the cause *before* replacing it. If he simply installs a new fuse, it will undoubtedly blow again. He now begins his systematic search for the source of the problem.

With the disconnect switch in the OFF position, John uses his multimeter, set to read resistance, to check for continuity between the power lines feeding the condensing unit and ground. When reading from the load side of the fuse on L2 to ground, he reads infinity (*Meter A* in Figure 7-30). When reading from the load side of the fuse on L1, his meter reads 0 ohms (*Meter B* in Figure 7-30), meaning there is a short to ground. He now knows that a short exists, but he still must locate it. He knows that the short is located within the condensing unit, so he evaluates each of the following possibilities. The short circuit is located:

- In the compressor
- In the wiring leading to the compressor
- In the condenser fan motor
- In the wiring leading to the condenser fan motor, or
- In the wiring from the disconnect box to the line side of the cooling contactor

By systematically checking each of the preceding scenarios, John should be able to locate the problem efficiently and accurately. He first locates any capacitors in the condensing unit and properly discharges them to avoid getting an electrical shock.

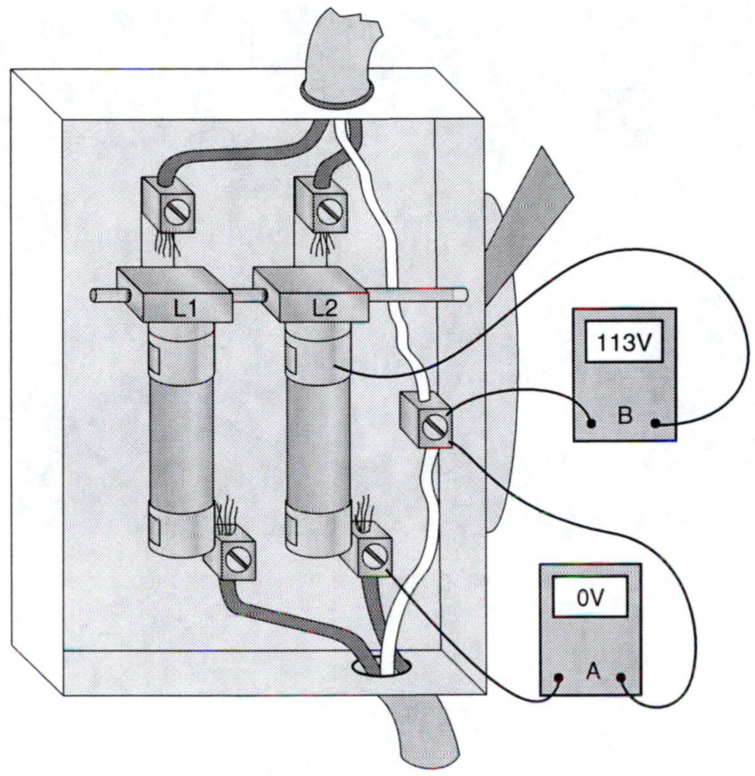

FIGURE 7-27 No power is passing through the fuse on the L2 side of the switch. This fuse is bad.

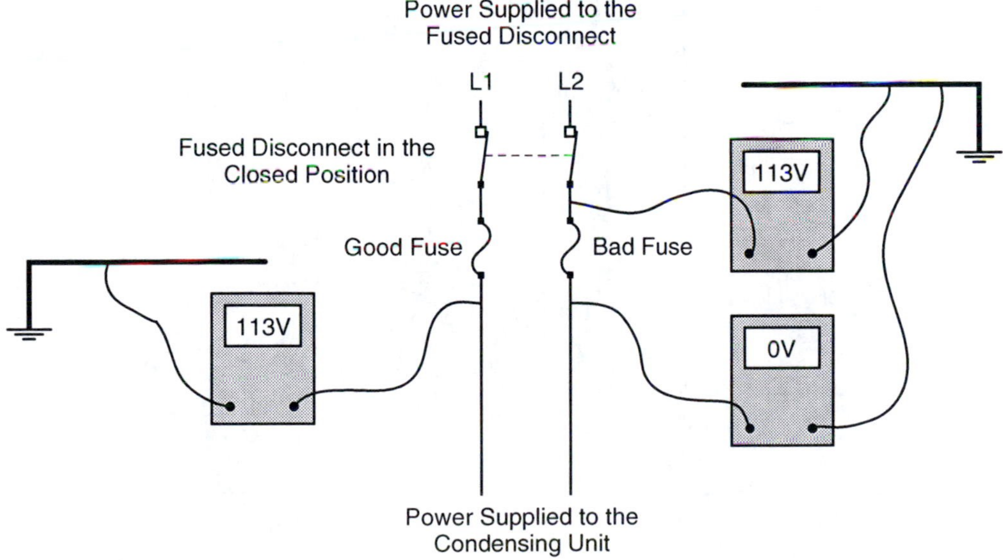

FIGURE 7-28 Schematic diagram showing the bad fuse

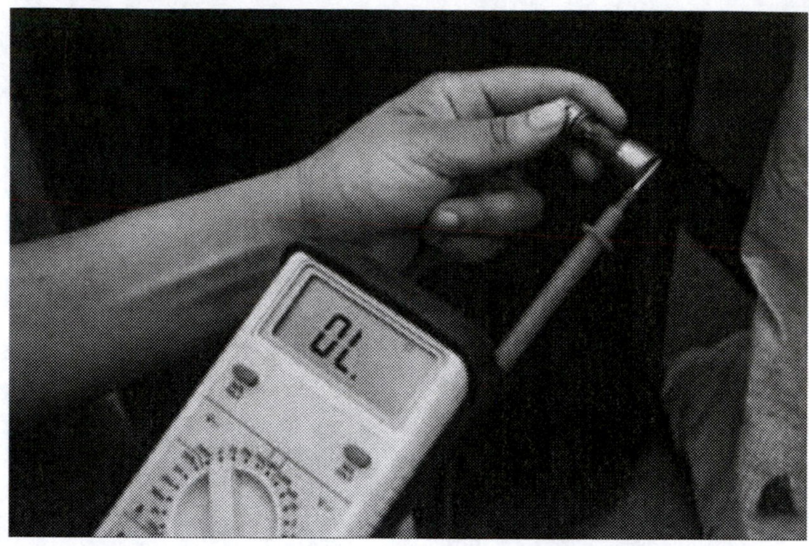

FIGURE 7-29 Checking a fuse using an ohmmeter. The OL display indicates that no continuity exists through the fuse. It is bad. *(Photo by Eugene Silberstein)*

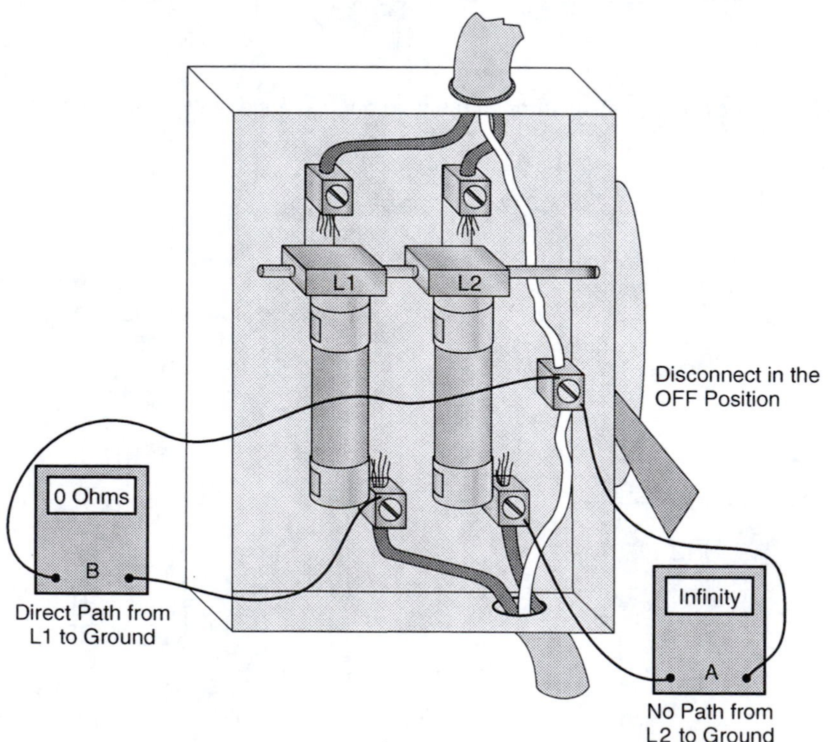

FIGURE 7-30 The load side of the disconnect switch has a short to ground on the L1 side.

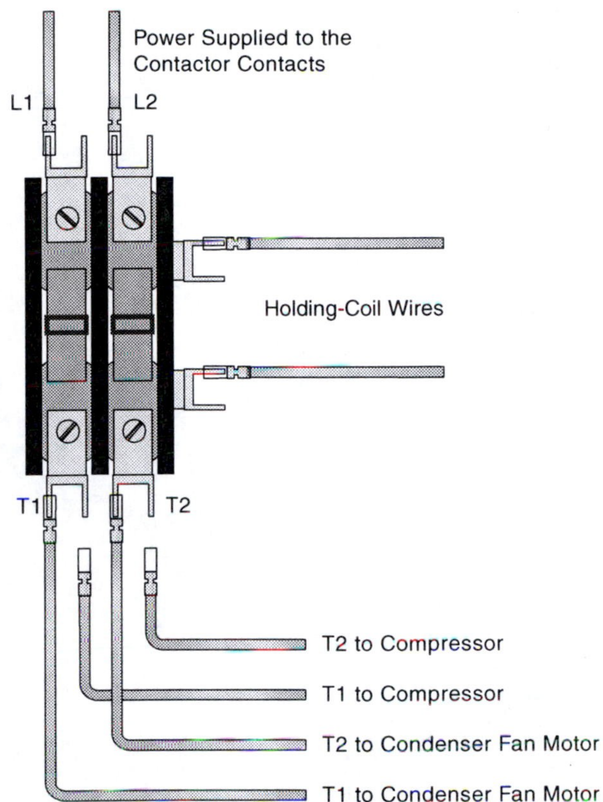

FIGURE 7-31 Removing the compressor wires from the load side of the contactor

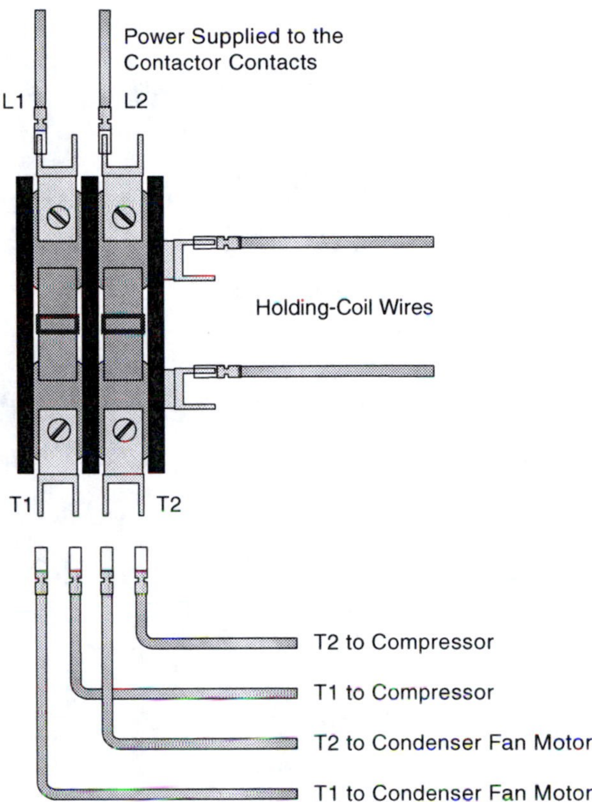

FIGURE 7-32 Removing the condenser fan motor wires from the load side of the contactor

---

SERVICE NOTE: *Even though the disconnect switch is in the OFF position, it is still possible to receive an electrical shock from the system. Remember that capacitors are energy-storing devices and can discharge even when the system is off. Always be sure to properly discharge these devices before equipment service begins.*

John locates the wires that supply power to the compressor. He labels the wires so he knows where they came from and disconnects them from the load side of the cooling contactor (Figure 7-31). He then again checks for continuity at the disconnect box. He still reads 0 ohms from L1 to ground. He can then eliminate the possibility that the compressor or the associated compressor wiring is at fault. He then locates the wires from the condenser fan motor, labels them, and disconnects them from the load side of the contactor as well (Figure 7-32). He still reads 0 ohms from L1 to ground. John is now able to establish that the problem lies in the wiring from the disconnect box to the line side of the contactor.

John now disconnects the two wires from the line side of the contactor and checks each for continuity to ground. He finds that the wire from L1 is making contact with ground (Figure 7-33). Upon closer inspection, he is able to see that the wire has rubbed against a sharp metal edge in the condensing unit. He goes to his truck and brings back a new length of wire and a new fuse and replaces the wire and the fuse. He reconnects all the wires from the condenser fan motor and compressor. As a final check, after all connections are made, John again checks for continuity from the lines feeding the condensing unit to ground and gets a reading of **infinite resistance**. This means that there is no electrical connection between the points.

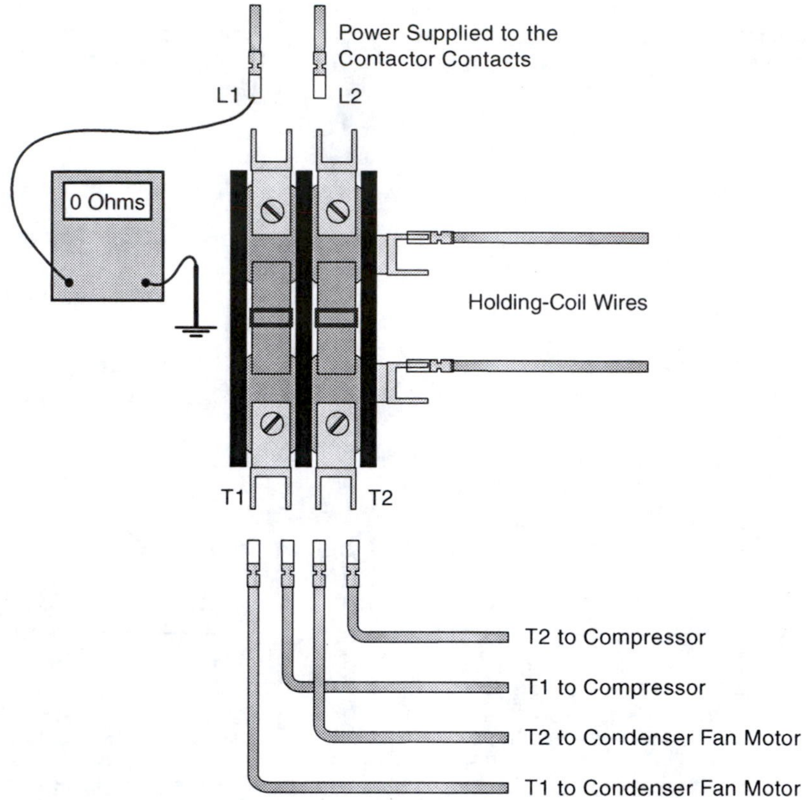

FIGURE 7-33 A short to ground exists within the conductor feeding the contactor.

John then closes the cover on the disconnect and turns the switch on. The condensing unit begins to operate. He touches the suction line entering the unit and finds it to be very cool. He feels the liquid line leaving the unit and finds it to be warm. He monitors the system for about 15 minutes while he completes his service ticket and explains the repair to the customer.

Service Call Discussion

Just as in the previous service calls, John listened to the customer complaint and, by simply feeling the air coming from the supply registers, was able to establish that the problem was most likely in the condensing unit or the associated wiring leading to it. He was able to establish this within the first 5 minutes of the service call. By applying basic circuit theory, he was able to quickly determine that the fuse was blown. John did not, however, simply replace the fuse. He first determined the cause. Also, notice how the service technician used two different methods to establish that the fuse was defective:

- He used a voltmeter to read the voltage across the fuse while it was in the circuit, just like reading voltage across a switch. A good fuse acts as a closed switch, while a defective fuse acts as an open switch.
- He also used an ohmmeter once the fuse was removed from the circuit. He knew that if no continuity existed through the fuse, the fuse was bad. Similarly, if continuity existed through the fuse, the fuse was good.

After he determined that there was a short to ground, he proceeded in a logical manner to isolate the location of the short. He did not simply check around haphazardly, hoping that the problem would surface on its own. By using his test instruments correctly and applying common sense, he was quickly able to isolate the trouble and, ultimately, resolve it.

SERVICE CALL 4: DEFECTIVE TRANSFORMER

Customer Complaint

A clothing store manager places a service call to the air-conditioning service company. She tells the dispatcher that the unit is not cooling and the paper ribbons hanging from the ceiling supply registers are not moving. She explains to the dispatcher that she turned the circuit breaker for the system off and then back on, but the system still did not come on. This unit is a package-type system (Figure 7-34) and is located in the alleyway behind the store. Package units are self-contained and do not require any field piping during installation. The clothing store manager is desperate and tells the dispatcher that her customers refuse to stay in the store because it is so hot. The dispatcher informs her that William, the company's lead man, is presently working on a construction job nearby and that he will pull him off the job to send him over right away.

Service Technician Evaluation

William arrives at the clothing store within 20 minutes. He immediately asks for a description of the problem. After listening to the details, he goes to the thermostat and sees that it is set to operate in the cooling mode and is set to maintain a temperature below the present temperature of the store. William then goes out back to the unit.

Since the unit gives the appearance of being off, William's first task is to check for line voltage. He removes a multimeter from his toolbox and puts his safety glasses on. He flips the service disconnect switch to the OFF position and opens the cover. He checks for voltage and finds 223 volts being supplied to the disconnect (between points *A* and *B* in Figure 7-35). He closes the disconnect box and turns the switch back on. William's next step is to open the service panel on the unit. Once the cover is removed, William is able to check the voltage on the line side of the contactor. Again, he reads 223 volts (*Meter A* in Figure 7-36). He then checks the voltage on the load side of the contactor and reads 0 volts, which indicates to him that the contactor's holding coil is not pulling the contacts closed (*Meter B* in Figure 7-36). He then checks the voltage being supplied to the holding coil and finds that the voltage at the coil is 0 volts (*Meter C* in Figure 7-36).

William now needs to establish that a low-voltage control circuit exists. He checks the voltage at the primary winding of the transformer and reads 223 volts. He then checks the voltage at the secondary winding and reads 0 volts. Since 223 volts are being applied to the primary and no voltage is present at the secondary,

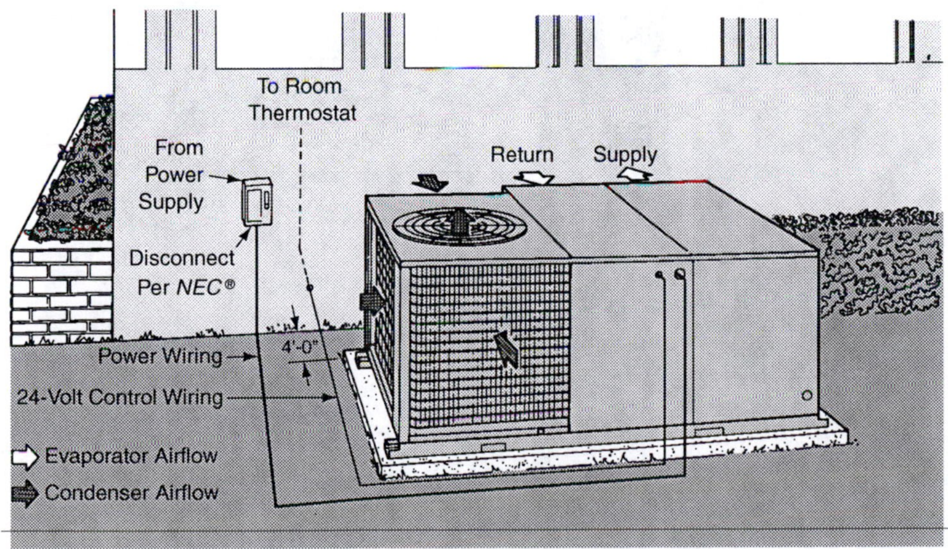

FIGURE 7-34 A typical package unit. This unit is self-contained. *(Reproduced courtesy of Carrier Corporation)*

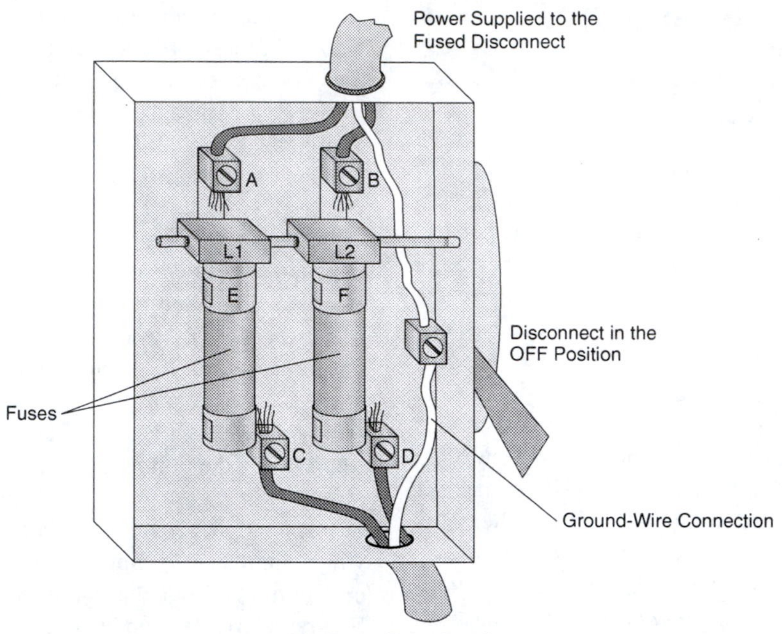

FIGURE 7-35 Even in the OFF position, a potential difference should still exist between points A and B.

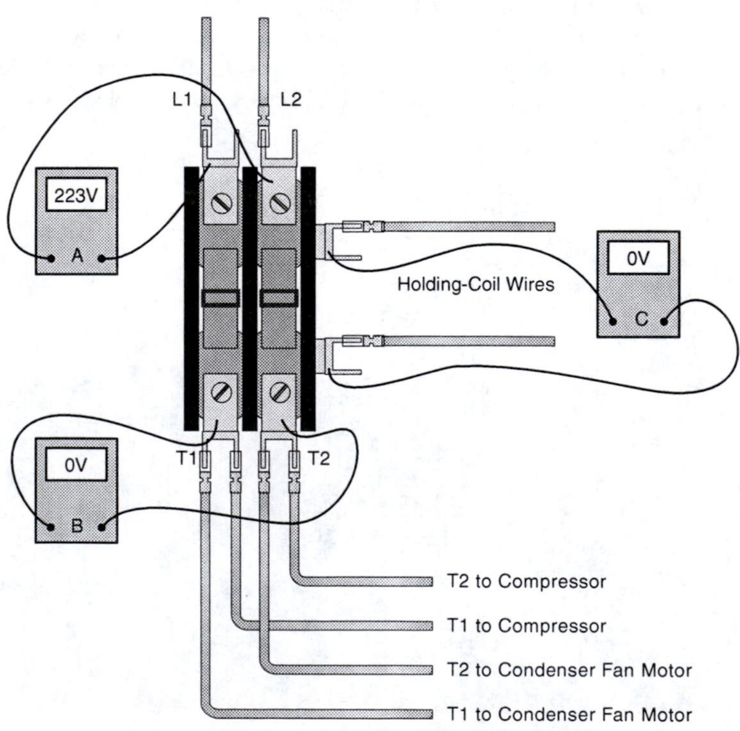

FIGURE 7-36 This contactor's contacts are open. No voltage is being supplied to the holding coil.

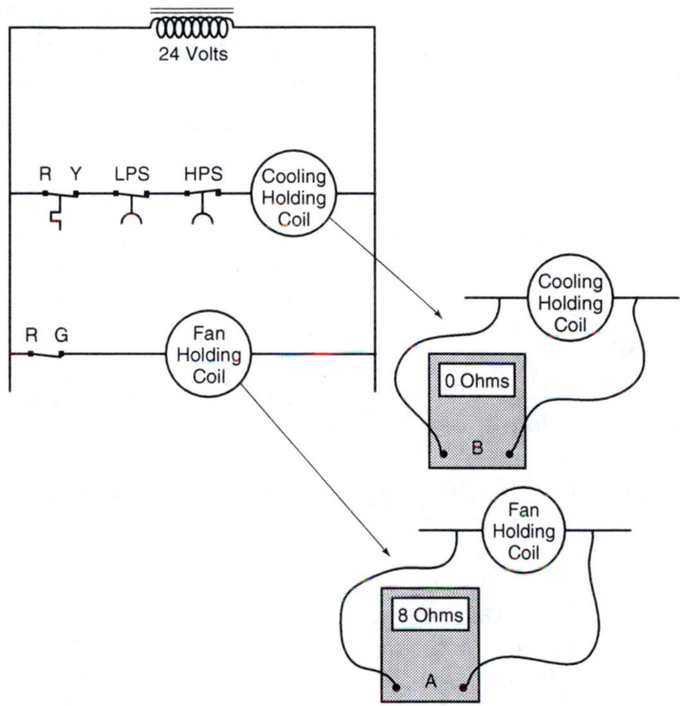

FIGURE 7-37 When removed from the circuit, the resistance of the individual holding coils can be checked. The fan relay coil is okay. The cooling-circuit holding coil is shorted and needs to be replaced.

William correctly concludes that the transformer is defective. He needs to know, though, which winding—the primary or secondary—is bad. He then shuts off the power to the unit, discharges the system capacitors, and removes the transformer from the unit.

With a multimeter set to read ohms, William finds no continuity through the secondary winding, indicating that this winding has burned. This indicates that there was, and probably still is, a short in the low-voltage control circuit. Before replacing the transformer, he takes a resistance reading of the control circuit and finds a resistance of 0 ohms, which indicates a direct short.

Since William notices that there are only two holding coils in the control circuit, he decides to check each one separately. By isolating the holding coil in the evaporator fan motor control circuit (*Meter A* in Figure 7-37), William establishes that the coil has a resistance of 8 ohms. He then replaces the wires on the coil. Doing the same for the cooling contactor's holding coil, he obtains a resistance reading of 0 ohms (*Meter B* in Figure 7-37). This coil is shorted, and the contactor needs to be replaced.

William goes to his truck and retrieves a replacement contactor, a transformer, and an in-line fuse for the low-voltage control circuit. He then tags all of the wires on the old contactor and removes it from the unit. He mounts the new contactor, after double-checking the amperage rating of the contacts and the voltage rating of the holding coil, and reconnects the wires. He then proceeds to remove the old transformer and installs the new one, being careful to cap any unused wires on the new transformer. To prevent the transformer from blowing in the future, William installs a 3-ampere fuse in the control circuit. If a short circuit develops in the future, the fuse will blow instead of the transformer. This will save the customer money on parts.

Before energizing the circuit, William checks the resistance of the control circuit to verify that the short circuit has been removed and the problem solved. After double-checking his wiring connections, he proceeds to energize the system.

The compressor, the condenser fan, and the evaporator fan all come on and the system begins to cool. William allows the system to operate while he checks the air filters and takes a temperature differential reading across the evaporator. After determining that the filters are clean and the temperature differential across the evaporator is 18 degrees, he completes his work order, making certain that he includes all parts used on the service call. He completes his cleanup, packs his tools, explains the repair to the store manager, gets his work order signed, and proceeds to his next service call.

Service Call Discussion

In this service call, the system appeared to be in the OFF position because there was no low-voltage control circuit. After verifying the complaint, William first checked for line voltage at the unit because the obvious first possible solution to the problem would be the absence of supply voltage. Once he determined that line voltage was being supplied to the unit, the next logical step was to check for low voltage; but, William first checked to make certain that voltage was present on the line side of the cooling contactor. This is not incorrect, and it is meant to bring light to the fact that each and every technician has his or her own mind and, therefore, a different way of troubleshooting and evaluating a system. No two technicians do exactly the same thing at the same time, and they all must do whatever works for them personally.

Once William was able to establish that the transformer was defective, he determined why the transformer went bad. He realized that components do not fail without cause. Since the circuit had only two low-voltage holding coils, William was able to determine which coil was shorted out. If neither coil were defective, he would have had to check the interconnecting low-voltage wiring to locate the short circuit.

The procedure William used to replace the contactor should serve as a model to the new field technician. Cooling contactors, in general, have at least eight wires connected to them. When replacing a contactor, good field practice includes labeling each wire so that each can be replaced in the proper position. Many technicians even draw simple wiring diagrams showing the contactor terminals and the wires that should be connected to each. *Do not* rely on memory! Even a small, unexpected distraction is enough to cause a technician to forget exactly which wire goes where. If this should occur, the time needed to complete the repair will obviously be increased.

SERVICE CALL 5: DEFECTIVE CONTACTOR HOLDING COIL

Customer Complaint

A residential customer has just returned from work and has found that her home is warm. She has a digital programmable thermostat that normally turns the air-conditioning system on about an hour before she arrives home. She looks at the thermostat and sees that it is set to the cooling mode. She then feels the supply air and finds it to be warm. She immediately places a call to her service company, even though it is late at night, because she is throwing a party the next day.

Service Technician Evaluation

Anthony arrives at the customer's home and listens attentively as the home owner vents her frustration and explains the situation with the system. To confirm her statements, Anthony goes to the thermostat and checks the settings. He then feels the supply air, which is, indeed, warm. He is able to make the following assertions:

- The transformer is providing low voltage.
- The evaporator fan is operating.
- The system is calling for cooling.

Anthony then decides to check the condensing unit, which is located in the back yard. Upon visual inspection, he can tell that the condensing unit is not operating. Making certain his safety glasses are on, he opens the service panel of the unit. With his voltmeter, he checks for voltage at the line side of the cooling contactor and finds 218 volts (*Meter A* in Figure 7-38). This system has a 220-volt rating, so this is fine. Upon checking the voltage on the load side of the cooling contactor, Anthony read 0 volts (*Meter B* in Figure 7-38). From this, he is able to establish that the contactor contacts are not closed. Changing the range on his voltmeter, he then checks the voltage being supplied to the contactor coil. He reads 24 volts (*Meter C* in Figure 7-38).

For all intents and purposes, the contactor contacts should be closed. The coil is receiving voltage, but the contacts are not closed. Anthony immediately suspects a defective contactor coil. After turning the service disconnect switch off and discharging the capacitors, Anthony removes the wires from the contactor coil

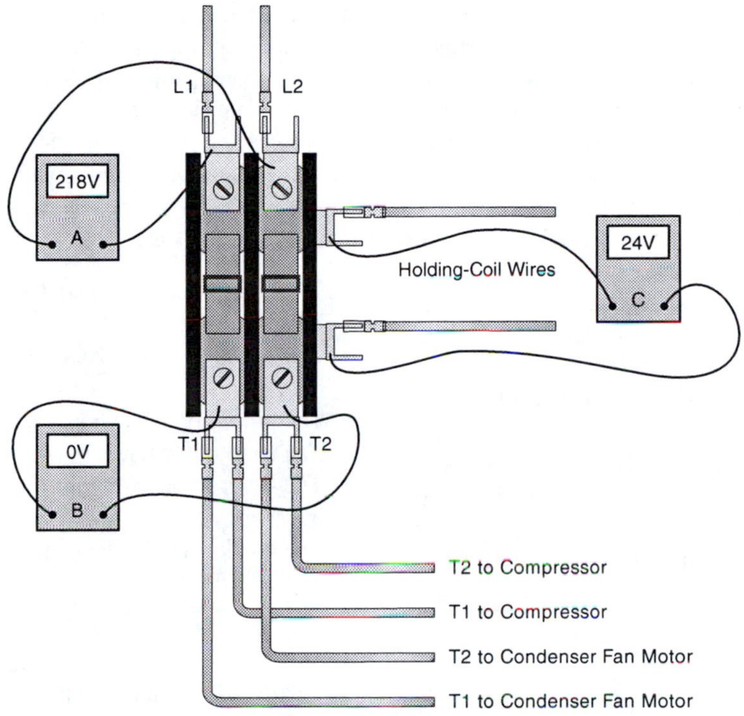

FIGURE 7-38 This contactor's contacts are open, and voltage is being supplied to the holding coil. A defective coil is immediately suspected.

and, with his multimeter set to read resistance, takes a reading of the contactor coil (Figure 7-39). The meter reads infinity, indicating that no continuity is present across the coil. The contactor needs to be replaced.

Anthony labels all of the wires on the old contactor and removes it from the unit. He checks the amperage rating and the coil voltage rating on the new contactor before installing the device. He mounts the new contactor securely and makes all necessary electrical connections. He then turns the service disconnect switch on, and the system begins to operate.

While monitoring the system, Anthony closes all service panels, packs his tools, completes his paperwork, explains the repair to the customer, and gets signed off the job.

Service Call Discussion

In this service call, Anthony was able to confirm the customer complaint and verify that the transformer was good and that the thermostat was calling for cooling. He knew that, based on the settings on the

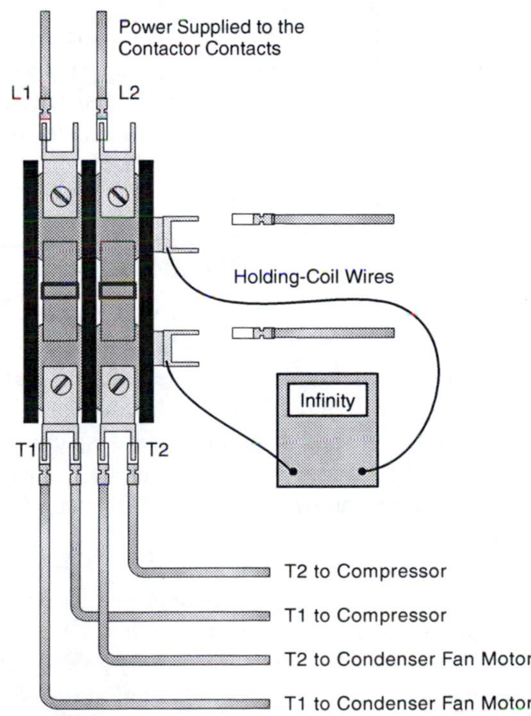

FIGURE 7-39 There is a break in the holding coil. It must be replaced.

thermostat, the unit should be operating in the cooling mode. He quickly went out to the condensing unit and saw that it was not operating. Anthony was able to isolate the problem by checking for and verifying the presence of both line and low voltage at the condensing unit. He then checked for line voltage at the line side of the cooling contactor and for low voltage at the contactor coil. Based on this gathered information, the contactor contacts should have been closed but they obviously were not. By checking the coil, he was able to conclude that the coil was defective.

Note that Anthony made it a point to check both the amperage rating and the coil voltage of the new contactor before installing the device. This is vital to ensure that the new component will function properly in the system. If the incorrect part is inadvertently installed, future failure will most likely result.

SERVICE CALL 6: BURNED MOTOR WINDING

Customer Complaint

A home owner who is concerned about her excessively high electric bill places a service call. Although the air-conditioning season has just started, she has noticed that her electric charges are higher than they normally have been for peak summer operation. She requests that a service technician come to her house to check out the system.

Service Technician Evaluation

Later that afternoon, Diane arrives at the home. While discussing the system problem, she asks the home owner whether or not the house is able to maintain a comfortable temperature. She is told that when the temperature outside is cool the house is comfortable but a little humid. As the temperature outside begins to rise, the house becomes a little uncomfortable but not unbearable.

Diane goes over to the thermostat and makes certain that it is set properly for cooling. She then feels the supply air, and it is not very cool but not excessively warm either. From this, she is only able to conclude that the indoor fan motor is operating. She needs to inspect the condensing unit.

As she walks around to the side of the house, she can hear a compressor running but it quickly shuts down. After 2 minutes or so, the compressor turns back on but the condenser fan motor does not. After running for about a minute, the compressor shuts down again. Diane immediately turns off the service disconnect switch. She is now able to evaluate the situation. The facts are:

- The customer is concerned about high electric bills.
- The system is not properly maintaining the house temperature.
- The humidity in the house is higher than normal.
- The evaporator fan is operating.
- The compressor is cycling on and off at 2-minute intervals.
- The condenser fan motor is not operating.

Diane is able to piece together the puzzle. She knows that the evaporator performs two functions. It cools, and it also removes humidity from the air. Since it is warm and humid inside the house, Diane is able to conclude that the evaporator is not functioning effectively. She then notices the condenser fan not operating and the compressor cycling on and off. This explains the high electric bills.

With the service disconnect switch in the OFF position, Diane opens the service panel on the condensing unit. Because the compressor is operating, she concludes the following:

- Line voltage is being supplied to the condensing unit.
- Low voltage is being supplied to the cooling contactor's holding coil, and the contacts are closed.
- The problem lies within the condenser fan motor or the wiring leading to it.

Diane discharges the capacitors and disconnects the condenser fan motor wires from the load side of the contactor. She also disconnects the wires from the motor's run capacitor, making certain to label the wires for future identification. With her multimeter set to read resistance, Diane checks the three condenser fan motor wires for continuity, two at a time. She obtains the following readings:

Common to run	8 ohms
Run to start	Infinity
Common to start	Infinity

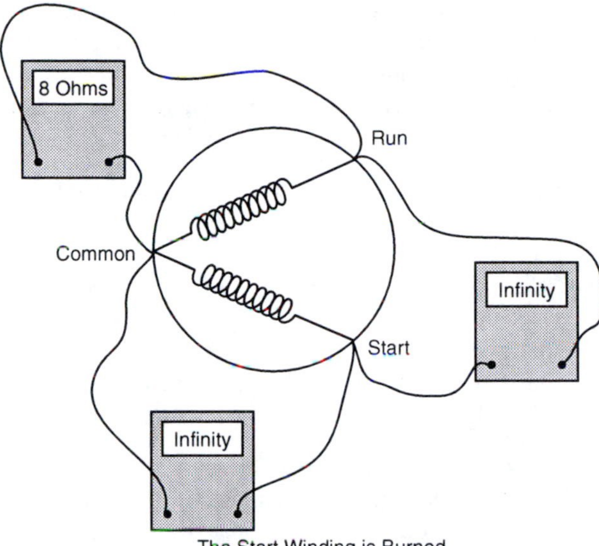

FIGURE 7-40 This motor has an open start winding. Note that infinite resistance exists across that winding.

Drawing a quick sketch (Figure 7-40), she notices that the start winding on the condenser fan motor is burned.

Since Diane does not have a replacement motor with her, she needs to have a new part ordered. From her experience, though, she knows that removing the condenser fan blade from the motor shaft without damaging it—especially on older units—is difficult, so she decides to try to remove the blade now. This way she will know if a new blade is also needed. (If she waits until the new motor is installed, she risks having to return to the home again to replace the blade. This would ultimately result in customer dissatisfaction. In addition to this, her boss would not be very happy either.) Fortunately, she is able to remove the blade without damaging it. She then sets the blade gently inside the condensing unit, takes all data tag information from the motor, and writes it on her work ticket.

To ensure that the new motor will be installed correctly, Diane replaces all of the wires in their proper places and leaves the old motor mounted. She leaves the service disconnect switch in the OFF position.

Since the repair cannot be completed today, Diane informs the customer what additional work needs to be done. She also informs the customer that the system fan can be used in the meantime to circulate air through the house. Diane calls her service manager with the news and gives him the motor information. Diane informs the customer that her service manager will call her back as soon as he knows how much the motor will cost to install and exactly when the job can be done.

Diane cleans up, packs her tools, completes her paperwork, gets signed off the job, and proceeds to her next service call.

Service Call Discussion

By seeing the compressor operating, although intermittently, Diane was able to quickly conclude that the problem was with either the condenser fan motor or its associated wiring. By checking the continuity in the motor windings, she was able to determine that the condenser fan motor was defective and in need of replacement. This service call was rather straightforward with the exception of her foresight to remove the fan blade from the motor shaft before leaving the job. Consider the following possible situation if she had not removed the blade.

Diane returns in a day or two with a brand new motor. She then begins to remove the old motor from the unit and finds that the *hub,* or portion of the blade that the motor shaft fits into, is rusted. While attempting to free the blade from the shaft, the blade becomes damaged. It now needs to be replaced along with the motor. Diane now places a phone call to her service manager and explains the situation. Obviously, the service manager is not at all pleased with the situation. Why?

- First, he must tell the customer that the repair cannot be completed as promised.
- Second, the customer must take another day off from work to meet the service technician.
- Third, the service manger must tell the customer that the repair will now cost more than the original estimate.
- Fourth, valuable time has been lost, costing the company money.

By taking the time to remove the blade before the estimate was given to the customer, Diane eliminated the possibility of this potentially awkward situation arising. She was able to tell her service manager exactly what part or parts she needed as well as approximately how long the repair would take.

SERVICE CALL 7: GROUNDED COMPRESSOR

Customer Complaint

A newlywed couple has just purchased their first home, a handyman's special. Upon turning the air-conditioning system on, they find that the air coming from the supply registers is warm. John, the new groom, having some technical knowledge, goes out back to the condensing unit. He opens the disconnect box and finds a 2-pole, 40-ampere circuit breaker that is tripped. He resets the breaker, and it immediately trips again. Realizing that he is getting in a little over his head, he calls a local air-conditioning contractor for help.

Service Technician Evaluation

Upon arriving at the home, David, the technician, is immediately told about the system. Since the circuit breaker at the condensing unit tripped immediately, he concludes that there must be a short circuit somewhere in the outdoor unit. With the disconnect switch in the OFF position and all capacitors discharged, he checks for continuity between the two power lines on the load side of the disconnect switch and ground. David reads 0 ohms from L2 to ground, indicating a direct short to ground within the condensing unit. The short must be in one of the following locations:

- The wiring from the disconnect box to the line side of the cooling contactor
- The compressor
- The wiring from the load side of the contactor to the compressor
- The condenser fan motor
- The wiring from the load side of the contactor to the condenser fan motor

To begin, David removes the wires from the line side of the contactor and again checks for continuity from the load side of the disconnect switch to ground. This time he reads infinity. This indicates that a short circuit is not present in the wires feeding the condensing unit and that the short lies within the condensing unit itself. He then checks for continuity between the load side of the contactor and ground and reads 0 ohms (Figure 7-41). David then removes

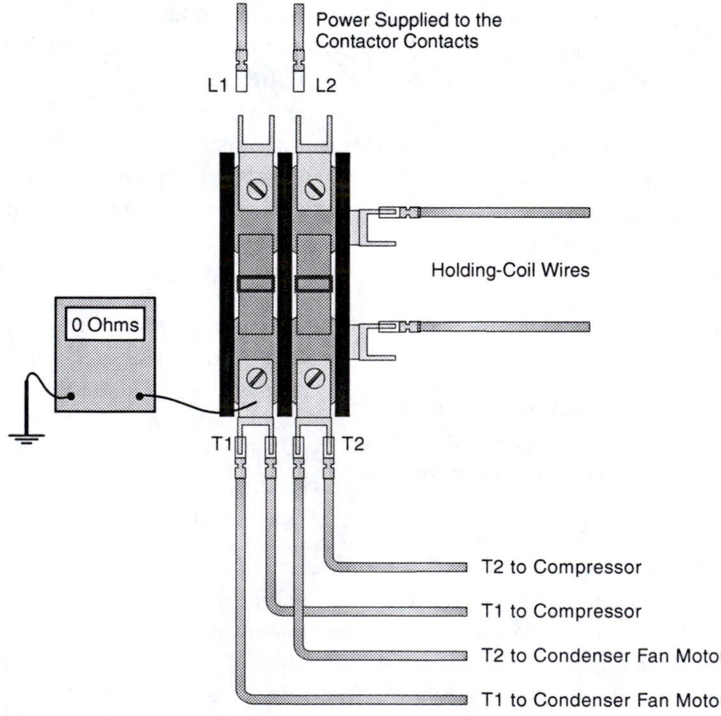

FIGURE 7-41 A short to ground exists on the load side of the contactor.

the compressor wires from the load side of the contactor and checks for continuity again. This time he reads infinity (Figure 7-42). Since he has removed the compressor and its associated wiring from the circuit, he has narrowed the search down to the compressor and its wiring components.

After replacing the wires leading to the compressor back onto the correct terminals of the contactor, David then removes the wires from the compressor itself, thereby isolating the compressor from the rest of the circuit. He then proceeds to check the compressor using a digital VOM set to read resistance. He scrapes a small section of the copper tubing on the compressor and holds one of the meter's test leads on this cleaned surface. This ensures that the meter will read properly. If the copper tubing has paint or other substance on it, the meter may not register continuity even if a short to ground exists. The other test lead is touched to each compressor terminal, one at a time, to check for continuity. David reads 0 ohms when checking from the start terminal to the copper tubing of the compressor (Figure 7-43). He concludes that the compressor is grounded and needs to be replaced.

Before placing a call to his service manager, David retrieves all the data tag information from the system and the compressor itself. He also visually inspects the area for any situation or condition that might possibly make the replacement of the compressor more difficult. He then calls in to his office and reports his findings. The service manager is able to properly price up the job and approach the customer with an estimate for the repair.

Service Call Discussion

In this service call, David was able to verify exactly what the customer was telling him by witnessing the tripping of the circuit breaker himself. The tripping of the breaker indicated either a short circuit within the wiring circuits or a short to ground. By using his VOM to check the circuit, he was able to determine that there was indeed a ground somewhere in the circuit.

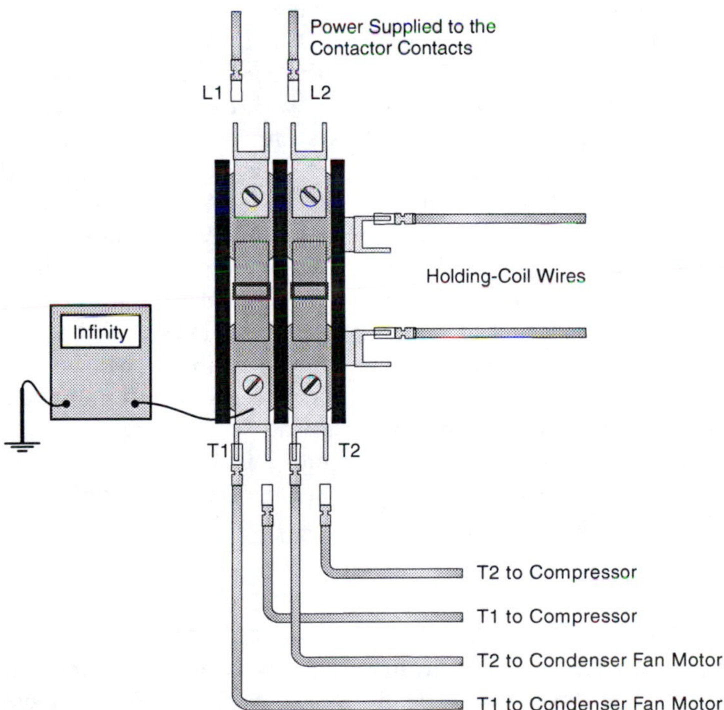

FIGURE 7-42 By removing the compressor from the circuit, the short is no longer present. The problem lies within the compressor or its associated wiring.

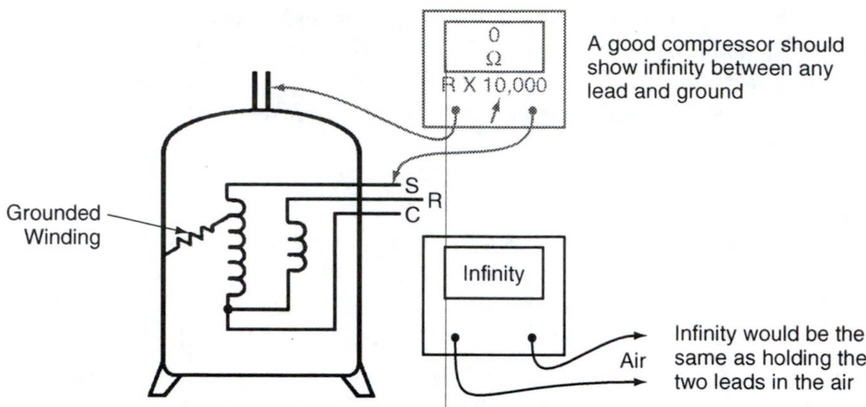

FIGURE 7-43 No continuity should exist between the compressor terminals and the shell of the device or its interconnecting piping.

When trying to isolate a problem in a system, knowing where the problem is *not* can be just as helpful as knowing where the problem *is*. By consistently and continually narrowing down the possibilities, the fault can be isolated in a concise and logical manner. By disconnecting the compressor wires from the load side of the contactor, David established a dividing line with which to work. If, after disconnecting the compressor wires, David still found continuity from the contactor to ground, he would have established that the ground did *not* exist in the compressor or its associated wiring. He would have then proceeded to check the condenser fan motor and its circuitry. This is a very systematic method of locating the system fault.

In any event, once the problem is isolated, the repair must still be performed. Replacing a compressor involves a number of different operations that, when performed in a logical manner, can reduce the overall repair time. For example, unless the system has a major refrigerant leak, the refrigerant must be properly recovered. **Refrigerant recovery** is the process of removing refrigerant from an air-conditioning or refrigeration system and storing it in a DOT-approved cylinder until it can be reintroduced into the customer's equipment or turned in for chemical reprocessing. Since the process of recovering refrigerant is generally time-consuming, good field practice includes beginning the recovery process first. While the refrigerant is being recovered, the old compressor can be disconnected electrically, its mounting hardware can be removed, and other preparation work can be done, as long as the recovery process is continually monitored. Once the recovery is complete, the old compressor will most likely be ready for removal from the system.

A similar situation arises when the new compressor is installed. The **evacuation** process, when air and moisture are removed from the system prior to system charging, is also time-consuming. For this reason, good field practice involves getting the new compressor in place and piped in as quickly as possible so that the evacuation process can begin. While the system is being evacuated, the compressor mounting hardware can be installed, the electrical connections can be made, and so forth. This way, once the system is properly evacuated, the charging process can begin.

SUMMARY

- A successful service call can be described by the following seven steps: verify complaint, gather information, inspect visually, isolate the problem, correct the problem, test the system, and complete the service call.
- Asking relevant questions of the customer can help the technician solve the problem.

- Performing a visual inspection of the equipment assists the technician in isolating the location of the system problem.
- The *cause* for component failure must be determined and corrected in addition to replacing the defective component.
- Always keep the customer informed as to what is being done and get permission before performing any additional work to the system.
- The system should be tested after the repair and be allowed to cycle on its own, if possible.
- Service calls are not complete until the paperwork has been completed.
- Service technicians should always be conscious of and respectful of the customer's property.
- Successful troubleshooting relies on the implementation of a logical and systematic approach to locating the system problem.

KEY TERMS

Continuity
Evacuation
Infinite resistance
Learning curve
Refrigerant recovery
Return plenum
Service panels
Supply plenum
Temperature differential
Ultraviolet solution

FOR DISCUSSION

1. A home owner calls and informs his service company that there is no air coming from the supply registers. The thermostat is set to the cooling mode, and the fan is switched to the ON position. It is the middle of July, and he can hear the indoor fan motor running and he can see that the outdoor unit is running. The house, though, is not getting cool. The home owner explains to the service manager that he has checked the return air filter and it is clean. The basement air handler is a vertical unit, and the fan/blower assembly is direct drive. Discuss the possible causes and remedies for this situation.

2. A home owner calls and informs his service company that no air is coming from the supply registers. The home owner tells the dispatcher that he cannot hear the indoor fan motor running even though the fan switch is set to the ON position. He can see that the outdoor unit is running. The house, though, is not getting cool. The home owner explains to the service manager that he has checked the return-air filter and it is clean. The basement air handler is a vertical unit, and the fan/blower assembly is direct drive. Discuss the possible causes and remedies for this situation.

3. Discuss why each of the seven steps for completing a successful service call is important. Discuss why none of these steps should be skipped or overlooked.

REVIEW QUESTIONS

1. Which of the following is an example of a technician verifying the customer's complaint?
   a. Checking the thermostat settings
   b. Feeling the air coming from the supply registers
   c. Taking a temperature reading of the occupied space
   d. All of the above

2. Correcting a system problem involves:
   a. Replacing a defective component.
   b. Eliminating the cause of the problem.
   c. Both *a* and *b* are correct.
   d. None of the above are correct.
3. True or False: The system problem must be corrected before information is gathered.
4. A multimeter is most likely used in which step of the troubleshooting process?
   a. Gathering information
   b. Isolating the problem
   c. Correcting the problem
   d. Visual inspection
5. True or False: A defective component can cause other components to go bad as well.
6. Which of the following would always be found on a completed work ticket?
   a. Number of hours spent on the job
   b. List of parts and materials used
   c. The customer's signature
   d. All of the above
7. What should Janet the technician do if a customer asks her a question about the service company that she is not qualified to answer?
   a. Be polite and answer the question anyway.
   b. Ignore the customer.
   c. Direct the customer to the service manager.
   d. Tell the customer, "I don't know," and leave it at that.
8. A refrigerant overcharge can be identified by which of the following?
   a. Low operating pressures
   b. High evaporator superheat
   c. Low condenser subcooling
   d. Increased system efficiency
9. A system undercharge can be identified by which symptom?
   a. Reduced cooling capacity
   b. Low discharge pressure
   c. A frosted evaporator coil
   d. All of the above
10. A fuse can be checked by:
    a. Taking a continuity reading across it.
    b. Taking a voltage reading across it.
    c. Both *a* and *b* are possible.
    d. None of the above are correct.
11. Which of the following can cause a line-voltage, power-circuit fuse to blow?
    a. A short circuit in a holding coil
    b. An open winding in a fan motor
    c. A grounded compressor
    d. All of the above

12. A defective control transformer will:
    a. Make the evaporator fan motor operate continuously.
    b. Make the system appear to be in the OFF mode.
    c. Cause the compressor to overheat.
    d. Cause the evaporator coil to freeze up.
13. If 24 volts are applied to a contactor's 24-volt holding coil:
    a. The normally closed contacts should open.
    b. The normally open contacts should close.
    c. The coil will burn out.
    d. Both *a* and *b* are correct.
14. If the evaporator fan motor is operating, it can be assumed that:
    a. The transformer is not defective.
    b. The compressor is operating.
    c. The control circuit is being powered.
    d. Both *a* and *c* are correct.
15. True or False: A short circuit causes the circuit current to increase.
16. While testing a split-phase motor with an ohmmeter, the technician reads 0 ohms between the common and the run terminals. The technician can conclude that the:
    a. Run winding is open.
    b. Run winding is shorted.
    c. Motor will operate normally.
    d. Motor is grounded.

CHAPTER 8

Vapor-Compression Heat-Pump Components and Accessories

OBJECTIVES After studying this chapter, the reader should be able to:

- Describe the basic heat-pump cooling cycle.
- Describe the basic heat-pump heating cycle.
- Explain how the heating/cooling changeover in a heat-pump system is accomplished.
- Describe the operation of the four-way reversing valve.
- Describe the function of a suction-line accumulator.
- Explain why heat exchangers are often found on heat-pump systems.
- List the various types of check valves and explain their importance in the heat-pump system.
- List two common uses for electric-strip heaters in heat-pump systems.
- Explain the concept of a bidirectional thermostatic expansion valve.
- Explain how the use of a bidirectional thermostatic expansion valve can simplify the refrigerant-piping circuit in a heat-pump system.
- Explain the construction of bidirectional filter driers.
- Explain the differences between standard heating/cooling thermostats and those designed to operate with heat-pump systems.
- Describe the operation of defrost timers used in heat-pump systems.

INTRODUCTION

Having reached this portion of the text, the reader should now possess the basic information and knowledge about the components that comprise the majority of the typical heat-pump system. With the addition of a relatively small number of components and concepts presented in this chapter, the reader will have a complete picture of the heat-pump system and its various modes of operation. These additional components are intended primarily to route refrigerant through the desired components at the desired time to achieve the desired result. Some of these components, such as the check valve, are strictly mechanical devices; others are electromechanical, as in the case of the four-way reversing valve; and still others are strictly electrical, such as the heat-pump thermostat. After completing this portion of the text, the reader will be able to apply the concepts discussed in earlier chapters to this exciting aspect of our industry.

OVERVIEW

The basic purpose of air conditioning and refrigeration is to transfer heat from a location where it is not desired to some other location where this heat is not objectionable. The heating system, on the other hand, transfers heat to a location in order to raise the temperature of that location. The basic vapor-compression refrigeration system that has been discussed thus far in the text is capable of producing a refrigeration effect but cannot produce the desired heating effect as described. By adding components to the basic air-conditioning system, a piece of equipment is created that performs both heating and cooling functions. This system, the **heat pump**, provides comfort cooling—or air conditioning—in the warmer months and heating in the cooler months. This is accomplished by altering the path that the refrigerant takes as it passes through the system.

As is known from basic refrigeration theory, refrigerant is discharged from the compressor and

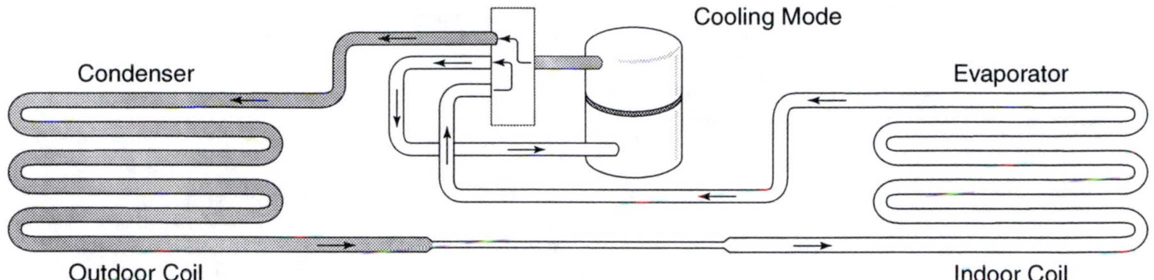

FIGURE 8-1(a) Heat-pump system operating in the cooling mode

travels on to the condenser in order to reject heat from the system. This heat is rejected to the medium passing over or through the condenser, typically air or water. As the system refrigerant transfers heat to the medium, the temperature of the medium increases. After leaving the condenser and passing through the expansion device, the refrigerant then enters the evaporator, where heat from the medium that is to be cooled is absorbed by the refrigerant. In an air-cooled, air-conditioning system, for example, the heat is removed from inside the conditioned space and transferred to the air located outside the structure. This works fine in the warmer months when the cooling effects of air conditioning are desired. But what happens in the cooler months when heating is needed?

In the colder months when heating is needed, heat transfer is still desired, but this time the direction of the transfer is opposite to that described in the previous paragraph. Now, heat needs to be transferred from outside the structure to the inside. This concept may be confusing because it is sometimes difficult to comprehend that 30°F outside air, for example, contains enough heat to actually have a warming effect on the area to be heated. In reality, air that is considered to be cold contains large amounts of heat that can be used effectively for this purpose.

It would therefore be desirable for the refrigerant, upon leaving the compressor, to travel first to the conditioned space in order to reject its heat into that location, thereby increasing the temperature of the space. The refrigerant would then travel through the expansion device and absorb heat from the air outside the structure before returning to the compressor. In essence, the locations of the evaporator and the condenser have been switched. The evaporator is now located outdoors to absorb heat, while the condenser is located indoors to reject heat into the conditioned space. For this reason, the coils in a heat-pump system are not referred to as the *evaporator* and the *condenser.* They are referred to as the **indoor coil** and the **outdoor coil**, referring to the coils' physical location in the system. When providing air conditioning, or cooling, the indoor coil acts as the evaporator and the outdoor coil acts as the condenser, as shown in Figure 8-1(a). This is the same as a typical air-conditioning system. In the heating mode, however, the indoor coil functions as the condenser, rejecting heat into the space, while the outdoor coil operates as the evaporator, as shown in Figure 8-1(b). The shaded areas in Figure 8-1(a) and Figure 8-1(b) represent the high-pressure side of the system. Notice how this area shifts from the outdoor coil to the indoor coil as the system switches over from the cooling to the heating mode, and vice versa.

Using just the four major system components, achieving this switchover from heating to cooling would be impossible without having to completely repipe the system each and every time a different mode of operation was desired. However, the addition of a few key components has made this transition as easy as pushing a button. The two key components that make this switchover possible are the **four-way reversing valve** (Figure 8-2) and the **check valve** (Figure 8-3).

The four-way reversing valve mechanically alters the refrigeration-piping configuration in order to route the compressor discharge refrigerant through the:

- Outdoor coil while operating in the cooling mode
- Indoor coil while operating in the heating mode

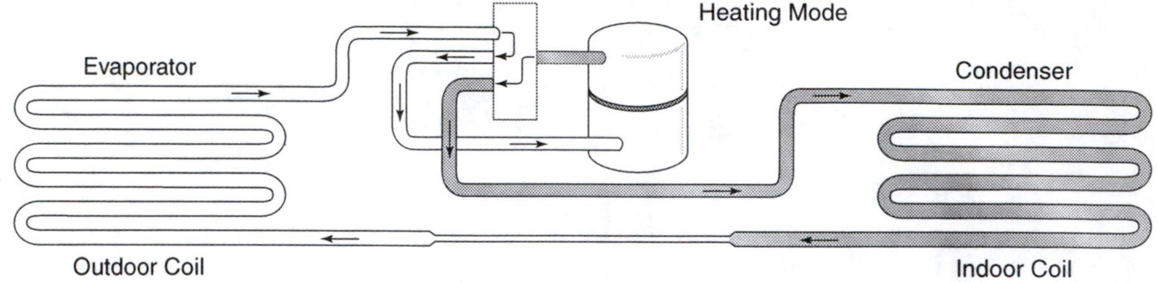

FIGURE 8-1(b) Heat-pump system operating in the heating mode

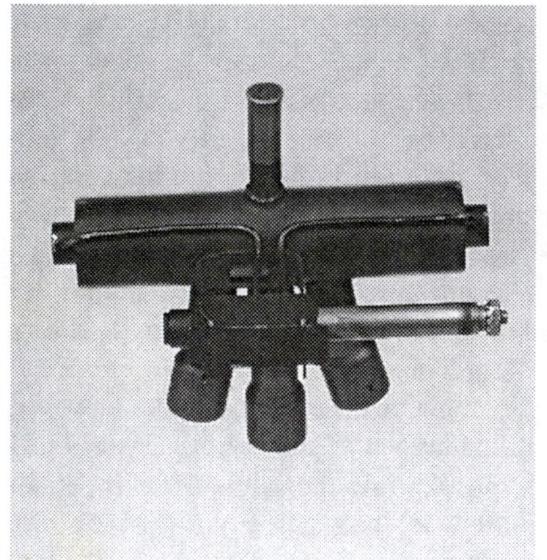

FIGURE 8-2 Four-way reversing valve. *(Photo by Bill Johnson)*

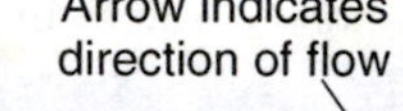

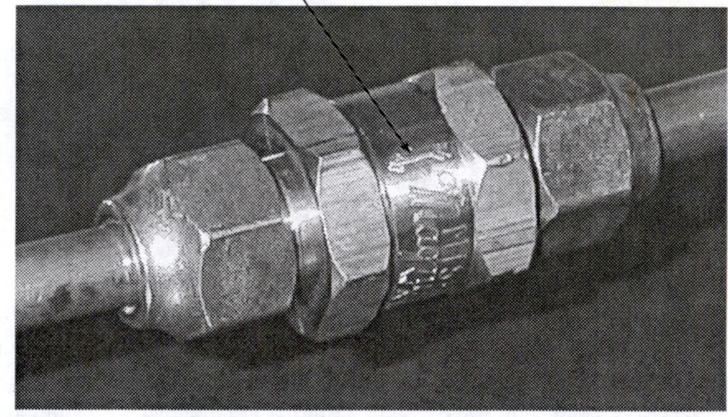

FIGURE 8-3 Flow-check valve. Direction of flow is indicated by the arrow on the device. *(Photo by Eugene Silberstein)*

The refrigerant is still discharged from, and returned to, the compressor in the identical fashion. The compressor's operation is exactly the same regardless of the mode in which the system is operating. Refer back to Figure 8-1(a) and Figure 8-1(b) and note that the refrigerant is still discharged from and returned to the compressor from the same ports.

The flow-check valve, commonly found on refrigeration systems with multiple evaporators, is commonly found on heat-pump systems as well. The purpose of this valve is to permit the flow of refrigerant in only one direction. This helps to reduce the possibility of refrigerant trying to push its way through an expansion device or filter drier in the wrong direction, which would result in an inefficient system as well as extremely high pressure drops across the devices. Since multiple expansion devices and filter driers are often found in heat pumps, the check valve also helps ensure that only the desired components are in the active refrigerant circuit at any given time.

In addition to the four-way reversing valve and the check valve, other components, when added to the heat-pump system, allow it to operate more effectively and efficiently. They include:

- Suction-line accumulators
- Auxiliary electric-strip heaters
- Bidirectional liquid-line filter driers
- Bidirectional thermostatic expansion valves

These devices are discussed in this chapter, as are thermostats that are specially designed to operate with heat-pump systems.

COMMON HEAT-PUMP-SYSTEM CONFIGURATIONS

Just as in standard vapor-compression air-conditioning and refrigeration systems, the condenser can generally be either air- or water-cooled and the evaporator can be used to remove heat from either air or liquid. Heat pumps are similar in the respect that they can facilitate the transfer of heat between a number of different mediums. Common classifications of heat pumps are:

- Air-to-air heat pump
- Air-to-liquid heat pump
- Liquid-to-air heat pump
- Liquid-to-liquid heat pump

The first portion of the classification denotes the source of heat for the system while operating in the heating mode. **Air-source heat pumps—air-to-air** and **air-to-liquid heat pumps**—transfer heat from air into the medium to be heated while operating in the heating mode. On the other hand, **liquid-source heat pumps—liquid-to-air** and **liquid-to-liquid heat pumps**—transfer heat from a liquid to the medium being heated while operating in the heating mode.

The second portion of the classification represents the medium that is ultimately being treated by the system. In air-to-air and liquid-to-air heat pumps, air is being treated at the indoor coil. Similarly, air-to-liquid and liquid-to-liquid heat pumps treat liquid at the indoor coil. Each of these configurations is discussed in detail later on in the text.

In addition to the types of heat pumps listed here, another classification stands alone and is covered in a separate chapter later in the text. This is the **geothermal heat pump**. The geothermal heat pump uses the earth as a source of heat in the colder months and as a heat sink in the warmer months. In the winter, heat is transferred from the earth to the refrigerant, which is then transferred to the medium to be treated. In the warmer months, the heat from the conditioned space is transferred to the refrigerant, which is then transferred to the earth. Two common types of geothermal heat pumps are the:

- Open-loop system
- Closed-loop system

The open-loop geothermal heat pump is a special type of liquid source heat pump. The **open-loop system** uses water as a heat source, but the source of this water is the earth. Common sources of water for use in this type of system are lakes and wells.

The **closed-loop system** does not use the water from the earth but, instead, uses the earth itself. Large coils of plastic piping, called **ground loops** or **ground coils**, are buried under the earth's surface, and the heat transfer takes place between the earth and the liquid in the piping network. The geothermal heat pump and its many different piping configurations are covered later on in the text.

FOUR-WAY REVERSING VALVES

As mentioned earlier in this chapter, it was desirable to have a system that could automatically switch between the heating and cooling modes of operation with the simple push of a button. The **four-way reversing valve** makes this changeover possible. This valve is controlled by a solenoid coil and changes the direction of refrigerant flow depending on whether or not the coil is energized. The valve is constructed of a shell, within which is a slide that has the ability to shift back and forth (Figure 8-4). When the solenoid is energized, the slide is situated at one end of the valve and when de-energized, the other. Valve manufacturers have established different criteria for determining if the system should be in the heating or cooling mode when the solenoid coil is energized. In this text, for the sake of clarity and uniformity, assume that the system is in the cooling mode when the solenoid valve is energized (Figure 8-5). Always check the individual system being worked on to determine which orientation is being used.

The four-way reversing valve has four ports to which the system refrigerant lines are connected. These ports are connected to:

1. The discharge line of the compressor
2. The suction line of the compressor
3. The outdoor coil (on the opposite side as the expansion device)
4. The indoor coil (on the opposite side as the expansion device)

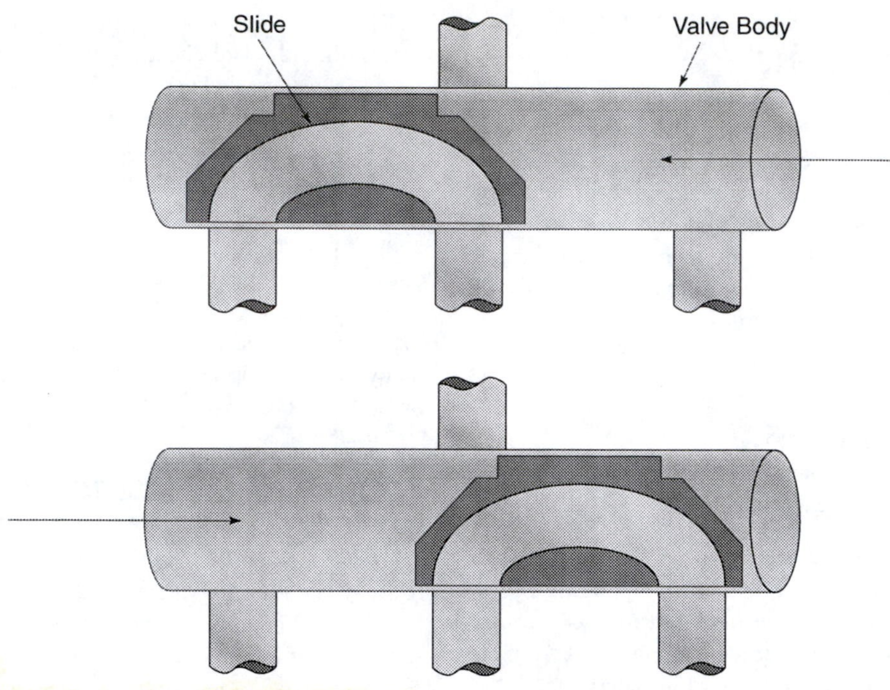

FIGURE 8-4 The internal slide on the reversing valve moves back and forth within the shell of the valve.

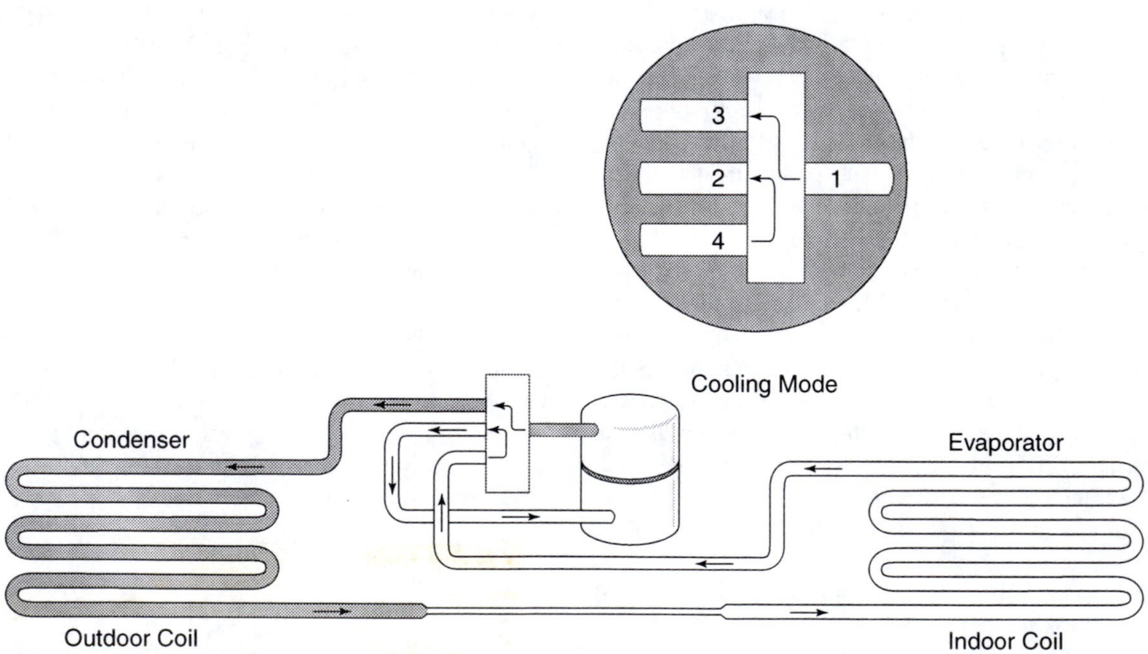

FIGURE 8-5 This valve is in the cooling mode when the solenoid coil is energized. 1. Connected to the compressor discharge line. 2. Connected to the compressor suction line. 3. Connected to the outdoor coil. 4. Connected to the indoor coil.

These locations are shown in Figure 8-5. One port is usually separate from the other three. This port is connected to the discharge line coming from the compressor. The compressor's discharge vapor enters the reversing valve at this point. The vapor refrigerant then leaves the valve through one of the two end ports on the other side of the valve. The refrigerant leaves the valve and travels to the condenser that is located either indoors or outdoors, depending on the desired mode of operation. In the case of Figure 8-5, the condenser is located outdoors, indicating that the system is operating in the cooling mode. Vapor refrigerant then flows back to the valve through the other end port, leaves the valve by way of the center port, and flows back to the compressor. Note that the discharge refrigerant always enters the valve at *Point 1* and returns to the compressor from *Point 2*.

Reversing-Valve Operation: The Cooling Cycle

When the solenoid valve is energized, in the cooling mode, the piping configuration is as shown in the bottom of Figure 8-6. Assume at this point that the system is equipped with a single metering device and that the same metering device is being used in both the heating and cooling modes. In this figure, note that the refrigerant leaves the compressor and first travels through the **reversing valve** and onto the **outdoor coil**, which in this case is the condenser. The refrigerant then flows through the metering device and then flows into the indoor coil. The indoor coil is acting as the evaporator and is absorbing heat from the conditioned space. This configuration is exactly the same as that of a system that operates as a cooling-only unit. Once the refrigerant flows through the evaporator, it then flows through the low-pressure side of the

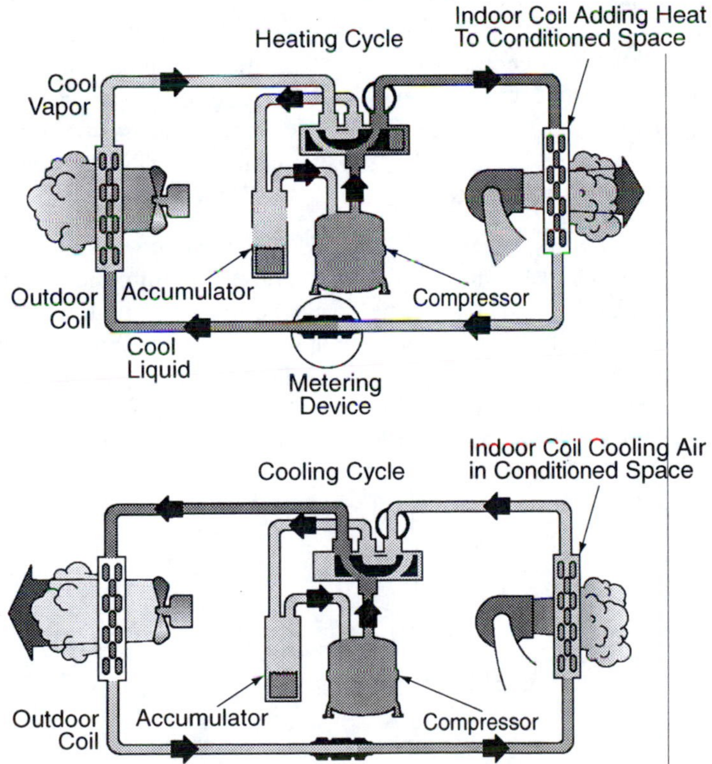

FIGURE 8-6 A refrigeration system showing the heating, *top*, and cooling, *bottom*, modes of operation

reversing valve. After leaving the reversing valve, the low-pressure refrigerant flows through the accumulator, which is discussed shortly, and then returns to the compressor to complete the cycle.

Reversing-Valve Operation: The Heating Cycle

When the solenoid is de-energized, putting the system in the heating mode, the flow through the coils is reversed, as shown in the top of Figure 8-6. The hot gas being discharged from the compressor still flows through the reversing valve and onto the condenser, but this time the condenser is the indoor coil. The refrigerant now gives up its heat to the occupied space, thereby heating the area. Once the refrigerant leaves the condenser and flows through the metering device, it then flows to the outdoor coil, which is acting as the evaporator. After flowing through the evaporator, the refrigerant travels back through the reversing valve and through the accumulator before returning to the compressor. Various key points in the system are identified in Figure 8-7; and the refrigerant flow through the system and the state of the refrigerant at these points, in both the heating and cooling modes of operation, are outlined in Figure 8-8.

Direct-Acting and Pilot-Operated Reversing Valves

When used on very small systems, the slide in the reversing valve can be controlled directly by the pull created by the magnetic field of the solenoid coil as described previously. When the solenoid coil is de-energized, a spring pressure pushes the slide to its normal position. This type of configuration is not found on larger valves for a number of reasons, including:

- The large distance that the slide must travel
- The large pressure differentials across the valve
- Erratic reversing-valve operation
- Premature solenoid coil failure

Larger systems are therefore equipped with **pilot-operated reversing valves.** This type of reversing valve uses the difference between the high- and low-side pressures to help push the valve slide from one position to the other. The direction of the force created by the high-side pressure is determined by the pilot ports (Figure 8-9), which in turn are controlled by the solenoid coil. The operation of the pilot-operated reversing valve is similar to that of a direct-acting valve in that the energizing and de-energizing of a solenoid coil cause the movement of a valve slide. The main difference is this:

- The solenoid-controlled slide on the pilot-operated valve controls the refrigerant flowing through the pilot ports of the reversing valve, not the refrigerant flowing through the main body of the valve.
- By controlling the refrigerant flow through the pilot ports, the system's high pressure is directed in such a way that it pushes the main valve slide to the desired position.

In operation, when the solenoid coil is de-energized, putting the system in the heating mode, the pilot valve is in the position as shown in the bottom of Figure 8-10(a). The compressor discharge gas at *"D"* flows into the pilot valve at *Port "D1"* and out of the pilot from *Port "B."* This high-pressure vapor then acts on the main valve slide, pushing it to

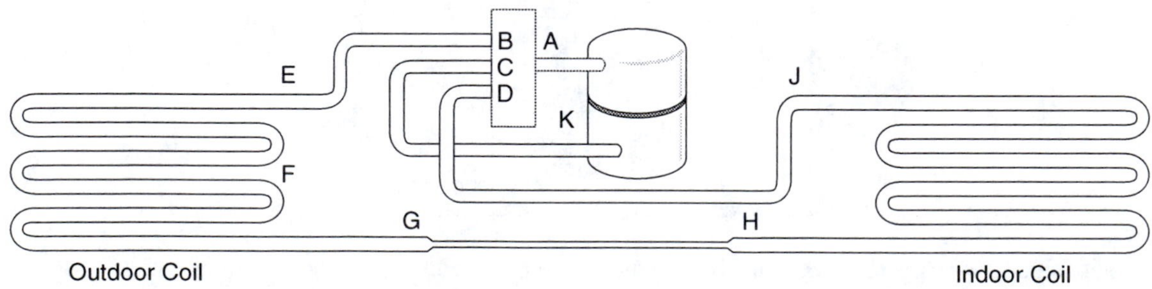

FIGURE 8-7 Heat-pump system with important locations identified. (See Figure 8-8)

Cooling Mode			Heating Mode		
Letter Designation	**Location in the System**	**State of the Refrigerant**	**Letter Designation**	**Location in the System**	**State of the Refrigerant**
A	Compressor discharge port	High-temp, high-pressure superheated vapor	A	Compressor discharge port	High-temp, high-pressure superheated vapor
B	Reversing-valve port leading to the outdoor coil	High-temp, high-pressure superheated vapor	D	Reversing-valve port leading to the indoor coil	High-temp, high-pressure superheated vapor
E	Inlet of the outdoor coil	High-temp, high-pressure superheated vapor	J	Inlet of the indoor coil	High-temp, high-pressure superheated vapor
F	Middle of outdoor coil	High-temp, high-pressure saturated vapor	H	Outlet of the indoor coil/inlet of the metering device	High-temp, high-pressure subcooled liquid
G	Outlet of outdoor coil/ Inlet of the metering device	High-temp, high-pressure subcooled liquid	G	Inlet of the outdoor coil/ Outlet of the metering device	Low-temp, low-pressure saturated liquid
H	Outlet of the metering device	Low-temp, low-pressure saturated liquid	F	Middle of the outdoor coil	Low-temp, low-pressure saturated liquid
J	Outlet of indoor coil	Low-temp, low-pressure superheated vapor	E	Outlet of the outdoor coil	Low-temp, low-pressure superheated vapor
D	Inlet of the reversing valve	Low-temp, low-pressure superheated vapor	B	Inlet of the reversing valve	Low-temp, low-pressure superheated vapor
C	Outlet of the reversing valve	Low-temp, low-pressure superheated vapor	C	Outlet of the reversing valve	Low-temp, low-pressure superheated vapor
K	Compressor	Low-temp, low-pressure superheated vapor	K	Compressor suction port	Low-temp, low-pressure superheated vapor

FIGURE 8-8 State of refrigerant at various points in the system

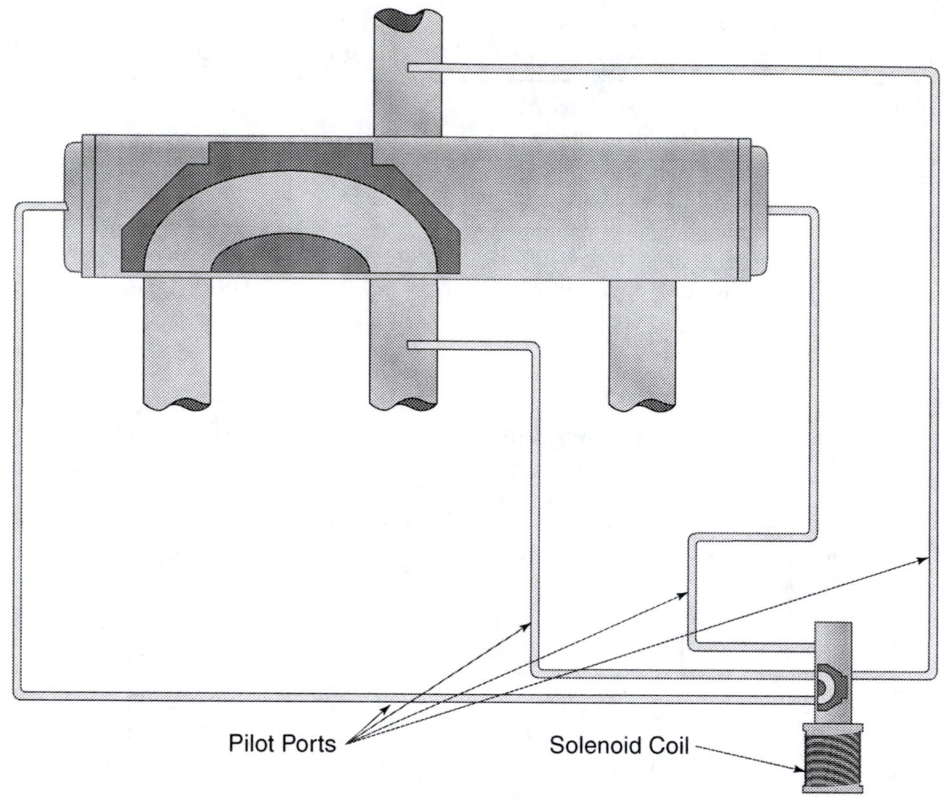

FIGURE 8-9 Pilot ports on a pilot-operated, four-way reversing valve

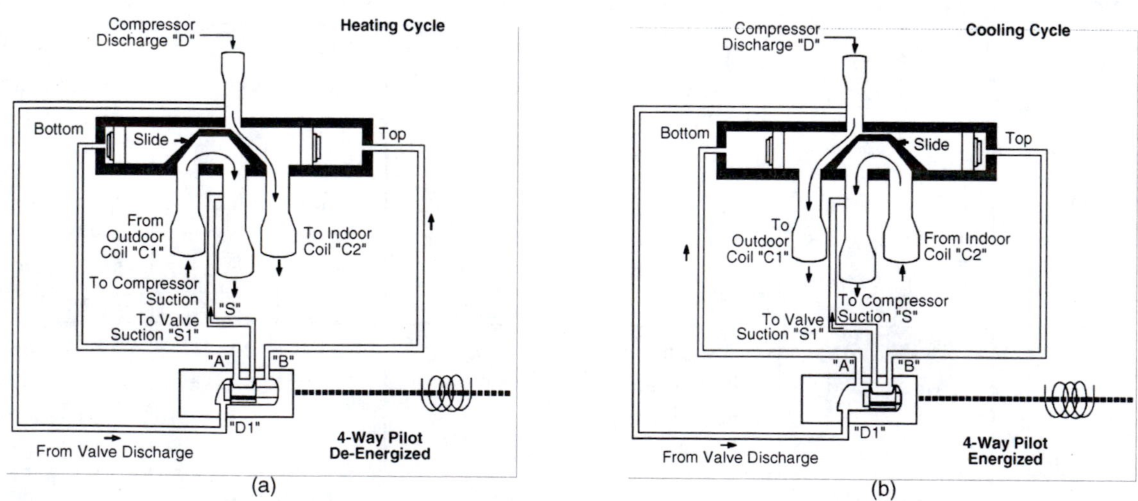

FIGURE 8-10 (a) Heating and (b) cooling positions of a pilot-operated, four-way reversing valve

	Heating Mode	Cooling Mode
Hot Gas Leaving the Compressor	Enters port "D1"	Enters port "D1"
Hot Gas Flowing to the Condenser	Leaves port "B"	Leaves port "A"
Refrigerant Leaving the Evaporator	Enters port "A"	Enters port "B"
Refrigerant Flowing Back to the Compressor	Leaves port "S1"	Leaves port "S1"

FIGURE 8-11 Summary of pilot-operated, reversing-valve port arrangement

the left. The refrigerant on the left side of the main valve slide is pushed out of the valve and enters the pilot valve through *Port "A."* It then leaves the pilot valve from *Port "S1"* and travels back to the compressor from *Port "S."* This valve is in the heating position since the discharge gas from the compressor, at *Point "D,"* is being circulated through the indoor coil, at location *"C2,"* first.

When the solenoid coil is energized, system calling for cooling, the pilot valve moves to the position shown in the bottom of Figure 10(b). The hot discharge gas from the compressor still enters the pilot valve through *Port "D1"* but leaves through *Port "A."* This sends the high-pressure gas to the left side of the slide in the main valve, pushing the main slide to the right. The vapor at the right side of the slide is pushed out of the valve and enters the pilot valve through *Port "B,"* leaves through *Port "S1,"* and returns to the compressor from *Port "S."* This system is in the cooling mode since the hot discharge gas from the compressor flows through the outdoor coil, denoted by *"C1,"* first.

You can see that the pilot-operated reversing valves are more desirable when used on larger applications. The solenoid coil is used to control the movement of a small plunger, which, in turn, operates the larger main slide on the valve itself. This eliminates the erratic operation often found on the direct-acting valves and also reduces the wear and tear on the solenoid coil. The purpose of both the pilot-operated and the direct-acting reversing valve is the same. They are both intended to direct the refrigerant to the correct coil in order to achieve the desired result. A summary of the refrigerant flow through the pilot-operated reversing valve, referring to Figure 8-10, is shown in Figure 8-11. Notice from Figure 8-11 that the hot gas from the compressor always enters the reversing valve at *Port "D1"* and the suction gas always returns to the compressor through *Port "S1."* Also note from the same figure that the reversing valve simply switches the internal connections to *Ports "A"* and *"B."*

CHECK VALVES

The purpose of the **check valve** is to permit refrigerant to flow in only one direction. As shown in Figure 8-3, the device is marked with an arrow that points in the direction of desired flow through the valve. As refrigerant flows through the valve in the direction indicated, there is very little resistance to flow. If, however, refrigerant attempts to flow through the valve in the direction opposite to that indicated on the valve body, the device will close and not permit the refrigerant to pass through. Several different types of check valves are commonly found in heat-pump systems and are differentiated by their construction and method of operation. The two common types are the:

- Ball-type check valve
- Disc-type check valve

The *ball-type check valve* is constructed of a hollow shell that is soldered or brazed into the refrigerant-piping circuit at the desired location. Inside the shell rests a ball that freely moves within the device. For this type of valve to operate correctly, it must be installed in the vertical position. When the system is off, the ball will fall to the bottom of the valve (Figure 8-12). When refrigerant flows in the direction indicated by the arrow on the device, the pressure of the refrigerant lifts the ball and the valve opens (Figure 8-13). When refrigerant attempts to flow through the valve in the wrong direction, the ball is pushed down and the valve closes as shown in Figure 8-12.

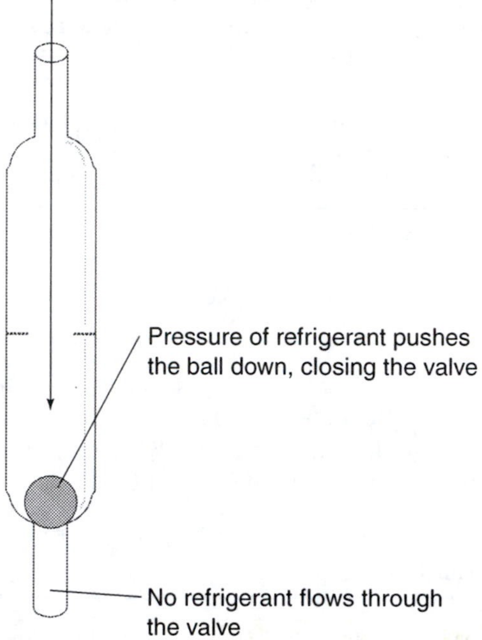
Pressure of refrigerant pushes the ball down, closing the valve
No refrigerant flows through the valve

FIGURE 8-12 This ball-type check valve is in the closed position.

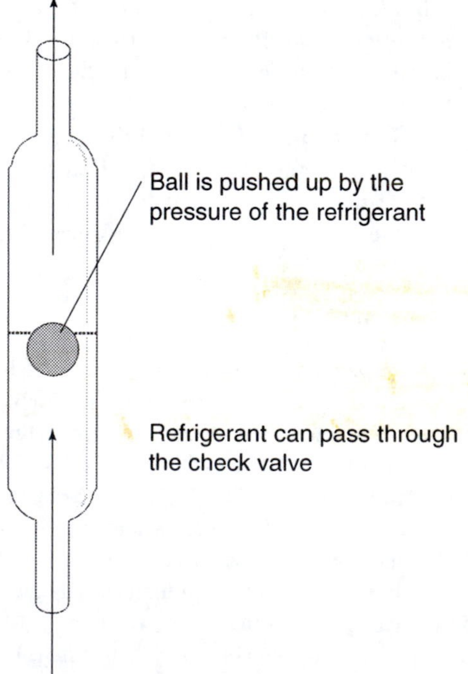
Ball is pushed up by the pressure of the refrigerant
Refrigerant can pass through the check valve

FIGURE 8-13 The ball is being pushed up, opening the check valve and allowing refrigerant to flow through it.

The *disc-type check valve* is constructed of a shell in which rests a disc that is held in place by magnetic attraction. The check valve in Figure 8-3 is a disc-type check valve. When refrigerant flows through the valve in the desired direction, the disc pivots and allows refrigerant to pass through (Figure 8-14). If refrigerant attempts to flow through the valve in the direction opposite to the desired flow, the disc is pushed against the valve opening, thereby closing the valve (Figure 8-15).

FIGURE 8-14 In the disc-type check valve, the valve is in the open position when the disc is parallel to the direction of refrigerant flow. *(Photo by Eugene Silberstein)*

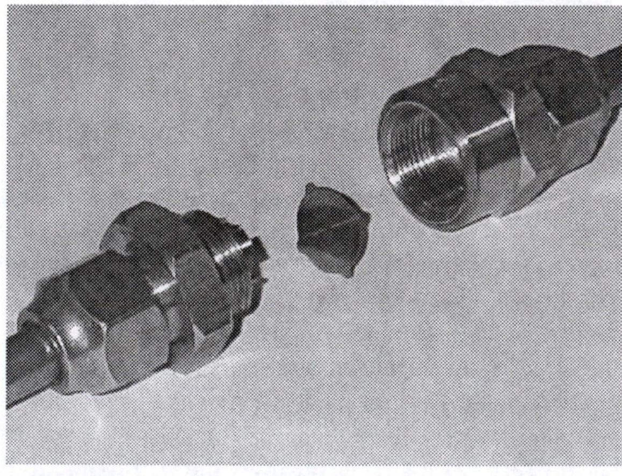

FIGURE 8-15 The check valve is closed when the disc is perpendicular to the direction of flow. *(Photo by Eugene Silberstein)*

Depending on the construction of the check valve, the device can be installed in the system in a number of different ways. Two of the most common methods are soldering or brazing and using flare connections. Some flow-check valves are designed so that they can be installed either way. These devices have inserts that can be removed to accommodate copper tubing, thereby allowing the device to be soldered or brazed (Figure 8-16). If the insert is left in place, the device is to be installed using flare connections (Figure 8-17). If needed or desired, one end of the flow-check valve can be connected with a flare nut while the other is soldered into the system.

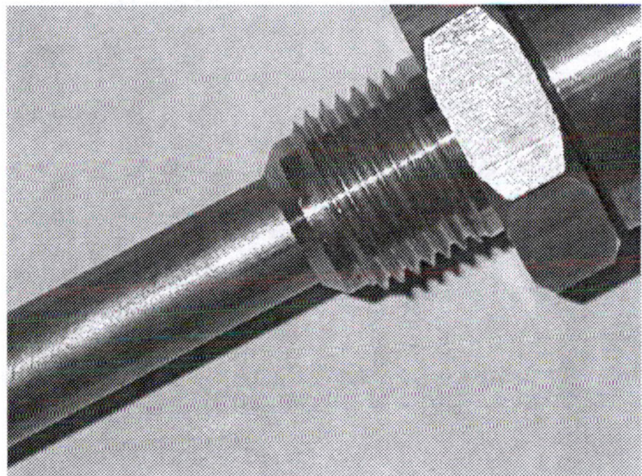

FIGURE 8-16 This type of valve can be connected by either flare or soldering. *(Photo by Eugene Silberstein)*

FIGURE 8-17 This insert is removed when the valve is soldered or left in place when installed using a flare connection. *(Photo by Eugene Silberstein)*

SERVICE NOTE: *When soldering check valves into the system, good field practice dictates wrapping the device with a wet cloth to prevent excessive heating, which could affect the operation of the valve.*

Check Valves in Capillary-Tube Systems

At the beginning of the discussion on the components used in the heat-pump system, the original assumption was made that the system used a capillary-tube metering device and that the same capillary tube was in the active circuit for both the heating and cooling cycles. Since the capillary tube is simply a fixed-bore metering device with no moving parts and no mechanical components, the change in direction of refrigerant flow did not pose an immediate problem. However, consider the strainer or filter drier that is located at the inlet of the capillary tube. During normal operation, the filter drier traps any particulate matter flowing in the refrigerant piping in order to prevent the clogging of the capillary tube (Figure 8-18). If refrigerant is permitted to flow through the capillary tube in the opposite direction, any debris or particulate matter that is trapped in the strainer will be pushed out of the device and back into the system (Figure 8-19). To alleviate this problem, it is desirable to allow refrigerant to flow in only one direction through the capillary tube. Therefore, two capillary tubes are commonly used—one for the indoor coil when it functions as the evaporator and one for the outdoor coil when it functions as the evaporator (Figure 8-20). To introduce the proper capillary tube to the active refrigerant circuit when needed, check valves are used. In practice, the filter driers are normally installed in series with the check valve as shown in Figure 8-20. The following is true in the heating mode:

- The flow-check valve at the indoor unit is in the open position.
- The refrigerant bypasses the capillary tube used in the cooling mode.
- The refrigerant passes through the indoor filter drier.
- The flow-check valve at the outdoor unit is in the closed position.
- The refrigerant bypasses the outdoor filter drier.
- The refrigerant passes through the outdoor capillary tube.

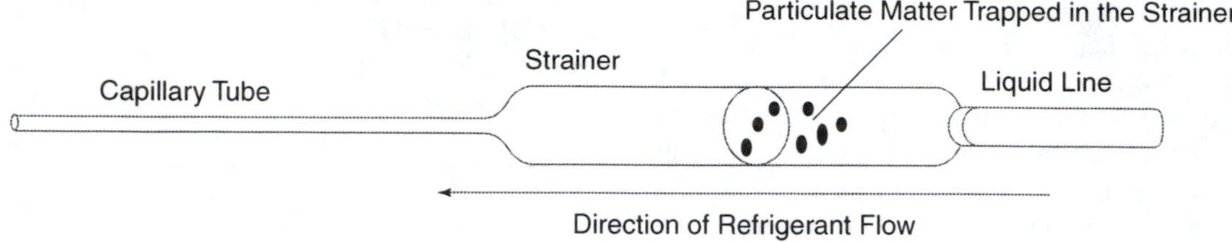

FIGURE 8-18 Particulate matter is trapped in the strainer.

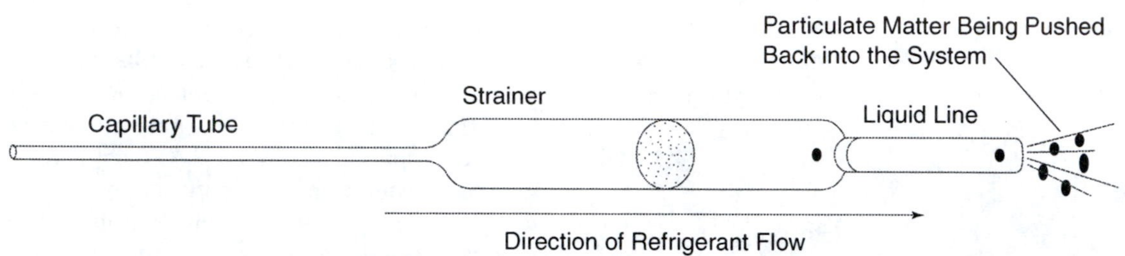

FIGURE 8-19 Particulate matter would be pushed back into the system if refrigerant flowed through the device in the opposite direction.

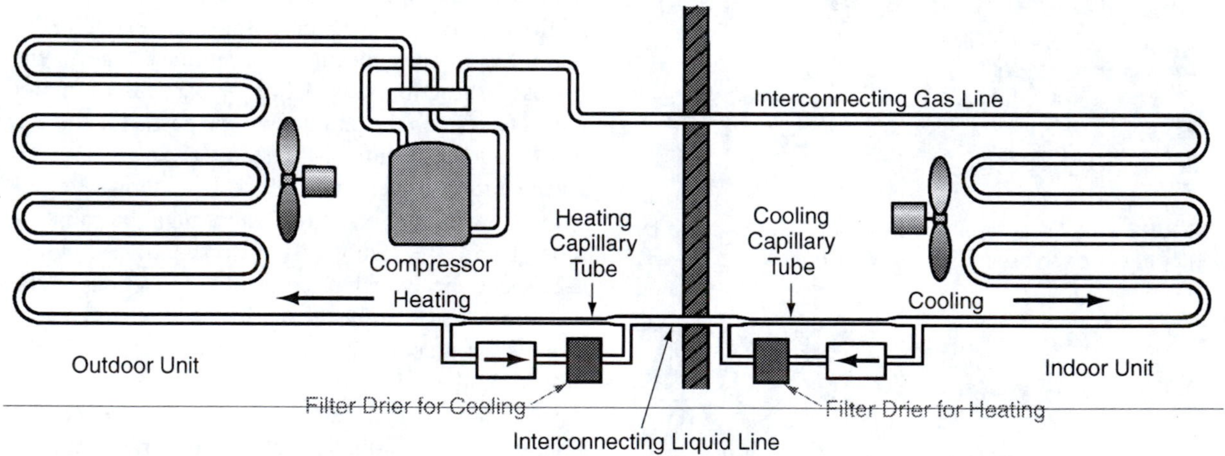

FIGURE 8-20 Filter driers are often installed in series with check valves.

In the cooling mode, the open check valve closes, and vice versa, as follows:

- The flow-check valve at the outdoor unit is in the open position.
- The refrigerant bypasses the capillary tube used in the heating mode.
- The refrigerant passes through the outdoor filter drier.
- The flow-check valve at the indoor unit is in the closed position.
- The refrigerant bypasses the indoor filter drier.
- The refrigerant passes through the indoor capillary tube.

Since the capillary tube provides a restriction to refrigerant flow, the installation of a check valve in parallel to the capillary tube will provide a low-resistance bypass for the refrigerant to take when the capillary tube is not intended to be in the active refrigerant circuit.

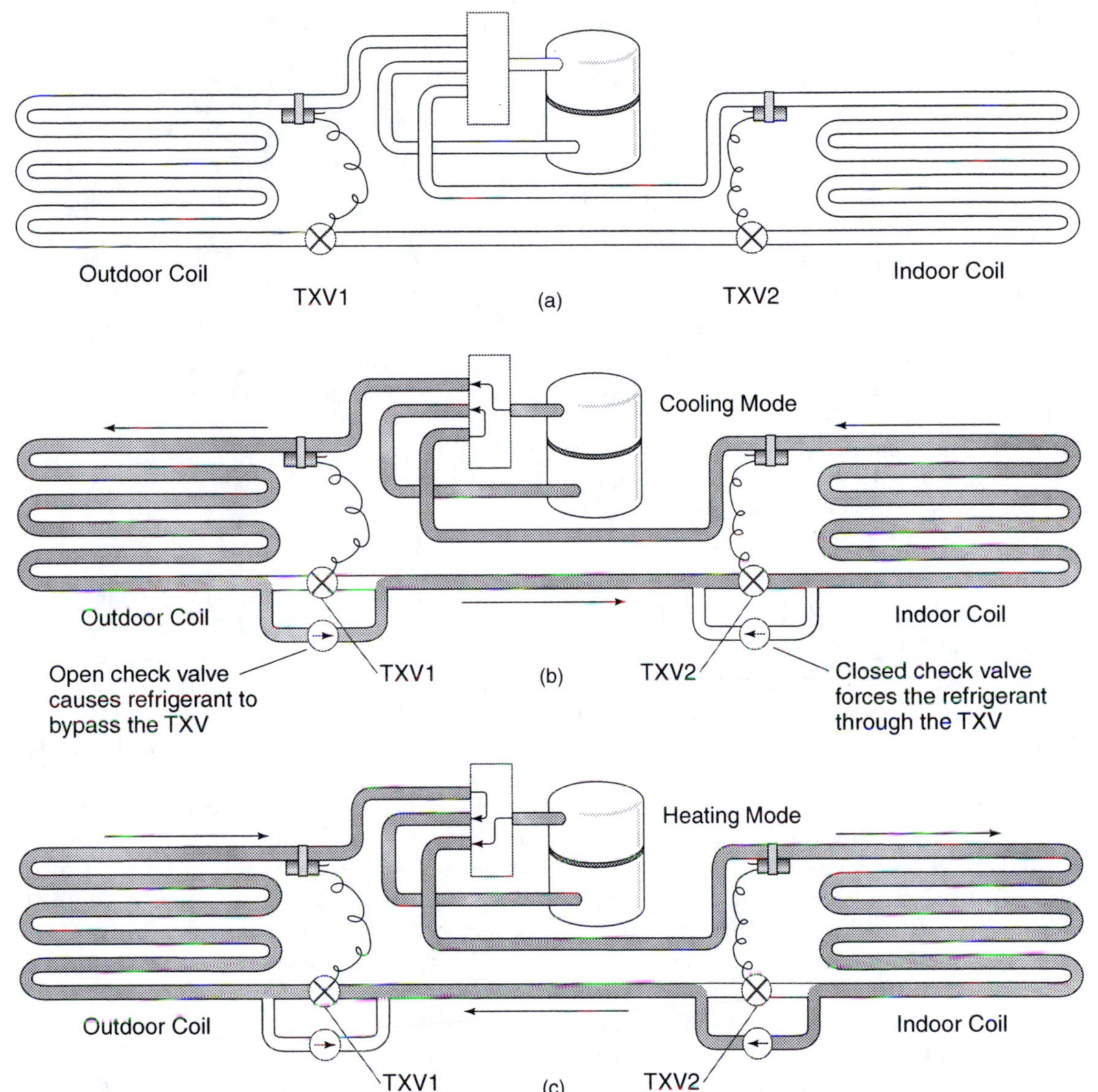

FIGURE 8-21 (a) Refrigerant would flow through the thermostatic expansion valve (TXV) in the wrong direction. (b) Refrigerant bypasses TXV 1 in the cooling mode. (c) Refrigerant bypasses TXV 2 in the heating mode.

Check Valves in Systems with Thermostatic Expansion Valves

Since thermostatic expansion valves (TXVs) must be able to sense the temperature at the outlet of the evaporator in order to maintain the proper superheat in the coil, TXVs must be located at both the indoor and outdoor coils of split-type systems in order to operate effectively. Without the use of check valves, refrigerant will be forced to travel through one TXV in the wrong direction whenever the system is operating, as shown in Figure 8-21(a). Refrigerant will be forced through *TXV 1* in the wrong direction during the cooling mode and through *TXV 2* in the heating mode. Problems quickly arise with a setup similar to this because of the large pressure drop created across the valve. To alleviate this situation, check valves are used in a similar manner to that described in the previous section. This type of configuration is shown in Figure 8-21(b) and Figure 8-21(c).

In the cooling mode, the refrigerant leaving the outdoor coil is able to bypass *TXV 1* because the direction of refrigerant flow is the same as the direction indicated on the check valve, shown in Figure 8-21(b). As the refrigerant reaches *TXV 2,* it is forced through the expansion device because the check valve piped in parallel with that valve is in the closed position. The shaded gray area on this piping diagram indicates the refrigerant flow pattern through the system. Figure 8-21(c) represents the piping circuit for the identical system operating in the heating mode. Some equipment manufacturers, in an attempt to simplify the piping configuration of such systems, have designed a special type of TXV that has a check valve built into it. This type of expansion valve is called *a bidirectional thermostatic expansion valve* and is discussed in the next section.

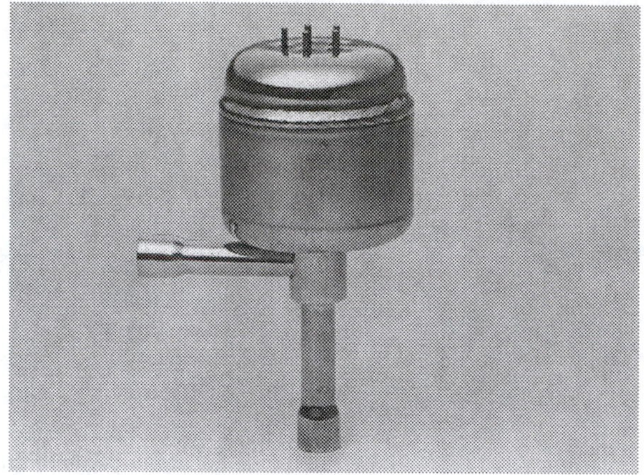

FIGURE 8-22 Electric expansion valve. *(Courtesy Alco Controls)*

BIDIRECTIONAL THERMOSTATIC EXPANSION VALVES

When standard thermostatic expansion valves are used on heat-pump systems, check valves are often used to ensure that the refrigerant flows through the correct expansion device at the correct time, while reducing the pressure drop in the liquid line. The addition of the check valves, however, adds to the complexity of the refrigerant-piping circuit. These additional components, piping and pipe connections, increase the possibility of system malfunction and refrigerant leak. For these reasons, manufacturers have introduced specially designed **bidirectional thermostatic expansion valves** that have the check valve built into the valve itself. This valve looks and operates in a manner similar to that of other TXVs, with the exception of the internal check valve.

In operation, when refrigerant flow is desired through the expansion device, the internal check valve is in the closed position and the refrigerant is directed through the needle-and-seat assembly of the expansion valve. When refrigerant flows through the valve in the opposite direction, when the bypass feature is desired, the check valve opens and the refrigerant flows around the needle-and-seat assembly, thereby reducing the pressure drop across the component. Note that the pressure drop across the valve in the bypass mode cannot be completely eliminated. A 5-ton air-conditioning system operating with R-22 as its refrigerant, for example, will experience a pressure drop of approximately 15 psig across the check valve, while a 2-ton air-conditioning system will experience a pressure drop in the range of 3 psig.

Normally, heat-pump systems are equipped with two such expansion valves, one at the inlet of the evaporator in both the heating and cooling modes of operation. In the cooling mode, the needle-and-seat assembly on the valve installed at the outdoor coil will be bypassed; while the valve at the indoor coil will be bypassed in the heating mode.

Another common type of bidirectional thermostatic expansion valve uses a stepper motor to control the movement of the valve's needle (Figure 8-22). The motor used to control this valve is either 12 or 24 volts DC and can be used in conjunction with electronically controlled systems. Configured just like a split-phase motor, the device is equipped with a stationary stator and a rotor that is connected directly to the needle assembly (Figure 8-23). In operation, it takes anywhere from 11 to 15 seconds for the valve to go from the full-open to the full-closed position. Similar to the construction of a hermetically sealed compressor, the electrical connections are made through a hermetic feed in the body of the device. This reduces the possibility of refrigerant leakage.

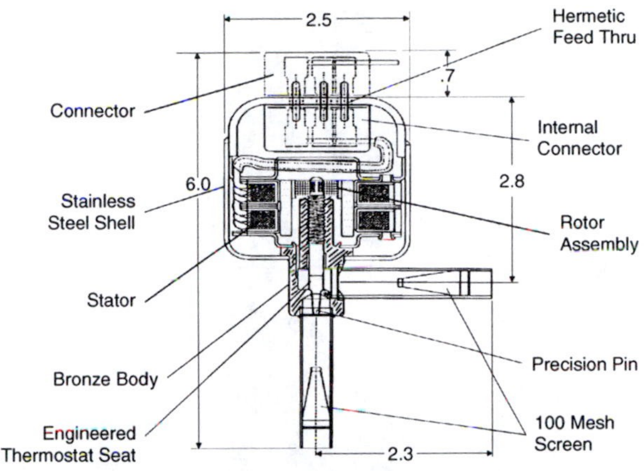

FIGURE 8-23 Cutaway of the electric expansion valve. *(Courtesy Alco Controls)*

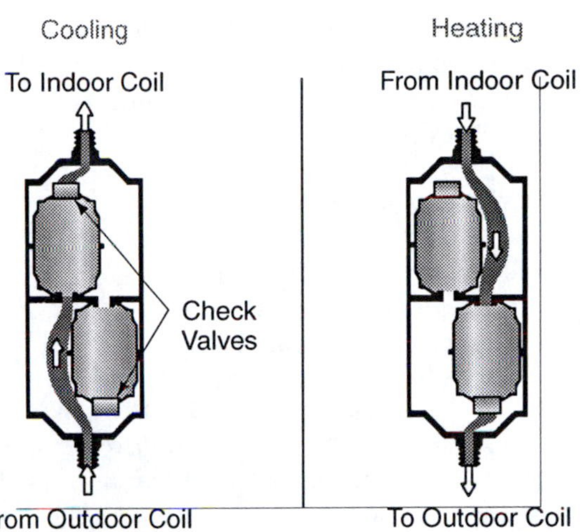

FIGURE 8-24 Bidirectional filter drier

BIDIRECTIONAL FILTER DRIERS

Just as it is common to find two expansion devices in a heat-pump system, it is also common to find two filter driers in the liquid line of the system. Such a piping configuration was shown in Figure 8-20. In another attempt to simplify the system installation and maintenance processes, specially designed filter driers allow refrigerant to flow through them in both directions, while still properly filtering and removing moisture from the refrigerant as it flows through. This device is called the **bidirectional liquid-line filter drier** (Figure 8-24).

Driers intended for single-directional flow have an indicator arrow on the shell that points in the direction that refrigerant should flow through it. Biflow driers have a double-headed arrow on the shell, indicating that the device can be installed in the piping circuit in either direction without affecting the operation of the device. The internal piping of the biflow filter drier, as shown in Figure 8-24, indicates that the device is made up of two smaller driers in a single shell and utilizes check valves to ensure the correct direction of flow. Referring to the figure, while the system is operating in the cooling mode, the drier on the top is being utilized, and, while operating in the heating mode, the drier at the bottom is in the active refrigerant circuit. This type of drier is a throwaway device, just like any other filter drier that is not equipped with replaceable cores.

SERVICE NOTE: *The formation of frost at the outlet of the liquid-line filter drier is an indication that the filter is blocked. In this case, the drier is acting as an expansion device, creating a substantial pressure drop. A blocked filter drier must be replaced if this occurs, unless of course the core of the drier is of the replaceable type. The installation of a sight glass with a moisture indicator at the outlet of the drier is also recommended. When replacing a filter drier, be sure to leave the protective seals on the device as long as possible to prevent the introduction of moisture or particulate matter into the shell of the device.*

SERVICE NOTE: *When soldering or brazing a filter drier into the system, good field practice dictates wrapping a damp cloth around the shell of the drier and directing the heat from the torch away from the drier.*

SUCTION-LINE ACCUMULATORS

The purpose of the evaporator in an air-conditioning system is to absorb heat from the medium to be cooled and, in an ideal system, completely vaporize the liquid refrigerant in the coil before allowing it to travel back to the compressor. An evaporator that functions correctly therefore helps protect the compressor from liquid floodback. However, certain system situations can result in the presence of liquid refrigerant in the suction line returning to the compressor. They include:

- Dirty or blocked evaporator coil
- Dirty or blocked air filters
- Defective evaporator fan motor
- Defective or dirty evaporator blower or fan
- Broken evaporator blower belt

Any of these situations will result in a reduction of air moving through the evaporator, thereby reducing the ability of the refrigerant to absorb heat and boil into a vapor. In addition to the situations mentioned here, heat-pump systems must also contend with the fact that, when operating in the heating mode, the outdoor coil, or evaporator, has the tendency to freeze. Freezing occurs when the surface temperature of the coil drops below 32°F and the condensing moisture from the air begins to freeze on the coil. Evaporator freezing prevents the refrigerant from absorbing heat from the outside air. The ice acts as an insulator between the refrigerant in the coil and the outside air. The refrigerant in the evaporator and suction line therefore remains in the liquid state, which can then travel back to the compressor, causing damage to the vapor pump.

To help prevent damage to the compressor, a **suction-line accumulator** (Figure 8-25) is installed in the suction line leading back to the compressor. On heat-pump systems, the accumulator is located in the suction line between the compressor and the reversing valve (Figure 8-26). The **accumulator** is a device that allows any liquid refrigerant that may be present in the suction line to boil into a vapor before returning to the compressor. The refrigerant leaving the evaporator flows into the top of the accumulator, and any liquid present will fall to the bottom of the shell. The outlet of the device is also at the top and is connected to a U tube that extends into the shell. The compressor's pumping action pulls the vapor refrigerant from the shell of the accumulator back to the compressor itself. Since the opening of the U tube is located near the top of the shell, any liquid refrigerant in the lower portion of the accumulator's shell must therefore boil into a vapor before returning to the compressor. The accumulator is usually located close to the compressor (Figure 8-27).

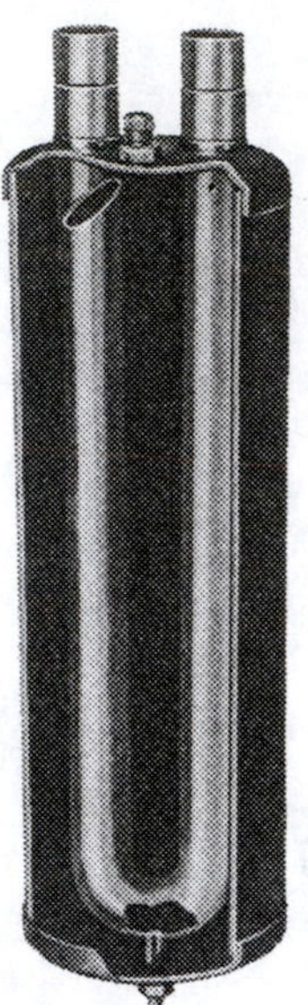

FIGURE 8-25 Suction-line accumulator. *(Courtesy AC&R Components, Inc.)*

Oil Return from the Accumulator to the Compressor

If the suction-line accumulator were constructed exactly as described in the previous section, the oil that normally travels with the refrigerant as it flows through the system would eventually migrate to the accumulator and cause the system to operate with less than the desired amount of oil. If oil is added to the system to compensate for the low oil level, it too will eventually wind up in the accumulator. To prevent this oil accumulation from occurring, the U tube in the accumulator has a small hole at the bottom of the bend (Figure 8-25). This oil-return hole is provided for the sole purpose of allowing the refrigeration oil to return to the compressor crankcase. This small hole helps ensure that when the amount of oil trapped in the accumulator exceeds the desirable level, it is able to slowly return to the compressor in a manner that will not damage the pump. Since the opening is not located at the very bottom of the shell, a small amount of oil is normally in the accumulator at all times.

Heat Exchangers in the Accumulator

Quite often, suction lines with four piping connections are encountered in the field. This type of accumulator is equipped with a **heat exchanger**. The heat

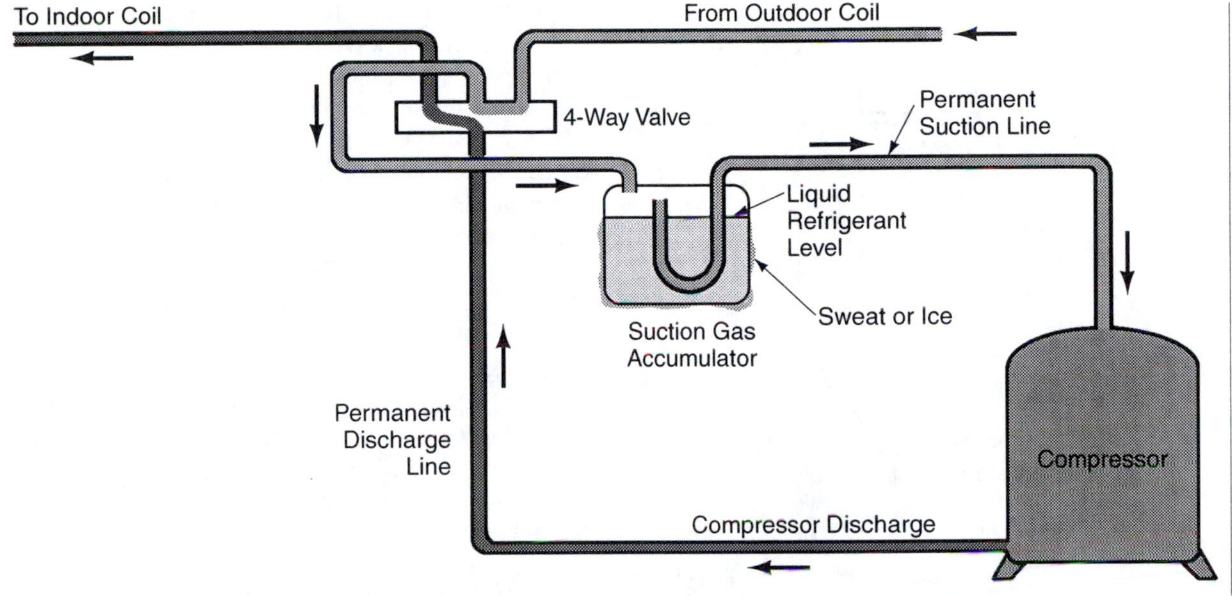

FIGURE 8-26 Accumulators are located in the suction line between the compressor and the reversing valve.

FIGURE 8-27 Suction-line accumulator located next to the compressor. *(Photo by Eugene Silberstein)*

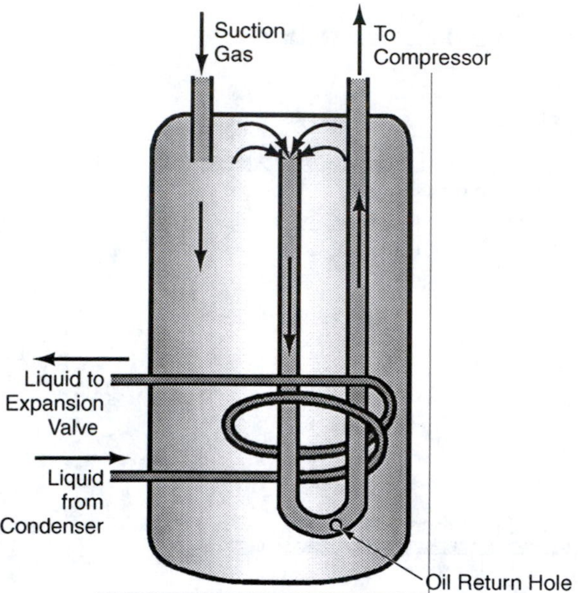

FIGURE 8-28 Suction-line accumulator with an internal heat exchanger

exchanger is an internal coil of tubing that is part of the system's liquid line (Figure 8-28). The purpose of this heat exchanger is to help increase the system's efficiency by transferring heat from the high-temperature liquid refrigerant leaving the condenser to the low-temperature refrigerant entering the compressor. This benefits the system in three ways:

- By removing heat from the liquid refrigerant, increasing the subcooling, before it enters the expansion device, the efficiency of the expansion device is increased.
- Increased subcooling increases both the system and evaporator efficiency.
- The heat added to the refrigerant in the suction line helps to boil any liquid refrigerant in the accumulator.

Why Do Suction-Line Accumulators Sweat?

The purpose of the suction-line accumulator is to allow any liquid refrigerant to vaporize before entering the compressor. For this to occur, heat must be absorbed by the refrigerant. The source of this heat is the air that surrounds the shell of the accumulator. If the accumulator were insulated, the rate of heat exchange between the refrigerant and the heat from the surface of the accumulator shell would be greatly reduced. Therefore, the accumulator is not insulated. Since the accumulator is located in the suction line and contains low-temperature refrigerant that is well below the dew-point temperature of the surrounding air, the device sweats. Therefore, observing puddles of water under the device and even traces of rust on the exterior of the shell itself is quite common.

SERVICE NOTE: *Accumulators are normally constructed of steel and are coated with a rust inhibitor to prevent them from rusting. As the component ages, or as the painted surface gets scratched or damaged, the accumulator is bound to rust and eventually cause the system to leak. Rusted accumulators must be carefully inspected periodically and should be replaced if the rust has compromised the integrity of the system.*

HEAT-PUMP-SYSTEM DEFROST METHODS

When operating in the heating mode, the outdoor coil, the evaporator, often freezes up. Since the evaporator-coil temperature is generally 20 to 25 degrees colder than the air passing over it, frost can begin to form on the surface of the coil when the outside ambient temperature is as high as 50°F. Ice buildup on the coil reduces the refrigerant's ability to absorb heat from the outside air, which in turn causes the coil temperature to drop even further, leading to even more ice buildup on the coil. Since the ice on the coil acts as an insulator between the refrigerant and the outside air, this ice must be removed for the system to operate properly. The process of removing this ice, called the **defrost cycle**, is not difficult but must be geared toward the ambient conditions that surround the outdoor unit. The following should be considered:

- Units operating in milder climates will require little, if any, **defrost**.
- Units operating during periods of high humidity will experience more frost buildup on the outdoor coil.

- Units operating during cold and wet conditions will accumulate ice at an accelerated rate.
- Faster ice formation on the outdoor coil will require more frequent defrost cycles.

The fact that the outdoor coil can operate as either the condenser or the evaporator, depending on the mode of operation, is the basis on which the defrosting of the outdoor coil is achieved. By switching the unit over to the cooling mode, the cold evaporator now becomes the hot condenser. The heat-laden refrigerant vapor from the compressor will flow first to the outdoor coil, transferring its heat to the ice that has formed on the coil, allowing it to melt. During the defrost cycle, the condenser fan motor is often de-energized in order to help concentrate the heat in the coil, helping to speed up the defrost process even when the outside ambient temperature is very low.

The defrost cycle must be initiated and terminated in a manner that that will best suit the design and operating conditions of the system. Decisions about when to start and stop defrost, as well as how to start and stop the cycle, vary from one manufacturer to the next. However, all defrost systems rely on at least one of the following to initiate and/or terminate defrost:

- Time
- Temperature
- Pressure

Each of these is discussed in detail in the following sections; but, regardless of the methods used to initiate and terminate defrost, all are designed and intended to achieve the same end result. During the defrost cycle, the following operations are completed:

- The reversing valve switches over to the cooling mode of operation.
- The outdoor fan motor is de-energized (most of the time).
- The air supplied to the occupied space is heated slightly to prevent cold air from entering the space during defrost.
- The cycle is long enough to ensure proper defrosting of the outdoor coil.
- The cycle is short enough so as not to have a large effect on the normal system operation.

Some of the most common defrost methods are described in the following sections. Defrost cycles that are initiated at predetermined time intervals are not always the most efficient, as will be seen shortly. Other methods that are only initiated when needed, called **demand defrost**, allow the system to operate more efficiently by reducing the defrost time, thereby allowing the system to heat the occupied space more evenly.

Time-Initiated–Time-Terminated Defrost

In this type of defrost, a **timer** (Figure 8-29) is used to both initiate and terminate the defrost cycle. In operation, the time-initiated–time-terminated method of defrost places the system in defrost mode at predetermined time intervals. A typical defrost cycle could be that for every 90 minutes of compressor run time the defrost cycle would be initiated for 10 minutes. As with other methods of defrost, this method will:

- Switch the system over to the cooling mode
- De-energize the outdoor fan motor
- Allow for the tempering of the air supplying the occupied space

There are, however, many drawbacks to this type of defrost method, making it a very unpopular method

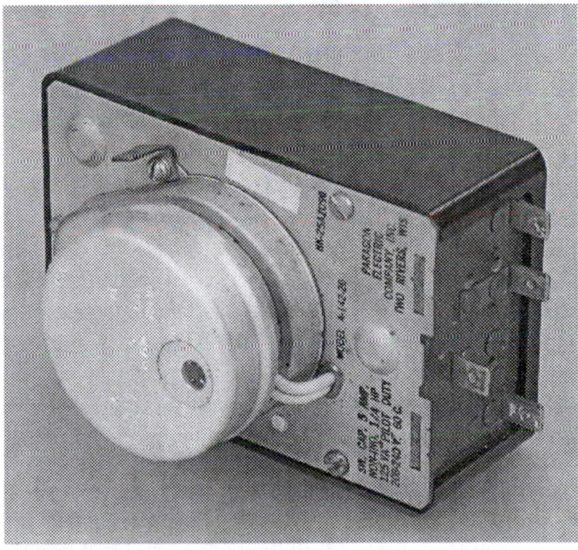

FIGURE 8-29 A typical defrost timer. *(Photo by Bill Johnson)*

to remove ice accumulation on the outdoor coil. These drawbacks include the following:

- The unit may go into defrost when no ice is present on the outdoor coil.
- The unit may come out of defrost before all of the ice on the coil has melted.
- The unit may stay in defrost much longer than needed.
- Additional defrost cycles reduce system effectiveness and efficiency.

If a heat-pump system goes into defrost when no ice is on the coil or if the system stays in defrost too long, unnecessary cool air may be introduced into the occupied space, reducing the effectiveness of the system. If the system comes out of defrost before all of the ice has been removed, ice will continue to form on the coil until, after a number of defrost cycles, the coil will remain frozen and the system performance will be greatly reduced. For these reasons, the time-initiated–time-terminated defrost cycle is no longer found on heat-pump systems but provides food for thought in the development of newer, more efficient defrost methods.

Time-and-Temperature-Initiation–Temperature-Termination Defrost

To help eliminate some of the problems with the time-initiated–time-terminated method of defrost, temperature was introduced as a factor in determining when the defrost cycle would be initiated and terminated. The temperature of the outdoor coil is measured by a device similar to the one in Figure 8-30. As stated in the previous section, the time-initiated defrost would begin continually at predetermined time intervals. This did not take into account the possibility that the coil may not be frosted and may not need to go into defrost.

By adding the temperature factor into the mix, two conditions must be met for defrost to start:

- The proper time period must have elapsed since the previous defrost attempt. This time period is normally set by the field technician and is typically 30, 45, or 90 minutes.
- There must be frost on the coil. This is determined by the coil temperature. If the coil temperature is low enough, typically lower than 26 or 28°F, the defrost cycle can begin as long as the time condition is also met.

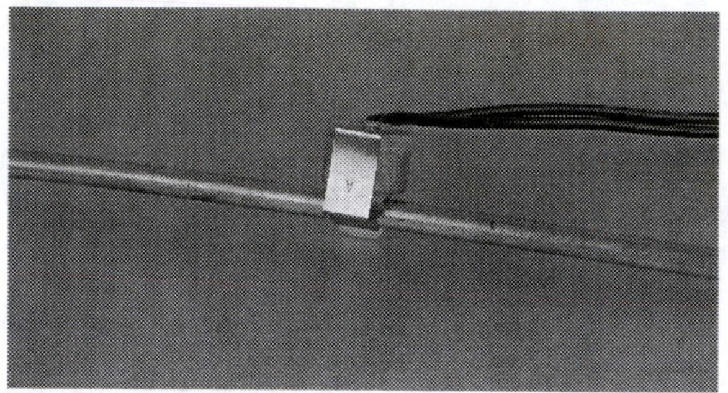

FIGURE 8-30 A defrost-termination thermostat. *(Photo by Bill Johnson)*

As can be seen from Figure 8-31, defrost is initiated only when both the time and temperature constraints are satisfied. Electrically speaking, these constraints can be represented by two switches wired in series with each other. The signal to defrost the outdoor coil can only pass when both switches are in the closed position. If the proper amount of time between defrost attempts has elapsed, the normally closed contacts in the defrost timer will close. If, however, the outdoor coil temperature sensor establishes that the outdoor coil is too warm to warrant the initiation of a defrost cycle, no defrost will take place. Additionally, if the outdoor coil temperature sensor closes, indicating that sufficient ice has formed on the coil, and the defrost timer contacts are open, indicating that the preset time period between defrosts has not passed, no defrost will be initiated. If, however, the coil temperature sensor contacts are closed, as soon as the defrost timer contacts close, defrost will be initiated.

Once these conditions are met and the defrost cycle has begun, it is terminated solely by the temperature of the outdoor coil. A sensing bulb or element mounted on the outdoor coil determines when the system will be taken out of defrost and placed back into the heating mode. The temperature that triggers the termination of defrost is typically in the range of from 55 to 60°F. Under normal conditions, a complete defrost cycle should last approximately 5 to 7 minutes with moderate frost, at which time the outdoor coil should be completely frost free. If there is

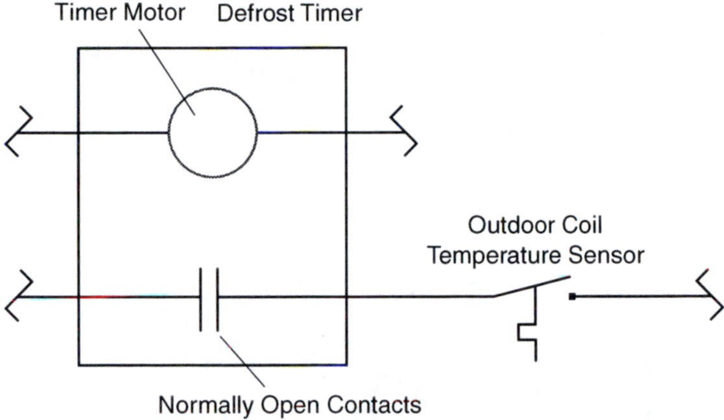

FIGURE 8-31 The timer contacts and outdoor coil, temperature-sensor contacts are wired in series with each other.

excessive frost or there is a system malfunction and the system fails to defrost the coil in a reasonable amount of time, usually 10 to 12 minutes, the system will automatically switch back to the heating mode. If this situation occurs, and the system is otherwise functioning properly, the time period that elapses between defrost attempts should be shortened.

Temperature-and-Air-Pressure-Initiation–Temperature-Termination Defrost

This method of defrosting the outdoor coil of a heat-pump system is very popular because it provides an effective and very efficient way to defrost the coil. In contrast to the method just described, time is not a factor that is taken into consideration when the defrost cycle is initiated. This method relies solely on:

- The temperature of the outdoor coil
- The pressure differential across the outdoor coil

For the unit to go into defrost, the temperature of the outdoor coil must be lower than the predetermined set point, as determined by either the factory setting in the control or the field setting as adjusted by the service technician. This temperature setting is generally in the range of 26 to 28°F and is monitored by a defrost thermostat, which closes its contacts when the coil temperature drops to the desired point.

In addition to the defrost thermostat, another component, the air-pressure switch, is also used to initiate the defrost cycle. The air-pressure switch (Figure 8-32) has the ability to measure the pressure both inside and outside the outdoor coil casing. A bellows located within the pressure-sensing device determines the difference in pressure between the inside and outside of the condenser casing. Since the pressure outside the casing and the pressure inside the casing push against opposite sides of the diaphragm within the device, the net pressure that is sensed is the difference between the two. For example, if the pressure inside the casing is the same as the pressure outside the casing, the net pressure difference will be 0 or, more realistically, very close to 0. These pressures are both important because the control, by taking the difference between them, can determine how much frost has formed on the coil's surface. This method of defrost is successful mainly because the pressure differential across the outdoor coil increases as the amount of ice on the coil increases. As more ice forms on the outdoor coil, the air pressure inside the casing drops to a pressure lower than that outside the casing, increasing the difference between the two pressures. Once the pressure differential has reached the predetermined set point, the contacts of the device will close in order to initiate defrost, as long as the temperature requirement for defrost is also met. This pressure differential is relatively small and is measured in units called **inches of water column** or, simply, inches.

Even when no frost is present on the outdoor coil, a small pressure differential still is caused by the fin-and-tube construction of the coil itself. For this reason, the contacts on the air-pressure switch do not

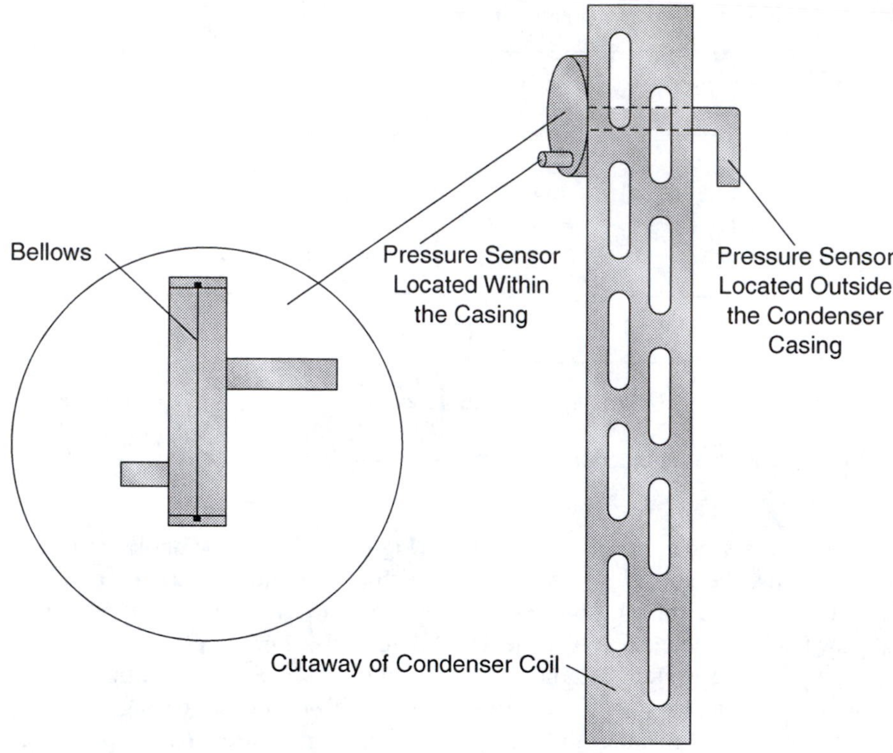

FIGURE 8-32 The pressure sensor measures the pressure both inside and outside the casing.

normally close until the pressure differential across the coil reaches approximately 0.5 inches of water column, depending on the specific control used and the construction of the coil.

Since both the temperature and the pressure differential conditions must be satisfied in order for the system to go into defrost, a number of situations that could cause the system to go into unnecessary defrost have been eliminated. For example, if the defrost initiation was only controlled by the pressure differential across the coil, the following scenarios could cause the unit to go into defrost prematurely:

- Dirty or blocked outdoor coil
- Erratic winds blowing through the outdoor coil

The addition of the temperature constraint eliminates this possibility. Just as with the defrost method described before, once the defrost cycle has been initiated, the cycle is terminated based on the temperature of the outdoor coil.

Solid-State Defrost

The defrost methods already described relied on a mechanical timer as well as other mechanical devices such as pressure switches and temperature sensors. Newer heat-pump systems do not employ mechanical defrost systems but, instead, have turned to solid-state technology. Instead of using bimetal temperature sensors, solid-state systems often utilize thermistors that sense changes in temperature and change their internal resistance according to these temperature changes. In many cases, the defrost system employs two thermistors (Figure 8-33). One senses the temperature of the outdoor coil, while the other senses the temperature of the outside ambient air. The differential between the resistances of the two thermistors is what initiates the defrost cycle. As frost accumulates on the outdoor coil, the temperature sensed by the thermistor will drop, since the ice acts as an insulator between the coil and the air passing through it. This will cause the resistance of the

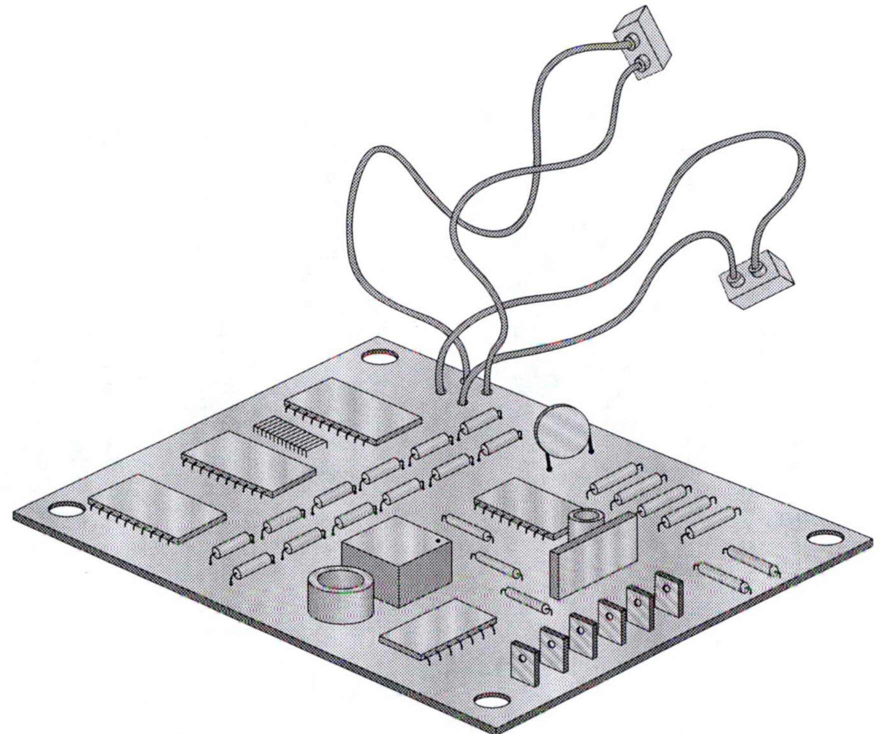

FIGURE 8-33 Solid-state defrost control board with two thermistors

thermistor to drop. A temperature differential in the range of 15 to 20 degrees will normally initiate defrost. Once the desired differential is achieved, the printed circuit board will energize the necessary contacts and holding coils to bring the system into defrost. Defrost is often terminated by a high-side pressure switch that is connected to the solid-state defrost control board.

Solid-state defrost can also be initiated by temperature; pressure; or a combination of pressure, temperature, and time. When time is used as a method to initiate defrost, solid-state timers are used, which are more reliable than mechanical, cam-type clocks. The time interval between defrost attempts can often be altered in the field by repositioning a jumper wire on the defrost circuit board itself (Figure 8-34). Many solid-state control boards give the field technician the choice of 30-, 60-, or 90-minute time intervals between defrost attempts. When pressure-initiated defrost is utilized, a dual-pressure-sensing switch similar to that used in the mechanical-type defrost control is connected to the circuit board and signals the system to enter defrost when the pressure differential rises to a predetermined point. Just as with the mechanical-type defrost systems described earlier, the defrost method that is ultimately used should be chosen based on the geographic location and requirements of the system.

AUXILIARY ELECTRIC HEAT STRIPS

Upon initiation of the defrost cycle, the system is switched over to the cooling mode of operation in order to effectively remove the ice that has formed on the outdoor coil. Therefore, during the defrost cycle, cool air is introduced to the occupied space. This seems to be counterproductive since you are now cooling the space that you have just heated. If left alone, this is exactly what will happen. For this reason, keep the defrost cycles as short as possible and minimize the number of cycles that are initiated.

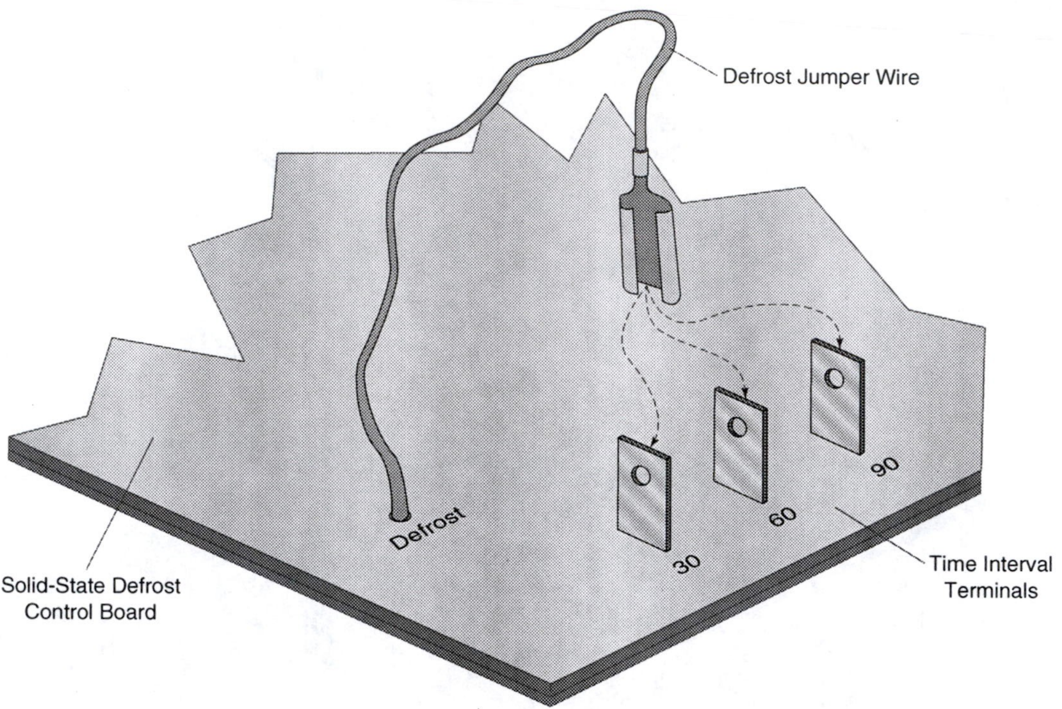

FIGURE 8-34 The time interval can be changed by repositioning the jumper wire on the circuit board.

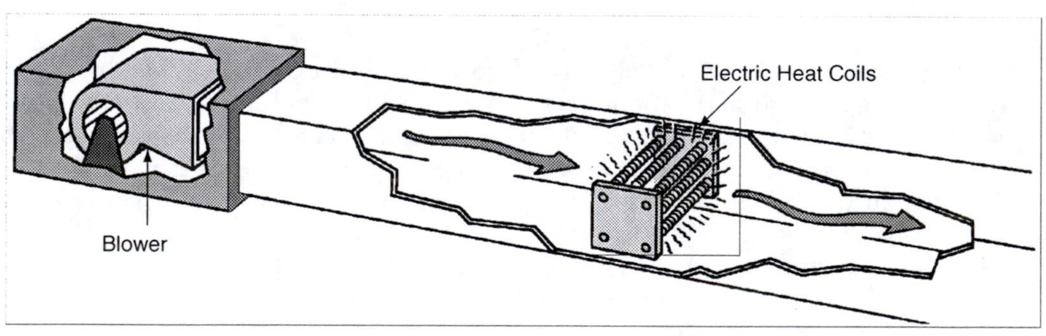

FIGURE 8-35 Electric heaters located in the supply duct

To reduce the amount of cool air being supplied to the occupied space during defrost, the air is heated slightly before entering the space to decrease the cooling effect that this air will have. This heating of the air is accomplished by supplementary, electric-strip heaters (Figure 8-35). These heaters are located within the supply ductwork of the system and are energized when the system goes into defrost. These heaters are resistive-type devices and will overheat if no air is passing over them. If this occurs, a fire hazard is created, as well as the possibility of component damage. To alleviate this potentially dangerous situation, a **sail**, or **flow switch** is installed in series with the heater's electric circuit. This will prevent the heaters from being energized if there is no airflow in the ductwork. The heaters are also equipped with high-limit controls that will de-energize the heaters if the temperature rises too high.

These electric heaters are also used to provide **supplementary heat** in the event that the heat-pump system cannot effectively satisfy the heating requirements of the occupied space. The control of this supplementary heat is controlled by the room thermostat.

HEAT-PUMP THERMOSTATS

Thermostats in general were covered earlier in the text. Heat-pump thermostats operate in a manner very similar to those discussed earlier in that they control the overall operation of the system. To briefly recap, the thermostat functions to:

- Turn the entire system on and off.
- Allow for the desired mode of operation, namely, heating, cooling, or simply air circulation.
- Allow for proper space temperature to be maintained.
- Automatically change the room temperature settings on devices so equipped, namely, electronic and programmable thermostats.

In addition to these functions, heat-pump thermostats are designed to perform a more varied range of functions, since the heat-pump system itself is more complex than a standard heating/cooling system. Heat-pump thermostats have these additional capacities:

- They are equipped to control two stages of heat.
- They are sometimes equipped with an automatic-changeover setting.
- They provide for emergency heat in the event of system malfunction.
- They provide a visible indication to the home owner or system operator when there is a trouble condition within the system.

Since the effectiveness of a heat-pump system varies depending on the ambient temperature in which it is operating, additional or supplementary heat may be needed. This heat source, in most cases, is electricity. If the heat-pump system is unable to maintain the space temperature effectively, this supplementary heating element will be energized in an attempt to aid the system in satisfying the heating load on the occupied space. For this reason, the heat-pump thermostat must be designed to control the two stages of heat that are present in the system, namely, the primary heat source, the heat pump itself, and the supplementary heat source.

Some heat-pump systems, as well as conventional heating/cooling systems, are equipped with automatic-changeover settings. This allows the system to automatically change over from heating to cooling and vice versa in order to maintain the desired temperature range within the space. In the case of the heat pump, it simply energizes and de-energizes the solenoid coil that controls the four-way reversing valve to accomplish this changeover.

Heat-pump systems also provide for the use of **emergency heat** in case of a system malfunction. This mode of operation is referred to as *emergency heat* and is initiated by the system operator by a switch on the face of the thermostat itself. If for any reason the heat pump is rendered inoperative, the supplementary heat will be energized and perform the function of the primary heat source. When the emergency heat mode is used, an indicator light on the face of the thermostat will illuminate, alerting the occupants that this mode of operation is in use. Since electric heat is much more costly to operate than the normal heat pump, the cause for the malfunction should be identified and remedied as quickly as possible.

In addition to the features of the heat-pump thermostat already mentioned, heat pumps are required to have manually reset, high-pressure cutouts in case the system operates outside of its designed pressure ranges. These devices are similar to those discussed earlier in the text. When these high-pressure cutouts open, the system is locked out and cannot operate until the control is manually reset. Many heat-pump thermostats are equipped with trouble lights that will alert the occupants of the lockout condition.

SUMMARY

- The four-way reversing valve permits the automatic changeover from heating to cooling, and vice versa.
- The outdoor coil functions as the condenser in the cooling mode and the evaporator in the heating mode.
- The indoor coil functions as the condenser in the heating mode and the evaporator in the cooling mode.
- Heat pumps can be either air- or liquid-source units and can be either air or liquid cooled.
- Geothermal heat pumps use the earth as a heat source, as well as a heat sink.
- Reversing valves can be either direct acting or pilot operated.
- Check valves ensure that refrigerant flows in the proper direction through the piping circuit.
- Bidirectional thermostatic expansion valves have built-in check valves and allow refrigerant to flow through them in either direction.
- Bidirectional filter driers are designed to operate effectively when refrigerant flows through them in either direction.
- Accumulators are designed to reduce the possibility of liquid refrigerant returning to the compressor.
- Accumulators equipped with heat exchangers help increase the efficiency of the system.
- System defrost is accomplished by putting the system in the cooling mode temporarily and is controlled by a combination of time, temperature, and pressure.
- Heat-pump thermostats are specially designed to control the additional components found on heat-pump systems.

KEY TERMS

Accumulator
Air-source heat pumps
Air-to-air heat pump
Air-to-liquid heat pump
Bidirectional liquid-line filter drier
Bidirectional thermostatic expansion valve
Check valve
Closed-loop system
Defrost
Defrost cycle
Defrost timer
Demand defrost
Disk-type check valve
Emergency heat
Flow switch
Four-way reversing valve
Geothermal heat pump
Ground coils
Ground loops
Heat exchanger
Heat pump
Inches of water column
Indoor coil
Liquid-source heat pumps
Liquid-to-air heat pump
Liquid-to-liquid heat pump
Open-loop system
Outdoor coil
Pilot-operated reversing valve
Reversing valve
Sail switch
Suction-line accumulator
Supplementary heat
Timer

FOR DISCUSSION

1. How can a service technician, faced with a system that constantly has ice buildup on the outdoor coil, remedy the situation?
2. A customer calls her service company and complains about unusually high heating bills during winter-month operation. What are the possible causes of the high utility bills, and what can be done to remedy the problem?

REVIEW QUESTIONS

1. During the cooling cycle of a heat-pump system, the indoor coil operates as the __________ and the outdoor coil operates as the __________.
2. During the heating cycle of a heat-pump system, the hot gas leaving the compressor travels first to the:
   a. Expansion device.
   b. Evaporator coil.
   c. Condenser coil.
   d. Four-way reversing valve.
3. All of the following are true about heat-pump systems except:
   a. The condenser fan is normally de-energized during the defrost cycle.
   b. The outdoor coil operates as the evaporator during the heating cycle.
   c. The compressor continues to operate during the defrost cycle.
   d. The indoor fan cycles off during system defrost.
4. An air-to-liquid heat-pump system:
   a. Uses air as the heat source while operating in the heating mode.
   b. Uses liquid as the heat source while operating in the heating mode.
   c. Uses air as the heat source while operating in the cooling mode.
   d. None of the above are correct.
5. The type of system that utilizes open loops is the:
   a. Air-to-air heat pump.
   b. Geothermal heat pump.
   c. Four-way heat pump.
   d. Liquid-to-liquid heat pump.
6. The two key components that make the automatic switchover from heating to cooling possible are the:
   a. Four-way reversing valve and the solenoid valve.
   b. Four-way reversing valve and the check valve.
   c. Check valve and the thermostatic expansion valve.
   d. Check valve and the solenoid coil.
7. Which of the following can have a negative effect on the operation of a heat-pump system?
   a. Excessive ice formation on the outdoor coil while operating in the cooling mode
   b. Excessive ice formation on the indoor coil when operating in the heating mode
   c. Excessive ice formation on the outdoor coil when operating in the heating mode
   d. Excessive ice formation on the discharge line while operating in the heating mode
8. A disc-type check valve is installed correctly when:
   a. It is in the vertical position.
   b. It is in the horizontal position.
   c. The indicator arrow on the device points in the direction of refrigerant flow.
   d. All of the above are correct.

9. A ball-type check valve is installed correctly when:
    a. It is in the vertical position.
    b. It is in the horizontal position.
    c. The indicator arrow on the device points in the direction of refrigerant flow.
    d. Both *a* and *c* are correct.
10. Excessive ice formation on the outdoor coils can result in:
    a. Increased system efficiency in the heating mode.
    b. Liquid refrigerant returning to the compressor.
    c. Improper heating of the conditioned space.
    d. Both *b* and *c* are correct.
11. The __________ is used to prevent liquid refrigerant from flowing back to the compressor.
12. Explain why defrost is necessary on heat-pump systems.
13. In which of the following climates would a heat-pump system tend to have more frequent defrost cycles?
    a. A warm, humid climate
    b. A cold, dry climate
    c. A cold, humid climate
    d. A warm, dry climate
14. A defrost cycle that is initiated by temperature will most likely use which of the following devices?
    a. A thermistor
    b. A pressure-sensing switch
    c. A defrost timer
    d. Both *a* and *b*
15. Which of the following will increase the subcooling in a system and increase its efficiency?
    a. A liquid-line solenoid
    b. A heat exchanger in the accumulator
    c. A thermostatic expansion valve
    d. A suction-line filter drier
16. Filter driers designed for use in heat-pump systems:
    a. Are often bidirectional.
    b. Are designed as two driers contained in a single shell.
    c. Must be installed in the vertical position.
    d. Both *a* and *b* are correct.
17. Which of the following defrost methods is the least efficient?
    a. Time-initiated–time-terminated
    b. Time-initiated–temperature-terminated
    c. Time/temperature-initiated–temperature-terminated
    d. Temperature/pressure-initiated–temperature terminated
18. Very cold air is being discharged into the occupied space during defrost. Which of the following is a likely cause for this situation?
    a. The indoor fan motor is not running.
    b. The outdoor fan motor is not running.
    c. The compressor is not running.
    d. The sail switch in the supply duct is defective.

19. Explain how heat-pump thermostats differ from conventional heating/cooling thermostats.

20. What thermostat feature enables the system to maintain the desired range of temperatures in the occupied space?

    a. Automatic changeover
    b. Emergency heat
    c. Automatic fan operation
    d. Two-stage cooling operation

CHAPTER 9

Troubleshooting Vapor-Compression Heat-Pump Systems

OBJECTIVES After studying this chapter, the reader should be able to:

- Explain the effects of a defective check valve on system operation.
- Explain how to determine if a check valve is malfunctioning.
- Explain how a leaking four-way reversing valve affects system operation.
- Explain how to determine if a four-way reversing valve is operating properly.
- Explain the effects of a blocked or clogged filter drier.
- Explain how to determine if a filter drier is blocked or clogged.
- Explain how a defective expansion valve affects system operation.
- Explain how to determine if an expansion valve is operating properly.

INTRODUCTION

As mentioned in Chapter 3, two fundamental issues separate an average service technician from an excellent one. These two factors are the speed and the accuracy with which the technician is able to perform his duties. These two elements play a major role, no matter what type of system is being serviced. In Chapter 3, troubleshooting of the major system components of a straight heating/cooling air-conditioning system was discussed. On heat-pump systems, however, additional components must be evaluated in order to properly troubleshoot and diagnose these systems. This chapter concentrates on these devices. From the mechanical standpoint, heat-pump systems differ from straight heating/cooling systems in that they are equipped with additional components that include:

- Check valves
- Four-way reversing valves

In addition to the check valves and reversing valves, expansion valves and filter driers on heat-pump systems also need to be properly evaluated. Although driers and expansion valves were discussed earlier on, these components, when installed and designed for heat-pump applications, tend to be constructed differently and operate differently than components designed for non-heat-pump-system applications. For this reason, these devices are examined again here, paying particular attention to the heat-pump applications for which they are intended.

OVERVIEW

Unlike straight heating/cooling systems, heat pumps tend to be somewhat more complex and, thereby, more difficult to troubleshoot. In a standard air-conditioning system, for example, the refrigerant flows in one and only one direction whenever the system is operating. In a heat-pump system, however, the refrigerant flows in different directions depending on the desired mode of operation. The direction of refrigerant flow is controlled by the flow-check valves, as well as the four-way reversing valves. As long as these devices function properly, the refrigerant flow can be properly controlled and the system will function as intended. If, however, these devices fail to perform as desired, system operation will be affected and major system component damage could occur. For example, a four-way reversing valve that develops an internal leak could allow the hot discharge gas from the compressor to be immediately returned to the suction port of the compressor, causing the compressor to burn out. An unsuspecting service technician may arrive on the job, determine that the compressor has burned out, and replace it without determining why the compressor became defective in the first place. After replacing the compressor and starting the system up, the technician will mostly likely to be shocked to find that the new compressor is defective as well! Being able to answer the question "Why did the compressor go bad?" will help the technician locate the cause of the problem, which is much more important than simply identifying the effect, or end result, of the problem.

A much more obvious scenario might look something like this. A home owner notices that her central air-conditioning system is not working, so she calls in her service contractor. The service contractor arrives at her home and notices that the line-voltage fuses at the condensing unit are blown. The technician replaces the fuses, and they blow again. Once again, the technician replaces the fuses and—you guessed it—they blow again . . . and again. This technician is attempting to remedy the effect, the blown fuses, without attempting to find the cause of the problem. Effective troubleshooting involves identifying not only the effect but the cause as well. So, to add to the initial examination of what distinguishes an average technician from an excellent technician, a more complete picture can now be painted. An excellent service technician with excellent troubleshooting skills must be able to:

- Identify the cause of the problem.
- Identify the effect(s) of the problem.
- Identify the causes and effects quickly and accurately.
- Convey findings to the office and to the customer in a courteous, professional manner.
- Conduct all service functions neatly and completely, always keeping the customer's concerns first and foremost. This includes paperwork and service tickets, fingerprints on the walls, and dirt marks on the carpets.

This chapter examines several components and provides the reader with an insight as to how these components should operate and what to look for when attempting to troubleshoot a system that is not operating correctly. Before getting into the heat-pump-system components, a hypothetical situation sets the stage for the material to follow.

Consider the condenser in Figure 9-1. This condenser has a mechanical hand valve connected between the inlet and the outlet of the condenser. This valve is intended to be in the closed position all the time, meaning that refrigerant never flows through it. Note in Figure 9-1(a) that the temperature of the refrigerant leaving the condenser is the same as the temperature of the refrigerant in the line flowing to the expansion device, namely, 100°F. The valve connected across the condenser is therefore in the closed position. Now consider the situation in Figure 9-1(b). The temperature of the refrigerant leaving the condenser is 100°F, but the temperature of the refrigerant increases to 120°F as it passes the bypass valve. Why is the temperature of the refrigerant suddenly increasing? It can be readily concluded that the hot gas from the compressor is bleeding through the supposedly closed valve and is then mixing with the refrigerant leaving the condenser. This leaking refrigerant will lead to a reduction in system efficiency, since the temperature of the refrigerant entering the expansion device will be higher than desired.

From this example, you can see that it is possible to troubleshoot and evaluate certain system components without having to physically remove them from the system. This is the main concern of this chapter. By knowing the difference between what *should be* and what *is,* the service technician can effectively perform his duties, thereby creating value for the customer as well as for the employer.

DEFECTIVE CHECK VALVE

In a typical heat-pump system utilizing two metering devices or expansion valves, two check valves frequently are connected in piping circuits that are connected parallel to the expansion devices. As discussed earlier, the purpose of the check valve is to divert refrigerant around the expansion valve that is not intended to be a part of the active refrigerant circuit. Figure 8-21 in the previous chapter shows the refrigerant path in both the heating and cooling modes of operation for such a system. If the check valve is not operating properly, the system's operating efficiency and effectiveness can be greatly reduced. The effects are different for the heating and cooling modes of operation and are outlined here. In

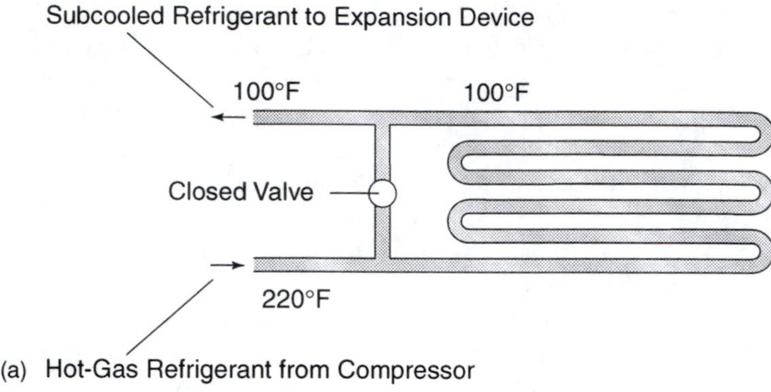

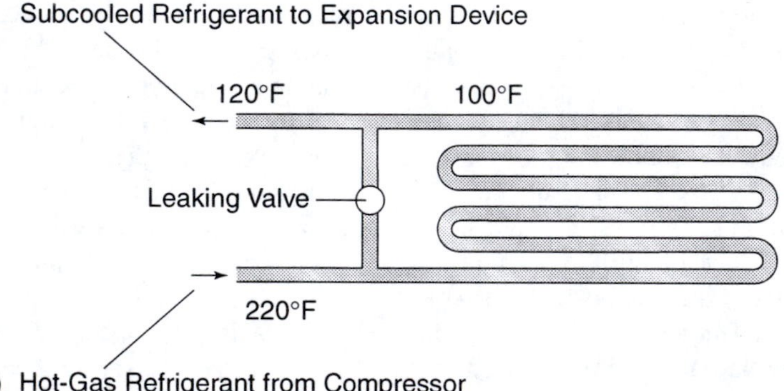

FIGURE 9-1 (a) This valve is in the closed position. The temperature of the refrigerant remains constant at 100°F. (b) This valve is leaking. Note how the temperature of the exiting refrigerant has increased.

the cooling mode, a malfunctioning check valve at the indoor coil can result in:

- High suction pressure
- Low or no superheat at the indoor coil
- Reduced refrigeration effect
- Liquid floodback to the compressor
- Major system component damage

In the cooling mode, a malfunctioning check valve at the outdoor coil can result in:

- High head pressure
- Low suction pressure
- High superheat at the indoor coil
- Greatly reduced refrigeration effect
- High compression ratios
- Reduced suction gas cooling to compressor

In the heating mode, a malfunctioning valve at the indoor coil can result in:

- High head pressure
- Low suction pressure
- High superheat at the outdoor coil
- Reduced heating of occupied space
- High compression ratios

In the heating mode, a malfunctioning check valve at the outdoor coil can result in:

- High suction pressure
- Low superheat at the outdoor coil
- Reduced ice formation on the outdoor coil
- Reduced suction gas compressor cooling
- Liquid floodback to the compressor

So how can the technician determine if the check valves are functioning properly without having to resort to removing the device from the system and physically inspecting the valve?

Is the Check Valve Defective?

As already known, the check valve is designed to be in either the open or closed position, depending on the mode of operation in which the heat-pump system is operating. If the system is operating in the cooling mode, the check valve at the outdoor unit will be in the open position, while the check valve at the indoor coil will be closed (Figure 9-2). When operating in this fashion, the temperature and pressure drops across the check valves will be either high or low, according to Figure 9-3.

If the check valve is in the open position, it is theoretically removed from the active refrigerant circuit.

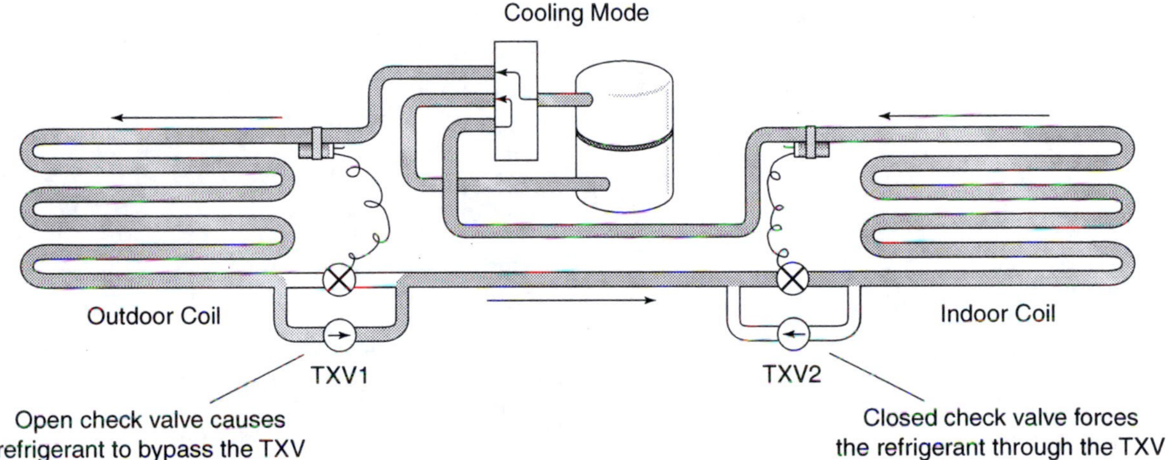

FIGURE 9-2 Heat-pump system operating in the cooling mode. The indoor check valve is in the closed position, while the outdoor check valve is open.

Cooling Mode		
	Outdoor Check Valve	**Indoor Check Valve**
Valve Position	Open	Closed
Temperature Drop Across the Valve	Low	High
Pressure Drop Across the Valve	Low	High
Heating Mode		
	Outdoor Check Valve	**Indoor Check Valve**
Valve Position	Closed	Open
Temperature Drop Across the Valve	High	Low
Pressure Drop Across the Valve	High	Low

FIGURE 9-3 Summary of pressure and temperature drops across check valves

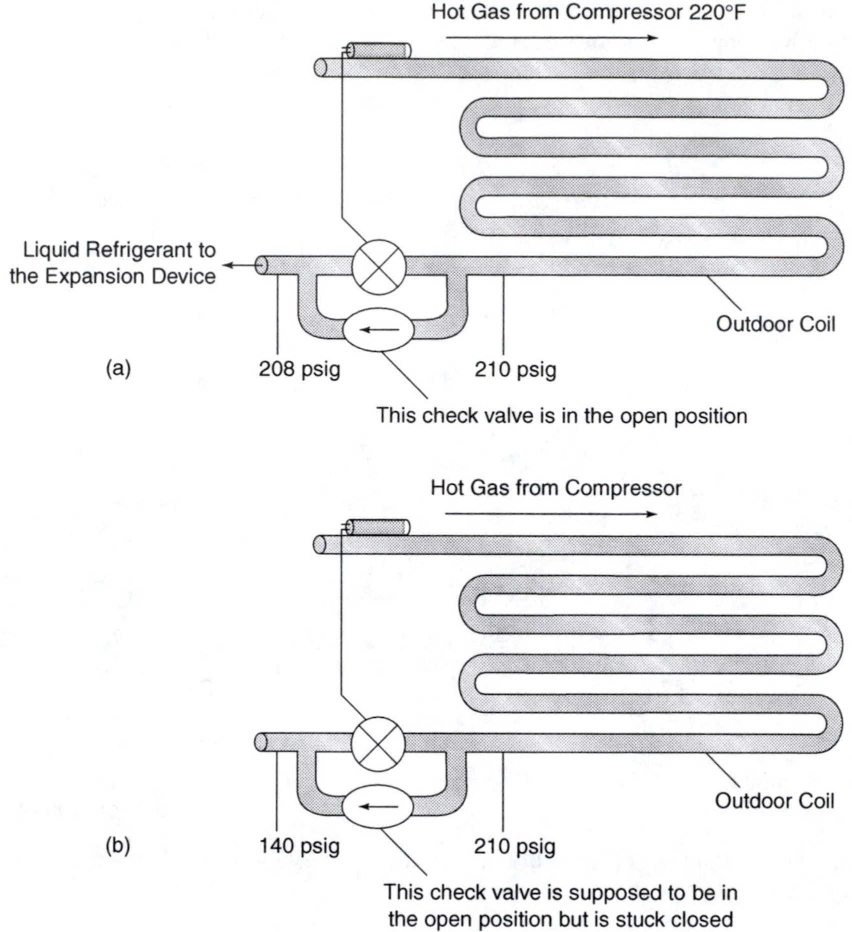

FIGURE 9-4 (a) This valve is in the open position allowing the subcooled liquid to pass through with only a small decrease in pressure. (b) This valve is stuck in the closed position. Note the large pressure drop across the valve.

Refrigerant should flow freely through the valve as if it were simply a small section of refrigerant piping. The device is therefore acting as a **pass-through device,** meaning that the refrigerant simply passes through the valve without experiencing any changes in temperature, pressure, or state. This would apply to the check valve located across the expansion valve at the outdoor coil while the system is operating in the cooling mode. However, a small pressure drop occurs across the valve even when in the open position. This is because of the internal structure of the valve, as well as the ball or disc that is located within the device. An open check valve is depicted in Figure 9-4(a). If the check valve is supposed to be in the open position and the pressure or temperature drop across the valve is high, the valve is closed, as in Figure 9-4(b).

If the check valve is in the closed position, no refrigerant should be flowing through the device. The closing of this valve will effectively direct all of the refrigerant through the expansion device. This would apply to the check valve located across the expansion device at the indoor coil while operating in the cooling mode. During this mode of operation, a large pressure and temperature drop should occur across the check valve, as in Figure 9-5(a). If the pressure and/or temperature drops across the valve are low when the device should be closed, the valve is in the open position, as in Figure 9-5(b).

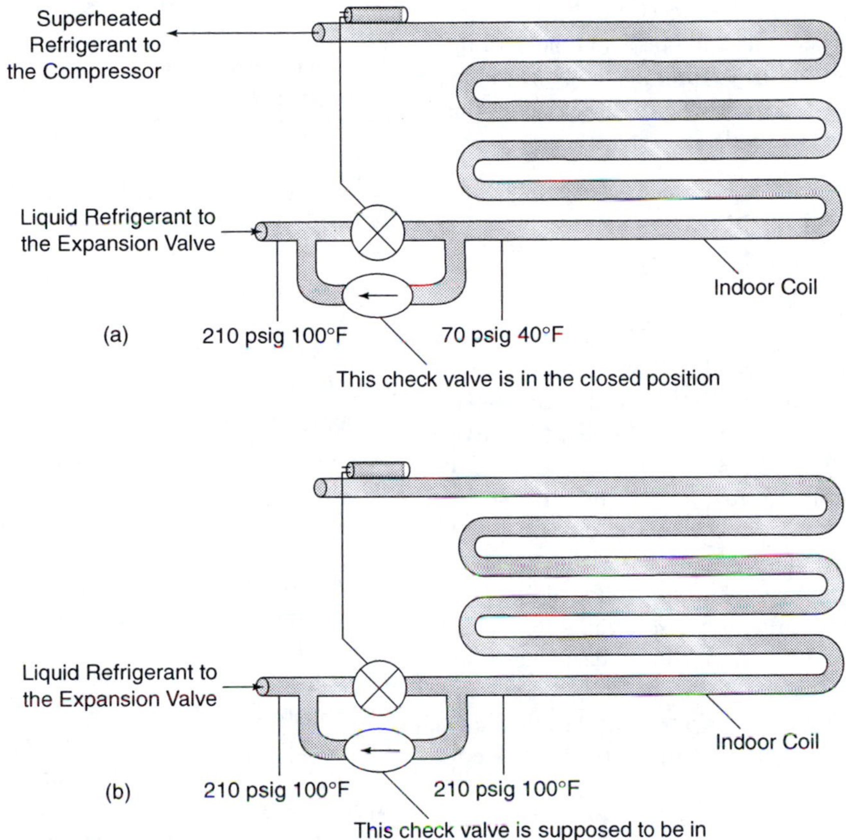

FIGURE 9-5 (a) This valve is in the closed position, forcing the refrigerant through the TXV. (b) This valve is open, allowing the subcooled liquid to enter the evaporator. The refrigerant will not be able to absorb any heat from the conditioned space.

Although measuring the pressure both before and after the check valve may be difficult due to piping restrictions in the system, measuring the temperature of the refrigerant line both before and after the device is always possible. A good-quality, digital thermometer should be used to ensure that accurate temperature readings are obtained.

SERVICE NOTE: *When using a digital thermometer to take readings of a refrigerant line, be sure to wrap the sensor, or probe, with insulating material. This will ensure that the surrounding air does not affect the thermometer's readings. Remember that the sensing element is round and, therefore, only a small portion of the element comes in physical contact with the surface of which you are measuring the temperature.*

Bad Check Valve or System Undercharge?

As just discussed, a defective check valve may give the servicing technician the impression that the refrigeration system is short of refrigerant. This will be true when the system is:

- Operating in the cooling mode and the outdoor check valve sticks in the closed position
- Operating in the heating mode and the indoor check valve sticks in the closed position

If a heat-pump system is suspected of operating with a low refrigerant charge, good field practice is to attempt to operate the system in its other mode. If the system fails to operate properly in the heating mode but functions properly in the cooling mode, the refrigerant charge in the system is not the problem. Similarly, if the system fails to operate in the cooling mode but functions in the heating mode, the refrigerant charge is not the problem. In either case, the check valve is most likely the culprit.

WHAT CAN BE DONE WITH A MALFUNCTIONING CHECK VALVE?

If the check valve is stuck closed when it should be open, or vice versa, the ball or disc within the valve is stuck and needs to be freed. To do this, the valve must be physically removed from the system. If a technician is going to go through the trouble of removing the valve from the piping circuit, good field practice is to simply replace the check valve. The logic behind this is simple. If the valve got stuck once, chances are good that it will stick again at some point in the future. Since the process of removing the check valve in the first place may involve pump-down, recovery, and evacuation, good field practice requires opening the system only once.

If, on the other hand, the suspicion is that the valve got stuck because of other system component failure or external events, the technician can try to free the valve externally by using a large magnet. The valve's body is typically made of brass, and the disc or ball is usually steel. By running a large magnet back and forth over the surface of the valve, the mechanism can possibly be freed without having to access the piping circuit. If this does not work, or if the valve is freed but then sticks again, it should definitely be replaced.

SERVICE NOTE: *When removing a defective flow-check valve from the system, make certain that the refrigerant has been properly recovered or the system has been properly pumped down. Removing a check valve or any other system component while there is pressure in the refrigerant lines is both a violation of EPA laws and a potentially hazardous situation that can cause severe physical injury to the technician, as well as to those in the surrounding areas.*

SERVICE NOTE: *When replacing flow-check valves, make certain the indicator arrow on the device points in the direction of desired refrigerant flow.*

MALFUNCTIONING FOUR-WAY REVERSING VALVES

Since the four-way reversing valve is intended to simply be a pass-through device, just as an open check valve, very little temperature or pressure drop should be experienced by the refrigerant passing through it. Hot discharge gas entering the reversing valve at 220°F should leave the valve at a temperature close to 220°F. Similarly, suction gas entering the valve at 40°F should leave the valve at a temperature close to 40°F. Any large discrepancies in temperature and/or pressure drop should be suspect, and the operation of the valve itself should be examined more closely. For the pilot-operated reversing valve to operate correctly, though, there must be a minimum **pressure differential** of 75 psig between the high- and low-pressure sides of the system. If the pressure differential is adequate, the integrity of the valve itself should be inspected. This means visually inspecting the valve body. Deep scratches, dents, or other imperfections in the body can be indications that the internal valve slide may not be able to move freely within the shell of the device.

Because the reversing valve is an electromagnetically controlled device, any failure of the device can be classified as either electrical or mechanical. An electrical malfunction of a four-way reversing valve is relatively straightforward and easy to determine. The technician must be able to answer the following two questions when determining whether an electrical failure is the problem:

- Is electrical current flowing to the solenoid coil of the valve?
- Is the coil creating the magnetic field necessary to cause the valve to switch positions?

If the answer to both of these questions is *yes,* then the failure of the reversing valve is of the mechanical nature. Probably the easiest way to determine if the solenoid coil is generating a magnetic field is to place a screwdriver or similar object next to the coil while it is energized, or supposed to be energized. For a system that fails in heating, for example, the solenoid coil will be energized in the cooling mode, and vice versa. The **fail position** for this valve would, therefore, be heating. If the valve is in the cooling mode when de-energized, the fail position is cooling. When energized, the pull of the magnetic field can be felt by placing the screwdriver next to the coil. If the coil is supposed to be energized and no magnetic field can be felt, the technician can then check to make certain that:

- Voltage is being supplied to the coil.
- Electrical continuity exists through the coil.

By using a voltmeter, the technician can determine if the proper voltage is being supplied to the coil. Also important is checking the voltage rating of the coil and making certain that the voltage being supplied is the same. If proper voltage is indeed being supplied to the coil, the power must be turned off and the coil removed from the circuit and checked for continuity. If no continuity exists through the coil, it is defective and should be replaced. Mechanical failure of the valve could have caused the coil to become defective. Further inspection should be made to ensure that the valve is functioning mechanically. If continuity exists through the coil, mechanical malfunction should be suspected. Mechanical damage is likely to be preventing the valve slide from moving within the valve body even though the magnetic field is being exerted on the valve. Physical damage to the valve may cause internal leaks or other mechanical failure.

LEAKING REVERSING VALVES

Since the refrigerant from the compressor's discharge port and the refrigerant flowing to the compressor's suction port must pass through the four-way reversing valve, internal leaks in the valve can have pronounced effects on system operation and performance. Internal leaks within the body of the reversing valves may or may not be evident, depending on the size of the leak. Small leaks might very well be present; but, if the system is still able to satisfy the heating or cooling requirements of the occupied space, a service call may never be initiated by the equipment owner. Larger leaks, however, can result in a number of problems, including:

- Compressor overheating and premature failure
- Excessive head pressure
- Higher than normal suction pressure
- Reduced suction gas cooling
- Reduced effectiveness in the heating mode
- Reduced cooling capacity in the cooling mode

The refrigerant charge within the system, however, is not lost when the valve develops an internal leak due to the fact that an internal leak simply transfers refrigerant from one section of the system to another, without escaping to the atmosphere.

An internal leak in the reversing valve can be detected easily by taking temperature readings at strategic locations in the system, primarily at the port connections on the reversing valve itself. This is often referred to as a **touch test.** Figure 9-6 shows a reversing valve with the four port connections labeled. In the cooling mode of operation, Port A should be hot, while Port B should be cold. Ports A and B are exposed to the discharge refrigerant leaving the compressor and the suction gas returning to the compressor, respectively. Port C should be hot as well since the discharge gas is leaving the valve from this port. Port D should be cold, since the suction gas from the evaporator is entering the reversing valve at this point. In the heating mode of operation, Ports A and D should be hot because the hot discharge gas flows through both of these ports. Ports B and C should be cold while operating in the heating mode. The relative temperatures of these ports are summarized in Figure 9-7. On pilot-operated reversing valves, the pilot tubes should be approximately the same temperature as the valve body itself.

By evaluating the temperatures at these points on the valve, the technician can quickly determine if the four-way reversing valve is operating correctly. Following are various scenarios where internal valve leakage or damage has occurred.

Units Operating in the Cooling Mode

Consider a typical air-conditioning system operating with R-22 as its refrigerant. While operating in the cooling mode, Ports A and C should be hot and Ports B and D should be cold. On this system,

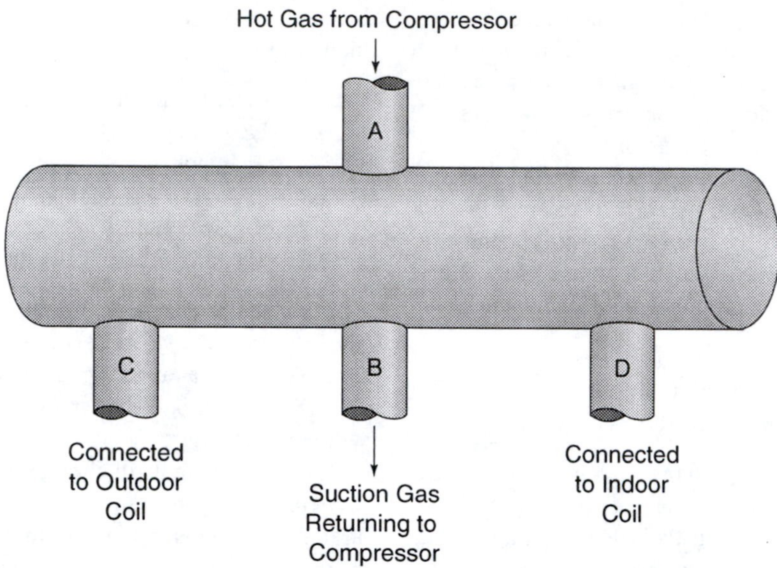

FIGURE 9-6 Labeled port connections on the reversing valve

Port	Cooling Mode	Heating Mode
A	Hot	Hot
B	Cold	Cold
C	Hot	Cold
D	Cold	Hot

FIGURE 9-7 Temperatures of the valve ports during the heating and cooling modes of operation

however, Port B is much warmer than Port D and the suction pressure is higher than normal. Assume that the temperature at Port D is 55°F but the temperature at Port B is 80°F. How could the refrigerant have picked up 25 degrees after traveling such a short distance? The answer is relatively straightforward. A leak in the body of the reversing valve has allowed the discharge gas from the compressor to leak into the low-pressure side of the valve. This can result in serious compressor damage, since the amount of compressor cooling will be reduced because of the increased suction-gas temperature. This reversing valve should be replaced.

Another scenario that shows how the reversing valve can be used to help evaluate the system involves measuring the relative temperatures of the pilot tubes. If, while operating in the cooling mode, Ports A and C are warm instead of hot and the pilot tube on the same side of the valve as Port C is also warmer than it should be, the compressor is defective and should be replaced. The pilot tube on the other end of the valve will most likely be at the same temperature as the valve body.

Still another scenario involves a valve that will not switch the system over from cooling to heating. If the ports are at their proper temperatures but the pilot ports are both hot, this indicates that both of the pilot ports are open, preventing the valve from shifting from one mode of operation to another. If the situation cannot be remedied by de-energizing and re-energizing the valve's solenoid coil, the valve should be replaced.

Units Operating in the Heating Mode

Consider another R-22 heat-pump system operating in the heating mode. During heating operation, Ports A and D should be hot, while Ports B and C should be cold. The pilot ports should also be approximately the same temperature as the valve body. If the pilot tubes are hot instead of warm and the system will not shift over from the heating to the cooling mode, it is an indication that the pilot is defective and the valve should be replaced.

Another situation, just as with the cooling operation, involves the temperatures of the pilot tubes. If Ports A and D are warm instead of hot and the pilot tube on the same side of the valve as Port D is warmer than the valve body, it is an indication that the compressor is defective and should be replaced.

One final scenario is when all of the ports and at least one of the pilot tubes are all hot. This is a very severe case and could easily lead to major system component damage. If this situation arises, the possible causes include:

- Major valve body damage exists.
- Valve slide is unable to complete its stroke.
- Both pilot ports are open.

If the valve body has become damaged, the entire valve must be replaced. If both pilot ports are open or if the slide cannot make its complete pass with the valve body, increasing the head pressure in the system may help push the slide in the valve or in the pilot and resolve the situation. If not, the valve must be replaced as well.

REMOVING DEFECTIVE FOUR-WAY REVERSING VALVES FROM THE SYSTEM

Once the technician has established that the reversing valve is defective, the valve must be removed from the system in preparation for the installation of the new component. Before the valve is physically removed, the refrigerant must be properly recovered by methods set forth in EPA laws and guidelines. Two methods are commonly used to remove the valve from the refrigeration circuit.

The first method of removing the four-way reversing valve is to simply cut the four refrigerant lines that are connected to the valve body. A refrigerant-tubing cutter should be used to cut the lines, which may prove difficult since, in most cases, three of the lines are close to each other. This makes the cutting of the lines much more difficult (Figure 9-8). Another drawback to this method is that, once the valve is removed, the installation of the new valve will require alterations to the piping circuit. The four lines that originally connected the reversing valve to the refrigeration circuit are now shorter than before, making it necessary to extend these lines with additional piping sections. This will increase the number of solder joints, thereby increasing the possibility of future refrigerant leaks.

Another method that can be used to remove the reversing valve from the system is to use a torch. When using a torch, the soldered or brazed joints at each of the four refrigerant lines must be heated in order for the valve to come loose. The one main drawback with this method is the lack of flexibility of the refrigerant tubing connecting the valve to the system. Disconnecting the single line from the valve is usually the easiest (Figure 9-9). The problem arises when the other solder joints are heated. Since the valve is still being held in place by the three remaining lines, it becomes difficult to disconnect the lines when they are heated individually. Insufficient play may be available in the refrigerant lines or the valve itself to facilitate the disconnecting of the lines. For these reasons, replacing a four-way reversing valve has always been a time-consuming task that no technician looks forward to.

The statement "Necessity is the mother of invention" definitely holds true in the HVAC industry. The problems stated in the preceding sections have created the demand for specialty tools in the industry, which help make the tasks that technicians perform much easier. One such tool is a torch tip designed specifically for reversing-valve removal (Figure 9-10). This tip, referred to as the *tuning fork* because of its unique shape, enables a technician to heat up all of the remaining three fittings on the reversing valve at once (Figure 9-11). Once the first line has been removed using a standard torch tip, this specialty tip can then be used to heat the remaining solder joints. When the filler material

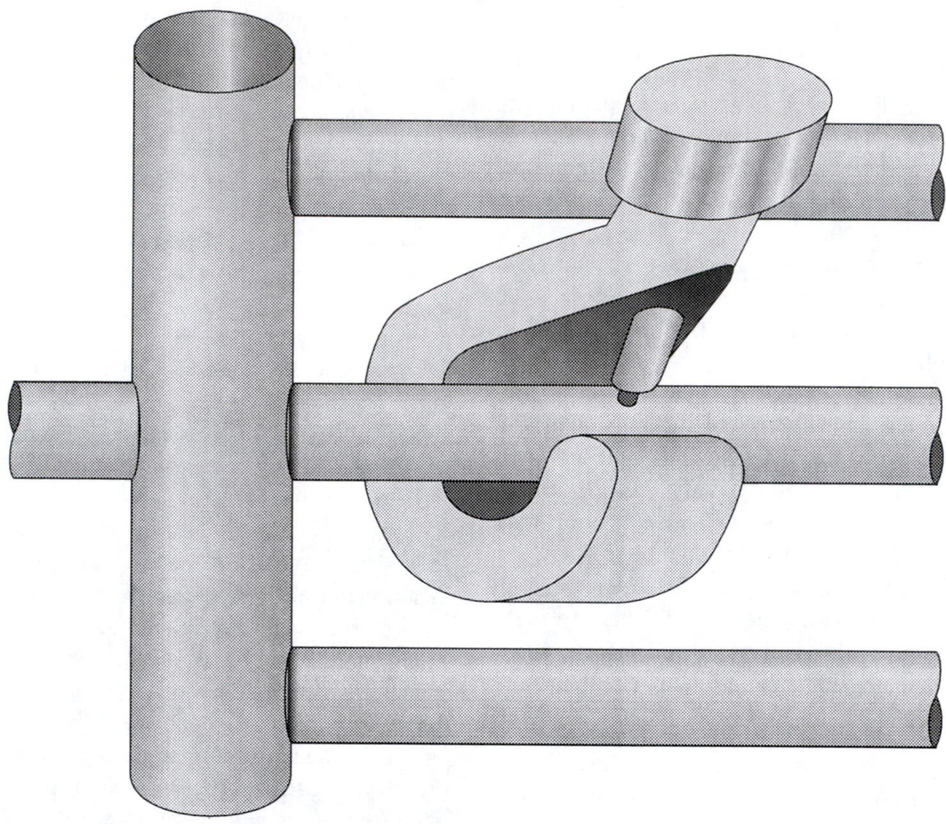

FIGURE 9-8 Cutting refrigerant lines on a four-way reversing valve is difficult with a standard tubing cutter.

FIGURE 9-9 The lone piping connection on a reversing valve is easy to remove with a standard torch tip. *(Courtesy Uniweld Products, Inc.)*

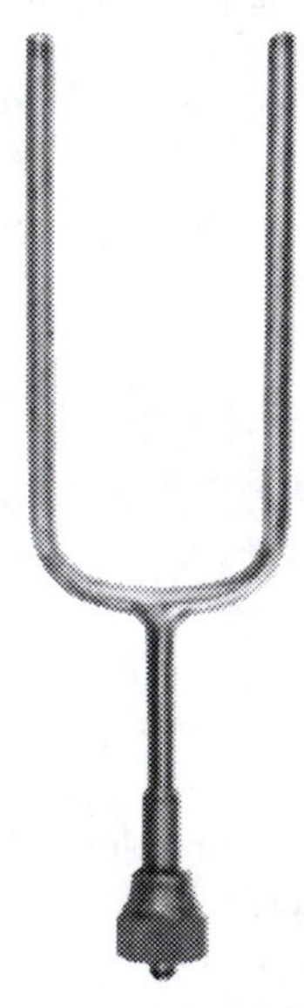

FIGURE 9-10 Specialized torch tip designed for removing reversing valves. *(Courtesy Uniweld Products, Inc.)*

FIGURE 9-11 The tip has the ability to heat all three lines at once, making valve removal easier. *(Courtesy Uniweld Products, Inc.)*

begins to melt, the entire valve can be removed from the system. One of the main benefits of using this type of tip is that the refrigerant piping does not need to be altered, as long as an exact replacement for the original reversing valve is used.

SERVICE NOTE: *When using any torch, always be sure to wear protective eye covering and have a fire extinguisher within reach.*

REPLACING FOUR-WAY REVERSING VALVES

If the technician has established that the reversing valve needs to be replaced, a few things must be kept in mind to ensure that the new valve functions correctly. Always be sure that:

- The new valve is thoroughly inspected before installation. Look for deep scratches, dents, or other imperfections. If at all possible, inspect the valve while you are still at the parts-supply house. Be wary of the valve if the box it comes in looks like it was dropped.
- The new valve is not dropped, banged, or otherwise mishandled. This will ensure that the slide can move freely within the valve body.
- When soldering or brazing the new valve in place, make certain that *heat paste* or a similar product is used. This will prevent the heat from the torch from reaching the plastic and neoprene components of the valve. They can melt if overheated! The valve mechanism should not be exposed to temperatures over 250°F. If no heat paste is available, a wet rag can be used, but care must be taken so that the water or its vapor is not permitted to enter the piping circuit.
- Make certain that the new valve fails in the same mode of operation as the old valve, and pipe it in accordingly. Normally, diagrams are inscribed directly on the surface of the valve body indicating which port will carry the discharge refrigerant from the valve in the de-energized position. Make certain that the piping corresponds correctly. Otherwise, the system will operate in the cooling mode when heating is desired, and vice versa.
- Make certain that the new valve is mounted horizontally, with the pilot ports and tubes on top. If the pilot tubes are underneath the valve, oil can accumulate in them and cause the reversing valve to malfunction.
- When choosing a replacement reversing valve, make certain the line sizes are the same as those on the old valve. Always try to obtain the exact replacement if at all possible. If the ports of the new valve are too small, an excessive pressure drop will occur across the valve, compromising system operation. If a larger valve is used, the time it takes for the valve to switch positions will be increased, thereby resulting in sluggish system operation.

BIDIRECTIONAL FILTER DRIERS

Bidirectional filter driers are commonly found on heat-pump systems instead of two, single-directional driers. The main reason for this is that the installation and servicing of the system are made easier. Just as with other types of filter driers, the bidirectional valves are designed to remove moisture, acid, and particulate matter that may find their way into the refrigerant circuit. Troubleshooting bidirectional filter driers is basically the same as with other types of filter driers, and the following points should always be recognized:

- Frost should never appear on a liquid-line filter drier. Frost is an indication that the drier is clogged and must be replaced. Remember, liquid-line filter driers are located in the liquid line and frost is never present on the liquid line of a properly operating system.
- A bidirectional filter drier may function properly in one mode and not in the other. One section of the filter drier may become blocked or clogged. For example, both the inlet and the outlet of the driers may be warm to the touch in the heating mode, but a substantial temperature and/or pressure drop may occur across the drier in the cooling mode. This drier must be replaced.
- Remember that the bidirectional filter drier is effectively two driers in one shell (Figure 8-24) and must be treated as such. These driers contain check valves similar to those described earlier in this chapter and in the last. Damage to these check valves can result in reduced system efficiency and effectiveness. If you have any doubt as to whether or not the drier is functioning correctly, it should be replaced immediately. Allowing particulate matter to flow freely through the system can result in clogged expansion valves and compressor damage.
- Good field practice also dictates replacing filter driers after a compressor motor burnout to help clean up the system and reduce the possibility of damage to the new compressor. Remember, compressor burnouts often produce acid within the system and the filter drier helps to absorb this acid.

SERVICE NOTE: *When replacing sweat-type filter driers, always avoid using excessive heat. Other specialty torch tips, such as the ones shown in Figure 9-12 and Figure 9-13, will help concentrate the heat and thereby reduce the amount of time needed to solder or braze the drier in place.*

FIGURE 9-12 Specialty torch tip reduces the time required to braze system components and reduces the chance of overheating the device being installed. *(Courtesy Uniweld Products, Inc.)*

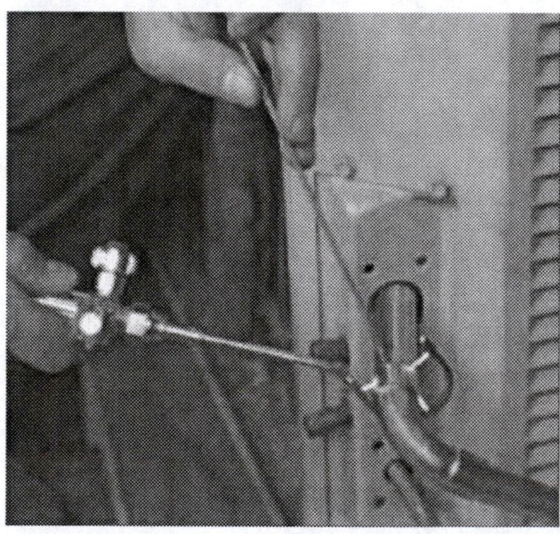

FIGURE 9-13 Specialty torch tip in action. *(Courtesy Uniweld Products, Inc.)*

TROUBLESHOOTING BIDIRECTIONAL THERMOSTATIC EXPANSION VALVES

Bidirectional thermostatic expansion valves are specially designed for heat-pump applications and are primarily intended to simplify the system piping. The bidirectional TXV is in effect a normal TXV that has a check valve built into it. This construction allows for refrigerant to bypass the valve's internal mechanisms—namely, the needle-and-seat assembly—when flowing in the opposite direction. The operation of the internal check valve is exactly the same as for that of the external check valve and is briefly summarized here.

The internal check valve in the bidirectional TXV will be in the *open* position when:

- The valve is located at the indoor coil and the system is operating in the heating mode.
- The valve is located at the outdoor coil and the system is operating in the cooling mode.

The internal check valve in the bidirectional TXV will be in the *closed* position when:

- The valve is located at the indoor coil and the system is operating in the cooling mode.
- The valve is located at the outdoor coil and the system is operating in the heating mode.

The bidirectional TXV is in effect two components in one; but, if one fails, the entire device needs to be replaced. On larger systems, however, instances occur in which valve components can be replaced without having to replace the entire valve.

In operation, the valve should function as a normal TXV when located at the indoor coil while the system operates in the cooling mode. The inlet of the valve should be warm, while the outlet of the valve should be cool. This indicates that the internal check valve is in the closed position. To make certain that the TXV is functioning properly, the evaporator superheat should be measured. The method for measuring evaporator superheat was discussed earlier in the text. If both the inlet and the outlet of the valve are warm, the check valve is in the open position and should be replaced.

In the cooling mode, however, the check valve in the TXV located at the outdoor coil should be in the open position when the system is operating in the cooling mode. Both the inlet and the outlet of the valve should be warm, indicating that the valve is functioning as a pass-through device and not affecting the state or temperature of the refrigerant. If the inlet is warm and the outlet is cool, the check valve is in the closed position and the valve should be replaced.

In the heating mode, the check valve at the indoor coil should be in the open position and both the inlet and the outlet of the valve should be warm to the touch. The check valve at the outdoor coil should be in the closed position, forcing the refrigerant through the needle-and-seat assembly. The inlet of the valve should be warm, while the outlet should be cool. Any deviations from these guidelines indicate that the valve is sticking and should be replaced.

THE THERMAL BULB ON THE BIDIRECTIONAL THERMOSTATIC EXPANSION VALVE

Just as with the thermostatic expansion valve intended for non-heat-pump use, the thermal bulb plays an integral role in the proper operation of the entire valve. For the TXV to properly measure and maintain the desired amount of superheat in the evaporator, the thermal bulb must:

- Be mounted securely to the suction line at the evaporator's outlet
- Be properly insulated to ensure proper temperature sensing
- Contain the proper refrigerant charge

Recall that the thermal bulb of the TXV must be mounted securely to the suction line at the outlet of the evaporator for the bulb to accurately measure the temperature of the line. The bulb should be secured to the line according to Figure 9-14. The mounting kit provided with the valve by the manufacturer should always be used. The majority of these kits supply two straps and two sets of nuts and bolts. They should both be used to prevent the bulb from turning or twisting. The bulb should never be mounted on the bottom of the suction line. The reason for this is that oil travels with the refrigerant as it flows through the refrigerant circuit. The refrigerant oil is typically heavier than the refrigerant and tends to occupy the lower portion of the piping. If the bulb is mounted at the bottom, the temperature of the oil, which may be different than the temperature of the refrigerant, may be measured instead.

Due to the round shape of both the thermal bulb and the suction line, the surface area of the bulb that

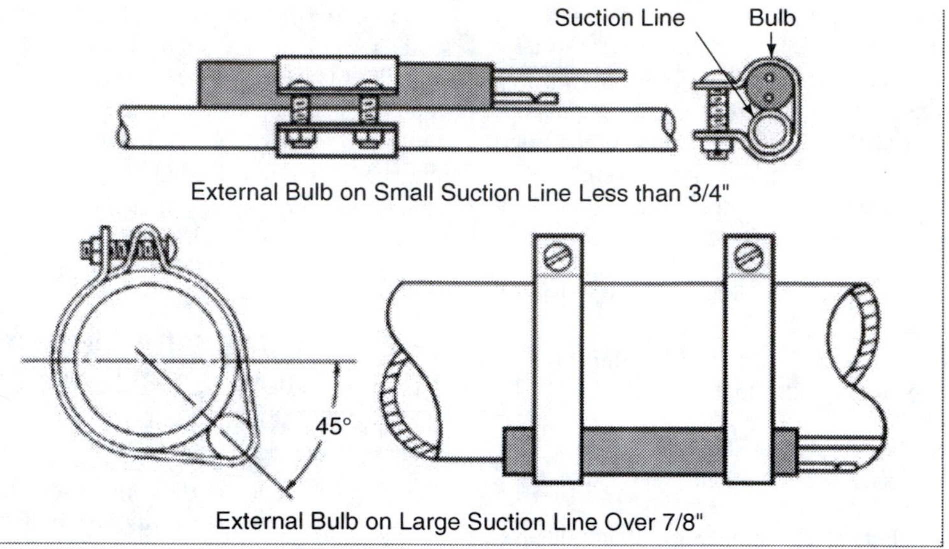

FIGURE 9-14 Proper positions for TXV thermal bulbs. *(Courtesy Alco Controls)*

actually comes in physical contact with the suction line is very small. For this reason, the technician must install the bulb on a straight section of suction-line piping and make certain that the bulb is wrapped with insulation to prevent inaccurate readings. Also, the bulb should never be mounted on pipe fittings, refrigerant traps, or other locations where the refrigerant is not free-flowing.

If the thermal bulb is not properly mounted to the suction line, it will sense a temperature that is higher than the actual suction-line temperature. This is because the evaporator is cooler than the space being conditioned as well as the area surrounding the coil, so the bulb will be exposed to a warmer temperature. This will give the valve the impression that the superheat is too high, and the valve will open accordingly. This could have potentially disastrous results, including liquid floodback to the compressor.

Finally, the technician must make certain that the thermal-bulb assembly does not lose its refrigerant charge. Since piping tends to vibrate during system operation, good field practice requires situating the transmission line on the thermal-bulb assembly in a manner that prevents pipe-to-pipe rubbing from occurring. If the thermal-bulb charge is lost, the thermostatic expansion valve will close. A summary of the effects of thermal-bulb problems that can arise is shown in Figure 9-15.

SUMMARY

- Closed check valves should have a large pressure and temperature drop across them.
- Open check valves should have low pressure and temperature drops across them.
- Open check valves should perform as pass-through devices.
- Depending on its location, a defective check valve can lead to reduced system efficiency, ineffective compressor operation, liquid floodback to the compressor, and/or improper conditioning of the occupied space.
- A defective check valve often can be located if the system functions in one mode but not the other.
- Defective check valves often give the impression of a system undercharge.
- Deep scratches and dents can have adverse effects on the operation of a reversing valve.
- Reversing valves can be evaluated by performing a touch test on the connecting ports.

Thermal Bulb Problem	System Head Pressure	System Suction Pressure	Evaporator Superheat	Condenser Subcooling	Δ T Across the Evaporator Coil (Air Temperature)	Major Effects on System
Loss of bulb charge	Lower	Lower	Higher	Higher	Lower	Reduction of refrigerant flow to the evaporator (reduced heat transfer to the refrigerant)
Loosely mounted thermal bulb	Higher	Higher	Lower	Lower	Higher	Possible liquid floodback to the compressor

FIGURE 9-15 Effects of thermal bulb problems

KEY TERMS

Fail position
Pass-through device
Pressure differential
Touch test

FOR DISCUSSION

1. Why should the reversing valve, unlike some other system components, be handled with extreme care? How could the carelessness of valve handling ultimately affect the performance of the system, company profitability, and customer satisfaction?
2. Why do most heat-pump systems fail in the heating mode? How does this benefit the customer?

REVIEW QUESTIONS

1. When a check valve is in the open position:
   a. The pressure drop across the device should be very low.
   b. The temperature drop across the device should be high.
   c. The temperature drop across the device should be low.
   d. Both *a* and *c* are correct.

2. If the check valve at the outdoor coil is stuck in the open position:
   a. The system will operate correctly in the cooling mode.
   b. The system will operate correctly in the heating mode.
   c. The system will operate correctly in both modes.
   d. The system will not operate correctly in either mode.
3. Which of the following is a possible result if the check valve at the indoor coil is stuck open while the system is operating in the cooling mode?
   a. Liquid floodback to the compressor
   b. Reduced cooling of the occupied space
   c. Higher suction pressure
   d. All of the above
4. If the check valves at both the indoor and the outdoor coil are in the closed position:
   a. The suction pressure will be excessively low.
   b. The operating pressures will not be affected.
   c. The suction pressure will be excessively high.
   d. The suction pressure will be low in the cooling mode but high in the heating mode.
5. If all ports on a four-way reversing valve are hot to the touch:
   a. The system is operating in the cooling mode.
   b. The system is in the heating mode.
   c. The system is operating in defrost mode.
   d. The system is malfunctioning.
6. If the suction line at the outlet of the indoor coil is cool to the touch and the suction line at the inlet of the compressor is warm:
   a. The system is operating in the heating mode.
   b. The reversing valve is likely to have an internal leak.
   c. The check valve at the indoor coil is in the open position.
   d. Both *a* and *c* are correct.
7. A technician initially suspects that a heat-pump system, while operating in the cooling mode, is operating with less than a full charge of refrigerant. What can be done to confirm this?
   a. Check to see if the system operates correctly in the heating mode.
   b. Check to make certain that the check valve at the outdoor coil is in the open position.
   c. Check to make certain that the check valve at the outdoor coil is in the closed position.
   d. Both *a* and *b* are correct.
8. If the suction pressure of a heat-pump system is 70 psig, which of the following pressures will most likely result in a malfunctioning pilot-operated reversing valve?
   a. 125 psig
   b. 170 psig
   c. 200 psig
   d. 225 psig
9. What is the main purpose for using heat paste on a reversing valve during installation?
   a. It heats the piping surrounding the reversing valve.
   b. It protects the solenoid coil from overheating.
   c. It allows the technician to use less heat during brazing.
   d. It protects the valve's internal components.

10. A bidirectional liquid-line filter drier that has frost on it in the cooling mode but not in the heating mode:
    a. Is operating according to system design.
    b. Is partially blocked and should be replaced.
    c. Is acting as a pressure-reducing device in the cooling mode.
    d. Both *b* and *c* are correct.
11. Defective check valves in a bidirectional filter drier can lead to:
    a. Clogged expansion valves.
    b. Compressor damage.
    c. Reduced operating efficiency.
    d. All of the above are possible.
12. An improperly mounted thermostatic expansion valve thermal bulb can result in which of the following?
    a. High superheat
    b. Liquid floodback to the compressor
    c. Reduced temperature differential across the evaporator
    d. Increased condenser subcooling
13. A system that sends the discharge gas to the outdoor coil first when the reversing valve's coil is energized fails in the ____________________ mode.
14. Explain what precautions must be taken when replacing a defective reversing valve.
15. If all four ports and at least one pilot tube on the reversing valves are hot, which of the following is a possible cause?
    a. Both pilot ports are open at the same time.
    b. The valve body has been damaged.
    c. The valve slide is unable to complete its stroke.
    d. All of the above are possible causes.

CHAPTER 10

Wiring Diagrams: Vapor-Compression Heat-Pump Systems

OBJECTIVES After studying this chapter, the reader should be able to:

- List the components that are controlled during normal heat-pump-system operation.
- Explain the function and operation of the lockout relay.
- Explain the function and operation of the holdback thermostat.
- Explain the sequence of operations of the heat-pump cooling cycle.
- Trace out and follow the electrical circuits of the heat-pump cooling cycle.
- Explain the sequence of operations of the heat-pump heating cycle.
- Trace out and follow the electrical circuits of the heat-pump heating cycle.
- Explain the sequence of operations of the heat-pump defrost cycle.
- Trace out and follow the electrical circuits of the heat-pump defrost cycle.
- Explain the sequence of operations of the heat-pump emergency-heat mode.
- Trace out and follow the electrical circuits of the heat-pump emergency-heat mode.

INTRODUCTION

In a conventional air-conditioning system, the main electrical components that are controlled are the compressor, the condenser fan motor, and the evaporator fan motor. The condenser fan motor and the compressor are usually connected in a single parallel circuit and are energized and de-energized as a single device. This makes the controlling of the air-conditioning system relatively easy since the system includes only an indoor fan circuit and a cooling circuit, both of which are ultimately controlled by the system thermostat. The heat pump, on the other hand, is a more complex system and, as can be expected, its controls are more numerous and complex. This chapter examines the control circuitry necessary to facilitate proper heat-pump-system operation.

OVERVIEW

Imagine placing coins in an automatic coffee vending machine and watching in sheer disbelief as the coffee, cream, and sugar are dispensed before the cup has been set in place. For the machine to operate as desired, the proper sequence of operations must be followed. This sequence is determined by the system's control and logic circuits. These circuits, sensors, and controls ensure that the previous function has been completed before the next can begin. In the coffee example, if the cup is never placed in its proper position, the coffee, cream, and sugar should never be released. Air-conditioning systems and, more specifically, heat pumps also rely on a certain sequence to ensure proper operation during all modes of operation.

In addition to the components found on traditional air-conditioning systems, heat pumps are equipped with additional components that must be properly controlled as well. These additional components include:

- The solenoid coil on the reversing valve
- The relays and controls for the supplementary electric heat
- Emergency-heat relays
- Relays that control the outdoor fan operation (this fan operates in the cooling cycle but is de-energized during the defrost cycle)

- Defrost components, timers, and sensors
- Various freeze controls, thermostats, and sensors

The majority of these switches, devices, and control circuits are monitored internally by the system thermostat. These external devices are intended to monitor the conditions under which the system is operating in order to adjust or modulate the operating parameters as needed. Proper wiring of the heat-pump system is vital to its overall performance. Therefore, the aspiring servicing technician must have a solid understanding of how and why certain system components are connected in the various circuits the way they are. The service technician also needs to feel comfortable working with wiring diagrams and schematics, as they are the road maps that can help untangle the seemingly endless lengths of wire that are present in even the simplest of heat-pump systems. A typical heat-pump wiring diagram is shown in Figure 10-1. By becoming familiar with the basic heat-pump system and its wiring diagrams, any differences that exist in each manufacturer's independent design can be understood without much difficulty. Remember that all heat-pump systems operate in basically the same manner, so understanding the components that are commonly found on them is the main obstacle that must be overcome in order to successfully service them.

The wiring of the heat-pump system takes place between three primary locations in the system: the outdoor unit, the indoor unit, and the thermostat. The outdoor unit contains both the control and the power circuits for the components located there, namely:

- The compressor
- The compressor contactor
- The outdoor fan motor
- The reversing valve
- The defrost relay
- The holdback thermostat
- Other defrost components (pressure switches, for example)

The indoor unit also contains control and power circuits for the components contained within that unit. These components include:

- The indoor fan motor
- The indoor fan relay
- Electric-strip heaters
- Fusible links
- High-limit switches
- Sail switches
- Time-delay relays
- The transformer

All of these components are ultimately controlled by the end user via the **thermostat.** The thermostat is a single component that controls all of the main circuits in the system. It indicates the mode in which the system is operating and can also alert the equipment owner of potential problems that may exist within the system itself. The thermostat also helps ensure that the individual system components are energized at the proper time. The examination of heat-pump wiring diagrams begins with the determination of when the individual system components are to be energized.

INDIVIDUAL COMPONENT OPERATION

For the heat-pump system to properly provide heating and cooling when desired, the system components must be energized at the proper time. The following sections examine the major heat-pump-system components and when they should be energized during normal system operation.

The Reversing Valve

Throughout this text, the assumption has been and will continue to be that the heat-pump system will fail in the heating mode of operation. Once again, this means that the system will be in the heating mode when the reversing valve is de-energized. Therefore, the reversing valve should be energized when:

- The system is in the cooling mode.
- The system is in defrost.

The interconnecting wiring, therefore, should be configured in such a manner that would facilitate the energizing of the valve's solenoid under these conditions. Since defrost is initiated when the system is in the heating mode, different sets of electrical contacts will cause the coil on the reversing valve to be energized.

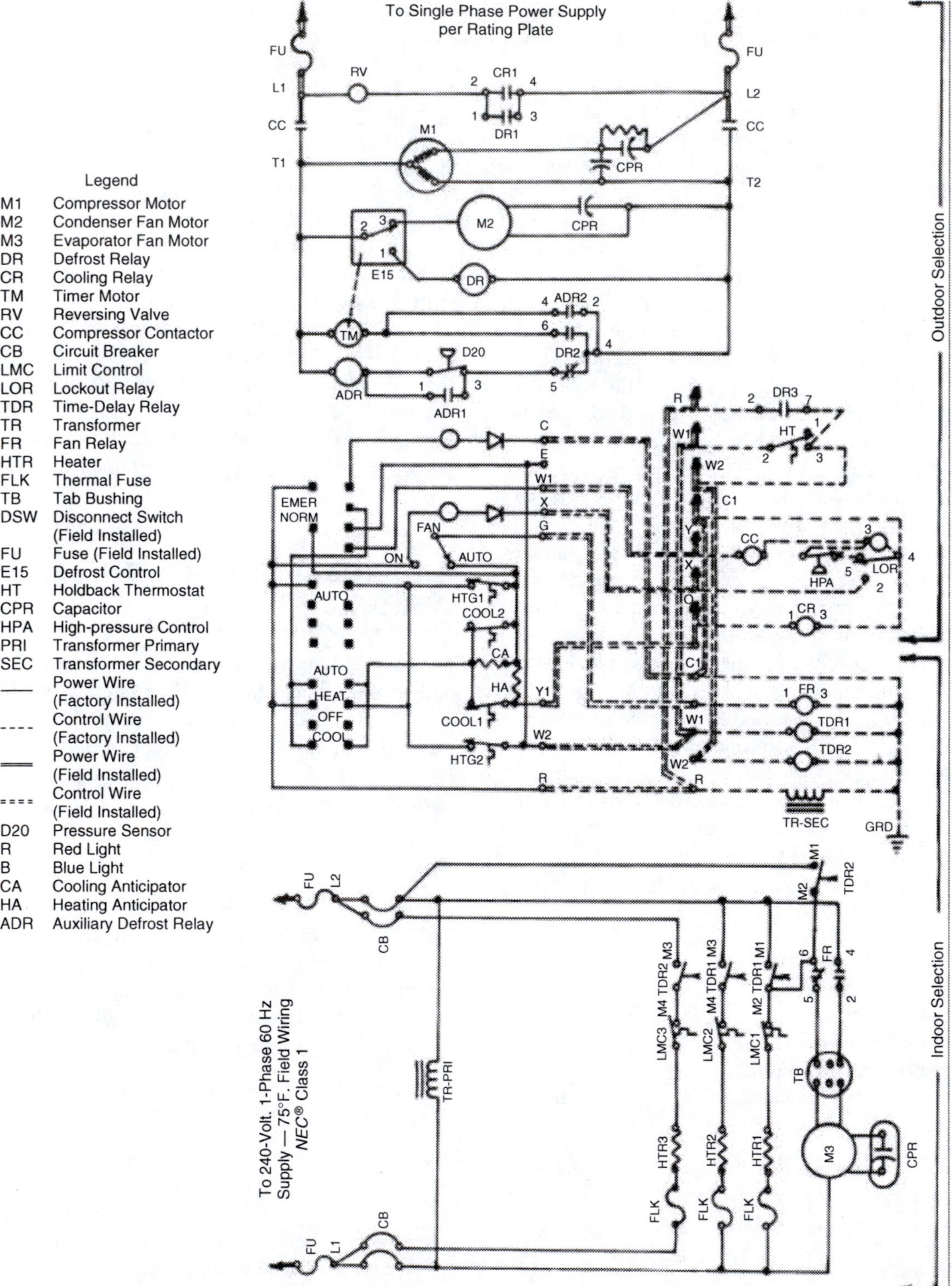
To Single Phase Power Supply
per Rating Plate
Legend
M1 Compressor Motor
M2 Condenser Fan Motor
M3 Evaporator Fan Motor
DR Defrost Relay
CR Cooling Relay
TM Timer Motor
RV Reversing Valve
CC Compressor Contactor
CB Circuit Breaker
LMC Limit Control
LOR Lockout Relay
TDR Time-Delay Relay
TR Transformer
FR Fan Relay
HTR Heater
FLK Thermal Fuse
TB Tab Bushing
DSW Disconnect Switch
(Field Installed)
FU Fuse (Field Installed)
E15 Defrost Control
HT Holdback Thermostat
CPR Capacitor
HPA High-pressure Control
PRI Transformer Primary
SEC Transformer Secondary
Power Wire
(Factory Installed)
Control Wire
(Factory Installed)
Power Wire
(Field Installed)
Control Wire
(Field Installed)
D20 Pressure Sensor
R Red Light
B Blue Light
CA Cooling Anticipator
HA Heating Anticipator
ADR Auxiliary Defrost Relay
Outdoor Selection
Indoor Selection
To 240-Volt. 1-Phase 60 Hz
Supply — 75°F. Field Wiring
NEC® Class 1
EMER
NORM
FAN
ON
AUTO
HEAT
OFF
COOL
TR-SEC
TR-PRI
GRD

FIGURE 10-1 Heat-pump system wiring diagram. *(Courtesy Addison Products Co.)*

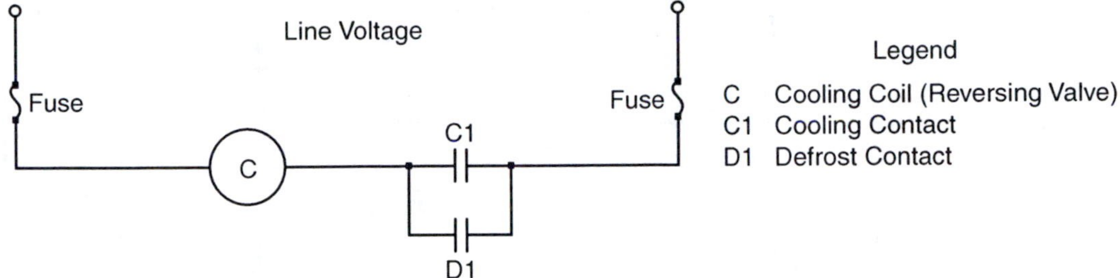

FIGURE 10-2 Simplified cooling coil and reversing-valve control circuit

These two sets of contacts should be connected in parallel with each other so that, if either set is closed, the desired outcome will be achieved. One set of contacts will close when the system is in the cooling mode, and the other set will close when defrost is initiated (Figure 10-2). Contacts C1 are controlled by the cooling terminal on the thermostat, and Contacts D1 are controlled by the defrost relay coil. A schematic diagram of a heat-pump system operating in the cooling mode is shown in Figure 10-3. The bold lines indicate the path of current flow. Note at the top of the diagram that Contacts CR1 are in the closed position, completing the circuit through the reversing valve. The holding coil that closed Contacts CR1 was energized because the thermostat was set to the cooling mode. The holding coil can be seen in the center portion of Figure 10-3 and is labeled CR. When the heat-pump system goes into defrost, the reversing valve is energized not by the CR1 contacts but by the DR1 contacts as shown in the top portion of Figure 10-4.

Many thermostats are equipped with terminals that are energized whenever the selector switch is set to the cooling mode. Connecting the cooling-relay coil or the reversing-valve circuit to this terminal will prevent the constant energizing and de-energizing of the reversing valve whenever the occupied space reaches the temperature set point. Although the compressor will cycle on and off with the temperature-controlled contacts, the reversing valve will remain energized, thereby reducing wear and tear on the valve itself. Note that the cooling relay and the compressor contactor are not the same component. The compressor contactor energizes and de-energizes the compressor in both the heating and cooling modes of operation. The cooling relay energizes and de-energizes the coil on the four-way reversing valve only in the cooling mode of operation.

Electric-Strip Heaters

Electric-strip heaters are used primarily for three purposes in the heat-pump system:

1. As a supplementary heat source
2. As a method to temper the air during defrost mode
3. As a primary heat source

When operating as a supplementary heat source, the electric-strip heaters are energized when the second-stage heating contacts of the thermostat close, indicating that the primary source of heat—namely, the heat pump itself—is not able to satisfy the load in the space. In addition, as a supplementary heat source, the electric-strip heaters are energized at the same time as and work in conjunction with the vapor-compression heat-pump system. When operating during the defrost cycle, the purpose of the electric-strip heaters is to temper the air being introduced to the occupied space. Finally, when operating as a primary heat source, the electric-strip heaters are functioning in place of the vapor-compression system. This is usually the case when there is a partial system failure.

No matter which role the electric-strip heaters are playing, they are controlled by a series of **time-delay relays.** Time-delay relays are commonly used when a number of high-amperage-drawing devices are to be energized at the same time. Time-delay relays operate in a manner similar to standard relays, but a specified delay occurs between the time that the coil is energized and the time that the controlled contacts open or close. The time-delay feature staggers the closing of the electric contacts. Such is the case with the electric-strip heaters. The time-delay relay coils are ultimately controlled by the heat-pump thermostat, which is discussed shortly. A typical power circuit

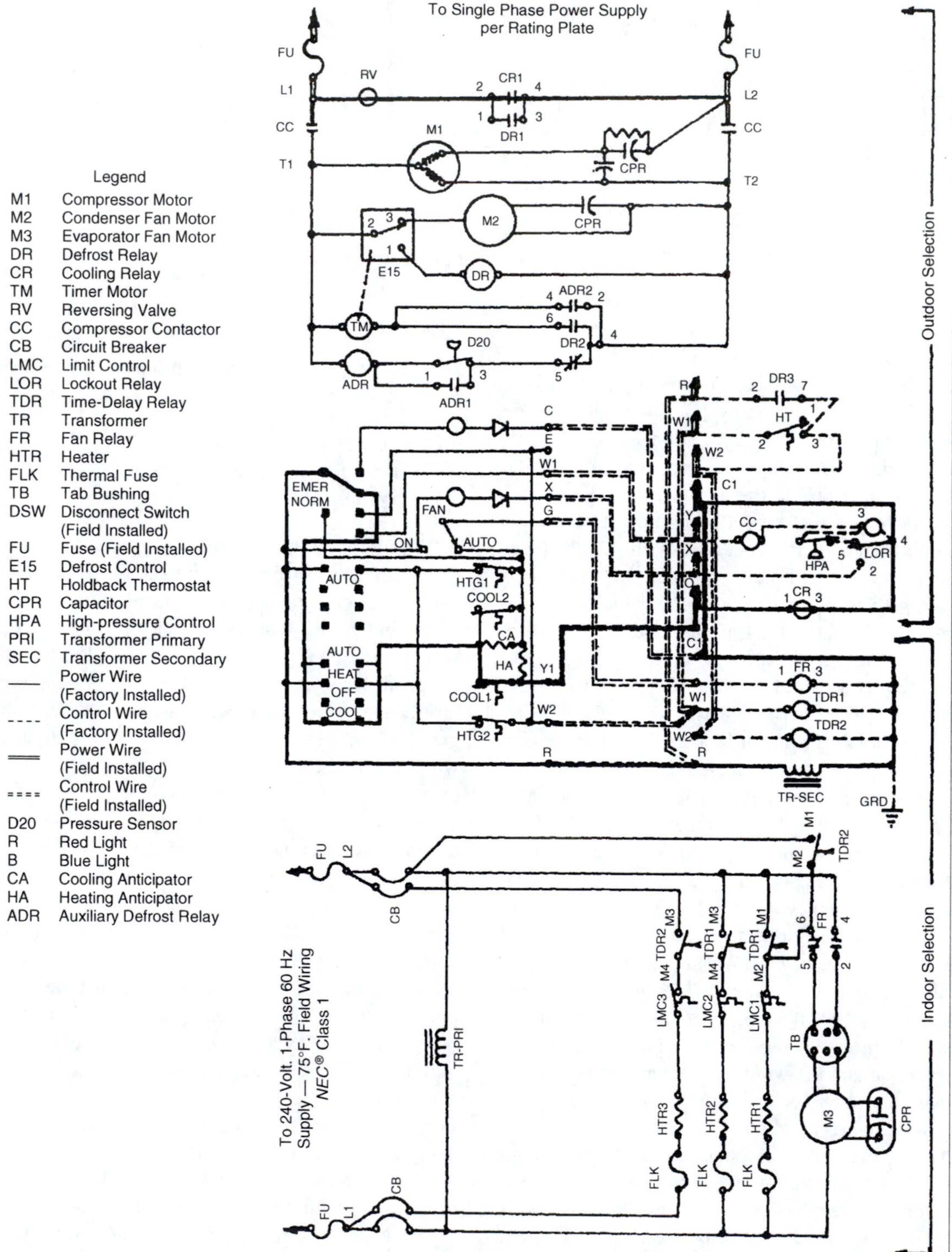

FIGURE 10-3 Reversing-valve circuit in the cooling mode. *(Courtesy Addison Products Co.)*

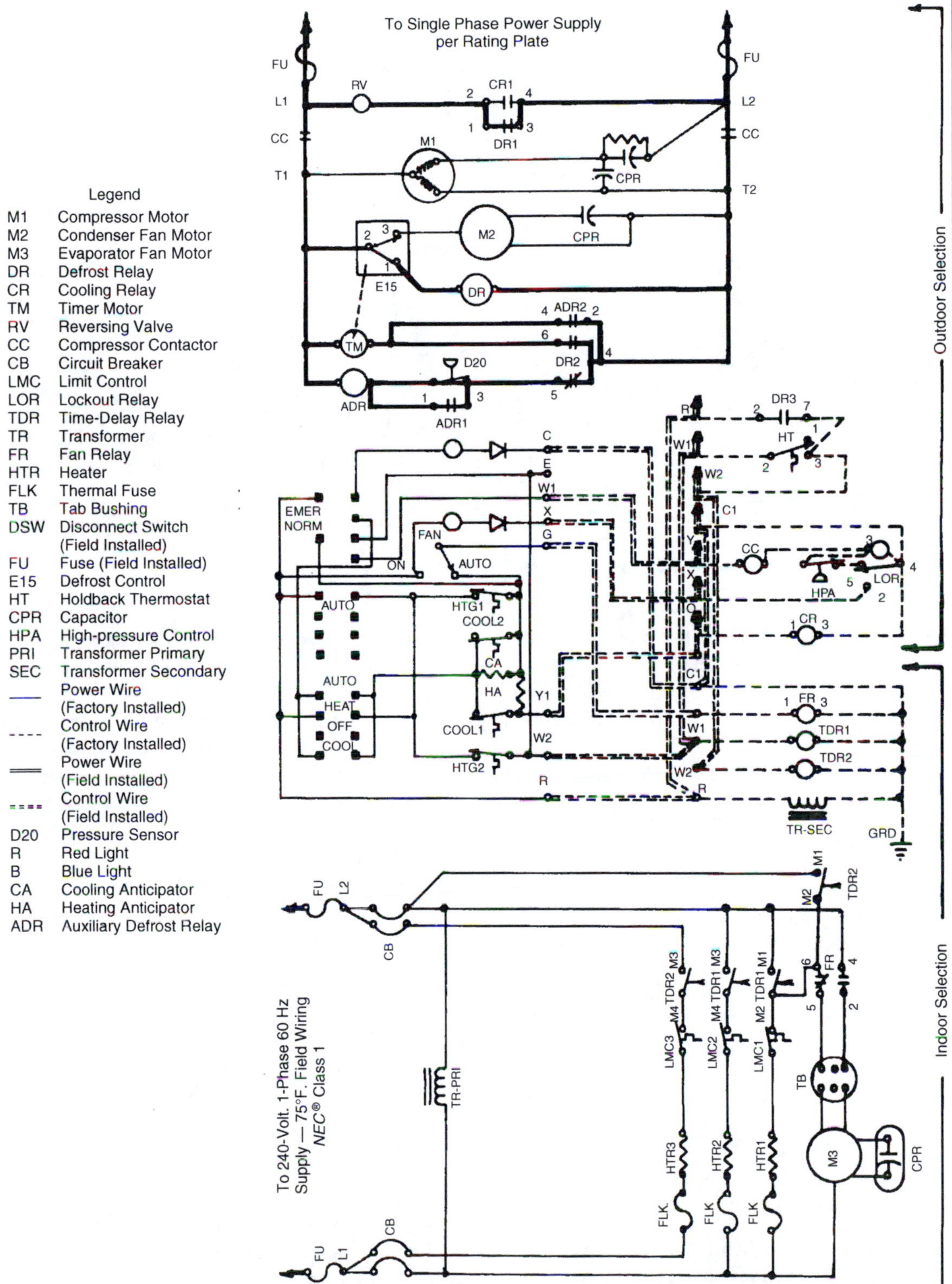

FIGURE 10-4 Reversing-valve circuit in the defrost mode. *(Courtesy Addison Products Co.)*

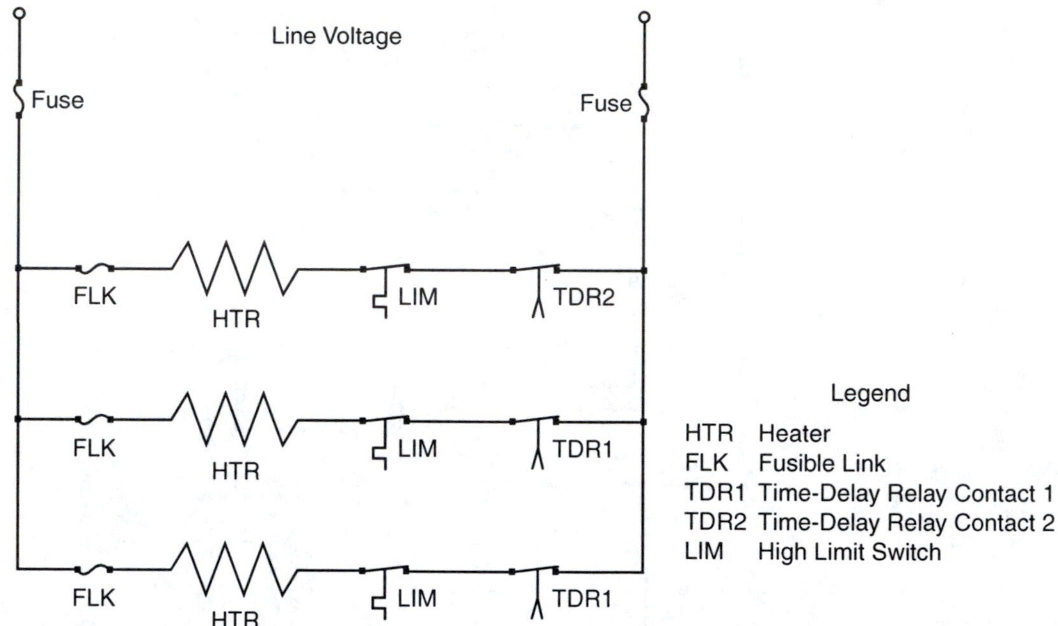

FIGURE 10-5 Typical electric-strip heater circuit, including safety devices

for the electric-strip heaters is shown in Figure 10-5 (which can also be seen in the bottom portion of Figure 10-1). Notice that these circuits have a number of safety features built into them. These features will de-energize the heating elements if unsafe conditions should arise in the system, which could include:

- An overheating condition in the duct
- A loss of airflow through the duct

Most heat-pump systems have backup safety devices to ensure that the system operates in a safe manner. For example, the circuit may have both a high-limit control and an airflow switch. The devices in the circuit shown in Figure 10-5 are:

- *Fuse:* Protects the entire indoor unit from excessive amperage draw
- *FLK:* Fusible link that protects each individual heater circuit
- *LIM:* High-limit switch that opens on a rise in temperature
- *TDR:* The time-delay relay contacts

Note that the fuse and the fusible link are both one-time devices, meaning that when they open, they must be replaced. The fusible link and the fuse are both sensing circuit current, which may seem redundant. As a matter of fact, they are not the same. Note that the fuse is protecting the entire circuit, while the fusible links are located in each separate heater-element circuit, thereby providing individual circuit protection.

Heat-pump systems are typically equipped with several strip heaters. Varying amounts of heat can then be generated based on need by energizing only a fraction of the system's heaters. This is accomplished with another system component, the **holdback thermostat.**

Holdback Thermostat

When the heat-pump system goes into defrost, it is in effect going into the cooling mode. Therefore, during the defrost cycle, cooled air will be introduced to the conditioned space. To reduce the amount of cooling that takes place, the heating elements used for second-stage heating and/or emergency heating are energized. Energizing these heating elements will temper the air being supplied to the conditioned space. If, however, all of the heating elements are energized during defrost, they may completely counteract the cooling effects of defrost and continue to increase the space temperature. Although this may seem desirable, it may have negative effects on the system. If, for

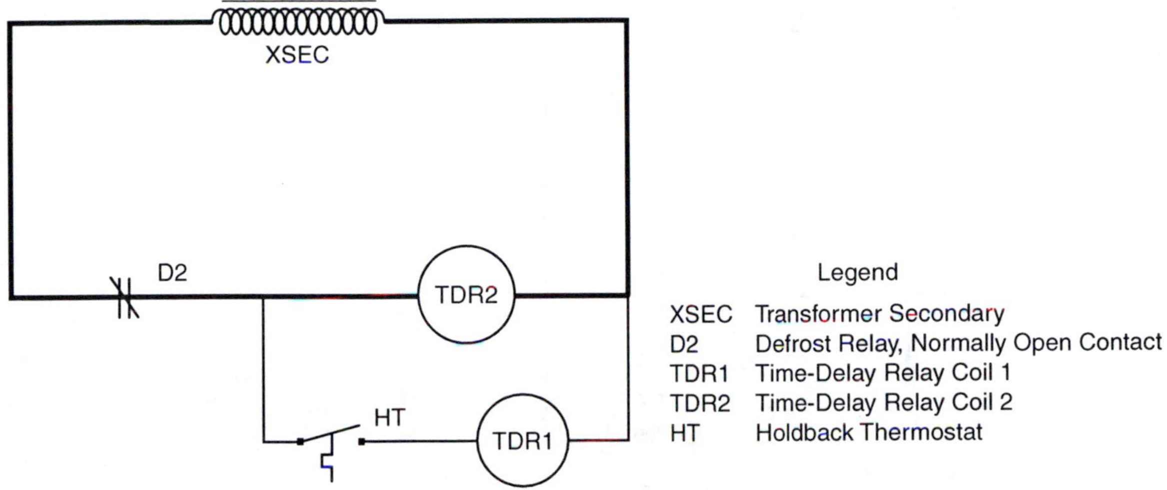

FIGURE 10-6 Time-delay relay circuit in defrost mode with a warm, outdoor ambient temperature

example, the space temperature continues to rise, the temperature setting in the thermostat may be reached, causing both the heating and defrost cycles to terminate. This may result in excess frost buildup on the outdoor coil. If this happens repeatedly, damage to the outdoor coil, including bursting refrigerant tubing, could result. Enter the holdback thermostat.

The purpose of the holdback thermostat is to make certain that the defrost cycle is able to terminate on its own, without any intervention from the space thermostat. It accomplishes this by sensing the outside ambient temperature. It is, therefore, in effect, an outdoor ambient thermostat. It then determines how many of the resistive heating elements should be energized, based on the sensible-heat capacity of the system, which is normally about 80 percent of the system's cooling capacity. The steps for calculating the approximate sensible-heat capacity are as follows:

- Multiply the tonnage of the system by 12,000 Btu/ton.
- Multiply this result by 0.80.

For a 4-ton system, the sensible heat capacity is approximately 38,400 Btu.

$$4 \text{ tons} \times 12{,}000 \text{ Btu/ton} \times 80\%$$

The maximum amount of electric heat that should be used can be found by dividing the sensible heat capacity by 3413. This figure is the number of British thermal units per kilowatt-hour. So, for the 4-ton system, a maximum of 11.25 kilowatt-hours (kWh) of heat (38,400/3413). If the required 4-ton heat-pump system was equipped with three 5-kW heaters, only two of them would be energized during the defrost mode. Energizing all three heaters would exceed the sensible-heat capacity and could result in continued space heating. Note that, in the case of tempering the air during defrost, less is better.

As the outside temperature drops, though, more heat can be supplied to the space during defrost. The holdback thermostat located in the outdoor section of the system monitors changes in the ambient temperature and either increases or decreases the number of heating elements in the active circuit. If the outdoor ambient is higher than the set point on the holdback thermostat, only a portion of the heating elements will be energized in the defrost mode (Figure 10-6). In this simplified circuit, only TDR2 is energized, since the holdback thermostat (denoted HT in the schematic diagram) is in the open position. The coil labeled TDR2 is the coil of a time-delay relay that ultimately energizes the electric-strip-heater circuits. If the outside ambient temperature drops below the holdback thermostat set point, its contacts will close, energizing the TDR1 coil in addition to the TDR2 coil (Figure 10-7). Having more strip heaters energized increases the amount of heat introduced to the conditioned space. The holdback thermostat is circled in Figure 10-8.

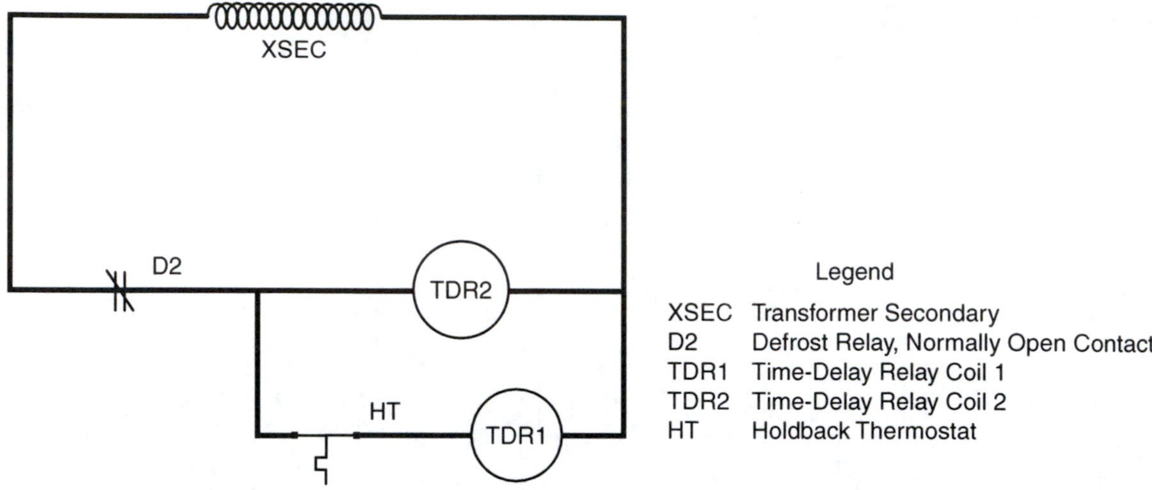

FIGURE 10-7 Time-delay relay circuit in defrost mode with a cool, outdoor ambient temperature

The holdback thermostat also functions to determine the number of heating elements that are energized when the system is operating in the emergency-heat mode as well as when supplementary heat is needed.

Compressor

The compressor in the vapor-compression heat pump is responsible for providing both first-stage heat and first-stage cooling. Therefore, it must be energized in both modes of operation. In the first-stage heating mode, the compressor must be energized when the temperature of the occupied space drops below that desired in the occupied space. In the cooling mode, however, the compressor must operate when the temperature of the occupied space rises above that desired in the space. The associated circuitry for achieving this desired outcome could be somewhat cumbersome and complicated were it not that these electric changeovers are achieved internally in the thermostat. The mode of operation is determined by the occupant in the event that the **manual-changeover** operation is chosen. In the manual-changeover mode, the system operator must choose between the heating and cooling mode of operation by moving the selector switch on the thermostat to the desired position. In the case of the **automatic changeover,** the system automatically changes over from the heating to the cooling mode in order to maintain the space at the desired temperature. This is accomplished by energizing and de-energizing the solenoid coil on the four-way reversing valve. Since the compressor is intended to operate under a wide range of conditions, the operating pressures must always be within the acceptable ranges. The compressor lockout relay plays an extremely important role in ensuring that the system and those around it are kept safe.

Compressor Lockout Relays

Probably one of the most important safety components on the typical heat-pump system, and probably for any air-conditioning or refrigeration system as well, is the lockout feature. The **lockout relay** prevents the compressor from operating if the head pressure reaches unsafe levels. Unlike some other types of pressure controls that automatically reset, the lockout relay must be reset manually. This is accomplished by disconnecting power to the system. A simplified circuit involving the lockout relay is shown in Figure 10-9. The contact terminals on the lockout relay have been arbitrarily numbered for ease in identification and explanation. The wiring diagrams for the specific system being worked on should be consulted for the proper terminal identification. The thermostat has also been eliminated from this diagram so the relay's operation can be clearly seen and understood but would normally be located in the circuit at Position A, as indicated in Figure 10-9. On a heat-pump wiring diagram, the lockout relay would look as it does in the circled area of Figure 10-10.

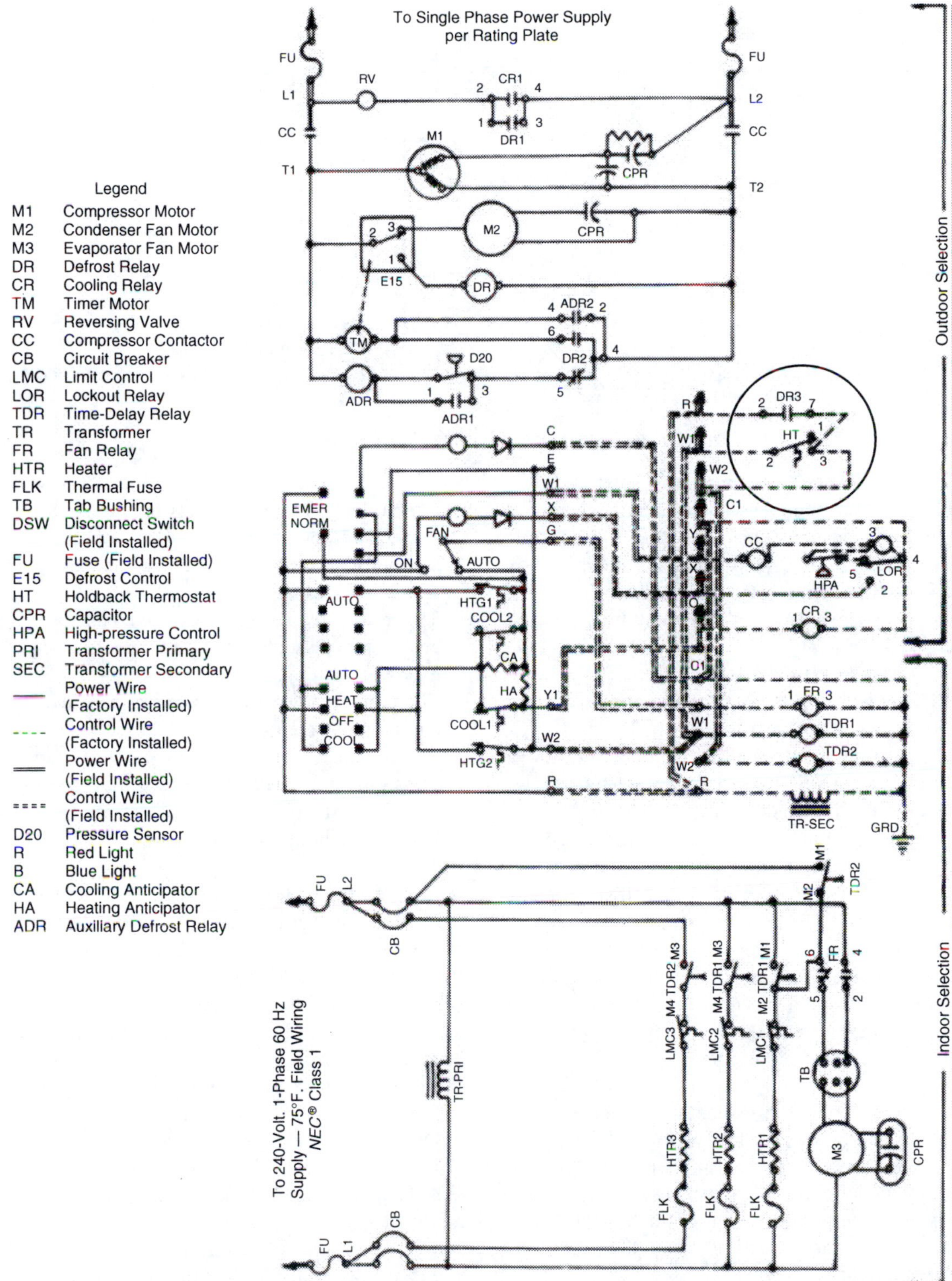

FIGURE 10-8 Holdback thermostat in heat-pump wiring diagram. *(Courtesy Addison Products Co.)*

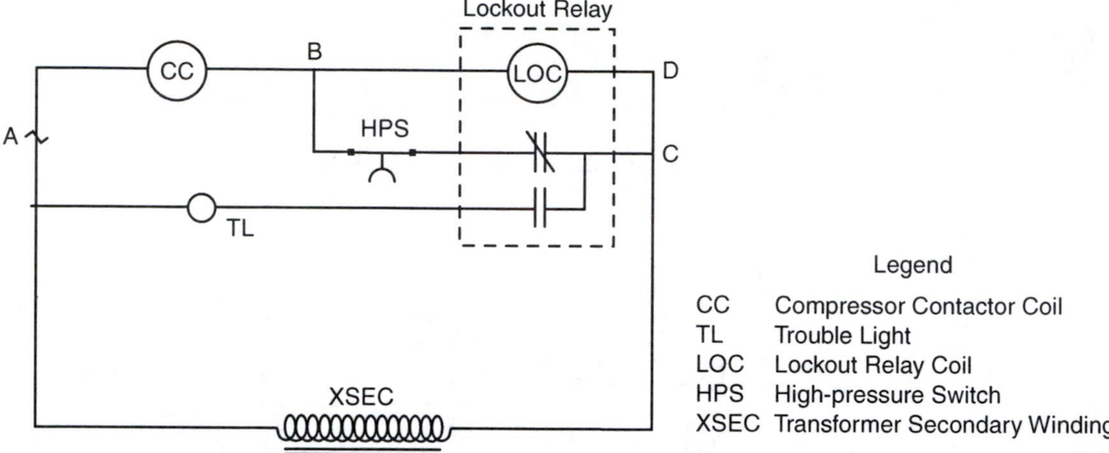

FIGURE 10-9 Typical lockout relay circuit with trouble light

During normal compressor operation, the compressor contactor coil is energized and the high-pressure switch is in the closed position. Current flows from the secondary winding of the transformer, through the compressor contactor coil, through the high-pressure switch, and then through the normally closed contacts, 3 and 5 (Figure 10-10), on the lockout relay. The circuit is completed by the connection from Terminal 5 back to the transformer (Figure 10-11). The current flow is indicated by the bold lines in the schematic diagram. Note that no current flows through the trouble light during normal compressor operation because of the open contacts, 4 and 5, on the lockout relay. In addition, no current flows through the coil on the lockout relay. This is because the path between point B and point D has a much higher resistance than the path between points on B and C. Since the high-pressure switch and Contacts 3 and 5 are closed, the resistance is effectively 0. All of the current, therefore, takes this path through the normally closed contacts, avoiding the coil on the lockout relay.

If the system pressure reaches unsafe levels, however, the high-pressure switch will open. This causes the electrical current to flow through both the compressor contactor coil and the lockout relay coil, since there is only one possible path for the current to take (Figure 10-12). This will trigger a number of actions within the circuit:

- The normally closed contacts on the lockout relay, *3* and *5,* will open.
- The normally opened contacts on the lockout relay, *4* and *5,* will close.
- The resistance of the compressor contactor coil circuit will increase due to the added resistance of the lockout relay coil.
- The current flow in the compressor contactor coil circuit will decrease.
- The compressor contactor contacts will open.
- The compressor will shut down until the lockout relay is manually reset.
- The trouble light will illuminate.

Because the high-pressure switch opened, the current in the control circuit is forced to flow through the lockout relay coil as well. This causes an increase in the resistance of this circuit. From Ohm's law, an increase in resistance causes a reduction in the amount of current flow in the circuit. This reduction in amperage prevents the compressor contactor coil from being able to keep its contacts closed because the magnetic field generated by the current flow through the contactor coil is weaker. The compressor is therefore de-energized. Since the lockout relay must be reset manually, the compressor will not operate even when the system pressure drops down to safe levels. Power to the circuit must be interrupted and then restored for the lockout relay to reset.

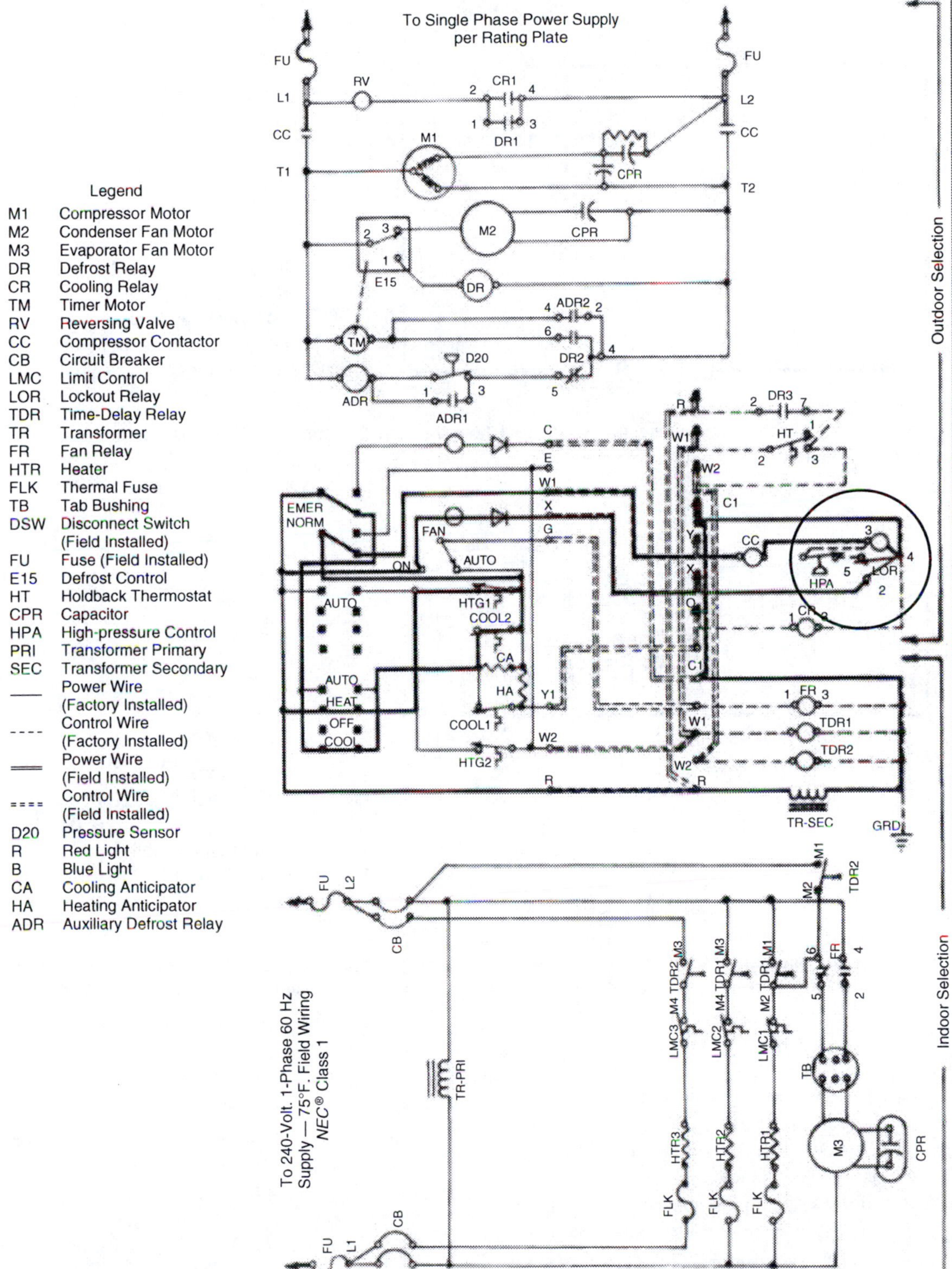

FIGURE 10-10 Lockout relay circuit. *(Courtesy Addison Products Co.)*

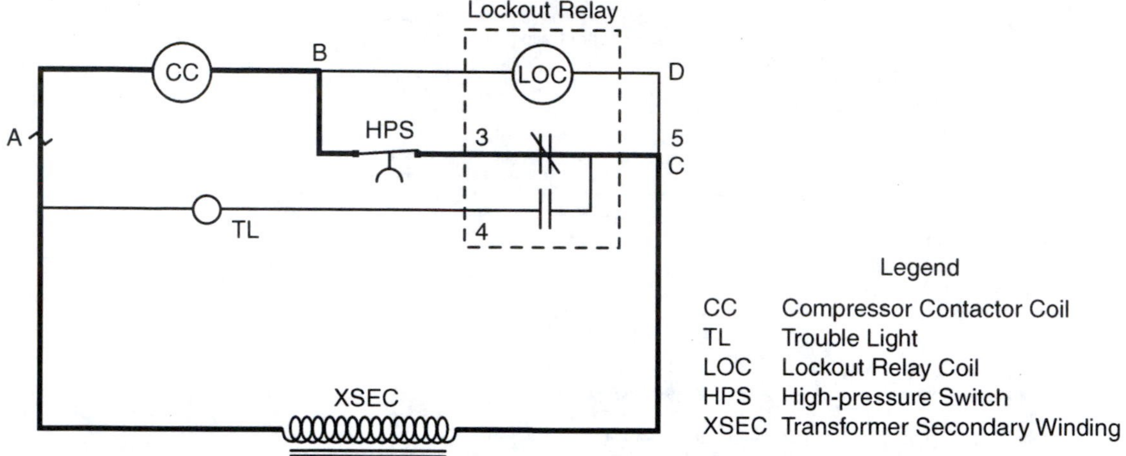

FIGURE 10-11 Bold lines indicate direction of current flow through the compressor contactor coil and the normally closed contacts on the lockout relay

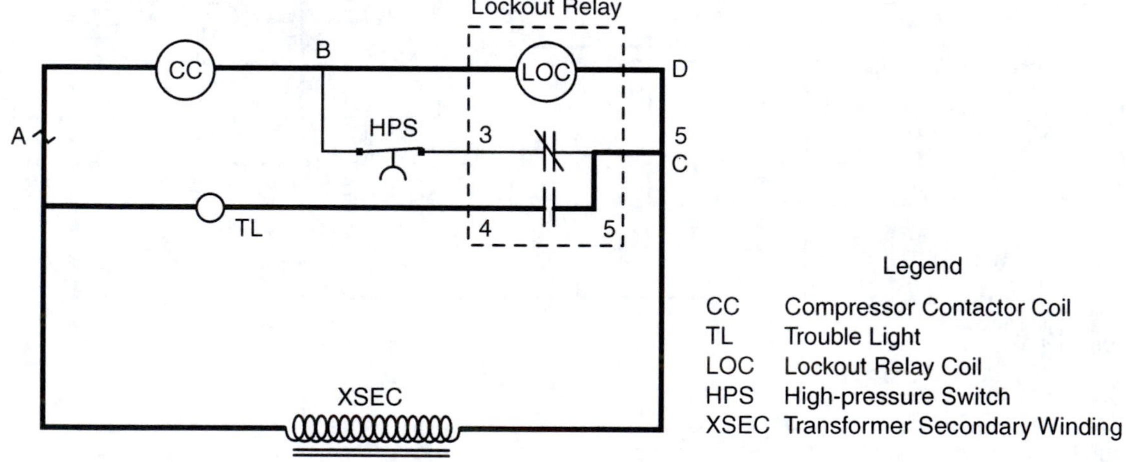

FIGURE 10-12 Bold lines indicate direction of current flow through the compressor contactor coil and the coil of the lockout relay. Notice how the high-pressure switch is now in the open position. The trouble-light circuit is also energized.

By forcing current through the lockout relay coil, the normally open contacts close and the normally closed contacts open. This energizes the trouble-light circuit and also makes it impossible for the compressor to restart if the system pressure drops. The trouble light, located on the thermostat itself, alerts the occupant of the lockout condition.

Condenser Fan Motor

On traditional, cooling-only systems, the outdoor fan motor is wired in parallel to the compressor so one is operating whenever the other is energized. This is fine in the cooling and heating modes of heat-pump operation but not when the system goes into defrost.

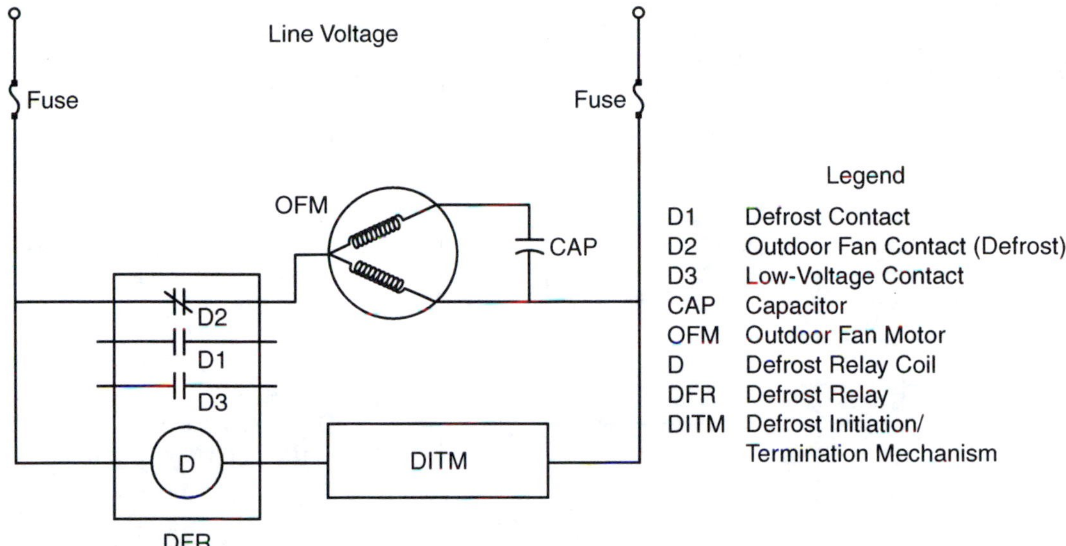

FIGURE 10-13 Outdoor fan motor circuit connected through the defrost relay

When the system, while operating in the heating mode, goes into defrost:

- The reversing valve switches the system over to the cooling mode.
- The outdoor fan motor is de-energized.
- Electric-strip heaters are energized.
- The compressor remains running.

Therefore, the outdoor fan motor and the compressor cannot be wired in the same circuit because they are not being energized at the same time. The outdoor fan must therefore be connected through a relay that will de-energize it when the system goes into the defrost mode. A sample circuit illustrating this is shown in Figure 10-13. The mechanisms that initiate and/or terminate defrost can rely on time, temperature, pressure, or a combination of the three. For this reason, the mechanism is labeled DITM, defrost initiation/termination mechanism, on the schematic diagram. When the system is in cooling mode and calling for cooling, the compressor contactor coil is energized and the normally closed contacts on the defrost relay, D2, are closed as well. This is because the DITM is in the open position. The outdoor fan motor is energized. The same holds true for the heating mode. In the defrost mode, however, the compressor contactor coil is energized, permitting the compressor to operate but the normally closed contacts on the defrost relay are now open, de-energizing the outdoor fan motor. The operation of the outdoor fan motor in the various modes of operation is summarized in Figure 10-14 and Figure 10-15. The actual circuitry for this portion of the system is shown in Figure 10-16.

Indoor Fan Motor

The indoor fan motor on a heat-pump system operates in a manner similar to the indoor fan motor on a cooling-only system. The operator of the system has the choice of the following modes of fan operation:

- Constant fan operation whenever the system is in operation
- Constant fan operation even when the system is not in operation
- Automatic fan operation

The mode of indoor fan operation is chosen by the system operator by setting the selector switch to the appropriate position on the thermostat. As can be readily seen from this discussion, a great deal of the heat-pump system's circuitry is built into the

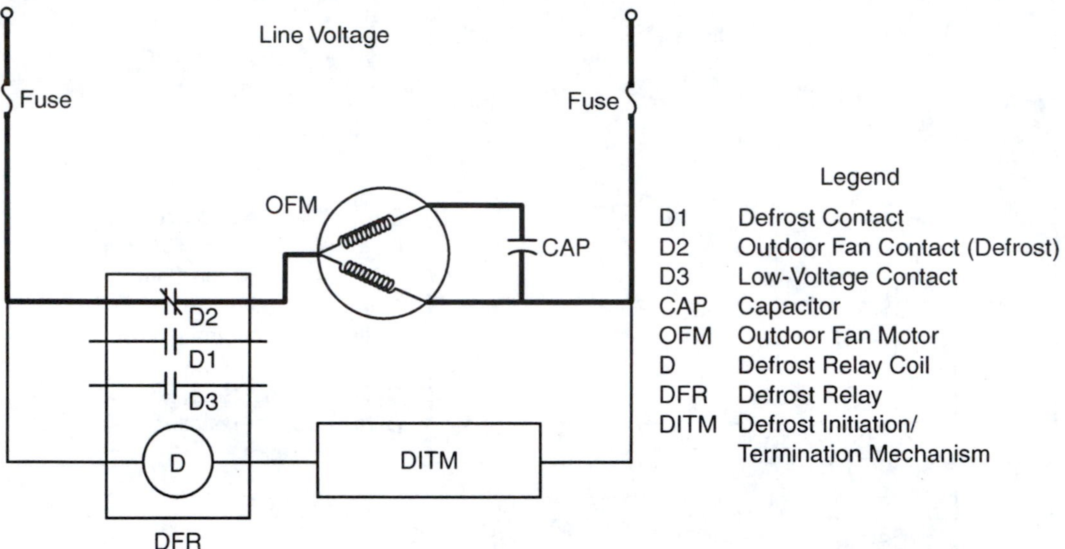

FIGURE 10-14 During both heating and cooling modes, the outdoor fan motor operates.

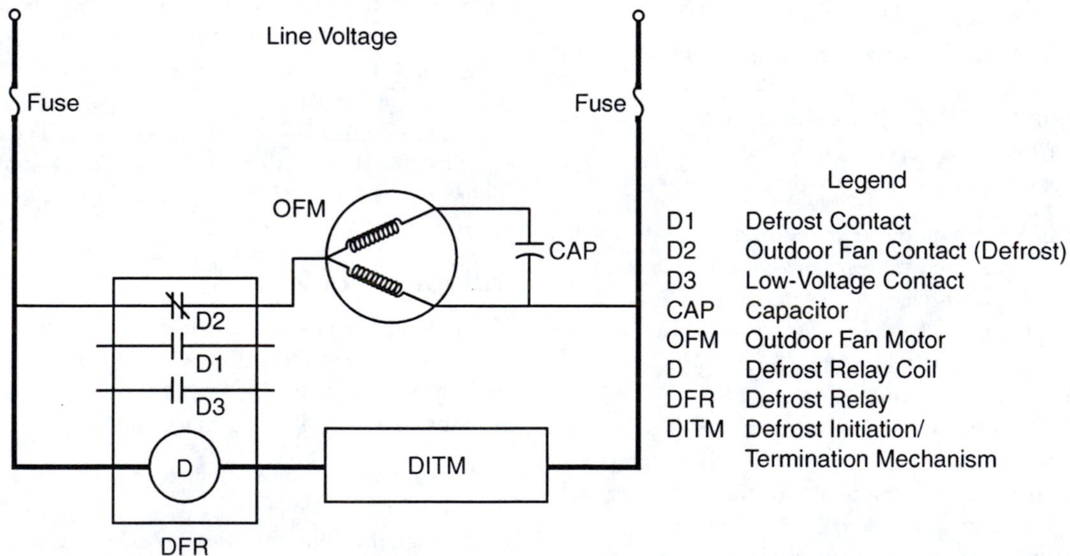

FIGURE 10-15 During defrost mode, the normally closed contacts open, de-energizing the outdoor fan motor.

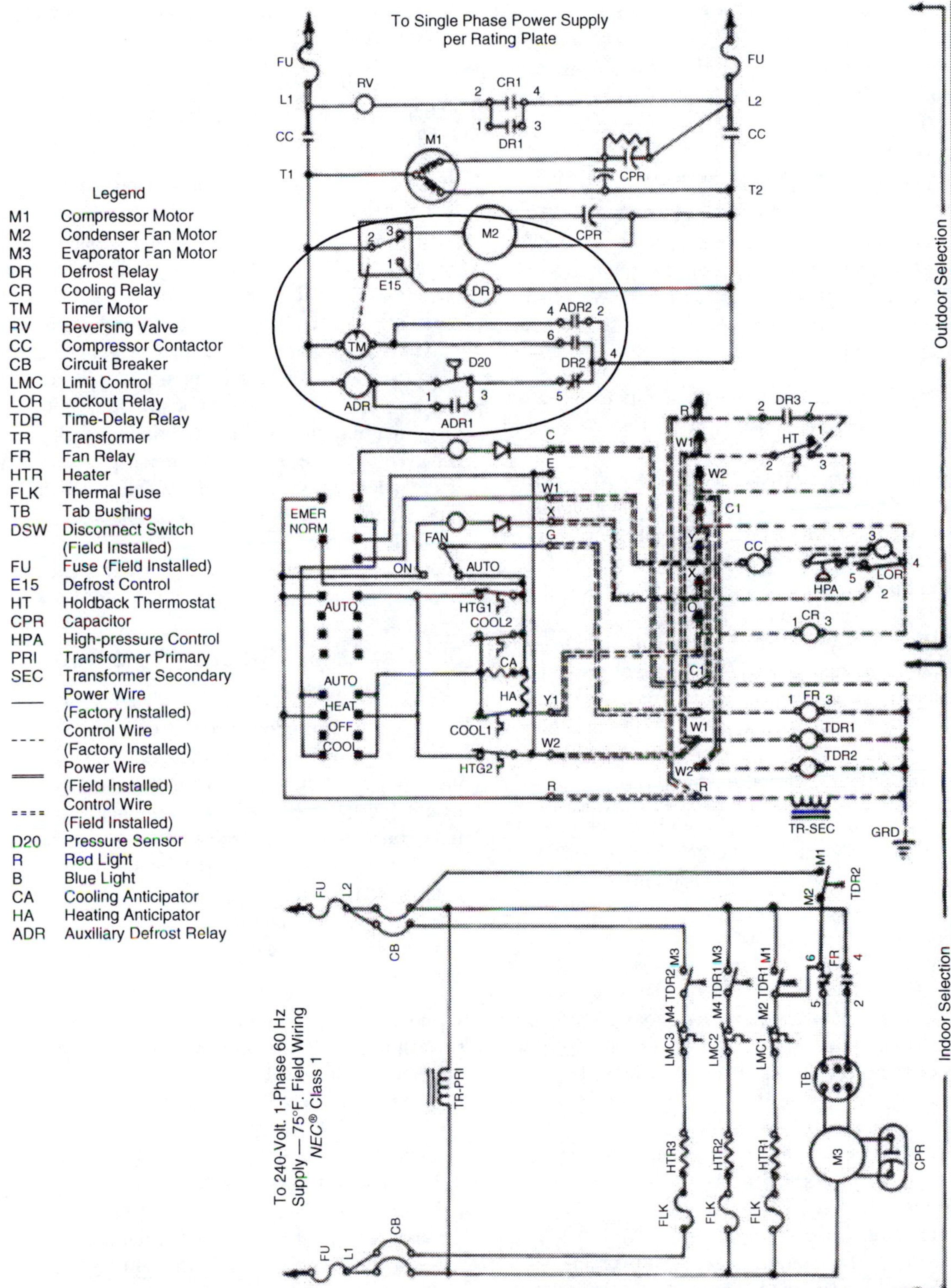

FIGURE 10-16 Circuitry controlling the outdoor fan motor. *(Courtesy Addison Products Co.)*

thermostat itself, making the system somewhat easier to *field wire* and to troubleshoot. **Field wiring** is the wiring that must be installed by the installing technician when the system is put into operation for the first time.

HEAT-PUMP THERMOSTATS

The preceding discussion on the operation of the various system components mentioned on more than one occasion the importance of the heat-pump-system thermostat. The heat-pump thermostat serves many purposes, including:

- Giving the system operator the ability to select the desired mode of operation as well as the desired space temperature
- Providing a convenient location to perform troubleshooting tasks
- Alerting the system operator if the system goes into lockout, on those thermostats so equipped
- Providing accurate switching and control of the major system components
- Alerting the system operator when the system is operating in emergency-heat mode, on those thermostats so equipped
- Allowing the space to be automatically maintained at preprogrammed temperatures at different times, on programmable digital models

Unlike the thermostats that are commonly found on cooling-only systems, the heat-pump thermostat is equipped with more contacts and terminals and are generally more complex. Note also that different thermostats are configured differently and the accompanying paperwork for the specific thermostat being used should be consulted before wiring the system. These configuration variations include different construction and, more important, different methodology of electrical terminal identification. Simply stated, Terminal X on one manufacturer's thermostat may be the equivalent of Terminal C on another. This is why referring to the manufacturer's paperwork for the specific thermostat is of utmost importance.

Before examining the individual thermostats, note that, for the thermostat to properly maintain the occupied space at the desired temperature, the device must be located in a manner that will enable it to measure the space temperature accurately. The thermostat should never be located:

- In direct sunlight
- Behind doors
- Close to supply registers

The thermostat should be located approximately 5 feet, or 1.5 meters, off the floor in an area that is at a temperature that is representative of the actual space temperature, both in the heating and cooling modes of operation. Refer to Figure 10-17 for a visual representation of proper and improper thermostat location.

The most common type of heat-pump system is comprised of a single cooling stage and two heating stages. Compressor operation provides both single stage heating and cooling. Second-stage heating is achieved by energizing the electric-strip heaters located in the indoor fan section of the system. Since this is the most common system configuration, discussion follows of two thermostat variations based on this configuration and one example of a thermostat that is able to control two heating stages as well as two cooling stages.

Single-Stage Cooling, Two-Stage Heating: Example 1

A typical heat-pump thermostat wiring diagram is shown in Figure 10-18. This diagram shows the internal electrical connections of the device as well as the external, or field, connections that must be made. The field connections and associated relays are shown on the right-hand side of the schematic. Next to these external components are the thermostat terminals. This is where the electrical connections are made. The terminal identifications are as follows:

- Terminal R is connected directly to the transformer and brings power to the thermostat.
- Terminal W2 is where the second-stage heating circuit is connected.
- Terminal L is used to control additional electrical contacts while the system is operating in the emergency-heat mode.
- Terminal B is where the heating changeover valve is connected. This terminal is where the reversing valve coil will be connected if the heat pump system is designed to fail in the cooling mode, or if the reversing valve must be energized when the system operates in the heating mode.

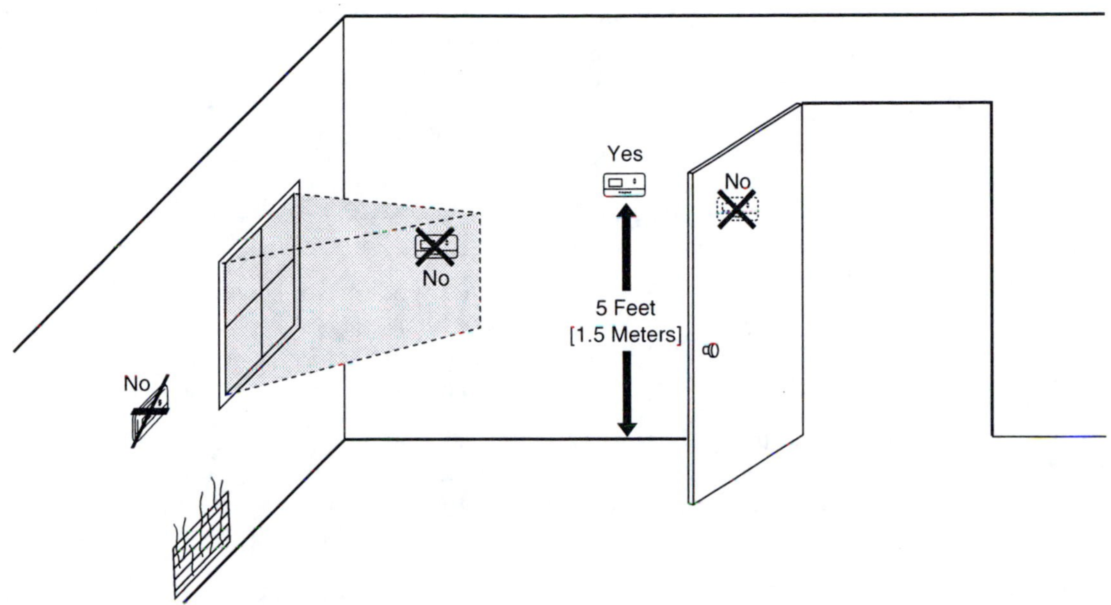

FIGURE 10-17 Typical thermostat location. *(Reprinted with permission of Honeywell Inc.)*

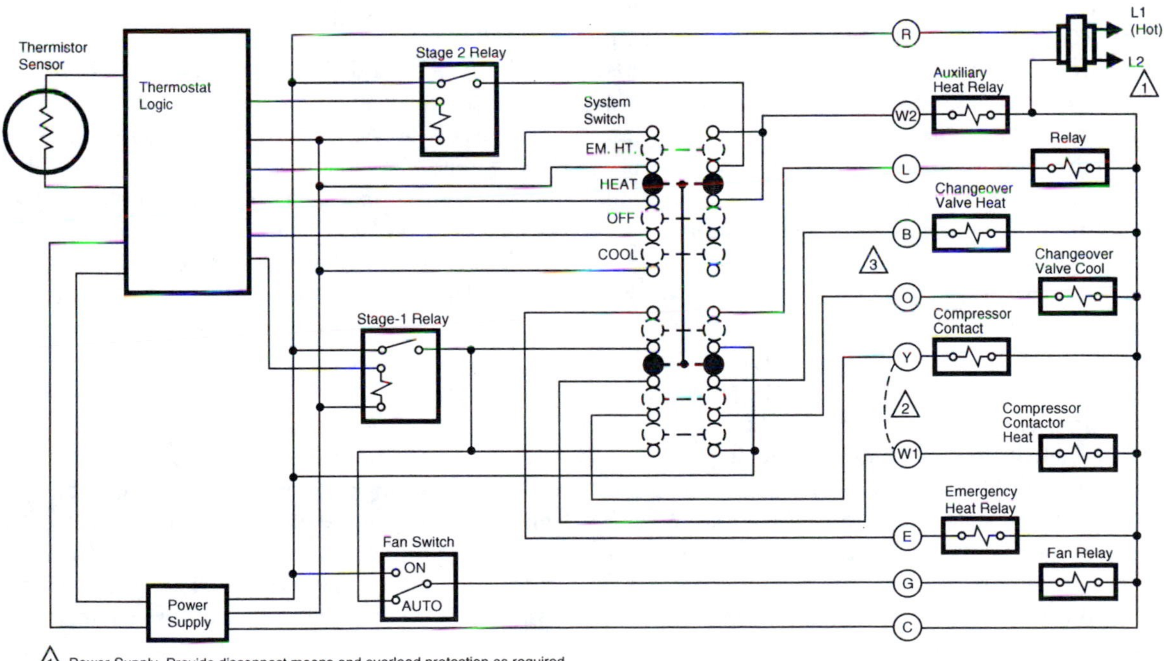

FIGURE 10-18 Electronic, two-stage-heat, one-stage-cool, wiring diagram with manual changeover. *(Reprinted with permission of Honeywell Inc.)*

- Terminal O is where the cooling changeover valve is connected. This terminal is where the reversing-valve coil will be connected if the heat-pump system is designed to fail in the heating mode, or if the reversing valve must be energized when the system operates in the cooling mode.
- Terminal Y is connected to a compressor contactor.
- Terminal W1 is also connected to a compressor contactor. If the same compressor is used for both first-stage heating and cooling, a jumper wire is connected between Terminal Y and Terminal W1.
- Terminal E is where the emergency-heating circuit is connected to the thermostat.
- Terminal G is where the indoor fan circuit is connected.
- Terminal O is the common terminal and is used to carry power back to the transformer.

When wiring a heat pump or any other system, it is quite common that all of the terminals on the thermostat will not be used. In the example of the thermostat just mentioned, Terminals B and Terminal O will never both be used. Terminal B is designed to be used for systems that require the reversing valve's solenoid coil to be energized in the heating mode, while Terminal O is used on systems that require the reversing valve's coil to be energized in the cooling mode. Since heat-pump-system manufacturers determine their own specific system configuration, thermostats that have both the heating and cooling changeover valve terminals are preferred, since they can be used effectively on either system layout. In addition, on systems that utilize the same compressor for both first-stage heating and cooling, a field wire will be connected to either Terminal Y or Terminal W1, but not both. A jumper wire is installed on the thermostat between these two terminals. If the sources for single-stage heating and cooling are different, a field wire will be connected to both Terminal Y and Terminal W1, in which case the jumper wire between these terminals should be removed.

The thermostat, whose schematic is shown in Figure 10-18, is of the electronic type and does not need to be perfectly level upon installation. It will still function properly, because the temperature of the space is sensed by a thermistor sensor located within the device. However, for aesthetic purposes, all thermostats should be mounted perfectly level.

SERVICE NOTE: *Every service technician should have a small, torpedo-type level to ensure that thermostats are properly leveled. Never rely on the method of leveling by eye.*

Single-Stage Cooling, Two-Stage Heating: Example 2

Another type of two-stage-heating, single-stage-cooling thermostat is shown in. Figure 10-19. Although very similar to the one just discussed, this thermostat has some features that are not shown in the previous example. These differences include:

- The temperature sensing is achieved by bimetal. coils, onto which are attached mercury-filled bulbs. The operation of this type of thermostat is mechanical.
- An indicator light alerts the system operator when the system is operating in the emergency heat mode.
- An indicator light alerts the system operator when the system is operating in the auxiliary, or supplementary, heat mode.

The mercury-filled bulbs on this type of heat-pump thermostat are similar to those found on cooling-only thermostats, as well as those found on single-stage-heating and single-stage-cooling systems. The mercury bulb that controls the first-stage heating, as well as the only stage for cooling operation, is acting as a single-pole, double-throw switch. Referring to the top left of Figure 10-19, note that the wiring connection at the center of the bulb is the common terminal while the left- and right-hand connections are for the heating and cooling modes, respectively. At the bottom left of the same schematic is the mercury bulb that controls the second heating stage. Notice that this bulb is a single-pole, single-throw switch. Because the opening and closing of the electrical contacts on the thermostat rely solely on the position of the mercury within the bulbs, this type of thermostat must be installed perfectly level to ensure proper thermostat and system operation.

The indicator lights for both the emergency and auxiliary heat modes of operation are connected to another terminal on the thermostat, namely, Terminal X. This terminal is connected directly back to the transformer. The main reason for illuminating a

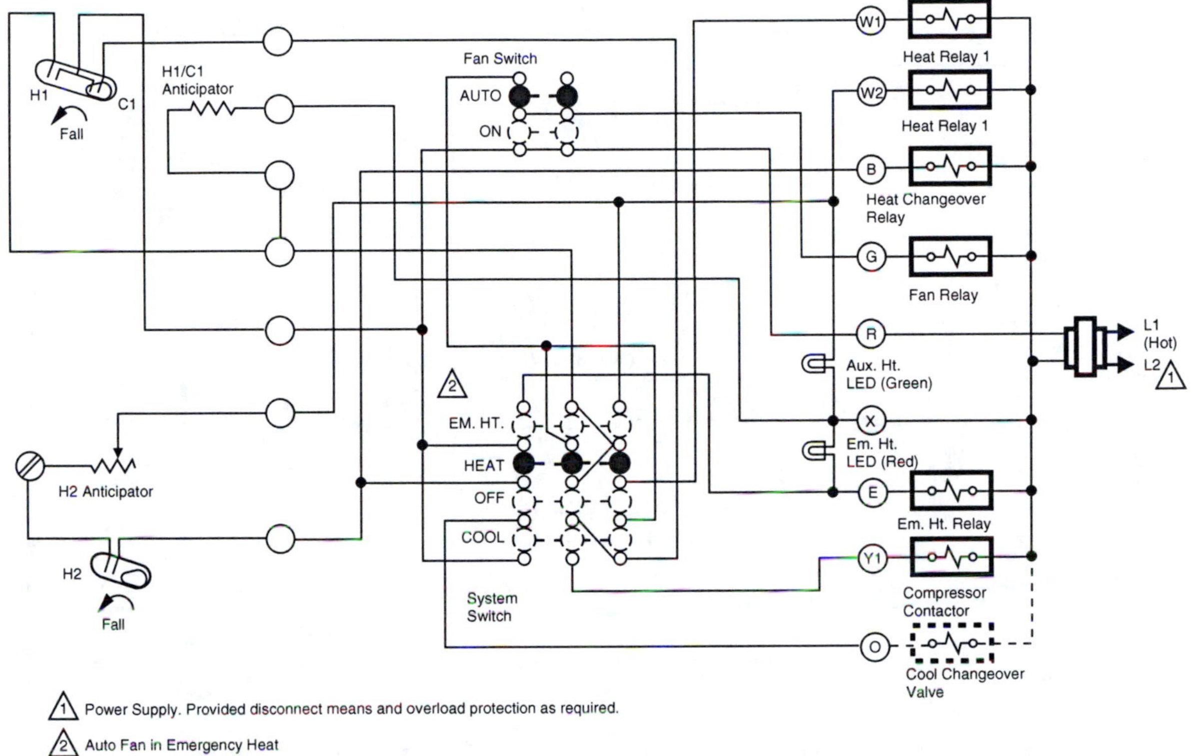

FIGURE 10-19 Mechanical, two-stage-heat, one-stage-cool, wiring diagram with manual changeover. *(Reprinted with permission of Honeywell Inc.)*

warning light is to alert the system operator that a typically more expensive means of conditioning is being used. If this light is illuminated on a regular and continuous basis, even in relatively mild weather, the customer should place a call to the service company.

Once again, this thermostat has separate terminals for first-stage heating and cooling. If the same source is being used for the first stages of heating and cooling, only one of these terminals will have a wire connected to it and a jumper wire will likely be connected between the two terminals on the thermostat's subbase.

Two-Stage Cooling, Two-Stage Heating: Example 1

Larger air-conditioning systems are often equipped with two cooling stages. This enables the system to operate at different cooling capacities, ultimately reducing the energy bills associated with running the equipment. Two-stage cooling systems generally employ one of the following strategies to accomplish the desired result:

- Two separate refrigeration circuits, each with its own piping and compressor
- One refrigeration circuit with a two-speed compressor
- One refrigeration circuit and one compressor equipped with unloaders

Systems with a second cooling stage require an additional set of electrical contacts, usually labeled Y2. A schematic for a two-stage heating and cooling system is shown in Figure 10-20. On systems that are equipped with two separate refrigeration circuits and two compressors, the second-stage cooling terminal, labeled Y2, will energize the compressor contactor coil on the second compressor. During single-stage cooling, only the first compressor is energized. On these systems, the evaporator is actually made up of two independent circuits so, when operating on stage-one cooling, refrigerant only flows through one half of the coil.

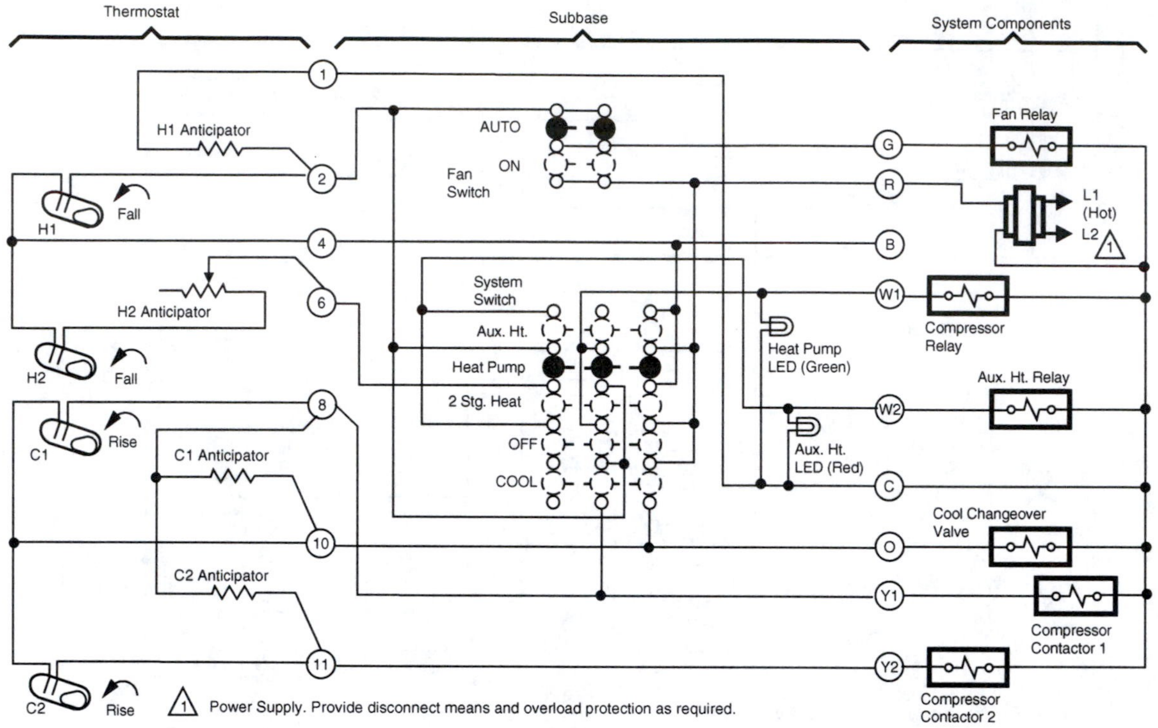

FIGURE 10-20 Mechanical, two-stage heat, two-stage-cool, wiring diagram with manual changeover. *(Reprinted with permission of Honeywell Inc.)*

On two-stage cooling systems that are equipped with two-speed compressors, the Y2 terminal will energize a starter or contactor circuit that will change the internal wiring configuration of the compressor motor windings, thereby changing the actual speed of the compressor motor. Equipment manufacturers, such as Lennox, have used two-speed compressors in a number of their air-conditioning systems with great success.

On larger compressors that are equipped with **unloaders,** the Y2 terminal will cause the system to operate under a loaded condition. Unloaders are intended to allow refrigerant to bypass the compression process, thereby reducing the system capacity as well as the amount of energy used to operate the compressor. Unloaders can be either mechanically or electrically controlled. Note that only electrically operated unloaders will be a part of the system's control circuit.

Also note from the schematic in Figure 10-20 that, with the exception of the additional cooling terminal Y2, the electrical connections are basically the same as with the two-stage heating and single-stage cooling thermostats that were examined earlier. This thermostat is equipped with four mercury bulbs, each of which controls one stage of either heating or cooling. Just as with the thermostat just discussed, this thermostat must be installed perfectly level because of the mercury bulbs contained within the device.

MODES OF OPERATION

This chapter has thus far examined the pieces that must all fit together in order to complete the heat-pump wiring diagram puzzle. It now examines both the power and control circuits for various modes of operation encountered in the typical heat-pump system. Although examining all possible combinations of components that are found on heat-pump systems and examining all of the different manufacturers' equipment would be impossible, what is presented is

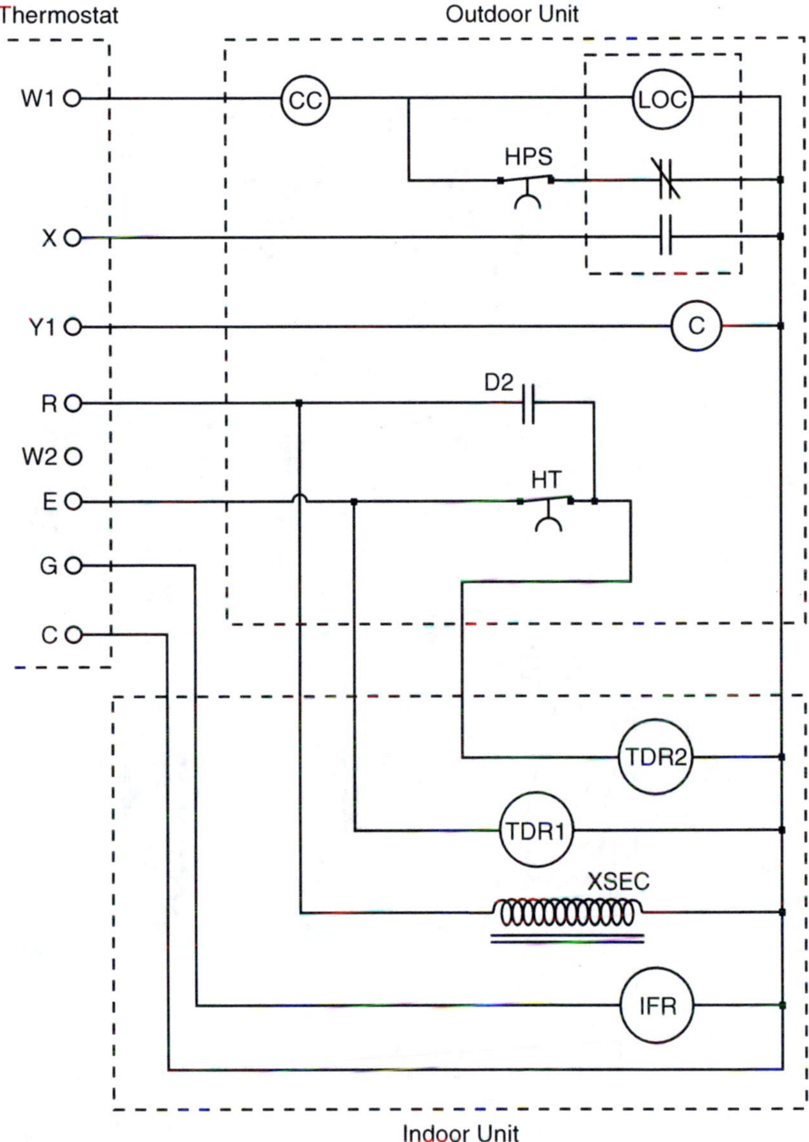

FIGURE 10-21 Sample low-voltage control circuit indicating thermostat, outdoor unit, and indoor unit

a generic heat-pump system, incorporating a variety of commonly found heat-pump components. A solid understanding of the material presented here will serve as a strong foundation for the understanding of even the most complicated heat-pump wiring diagrams. Figure 10-21 represents a simplified, typical, heat-pump, low-voltage control circuit. An actual low-voltage wiring diagram is shown in the middle section of Figure 10-22. Since the thermostat functions as an independent component, the terminals are shown but not the internal workings of the device, as were shown in the sections that dealt with them. However, the terminals and contacts that are closed during each mode of operation are shown. In addition, the wiring diagram is divided into three sections representing the thermostat, the outdoor unit, and the indoor unit. This will help identify where certain components are located within the control circuit.

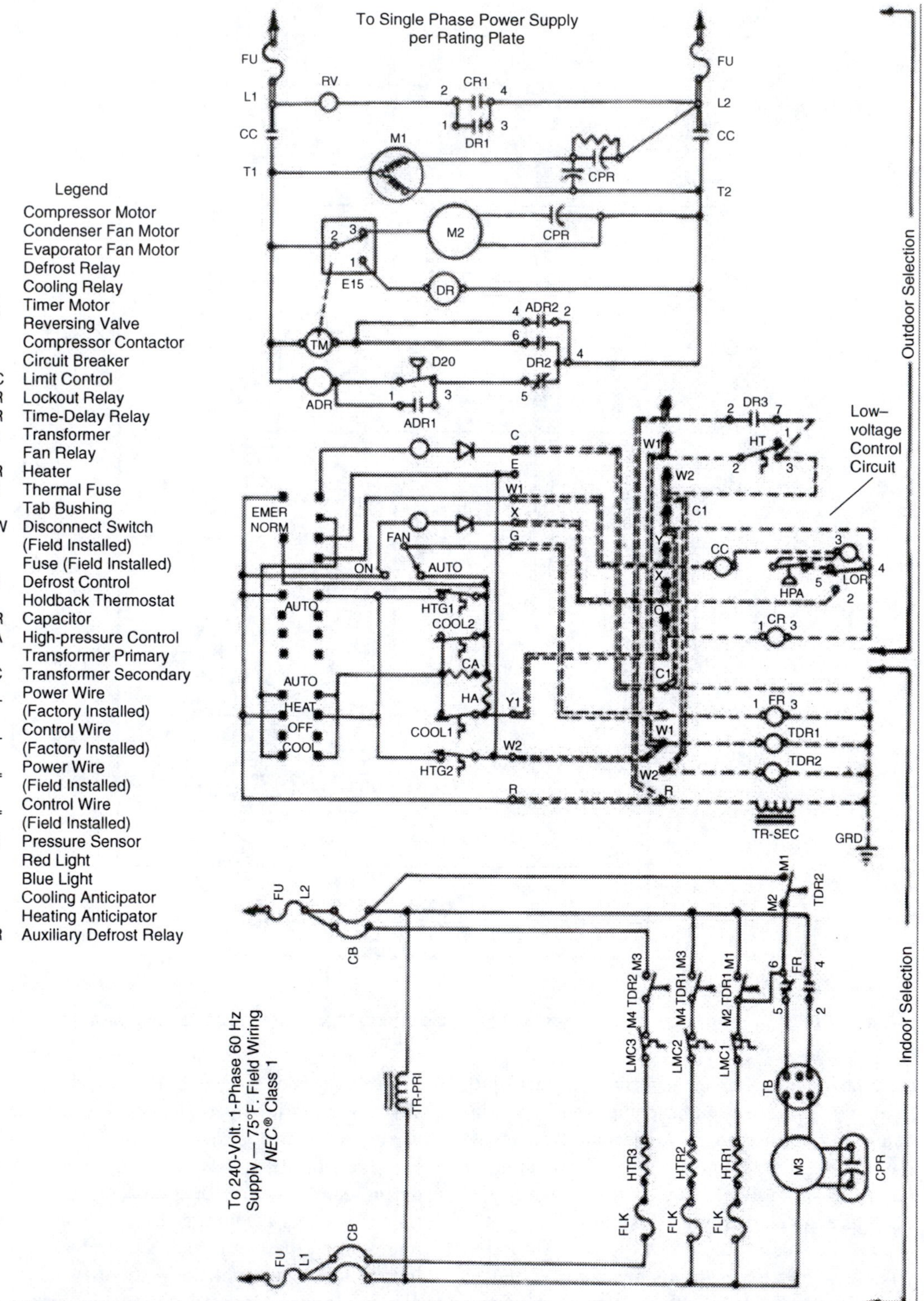

FIGURE 10-22 Heat-pump-system, low-voltage control circuit. *(Courtesy Addison Products Co.)*

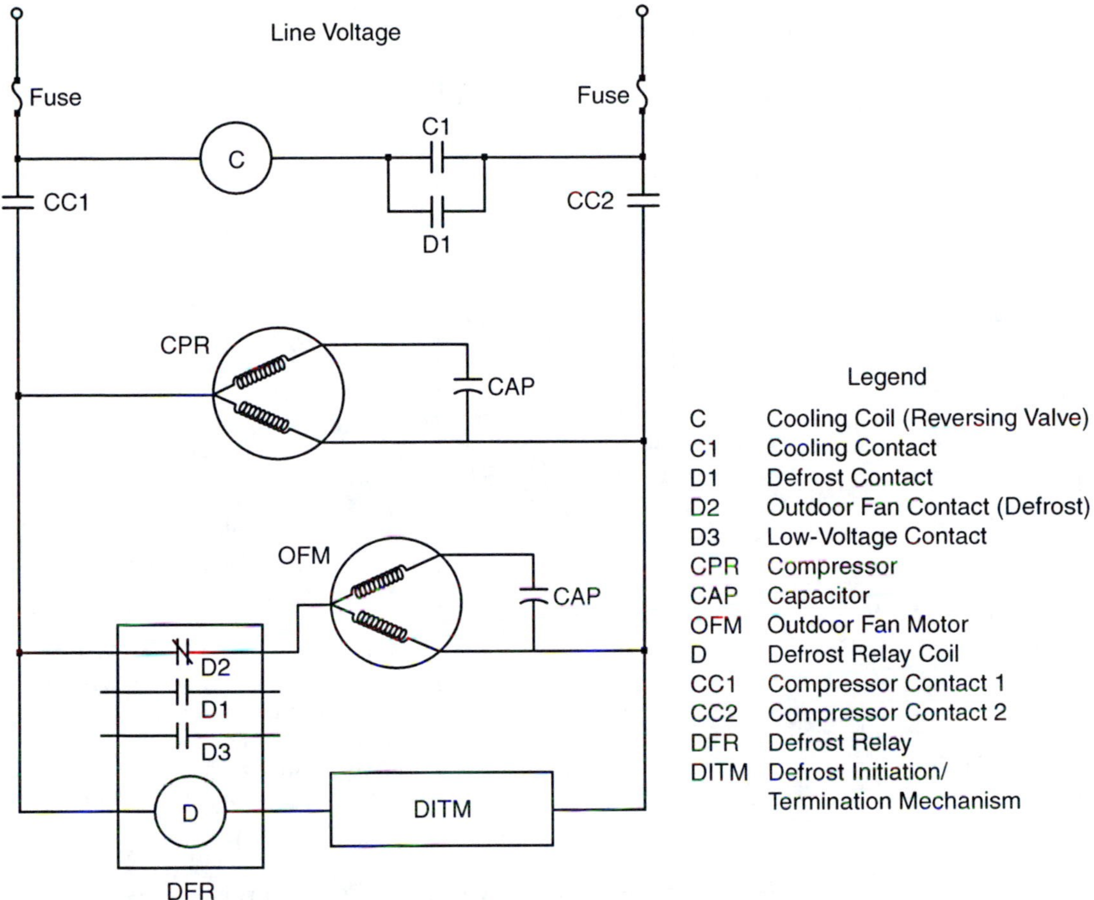

FIGURE 10-23 Ladder diagram of the line-voltage power circuit for the outdoor unit

During the evaluation of the control circuits for the various modes of operation, repeated reference is made to the ladder diagram for the outdoor unit (Figure 10-23) and the indoor unit (Figure 10-24). The low-voltage thermostat connections in the sample system have the following terminal designations:

- W1: Compressor circuit terminal
- X: Used for the trouble-light circuit (lockout mode)
- Y1: Reversing-valve circuit
- R: Power from the transformer
- W2: Second-stage heating circuit
- E: Emergency-heat circuit
- G: Fan circuit
- C: Common terminal bringing power back to the transformer

Cooling Mode

While operating in the cooling mode, the thermostat closes the electrical connections between Terminals W1, Y1, R, and G. These internal connections are shown on the left side of Figure 10-25. Since the power from the secondary winding of the transformer feeds into the thermostat at Terminal R, the investigation begins there. From Terminal R the current will flow through three separate parallel circuits, as denoted by the bold lines in the same diagram. The three circuits are:

- The circuit through the compressor contactor coil, CC
- The circuit through the cooling coil, C
- The circuit through the indoor fan relay coil, IFR

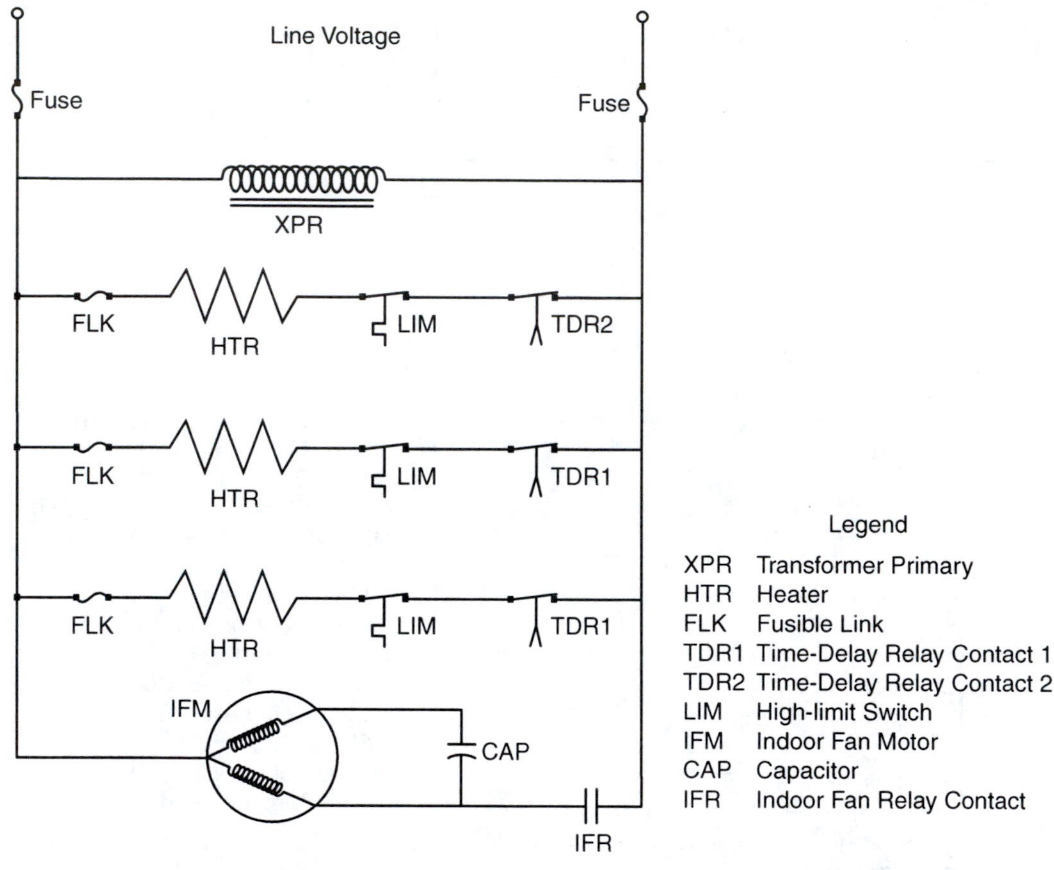

FIGURE 10-24 Ladder diagram of the line-voltage power circuit for the indoor unit

Assuming that the thermostat's contacts remain closed, the compressor will continue to operate as long as the contacts on the high-pressure switch remain closed. The closed high-pressure switch contacts also permit the current in the compressor contactor coil circuit to bypass the lockout relay coil. Once the current passes through the high-pressure switch and the lockout relay's normally closed contacts, it then flows back to the transformer to complete the circuit. The entire wiring diagram for a heat-pump system operating in the cooling mode can be seen in Figure 10-26.

The circuits for the cooling coil and the indoor fan motor relay coil are simple series circuits. These coils are controlled exclusively by the thermostat. As mentioned earlier, the cooling contactor coil is the solenoid that controls the position of the four-way reversing valve. Referring back to the power circuit for the outdoor fan motor, the following can be established:

- Contacts CC1 and CC2 are closed, energizing both the compressor and the outdoor fan motor.
- Contacts C1 are closed, energizing the cooling coil.

Referring to the power circuit for the indoor unit, note that the normally open Contacts IFR will be closed, energizing the indoor fan, permitting cooled air to be introduced to the conditioned space.

Lockout Mode

Because the heat-pump system operates within a wide range of conditions, ensuring the safety of those around the equipment is an important factor that must

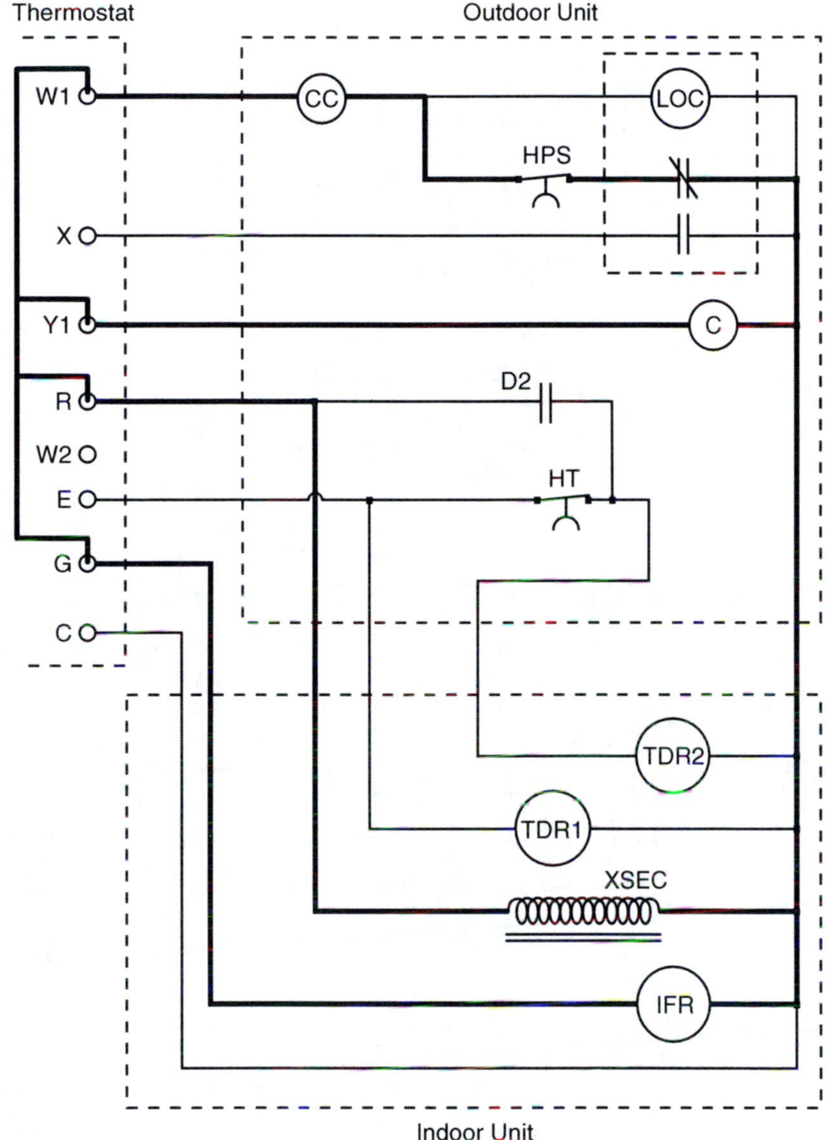

FIGURE 10-25 Control circuit for a heat-pump system operating in the cooling mode

be addressed. As discussed earlier, the lockout relay prevents the compressor from operating if the pressure in the system reaches unsafe and dangerous levels. A typical lockout condition is shown in Figure 10-27. (The lockout circuit was also shown in Figure 10-10.) The cooling-coil circuit has not been bolded in this circuit in order to concentrate on the lockout condition of the system. It is, however, quite common for the system to be operating in the cooling mode when the lockout mode is initiated. In the lockout mode:

- The high-pressure switch has opened its contacts in response to a high-pressure condition.
- The current flowing through the compressor contactor coil is now forced through the lockout relay coil.

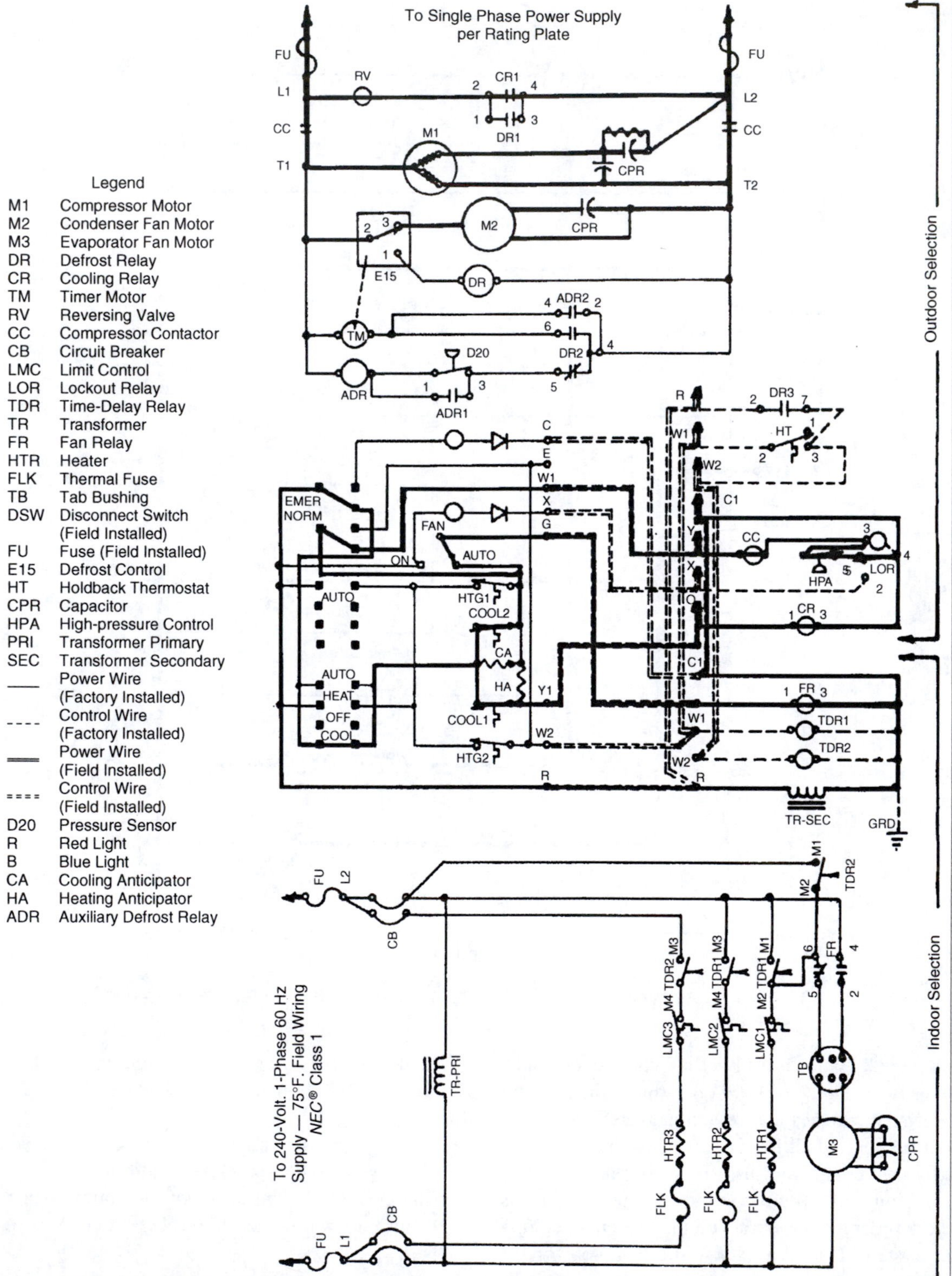

FIGURE 10-26 Heat-pump cooling mode wiring diagram. *(Courtesy Addison Products Co.)*

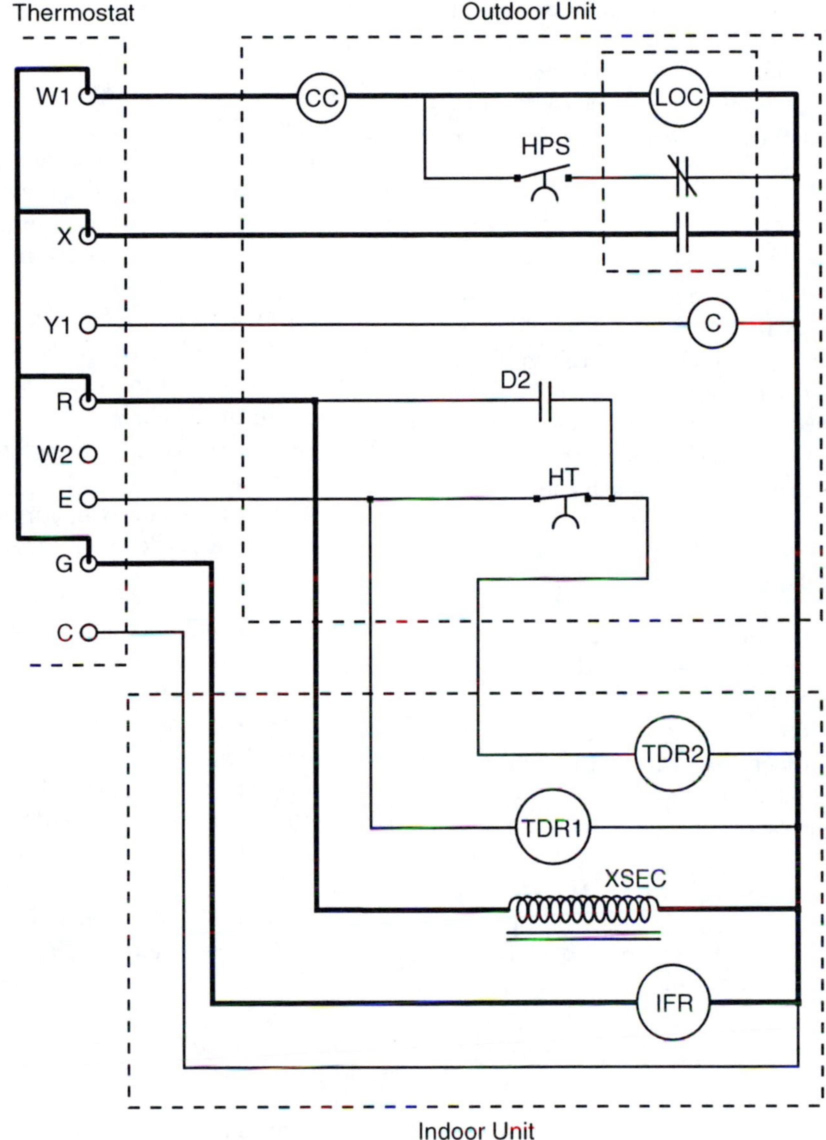

FIGURE 10-27 Control circuit for a heat-pump system in lockout mode

In Figure 10-27, the current leaving Terminal W1 must now flow through the coils of both the compressor contactor and the lockout relay. This causes the current in the circuit to decrease, thereby decreasing the strength of the magnetic field created in the compressor contactor coil. This reduction in magnetic field makes it impossible for the compressor contactor contacts to remain closed. The compressor is therefore de-energized.

In addition to de-energizing the compressor, the lockout relay also energizes the circuit containing the trouble light, indicated by Terminal X in the same figure. The current flowing through the lockout relay coil is sufficient to cause the normally open contacts on the relay to close, causing the trouble light to illuminate. This light alerts the occupant of the lockout condition so that a service call can be placed. The indoor fan circuit is not affected by the compressor lockout and will continue to operate, as indicated by the control circuit schematic. The fact that the fan can continue to operate will benefit the system operator if the emergency-heat mode needs to be used until the service technician is able to repair the problem.

Heating: Stage 1

Because the compressor is responsible for providing first-stage heating as well as first-stage cooling, the compressor contactor coil must be energized during the heating mode. The indoor fan circuit is also energized, allowing heated air to be introduced to the occupied space. This control circuit is shown in Figure 10-28. In the first-stage heating mode:

- The compressor contactor coil is energized.
- The indoor fan circuit coil is energized.

Comparing Figure 10-28 with Figure 10-25, the only operational difference is that the reversing-valve circuit is not energized. The main differences in the circuitry occur within the thermostat itself. In the heating mode, the compressor is energized as the temperature of the occupied space falls, while in the cooling mode the compressor is energized when the space temperature rises. Referring back to the ladder diagram of the outdoor unit (Figure 10-23), Contacts CC1 and CC2 are closed, energizing the compressor and the outdoor fan motor. Contacts C1 and D1 remain open, thereby keeping the cooling coil out of the active power circuit. A complete wiring diagram showing first-stage heating operation is illustrated in Figure 10-29.

Heating: Stage 2

If the heat-pump system is not able to satisfy the heating requirements of the occupied space, the second heating stage will be energized. This normally occurs on colder than normal days, when there is a malfunction in the heat-pump system or simply when the system has been off for a period of time. Assuming that the outdoor ambient temperature is very low, the low-voltage control circuit will resemble that in Figure 10-30. The conditions giving rise to second-stage heat are as follows:

- The first heating stage is still energized and operating.
- Terminal W2 on the thermostat is powered.
- Time-delay relay coil TDR1 is energized.
- If the outdoor ambient temperature is low enough, the holdback thermostat is in the closed position.
- If the holdback thermostat is closed, the time-delay relay coil TDR2 is energized as well.
- The electric-strip heaters located in the indoor fan section are energized.
- The compressor is still operating to provide first-stage heating.
- The outdoor fan motor is still operating.
- The cooling coil is de-energized.

The second-stage heating and indoor fan circuits are the same as in the first-stage heating example, so here the focus is on the supplementary electric-strip heaters and their associated wiring.

When second-stage heating is called for, Terminal W2 on the thermostat is energized in addition to the first-stage heating terminal, W1. As shown in Figure 10-30, the current flows from Terminal W2 and feeds two additional parallel circuits containing TDR1 and TDR2. The TDR1 coil will be energized whenever the W2 terminal is energized because there are no other controlling factors in this circuit. The TDR2 coil, however, will be energized only when W2 is energized and the holdback thermostat contacts are closed. This follows logically because in the warmer weather:

- Less supplementary heat should be required.
- The contacts on the holdback thermostat are open, de-energizing TDR2.

Therefore, when the outside temperature is warmer and the second-stage heating circuit is energized, only a portion of the heating elements will be activated; whereas, in the event of a colder outside ambient, all of the supplementary electric-strip heaters will be in the active circuit. The complete second-stage wiring diagram can be seen in Figure 10-31 on page 296.

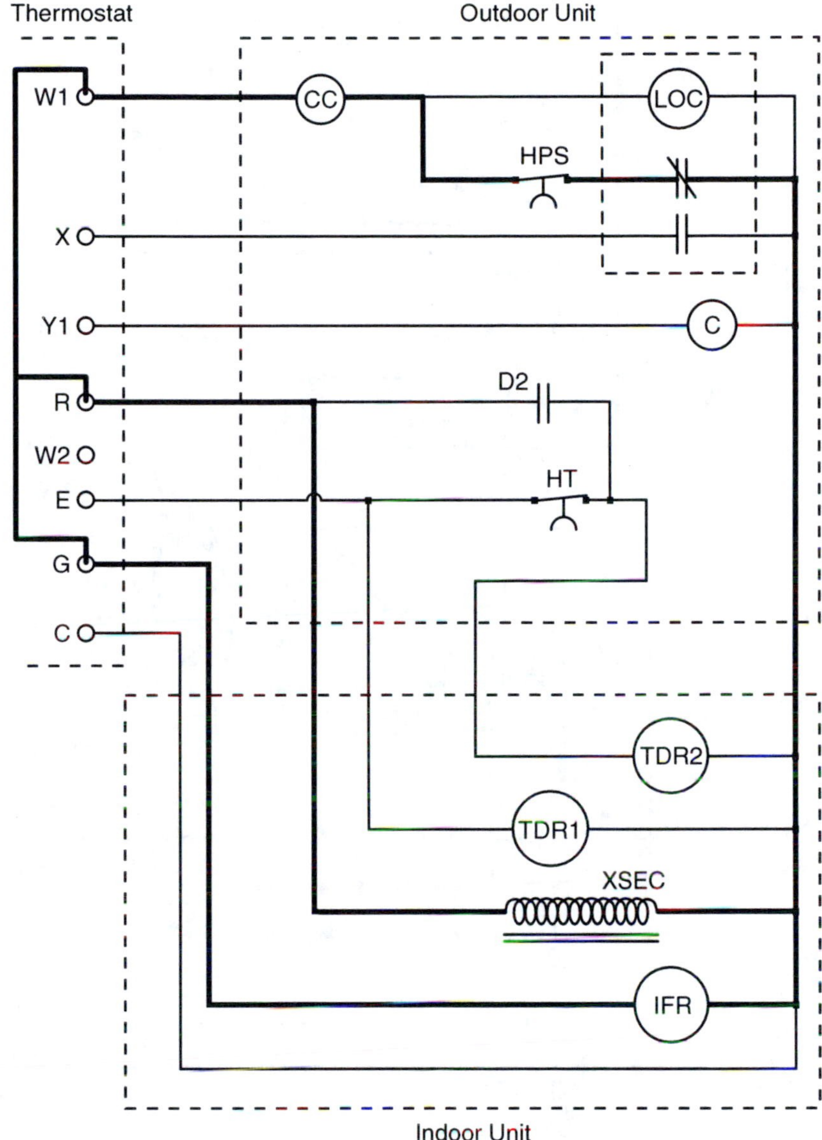

FIGURE 10-28 Control circuit for a heat-pump system operating in first-stage heating

Defrost Mode: Warm Outside Ambient Temperature

The sample heat-pump system is equipped with a line-voltage defrost relay, which is located in the outdoor unit. The coil of this relay is energized whenever the system is calling for the system to go into the defrost mode.

Earlier in the text, the various ways that a system can be put into or taken out of defrost were addressed and they included either one or more of the following factors:

- Time
- Temperature
- Pressure

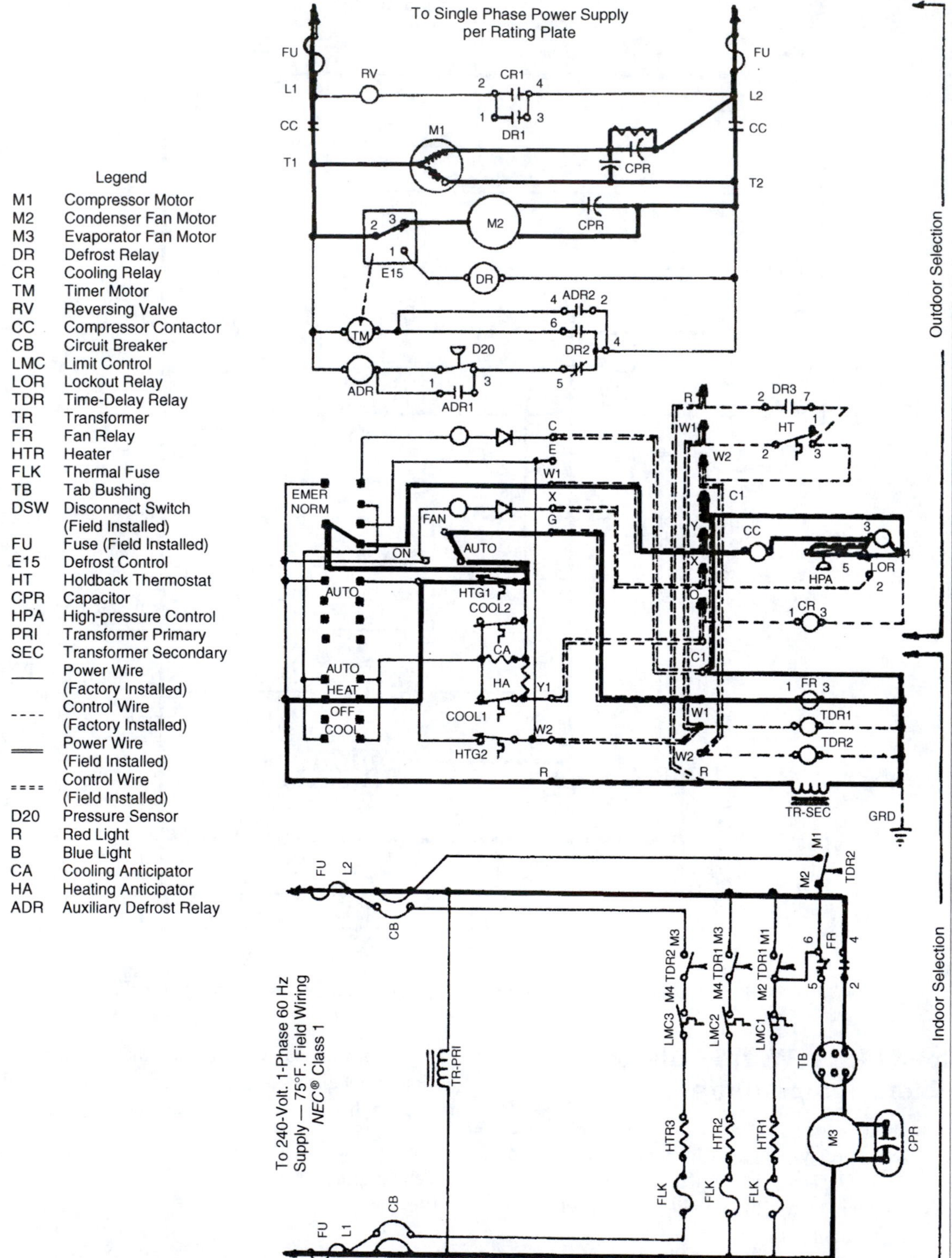

FIGURE 10-29 Heat-pump first-stage heating-mode wiring diagram. *(Courtesy Addison Products Co.)*

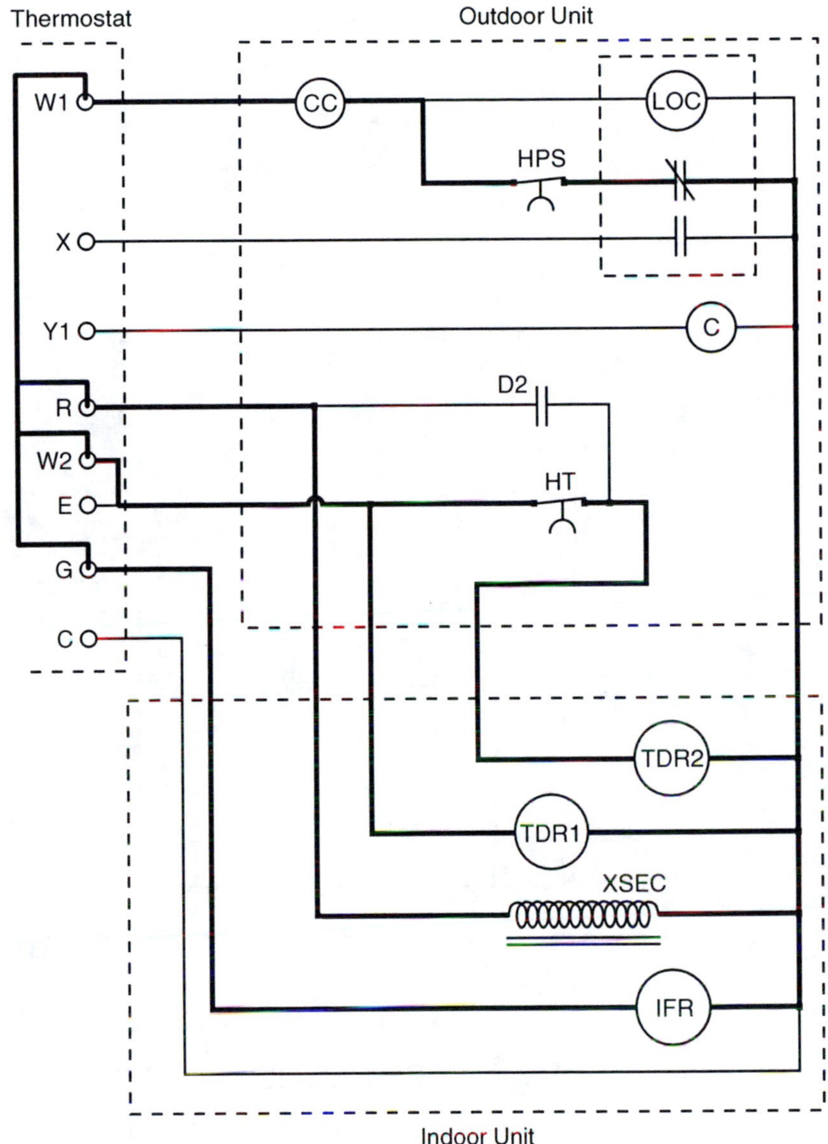

FIGURE 10-30 Control circuit for a heat-pump system operating in second-stage heating

Regardless of which method is used to initiate or terminate defrost, the end result is the same. For this reason, the defrost-controlling factors were labeled as DITM, or defrost initiation/termination mechanism, on the schematic shown in Figure 10-23. One interesting feature of this defrost relay is that the normally closed set of contacts, D2, is controlling a line-voltage circuit for the outdoor fan, while a normally open set of contacts, D3, is controlling a low-voltage circuit for the TDR2 coil. A third set of contacts, D1, is used to energize the solenoid coil on the four-way reversing valve, which, in this case, is located in the line-voltage circuit. Once defrost is initiated, a number of things occur. At the outdoor unit:

- The compressor will continue to operate.
- The defrost relay coil will be energized.

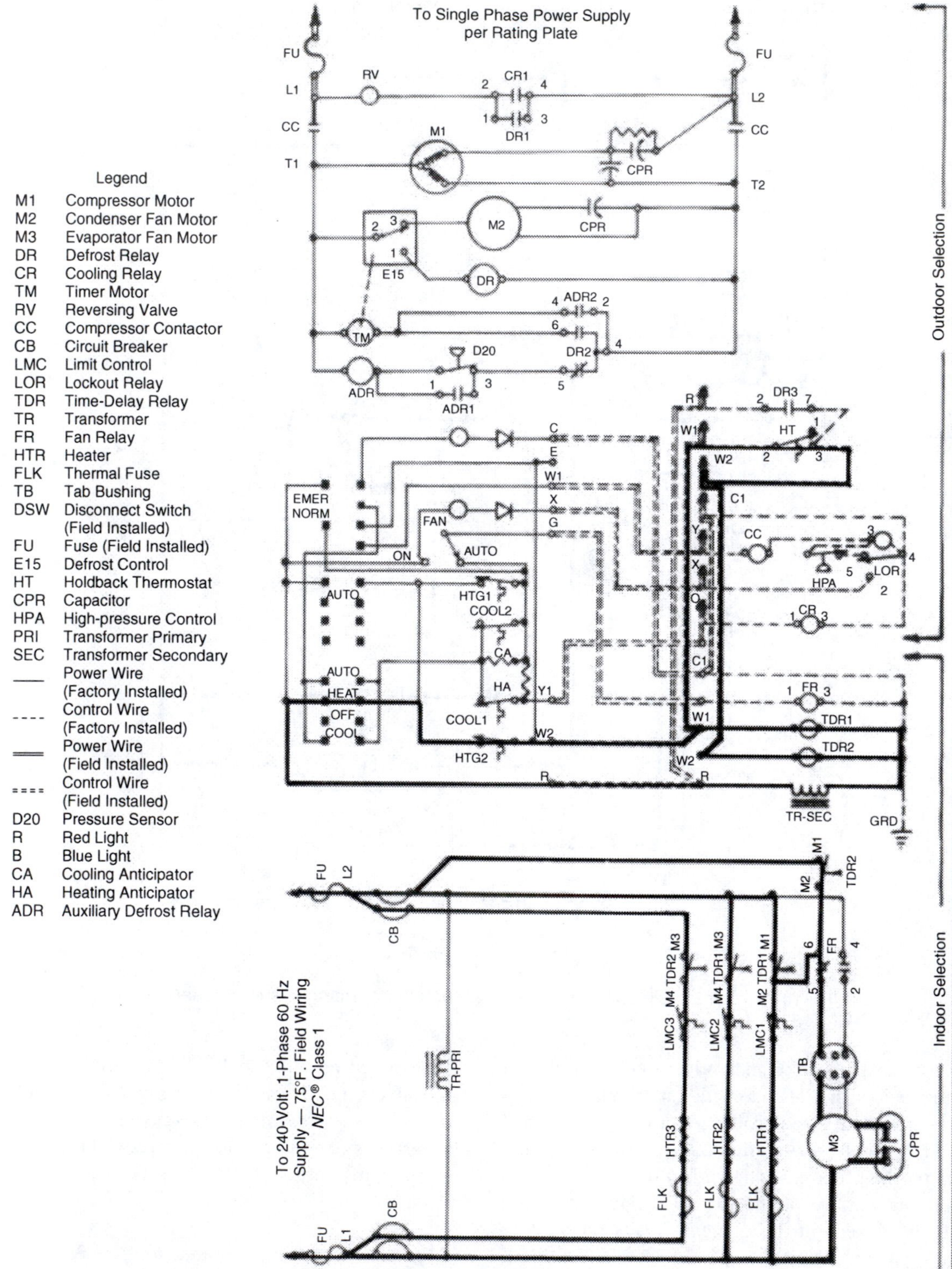

FIGURE 10-31 Heat-pump second-stage heating wiring diagram. *(Courtesy Addison Products Co.)*

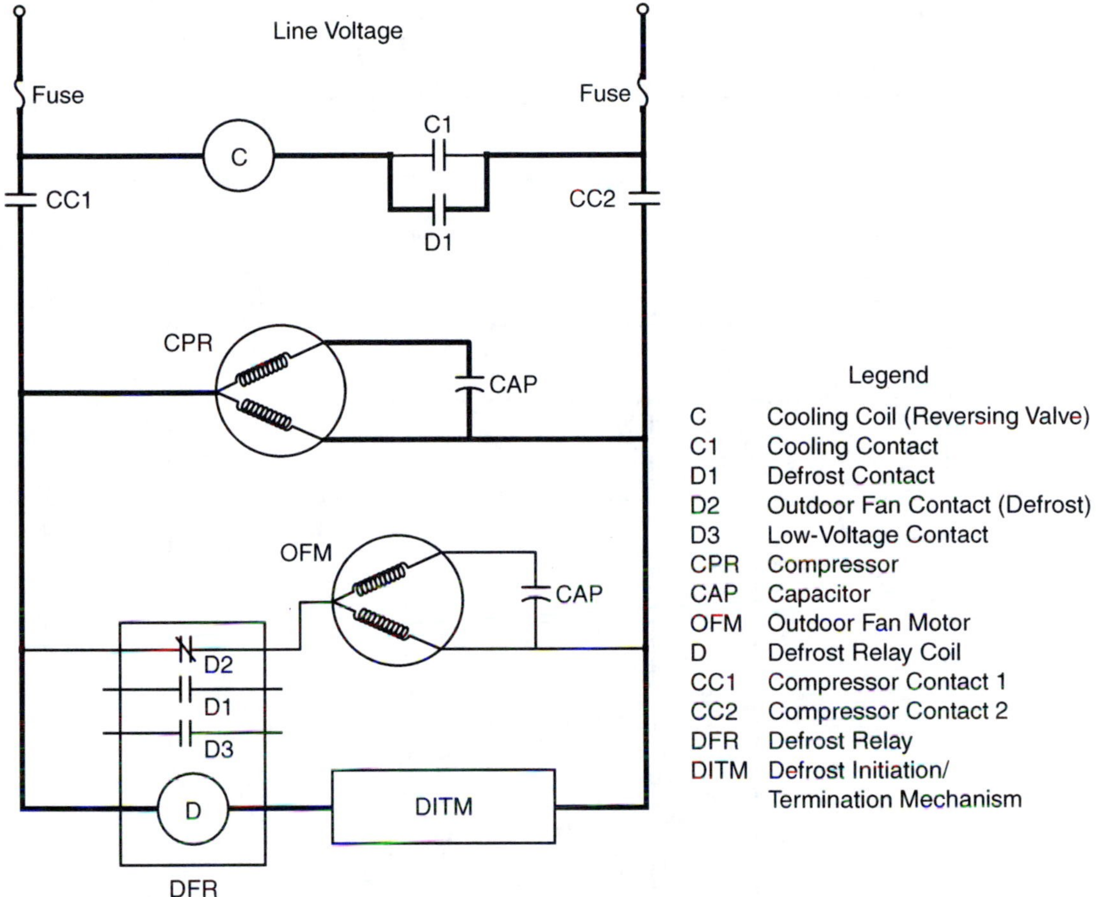

FIGURE 10-32 Outdoor unit power circuit while operating in the defrost mode

- Contacts D2 on the defrost relay will open, de-energizing the outdoor fan motor.
- Contacts D3 will close, energizing the TDR2 circuit.
- Contacts D1 will close, energizing the cooling coil, switching the system into cooling mode.

The current flow through the outdoor unit is shown in Figure 10-32. At the indoor unit, the following conditions are present:

- The indoor fan will continue to operate.
- The TDR2 contacts will be closed, energizing the first electric-strip heater, assuming that the holdback thermostat and all safety switches in the heater circuit are closed.

The low-voltage control circuit for this scenario is shown in Figure 10-33.

Defrost Mode: Colder Outside Ambient Temperature

The defrost that occurs when the ambient is low differs from the defrost under warmer ambient conditions and varies only with respect to the number of electric-strip heaters that are energized. This, of course, is determined by the holdback thermostat. The scenario is exactly the same as the one just addressed, with the exception that now all of the strip heaters are in the active electrical

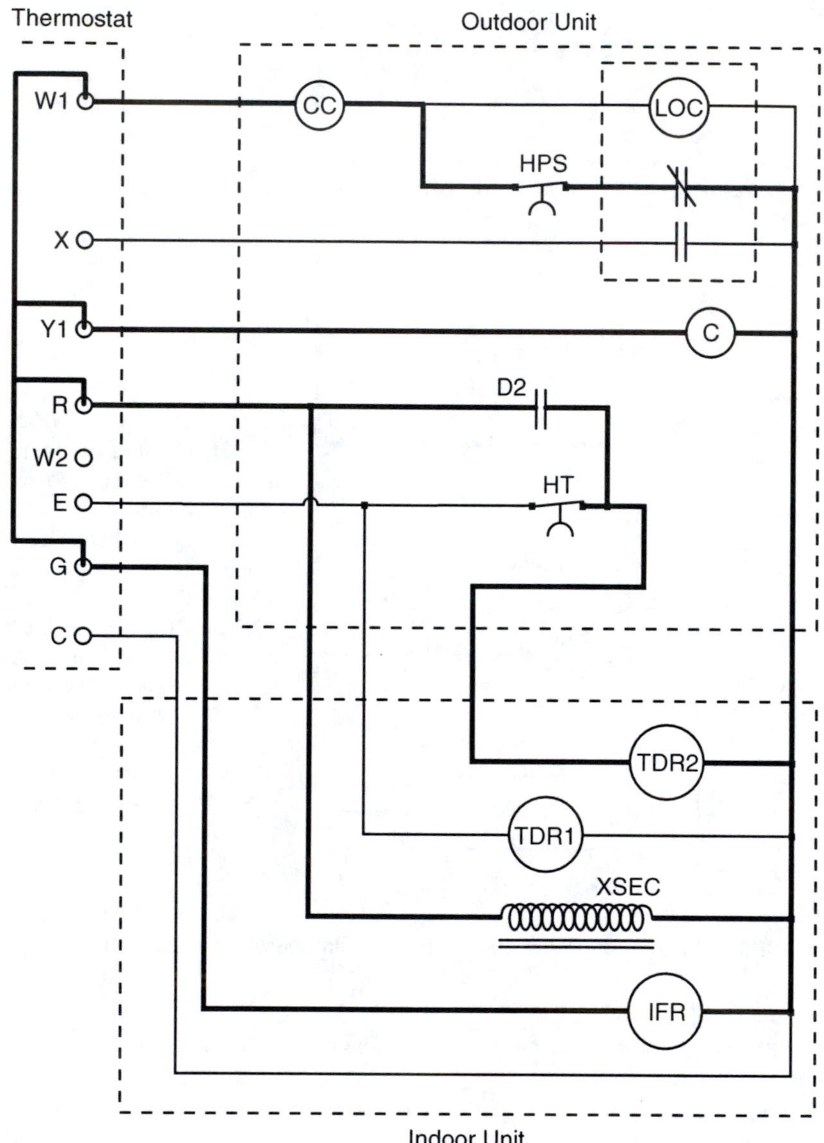

FIGURE 10-33 Control circuit for a heat-pump system operating in the defrost mode with a warm outdoor ambient temperature

circuit. This situation is shown in Figure 10-34. So, at the indoor unit:

- The indoor fan will continue to operate.
- The TDR1 and TDR2 contacts will be closed, energizing all three electric-strip heaters, assuming that the holdback thermostat and all safety switches in the heater circuits are closed.

Emergency-Heat Mode

In the event of a failure in the vapor-compression heat-pump cycle, the system operator can initiate the emergency-heat mode. Having the system operate in this mode for an extended period of time is not desirable since the source of this heat is electricity, which tends to be a much more expensive means of providing the

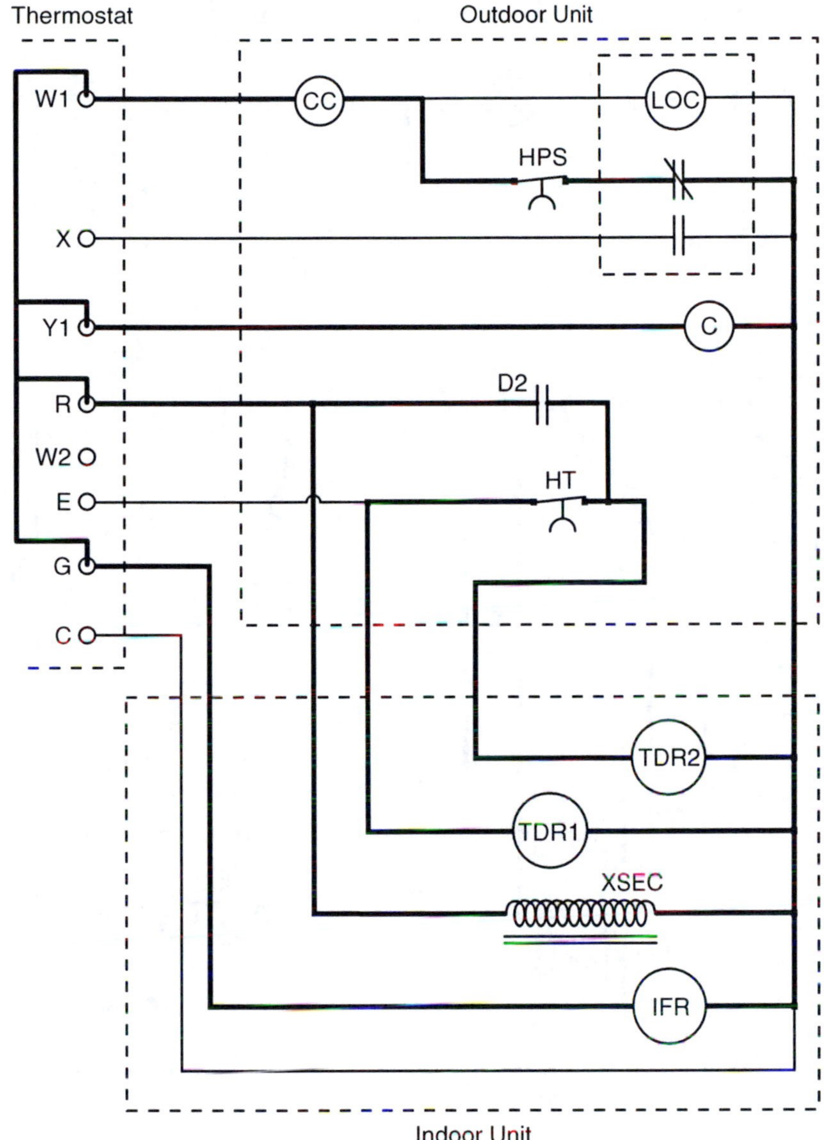

FIGURE 10-34 Control circuit for a heat-pump system operating in the defrost mode with a cold outdoor ambient temperature

required heat. A typical schematic for a system operating in the emergency-heat mode is shown in Figure 10-35. In the emergency-heat mode, the following conditions are satisfied:

- The compressor contactor coil is de-energized.
- The indoor fan is energized via the G terminal on the thermostat.
- The TDR1 coil is energized through the E terminal on the thermostat.
- The emergency-heat light is illuminated through the C terminal on the thermostat.

In the emergency-heat mode, the outdoor unit is not operating at all, while the indoor unit has both the fan and at least a portion of the electric-strip heaters

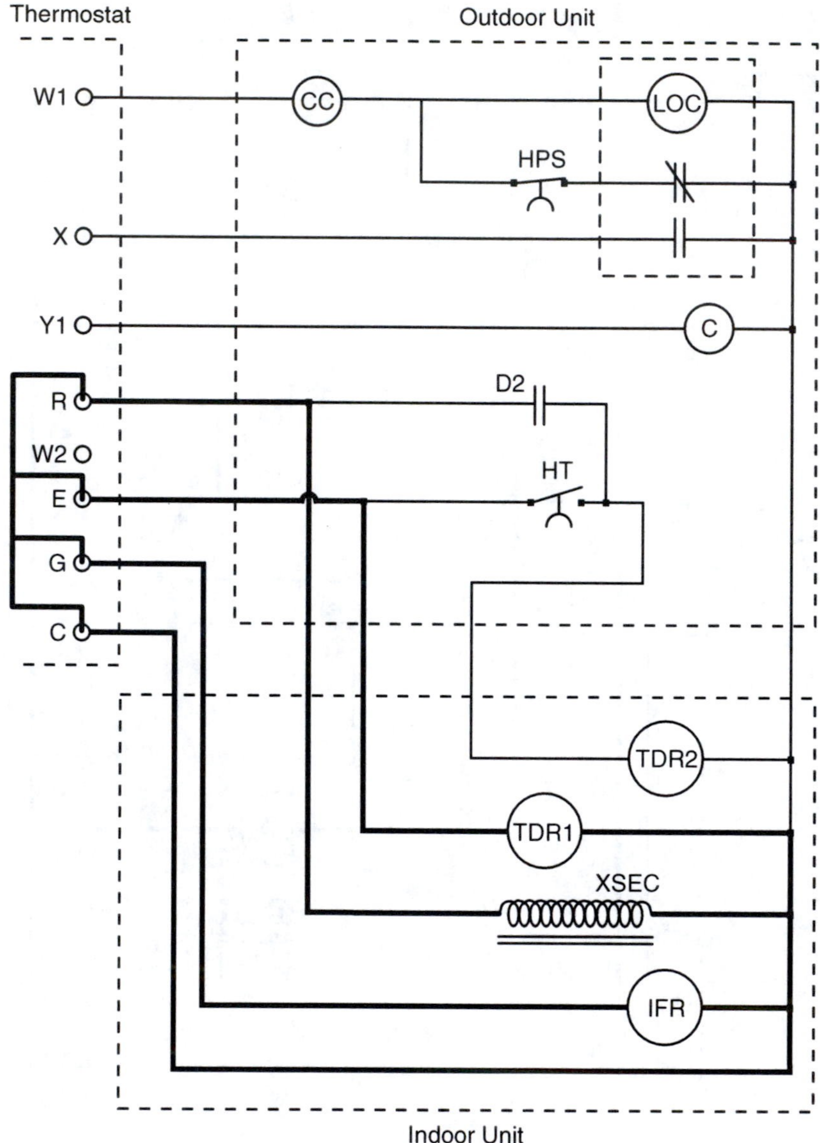

FIGURE 10-35 Control circuit for a heat-pump system operating in the emergency-heat mode with a warm outdoor ambient temperature

energized. The wiring diagram in Figure 10-35 shows the circuit when the outdoor ambient is above the set point on the holdback thermostat. Note that the contacts on the holdback thermostat are open and only TDR1 is energized. Figure 10-36 shows what the circuit will look like when the contacts on the holdback thermostat close, thereby allowing all of the heater circuits to be energized, maximizing the amount of heat being introduced to the conditioned space.

Finally, note that even though the electric-strip heaters are energized in both the emergency-heat mode and in the second-stage heating mode, the operation of the strips in the emergency mode is initiated manually by the system operator. Second-stage heating is initiated automatically by the thermostat.

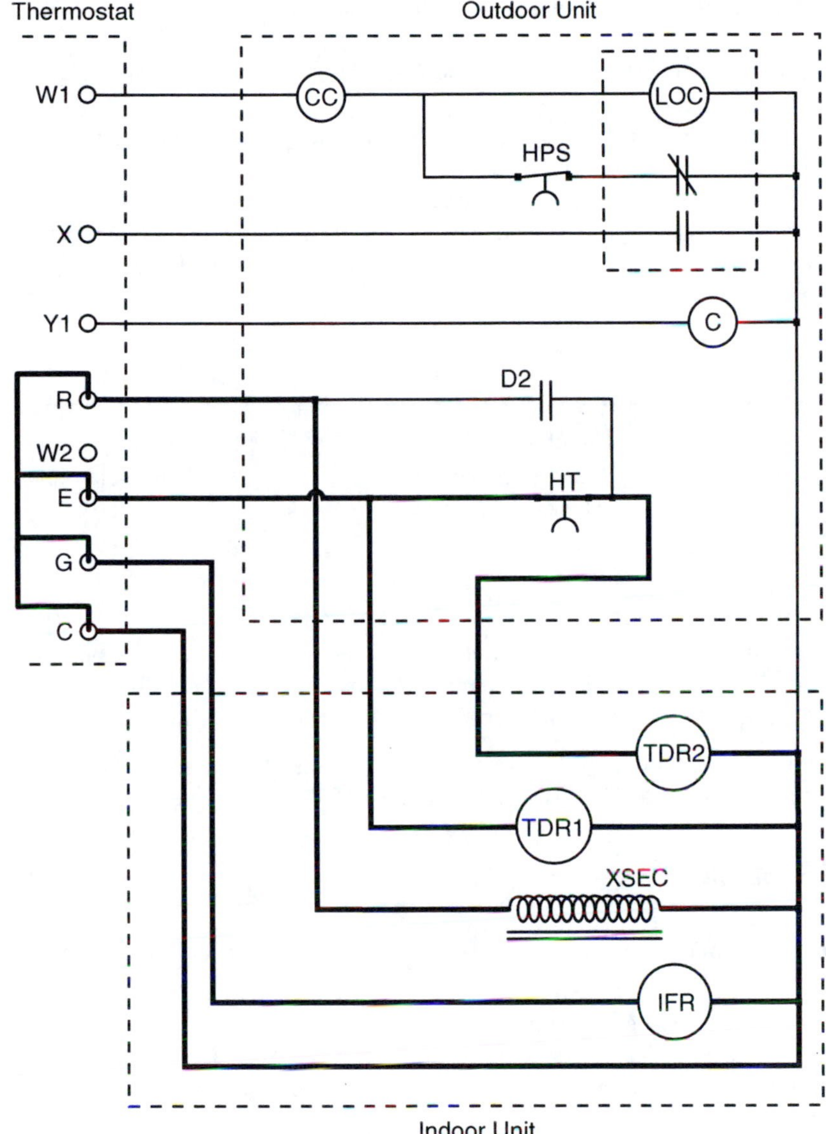

FIGURE 10-36 Control circuit for a heat-pump system operating in the emergency-heat mode with a cold outdoor ambient temperature

SUMMARY

- The majority of the low-voltage electrical circuits are controlled by the thermostat.
- For the system to operate as desired, the proper sequence of operations must be followed.
- The compressor provides first-stage heating and cooling and is often controlled by the same thermostat terminal in both modes.
- The reversing valve must switch the system over to the cooling mode in both the cooling and defrost modes.

- Electric-strip heaters are energized during second-stage heating, during defrost, and during the emergency-heat mode.
- The outdoor fan motor is connected through the defrost relay to prevent it from operating during the defrost mode.
- The holdback thermostat determines the number of heating elements that will be in the active electrical circuit depending on the outside ambient temperature.
- The lockout relay prevents the compressor from operating if system pressures reach unsafe levels.
- The lockout relay must be manually reset by disconnecting and restoring power to the circuit.
- Typical heat-pump systems have two heating stages and one cooling stage, but larger systems can have multiple heating and cooling stages.

KEY TERMS

Automatic changeover
Cylinder unloaders
Field wiring
Holdback thermostat
Lockout relay
Manual changeover
Thermostat
Time-delay relays
Unloaders

FOR DISCUSSION

1. Consider a customer who must constantly use the emergency-heat mode on the heat-pump system in order to obtain the desired space temperature. Why is this not desirable and what are some possible system problems that can lead to this situation?
2. Discuss the possible results of a system that tempers, or heats, the air supplied to the occupied space too much during the defrost mode. How is this situation avoided?

REVIEW QUESTIONS

1. For a heat-pump system that fails in the heating mode, when will the cooling coil on the reversing valve be energized?
   a. When a normally open set of heating contacts closes
   b. When the first-stage cooling terminal on the thermostat is energized
   c. When a normally open set of contacts on the defrost relay closes
   d. Both *b* and *c*
2. In which of the following cases are the electric-strip heaters energized?
   a. When the first-stage heating terminal on the thermostat is energized
   b. When the system is operating in the cooling mode in order to temper the air being supplied to the occupied space
   c. When the defrost mode is initiated
   d. When the time-delay relay coils are de-energized
3. Which of the following is intended to protect the individual electric-strip heater circuit from an overcurrent condition?
   a. The time-delay relay coil
   b. The fusible link
   c. The main fuse at the indoor unit
   d. The high-limit switch

4. The device that helps ensure that the space thermostat does not terminate the defrost cycle prematurely is the __________.

5. Under which of the following conditions are all of the electric-strip heaters energized?
   a. In defrost mode with a warm outdoor ambient temperature
   b. In the cooling mode with a cold outdoor ambient temperature
   c. In the defrost mode with a cold outdoor ambient temperature
   d. In the emergency-heat mode with a warm outdoor ambient temperature

6. Which of the following could cause the electric-strip heaters to become de-energized?
   a. The cycling of the system from first-stage heating to second-stage heating
   b. The cycling of the system from first-stage heating into defrost
   c. The cycling of the system from defrost to first-stage heating
   d. The cycling of the system from defrost to second-stage heating

7. In a heat-pump system operating in the automatic changeover mode:
   a. The compressor is energized and de-energized as the system switches over from heating to cooling, and vice versa.
   b. The compressor continues to operate, but the reversing-valve coil is energized and de-energized as the system switches over from heating to cooling, and vice versa.
   c. The outdoor fan motor cycles on and off as the system switches over from heating to cooling, and vice versa.
   d. Both *a* and *c* are correct.

8. Energizing the lockout relay coil:
   a. Increases the resistance of the compressor contactor circuit.
   b. Causes a reduction in the current flow through the compressor contactor circuit.
   c. Causes the system trouble light on the thermostat to illuminate.
   d. All of the above are correct.

9. The lockout relay:
   a. Must be reset manually by disconnecting and then restoring power to the circuit.
   b. Will reset itself automatically when the system pressure falls to a safe level.
   c. Can be jumped out and removed from the system indefinitely without sacrificing the safety feature.
   d. None of the above are correct.

10. Heat-pump thermostats that have two cooling stages can be found on systems:
    a. That are equipped with a two-speed compressor.
    b. That have two separate refrigeration circuits and two compressors.
    c. That have compressors that are equipped with electrical unloaders.
    d. All of the above are correct.

11. A heat-pump system that cannot properly maintain the desired space temperature while operating in the first-stage heating mode but provides adequate cooling:
    a. May have an improperly mounted thermostat.
    b. May have a refrigerant undercharge.
    c. Neither *a* nor *b* is possible.
    d. Both *a* and *b* are possible.

12. When the system goes into lockout mode:
    a. The indoor fan will still operate and the emergency heat mode can be used.
    b. The indoor fan cannot operate until the lockout relay is reset.
    c. The compressor can still operate, but only in the heating mode.
    d. The outdoor fan motor can still operate.
13. A two-stage-heating/two-stage-cooling thermostat will have __________ single-pole, single-throw, mercury-filled bulb(s).
    a. One
    b. Two
    c. Three
    d. Four
14. Explain why the outdoor fan motor and the compressor cannot be wired together in the same circuit as in the case of a cooling-only system.
15. Explain why the compressor remains energized when the heat-pump system switches over from first-stage heat to second-stage heat.

CHAPTER 11

Troubleshooting Basic Vapor-Compression Heat-Pump Control Circuits and Components

OBJECTIVES After studying this chapter, the reader should be able to:

- Explain how to determine if a heat-pump thermostat is making the correct electrical connections.
- Explain how to evaluate a heat-pump thermostat from the thermostat location as well as from the indoor fan section.
- Explain how to determine if there is a break in the interconnecting wiring between the indoor unit, the outdoor unit, and the thermostat.
- Explain how to determine if the defrost timer in a heat-pump system is operating properly.
- Explain how to determine if the solid-state defrost circuit board is operating properly.
- Explain how to determine if the electric-strip heater circuits are operating properly.
- Explain how to isolate faulty components in the electric-strip heater circuit.
- Explain how to determine if the holdback thermostat is operating properly.
- List the procedures to effectively troubleshoot an outside ambient thermostat.

INTRODUCTION

As discussed in earlier troubleshooting chapters, the ability to recognize and understand the various wiring diagrams found on different systems is an integral part of becoming an effective system troubleshooter. Although the troubleshooting of basic electrical components was discussed earlier in the text, this chapter deals with those components found in the heat-pump system. The techniques and test instruments used earlier in the text are the same ones that are employed here. The reader should already have a basic understanding of the fundamentals of electrical circuits and troubleshooting before proceeding to this chapter. A number of areas are discussed in the following pages, thereby strengthening the ability of the reader to effectively evaluate the electrical circuits that comprise such an important part of the heat-pump system as a whole. This chapter is, in effect, a continuation of Chapter 6, which dealt with the troubleshooting of electrical components in the vapor-compression refrigeration system.

OVERVIEW

Troubleshooting the electrical circuits in a heat-pump system requires the same set of skills that are used when evaluating any other electrical circuit. To be effective when attempting to determine the cause of an electrical system failure, the technician should be able to answer the following questions before actually making a system repair:

- What electrical controls and loads are located in the circuit in question?
- Are the associated controls and switches in the circuit normally open or normally closed devices?
- Is the intended operation of these controls clearly understood?

Knowing the answers to these questions will help establish the baseline by which the technician can begin the evaluation. In addition to the preceding questions, the technician should also have the skills required to properly utilize and operate the test equipment that is vital to troubleshooting success.

Technicians should be well versed in taking voltage, current, and resistance readings in an electrical circuit and be able to properly interpret those readings. For example, if the voltage reading across a set of contacts is 24 volts, does this mean that the contacts are open when they should in reality be closed or are the contacts open because they should indeed be open? Or could it be that the contacts are pushing closed as they should but dust, dirt, or some other object is preventing the contacts from physically closing?

Since many electrical devices in the heat-pump circuitry are ultimately controlled by either temperatures or pressures both within and external to the system, the apparent failure of an electrical system component may indeed be a deficiency somewhere else in the mechanical arena. The technician must determine where mechanical problems end and electrical problems begin. Often, a very fine line exists between the two. The technician must understand the importance, for example, of checking the system's refrigerant charge when faced with a low-pressure control that has opened its contacts. A complete or partial loss of refrigerant may lead to the apparent failure of the pressure control. This relates to the cause-and-effect issue that was addressed at the beginning of this text. By replacing the seemingly defective pressure control, the cause of the problem—namely, the loss of refrigerant—has not been resolved.

HEAT-PUMP THERMOSTATS

A great deal of the heat-pump system's operations are controlled by the thermostat. Therefore, technicians very easily, and often incorrectly, conclude that the thermostat is to blame when the system completely or partially fails. A few key issues should be addressed with respect to the thermostat:

- The thermostat acts as a pass-through device.
- With the exception of the emergency-heat light and the trouble light, the thermostat has no electrical loads.
- The thermostat is effectively comprised of multiple switches that open and close according to the desired mode of operation.

The methods employed for determining if the thermostat is functioning properly will vary slightly depending on the construction and design of the device. Electronic thermostats that are equipped with a ribbon-wire connection or a molded plastic plug (Figure 11-1) connecting the thermostat and the subbase are among the easiest to troubleshoot. Thermostats that have no physical connection to the subbase when removed from it are somewhat more difficult to evaluate, for reasons that are discussed shortly. A discussion of the digital thermostat follows.

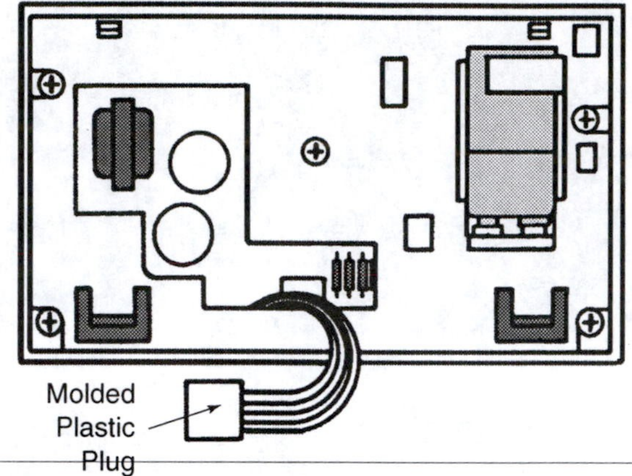

FIGURE 11-1 Molded plastic plug connecting the thermostat to the subbase

Digital Thermostats (Ribbon-Wire or Molded Plastic Connection)

The discussion of digital thermostats makes the following assumptions about the electrical terminals on the thermostat and about the system itself:

- The system has two heating stages and a single cooling stage.
- The system fails in the heating mode.
- The thermostat terminals are labeled as follows:
  R: Power from the transformer
  G: Indoor fan circuit
  W1: First-stage heating/first-stage cooling/defrost (compressor)
  Y1: First-stage cooling
  C: Power returning to transformer
  W2: Second-stage heating
  E: Emergency heat

This type of thermostat is the easiest to evaluate since the low-voltage wiring and the associated electrical terminals can be easily accessed without having to disturb or disconnect the actual wiring connections.

	Contacts					
Mode	**R & G**	**R & W1**	**R & W2**	**R & E**	**R & C**	**R & Y1**
Cooling	0 volts	0 volts	24 volts	24 volts	24 volts	0 volts
First-Stage Heating	0 volts	0 volts	24 volts	24 volts	24 volts	24 volts
Second-Stage Heating	0 volts	0 volts	0 volts	24 volts	24 volts	24 volts
Defrost	0 volts	0 volts	24 volts	24 volts	24 volts	0 volts
Emergency Heat	0 volts	24 volts	0 volts	0 volts	24 volts	24 volts

FIGURE 11-2 Voltage readings through a properly functioning heat-pump thermostat

Because physical contact occurs between the thermostat and the subbase, the subbase terminals can be accessed while leaving the thermostat in operation. With the securing screws on the thermostat loosened, the technician can relatively easily take the proper test readings between the appropriate terminals on the subbase to determine if the position of the thermostat's internal switches corresponds to the desired mode of operation. A summary of the voltages that should be obtained from reading between the given terminals is given in Figure 11-2. Note that since the temperature sensing on digital thermostats is performed by system logic circuits, as shown in Figure 10-12, the removal of the thermostat from the subbase will not affect the accuracy of the temperature being sensed by the device.

The voltages outlined in Figure 11-2 assume that the thermostat is actually functioning in the mode to which it is set. For example, when examining the cooling-mode portion of the chart, the technician can assume that the indoor fan is operating and the thermostat is set to a temperature that is well below the actual space temperature. Similarly, for heating, the technician can assume that the indoor fan is operating and the thermostat is set to a temperature that is above the temperature of the occupied space.

The information presented in Figure 11-2 can be extremely useful to the troubleshooting technician. For example, if the settings on the thermostat indicate that the fan should be operating and a voltage reading exists between Terminal G and Terminal R on the subbase, the internal switch or switches in the thermostat are not functioning properly and the device should be replaced. The same rules apply to other modes of operation that are ultimately controlled by the thermostat.

SERVICE NOTE: *Many digital/programmable thermostats are equipped with built-in time delays that can be as long as 5 minutes. The time delay that exists between the time that a mode of operation is selected and the time that operation is initiated varies from manufacturer to manufacturer. Be sure to check the manufacturer's specifications with regard to the specific thermostat being serviced. Always make certain that the time-delay period has been allowed to elapse before concluding that the thermostat is defective.*

If, while checking the thermostat, the technician finds a 24-volt reading between Terminal R and Terminal W1 when the thermostat is set to operate in the first-stage heating mode, the conclusion may be that the thermostat is not operating correctly. After allowing the time-delay period to elapse, the technician can safely conclude that the device is not functioning properly. To be absolutely sure, the technician can place a temporary jumper wire between Terminal R and Terminal W1 to verify that the compressor will operate. Always be sure to de-energize the circuits before making any changes to them.

Another method that can be used effectively to troubleshoot a thermostat is to take resistance or continuity readings through the device itself. To do this, however, the low-voltage wiring connections on the thermostat's subbase must first be disconnected. This will ensure that any readings taken will be those taken through the thermostat and not through any feedback loop that may result from the external wiring connections in the system.

SERVICE NOTE: *To prevent damage to the system components, good field practice requires disconnecting the power to the system before performing any sort of work on the electrical circuitry. This includes low-voltage circuits as well as line-voltage circuits. Short circuits in a low-voltage circuit may result in damage to the system transformer.*

When taking resistance or continuity readings through the thermostat, or across any other device, note that all power sources should be disconnected to:

- Prevent system component damage
- Prevent damage to the test instruments being used

Safety Tips:

Taking continuity readings on live circuits can cause personal injury and/or damage to the test instrument.

When taking continuity readings in a circuit, recall that the reading:

- Will be low when an electrical path exists between the points being evaluated
- Will be high when no path exists for the electrical current to take

If the technician is using an analog test instrument to take continuity readings, the meter's needle will not move if no continuity exists between the tested points. If the needle does move, a path is available for electrical current to take if it is applied to the circuit.

Resistance readings are not as cut-and-dry as the readings obtained by simple continuity testing. Resistance readings are designed not only to tell whether or not a path is available for the current to take but also to give the actual value of the resistance between the points being tested. Because what the actual resistance of a circuit should be is not often known, the technician can simply use continuity as a means of determining whether or not a particular set of contacts is in the open or closed position. Figure 11-3 presents a summary of the readings that should be expected between the various terminals on the heat-pump thermostat during different modes of operation. In the diagram, low resistance implies that continuity exists between the tested points, while high resistance implies that no continuity exists between the tested points.

DIGITAL AND ANALOG THERMOSTATS (WITHOUT RIBBON-WIRE OR MOLDED PLASTIC CONNECTIONS)

As already noted, thermostats that are equipped with some physical, electrical connection to the subbase when removed from the base are relatively easy to evaluate. However, when no physical contact exists between the thermostat and subbase, the evaluation process should ideally be performed at the terminal board located in the indoor unit. Here is where the wires from the thermostat and the wires from the outdoor unit come together. Troubleshooting a thermostat from the indoor unit location has advantages as well as disadvantages.

Disadvantages:

- Possible access problems to the equipment location
- Possibility of faulty wiring connecting the thermostat to the indoor unit
- Constantly having to move from the thermostat to the indoor unit to change thermostat settings

Advantages:

- Being able to isolate the electrical circuits for the indoor unit, the outdoor unit, and the thermostat
- Being able to evaluate the indoor unit and the outdoor unit from the same location
- Often having the system transformer located in the indoor unit as well as any solid state circuit boards that control system operation in conjunction with the thermostat

Mode	Contacts					
	R & G	R & W1	R & W2	R & E	R & C	R & Y1
Cooling	Low resistance	Low resistance	High resistance	High resistance	High resistance	Low resistance
First-Stage Heating	Low resistance	Low resistance	High resistance	High resistance	High resistance	High resistance
Second-Stage Heating	Low resistance	Low resistance	Low resistance	High resistance	High resistance	High resistance
Defrost	Low resistance	Low resistance	High resistance	High resistance	High resistance	High resistance
Emergency Heat	Low resistance	High resistance	Low resistance	Low resistance	High resistance	High resistance

FIGURE 11-3 Resistance readings through a properly functioning heat-pump thermostat

The terminal board in the indoor section of the heat-pump system is labeled with many of the same terminals as the thermostat subbase. Note, however, that since the same company often does not manufacture both the thermostat and the actual equipment, the method of terminal identification may be different. One saving grace in this situation is that a low-voltage wiring diagram is usually on one of the service panels on the indoor unit. The legend on the schematic will identify each terminal on the board. The summary charts presented in Figure 11-2 and Figure 11-3 can be used to troubleshoot the thermostat in this situation as well, but the terminal labels may be somewhat different.

DEFECTIVE WIRING BETWEEN THE INDOOR UNIT, THE OUTDOOR UNIT, AND THE THERMOSTAT

Now consider the situation in which the wiring between the thermostat and the indoor unit is defective. Having set the thermostat to the cooling mode of operation and having made certain that the thermostat was set to a temperature well below the occupied space temperature, the technician found that the compressor failed to operate. The technician then placed a temporary jumper between Terminal R and Terminal W1 on the indoor unit's terminal block, and the compressor began to function. To make certain that the thermostat was the cause of the problem, the technician removed the jumper wire from the terminal block and then placed the jumper across the same terminals on the thermostat subbase. The compressor failed to operate (Figure 11-4). The technician was then able to conclude that there was a problem with the wiring that connected the thermostat to the indoor unit.

To establish that a break actually exists in the low-voltage wiring, simply taking a continuity reading from one end of the wire to the other would be ideal. Unfortunately, the ends of the wire can be many feet from each other or even on different floors of the structure. In the example just presented, determining whether or not the wire has a break in it can be achieved using the following method:

- Disconnect the power to the system.
- Disconnect the wires from the R and W1 terminals on the thermostat subbase.
- Twist these two wires together and tighten with a wire nut.
- Disconnect the wires from the R and W1 terminals on the indoor unit's terminal board.
- Using a continuity tester, check for continuity between the two wires that were removed from the terminal board.

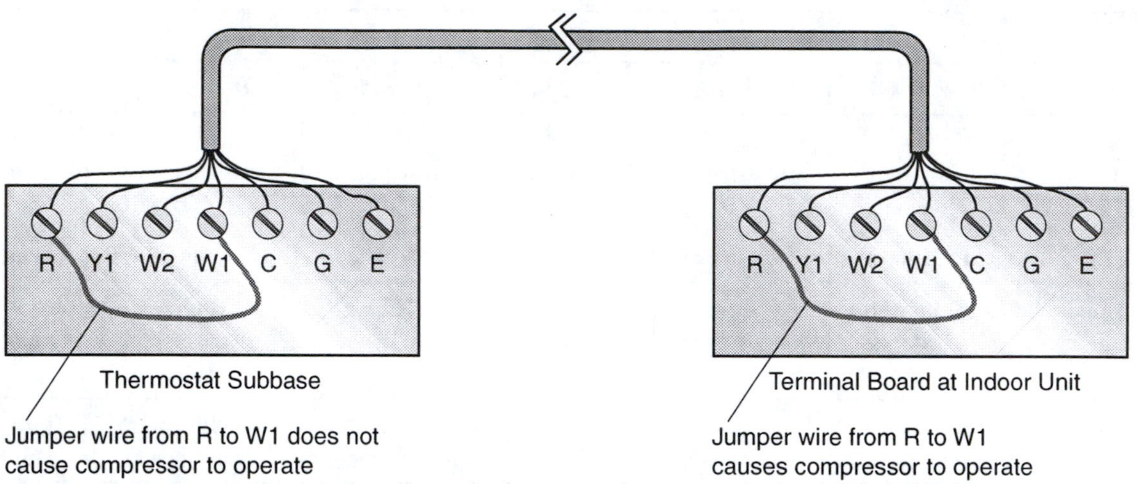

FIGURE 11-4 The jumper wire at the indoor unit's terminal board causes the compressor to operate. The jumper at the thermostat does not.

- If no continuity exists between the two wires, the wire has a break in it.
- If continuity exists between the two wires, the wires are intact.

This method is shown in Figure 11-5. Multiple wires can be check with this method, but they must be checked one set at a time. The same method also can be used to check for possible breaks in wiring that runs between the indoor unit and the outdoor unit.

MECHANICAL DEFROST TIMERS

The defrost timer, just as any other electrical component, can be effectively evaluated with the help of a good-quality multimeter. A multimeter is a piece of troubleshooting equipment used by field technicians to obtain voltage, current and resistance readings of electric circuits. The defrost timer is made up of several different internal components including the motor itself as well as the normally open and normally closed contacts contained within the device. Evaluating the motor usually can be done visually. Most defrost timers are equipped with plastic-enclosed motors (Figure 11-6), so that the operation of the motor can be visually determined. Other defrost timers are equipped with small windows in the device that enable the technician to inspect the motor. These windows are usually covered with clear plastic film to prevent particulate matter from entering the motor compartment of the timer. Having the opportunity to inspect the motor gives the technician confirmation that the motor is indeed turning.

Unfortunately, knowing that the motor is turning is not enough to determine that the defrost timer is operating properly. The technician must also establish that the contacts in the device are opening and closing as desired. For this reason, most defrost timers are equipped with knobs that allow the servicing technician to manually advance the timer either into or out of the defrost mode. These knobs will turn the cam within the timer (Figure 11-7), causing the timer to advance. The advancing of the timer saves the technician valuable time in determining if the contacts are opening and closing as desired.

Defective Timer Motor

Visually inspecting the timer motor in order to determine if it is turning or not is an easy task, as already established. If the motor is not operating, a few questions must be answered before jumping to any conclusions:

- Is voltage being supplied to the motor?
- Is the timer designed for continuous operation?
- Does the timer only operate when the compressor operates?

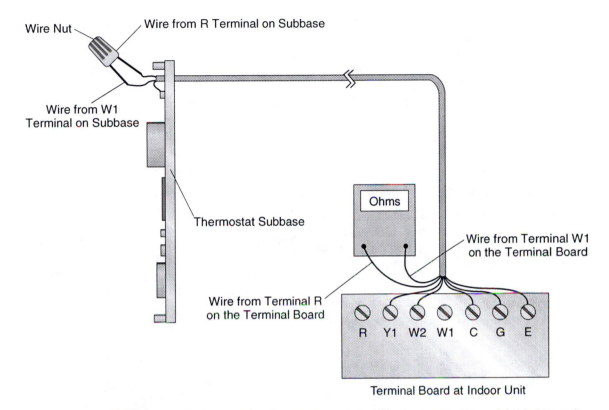

FIGURE 11-5 Taking a continuity reading between wires connecting the thermostat and the indoor unit

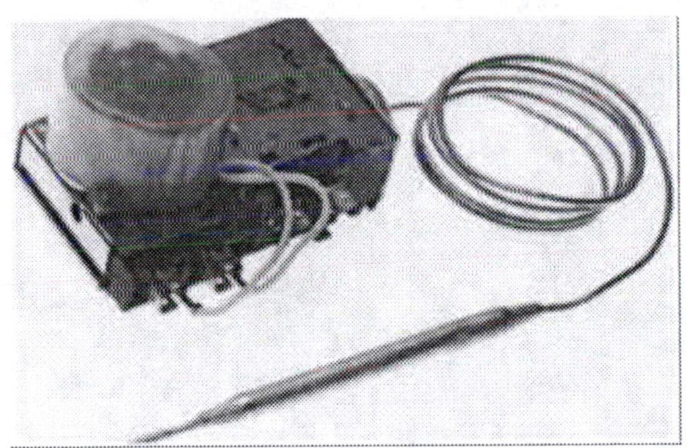

FIGURE 11-6 The motor can be visually inspected in this E15 time/temperature defrost control. *(Courtesy Ranco Inc.)*

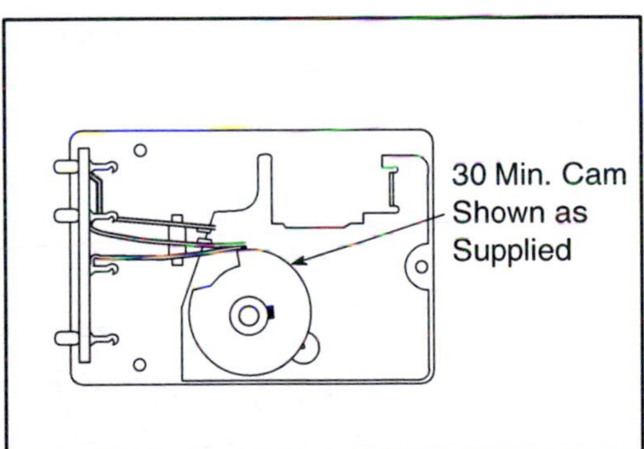

FIGURE 11-7 The cam in the defrost timer can be manually advanced. *(Courtesy Addison Products Co.)*

After answering these questions, the technician will be able to determine if the defrost timer motor is defective or not. The first question is relatively easy to answer. By taking a voltage reading at the timer motor terminals, the technician can determine if voltage is being supplied to the device. If the proper voltage is present and the motor does not function, it is defective and must be replaced. If no voltage is being supplied to the motor, the technician must determine exactly how the timer motor is wired into the circuit. Some heat-pump systems are designed to have the timer motor energized continuously, while others are designed to run only when the compressor is operating. In either case, if the compressor is operating, the timer motor should be operating as well. If no voltage is being supplied to the timer motor when the compressor is operating, the interconnecting wiring should be checked. The symptoms of a defective defrost timer include excessive ice buildup on the outdoor coil and reduced heating efficiency.

Mechanical Failure of the Timer Contacts

Once the timer motor is determined to be functioning correctly, the timer contacts must be evaluated. Before the contacts can be evaluated, their proper position must be determined. In most cases:

- Normally open contacts are open when the system is not in defrost mode.
- Normally closed contacts are closed when the system is not in defrost mode.

The easiest way to check the contacts on the defrost timer is with a voltmeter. If the system is operating in the heating mode, a reading of 0 volts should be obtained across the normally closed contacts, while a substantial voltage reading should be obtained across the normally open contacts. Figure 11-8 shows a cutaway view of a defrost timer in which both sets of contacts are closed and the voltage reading from the common terminal to both Terminal 1 and Terminal 2 is 0 volts, indicating that both sets of contacts are closed. In Figure 11-9, the cam has moved, causing both sets of contacts to open. The voltage readings obtained both indicate that the internal contacts are now in the open position. The wiring schematic for the specific timer that is being worked on should be referred to since each manufacturer's devices are configured differently.

Specific Defrost Example

In the case of one manufacturer's control circuit shown in Figure 11-10, the defrost timer has a normally closed set of contacts between Terminal 2 and Terminal 3 and a normally open set of contacts between Terminal 1 and Terminal 2. This system uses a combination of time, temperature, and pressure to achieve system defrost. It is also designed so that the entire control circuit is energized only when the compressor is operating. In this specific example:

- A reading of 0 volts should be obtained between Terminal 2 and Terminal 3.
- A line-voltage reading should be obtained between Terminal 1 and Terminal 2.
- A line-voltage reading should be obtained across the D20 pressure control.
- The outdoor fan should be operating.
- No voltage is being supplied to the timer motor.
- No voltage is being supplied to the defrost relay coil.

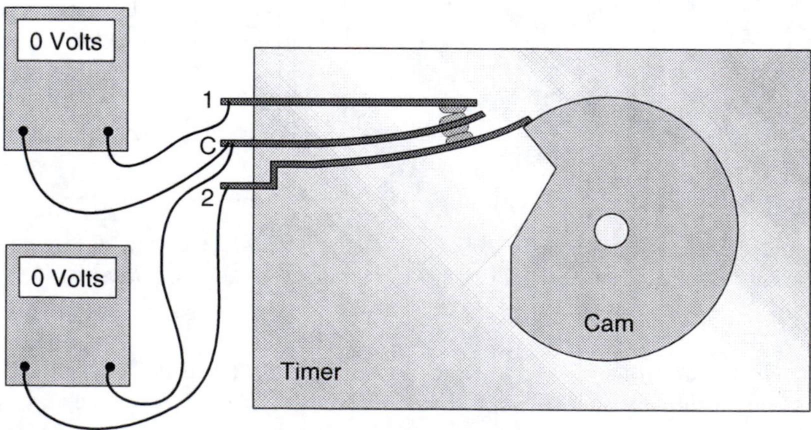

FIGURE 11-8 The position of the cam determines the position of the internal contacts. These contacts are in the closed position.

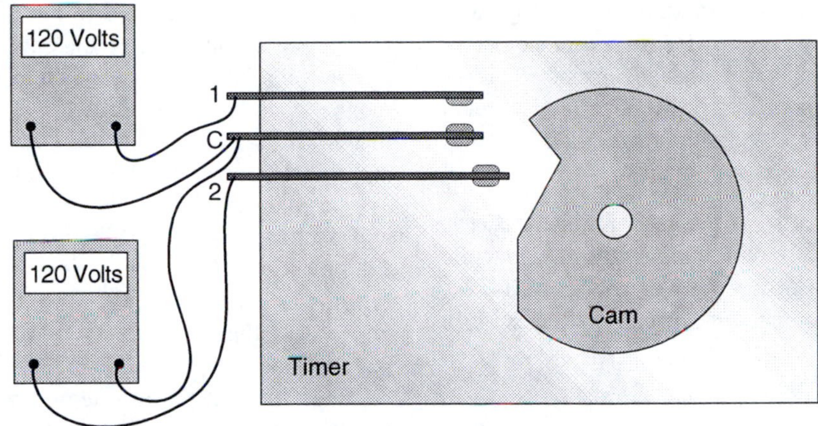

FIGURE 11-9 After turning slightly, the contacts have now opened.

L1
L2
Defrost Relay
E15
Supplemental Heat
Reversing Valve Solenoid
1
2
3
Timer Motor
D20
Outdoor Fan
Outdoor Coil Temperature-Sensing Bulb

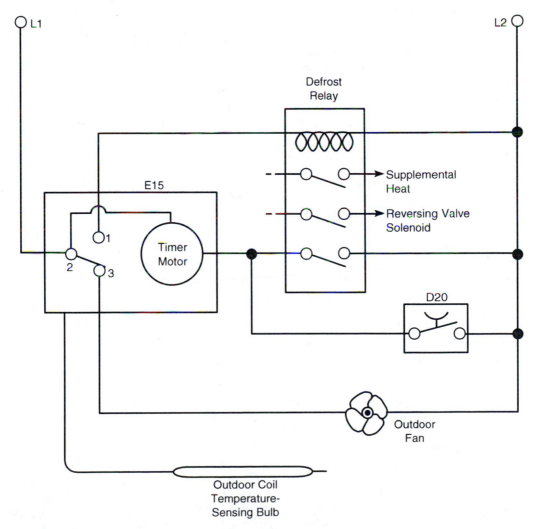

FIGURE 11-10 E15/D20 defrost-control wiring circuit. *(Courtesy Addison Products Co.)*

As mentioned earlier in the text, this type of defrost system only permits defrost to initiate once a pressure differential is established across the outdoor coil as well an appropriate temperature at the outdoor coil itself. Therefore, the following conditions must be met in order for defrost to begin:

- The D20 pressure switch must sense the predetermined pressure differential across the outdoor coil.
- The temperature of the outdoor coil must fall to a predetermined temperature, usually in the range of 25 to 27°F.

In this example, if the timer motor is not energized, there are two possible reasons:

- Compressor operation is not being called for.
- The D20 pressure control is in the open position.

If the pressure switch is in the closed position and the compressor is operating and the timer motor is still not operating, it is defective and must be replaced. If the timer motor is operating and the system will not go into defrost, there are two possible reasons:

- The temperature of the outdoor coil is not low enough to initiate defrost.
- The internal contacts of the timer are not functioning.

Remember that both conditions, pressure and temperature, must be met in order for this system to go into defrost. The location of the outdoor coil sensing bulb should be checked to ensure that sufficient frost is present to actually initiate a defrost cycle. If not, the system will continue to operate in the heating mode without initiating a defrost. Chances are, though, that if no excessive frost buildup appears on the outdoor coil, the system is defrosting properly.

Once defrost is initiated, the contacts between Terminal 2 and Terminal 3 will open and the contacts between Terminal 1 and Terminal 2 will close. The following conditions should hold:

- A line-voltage reading should be obtained between Terminal 2 and Terminal 3.
- A reading of 0 volts should be obtained between Terminal 1 and Terminal 2.
- A line-voltage reading should be obtained across the defrost relay coil.
- A reading of 0 volts should be obtained across each set of normally open contacts on the defrost relay.

As the frost is removed from the outdoor coil, the pressure differential across the outdoor coil will drop, causing the D20 pressure switch to open its contacts. However, a closed set of contacts is connected in parallel with the pressure switch. This set of contacts is controlled by the defrost relay coil and will keep the system in defrost until the temperature of the outdoor coil rises to the predetermined set point. Even though the pressure switch opens its contacts, the voltage reading across the device should still be 0 volts because of the closed contacts in parallel with the pressure control.

Since many defrost systems operate on concepts similar to the one just discussed, an understanding of this scenario will help make the troubleshooting of this and other types of defrost systems much easier.

SOLID-STATE DEFROST CIRCUIT BOARDS

Later model heat-pump systems are often manufactured with solid-state defrost control boards, as opposed to the mechanical timers discussed in the previous section. The main differences between mechanically and electronically controlled defrost methods are:

- No physical timer motor is on the solid-state boards.
- No means exist by which the timer can physically be advanced into defrost.
- No physical means exist by which the technician can check the temperature sensor on the timer.

These items may seem to imply that the solid-state defrost control board cannot be evaluated in the field by a technician, but such is not the case. Manufacturers have provided field technicians with methods by which these boards can be effectively field-tested. In the earlier days of solid-state defrost controls, field technicians would quite commonly—and improperly—condemn the board simply because the system was not operating properly. This resulted in lost time and profit for the service company as well as increased system downtime, which often led to an unhappy customer. The following sections describe methods that can be used to determine if the solid-state control board is functioning properly.

Since each equipment manufacturer utilizes different means by which these boards are checked, the technician should refer to the manufacturer's guidelines for specific troubleshooting procedures.

Checking Normal System Operation

The first step in determining whether or not the solid-state defrost control is operating is to determine whether or not the system operates properly in both the heating and cooling modes. If the system works properly in the heating mode but provides heat when set to operate in the cooling mode, the technician should suspect that a problem exists with the reversing valve or its associated wiring. The problem most likely does not lie within the defrost board. The technician must evaluate the system in this manner to avoid making unwarranted assumptions that would lead to an incorrect diagnosis of the system. Stating that the defrost board must be defective because the system does not defrost is not a valid conclusion. That would be just like stating that the compressor must be defective because it is not running. Once the technician has established that the system operates correctly in both the heating and cooling modes, the inspection of the board should begin.

Troubleshooting the Solid-State Defrost Board: Frost Buildup on the Outdoor Coil

If substantial frost buildup appears on the outdoor coil, the system should ideally go into the defrost mode to melt the ice from the coil's surface. If the system does not defrost, the first order of business is to check the outdoor coil thermostat or thermistor. The following steps will determine if the outdoor coil temperature sensor is operating:

1. Turn the space thermostat to the OFF position.
2. Disconnect power to the outdoor unit.
3. Remove the wires for the outdoor thermostat from the circuit board.
4. Check the thermostat with an ohmmeter.
5. A reading of infinite resistance is an indication that the thermostat's contacts are open and not calling for defrost. If sufficient frost appears on the coil and the sensor is mounted properly, the sensor should be replaced. To confirm that the sensor is defective, place a jumper wire across the terminals on the board that the sensor was connected to and restart the system. If the system goes into defrost, the sensor is indeed in need of replacement. Once the new sensor is installed, the system should defrost properly. If not, proceed to Step 7.
6. A reading of 0 ohms indicates that the temperature sensor is in the closed position and calling for defrost.
7. Remove the jumper wire across the sensor terminals and reconnect the sensor to the board.
8. The board should now be advanced into the defrost mode.
9. Most solid-state boards are equipped with speed-up terminals that will cause the solid-state circuitry to reduce the time interval between defrost attempts. When sped up, the time interval is often reduced to 1/60 of the original interval. A 60-minute interval will therefore be reduced to 60 seconds.
10. Place a jumper wire across the speed-up thermals and wait for the system to switch over into defrost. If the system does not switch over to defrost after 60 seconds, assuming the original time interval was 60 minutes, the circuit board is defective.
11. If the board switches the system into defrost with the outdoor temperature sensor in place, the circuit board is operating properly.

SERVICE NOTE: *Be sure to disconnect all power sources before making any changes to system wiring. Also be sure to check the manufacturer's specific procedures for checking and evaluating a solid-state defrost board so as to prevent damage to the component.*

ELECTRIC-STRIP HEATER CIRCUITS

The electric-strip heaters are vital system components that provide heat in the defrost, emergency-heat, and second-stage-heating modes. These heaters, though, can become fire hazards if they are permitted to operate when insufficient air is moving through the associated ductwork. For this reason, the strip heater circuits are equipped with a number of safety switches that prevent them from operating if the system conditions are not perfect. To briefly recap, the heater circuits are controlled by a number of switches and safeties, including:

- Time-delay relay contacts
- High-limit switches
- Fusible links
- Main circuit fuses

The time-delay relay contacts are intended to prevent all of the strip heaters from energizing at the same time. These are essentially control switches and are not intended to act as safety devices. The high-limit switches are usually automatically resetting devices and sense excessive temperatures in the ductwork or heater cabinet. The fusible links are one-time devices that will melt if the temperature in the heater cabinet reaches unsafe levels. The temperature at which the fusible link will melt is higher than the temperature at which the high-limit switch will open. The fusible link is therefore acting as a backup for the high-limit switch. Once the fusible link has melted, the device must be replaced. Of course, this will take place after the reason for the overheating condition has been determined and rectified. The main fuses are intended to protect the whole circuit but do not protect the individual heater circuits.

In addition to the items just mentioned, heater circuits are sometimes equipped with **sail switches** (Figure 11-11), the purpose of which is to prevent the heaters from energizing if there is no air movement in the air duct. The contacts on the sail switch are normally open. These contacts will close when the air passing through the duct pushes the sail, causing the internal contacts to close. If the airflow is stopped, the contacts open again.

Just as with any other electrical circuit, the electric-strip heater circuits are made up of switches and loads that can be evaluated with the aid of a good-quality multimeter. The ladder diagram in Figure 11-12 shows a typical electric-strip heater circuit. This diagram has key troubleshooting points labeled A through K in order to facilitate discussion of this type of circuit.

When the electric-strip heaters are energized and operating properly, the voltage reading taken at the various points in the circuit should be as follows:

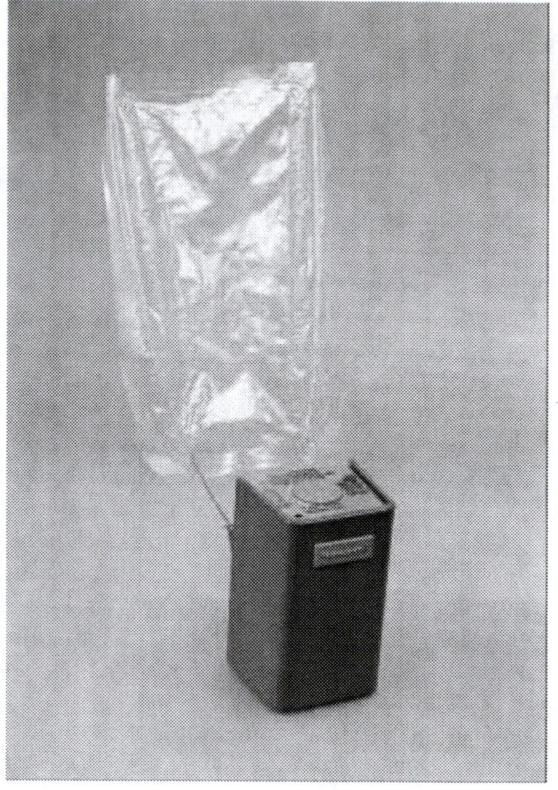

FIGURE 11-11 Sail switches monitor air movement in a duct. *(Photo by Bill Johnson)*

- Line voltage between point A and point E
- Line voltage between point K and point F
- Line voltage between point B and point C
- Line voltage between point J and point H

These conditions indicate that line voltage is being supplied to the heaters and that all of the safety and control switches are in the closed position. In the event that the electric-strip heaters are not operating, voltage readings taken at strategic points in the circuit will lead the technician to the problem. Now examine the top rung in the ladder diagram in Figure 11-12.

Consider the situation in which the electric heater is not operating. The technician takes voltage readings between the indicated points (Figure 11-13):

- Voltage between point A and point E is 220 volts.
- Voltage between point A and point D is 220 volts.
- Voltage between point A and point C is 0 volts.

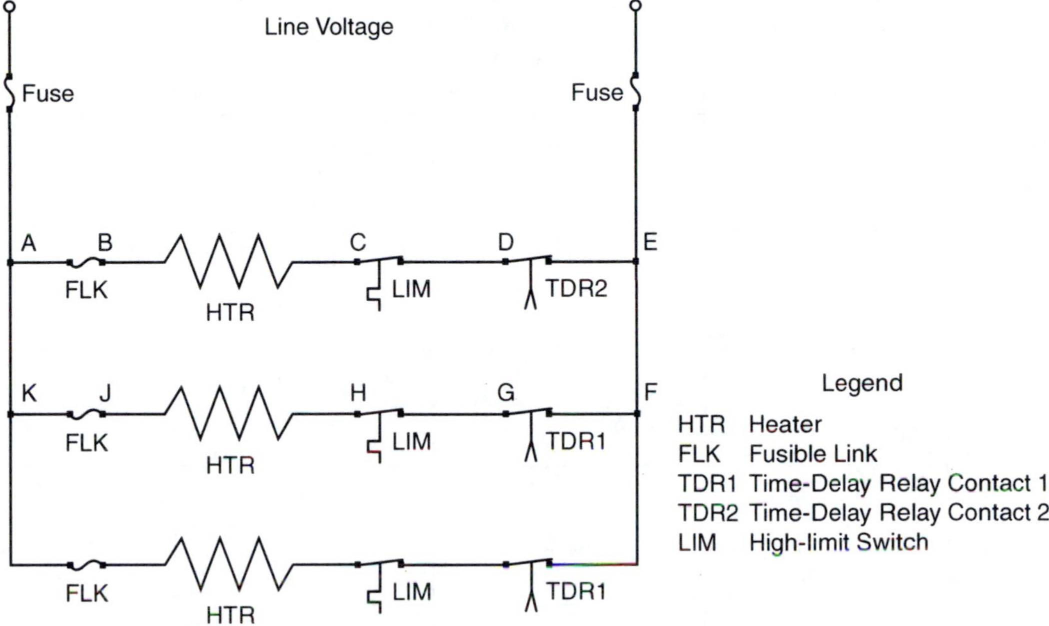

FIGURE 11-12 Typical electric-strip heater circuit

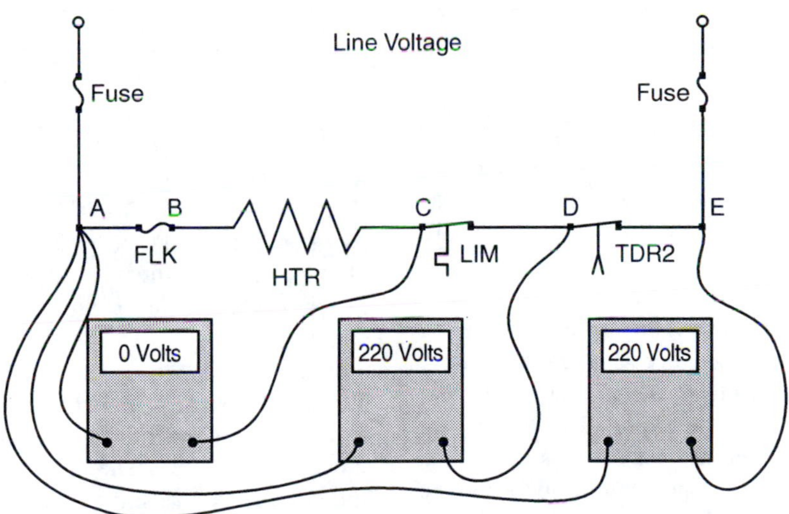

FIGURE 11-13 The voltage readings indicate that the high-limit switch is in the open position.

From these measurements, the technician can establish that both line-voltage fuses are good since line voltage occurs at point A and point E. The technician can also conclude that the TDR2 contacts are closed since line voltage occurs between point A and point D. Since a reading of 0 volts is obtained between point A and point C, the technician can conclude that the high-limit switch between point C and point D is in the open position. A voltage reading of 220 volts between these points will confirm that fact.

	Voltage between Points A and B	Voltage between Points B and C	Voltage between Points C and D	Voltage between Points D and E
Properly Operating Heater	0 volts	220 volts	0 volts	0 volts
Defective Fusible Link	220 volts	0 volts	0 volts	0 volts
Defective High-limit Switch	0 volts	0 volts	220 volts	0 volts
Defective Heater	0 volts	220 volts	0 volts	0 volts
Open TDR2 Contacts	0 volts	0 volts	0 volts	220 volts
Defective Line Fuse	0 volts	0 volts	0 volts	0 volts

FIGURE 11-14 Voltages across components of the electric-strip heater circuit

Still referring to the circuits in Figure 11-12 and Figure 11-13, a similar method can be presented for evaluating the fusible links and the heater itself. The chart in Figure 11-14 provides a summary of the voltage readings that will be obtained under a variety of circumstances.

Note that the voltage reading across the heater when it is operating is the same as when the electric heater is defective. Just as in the case of a defective lightbulb, voltage is being supplied to the device but the device is not operational. The easiest way to determine if the heater is operating or not is to take an amperage reading of the circuit with a clamp-on ammeter. If the circuit is not drawing any current and line voltage is being supplied to the heater itself, the heater is defective and needs to be replaced. On the other hand, if line voltage is supplied to the heater and the circuit has current flowing through it, the heater is energized and operating.

DEFECTIVE HOLDBACK THERMOSTAT

As mentioned in the preceding chapter, the purpose of the holdback thermostat is to ensure that the defrost cycle is able to run its course without being prematurely terminated. This could occur if too much heat is introduced to the occupied space while the air is being tempered. The holdback thermostat, or **outdoor ambient thermostat,** is a simple, single-pole, single-throw switch. The switch is designed to open and close its contacts at a predetermined temperature as set on the device itself (Figure 11-15). The device is relatively easy to troubleshoot since the temperature at which the contacts open and close can be adjusted to facilitate the evaluating of the control. Recall that this device will close its contacts on a drop in temperature, meaning that the contacts are in

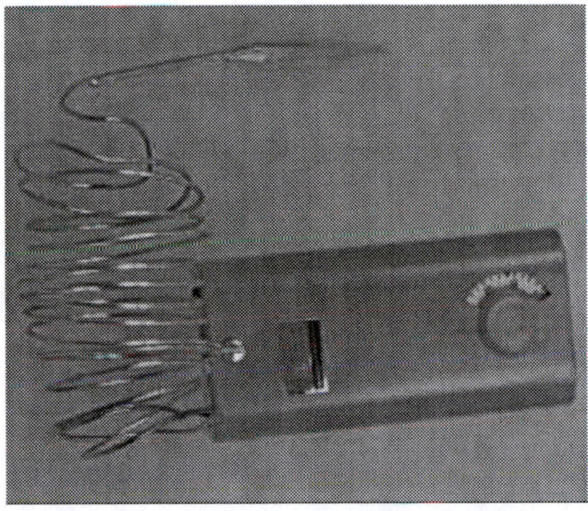

FIGURE 11-15 Line-voltage temperature control with adjustable set point. *(Photo by Bill Johnson)*

the open position when the ambient temperature is mild. To check the device using a voltmeter:

- Disconnect power to the system.
- Remove the cover from the device.
- Place one test lead of the voltmeter on each of the two electrical contacts on the thermostat.
- Restore power to the system.
- Advance the system to the defrost mode of operation.
- If 0 volts is read between the contacts, the thermostat is in the closed position. Turn the knob on the thermostat to a lower temperature setting until the device produces an audible click. The reading across the contacts should now be equal to the voltage supplied to the circuit. If the reading is still 0 volts, the thermostat is defective and needs to be replaced.
- If a line-voltage reading is obtained between the contacts, the thermostat is in the open position. Turn the knob on the thermostat to a higher temperature setting until the device produces an audible click. The reading between the contacts should now be 0 volts. If the voltage reading remains high, the contacts are not closing and the device should be replaced.

The holdback thermostat can also be checked with an ohmmeter, as follows:

- Disconnect power to the system.
- Remove the cover from the thermostat.
- Disconnect the wires from the thermostat.
- Place one test lead from the ohmmeter on each of the contacts on the thermostat.
- If the meter registers continuity, the thermostat contacts are in the closed position. Turn the knob on the thermostat to a lower temperature setting until the device produces an audible click. The reading on the meter should now indicate that no continuity exists between the contacts. If the meter still indicates continuity, the thermostat is defective and should be replaced.
- If the meter registers no continuity between the contacts, the thermostat is in the open position. Turn the knob on the thermostat to a higher temperature setting until an audible click is heard. The meter should now register continuity between the contacts. If the meter still registers no continuity between the contacts, the device is defective and should be replaced.

SUMMARY

- Thermostats with ribbon-wire or molded plastic connectors can be easily evaluated at the thermostat location.
- Thermostats that have no physical connection to the subbase should be evaluated from the indoor unit location.
- Jumper wires can be used to confirm that a thermostat is not operating as desired.
- Faulty wiring between pieces of equipment can be found by taking continuity readings through individual pairs of wires.
- Resistance readings must be taken on circuits that are de-energized.
- Voltage readings must be taken when the circuit is energized.
- Defrost timers can be manually advanced in order to make evaluation easier.

- The defrost timer has two major components, the motor and the contacts, which must each be evaluated to determine if the device is operating properly.
- Safety devices in the electric-strip heater circuit include sail switches, fusible links, and high-limit switches.
- Electric-strip heater circuits are best evaluated with a voltmeter.
- Holdback thermostats are simple, single-pole, single-throw switches that can be easily evaluated with either a voltmeter or an ohmmeter.

KEY TERMS

Multimeter
Outdoor ambient thermostat
Sail switch

FOR DISCUSSION

1. How can a field technician confirm that a break has occurred in a low-voltage wire running between the thermostat and the indoor unit? If it is established that a break has occurred in the wire, what can a technician do to rectify the situation?
2. The fusible link in an electric-strip heater circuit melts on a regular basis. Discuss the safety issues that should be addressed and make a plan as to how the system should be checked to rectify the situation.

REVIEW QUESTIONS

1. A jumper is placed across the R and G terminals on the thermostat's subbase, but the indoor fan motor does not operate. Which of the following is a possible reason for the fan's failure to function?
   a. The transformer is defective.
   b. The thermostat is defective.
   c. A break has occurred in the wiring from the thermostat to the indoor fan section.
   d. Both *a* and *c* are correct.
2. Which of the following is correct with respect to readings taken across a closed set of contacts?
   a. A voltage reading of 0 volts should be obtained across the contacts.
   b. A low-resistance reading should be obtained across the contacts.
   c. Continuity should register between the contacts.
   d. All of the above are correct.
3. The best way for a technician to evaluate a control circuit is to
   a. Use a voltmeter with the circuit power off.
   b. Use an ammeter with the circuit power off.
   c. Use an ohmmeter with the circuit power off.
   d. Use a continuity tester with the circuit power on.
4. Digital thermostats are often equipped with __________ that protects the system from being cycled on and off rapidly by the system operator.

5. Determination that a defrost timer motor is operating can be accomplished by
   a. Determining that voltage is being supplied to the motor.
   b. Visually inspecting the motor.
   c. Establishing that continuity exists through the motor.
   d. Both *a* and *b* are correct.
6. One advantage of troubleshooting a control circuit from the terminal block at the indoor unit is:
   a. The ease and speed at which the system thermostat can be switched over to the various modes of operation.
   b. The ability to isolate the outdoor unit, the indoor unit, and the thermostat circuits.
   c. The ability to rule out faulty wiring between the thermostat and the indoor unit.
   d. All of the above are advantages.
7. What conclusion can be drawn if proper voltage is supplied to a defrost timer motor and the current reading of the timer motor circuit is 0 volts?
   a. The timer motor is defective and should be replaced.
   b. The timer motor is switching the system back into the heating mode.
   c. The timer motor is operating properly.
   d. The normally open contacts in the timer are in the closed position.
8. If the defrost timer motor is operating, which of the following is an incorrect assumption?
   a. Continuity exists through the motor.
   b. Voltage is being supplied to the motor.
   c. The defrost timer is operating correctly.
   d. All of the above are correct assumptions.
9. How can a service technician determine if the contacts in a defrost timer are opening and closing properly?
   a. The technician can check for continuity across the terminals while the circuit is energized.
   b. The technician can wait for the system to go into defrost on its own and then check for continuity through the contacts.
   c. The technician can check the contacts with a voltmeter, advance the timer manually, and then check the contacts again to see if they switched position.
   d. The technician can disconnect the wiring from the timer terminal and take voltage readings between the terminals.
10. Which of the following components does not protect the individual electric-strip heater circuits?
    a. The high-limit switch
    b. The fusible link
    c. The main circuit fuse
    d. None of the above
11. Which of the following is most likely if only two out of three electric-strip heaters are energized?
    a. The main circuit fuse is blown.
    b. The sail switch is in the open position.
    c. Time-delay relay contacts are open.
    d. The outdoor ambient temperature is extremely low.

12. Line voltage supplied to an electric-strip heater is an indication that:
    a. The heater is energized.
    b. The safety devices in the circuit are all in the closed position.
    c. The main circuit fuse has blown.
    d. The system is operating in the cooling mode.
13. What is the best way for a technician to determine if an electric-strip heater is energized?
    a. Check for amperage in the heater circuit.
    b. Check for voltage to the heater circuit.
    c. Touch the heater to determine if it is hot.
    d. All of the above are acceptable methods for checking an electric-strip heater.
14. A holdback thermostat's contacts are currently in the open position. To confirm that the contacts will close, the technician should:
    a. Turn the adjusting knob to a higher temperature setting.
    b. Turn the adjusting knob to a lower temperature setting.
    c. Never turn the adjusting knob on the thermostat.
    d. Heat up the remote bulb on the thermostat.
15. Before working on or making any changes to any electrical circuit, the technician should:
    a. Understand the circuits and their components.
    b. Disconnect power sources.
    c. Work carefully and systematically.
    d. All of the above are correct.

CHAPTER 12

Servicing Vapor-Compression Heat-Pump Controls and Control Circuits

OBJECTIVES After studying this chapter, the reader should be able to:

- Review the seven steps in performing a successful service call.
- Explain how a technician can determine if the system thermostat is defective.
- Explain how a technician can determine if the interconnecting wiring between the thermostat, indoor unit, and outdoor unit is faulty.
- Explain how a technician can establish if the holdback thermostat is faulty.
- Explain how to determine if the solenoid coil on a four-way reversing valve has an open winding.
- Explain how to determine if the solenoid coil on a four-way reversing valve is shorted.
- Explain how a technician can determine if the electric-strip heaters are defective.
- Explain how to determine if the electric heater control circuit components are defective.

INTRODUCTION

The various chapters on components, troubleshooting, and servicing in this text are filling the reader's mental toolbox with more and more information regarding the operation and troubleshooting of the heat-pump system. This chapter expands on the information presented in Chapter 11 in the form of sample service calls that technicians are likely to encounter in the field. Just as with earlier service calls in this text, each example provides the reader with the original customer complaint; the technician's evaluation of the system and its components; and, finally, a brief discussion of the service call and the methods used by the technician. The service calls that are presented involve the failure of the system components discussed in the previous chapter. This chapter, however, presents situations in which the technician may not have proceeded in the most efficient manner possible. With experience, service technicians can achieve maximum efficiency by taking control of the situation as opposed to letting the system take control of them.

OVERVIEW

Electrical troubleshooting of a heat-pump system requires the servicing technician to have a good understanding of the components that make up the system. This understanding includes knowledge of the proper sequence of operations of the system as well as the approximate voltage, current, and resistance readings that should be obtained across or through the various contacts, terminals, and components in the system. In addition to the basic system knowledge, the technician must also be able to work through a service call in a manner that will ultimately benefit the customer and the company as well. For example, a technician's spending 3 hours diagnosing a system with a defective thermostat would not be beneficial. The customer would most definitely be faced with a very large repair bill, and the service manager and the company owner would both be less than happy to hear that it took a technician 3 hours to perform a 1-hour task.

In addition to the technical knowledge that the technician must have in order to be successful in the field, a plan is also required that will enable the technician to complete each and every service call successfully. As outlined in Chapter 7, seven steps should be followed in order to complete a successful service call:

- Verify the complaint.
- Gather information.
- Perform a visual inspection.
- Isolate the problem.
- Correct the problem.
- Test the system operation.
- Complete the service call.

Each of these steps is explained in detail in Chapter 7; the reader should refer back to that section in the text to review this information. By proceeding in a logical manner, the chances of making potentially costly mistakes in the field are greatly reduced. The service technician should always ask himself the following question after diagnosing a system with a defective component:

Why did the component fail?

Being able to properly answer this question will help eliminate the possibility of damaging newly installed replacement parts.

SERVICE CALL 1: DEFECTIVE THERMOSTAT

Customer Complaint

John, a home owner, notices that his heat-pump system is not able to provide adequate heat to the occupied space. He checks the outdoor unit and notes that the outdoor fan is not operating. He switches the system over to the emergency-heat mode to provide temporary relief from the cold space temperature. He then places a call to his service company. Meanwhile, the system, operating in the emergency-heat mode, is able to properly heat the space. A short time later, the technician—Steve—arrives at the residence. John, while explaining the situation to Steve, also mentions that his living room and hallway were recently painted and that the painters removed the thermostat from the wall. He also states that, prior to the paint job, the system seemed to have been operating properly.

Service Technician Evaluation

After listening to the customer complaint, Steve goes over to the thermostat and confirms that it is, indeed, set to the emergency-heat mode and that warm air is being introduced to the occupied space. To check the first-stage heating mode, Steve switches the system to the normal heating mode and makes certain that the thermostat is set to a temperature higher than that of the occupied space. He then goes to the outdoor unit, removes the service panel, and checks to see if power is being supplied to the unit. Steve obtains the following readings, which are shown in Figure 12-1:

- 220 volts on the line side of the compressor contactor
- 0 volts across the holding coil on the four-way reversing valve
- 0 volts being supplied to the outdoor unit through the low-voltage wiring

The 220-volt reading on the line side of the compressor contactor indicates to Steve that the problem is most likely in the low-voltage circuit. After checking the voltage at the holding coil of the four-way reversing valve and inspecting the piping configuration as well as the valve itself, Steve determines that the system fails in the heating mode and that the reading of 0 volts across the solenoid coil is correct for heating operation. The reading of 0 volts between the low-voltage wires feeding the outdoor unit indicates that the problem does not exist within the outdoor unit. The results from Steve's information gathering thus far yield the following:

- Because the indoor fan is operating, the system transformer is good.
- The problem lies in one of the following areas:
  - The interconnecting wiring between the thermostat and the indoor unit
  - The interconnecting wiring between the indoor unit and the outdoor unit
  - The thermostat itself
  - The wiring in the indoor unit

To narrow down his search, Steve decides to check the indoor fan section next, using the terminal block within that section to check the system's low-voltage circuits. Locating the air handler in the

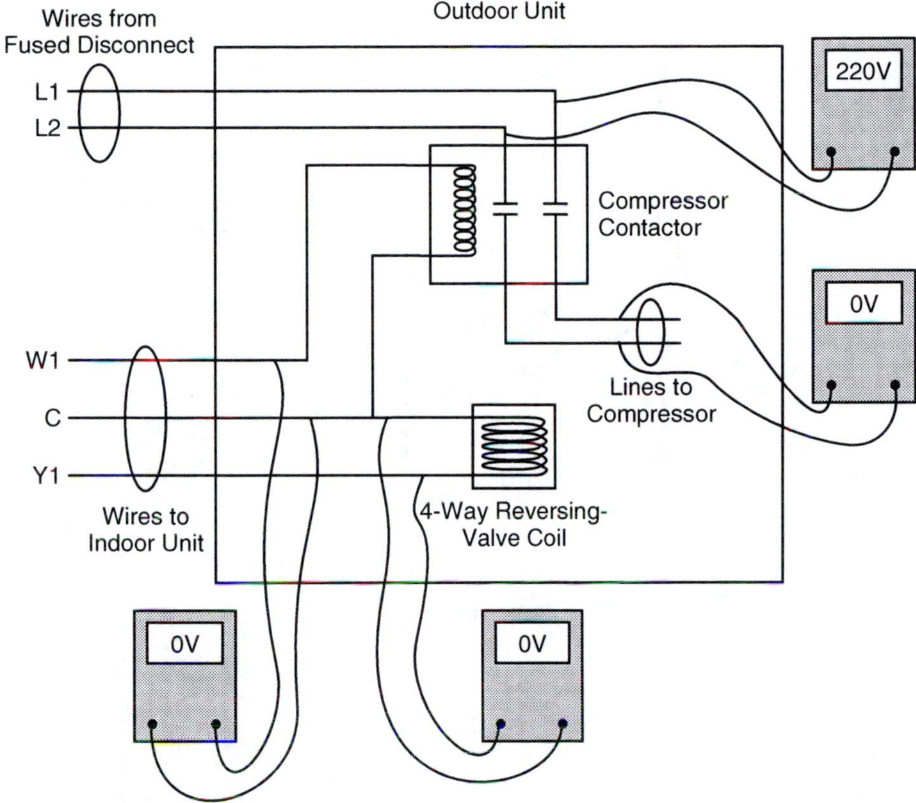

FIGURE 12-1 Voltage measurements across a compressor contactor with a de-energized holding coil

basement, he removes the front panel on the unit to gain access to the terminal board. Using his voltmeter, Steve obtains the following voltage readings (Figure 12-2) from the terminal block:

- 24 volts between Terminal R and Terminal C
- 24 volts between Terminal R and Terminal W1
- 0 volts between Terminal R and Terminal G

The 24-volt reading between Terminal R and Terminal C simply confirms that power is being supplied by the transformer. In actuality, Steve already knows this, since the indoor fan is operational. Similarly, the 0-volt reading between Terminal R and Terminal G tells him that the indoor fan circuit is closed. Again, Steve already knows this since the fan is operating. The 24-volt reading between Terminal R and Terminal W1, however, indicates that the circuit between those two terminals is in the open position when it should be closed.

To confirm that the outdoor unit is operational, Steve takes an insulated jumper wire and places it across Terminal R and Terminal W1 (Figure 12-3). The outdoor unit begins to operate. Steve is now able to determine that the problem is situated in one of the following two locations:

- The thermostat
- The interconnecting wiring between the thermostat and the indoor unit

Steve removes the jumper wire from the terminal board and then proceeds to the thermostat. Removing the thermostat from the subbase, he places the insulated jumper between Terminal R and Terminal W1 on the subbase and again checks the outdoor unit. Again, the unit begins to operate. Since this thermostat is the mechanical type, Steve does not need to concern himself with any internal time delays that are normally integrated in electronic-type thermostats.

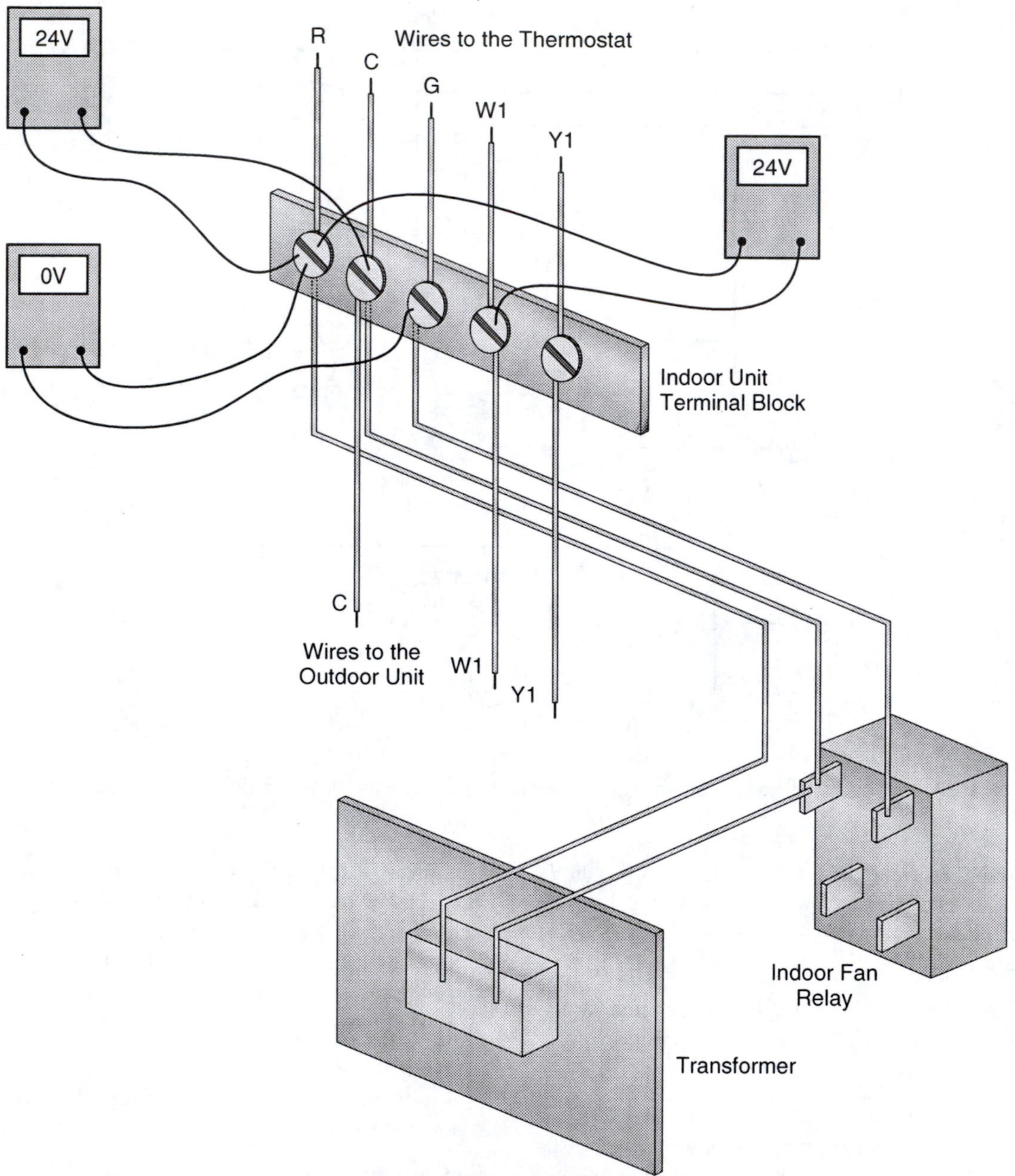

FIGURE 12-2 Voltage readings taken at the terminal block on an indoor unit

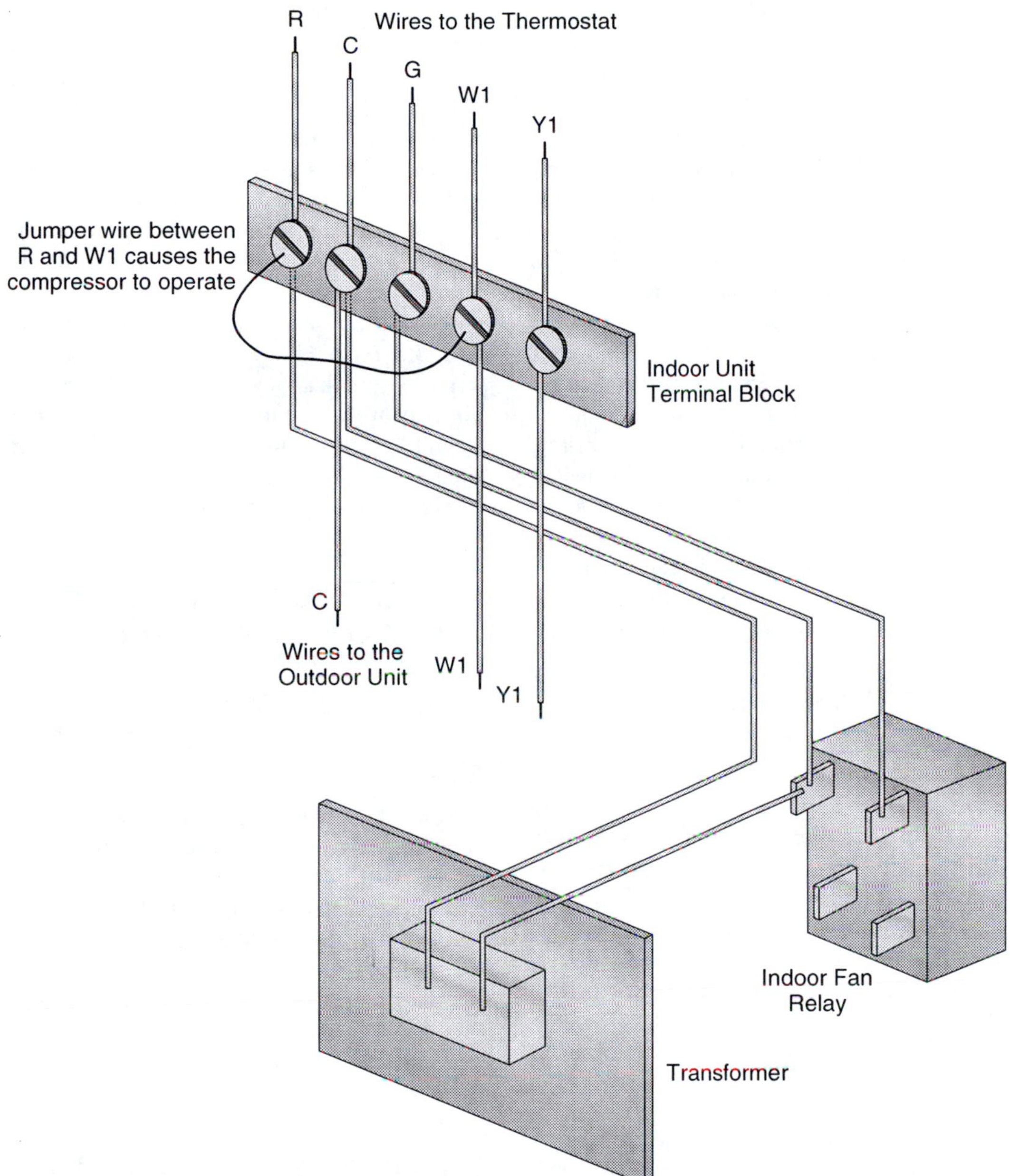

FIGURE 12-3 A jumper placed between Terminals R and Terminal W1 causes the system compressor to operate.

He then concludes that the thermostat is defective. Steve goes to his truck and retrieves a replacement thermostat, making certain that the new thermostat is configured exactly the same as the defective device to ensure that all control circuits operate properly. Before actually disconnecting the wiring from the existing device, he makes certain to disconnect the power to the system. After doing so, he removes the existing thermostat and replaces it with the new device. After securing the thermostat, Steve makes certain that it is mounted perfectly level to ensure that it operates properly.

Once the thermostat is replaced, Steve restores power to the system. He then starts up the unit and makes certain that the system operates. While writing up his work order, he asks the home owner if he has any questions regarding the new thermostat. After he completes writing up the work order, he gives John a quick lesson on how to operate and properly set his new thermostat and also makes certain that he leaves the information booklet with the home owner.

Service Call Discussion

In this service call, Steve utilized the seven-step procedure that was outlined at the beginning of this chapter and also earlier in the text. He was quickly able to verify the customer's complaint, and he then proceeded to gather information regarding the system. He proceeded in a logical manner to eventually determine the system problem and ultimately resolve it. However, what could Steve have done differently? Did he effectively use the information that he gathered?

When Steve arrived at the customer's house, he was told that the heating was not operating properly. He was also told the following:

- The customer had just had his living room and hallway painted.
- The painter had removed the thermostat from the wall.
- The system had worked fine prior to the paint job.

Although he eventually determined the cause for the system failure, he could have saved valuable time if he had made the initial assumption that the painters were in some way responsible for the problem. If he had done this, his initial inspection might have been at the thermostat location and not at the outdoor unit. If this were indeed the first place he checked, he would have established almost immediately that the thermostat was defective. After placing a jumper wire between Terminal R and Terminal W1 on the subbase and observing the outdoor unit come on, he would have also saved himself a trip to the basement to take voltage readings at the terminal board within the indoor fan section.

In addition, once the unit was put back into operation, Steve should have checked the system's air filters to ensure that they were clean. Although this was not part of the original service call or complaint, it should be part of all service calls. Checking the air filters only takes a short time, and the customer will be pleased to see that it is done.

SERVICE CALL 2: DEFECTIVE INTERCONNECTING WIRING BETWEEN THE INDOOR AND OUTDOOR UNITS

Customer Complaint

It is mid-July, and a residential customer phones in a complaint about her heat-pump system. She set the thermostat in the cooling mode, but the air coming from the registers is not cold. She informs the dispatcher that the air is not extremely hot but that the system is definitely not cooling. She also tells the dispatcher that she can normally see and hear the outdoor unit operating, but now it is not operating at all. She sets up an appointment for 8:00 A.M. the next day.

Service Technician Evaluation

At 7:55 A.M. the next day, the technician, Kevin, arrives at the customer's home. After introducing himself to the home owner, he asks her what kind of problem she is experiencing with the unit. She informs him that the outdoor unit is not operating, even after she has set the thermostat to operate in the cooling mode. Kevin sets the thermostat to cooling, and he can hear the indoor fan motor begin to operate.

Setting the thermostat to its coldest setting, he then goes to the outdoor unit located in the backyard. He disconnects the power, opens the access panel on the unit, and takes a low-voltage reading between the wires feeding control voltage to the unit. The reading is 0 volts. He also notes that the two wires that are supposed to be feeding 24 volts to the outdoor unit are color-coded yellow and white.

Next, Kevin goes to the air handler in the basement and checks the voltage between the terminals, C and W1, to which the yellow and white wires are connected. The reading is 24 volts. Since these two wires are running directly between the indoor unit and the outdoor unit, Kevin suspects that the problem is with the wiring. After disconnecting the power to the indoor unit, he proceeds to remove the yellow and white wires from their respective terminals. He then twists these wires together and goes back to the outdoor unit.

Once outside, Kevin loosens the connections between the yellow and white wires and the outdoor unit. Using his continuity tester, he then checks the loop that he has created with the two wires. He knows that, if continuity exists in the loop, the wires are in good shape. If not, one or possibly both of the wires are damaged. His readings indicate that no continuity exists in the loop. Luckily for him, the low-voltage wire running between the indoor and outdoor unit has five conductors within the vinyl casing. Removing the insulation from one of the extra wires, the red one, he then ties the red wire and the yellow wire together.

Back at the indoor unit, Kevin removes the connection that was made between the yellow and white wires and now checks for continuity between the red and yellow wires. This time, he finds continuity in the loop. He then concludes that the white wire must have been damaged. To complete the repair, Kevin replaces the yellow wire on the proper terminal and then places the red wire on the terminal to which the white wire was originally connected. He then closes up the indoor unit and, once again, makes his way outside. He looses the connection between the red and yellow wires and then connects them back to the outdoor unit. After restoring power to the system, the outdoor unit begins to operate. He completes his paperwork and leaves the customer's house at 8:35 A.M.

Service Call Discussion

In this call, Kevin was able to quickly diagnose the problem. Voltage was present at the indoor unit, but it was not making its way to the outdoor unit. Had Kevin not been able to use the spare wires that were available to him, he would have had to run a new set of control wires between the indoor unit and the outdoor unit. This would have taken much more time and would have ultimately cost the customer more money. For this reason, good practice includes providing spare wires when the system is initially installed for possible future use. Kevin took advantage of the extra wires that were there, allowing him to complete the service call in a relatively short amount of time. If he had determined that the low-voltage line had been completely severed, though, he would have had no alternative but to run a completely new line between the two pieces of equipment.

SERVICE CALL 3: IMPROPER HOLDBACK THERMOSTAT SETTING

Customer Complaint

Mary observes that the outdoor unit on her heat-pump system has a large amount of frost buildup on the coil. The system seems to be heating the space, but Mary is concerned about the excessive ice. She has seen ice formation on the coil before, but she thinks that a little more than normal is there now. She places a call to her service company and explains the situation to the service manager.

Service Technician Evaluation

Upon arriving at the customer's home, the service technician, John, asks Mary about her system. She gives him the same information that she has given to the service manager. John makes his way to the outdoor unit and confirms that a large amount of frost is on the coil. John immediately assumes that a defrost problem must exist, since the system appears to be providing adequate heat to the occupied space. He notices that the outdoor fan is operating, and he is also able to hear the compressor operating. He removes the service panel from the outdoor unit so

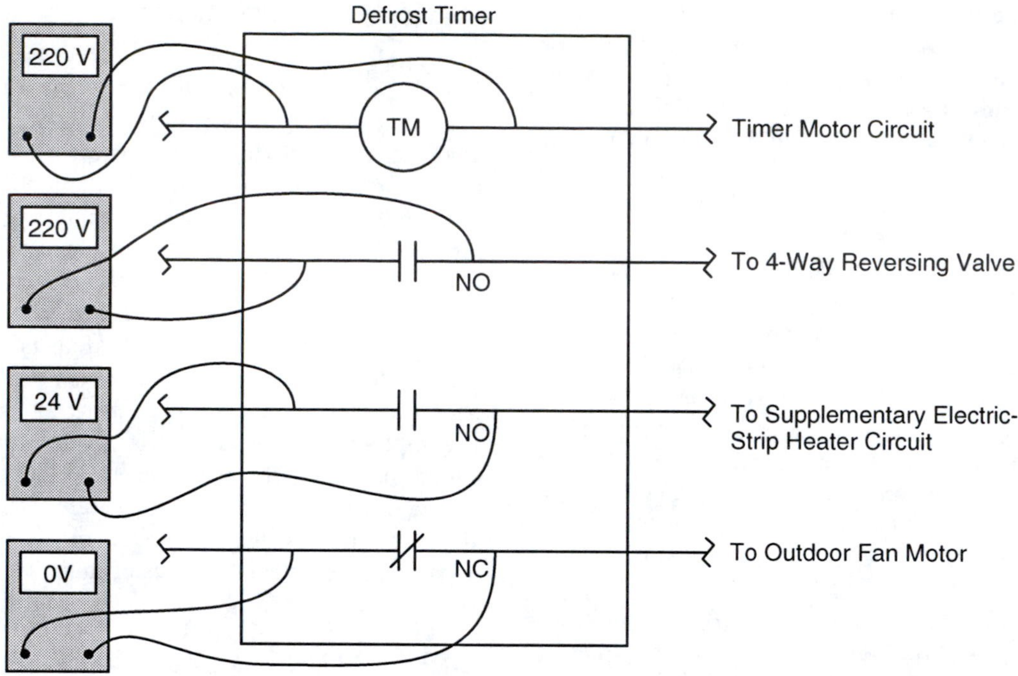

FIGURE 12-4 Voltage readings across a defrost timer with the system operating in the heating mode

that he can inspect the defrost circuitry. John's initial thoughts are that the problem could be with:

- The defrost timer motor
- The defrost timer contacts
- The interconnecting wiring in the defrost circuits

After removing the access panel, John notices that this system is equipped with a defrost cycle that is initiated by time and pressure and terminated by a combination of pressure and temperature. He also notes that the system is designed to fail in the heating mode. He first visually inspects the defrost timer motor and observes that the motor is indeed operating. Following the wiring diagram on the defrost control, he then checks the voltage across the contacts. He obtains the following readings:

- 0 volts across all of the normally closed contacts
- 220 volts across the normally open contacts that control the coil of the four-way reversing valve
- 24 volts across the normally open contacts that control the supplementary heater circuits in the defrost mode

From these readings (Figure 12–4), John is able to determine that the system is operating in the heating mode. He is able to conclude this because the four-way reversing-valve coil is not energized and the supplementary electric-strip heater circuit interlock is not closed. Up to this point, the system appears to be operating properly in the heating mode. He then decides to check the defrost mode of operation. Slowly advancing the knob on the defrost timer, he hears an audible click, signaling the initiation of defrost.

John is able to observe the outdoor fan motor cycle off and also hears the swish of the refrigerant in the system as the reversing valve changes over. With his voltmeter, he takes readings across the contacts on the defrost timer. His readings (Figure 12–5) are as follows:

- 0 volts across the contacts that control the four-way reversing-valve coil
- 0 volts across the contacts that control the supplementary electric-strip heater circuit
- Line voltage across the normally closed contacts

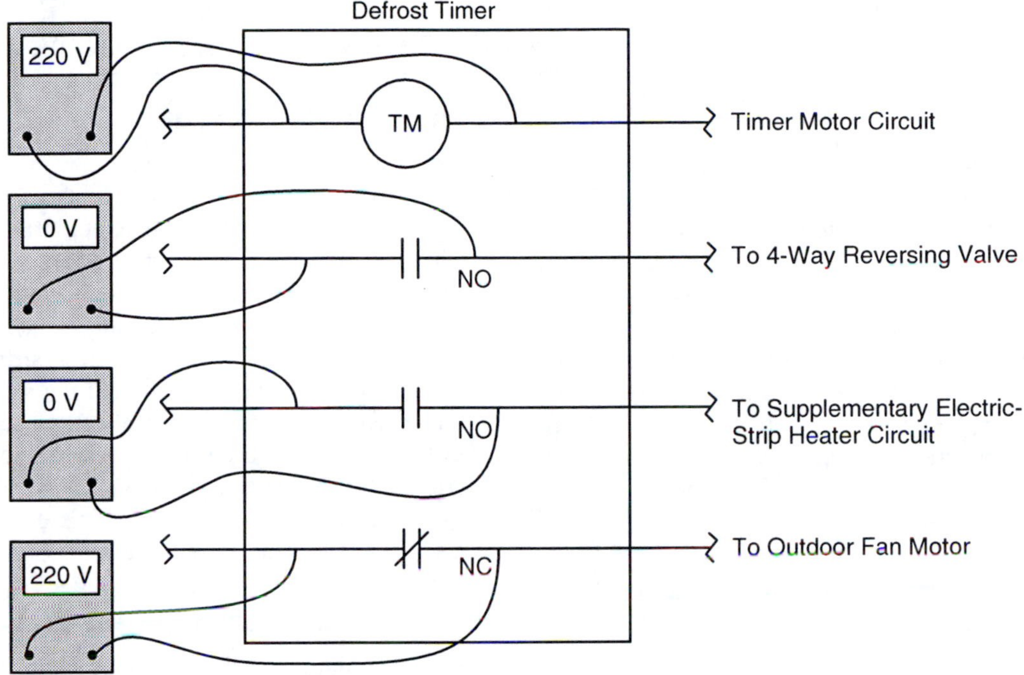

FIGURE 12-5 Voltage readings across a defrost timer with the system operating in the defrost mode

At this point, the system seems to be defrosting properly. Since John is aware that, in this system, defrost can be terminated by either a decrease in the pressure drop across the outdoor coil or an increase in the outdoor coil temperature, he decides to monitor the defrost cycle. After about 10 minutes, he hears the compressor cycle off. He does not, however, hear the swish of the refrigerant in the system, indicating that the reversing-valve coil is still energized. Taking a voltage reading across the compressor contactor coil, John finds that no voltage is being supplied to the coil.

John makes his way into the house and goes to the thermostat location. He notices that the temperature of the occupied space is the same as the set point on the thermostat. He increases the set point on the thermostat to its highest setting and then goes back outdoors. The compressor is operating again. John then concludes that the air being supplied to the occupied space during defrost is warm enough to continue to increase the temperature of the space. This temperature increase is sufficient to satisfy the thermostat and to cause the compressor to cycle off and, in effect, terminate the defrost cycle. John then turns his attention to the supplementary electric-strip heaters, thinking that, possibly, too many of the heaters are being energized during defrost. Knowing that the number of strip heaters that are energized in the defrost mode is controlled by the holdback thermostat, he goes outside to inspect the control.

Disconnecting the power to the outdoor unit, John removes the wires from the holdback thermostat and checks the device for continuity. He reads continuity through the device, indicating that the contacts in the control are in the closed position. Since this device closes its contacts on a drop in temperature, John increases the temperature setting on the thermostat to see if the contacts will open. He hears an audible click and again checks the control for continuity. This time, he reads no continuity. To remedy the situation, John resets the setting on the control to a temperature that is approximately 5 degrees lower than the original setting on the control—in an effort to prevent all of the defrost heaters from being energized in defrost unless the outside ambient temperature were considerably lower than it presently is. He then reconnects the wires on the holdback thermostat and restores power to the system. He then advances the system into defrost again.

Once again, the outdoor fan cycles off and the reversing valve switches over. This time, the system stays in defrost and the ice begins to melt. After the coil is defrosted, the system switches back over to the heating mode. Taking a voltage reading across the air-pressure switch, John determines that the pressure switch is in the open position, indicating that defrost is terminated by the reduction of the pressure drop across the outdoor coil. Having properly adjusted the holdback thermostat, John continues to monitor system operation while he fills out his job order. Making his way into the house, he checks the condition of the air filters located in the hallway and obtains the customer's signature on the work order.

Service Call Discussion

Once again, the service technician utilized the seven steps to completing a successful service call. In this service call, the technician determined that the system was operating properly in the heating mode and that it also was able to enter the defrost mode. The contacts on the defrost timer were switching over properly, which gave the impression that the system was in good working order; but John knew that something was definitely wrong with it. He was persistent and was able to ultimately find the problem with the system. All too often, technicians decide that an "intermittent problem" must have been causing the system to malfunction but that the system was working fine when the technician checked it out. In more instances than not, finding nothing wrong with a system will result in a callback to the job for the same problem at some point in the not-too-distant future. Therefore, the technician must exhaust all possibilities before resorting to the intermittent-problem defense. Even though a holdback thermostat is rarely in need of adjustment and/or calibration, John found it necessary to investigate, since the system seemed to have been operating fine otherwise.

SERVICE CALL 4: DEFECTIVE REVERSING-VALVE SOLENOID COIL

Customer Complaint

A home owner calls her service company and explains that when she turned her heat-pump system on to operate in the cooling mode, hot air was being discharged into the occupied space. She explains to the service manager that the system operated fine in the colder winter months and that this is the first time that she has attempted to operate the system in the cooling mode. She has since turned the system off and is patiently waiting the arrival of the technician. Shortly after placing the call, the service technician, Sal, arrived at the customer's home. Knowing that his truck has a slow oil leak, Sal parks his truck on the street to avoid getting oil on the customer's driveway.

Service Technician Evaluation

After being told about the system problem, Sal goes over to the system thermostat and sets it to operate in the cooling mode. He makes certain that the thermostat is set at a temperature well below the actual temperature of the space. He then checks the air being supplied from the registers and confirms that heated air is entering the room. From this, he is able to establish the following:

- The low-voltage transformer is operational.
- The indoor fan is operating.

He is not, however, able to determine the source of the heat. The heat could be supplied by the compressor operating in the heating mode or by the supplementary electric-strip heaters.

Sal decides to visually inspect the outdoor unit before checking the electric heaters. Going outside, he can hear the compressor operating and can see that the outdoor fan is also operating. Placing his hand on the outdoor coil, he determines that the system is operating in the heating mode—because the coil is cool to the touch. From this, he concludes that the system problem is likely in one of the following areas:

- A problem exists with the system thermostat.
- A problem exists with the low-voltage control circuits and/or wiring.

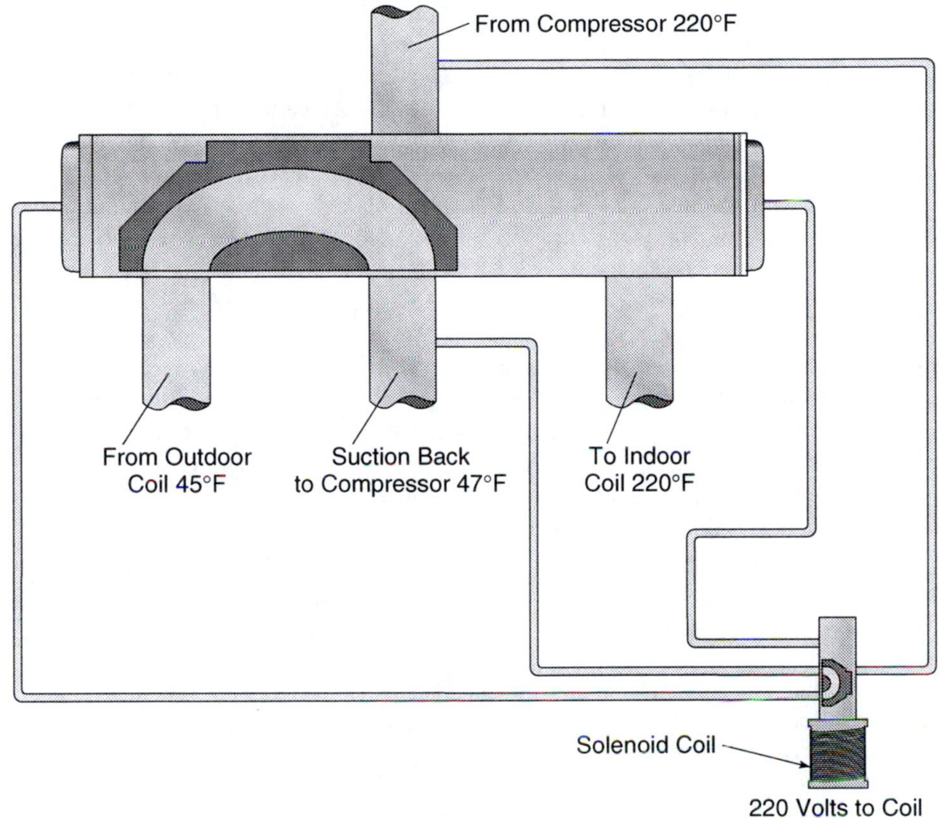

FIGURE 12-6 System temperature readings taken at the four-way reversing valve

- A mechanical problem exists with the four-way reversing valve.
- A problem exists with the solenoid coil on the four-way reversing valve.

Since he is already outside, Sal decides to first check for problems with the outdoor unit. After turning the power off, he removes the access panel from the unit. Turning the system back on, he takes temperature readings on the lines connected to the four-way reversing valve. He obtains the following readings:

- 220°F on the discharge line leaving the compressor entering the valve
- 220°F on the line leading to the indoor section
- 45°F on the line connected to the outdoor coil
- 47°F on the suction line returning to the compressor

> ***Safety Tips:***
>
> When taking temperature readings, always be careful to not touch high-temperature refrigerant lines. They can be extremely hot and can cause burns.

These temperature readings are shown in Figure 12-6. From this, Sal is able to determine that the system is indeed operating in the heating mode and that the four-way valve is not stuck midway between the heating and cooling position. He then chooses to check the solenoid coil on the valve itself.

After inspecting the valve, Sal determines that this system fails in the heating mode; so, in order for the system to provide cooling, the solenoid coil must be energized. Using his voltmeter, he determines that

220 volts is being supplied to the solenoid coil. Disconnecting the power to the unit, Sal removes the solenoid coil from the circuit and, with the aid of an ohmmeter, checks the coil. He reads infinite resistance, indicating that the coil is open and therefore in need of replacement.

At his truck, Sal finds a replacement coil and checks to make certain that both the configuration and the voltage rating of the new coil match those of the defective coil. Returning to the outdoor unit, he installs the new coil, making proper electrical connections. After restoring power to the system, he can hear both the compressor and the outdoor fan come on. He can also hear the swish associated with the changing over of the reversing valve. Taking temperature readings at the reversing valve, he determines that the valve is now in the cooling position; and he can feel the heat being discharged by the outdoor coil and fan. While outside, he takes temperature and pressure readings at the outdoor unit and also checks the amperage draw of the compressor and the outdoor fan motor.

Going down to the basement, the location of the system air handler, Sal proceeds to take temperature readings of both the supply and the return air. The air readings are:

- 80°F in the return duct
- 61°F in the supply duct coming off the system's supply plenum

From these, he is able to establish a temperature differential across the coil of 19 degrees, which is well within the ideal range of 17 to 20 degrees. While in the basement, he checks the air filters and, since they are very dirty, replaces them. He fills out his paperwork, obtains the necessary signature, and proceeds to his next service call.

Service Call Discussion

The preceding is a very good example of how a technician should complete a service call effectively. Sal, well aware that his truck had a slow oil leak, showed respect for the customer's property by parking his vehicle in the street. Remember, a little care and concern go a long way. While outside, Sal decided to exhaust those possible causes for the system failure that were associated with the outdoor unit. This reduced the number of times that he had to enter the home, limiting the potential for bringing dirt into the house. In addition, reducing the number of trips into and out of the house saved valuable time.

Sal made certain that he checked the operation of the system to be certain that it was in good working order before leaving the premises. While in the basement checking the system operation after the repair was made, Sal made it a point to check the air filters. Finding that they were dirty, he replaced them. While at his truck, he made certain that the new coil he intended to install on the system matched the defective component as far as the configuration and the voltage rating were concerned. This reduced the possibility of installing and possibly damaging a component that was not intended for that specific application.

SERVICE CALL 5: DEFECTIVE REVERSING-VALVE SOLENOID COIL

Customer Complaint

A store owner places a service call and tells the dispatcher that the air-conditioning system in his store will not come on. The owner informs the office that the system fuses seem to be blown because the indoor blower will not even operate. Two weeks prior, a spring start-up was performed on the system and it has been operating fine until now. The temperature in the store is 90°F, and the owner is concerned that he will lose customers due to the uncomfortable environment in the store. The dispatcher informs the customer that the technician who performed the start-up on the system is working on an installation job a few blocks away and that he will have him come right over to check out the system. Twenty minutes later, Frank arrives at the store.

Service Technician Evaluation

Frank arrives at the store and immediately goes to the space thermostat and turns the system on. The unit will not start. Apparently, no power is being supplied to the system at all. Frank goes to the stockroom at the back of the store where the air handler is located and checks the line-voltage fuses in the system disconnect box. Reading 220 volts between L1 and L2 on the load sides of the fuses, he establishes that

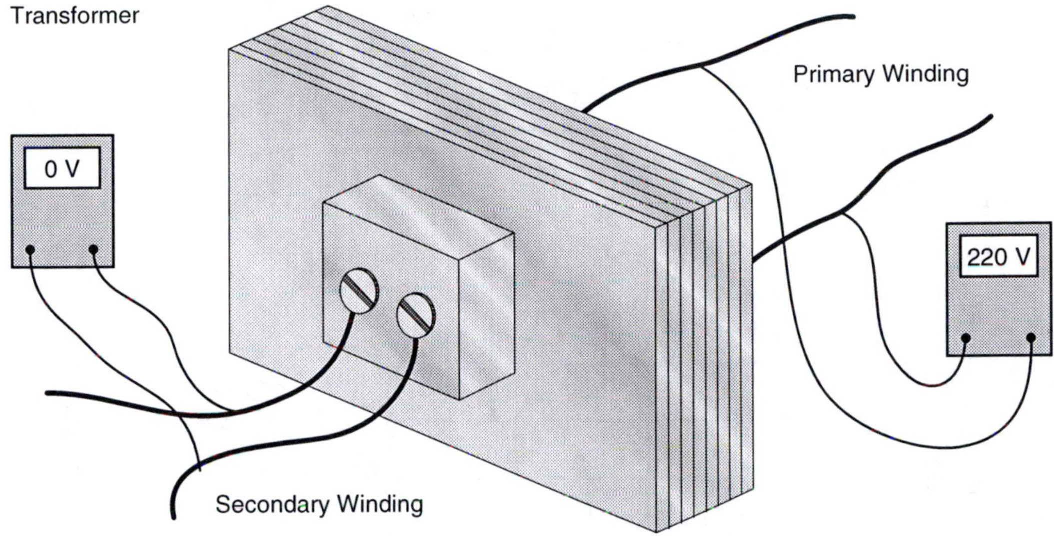

FIGURE 12-7 The absence of voltage at the secondary coil of the transformer indicates that the transformer is defective.

power is indeed being supplied to the system. He then suspects a low-voltage problem. He sets the thermostat to the OFF position and sets the fan switch to the AUTO position.

Opening up the air handler and accessing the terminal board in the air handler, Frank checks the voltage on the low-voltage side of the transformer. The reading is 0 volts. He then checks the voltage being supplied to the primary side of the transformer and finds that 220 volts are being supplied to that winding (Figure 12-7). He concludes that the transformer is defective and goes to his truck to get a replacement.

Making certain that the power to the unit is disconnected, Frank proceeds to label and remove the wires from the defective transformer. He mounts the new transformer in place and makes the necessary wiring connections. After this phase of the repair is complete, he restores power to the system via the fused disconnect at the air handler. He then goes over to the space thermostat and first turns the fan switch to the ON position. The indoor fan begins to operate. He then puts the system into the cooling mode and sets the thermostat to a temperature that is well below the temperature of the occupied space. Frank can hear the indoor fan motor stop running. Turning the system off again, he returns to the stockroom to see what happened. Checking the voltage at the transformer's secondary winding, he finds that 0 volts are being supplied by the new transformer. Checking the voltage at the primary winding, he finds that 220 volts are being supplied to the device. Once again, the transformer is defective. After a few minutes of careful thought, he makes the following observations:

- The indoor fan motor operates when the system is in the OFF position and the fan switch is in the ON position.
- When the system was put into the cooling mode, the transformer burned out.

SERVICE NOTE: *Always make certain that you establish why a component failed before installing the new one. This will help reduce the possibility of damaging brand new replacement parts.*

Frank then concludes that a short circuit exists in the low-voltage controls associated with the cooling mode of operation. He concludes that the short is not in the indoor fan motor circuit because the fan operated fine after the transformer was replaced. He disconnects the power to the air handler and then proceeds to remove the wires from the secondary winding of the transformer. He then sets the thermostat

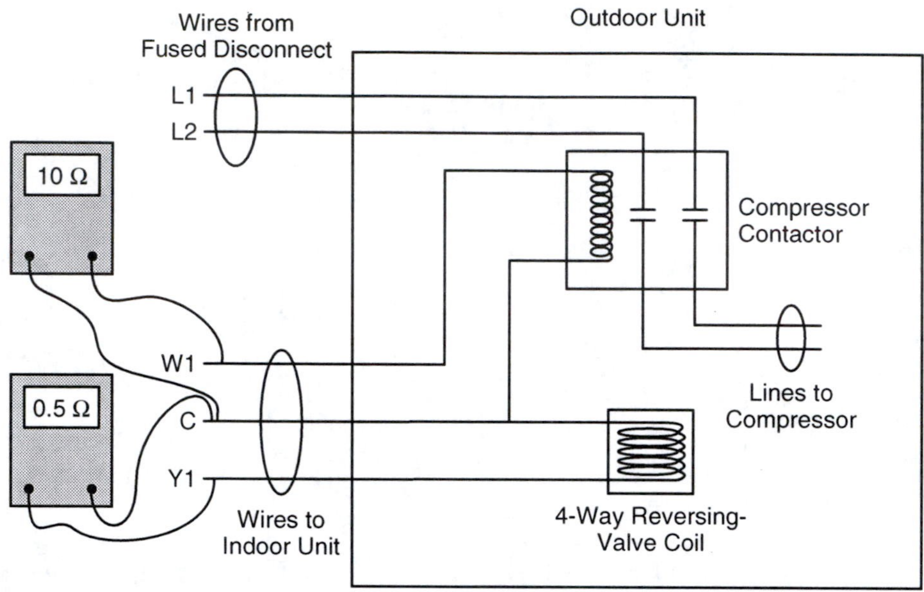

FIGURE 12-8 A short circuit is indicated by the low resistance in the solenoid coil branch of the control circuit.

to the cooling mode and makes certain that it is set to a temperature well below the actual space temperature. Using his ohmmeter, he takes a resistance reading of the low-voltage control circuit between the wires that have been removed from the transformer's secondary winding. The reading is 0.5 ohms. This reading is very low and is definitely the cause of a blown transformer. He now has to determine where the short circuit is located. He establishes that the short could be in one of the following locations:

- The interconnecting wiring between the indoor unit and the outdoor unit
- The compressor contactor coil
- The four-way reversing-valve coil
- The wiring within the outdoor unit itself

The problem is not likely to be a short circuit within the thermostat since the thermostat functions as a series of switches. A short circuit within the device would most likely cause system components to be energized at the wrong time. Going to the outdoor unit, Frank disconnects the power to the unit and removes the service access panel. To eliminate the possibility of a short circuit in the interconnecting wiring, Frank loosens the wire nuts on the wires carrying low voltage to the outdoor unit and disconnects them from the outdoor unit. Back at the indoor unit, an infinite resistance reading indicates that no short circuit exists in the interconnecting wiring between the thermostat and the outdoor unit. He is then able to conclude that the short circuit exists in the outdoor unit. To confirm this, Frank takes a resistance reading between the low voltage wires that he has just disconnected from the wires coming from the indoor unit. He obtains a reading of 0.5 ohms between Wire C and Wire Y1 and 10 ohms between Wire W1 and Wire C (Figure 12-8). This confirms that the short circuit is located within the outdoor unit, and Frank is now able to scratch the contactor coil off his list of possible locations for the short circuit. The problem now has to be in one of the following locations:

- The four-way reversing-valve coil
- The wiring within the outdoor unit itself

Leaving the wires disconnected from the contactor coil, Frank then disconnects the wires from the solenoid coil on the four-way reversing valve. A resistance reading of the outdoor unit indicates infinite resistance. This confirms that the short circuit is in the solenoid coil. Taking a resistance reading directly across the coil produces a reading of 0.5 ohms. Going back to his truck, Frank retrieves a new solenoid coil, another transformer, and an in-line fuse for installation in the control circuit.

After replacing the solenoid coil and replacing all of the wires on the compressor contactor, he proceeds to take another resistance reading of the control circuit in the outdoor unit. This time he reads 10 ohms between Wire W1 and Wire C and 9 ohms between Wire C and Wire Y1. He then goes inside to the air handler to once again replace the control transformer. This time, however, Frank installs an in-line fuse to protect the transformer from excessive amperage in the future. Once the repair is complete, Frank checks the operation of the system and writes out his work order and lists all of the parts he used, including a solenoid coil, an in-line fuse, and two transformers. After obtaining the required signature on the work order, he proceeds back to the installation job on which he had been working.

Service Call Discussion

Frank was a relatively new technician and had often been a part of the installation team within his company. To familiarize him with the field, his boss had begun sending him out on spring start-up service calls last year. Not until this year, after hours of in-house training, did his boss begin sending him out on service calls involving actual system malfunctions. On this service call, though, Frank had some difficulties and also made some mistakes that could have easily been avoided. Examining this service call will reveal Frank's mistakes and also suggest what he might do in the future to help him avoid making the same mistakes again.

When Frank first arrived at the store, he checked the system from the thermostat location to see if the system would come on. When it did not, Frank quickly checked for line voltage being supplied to the unit and then for the presence of low voltage. He then quickly and correctly concluded that the control transformer was defective. Up to this point, Frank seemed to be working quickly, effectively, and efficiently. Here, however, he made a crucial mistake. Instead of asking himself why the transformer went bad, he simply replaced the device, only to find that he had now damaged a new replacement transformer as well.

Only after the damage had already been done did Frank attempt to find the cause for the component failure. He proceeded to locate the problem in a systematic manner and ultimately zoned in on the problem. He made the repair and then proceeded to take another resistance reading of the outdoor unit just to make certain that the problem had been alleviated.

So, what was Frank's second mistake? It was in the preparation of his work order. When he filled out his paperwork, he listed two transformers. This ultimately indicated that the customer would be billed for two transformers, not one. It was not the customer's responsibility to pay for a component that was damaged by the servicing technician. Had Frank proceeded to locate the cause of the problem before replacing the transformer, this would have been avoided. Imagine what would have happened if, instead of replacing an inexpensive component like a transformer, Frank had replaced a compressor without initially determining the cause for its failure. Hours of labor and the cost of a new compressor would have had to be absorbed by the service company.

SERVICE CALL 6: DEFECTIVE ELECTRIC-STRIP HEATER

Customer Complaint

The owner of a small dress shop places a call to her service company. She informs the dispatcher that, while operating in the heating mode, her heat-pump system sometimes blows cold air into the store. She also tells the dispatcher that the system seems to be maintaining the space at the desired temperature but that the cold draft from the system is not very pleasant. A short time later, the service technician arrives at the store.

Service Technician Evaluation

Upon arriving at the store, the technician, Marc, asks the store owner about the problem and feels the air coming out of the discharge grills. The air is warm. Since the system seems to be heating correctly, Marc immediately suspects that the cold air being introduced to the space is the result of incorrect air tempering during defrost. Marc decides to put the system into defrost so that he can monitor the cycle. He goes out back to the outdoor unit and removes the service panel on the unit. He advances the knob on the defrost timer to initiate defrost, but, because no frost is present on the outdoor coil, the system immediately switches back into the heating mode. After inspecting the system, Marc establishes that the defrost cycle is initiated by a combination of time and pressure.

Marc then disconnects the power to the outdoor unit and carefully places a jumper wire across the terminals on the pressure switch that senses the pressure differential across the outdoor coil, thereby simulating ice formation on the coil. He then restores power to the unit and again advances the defrost timer until the system is again calling for defrost. This time, the system goes into defrost and remains there. Marc then goes into the store and checks the air being introduced to the space. The air feels very cool. He now has to check the air handler and the electric-strip heaters.

Marc goes to his truck and retrieves his ladder and droplight. After moving several display racks and covering the merchandise with a clean drop cloth, he sets up his ladder in preparation for gaining access to the air handler, which is located in the ceiling of the store. Once inside the ceiling, he can hear the indoor fan operating; he opens the access panel on the indoor unit, giving him access to the electric-strip heaters. Taking amperage readings of the two individual electric-strip heater circuits, he finds 8 amperes flowing through one and 0 amperes flowing through the other.

Safety Tip:

When choosing a ladder, wood or fiberglass ladders are a better choice than metal. Although heavier than metal, they do not conduct electricity. For this reason, they are safer, especially when working on electrical circuits.

Safety Tip:

When taking amperage readings of electric-strip heater circuits, be extremely careful not to touch any of the associated wiring. Remember, these circuits are live and can result in serious personal injury.

Setting his meter to read voltage, he takes a reading across the heater that is not drawing any current. The voltage reading is 220 volts. Since power is being supplied to the heater and the circuit is not drawing any current, Marc concludes that the heater itself is defective. To confirm this, he disconnects the power to the indoor unit and disconnects the wires from the heater. He then checks for continuity through the heater and finds none. The heater is defective. Marc then carefully removes the heater from the unit and tapes up the wires that were originally connected to it. He secures the wires within the unit to prevent them from coming in contact with any metallic surface. Once removed, he calls his service manager with the information from the heater. After making several phone calls, the service manager determines that this heater has to be ordered and it will take 3 days to get it.

Marc, upon getting this news from his manager, immediately informs the storeowner as to the status of the repair. She thanks him for keeping her informed, and he proceeds to close up the system in the ceiling. He makes certain that he re-secures the service panel so that the system can still be used until the final repair can be made. He puts his ladder back into the truck and offers to move the displays back to their original locations. The store owner insists that she can have her own staff relocate the displays.

SERVICE NOTE: *In stores, the merchandise on display represents the bread and butter for the shop owner. Do not touch or move any merchandise with dirty hands. Treat these items with the same respect that you would any property belonging to a customer.*

Marc then goes back to the outdoor unit and disconnects the power. He then removes the jumper wire that he has placed across the pressure control and then restores power to the system. He replaces the service panel on the unit and then restores power to it. The system comes on immediately and begins to operate in the heating mode.

SERVICE NOTE: *Always remember to remove any jumper wires that may have been placed across controls, switches, or other devices. Leaving them in place can adversely affect system operation.*

Three days later, Marc returns to the store with the new electric-strip heater. Since the service manager had called the store owner earlier in the morning, she has already moved the displays and cleared the area for him. He makes certain that the power to the unit is disconnected and sets up his ladder. Having removed the access panel from the air handler, he proceeds to mount the new heater in position. Before making the electrical connections, Marc again checks to be certain that the power to the unit is off. He then makes the necessary connections. Before closing the access panel, he restores power to the air handler, goes to the outdoor unit, and again places a jumper across the pressure switch. This energizes the outdoor unit and then advances the system into defrost. With his ammeter, Marc checks the amperage in both of the heater circuits. He now reads 17 amperes in each circuit, indicating that both heaters are now operational.

With the access panel on the air handler open, he decides to oil the fan motor. He then replaces the access panel on the air handler and closes the ceiling access panel. Noticing that the filter rack is located in the ceiling as well, Marc decides to perform a filter change on the system, knowing that the equipment is soon due for one. He goes outdoors and, after disconnecting the power to that section of the system, removes the pressure switch jumper wire. Finally, he turns the power back on and allows the system to operate. He completes his paperwork, making certain that he includes the new part that was supplied for the system. The store owner signs the work ticket, and Marc then leaves for his next service call.

Service Call Discussion

This service call was not completed in one visit. The system needed a part that was not immediately available. Marc made certain that he left the system operational so that the store would not be without heat. Upon leaving the store after the initial visit, he also made certain that all of the merchandise displays that were moved to gain access to the unit were placed back in their original locations. During the service call, he was also very careful to make certain that the power to the unit was disconnected before making any changes to the system's wiring. Upon returning to the job with the correct part, Marc made certain that the part was the correct replacement before installing it. After making the necessary repair, Marc took amperage readings of the heater circuits to ensure that they were operating properly.

In an effort to minimize the inconvenience to the store owner, Marc, while up in the ceiling, oiled the fan motor and replaced the air filters. This eliminated the need to disturb the displays again when another technician came to perform routine maintenance on the system. On Marc's job order, he made certain that he included the parts that he replaced and also made note that the air filters were replaced and the motor was lubricated.

SERVICE CALL 7: DEFECTIVE SAIL SWITCH

Customer Complaint

It is mid-January, and a residential customer places a call to her service company. She tells the dispatcher that her heat-pump system blows very cold air every so often while operating in the heating mode, cooling off the space that was just being heated. The call is given to Robert, the service technician. On his way over to the customer's house, he figures that, since the system provides adequate heating and only blows cold air occasionally, the problem is likely with the supplementary electric-strip heaters.

Service Technician Evaluation

Robert arrives at the customer's residence and listens to the customer's complaint about the system. Again, he makes the initial determination that the supplementary heaters, which are intended to temper the air during defrost, are not operating. First, he goes to the thermostat and makes certain that the thermostat is set to the heating mode and that the set point on the thermostat is well above the actual space temperature. He feels the air being discharged from the supply grills in the space. It is warm. He then goes to the backyard to check the outdoor unit. The unit is operating. He feels the outdoor coil and, since it is cool to the touch, determines that the system is indeed operating in the heating mode. Carefully removing the service panel from the outdoor unit, he advances the defrost control knob until the unit switches over to the defrost mode. The outdoor fan motor shuts down,

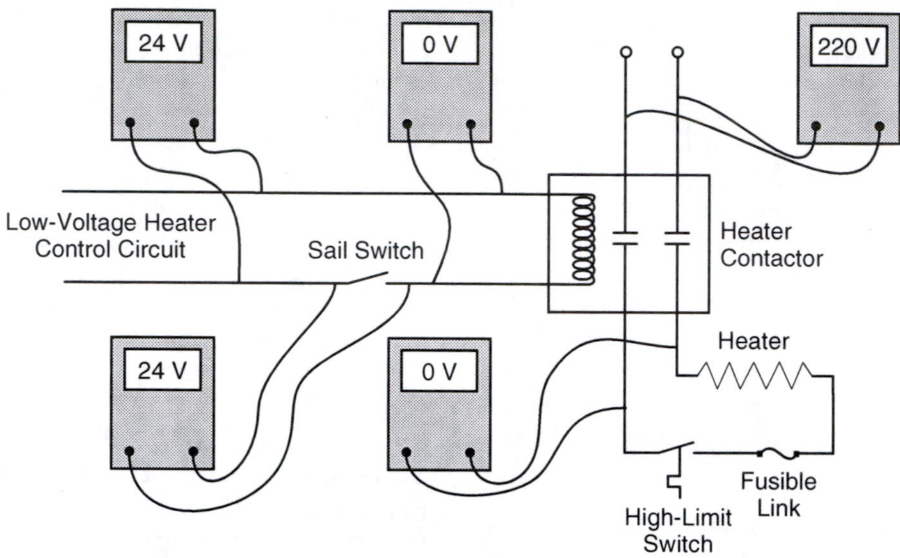

FIGURE 12-9 Voltage readings in the power and control circuits for the supplementary electric-strip heaters

and he hears the swish of the refrigerant as the four-way reversing valve switches over to the cooling position. He then makes his way into the attic so that he can check the supplementary electric-strip heater circuits.

Once in the attic, Robert removes the access panel from the air handler and, using his voltmeter, checks the voltage on the line side of the heater contactor. The reading on each of the two heater contactors is 220 volts. However, the voltage reading is 0 volts on the load side of the contactor (Figure 12-9). This indicates that voltage is being supplied to the heater circuit and that a problem exists in the control circuit. Checking for voltage at the contactor coils, Robert obtains readings of 0 volts across both coils. He makes the following observations:

- The fusible link in the heater circuit is in the power circuit, in series with the individual heaters.
- The high-limit switch is also installed in series with the individual heaters.
- The indoor fan motor is operating.
- The sail switch is located in the low-voltage control circuit.

Again, with his voltmeter, Robert checks for low voltage being supplied to the electric-strip heater control circuit. He reads 24 volts at the terminal block. Noting that the only control in the circuit is physically located between the terminal block and the holding coils on the heater contacts, he decides to check the sail switch. Checking the voltage across the sail switch, he reads 24 volts (Figure 12-9). After turning off the disconnect switch at the indoor unit, Robert locates the sail switch in the ductwork. He disconnects the two wires from the device and then proceeds to check the control with his ohmmeter. He reads infinite resistance across the control. He then removes the switch from the duct section. He pushes the sail, simulating airflow through the duct, and then again checks for continuity through the device. Again, he reads infinite resistance. He then determines that the sail switch is defective, since continuity should be read through the control when the sail is pushed by the air flowing through the duct. He obtains the data from the control and then replaces it in the duct to prevent air leakage into the attic.

After explaining the problem to the customer, Robert calls his service manager to report his findings. The service manager places a call to the supply house and confirms that a replacement part is available for immediate pickup. Robert then goes to the supply house, which is only a short distance away, and obtains the new switch. Before installing the new switch, he checks it with his ohmmeter. He obtains an infinite resistance reading when the sail is at rest and very low resistance when the sail is moved. He then mounts the switch in the duct section and makes the necessary electrical connections. He then restores power to the air handler.

Once the replacement of the switch is complete, Robert goes back to the outdoor unit and again finds the system operating in the heating mode. Once again, he advances the defrost control into defrost. He then goes into the house and takes a temperature reading of the air being supplied to the space, which he determines is indeed being tempered by the electric-strip heaters. He fills out his paperwork and leaves the customer's house for his next call.

Service Call Discussion

As in this service call, using the information provided by the customer to help determine the problem with the system is always important. Also important is understanding that the information that the customer provides may not always be accurate. The customer may have a feeling about what the problem may be, but technicians must remember that customers are not the experts—technicians are. In many instances, customers may be aware that something is not working right simply because they know how the system has performed in the past. In this case, however, the customer was very accurate in her determination that the system gave off cold air only some of the time. As was seen in this example, the information provided by the customer was extremely valuable. It actually gave Robert the ability to more or less troubleshoot the system even before he arrived at the home.

Robert did two very smart things during this call—above and beyond being careful when dealing with the system's electrical circuits. First, once he made the diagnosis of the system, he made the customer aware of his findings. Customers always appreciate the fact that the technician thought enough about them to keep them abreast of exactly what was going on with the system.

Second, after making the customer aware of the problem, Robert called his service manager. Letting the service manager know what is going on with a particular service call, especially when a part is needed, is always important. Robert allowed the service manager to decide whether or not the repair should be completed at that time and also allowed the manager to call the supply house to find out if the part was available. If Robert had taken the matter into his own hands, he might have told the customer that he had to go get a part and then might have returned empty-handed when he found out that no replacement was available. This would have been a major inconvenience for the home owner, because, quite often, the wait time at parts supply houses is very long. Acting in this manner also would have wasted valuable time that could have been spent on other service calls. Robert realized this and showed respect for both his customer and his boss.

SUMMARY

- Approach each service call in a logical manner.
- Only use the information gathered from the customer as a guideline.
- Utilize the seven-step method for completing a successful service call.
- Always respect the property of the customer.
- Before replacing a component, find out why it failed in the first place.
- Try to avoid unnecessary trips into, out of, and around the customer's home.
- Before replacing a component, make certain that the new part matches the old one with respect to voltage rating and physical configuration.
- Always provide a little extra service, such as oiling the motors and checking the air filters.
- Always keep the customer and the service manager informed as to the status of the service call.
- Make certain that all jumper wires that were used to troubleshoot the system are removed before leaving the site.

FOR DISCUSSION

1. How can a system with a defective solenoid coil on the four-way reversing valve experience excessive ice buildup on the outdoor coil?
2. Explain the conditions that could prevent a heat-pump system from completing its defrost cycle.

REVIEW QUESTIONS

1. A defective space thermostat can cause which of the following?
   a. The control transformer becomes defective.
   b. The outdoor fan motor stops operating while the compressor continues to operate.
   c. The indoor fan motor stops operating.
   d. Both *a* and *b* are possible.
2. What is the most likely cause if a heat-pump system provides heat to the occupied space when the thermostat is set to operate in the cooling mode?
   a. A defective solenoid coil on the four-way reversing valve
   b. A defective compressor
   c. A defective indoor fan motor
   d. A defective outdoor fan motor
3. Providing extra low-voltage conductors between the thermostat, the indoor unit, and the outdoor unit when the system is initially installed:
   a. Is a waste of time and money and should never be done.
   b. Is intended to confuse other companies that may come to service the equipment.
   c. Is very useful to the servicing technician in the event that one of the conductors becomes damaged.
   d. None of the above are correct.
4. Excessive ice formation on the outdoor coil of a heat-pump system can be caused by which of the following?
   a. A defective holdback thermostat
   b. A defective indoor fan motor
   c. A defective defrost control
   d. Both *a* and *c*
5. Explain the difference between a solenoid coil with an open winding and a coil with a shorted winding. What are the symptoms that a service technician might encounter in each case?
6. For a heat-pump system that fails in the heating mode, which of the following could be a possible result if the solenoid coil on a four-way reversing valve had an open winding?
   a. The system's control transformer would burn out.
   b. The system would operate in the cooling mode when set to operate in the heating mode.
   c. The system would operate in the heating mode when set to operate in the cooling mode.
   d. The system would continually operate in the cooling mode.

7. Consider a compressor contactor with a 24-volt holding coil. The field technician reads 24 volts across the holding coil, 220 volts on the line side of the contactor, and 0 volts on the load side of the contactor. Which of the following is the most likely scenario?
   a. The thermostat is not calling for compressor operation.
   b. The holding coil on the contactor is shorted.
   c. The holding coil on the contactor has an open winding.
   d. The system has no problem. The compressor should be operating under these conditions.
8. Consider a compressor contactor with a 24-volt holding coil. The field technician reads 24 volts across the holding coil, 220 volts on the line side of the contactor, and 220 volts on the load side of the contactor. Which of the following is the most likely scenario?
   a. The thermostat is not calling for compressor operation.
   b. The holding coil on the contactor is shorted.
   c. The holding coil on the contactor has an open winding.
   d. The system has no problem. The compressor should be operating under these conditions.
9. If the indoor fan motor is operating, which of the following *must* be true?
   a. The system is operating in the cooling mode.
   b. The fan switch is in the ON position.
   c. The control transformer is functioning.
   d. The supplementary electric-strip heaters are energized.
10. If the supplementary electric-strip heaters are energized, which of the following *must* be true?
    a. The indoor fan is energized.
    b. The system is operating in defrost.
    c. The system is calling for second-stage heat.
    d. All of the above must be true.
11. The outdoor unit on a split-type heat-pump system fails to operate. A jumper wire placed across the appropriate terminals on the indoor unit's terminal block causes the outdoor unit to begin operation. Where can the problem be located?
    a. The interconnecting wiring between the thermostat and the indoor unit
    b. The space thermostat
    c. The outdoor unit
    d. Both *a* and *b* possible
12. Excessive ice buildup on the outdoor unit may be a result of:
    a. A system that will not switch over to the heating mode.
    b. A system that has defective supplementary electric-strip heaters.
    c. A system that has a defective or improperly adjusted holdback thermostat.
    d. All of the above are possible causes for excessive ice buildup on the outdoor unit.

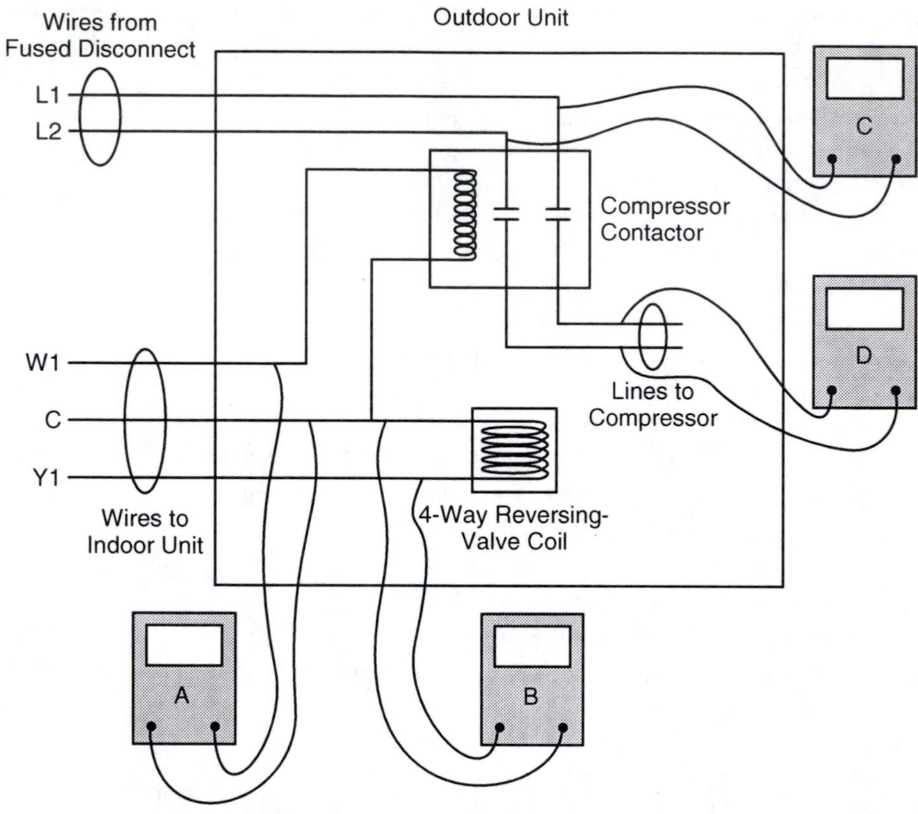

FIGURE 12-10 Use this wiring diagram for Review Question 13 through Review Question 15. This system fails in the heating mode and has a 220-volt power supply and a 24-volt control circuit.

Question 13 through Question 15 are based on the wiring diagram shown in Figure 12-10. This system fails in the heating mode, operates with a 220-volt power supply, and has a 24-volt control circuit.

13. If the system is operating in first-stage heating, the voltage reading on Meter A will be __________ volts, the reading on Meter B will be __________ volts, the reading on Meter C will be __________ volts, and the reading on Meter D will be __________ volts.

14. If the system is operating in the defrost mode, the voltage reading on Meter A will be __________ volts, the reading on Meter B will be __________ volts, the reading on Meter C will be __________ volts, and the reading on Meter D will be __________ volts.

15. If the system is operating in the cooling mode, the voltage reading on Meter A will be __________ volts, the reading on Meter B will be __________ volts, the reading on Meter C will be __________ volts, and the reading on Meter D will be __________ volts.

CHAPTER 13

Vapor-Compression Heat-Pump Refrigeration Circuits

OBJECTIVES After studying this chapter, the reader should be able to:

- Describe various types of heat-pump systems and their applications.
- Describe the air-to-air heat-pump refrigeration circuit.
- Describe the air-to-liquid heat-pump refrigeration circuit.
- Describe the liquid-to-air heat-pump refrigeration circuit.
- Describe the liquid-to-liquid heat-pump refrigeration circuit.

INTRODUCTION

Up to this point in the text, the focus has been on the air-to-air-type heat-pump system. In this type of heat pump, the condenser is the air-cooled type and the equipment is used to either heat or cool the air that passes through the air handler. The heat pumps discussed thus far are used primarily to provide either comfort cooling or heating to the occupied space. Heat pumps are, however, also used to alter the temperatures of other fluids—including water and polyethylene glycol, which is a brine solution with a high boiling and low freezing point. This chapter takes a closer look at different heat-pump systems and the mediums that are used as either the source or sink for the heat energy used to provide heating or cooling as required by the system.

OVERVIEW

As mentioned in Chapter 8, each of a number of different heat-pump-system configurations is geared toward the requirements of the system as well as the environmental and physical conditions under which the system is to operate. Heat-pump systems are designed to facilitate the exchange of heat between a number of different mediums, including air and liquid. The condenser on the heat-pump system can be either air- or liquid-cooled; and, similarly, the evaporator can absorb heat from either air or a liquid. The common configurations for heat-pump systems are as follows:

- Air-to-air systems
- Air-to-liquid systems
- Liquid-to-air systems
- Liquid-to-liquid systems

The medium that is listed first denotes the source of heat for the system while operating in the heating mode, while the second refers to the medium that is ultimately treated by the system. In the case of the air-to-liquid heat-pump system, the evaporator absorbs heat from heat-laden air passing over the coil and the system transfers this heat to the liquid that is passing over or through the condenser. In the case of the liquid-to-air heat-pump system, the evaporator absorbs heat from heat-laden liquid passing over or through the coil and the system transfers this heat to the air passing over the condenser.

Although the theory of operation of each type of heat pump is similar, regardless of the medium used as either the heat source or heat sink, the servicing technician must be aware of the differences. One of the major differences between heat pumps strictly utilizing air and those utilizing liquid is the need to ensure proper liquid flow through the condenser/evaporator. The technician must also be concerned with the operation of the cooling-tower equipment. If the cooling tower is not functioning properly, the operating effectiveness and efficiency of the heat-pump equipment will also be affected. The service technician must also be aware of other situations that can arise when dealing with liquid systems, including:

- Fouling of the cooling-tower equipment
- Scale formation in the water circuits
- Water-regulating-valve problems
- System water treatment

When servicing water-cooled systems, the technician must evaluate, in essence, two individual systems—the heat-pump system itself and the cooling tower, if there is one. Since cooling towers are often open to the atmosphere, the possibility exists of foreign matter making its way into the cooling tower's water circuit. A number of foreign objects can find their way into the cooling tower including:

- Leaves
- Bird waste
- Paper and plastic debris
- Earth, grass, twigs, and branches

In addition to these, the cooling tower provides a perfect breeding facility for bacteria and algae. The cooling-tower environment provides both moisture and heat, providing perfect breeding conditions. For this reason, the tower must be properly maintained and cleaned on a regular basis. A proper water-treatment program is also desirable to help reduce the rate of scale buildup within the water circuits and to slow the rate of algae growth. Scale formation within the water circuit acts as an insulator and thereby reduces the rate of heat transfer between the water circuit and the medium flowing over it.

In addition to the types of heat-pump systems already mentioned, a separate category of heat-pump systems utilizes the earth as both the source and the sink for heat. These systems are referred to as *geothermal heat pumps* and are covered in subsequent chapters of this text.

AIR-TO-AIR HEAT-PUMP REFRIGERATION CIRCUITS

As discussed earlier in the text, the air-to-air heat-pump circuit utilizes air as its **heat source** during the heating mode of operation and as the **heat sink** during the cooling mode of operation. In the cooling mode of operation, the hot discharge gas from the compressor is directed to the outdoor coil first so that the heat absorbed by the system can be rejected from the system (Figure 13-1). The piping configuration for this mode of operation is shown in Figure 13-2. For systems that fail in the heating mode of operation, the solenoid on the four-way reversing valve is energized, causing the valve slide to shift to the cooling position. In the cooling mode, the outdoor air is

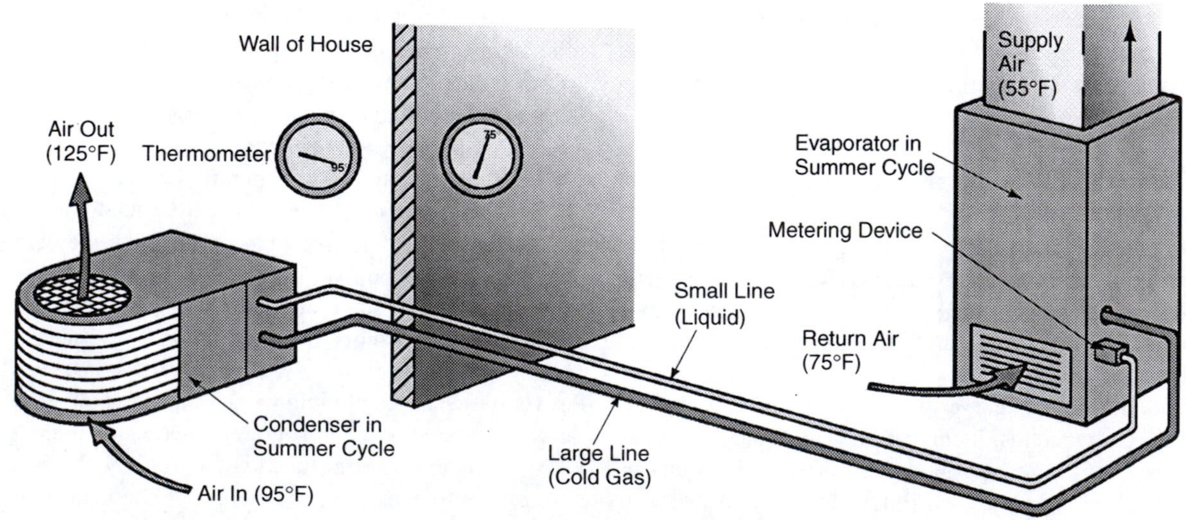

FIGURE 13-1 Air-to-air heat pump removing heat from the structure in the cooling mode

acting as the heat sink, absorbing the heat that is discharged by the system condenser.

In the heating mode, however, the outdoor air is acting as the heat source. The outdoor unit is acting as the evaporator, absorbing heat from the outside air. This heat is then transferred to the indoor unit, or the condenser. The hot discharge gas from the compressor is routed first to the indoor unit, now acting as the condenser. The heat is then introduced to the occupied space (Figure 13-3). The reversing valve, now in the de-energized position, facilitates this change in the direction of refrigerant flow.

As the outside ambient temperature drops, the efficiency and the effectiveness of the heat-pump system drop as well. Notice that, in Figure 13-3, the temperature differential across the indoor coil is only in the range of 10 degrees. This may not be enough to satisfy the heating requirements of the space. In this case, the second-stage heating mode, or supplementary heating strips, may be needed to help satisfy the heating requirements (Figure 13-4). The supplementary heating strips, as mentioned earlier, are intended to be utilized when the first-stage heating mode cannot increase the space temperature by itself.

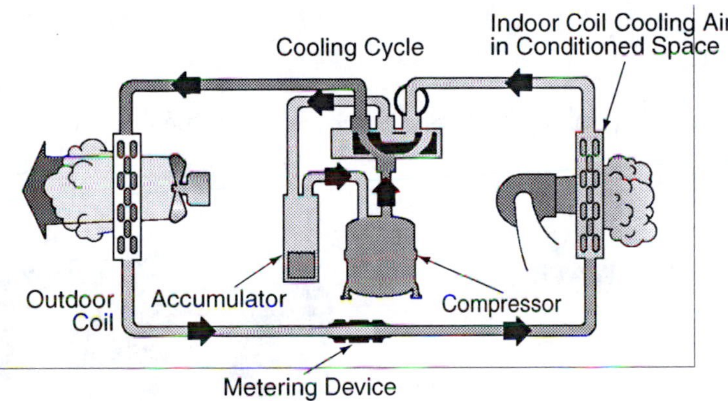

FIGURE 13-2 Piping diagram showing the direction of refrigerant flow in the cooling mode. *(Reproduced courtesy of Carrier Corporation)*

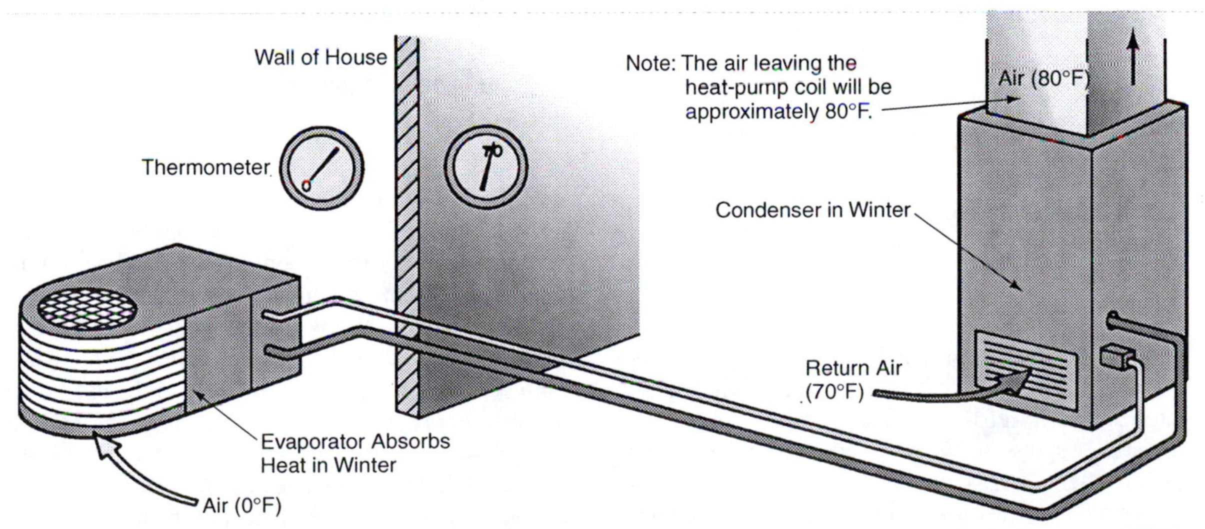

FIGURE 13-3 In the winter, the heat pump transfers heat into the structure.

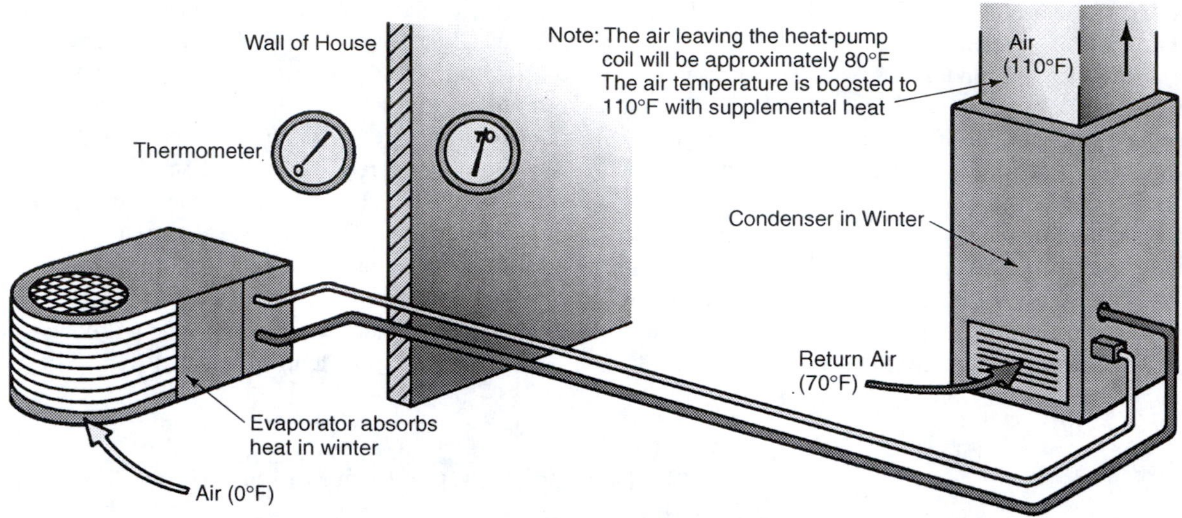

FIGURE 13-4 The use of supplementary electric-strip heaters increases the temperature of the air supplied to the space.

LIQUID-TO-AIR HEAT-PUMP REFRIGERATION AND WATER CIRCUITS

As mentioned in the chapter overview, the liquid-to-air heat-pump configuration uses liquid as the heat source when operating in the heating mode. Instead of utilizing a fin-and-tube coil located outdoors as the heat source, a tube-in-tube or coaxial heat exchanger is used (Figure 13-5). The liquid flowing through this heat exchanger can be either water or a mixture of water and antifreeze such as polyethylene glycol. A typical liquid-to-air piping configuration is shown in Figure 13-6. The source of liquid (water) can be one of the following:

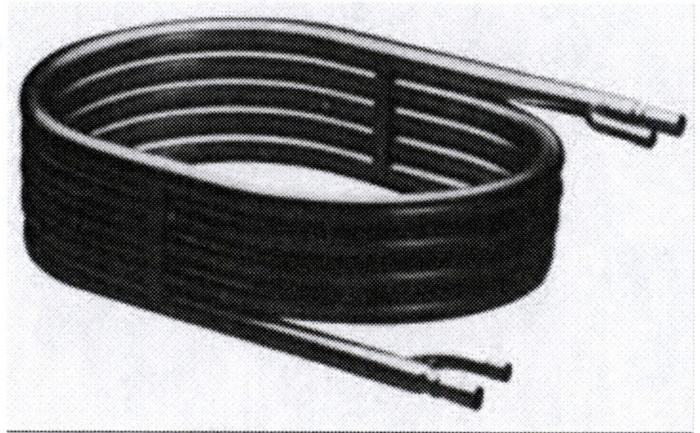

FIGURE 13-5 A tube-in-tube heat exchanger

- Local municipal water supply (tap water)
- Cooling tower
- The earth
- Lake or pond

When the local municipal water supply is used, the water is normally wasted down the drain. These systems are therefore more commonly known as **wastewater systems.** Wastewater systems are illegal in many localities, so local zoning laws and codes must be adhered to with regard to this type of equipment. The laws regarding wastewater systems reflect the population of the area, the availability of water, and the capacity of the system intended to be used. In urban areas, the most common source for the water is a cooling tower. Cooling towers were discussed earlier in the text, and their use in liquid-to-air heat-pump systems enables the same water to be used over and over, thereby reducing water waste. In certain areas of the country, using water from the earth itself is beneficial. These systems, referred to as *geothermal heat pumps,* are covered later in the text.

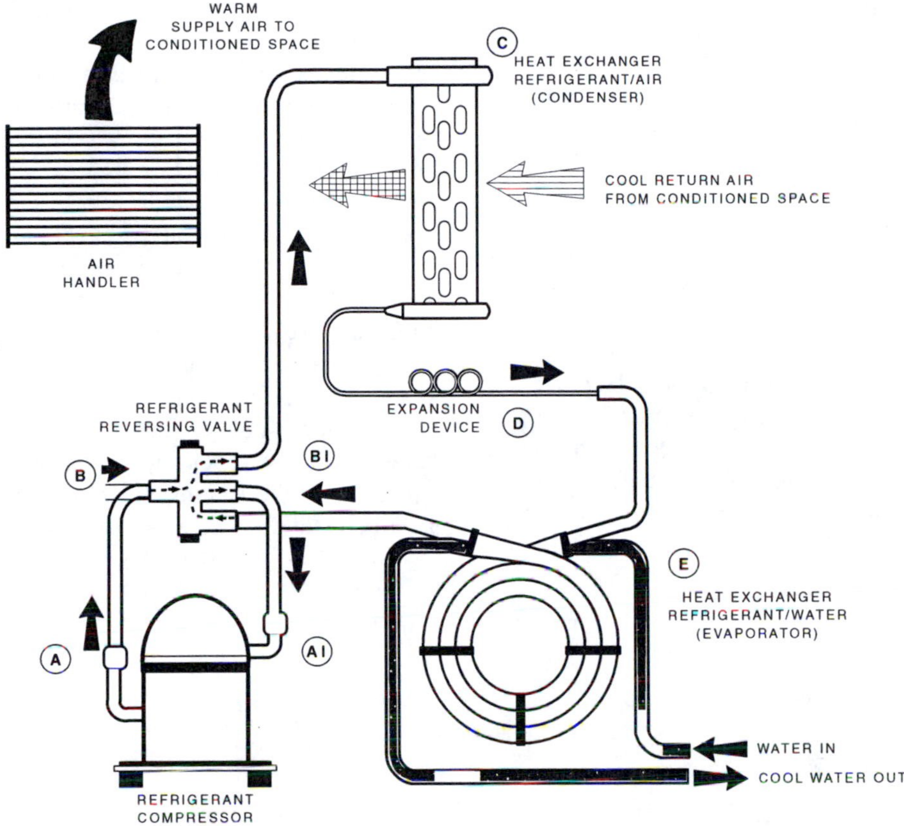

FIGURE 13-6 Liquid-to-air piping diagram. Heating mode. *(Courtesy Climate Master, Friedrich)*

Liquid-to-Air Heat-Pump Refrigeration Circuit: Heating Mode

In Figure 13-6, the system is operating in the heating mode. The hot discharge gas leaving the compressor at Point A flows into the four-way reversing valve at Point B. Upon leaving the reversing valve, the hot discharge gas flows first to the indoor finned coil at Point C, heating the air from the conditioned space as it flows through the coil. Once the refrigerant has condensed in the indoor coil, it then flows through the metering device at Point D, causing a reduction in both the temperature and the pressure of the refrigerant. The refrigerant then flows through the tube-in-tube heat exchanger at Point E, where the refrigerant absorbs heat from the liquid flowing through the exchanger. As explained earlier in the text, the refrigerant flows in the outer tube, while the liquid flows within the inner tube. After flowing through the heat exchanger, the low-pressure, low-temperature, vapor refrigerant returns to the reversing valve at Point B1 and back to the compressor at Point A1.

The temperature of the liquid leaving the heat exchanger is cooler than the temperature of the water entering the heat exchanger, since it has transferred some of its heat energy to the refrigerant. The amount of water or liquid flowing through the heat exchanger is regulated by a water-regulating valve located on the water supply, which opens as the system's low pressure drops. This water-regulating valve is set to maintain the desired low pressure on the suction side of the refrigeration system.

Liquid-to-Air Heat-Pump Refrigeration Circuit: Cooling Mode

In the cooling mode, the tube-in-tube heat exchanger functions as the heat sink for the heat that is removed from the occupied space. The piping configuration for a liquid-to-air heat-pump system operating in the cooling mode is shown in Figure 13-7.

Following the refrigerant flow through the liquid-to-air system operating in the cooling mode, the hot discharge refrigerant leaves the compressor at Point A and flows to the reversing valve at Point B. Upon leaving the reversing valve at Point B1, the refrigerant flows first to the heat exchanger at Point C. Here, the refrigerant gives up its heat to the cool water that is entering the heat exchanger. After condensing in the heat exchanger, the subcooled refrigerant flows to the metering device at Point D, where the pressure and temperature of the refrigerant are reduced. The low-pressure, low-temperature refrigerant then enters the evaporator where it absorbs heat from the air passing over it. The cooled air then returns to the occupied space, thereby reducing the temperature of the area. Finally, after flowing through the evaporator, the refrigerant flows through the four-way reversing valve and then back to the compressor at Point A1.

Just as in the heating mode, a water-regulating valve controls the liquid flow through the tube-in-tube heat exchanger. In this case, though, the regulating valve senses the high pressure of the system, opening wider as the pressure increases. A potentially difficult situation then arises. How can a water-regulating valve open on a drop in pressure when operating in the heating mode and open on a pressure rise while operating in the cooling mode? This question is answered in the next section.

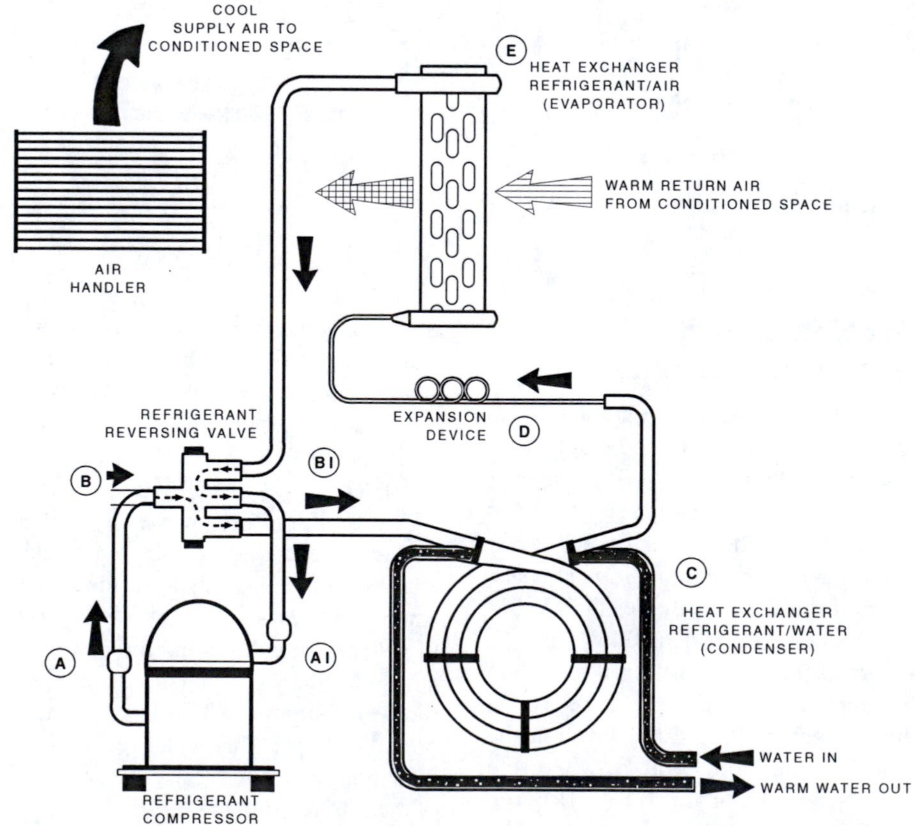

FIGURE 13-7 Liquid-to-air piping diagram. Cooling mode. *(Courtesy Climate Master, Friedrich)*

Liquid-to-Air Heat-Pump Refrigeration Circuit: Dual Water-Regulating Valves

On a liquid-to-air heat-pump system operating in the heating mode, the water-regulating valve tends to open as the low-side system pressure drops below acceptable levels. This is to ensure that the tube-in-tube heat exchanger does not freeze. In the cooling mode, however, the water-regulating valve is designed to open as the system's high pressure increases in an effort to keep the head pressure of the system within an acceptable range. The method commonly used to accommodate both scenarios is to utilize two water-regulating valves connected in parallel with each other. These water-regulating valves sense the pressure at the four-way reversing-valve port that feeds refrigerant to the tube-in-tube heat exchanger while operating in the cooling mode (Figure 13-8). The regulating valve arrangement can be seen at the right side of Figure 13-8.

In the cooling mode, the hot discharge gas from the compressor leaves the four-way reversing valve and flows to the tube-in-tube heat exchanger. Both water-regulating valves sense this pressure. The water-regulating valve that controls the water flow in the heating mode will be in the closed position because this valve is designed to open when the low-side pressure drops below a predetermined pressure, which for an R-22 air-conditioning system will be in the range of 70 psig. The high-pressure refrigerant leaving the compressor will be in the range of 210 psig, keeping the regulating valve in the closed position. Thus, all of the water flowing through the tube-in-tube heat exchanger will be directed through the other water-regulating valve. As the head pressure of the

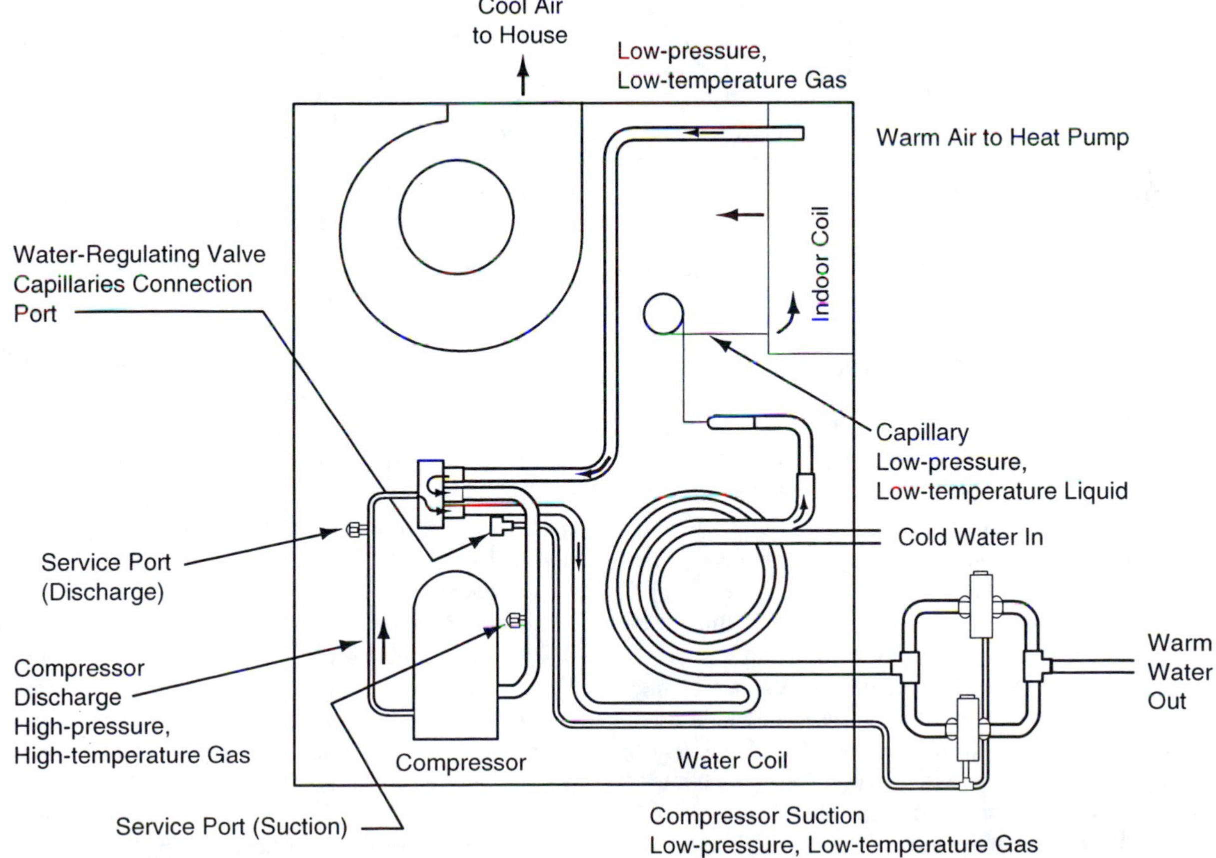

FIGURE 13-8 Dual water-regulating-valve arrangement. Cooling mode. *(Courtesy Bard Manufacturing Co.)*

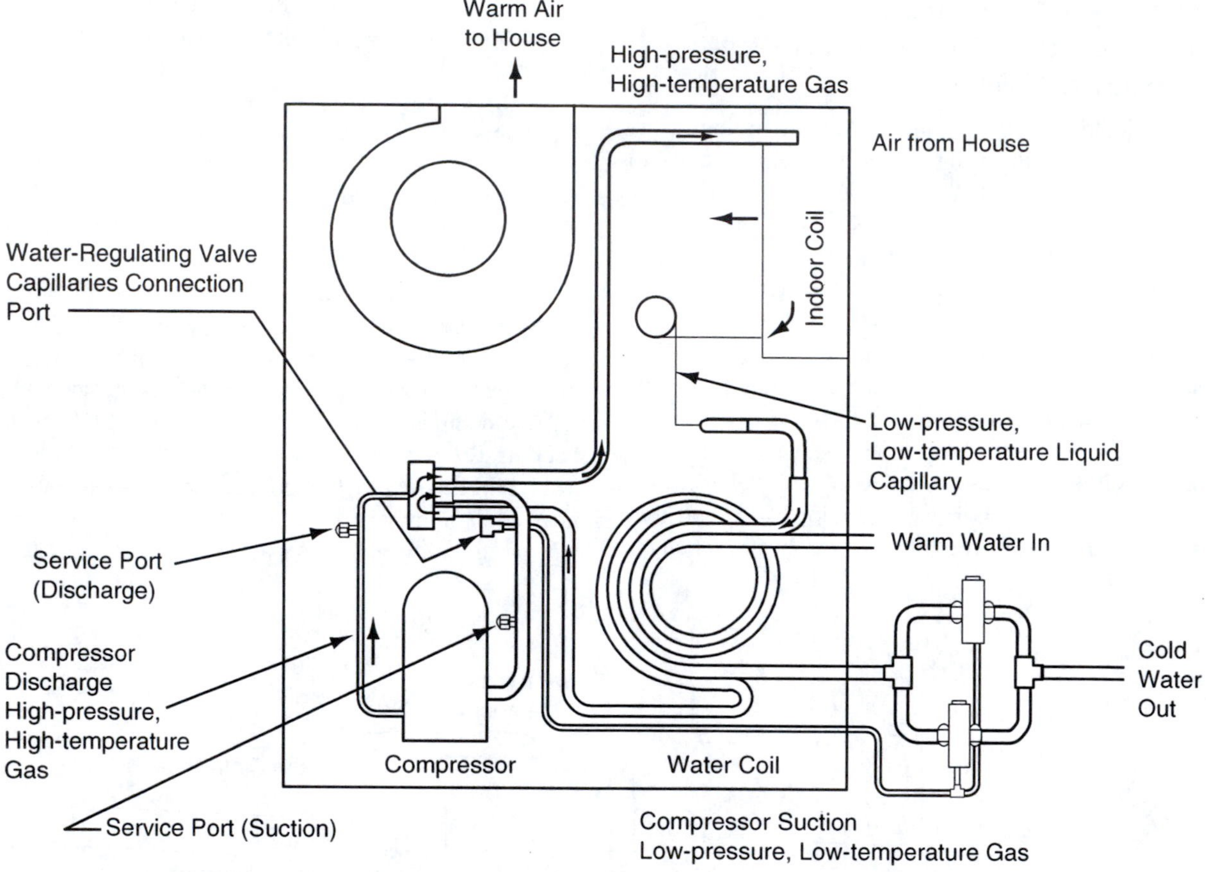

FIGURE 13-9 Dual water-regulating-valve arrangement. Heating mode. *(Courtesy Bard Manufacturing Co.)*

system rises above the predetermined set point, which will be in the 210 psig range, the valve will open allowing more water to flow through the exchanger, thereby reducing the head pressure.

In the heating mode (Figure 13-9), the hot discharge gas is first directed to the indoor coil where it gives up its heat to the air passing over the coil. The system is now providing heat to the occupied space. The refrigerant flowing through the tube-in-tube heat exchanger is now at a low temperature and low pressure, and it picks up heat from the water flowing through the exchanger. The heat exchanger is now functioning as the heat source. The refrigerant returning to the compressor is at a low temperature and low pressure, and this pressure is now being sensed by both water-regulating valves. Since this pressure is low, the water-regulating valve that controls the system while operating in the cooling mode will be in the closed position. Once again, the other valve will now control all the water flowing through the tube-in-tube heat exchanger. If the suction pressure of the system falls below the predetermined set point, which is in the range of 70 psig, the regulating valve will open, thereby increasing the system's suction pressure.

The operation of the two-water-regulating-valve configuration is outlined in Figure 13-10. This figure illustrates that, when the system is in the off position and the system is at **standing pressure,** both valves are in the closed position. Standing pressure is pressure that is present in the system when the compressor is not operating. The standing pressure at this point should correspond to the ambient temperature as indicated by the pressure/temperature chart.

	Heating Mode	Heating Mode	Cooling Mode	Cooling Mode
Pressure Sensed by the Regulating Valves	**Cooling Valve**	**Heating Valve**	**Cooling Valve**	**Heating Valve**
50 psig	Closed	**Open**	Closed	**Open**
60 psig	Closed	**Open**	Closed	**Open**
70 psig	Closed	**Open**	Closed	**Open**
80 psig	Closed	Closed	Closed	Closed
90 psig	Closed	Closed	Closed	Closed
100 psig	Closed	Closed	Closed	Closed
110 psig	Closed	Closed	Closed	Closed
120 psig	Closed	Closed	Closed	Closed
130 psig	Closed	Closed	Closed	Closed
140 psig	Closed	Closed	Closed	Closed
150 psig	Closed	Closed	Closed	Closed
160 psig	Closed	Closed	Closed	Closed
160 psig	Closed	Closed	Closed	Closed
170 psig	Closed	Closed	Closed	Closed
180 psig	Closed	Closed	Closed	Closed
190 psig	Closed	Closed	Closed	Closed
200 psig	Closed	Closed	Closed	Closed
210 psig	Closed	Closed	Closed	Closed
220 psig	**Open**	Closed	**Open**	Closed

FIGURE 13-10 Positions of water-regulating valves in both heating and cooling modes of operation

It can also be seen from Figure 13-10 that, when the system is operating in the heating mode, the cooling valve will be in the closed position. While operating in this mode, the pressure sensed by the valves will not be high enough to cause the cooling valve to open. Similarly, while operating in the cooling mode, the pressure sensed by the valves will never be low enough for the heating valve to be in the open position. Therefore, when operating in either mode, one water-regulating valve is effectively removed from the active circuit at any given point in time.

Service note: To properly evaluate the operation of the heat-pump system, the technician should be able to take temperature readings of the water entering and leaving the tube-in-tube heat exchanger. For this reason, thermometer wells should be installed in both the supply and return water piping circuits. Hose bibs should also be installed in the lines to facilitate the cleaning of the water loop (Figure 13-11).

Liquid-to-Air Heat-Pump Refrigeration Circuit: Single Water-Regulating Valve

The piping configuration for the liquid-to-air heat-pump system using two water-regulating valves can be somewhat simplified by the use of a single water-regulating valve that is specially designed for heat-pump application. This valve has a dual-adjusting

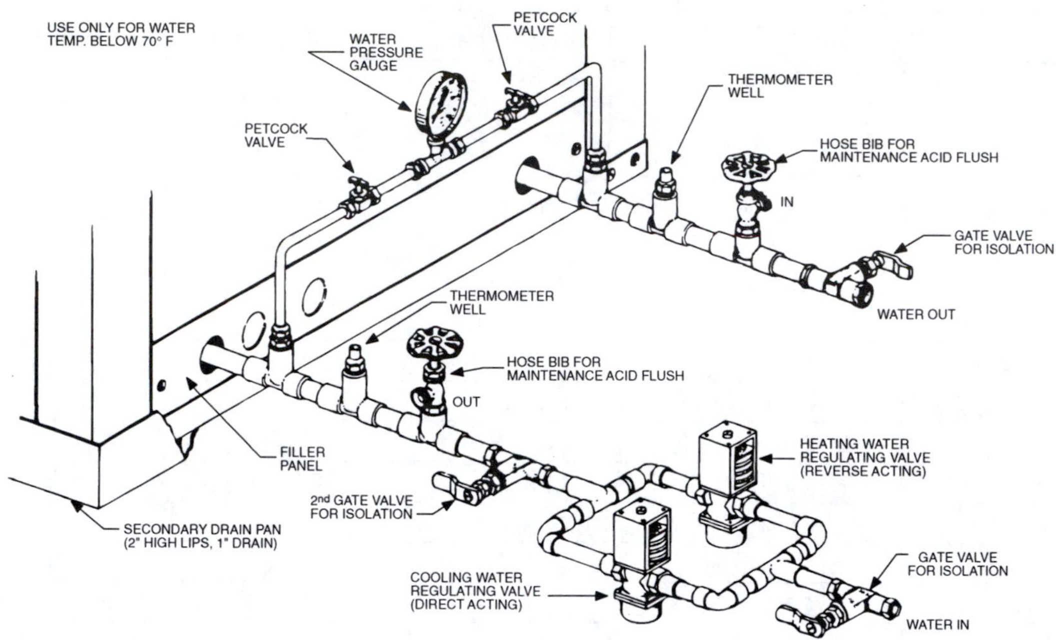

FIGURE 13-11 Typical water-regulating-valve piping arrangement. *(Courtesy Addison Products Co.)*

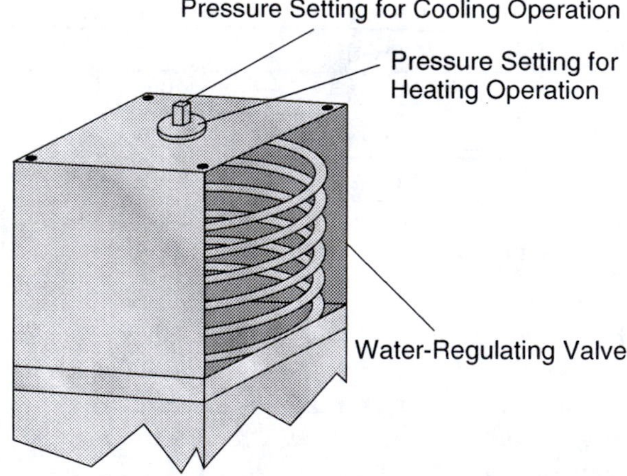

FIGURE 13-12 Pressure-adjustment screws on a dual-pressure water-regulating valve

screw on the top of the valve, which facilitates the setting of the pressures to be maintained in both the heating and the cooling modes of operation (Figure 13-12).

This valve is closed when the system is not operating. When the system is operating in the heating mode, the valve opens when the suction pressure drops below the predetermined set point. Similarly, in the cooling mode, the valve opens when the high-side pressure rises above the predetermined high-pressure set point. For an R-22 high-temperature system, assume that the low-side pressure and the high-side pressure settings are 70 psig and 210 psig, respectively. In this instance, the water-regulating valve is in the open position when it senses a pressure below 70 psig or above 210 psig. Otherwise, the valve is in the closed position.

Heat Reclaim Used with Liquid-to-Air Heat-Pump Systems

One popular feature of liquid-to-air heat-pump systems is the ease in utilizing the hot discharge vapor from the compressor as a source of additional heating capacity for use in domestic hot water systems. To accomplish this additional heat transfer, a tube-in-tube heat exchanger, referred to as a **desuperheater,** is located between the compressor discharge port and the four-way reversing valve (Figure 13-13). The discharge refrigerant is able to transfer heat to the domestic water flowing through the exchanger before

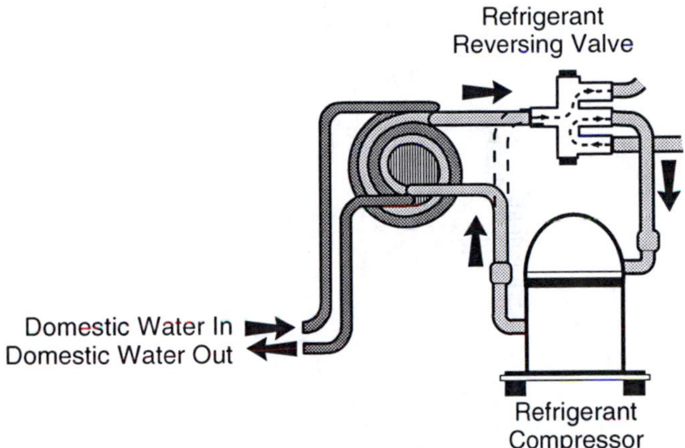

FIGURE 13-13 A typical desuperheater. *(Courtesy Friedrich Air Conditioning and Refrigeration Co., Climate Master Division)*

flowing to the condenser. In this case, the condenser is water-cooled during the cooling mode and air-cooled during the heating mode (Figure 13-14).

Desuperheaters are especially effective when operating in the cooling mode. Without the use of the desuperheater, the heat being discharged from the compressor would be wasted as it is transferred to the water-cooled condenser. With the heat exchanger, the heat is, in effect, reclaimed, as it can be used again. In this case, the heat is being transferred to water for domestic use. During the heating mode, a benefit still exists but it is not as great. Using the **heat reclaim** in the heating mode reduces the temperature of the refrigerant that flows through the condenser, which supplies heat to the occupied space. Therefore, the heat exchanger must be sized carefully so that it is not too efficient. An overly efficient heat exchanger will reduce the heating capacity of the heat pump.

AIR-TO-LIQUID HEAT-PUMP REFRIGERATION CIRCUITS

As the name implies, an air-to-liquid heat-pump system uses air as the source of heat when the system is operating in the heating mode. This type of heat pump functions to ultimately treat and maintain the temperature of a liquid. The most common uses for the air-to-liquid heat pumps are:

- Swimming pools
- Spas
- Hot water heaters

Since these very special heat-pump systems are used in areas that are typically warm, they are often designed without reversing valves. The absence of the reversing valve indicates that this type of system does not require a defrost cycle. Specific applications, as discussed later, may require the use of a four-way reversing valve.

Heat-Pump Swimming Pool Heaters

A heat-pump system designed for use as a pool heater functions much the same as a standard air-to-air heat pump in that the heat source for both system types is the ambient air. In appearance, they also closely resemble the air-to-air heat-pump system. Instead of refrigerant pipes connected to the unit, though, water pipes are connected to the pool water circuit (Figure 13-15). Other components such as chlorinators and check valves are also part of the installation (Figure 13-16). The heat-pump swimming pool heater is designed in a manner that is very different from a system designed to provide comfort cooling and/or heating. These differences include:

- They are equipped with a tube-in-tube heat exchanger.

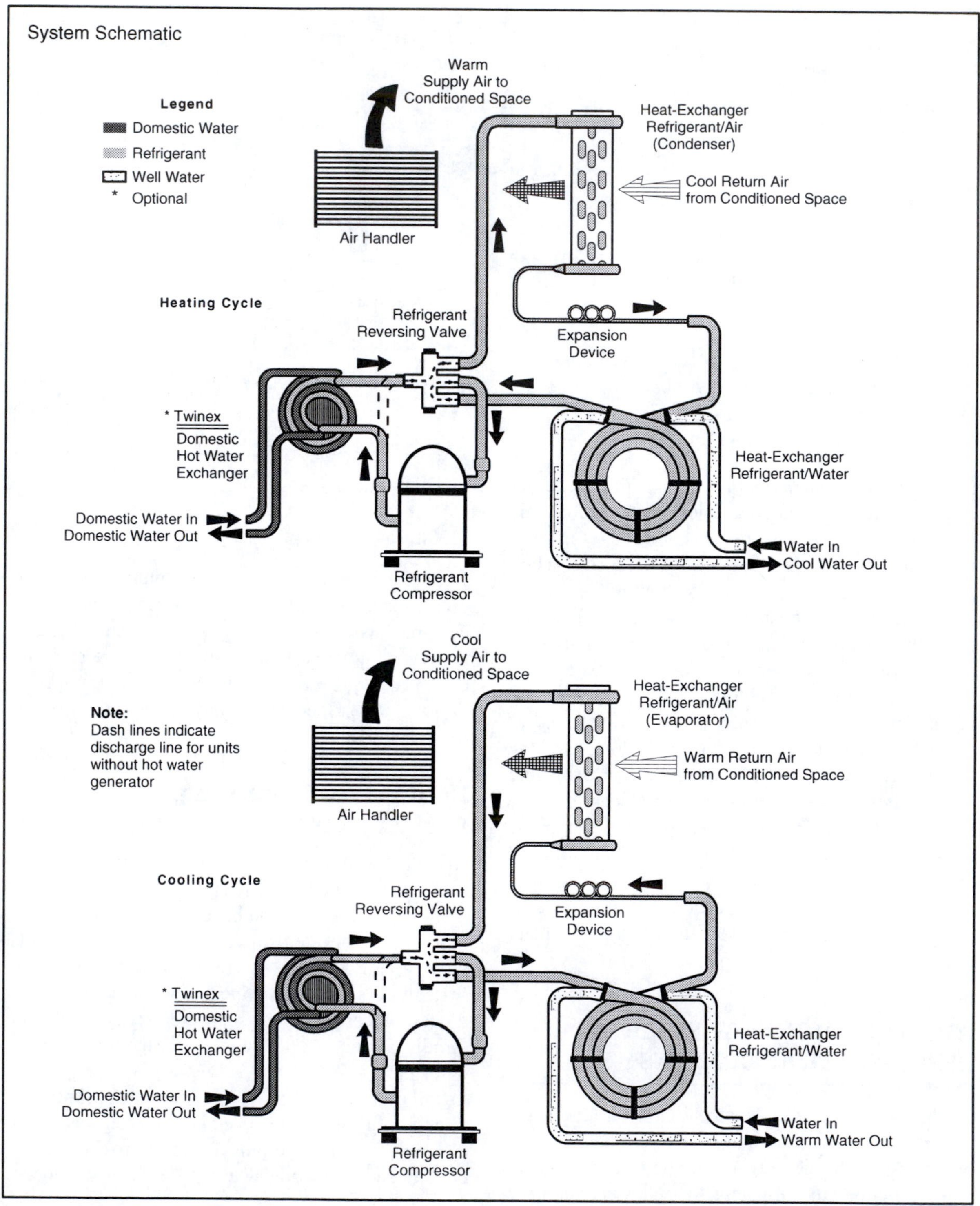

FIGURE 13-14 Refrigeration and water circuits. *(Courtesy Friedrich Air Conditioning and Refrigeration Co., Climate Master Division)*

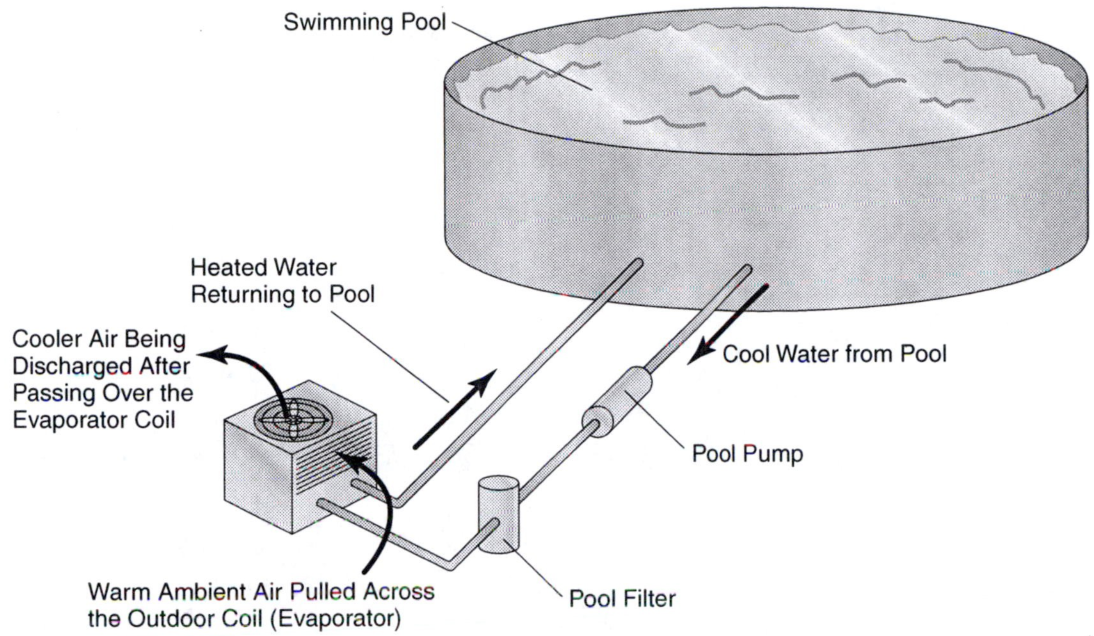

FIGURE 13-15 Heat-pump pool heater resembles an air-cooled condensing unit.

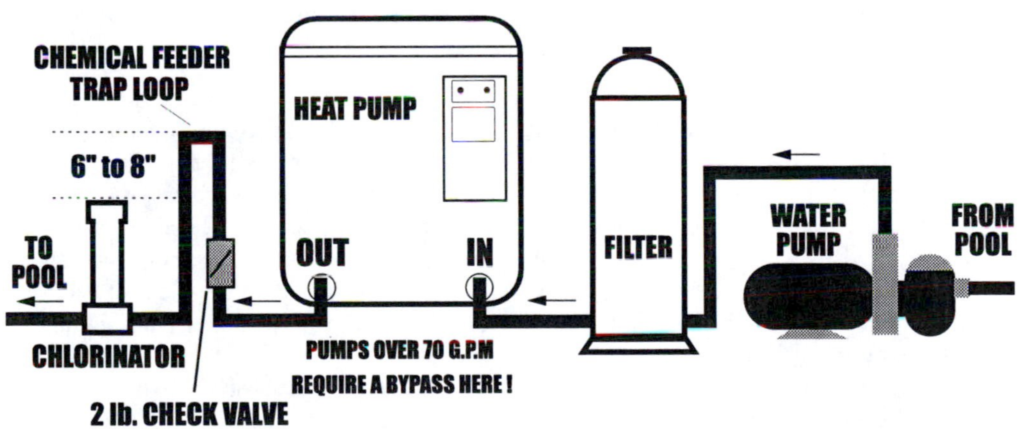

FIGURE 13-16 Location of heat-pump pool heater with respect to other pool components. *(Courtesy Calorex U.S.A.)*

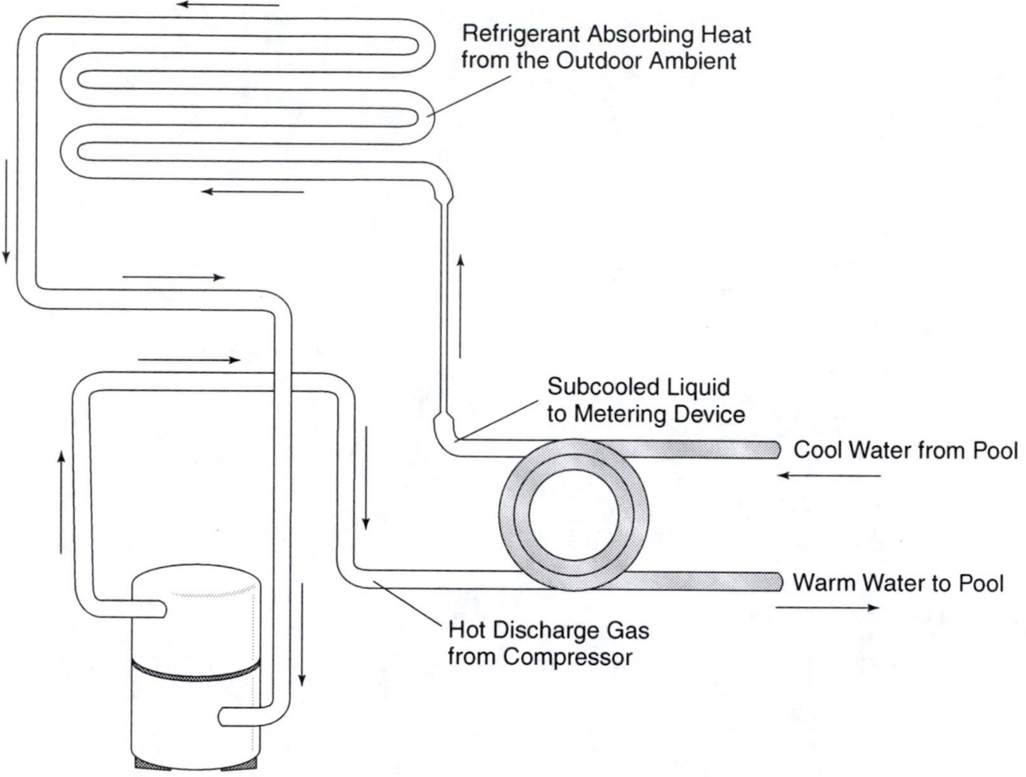

FIGURE 13-17 Typical refrigerant piping circuit of a heat-pump swimming pool heater

- They are not normally equipped with four-way reversing valves.
- The outdoor fan motor runs continuously.
- No supplementary electric-strip heaters are utilized.

As opposed to the air-to-air-type heat-pump system, the air-to-liquid system is intended to heat the liquid that is ultimately being treated. In this case, the liquid is swimming pool water. For this reason, the heat exchanger is of the tube-in-tube type. This heat exchanger is similar in construction to the tube-in-tube condenser discussed at the beginning of the text. In operation, the heat exchanger functions exclusively as the condenser since this system is not intended to operate in the cooling mode. A typical piping diagram for the refrigeration system is shown in Figure 13-17. The piping connections from the pool water circuit to the heat-pump system are normally made with polyvinyl chloride (PVC) pipe connections in the 2-inch range. As shown in Figure 13-15 and Figure 13-16, the piping comes directly from and returns to the pool, with the pool filter and pump connected in series with the tube-in-tube heat exchanger. The pool filter should be connected in the circuit before the heat exchanger to help prevent the heat exchanger from getting clogged or fouled.

Note that the heat-pump swimming pool heater is not normally equipped with a four-way reversing valve as shown in Figure 13-17. Because these units are typically operated in the summer months, no possibility exists of frost developing on the outdoor coil. During normal operation, the outdoor finned coil is functioning as the evaporator, absorbing heat from the warm ambient air. Thus, the outdoor fan motor operates continuously, maximizing the rate of heat transfer. This heat is then transferred to the pool water. Since no

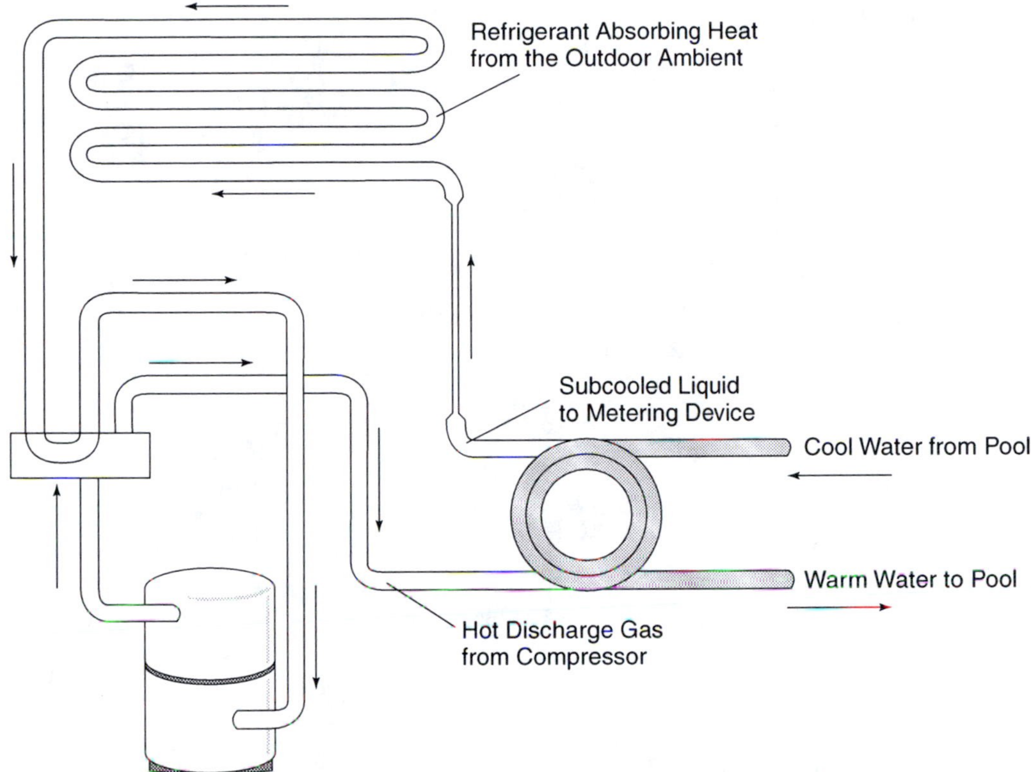

FIGURE 13-18 Refrigerant piping circuit for a heat-pump swimming pool heater equipped with a reversing valve

frost buildup occurs, the coil does not need to be defrosted. However, an exception to this does exist.

If the pool is enclosed and is intended to be used year-round and the heat-pump pool heater is located outdoors, the possibility of frost buildup on the outdoor coil exists in the cold winter months. If this is indeed the case, the manufacturer of the equipment can accommodate this special situation by installing a reversing valve in the system as well as a defrost control, which will permit conventional heat-pump operation (Figure 13-18). As mentioned, this is the exception and not the rule. Under these circumstances, backup gas heaters are commonly available that will heat the pool, since the heat pump will not operate effectively when the ambient temperature drops below 45° F.

Note that, in the case of the pool heater, no supplementary electric-strip heaters are used. Heat-pump pool heaters rely solely on the heat-pump operation to provide adequate pool heating in the absence of a backup. Because no second-stage heating is used, so to speak, when the ambient temperature drops to approximately 17° F, the heat-pump system will fail to operate. This is why backup systems are required if the system is located outdoors and the pool is to be used year-round.

Heat-Pump Spa Heaters

The same air-to-liquid technology is often used to heat spas. In many cases, depending on the size of the pool and spa, a single heat-pump system can be used to heat both. If the spa is connected to the pool where the spa has a spillover, or waterfall into the pool, the piping arrangement resembles that in Figure 13-19. If the spa and pool are separate units, a

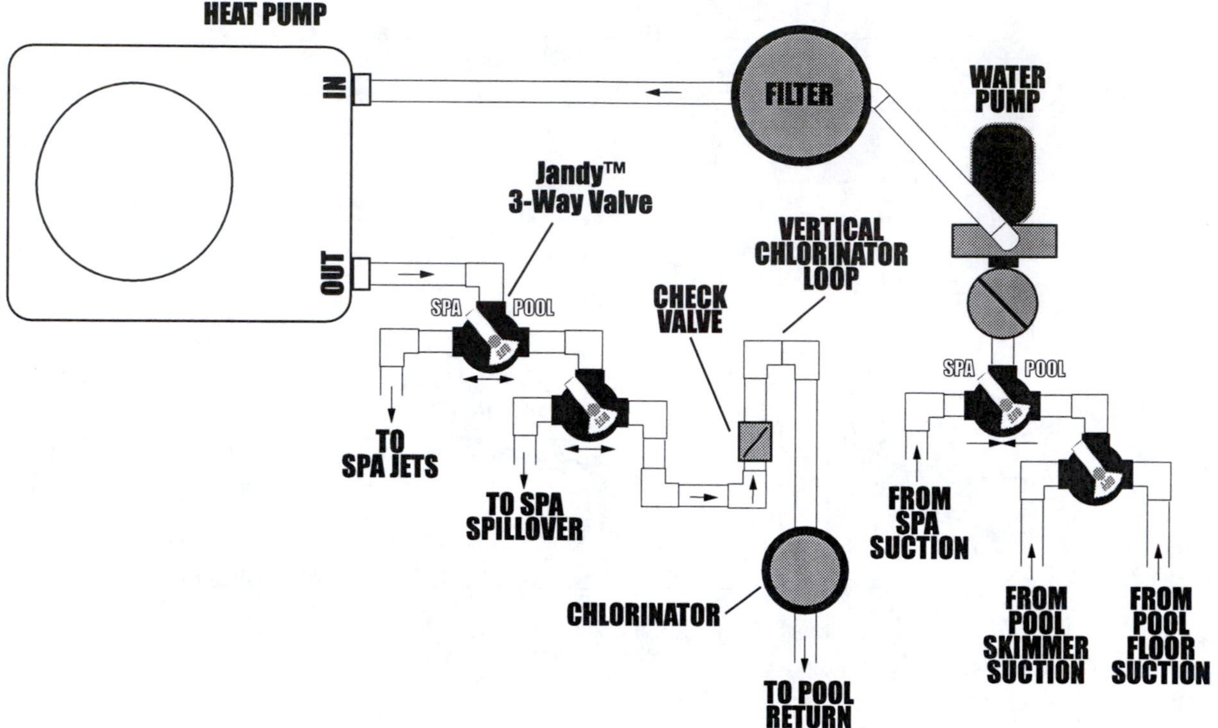

FIGURE 13-19 External piping arrangement for a combination pool/spa heating system. *(Courtesy Calorex U.S.A.)*

single unit can still be used to heat both the spa and the pool but the piping configuration (Figure 13-20) is somewhat different.

Heat-Pump Hot Water Heaters

When used for domestic hot water applications, a self-contained heat-pump system is tied into the domestic hot water piping system. The heat-pump water heater can be obtained without a tank (Figure 13-21) or as a complete unit (Figure 13-22). The tankless models are designed to be connected to an existing water tank, while the complete units are intended to replace existing tanks. When connected to an existing tank, the heat pump is located next to the tank and is connected to the tank via a hose arrangement (Figure 13-23). This hose arrangement is configured as a tube-in-tube plumbing tree that allows for colder water to enter the heat pump, while heated water is introduced back to the tank. The heated water from the heat pump is directed upward into the tank, as colder water from the bottom of the tank is permitted to flow into the heat pump. The heat pump is designed with a thermometer and a temperature control to maintain the desired water temperature in the tank. Self-contained units incorporate the heat pump and the tank in a single unit (Figure 13-22).

Installing Air-to-Liquid Heat Pumps

No matter which type of the air-to-liquid heat pumps is being used, the installation of the heat-pump end of the equipment is almost a no-brainer. The entire refrigeration system is self-contained. No field refrigerant piping is used. The unit often resembles a free-standing plug-in appliance. The water piping to the

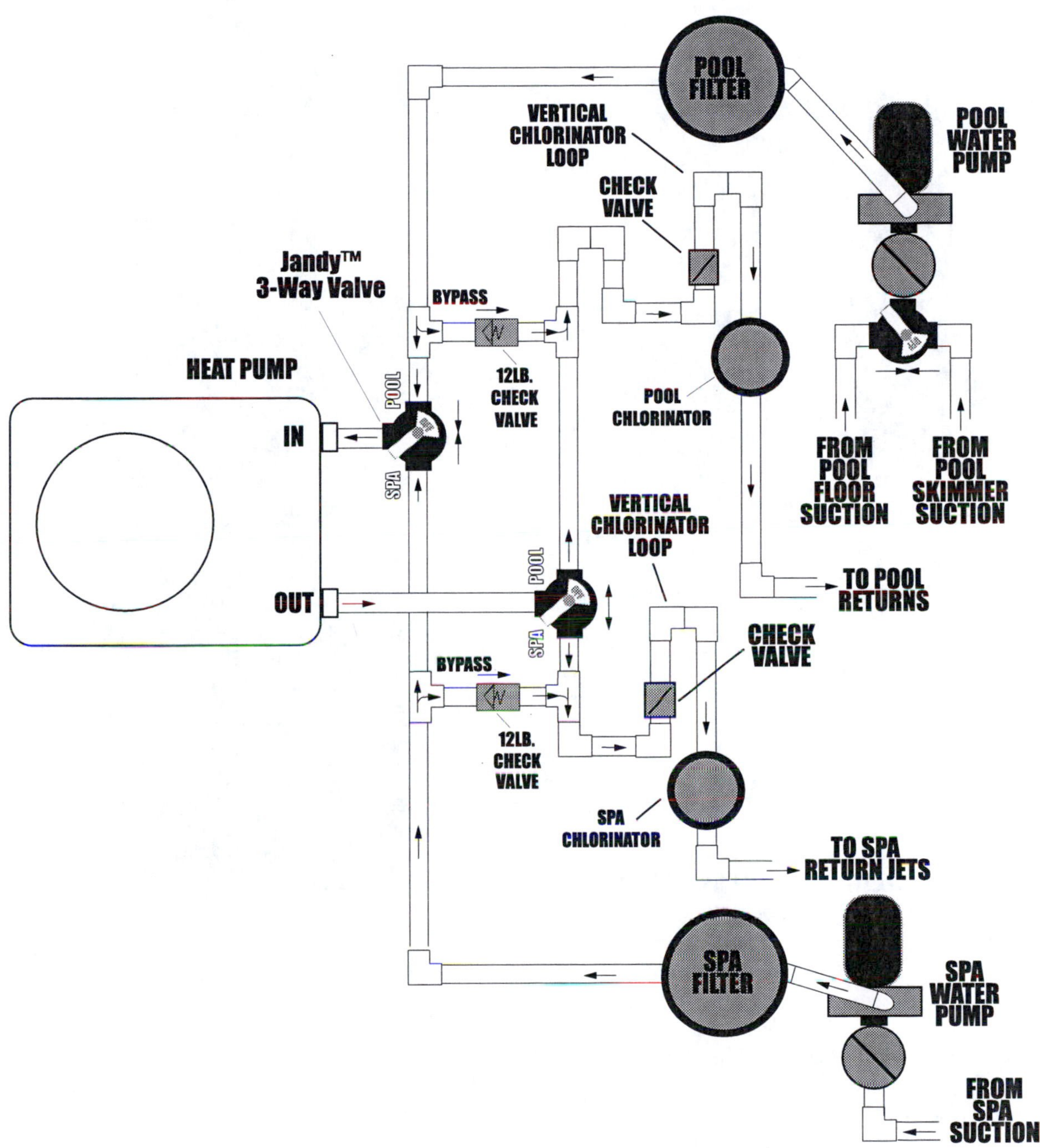

FIGURE 13-20 External piping arrangement for separate swimming pool and spa. *(Courtesy Calorex U.S.A.)*

FIGURE 13-21 Round heat-pump water heater without tank. *(Courtesy Energy Utilization Systems, Inc.)*

heat-pump unit is a different story, though. The heat-pump system must be tied into the existing water loops, be they domestic water or swimming pools and spas. Anyone installing a specialized piece of equipment, such as a pool heater, must follow all of the manufacturer's specifications closely.

LIQUID-TO-LIQUID HEAT-PUMP REFRIGERATION CIRCUITS

The liquid-to-liquid heat-pump system can be considered a cross between the liquid-to-air and the air-to-liquid systems. Like the liquid-to-air system, liquid is the source of heat for the system. Like the air-to-liquid system, the primary application of a liquid-to-liquid heat-pump system is to heat a liquid and is, therefore, generally not equipped with a reversing valve. The liquid-to-liquid system is used primarily to extract heat from a free source and transfer that heat to either a hydronic heating system or to domestic hot water.

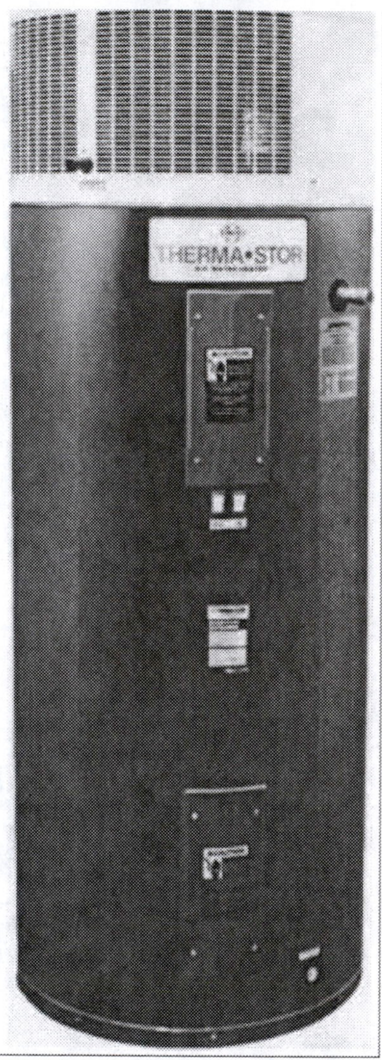

FIGURE 13-22 Self-contained heat-pump hot water heater. *(Courtesy DEC International, Therma-Stor Products Group)*

The most common application for the liquid-to-liquid heat pump is domestic hot water. Utilized in a manner similar to that of the air-to-liquid heat-pump water heater, the liquid-to-liquid variety is also piped to a water storage tank. The makeup water to the tank, along with cooler water from the bottom of the tank, is piped to the heat pump. After

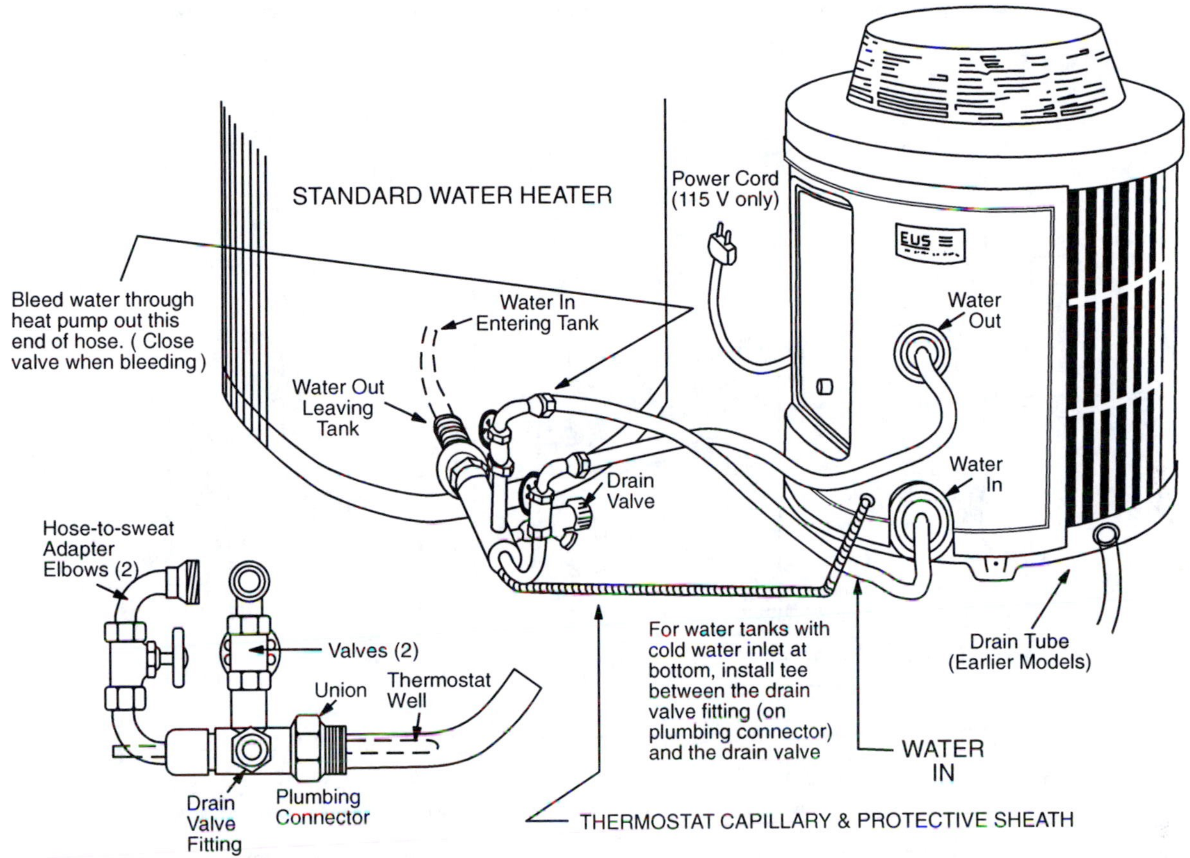

FIGURE 13-23 Piping diagram for an add-on heat-pump hot water heater

flowing through the high-side heat exchanger, the water is introduced to the water storage tank (Figure 13-24). The low-side heat exchanger is piped to the free heat source. Figure 13-24 also shows the refrigerant circuit, complete with solenoid valve, expansion valve, and two liquid-to-liquid heat exchangers.

Just as with the air-to-liquid hot water heat-pump system, manifolds have been specially designed to allow the use of a standard hot water heater tank in conjunction with the heat pump. These manifolds (Figure 13-25) reduce the amount of mixing between the cooler water at the bottom of the tank and the water just heated by the heat pump.

SUMMARY

- Heat-pump systems can be classified as air-to-air, air-to-liquid, liquid-to-liquid, or liquid-to-air systems.
- Liquid-source heat pumps utilize liquid as the heat source when operating in the heating mode.
- Air-source heat pumps utilize air as the heat source when operating in the heating mode.

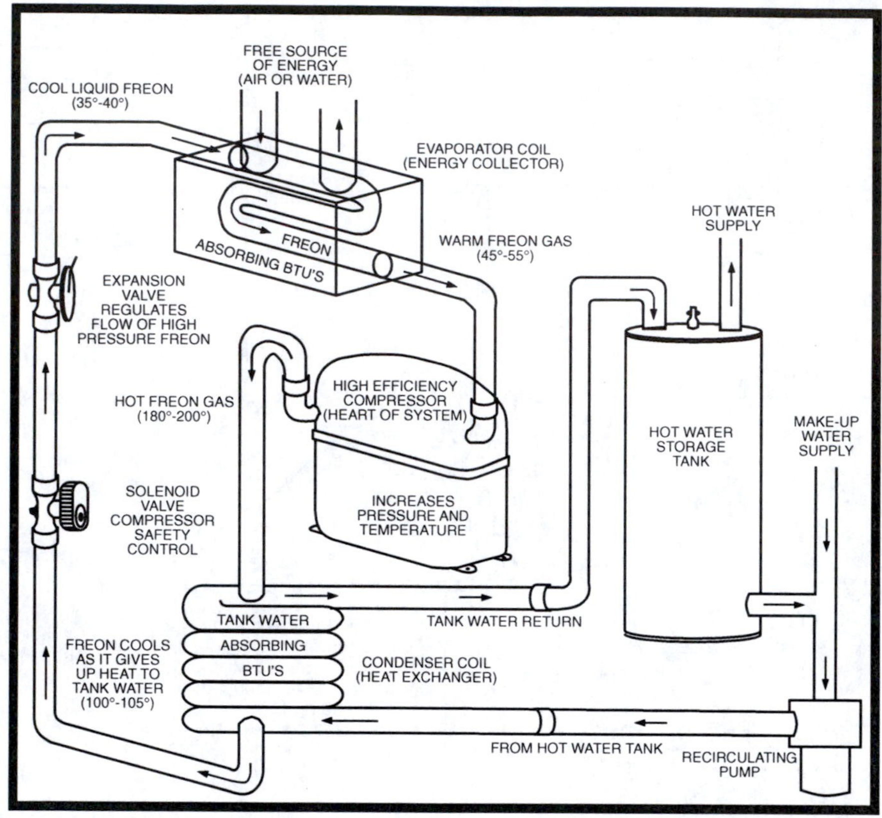

FIGURE 13-24 Piping diagram for a liquid-to-liquid heat-pump system

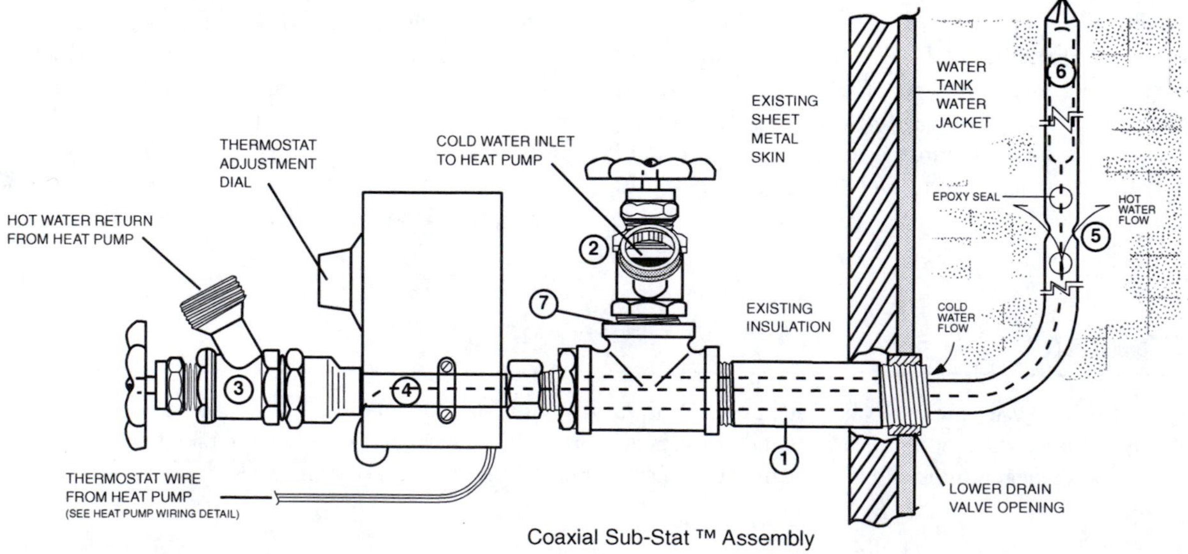

FIGURE 13-25 Coaxial thermostat and piping tree arrangement. *(Courtesy Friedrich Climate Master, Inc.)*

- Liquid-source heat pumps can use local municipal water supplies, cooling towers, lakes, ponds, or the earth as heat sources.
- Liquid-to-air and air-to-liquid heat pumps are usually configured as self-contained refrigeration systems.
- Liquid-to-liquid and air-to-air heat pumps are typically installed and configured in a manner similar to conventional air-cooled and water-cooled systems.
- Water-regulating valves are used to control the system pressures in both the heating and cooling modes of operation.
- Air-to-liquid heat-pump systems are most commonly used to heat pools, spas, and domestic water.
- Desuperheaters are often found on liquid-to-air heat-pump systems to help reclaim some of the heat that would normally be wasted.

KEY TERMS

Desuperheater
Heat reclaim
Heat sink
Heat source
Wastewater system

FOR DISCUSSION

1. What factors should be considered when the choice must be made between air-source and water-source heat-pump systems?
2. Discuss why proper cooling tower maintenance and water treatment are important to efficient system operation.

REVIEW QUESTIONS

1. Typically, a liquid-to-liquid heat-pump system utilizes:
   a. One tube-in-tube heat exchanger and one fin-and-tube heat exchanger.
   b. Two fin-and-tube heat exchangers.
   c. Two tube-in-tube heat exchangers.
   d. No heat exchangers are needed on a liquid-to-liquid heat pump.
2. Typically, an air-to-air heat-pump system utilizes:
   a. One tube-in-tube heat exchanger and one fin-and-tube heat exchanger.
   b. Two fin-and-tube heat exchangers.
   c. Two tube-in-tube heat exchangers.
   d. No heat exchangers are needed on an air-to-air heat pump.
3. Which of the following can be described as a liquid-source heat-pump system?
   a. One that uses water as the heat source during the heating mode
   b. One that uses water as the heat sink in the cooling mode
   c. One that uses water as the heat source in the cooling mode
   d. Both *a* and *b*
4. In a large urban city, the source of water for large liquid-to-air heat-pump systems is most likely:
   a. A cooling tower.
   b. The earth.
   c. The local municipal water supply.
   d. A pond or lake.

5. Leaves, bird waste, paper, and plastic debris are most likely to affect the operation of which of the following?
   a. A liquid-to-air heat-pump system
   b. An air-to-air heat-pump system
   c. An air-to-liquid heat-pump system
   d. A geothermal heat-pump system
6. Heat pumps used for heating pools, spas, and domestic water:
   a. Require the installation of a four-way reversing valve.
   b. Are self-contained refrigeration systems.
   c. Are never air-cooled.
   d. All of the above are correct.
7. On liquid-to-air heat-pump systems, a(n) __________ can be located between the compressor discharge port and the reversing valve to help reclaim heat that would otherwise be lost.
8. Explain why two water-regulating valves are often used on a liquid-to-air heat-pump system.
9. Water-regulating valves on a liquid-to-air heat-pump system:
   a. Prevent the system pressure from dropping below a predetermined set point.
   b. Prevent the system pressure from rising above a predetermined set point.
   c. Both *a* and *b* are correct.
   d. Neither *a* nor *b* is correct.
10. A liquid-to-liquid heat pump is equipped with a heat exchanger between the compressor and the reversing valve for heat-reclaim purposes. Technician A concludes that the installation of the heat exchanger decreases the efficiency of the system in the cooling mode. Technician B concludes that the installation of the heat exchanger will reduce the temperature of the refrigerant entering the condenser in the heating mode. Which of the following statements is most likely correct?
    a. Technician A is correct and Technician B is incorrect.
    b. Technician B is correct and Technician A is incorrect.
    c. Both Technician A and Technician B are correct.
    d. Both Technician A and Technician B are incorrect.
11. Consider a tube-in-tube refrigerant/liquid heat exchanger. Technician A states that, in order to maximize the heat-exchange rate, the refrigerant should be piped into the outer tube. Technician B states that, in order to maximize the heat-exchange rate, the refrigerant and liquid should flow through the heat exchanger in opposite directions. Which of the following statements most likely is correct?
    a. Technician A is correct and Technician B is incorrect.
    b. Technician B is correct and Technician A is incorrect.
    c. Both Technician A and Technician B are correct.
    d. Both Technician A and Technician B are incorrect.
12. Air that is discharged from an air-to-liquid heat-pump system is:
    a. Cooler than the temperature of the coil.
    b. Cooler than the temperature of the air entering the coil.
    c. Warmer than the temperature of the air entering the coil.
    d. The same temperature as the air entering the coil.

CHAPTER 14

Geothermal Heat Pumps

OBJECTIVES After studying this chapter, the reader should be able to:

- Describe a geothermal heat pump.
- Explain why geothermal heat pumps have the capability to operate efficiently year-round.
- Define a geothermal closed-loop system.
- Describe the vertical, horizontal, and parallel loop configurations found on closed-loop systems.
- List the advantages and disadvantages of using a closed-loop system.
- Define a geothermal open-loop system.
- Describe single-well, two-well, and dry-well configurations found on open-loop systems.
- List the advantages and disadvantages of using an open-loop system.

INTRODUCTION

When an adequate source of water or ground space is available, a geothermal heat-pump system may provide a viable alternative to air- or liquid-source heat pumps. In an air-source heat pump, the efficiency of the system depends directly on the temperature of the air that passes through the outdoor coil both in the heating and cooling modes of operation.

During periods of extreme temperatures, the air-source heat pumps fail to operate at their full potential—a time when peak efficiency is desired. Liquid-source heat-pump systems can also operate at reduced efficiency when the temperatures are out of the design range of the system. During periods of high temperature and humidity, the effectiveness of a cooling tower, for example, is greatly reduced. During the heating months, liquid must be heated in order to provide an adequate heat source for the system. This can be costly. This chapter concentrates on the geothermal heat-pump system and explains why the effectiveness and efficiency of this type of system remain relatively constant year-round.

OVERVIEW

Just as with the other types of heat-pump systems discussed in this text, the geothermal heat-pump system is a reverse-acting system that provides both heating and cooling, depending on the desired mode of operation. Configured in a manner similar to that of a liquid-to-air system, the geothermal heat pump uses a liquid/liquid heat exchanger as the source of heat during the winter months and as the heat sink during the warmer summer months. The main difference between the liquid-to-air heat pump and the geothermal system is that the temperature of the liquid flowing through the heat exchanger is maintained at a relatively constant temperature by the earth itself. Although the surface temperature of the ground changes with the seasons, the temperature of both the earth and the water some distance below the surface stays constant. This constant temperature allows the geothermal system to operate consistently during a wide range of weather conditions. This constant and unchanging operation results in an increase in the operating efficiency of the system. The term **energy efficiency ratio (EER)** is used to rate this efficiency with respect to other systems. The EER is calculated by dividing the Btu output of the system by the power input in watts. The higher this ratio is, the more efficient the system is. One drawback of the EER calculation is the fact that the energy used during system start-up and shutdown is not a factor in the calculation. To compensate for this, another ratio, the **seasonal energy efficiency ratio (SEER)** is used. The SEER takes the energy used during start-up and shutdown into consideration. Both the EER and the SEER are higher for heat-pump systems, especially geothermal systems, than for conventional heating systems.

The determination must be made as to whether the ground or the water will be used as both the heat source for winter operation and the heat sink for summer operation. These systems are classified as open-loop systems when water is used and closed-loop systems when the ground is used. A number of factors will determine which type of system is ultimately used. These factors include:

- Quantity of available water
- Temperature of available water
- Quality of available water
- Local building and environmental codes
- Available ground space

For an open-loop system to operate properly, an adequate supply of water must be available. The amount of water that is needed depends on the size of the system to be installed. Larger systems will require substantially more water than smaller systems. When an open-loop system is installed, a well must be installed so that the underground water source can be accessed. The required depth of the well, as well as the amount of water that will be available, can be estimated by a contractor that specializes in well drilling in that specific geographic area. An inadequate water supply will reduce the efficiency and heat-transfer capability of the system. Required flow rates typically range from 3 to 7 gallons per minute (GPM) per ton of refrigeration. Areas that experience extreme temperatures and climates will normally require a greater flow rate. Manufacturers supply required flow-rate information based on system capacity, geographic location, and supply water temperature. The temperature of the groundwater varies with the geographic location. Figure 14-1 shows a map of the groundwater temperatures at depths of 50 to 150 feet.

The quality of the water is also a major factor that must be considered. The water that can be used in a geothermal system does not necessarily have to meet the standards set forth for human consumption, but guidelines are established, nonetheless. Factors that must be considered include:

- Calcium content
- Magnesium content
- pH level
- Hydrogen content
- Chlorine content

A water test performed by the local health department can determine if the available water meets the manufacturer's requirements. If the water is safe for human consumption, it most certainly meets the requirements for heat-pump applications. Note that tying a geothermal heat pump to a potable water source will in no way affect the quality of the water.

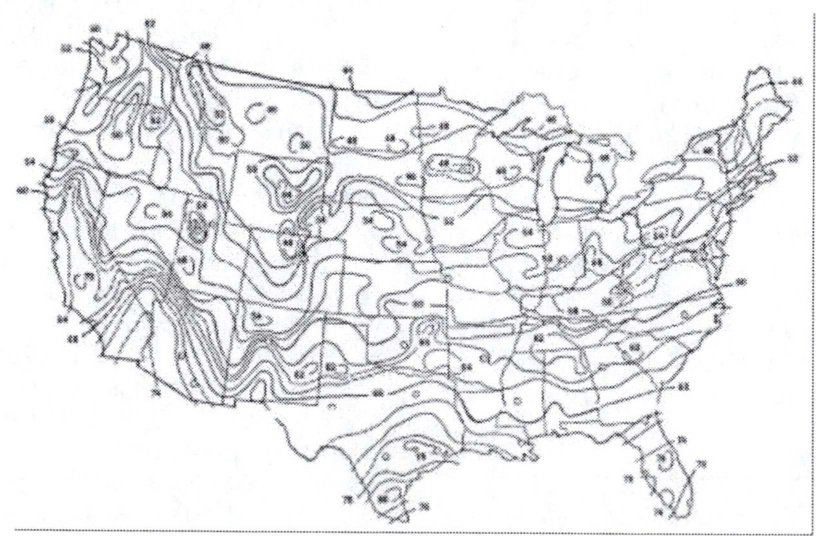

FIGURE 14-1 Groundwater temperatures (°F) in wells ranging from 50 to 150 feet deep. *(Courtesy Mammoth Corporation)*

The water that is reintroduced into the water source will vary only in temperature. The quality of the water is not changed. Even in the unlikely event that a leak should develop within the water/refrigerant heat exchanger, no health hazard will result. Refrigerants and oils used in heat-pump systems are stable, nontoxic, and noncorrosive and meet UL standards.

Two other factors that need to be considered are local codes and available ground area. Some localities do not allow the installation of an open-loop system. In these instances, a closed-loop system should be used. Closed-loop systems, however, require a sufficient amount of land space for the burial of the piping circuit.

CLOSED-LOOP SYSTEMS

Closed-loop systems, or **earth-coupled systems,** are designed so that the water or the antifreeze solution in the system does not come in contact with the earth or the water in the earth. Antifreeze should be used in closed-loop systems that have a groundwater temperature below 60° F. A closed-loop system is made up of a plastic piping arrangement that is buried in the ground. The piping is referred to as either a **ground loop** or a **water loop.** The water or antifreeze solution flows through this piping arrangement and either absorbs heat from or transfers heat to the surrounding earth. The liquid is circulated through the piping by means of a centrifugal pump. Ideally, once filled, the ground loop will remain filled. The liquid in the loop absorbs heat from the ground and transfers it to the refrigerant in the heating mode (Figure 14-2) and absorbs heat from the refrigerant and transfers it to the ground in the cooling mode (Figure 14-3). Depending on the available space and the depths that can be reached in the earth, these ground loops can take on a number of different configurations, including:

- Vertical
- Horizontal
- Parallel

No matter which ground loop configuration is used, good practice requires performing a standing pressure test on the piping circuit before covering the loops with soil.

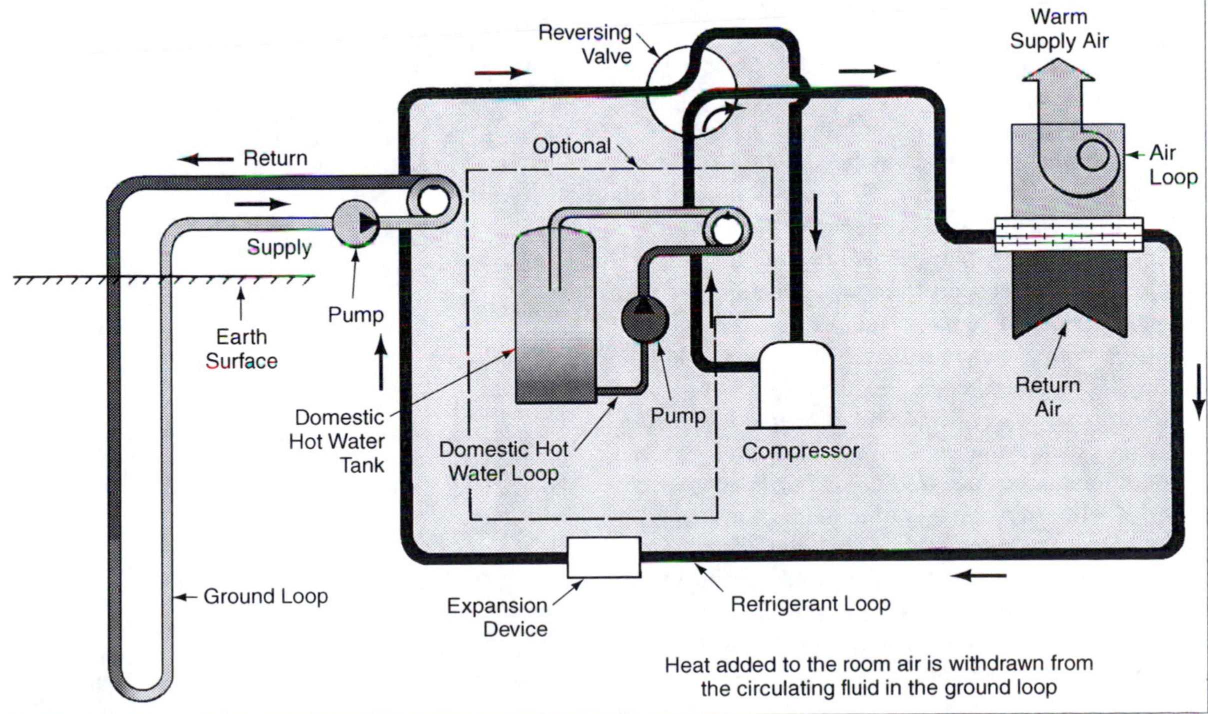

FIGURE 14-2 Closed-loop, water-source heat pump in the heating mode. *(Courtesy Oklahoma State University)*

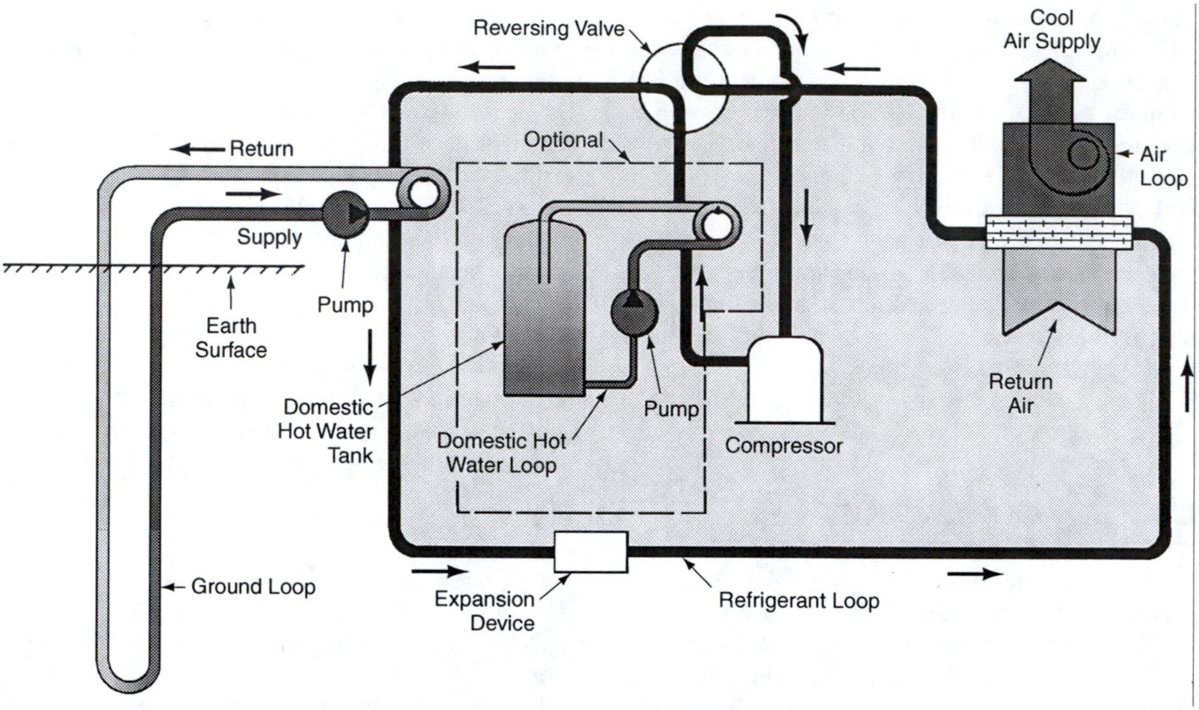

FIGURE 14-3 Closed-loop, water-source heat pump in the cooling mode. *(Courtesy Oklahoma State University)*

Vertical Configuration

In areas where the amount of ground space is small, **vertical ground loops** are desirable. Vertical holes are drilled into the ground to a depth of approximately 150 feet. The actual number of holes that need to be drilled is determined by the capacity of the system and the earth material encountered during the drilling process. The holes should be drilled approximately 10 to 15 feet apart from each other and should be approximately 5 to 6 times wider than the diameter of the piping material used. Generally speaking, 200 to 400 feet of piping are required for each ton of system capacity. A typical vertical ground loop configuration is shown in Figure 14-4. This arrangement is configured with three vertical U-bends connected in series with each other. During the heating season, the cool water solution enters the ground loop and absorbs heat from the soil surrounding the loop (Figure 14-5). This heat is then transferred to the refrigerant, which, in turn, is transferred to the air passing over the condenser.

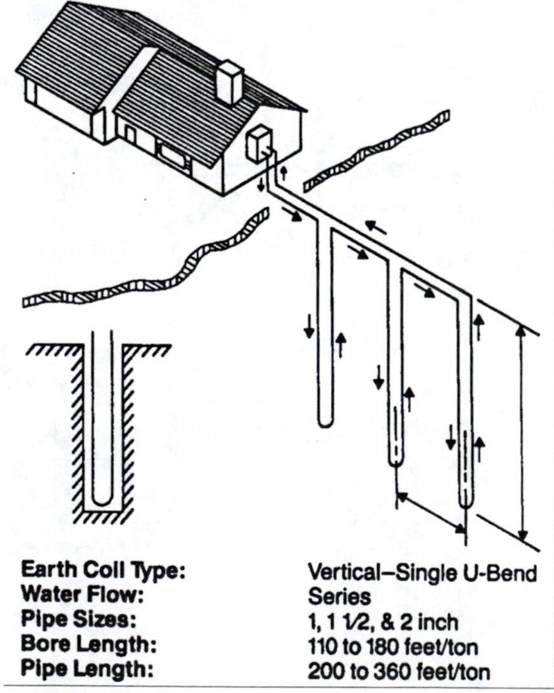

Earth Coil Type: Vertical–Single U-Bend
Water Flow: Series
Pipe Sizes: 1, 1 1/2, & 2 inch
Bore Length: 110 to 180 feet/ton
Pipe Length: 200 to 360 feet/ton

FIGURE 14-4 Vertical, series ground loop configuration. *(Courtesy Mammoth Corporation)*

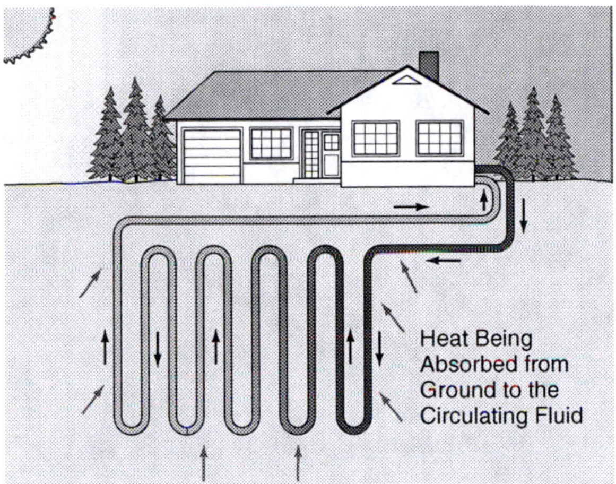

FIGURE 14-5 Vertical, series ground loop absorbing heat from the soil in the heating mode

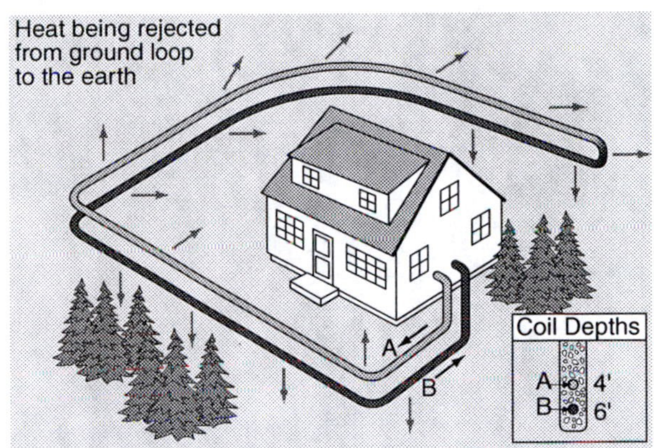

FIGURE 14-7 Two-layer, horizontal ground loop in the cooling mode

Horizontal Configuration

In areas where drilling deep holes is not possible because of rock and other obstructions, the **horizontal ground loop** configuration (Figure 14-6) is popular. Because these piping configurations are closer to the surface of the ground, longer piping runs are needed. In this case, roughly 300 to 500 feet of piping are required for each ton of system capacity. These systems are ideal where a lot of ground surface area is available.

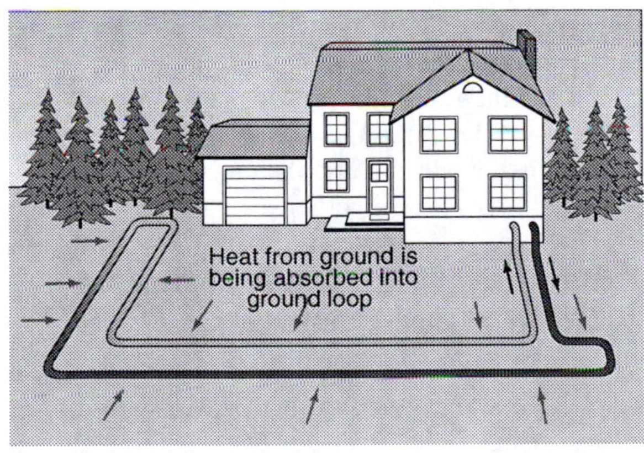

FIGURE 14-6 Single-layer, horizontal ground loop

The main drawback to this type of configuration is the fact that the length of the trench that must be dug is equal to the length of the required piping. This can become costly and is, therefore, recommended for use on smaller systems.

In an effort to reduce the cost of digging the trench and in areas where limited ground surface area is available, a modified horizontal loop configuration can be utilized. In this configuration, two layers of piping are installed, one on top of the other (Figure 14-7). This cuts the amount of trenching that must be done in half. These systems require approximately 20 percent more piping per ton than the single horizontal loop. Ideally, the lower pipe will be at least 6 feet below grade and the upper and lower pipes should be approximately 2 feet apart.

Parallel Loops

The configurations described so far are all configured as single-loop systems. In single-loop systems, the water flows through all of the piping material with each pass through the system. This can result in a large amount of resistance when larger systems are installed. In an effort to reduce this resistance, **parallel ground loops** are often used. These loops can be installed either vertically (Figure 14-8) or horizontally (Figure 14-9).

Because of the resulting reduced resistance, smaller pumps can be used to move the liquid

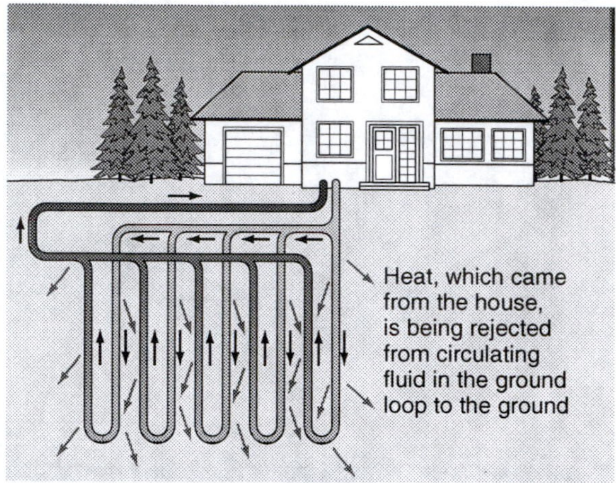

Heat, which came
from the house,
is being rejected
from circulating
fluid in the ground
loop to the ground

FIGURE 14-8 Vertical, parallel ground loop configuration in the cooling mode

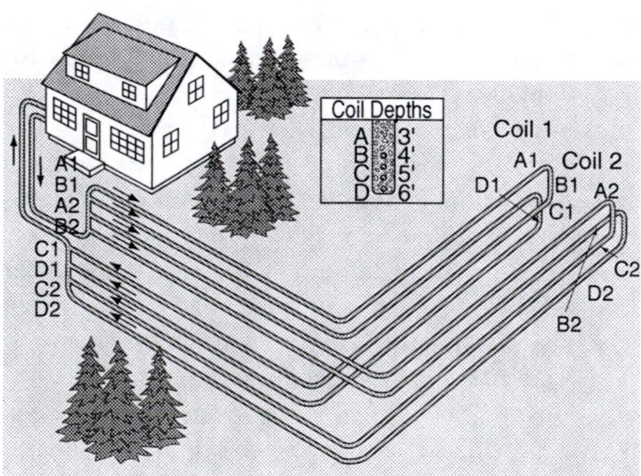

FIGURE 14-9 Four-pipe, horizontal ground loop configuration

through the piping circuit. The one main drawback with a parallel system is that, if one of the branches becomes blocked or clogged, it is extremely difficult to clear. On the four-layer horizontal configuration shown in Figure 14-9, the pipes should be 12 to 18 inches apart from each other to ensure proper heat transfer between the liquid in the pipe and the surrounding soil. One minor drawback of parallel loop systems is that they require somewhat more piping material than series ground loop configurations.

FIGURE 14-10 Spiral-type ground loop

Other Configurations

Depending on the restrictions posed by the ground material and the availability of a nearby water source, closed-loop systems may take on a configuration different from those just discussed. For example, if limited ground area is available for a horizontal loop system, as well as limited drilling capability, a spiral loop system (Figure 14-10) may be the answer. The spiral configuration requires that a large hole be dug—as opposed to a narrow trench. The hole should be as deep as possible to accommodate the required amount of piping. The pipes should also be as far apart as possible to maximize the rate of heat transfer between the liquid and the soil.

In areas where a nearby water source is available but open-loop systems are not permitted, a ground loop installed in the lake or pond may be the answer (Figure 14-11). Since the ground loop is sealed, no actual contact occurs between the liquid in the piping circuit and the water source. Although the water temperature will be slightly less than with an equivalent earth loop, the reduced installation cost will compensate for the small loss in efficiency.

Advantages and Disadvantages of Closed-Loop Systems

No matter what type of system is chosen, there are always pros and cons. Following are lists of the advantages and disadvantages of choosing a closed-loop system and a comparison of different closed-loop systems.

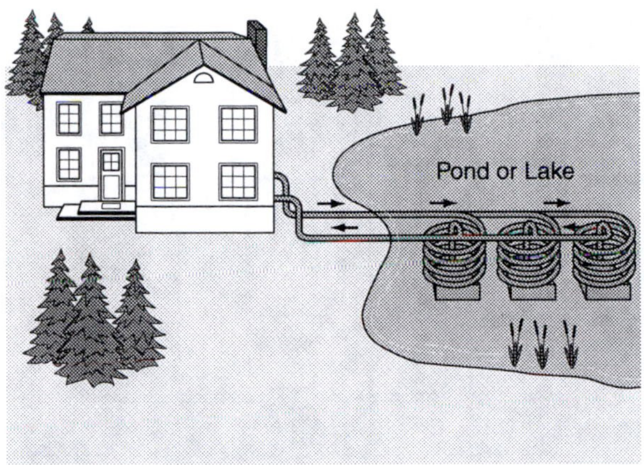

FIGURE 14-11 Ground loop installed in a pond or lake

Advantages of selecting a closed-loop system:

- Reduced scaling and corrosion occur.
- Systems are not affected by drought conditions.
- Fouling of the liquid piping circuit is eliminated.
- Maintenance of the liquid circuit is minimal.
- No chemical treatment of the water circuit is needed.
- Water tests are not required to ensure compliance with manufacturer's specifications.

Disadvantages of a closed-loop system:

- Water leaks are difficult to locate.
- Blockages in parallel loops are difficult to clear.
- Initial installations can be more time-consuming when compared to open-loop systems.
- Initial installation can result in more damage to finished lawns and yards when compared to open-loop systems.

Comparing different closed-loop systems:

- Horizontal loop systems require more trenching than vertical loop systems.
- Horizontal loops require more available ground area than vertical loops.
- Blockages are cleared more easily on series loop systems than on parallel loops.
- Parallel loop systems require smaller pumps than series loop systems.
- Parallel loop systems offer less resistance to water flow than series loop systems.
- Vertical loops require deeper penetrations into the earth than horizontal loops.
- Spiral loops can be installed in areas that have little ground surface area and that have a difficult earth material to penetrate.
- Loops installed closer to the ground surface require more linear feet of piping per ton than those installed deeper in the ground.
- Trapped air is easier to remove from a series loop system than from a parallel loop system.
- The heat-transfer rate per foot of piping is greater in a series loop system.
- Larger-diameter pipe is needed in series loop systems than in parallel loop systems.
- The installation costs are higher on series loop systems than on parallel loop systems.

OPEN-LOOP SYSTEMS

Unlike ground loop systems in which the liquid is constantly recirculated within the closed system, open-loop systems rely on a constant water source. This source is typically a well system. Just as with the ground loop system, heat is transferred from the water to the refrigerant in the heating mode, as shown in Figure 14-12(a), and from the refrigerant to the water in the cooling mode, as shown in Figure 14-12(b). As mentioned at the beginning of this chapter, the water that is used in an open-loop system must meet certain requirements as indicated by the equipment manufacturer. The main factors that apply to open-loop systems are:

- Water quantity
- Water quality
- Temperature of the water
- Local governmental and environmental codes

These were addressed earlier, and each should be evaluated carefully before a system is installed to ensure not only that the system will operate correctly but also that installing an open-loop system is legal. The three common types of open-loop systems are the:

- Dedicated, single-well system
- Dedicated, two-well system
- Dedicated, geothermal well system

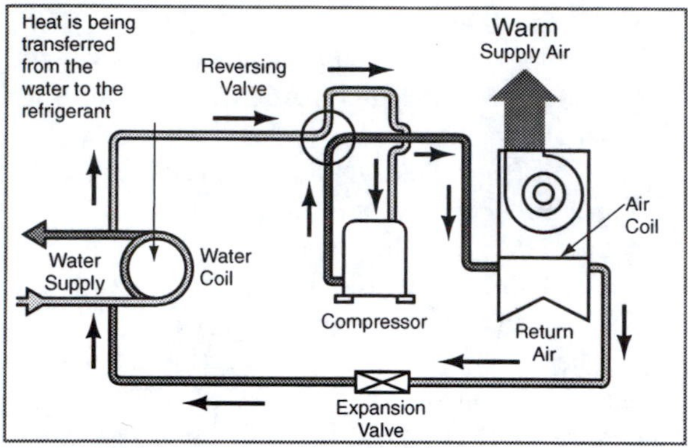

In the heating mode, hot refrigerant flows through the air coil supplying warm air to the conditioned space.

(a)

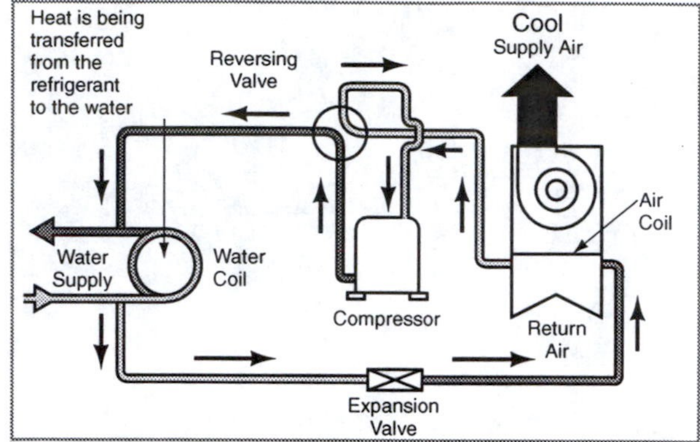

In the cooling mode, cold refrigerant flows through the air coil supplying cool air to the conditioned space

(b)

FIGURE 14-12 Open-loop, water-source heat pump in (a) the heating mode and (b) the cooling mode. *(Courtesy Mammoth Corporation)*

Dedicated, Single-Well Systems

In a dedicated, single-well system, the source of heat for winter operation, as well as the sink for heat in summer operation, is water that is pumped to the system from a well. The well is drilled to a depth required to locate groundwater. Most wells are in the range of 125 to 150 feet, although, in some locations, groundwater can be reached with a well only 75 feet deep. Once the well is drilled, it must have a lining installed to prevent it from caving in. This casing is often made of PVC pipe but can be made of steel. PVC is preferred in most cases due to the ease of joining sections and due to the fact that PVC is much lighter than steel. The well casing must extend below the natural level of the groundwater.

A submersible pump is located toward the bottom of the well, and the electrical connections are made at the top of the well, under the cap. Connected to the pump is a discharge pipe that carries the water being supplied to the heat-pump system. A cross-sectional view of a typical well is shown in Figure 14-13.

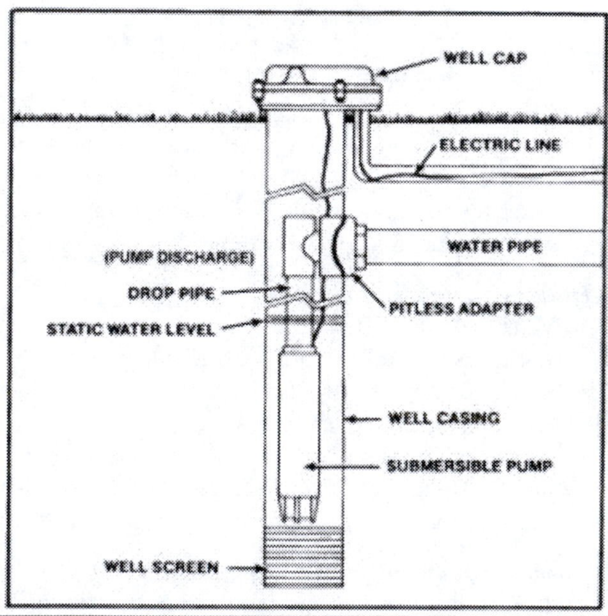

FIGURE 14-13 Basic drilled well. *(Courtesy Mammoth Corporation)*

If the pump operated to maintain constant water flow through the heat exchanger, it would never cycle off. A pump that operated continuously would require frequent service and repair. The location of the pump, unfortunately, does not make such frequent service an easy task. Retrieving the pump from the location would involve the removal of the connecting discharge pipe section by section as the pump was lifted from the well. After servicing, the piping would have to be reassembled as the pump was lowered again.

In an effort to reduce the run time of the pump, **pressure tanks** are often added to well systems. A pressure tank is simply a pressurized tank that stores water. The tank is equipped with an air bladder at the top that is pressurized. As the water is forced by the

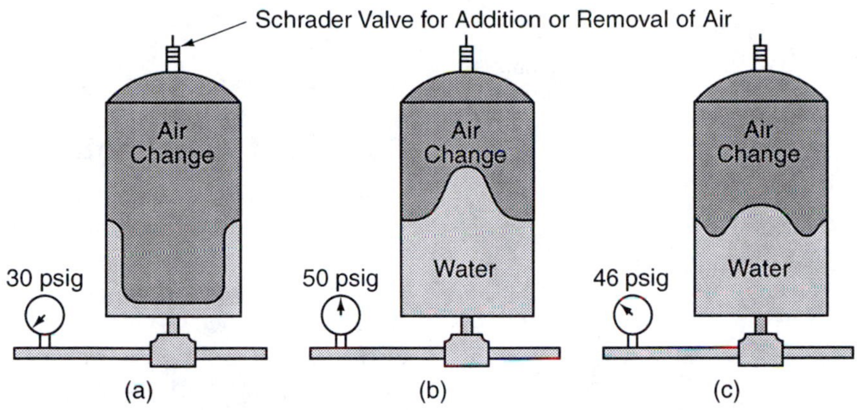

(a) Factory Air Charge

(b) Well pump has pressurized air charge and water pressure is 50 psig. The well pump will now shut off because pressure switch has opened on a rise in pressure.

(c) When water is used by the heat pump, the pressure in the air charge pushes water into the system. The pump stays off. The well pump only comes on when the pressure switch closes on a drop in pressure.

FIGURE 14-14 Cutaway view of a pressure tank

submersible pump up to the heat pump, it first flows into the pressure tank. As the amount of water in the tank increases, the pressure on the bladder increases (Figure 14-14). Once the pressure in the tank reaches a predetermined set point, a **pressure switch** de-energizes the pump. The water is then stored in the tank until needed by the heat-pump system. Once the pressure in the tank drops to a predetermined level, the pump is once again energized. The desired pressure level in the tank can be changed to either increase or decrease the cycle time of the pump.

The pressure tank also serves another important function in the heat-pump system. Since the tank is pressurized, the water in the heat exchanger is also kept under pressure. This is desirable because minerals that are present in water are more soluble under higher pressures. This will reduce the amount of scaling that occurs on the interior surfaces of the heat exchanger. Keeping the water pressurized also reduces the amount of air in the heat exchanger. A **slow-closing solenoid valve** installed in the water circuit at the outlet of the heat exchanger keeps the pressure in the heat exchanger at the same pressure as that in the pressure tank. The slow-closing solenoid is desired because it eliminates the **water hammer** created by standard solenoid valves. Water hammer is the loud hammerlike sound generated when a water valve closes immediately. Once the water flows through the heat exchanger, it is then discharged into a pond, lake, stream, or **dry well,** which is simply a large gravel-, sand-, and rock-filled hole in the ground (Figure 14-15). The discharge from the heat exchanger is pumped to the dry well where the water seeps through the sand and gravel to the **aquifer,** which is the underground water formation. A typical heat-pump setup with a dedicated single well is shown in Figure 14-16.

Dedicated, Two-Well Systems

Dedicated, two-well systems are similar to the dedicated, single-well system in that they use a submersible pump to supply water to the heat exchanger. The main difference between the one- and two-well systems is in the disposal of that water once it has left the heat exchanger. In the single-well system, the water is pumped to a lake, pond, or dry well. In a two-well system, the water is pumped into another drilled well, referred to as a return well (Figure 14-17). The return well should be at least as wide as the supply well to ensure that it can safely handle the water flowing into it. In addition, the return well and the supply

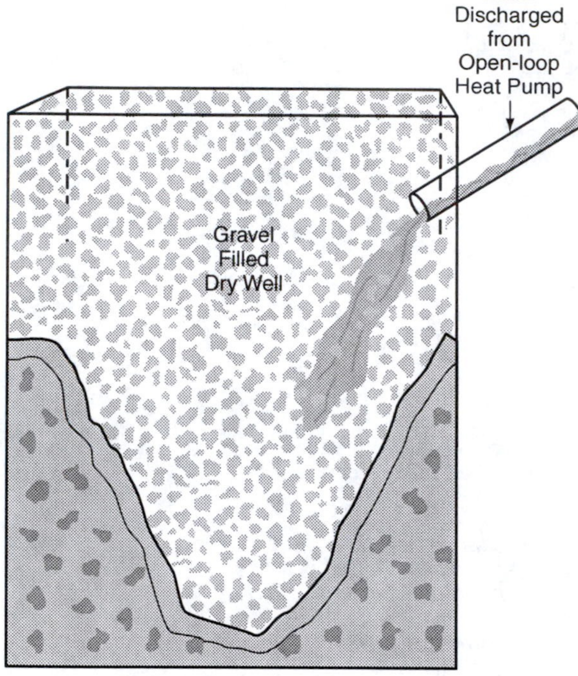

FIGURE 14-15 A dry well

well should be located as far apart as possible to prevent the **return water** from affecting the temperature of the **supply water.**

Dedicated, Geothermal Well Systems

Geothermal wells are commonly used when there is insufficient water to meet the system requirements. They are often used when the water source is in a **consolidated formation,** such as granite or sandstone. In a geothermal well, the return water that is discharged by the heat exchanger is sent back to the same well from which the supply water is taken. An ample supply of water should be in the well to reduce the possibility of the return water affecting the temperature of the supply water. The return water is returned to the bottom of the well, while the supply water is removed from the top portion of the aquifer but still below the **water table,** which is the level or surface of the water. A dedicated geothermal well is shown in Figure 14-18. Because limited water is available for these types of wells, dedicated geothermal wells are typically much deeper than standard dedicated, single- and two-well systems.

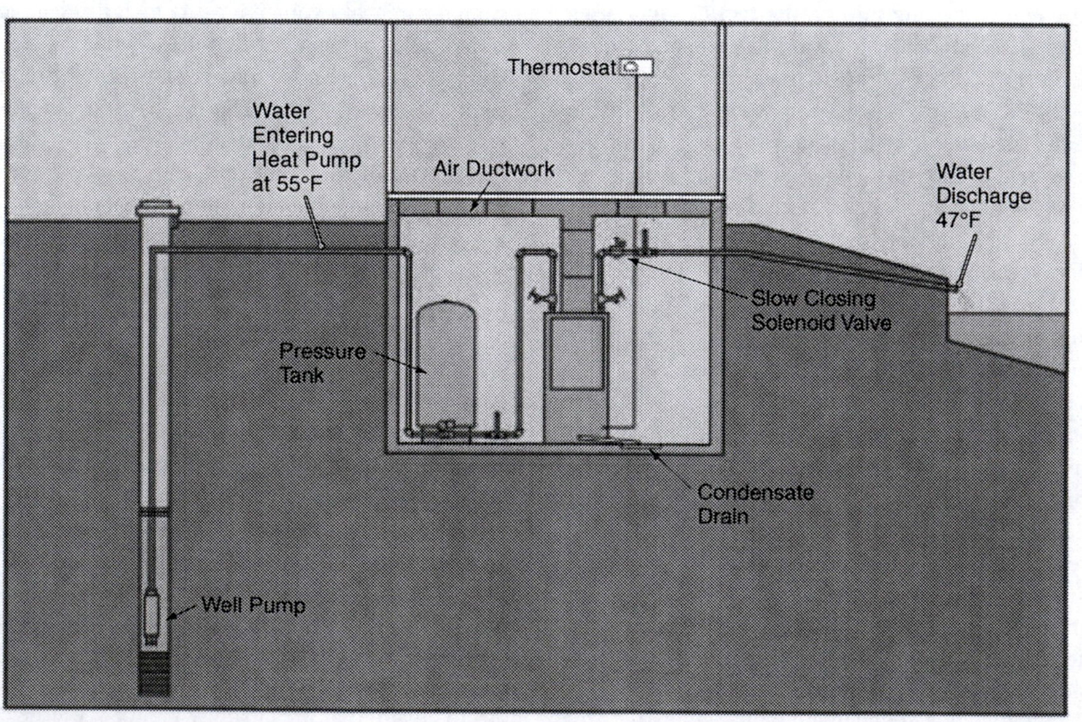

FIGURE 14-16 Dedicated, single-well system discharging return water into a pond or lake in the heating mode. *(Courtesy Mammoth Corporation)*

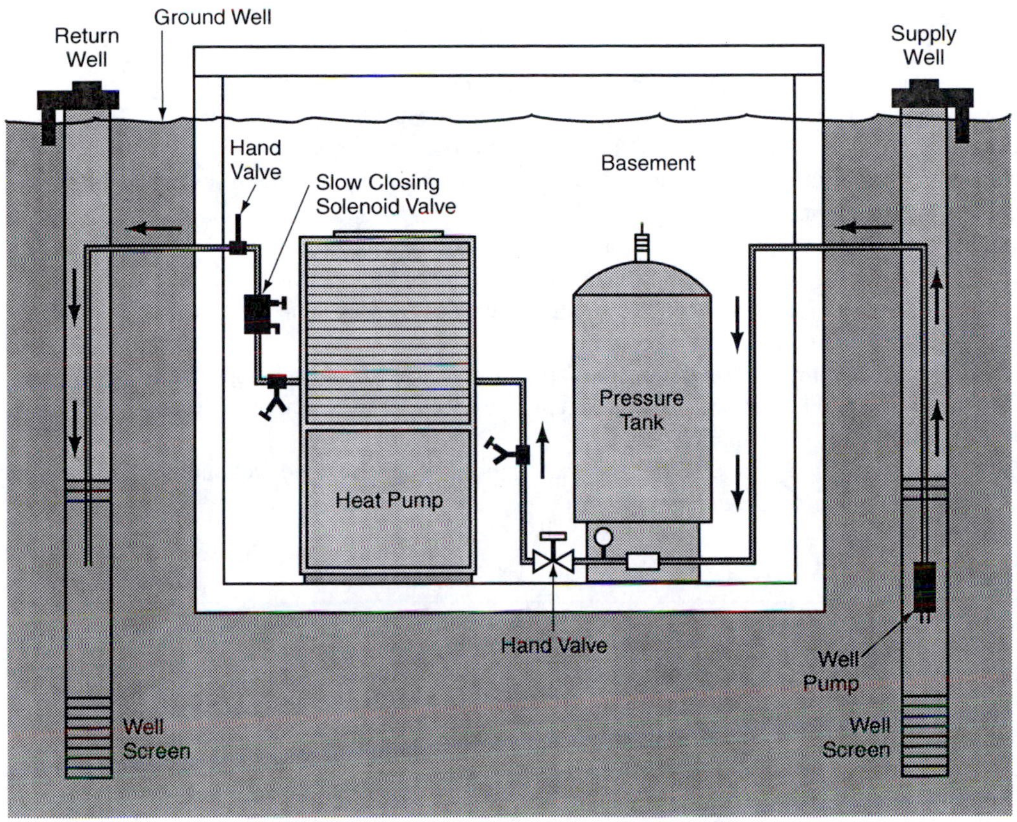

FIGURE 14-17 Dedicated, two-well system

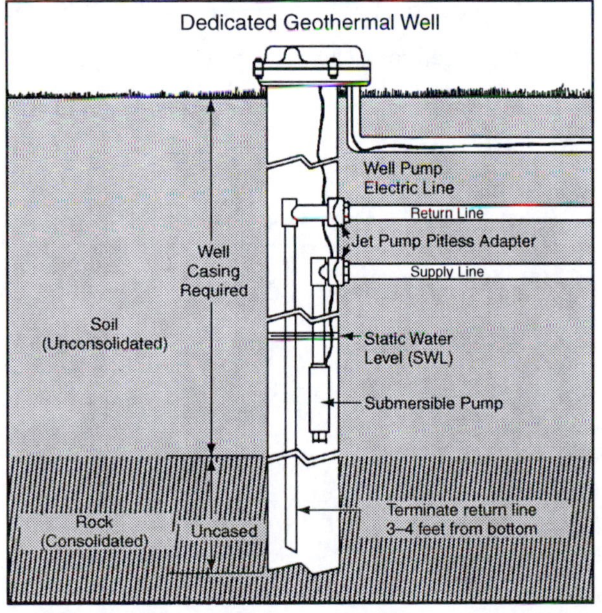

FIGURE 14-18 Dedicated geothermal well system. *(Courtesy Mammoth Corporation)*

SUMMARY

- Geothermal heat pumps provide consistent system performance year-round.
- Geothermal systems can be either closed-loop or open-loop systems.
- Closed-loop systems can be configured in horizontal or vertical patterns.
- Closed-loop systems recirculate the same water/antifreeze solution, eliminating water-quality problems that arise in open-loop systems.
- Ground loops can be piped in either series or parallel configurations.
- The liquid in a closed-loop system is completely isolated from the earth and the groundwater.
- Pressure tanks are commonly found on open-loop systems and are designed to store water for use in the heat exchanger.
- Open-loop systems require the use of a well to supply water to the heat exchanger.
- Water can be discharged to a pond, a lake, a stream, a dry well, or to another drilled well.
- Open-loop systems can be configured with either one or two wells.
- Geothermal wells have the supply and return water piped within the same well and are used primarily when the water source is in consolidated formations.

KEY TERMS

Aquifer
Consolidated formation
Dry wells
Earth-coupled systems
Energy efficiency ratio (EER)
Ground loops
Pressure switch
Pressure tanks
Return water
Seasonal energy efficiency ratio (SEER)
Slow-closing solenoid valve
Supply water
Water hammer
Water loops
Water table

FOR DISCUSSION

1. Under what conditions would a geothermal well system be ideal? A dedicated, two-well system?
2. Why might a particular installation call for a closed-loop system as opposed to an open-loop system?
3. From a service standpoint, which type of system, a ground loop or an open loop, would be more easily serviced in the event of a pump failure?

REVIEW QUESTIONS

1. In which of the following ground-loop systems would a blockage be most difficult to clear?
   a. Four-layer horizontal ground loop
   b. Vertical series loop
   c. Spiral loop
   d. Single-layer horizontal ground loop
2. Generally speaking, which state has the warmest groundwater temperature?
   a. Washington
   b. California
   c. New York
   d. Texas

3. How does the groundwater temperature relate to the rate of heat transfer?
   a. Low groundwater temperature results in a large heat-transfer rate in the heating mode.
   b. High groundwater temperature results in a large heat-transfer rate in the cooling mode.
   c. Low groundwater temperature results in a large heat-transfer rate in the cooling mode.
   d. High groundwater temperature results in a low heat-transfer rate in the heating mode.
4. A 3-ton heat-pump system is installed with a series vertical ground loop. Approximately how many feet of piping are required to achieve the desired rate of heat transfer?
   a. 200 feet
   b. 400 feet
   c. 900 feet
   d. 2000 feet
5. One of the main advantages of a parallel loop system is that:
   a. Blockages are easily cleared.
   b. Less piping material is used.
   c. A smaller pump is required.
   d. All of the above are advantages of a parallel loop system.
6. Antifreeze should be added to the water loop when the groundwater temperature is, on average, less than ________ °F.
7. Which type of system gets its supply water from the top of a water column, circulates this water through the heat exchanger, and then returns the water to the same water column?
   a. A dedicated, single-well system
   b. A dedicated, two-well system
   c. A dry well
   d. A dedicated, geothermal well system
8. Explain the operation and function of a pressure tank as used on an open-loop system.
9. Explain how heat is ultimately removed from a ground loop heat-pump system operating in the cooling mode. Include all three processes: air circuit, refrigerant circuit, and water circuit.
10. A dry well is typically filled with:
    a. Sand.
    b. Rocks.
    c. Gravel.
    d. All of the above are correct.
11. The return well on a dedicated, two-well system:
    a. Must be smaller in diameter than the supply well.
    b. Must be located as close to the supply well as possible to ensure adequate water flow between them.
    c. Both *a* and *b* are correct.
    d. Neither *a* nor *b* is correct.
12. When servicing a submersible pump in a dedicated, single-well system:
    a. The pump must be removed from the well for service.
    b. The discharge pipe on the pump must be removed and disassembled.
    c. Both *a* and *b* are correct.
    d. Neither *a* nor *b* is correct.

13. In a(n) ______________ system, the water that leaves the heat exchanger is discharged to a lake, pond, or stream.

14. When a severe drought occurs, which of the following system types will be affected least?
    a. A dedicated, geothermal well system
    b. A dedicated, two-well system
    c. A vertical series loop system
    d. None of the above systems will be affected by a severe drought.

15. Explain the purpose of a slow-closing solenoid valve at the outlet of the heat exchanger of an open-loop system.

CHAPTER 15

Troubleshooting Geothermal Heat Pumps

OBJECTIVES After studying this chapter, the reader should be able to:

- Explain how to identify water circuit problems in a geothermal heat-pump system.
- Explain how to identify a defective water pump.
- Explain how to determine if the water flow rate is sufficient.
- Explain how to identify airflow problems in a geothermal heat-pump system.
- Explain how to determine if the airflow through a geothermal heat-pump system is sufficient.
- Explain how to identify motor problems in a geothermal heat-pump system.

INTRODUCTION

A service technician who has had experience working on air-to-air and air-to-liquid heat-pump systems may be somewhat intimidated when encountering a geothermal heat-pump system for the first time. Rest assured that, even though the heat source/sink for the system may seem foreign, many of the same techniques used to troubleshoot those systems can be applied to geothermal systems as well. The refrigeration system on a geothermal system is identical to that of any other heat-pump system, including an air-to-air system, with the simple exception of the heat exchanger configuration on the outdoor section. The air side of the geothermal system is also identical to that of any other system. However, some differences exist that are addressed in this chapter.

OVERVIEW

Having covered a great deal of this text already, the reader has been exposed to a number of different troubleshooting techniques that can be used successfully in the proper evaluation and diagnosis of a wide variety of systems. The majority of these techniques are directly applicable in the case of the geothermal heat-pump system. Knowledge of refrigerant pressures, saturation temperatures, superheat, and subcooling measurements all come into play, as well as knowledge of mechanical and electrical theory. Troubleshooting a defective motor winding requires the same skills no matter where the motor is located. Identifying a grounded or shorted compressor is a skill that goes far beyond the bounds of air-to-air systems.

Troubleshooting geothermal heat-pump systems, although very similar in most respects to other heat-pump and conventional air-conditioning systems, entails some differences of which the servicing technician must be aware. These include:

- The technician should be able to take and interpret temperature readings at both the inlet and the outlet of the water/refrigerant heat exchanger.
- The technician should be able to take and interpret pressure readings at both the inlet and the outlet of the water/refrigerant heat exchanger.
- The technician should be able to take and interpret flow meter readings in the water circuit.
- The technician should be able to utilize flowcharts, similar to that in Figure 15-1, to estimate water flow in the event that a flow meter has not been installed on the system.

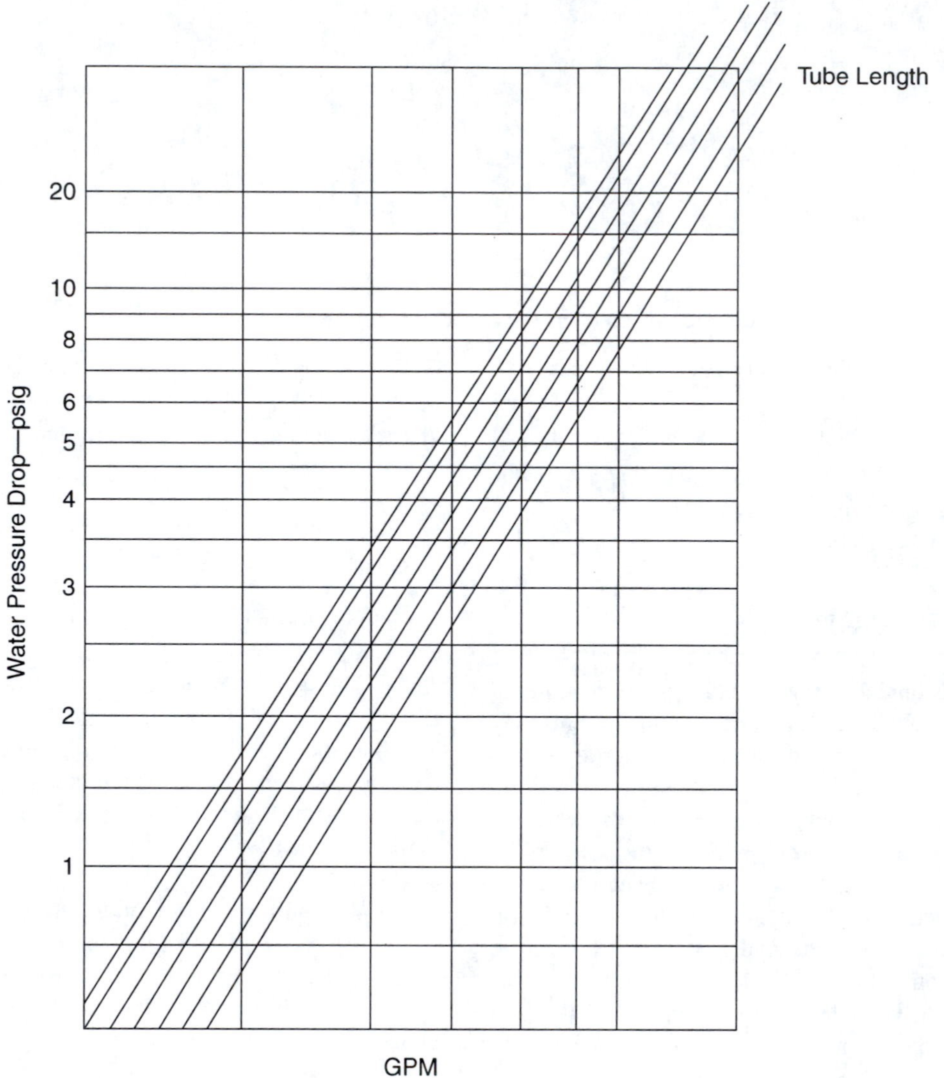

FIGURE 15-1 Representational chart showing the relationship between pressure drop and water flow in gallons per minute (GPM)

SERVICE NOTE: *Figure 15-1 is only representational and does not reflect actual and accurate data. Manufacturers supply these tables in the installation and service manuals that accompany new systems. Since heat exchangers vary from one manufacturer to the next, data also vary from system to system.*

- The technician should be able to interpret manufacturer's guidelines regarding proper water flow through the water/refrigerant heat exchanger.
- The technician should be able to determine approximate groundwater temperatures from a chart or table similar to Figure 14-1.

WATER CIRCUIT PROBLEMS

Just as with liquid-to-air heat-pump systems that are connected to cooling towers, the operating effectiveness and efficiency of geothermal heat-pump systems rely on a water circuit that provides an adequate flow rate through the heat exchanger. Improper water flow will result in reduced system efficiency heat transfer. On all liquid-to-air heat pumps, including geothermal systems, both a liquid/refrigerant heat transfer and a refrigerant/air heat transfer take place. The refrigerant/air heat transfer is exactly the same, no matter what type of air-conditioning system is involved. The water/refrigerant heat transfer, though, can be affected by four major factors:

- Water leaks
- Defective water pumps
- Mineral deposits
- Improper water flow

Water Leaks

Since both heat exchangers on geothermal heat-pump systems must operate properly in order for the system to function correctly, the water circuit must be properly maintained. However, on occasion, water leaks do occur and the technician's job is to quickly and accurately diagnose the system problem. A wrong diagnosis on a geothermal heat-pump system can result in a great deal of lost time and lost money. Having to unnecessarily dig up a ground loop, for example, would definitely not make the customer or the service company owner very happy.

OPEN-LOOP-SYSTEM WATER LEAKS. Water leaks on an open-loop system are generally not as big a problem as they are on ground loop systems. On open-loop systems, the supply water pump is submerged below the water table in the supply well. Leaks on the suction side of the pump are, therefore, almost nonexistent, if they exist at all. For the most part, all water leaks on the water circuit of an open-loop system are on the supply side of the pump. The possible locations of water leaks on the supply side of the system are:

- In the drop pipe connecting the pump to the underground water line
- In the water line leading to the pressure tank
- In the water line between the pressure tank and the heat exchanger
- In the heat exchanger itself
- In the line leaving the heat exchanger

When a water leak is present in the drop pipe or in the underground water line, chances are that the leak will never be noticed, let alone repaired. There is no way that the customer would know that a leak exists, and a service technician would not look for a leak here. There are no symptoms that would indicate a leak in those locations. The submersible pump will continue to operate until the pressure in the pressure tank has reached the preset cutout pressure. If, because of a small leak on the line leading to the pressure pump, the pump takes an extra minute or two to reach the desired pressure, nobody will notice. However, if the leak becomes larger, the pump will not be able to build up the desired pressure in the tank. Then, the lines and possibly the pump will need to be evaluated.

A small leak between the pressure tank and the heat exchanger can easily be spotted. This line is generally run directly between the two pieces of equipment in full view. Routine inspection should reveal any leaks in that relatively short run of piping.

Leaks within the heat exchanger are very difficult to diagnose. The first symptom that a leak exists is the loss of system refrigerant. A refrigerant leak can be located anywhere in the system, and a heat exchanger leak should only be suspected after the rest of the system has been checked thoroughly. If the rest of the system has been checked carefully for leaks, the best way to check the heat exchanger for leaks is to remove it from the system. Once the heat exchanger has been removed from the system, one of the refrigerant lines should be **brazed** closed. This is best accomplished by brazing a piece of copper tubing to the tube on the heat exchanger, crimping it closed, and then brazing it closed (Figure 15-2). A Schrader valve should be brazed to the other line on the heat exchanger, and nitrogen should be introduced to the heat exchanger

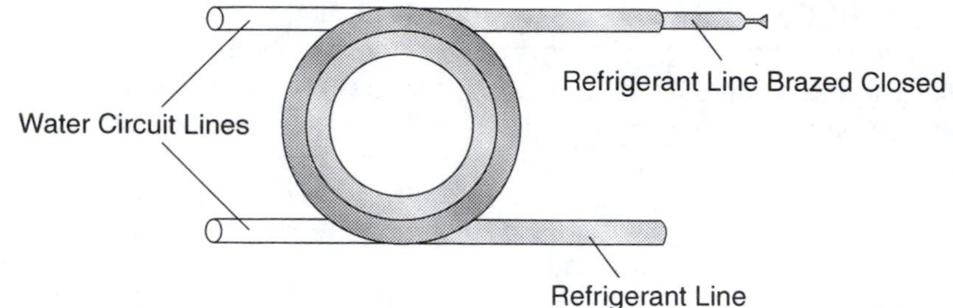

FIGURE 15-2 Heat exchanger removed from the system with one refrigerant port brazed closed

(Figure 15-3). A **standing-pressure test** is, in effect, being performed on the heat exchanger. If the pressure in the heat exchanger begins to drop, the exchanger has a leak and should be replaced. The water should also be tested to make certain that it is within the manufacturer's guidelines for pH, calcium, magnesium, and chloride levels. The water may have played a role in the deterioration of the heat exchanger.

If a water leak is present in the line leaving the heat exchanger, once again it will not affect the system operation but can cause a flood on the floor. As with the line leading to the heat exchanger, a leak at the outlet of the heat exchanger will become evident during routine maintenance. As can be seen, a leak on an open-loop heat-pump system will generally not affect system operation unless the leak is located in the heat exchanger. Leaks should

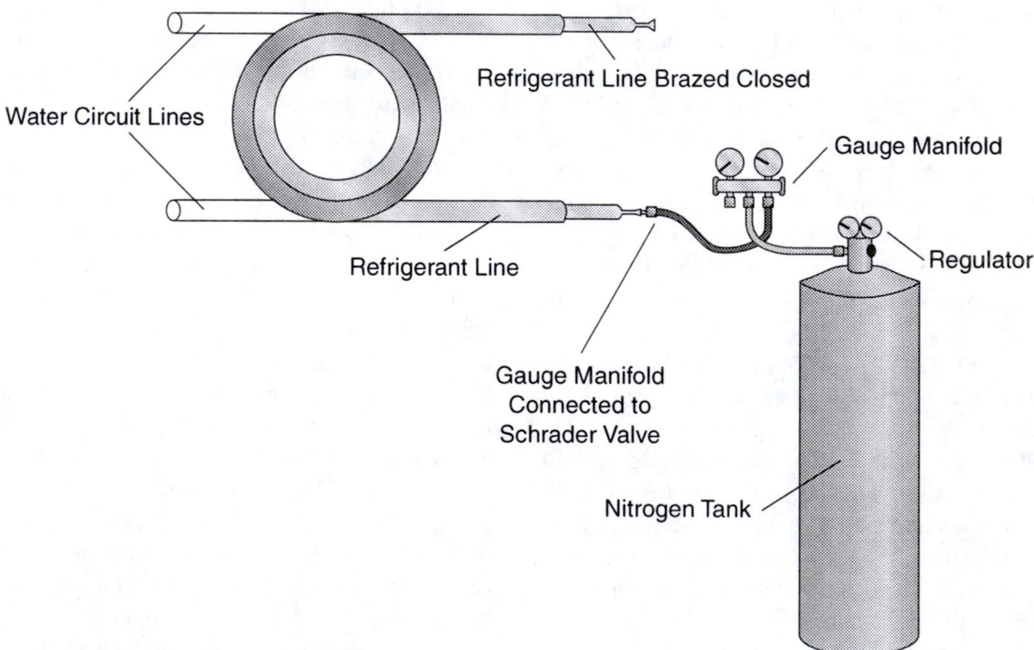

FIGURE 15-3 The heat exchanger is pressurized with dry nitrogen to locate a potential leak.

be treated and repaired just as any other plumbing leak would be.

CLOSED-LOOP-SYSTEM WATER LEAKS. Water leaks on closed-loop systems tend to pose much larger problems than those found on open-loop systems. The majority of the water piping circuit is buried deep underground, and a leak in the ground loop can lead to a costly repair. Luckily, though, the majority of water leaks on ground loop systems occur at the circulating pump or the flanges that connect it to the system. A number of telltale signs indicate that a water leak is present in the water circuit. They are:

- An audible crackling sound at the circulating pump
- A low temperature differential of the water across the heat exchanger in both the heating and cooling modes
- Low suction pressure in the heating mode
- High head pressure in the cooling mode
- Low pressure drop across the heat exchanger
- Low flow rate through the heat exchanger

Probably the sign that is most indicative of a water leak on a ground loop system is the audible crackling sound at the pump. This is a sure sign that air has entered the water circuit. A low temperature differential across the heat exchanger indicates that the heat-exchange rate is lower than it should be. This can result from:

- Water flowing too fast through the heat exchanger
- High water temperature in the summer
- Low water temperature in the winter

The first possibility can be ruled out immediately unless the pump has been recently replaced with a larger, higher-volume pump. In this case, the pump should be replaced with one that meets system requirements. The second and third possibilities go hand in hand. If the ground loop is shorter than it should be, the water will not have enough time to pick up heat in the winter and reject heat in the summer. This would result in lower suction pressure while operating in the heating mode and higher head pressures when operating in the cooling mode. Assuming that the system was designed properly, the ground loop should be the correct length. The impression of an undersized ground loop can easily occur on parallel ground loop systems where one of the parallel loops has become inoperable due to an air restriction. An air restriction in the loop indicates that a leak is present in the system. Air restrictions can be remedied relatively easily if a manifold-type valve arrangement is installed on the system. Strategically placed valves and hose bibs can help easily clear away an air restriction (Figure 15-4). Under normal operation, all valves are in the open position and the hose bibs are closed. Each parallel loop can be isolated and have water pushed through it to alleviate any air pockets that may be present (Figure 15-5). In Figure 15-5, all valves are closed with the exception of the one supplying water to the first loop. The hose bib on the first loop is open, and any air is pushed through the loop and out through the hose bib.

Another possible reason for high head pressure in the cooling mode and low suction pressure in the heating mode is the loss of system water. Less water in the loop means a lower heat-transfer rate in the heat exchanger. A loss of system water is indicated by reduced water flow through the exchanger and a low pressure drop across the heat exchanger. Technicians should be able to take both pressure and temperature readings at the inlet and outlet of the exchanger. The most commonly used method to obtain these readings is with the aid of a **Pete's port** (Figure 15-6). A Pete's port is a self-sealing access port that enables the technician to take pressure and temperature readings at both ends of the heat exchanger. Pressure gauges or thermometers can be inserted into the port without having to use any tools whatsoever. Simply pushing the pressure gauge or thermometer into the port causes the rubber gasket to shift, giving access to the water circuit (Figure 15-7). Taking pressure and temperature readings at both ends of the exchanger enables the technician to calculate **pressure drops,** temperature differentials, (previously discussed) and flow rates. The piping arrangement of a typical ground loop system is shown in Figure 15-8. This arrangement gives the technician access to the ground loop by means of hand valves installed between the ground loop and the heat exchanger.

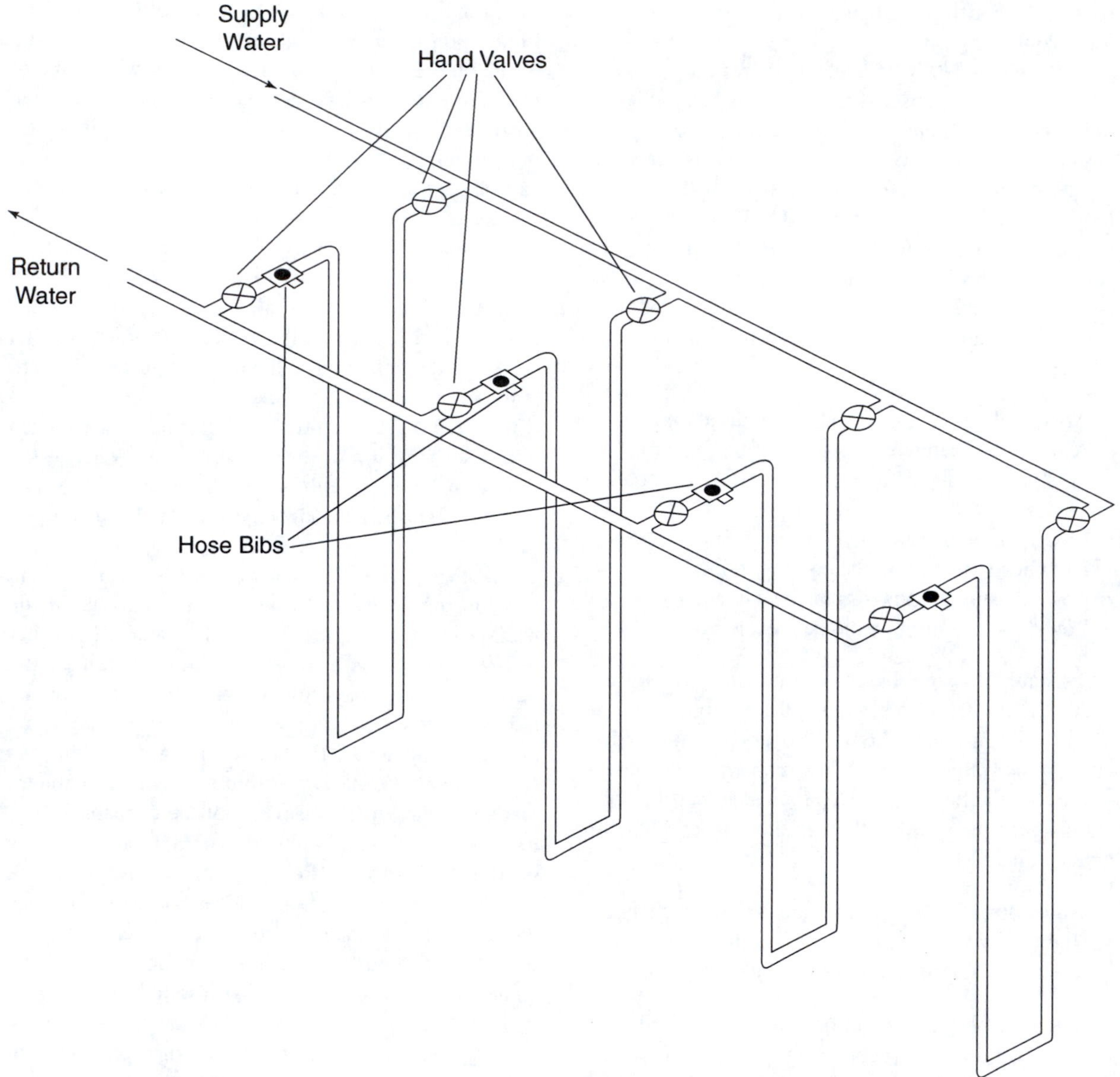

FIGURE 15-4 Manifold valving arrangement on a parallel loop system

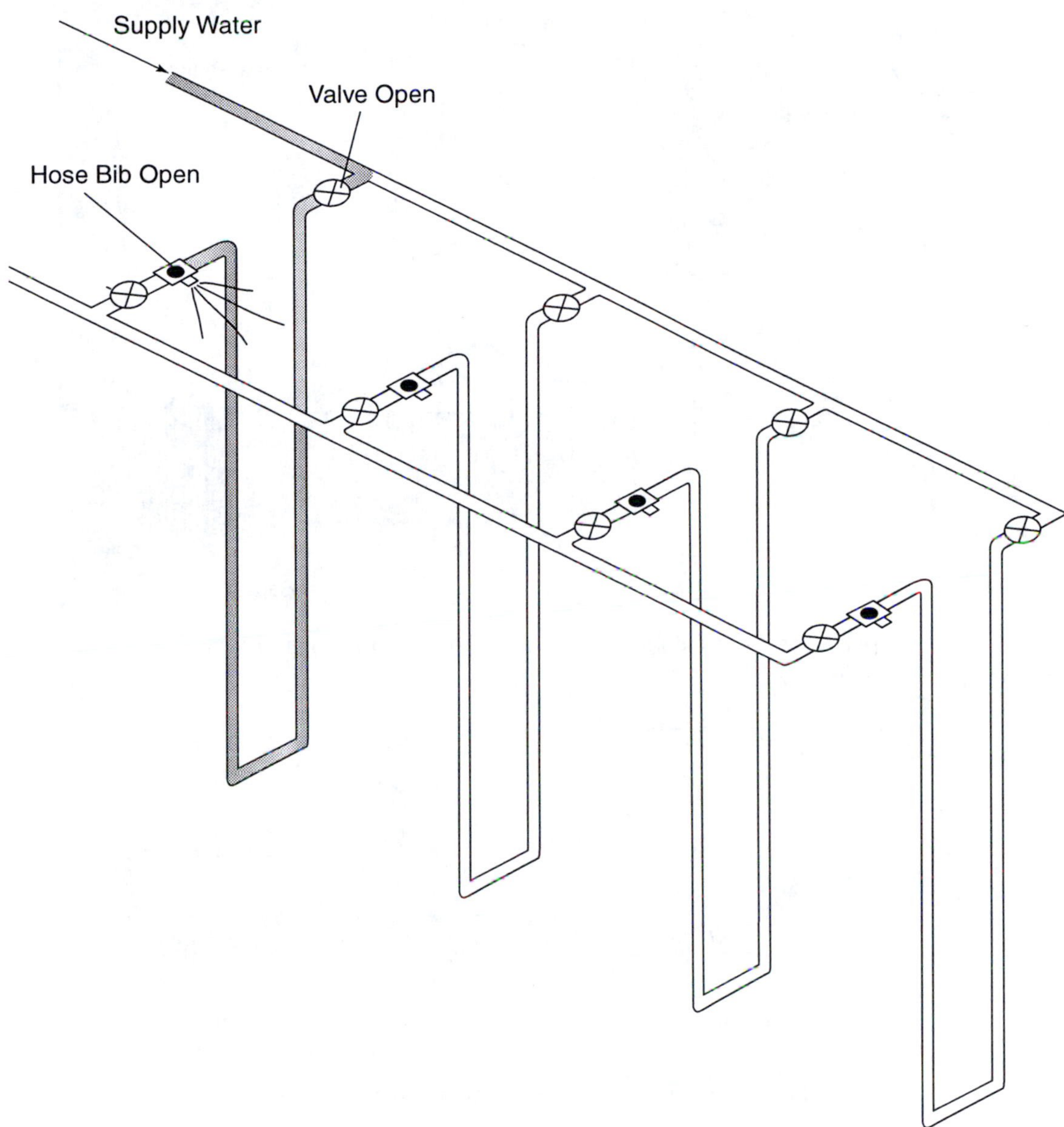

FIGURE 15-5 Water can be forced through any loop to remove air pockets.

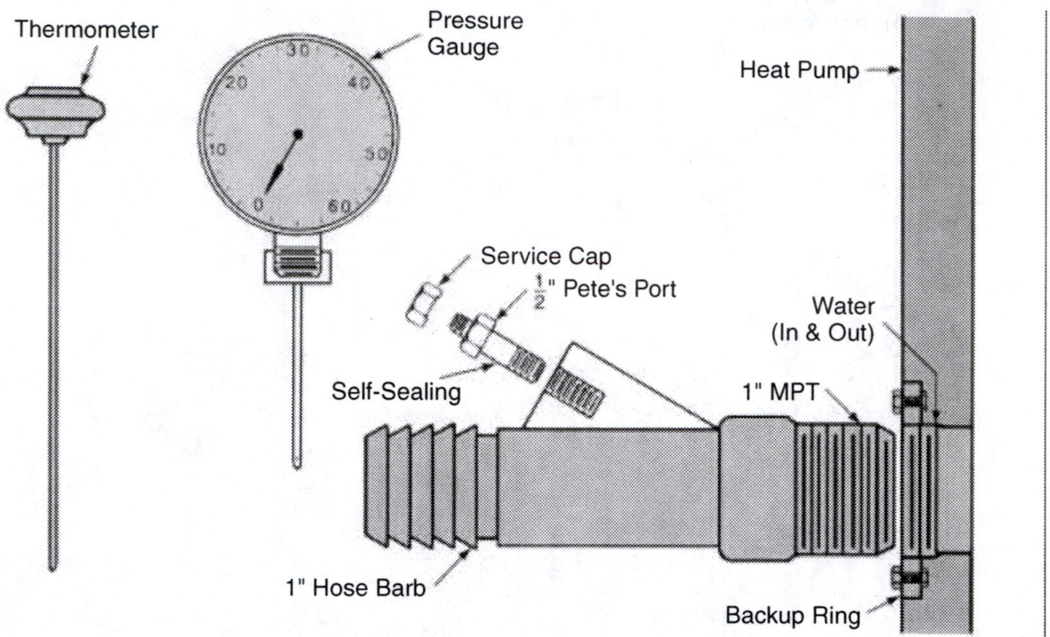

FIGURE 15-6 Pete's port used to obtain pressure and temperature readings in a ground loop. *(Courtesy Oklahoma State University)*

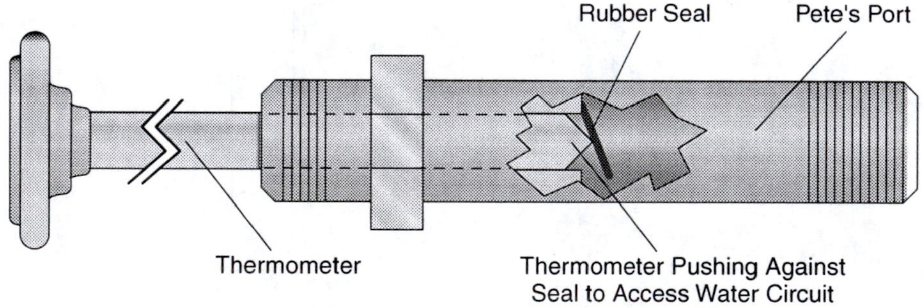

FIGURE 15-7 Cutaway view of a Pete's port access fitting

Note that any of the possible results of a leaking water circuit could also be the result of other system problems. A system operating with a low suction pressure in the heating mode does not automatically mean that the system has a water leak. When a water leak exists, more than one and often all of the symptoms appear. Before concluding that a leak exists, however, the technician must inspect the pump and make certain that it is operating properly.

Defective Water Pumps

Centrifugal pumps (Figure 15-9), commonly found on ground loop systems, are much the same type of pump as those found on domestic boilers. These pumps are comprised of the motor and the **impeller** (Figure 15-10). These two components are connected together with a spring-type linkage and a watertight seal. A defective motor does not necessarily mean

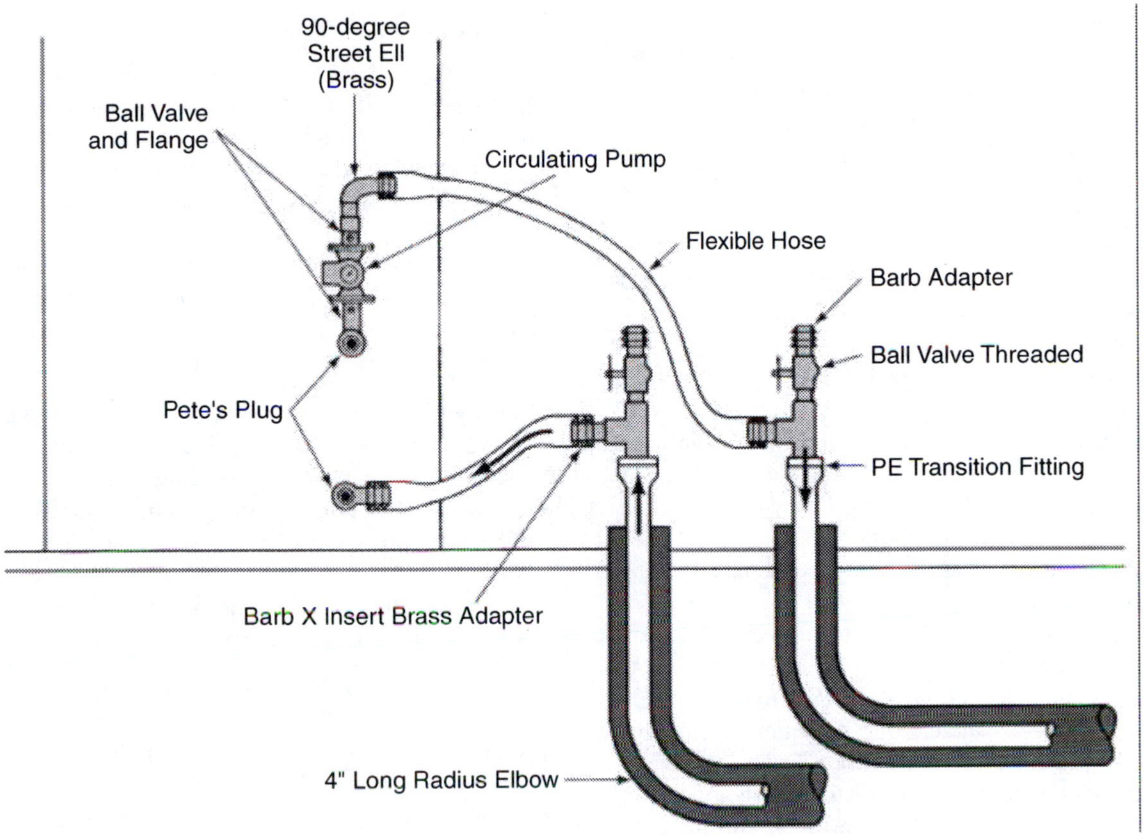

FIGURE 15-8 Piping hookup from a ground loop to the heat-pump system. *(Courtesy Oklahoma State University)*

FIGURE 15-9 Centrifugal pump. *(Courtesy ITT Fluid Handling Division)*

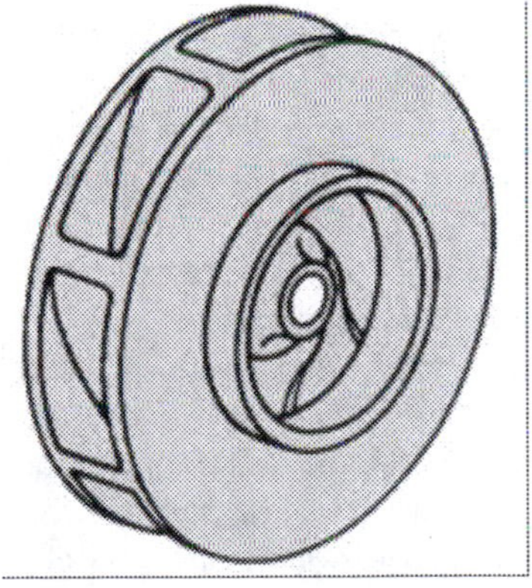

FIGURE 15-10 Pump impeller

that the impeller assembly is defective as well, and vice versa. The technician must be able to evaluate each of these components separately to avoid making an improper system diagnosis.

PUMP MOTORS. Just as with any other motor, pump motors can be evaluated both electrically and mechanically. Electrical problems usually involve open, grounded, and shorted motor windings. Equipped with a digital multimeter, a technician can check the motor for these conditions and diagnose accordingly. If the motor windings are found to be grounded, shorted, or open, the pump motor should be replaced. Mechanical problems include defective bearings and improper lubrication, which can lead to the motor jamming and not being able to start. When checking the motor, the technician should disconnect the impeller linkage from the motor shaft. This ensures that the motor is being evaluated and not the impeller or the linkage.

For example, consider the following scenario. A customer calls her service company and reports that her system is not cooling. Upon arrival, the technician notices that the circulating pump on the ground loop is humming but not turning. The technician concludes that the pump motor is defective and runs out to get a replacement. After replacing the motor and restarting the system, the new motor hums as well but does not turn. Talk about embarrassing! The problem is a jammed pump impeller, not the motor at all!

IMPELLER ASSEMBLIES. Impeller assemblies should also be evaluated with the motor removed from the linkage. The linkage and the impeller should turn freely with some resistance. An impeller that must be turned only with excessive force should be a prime candidate for replacement. The impeller and the linkage can be inspected and checked without having to access the water circuit directly.

If the pump motor operates correctly and the impeller seems to spin freely, ample water flow should occur through the water loop. If a flow meter registers no water flow at all, the impeller may indeed be broken inside the casing of the pump or the flow meter may be bad. How can the technician tell? If no pressure drop occurs across the heat exchanger, no flow is occurring through it and the pump is not moving any water through the piping circuit. If a pressure drop occurs across the heat exchanger and the flow meter indicates no flow, it is time to get a new flow meter.

Mineral Deposits

Mineral deposits on the interior of water system piping are practically nonexistent on ground loop systems due to their sealed nature. On the other hand, mineral deposits are a major concern on open-loop systems. Mineral deposits act as an insulator between the refrigerant and the water in the heat exchanger. The presence of mineral deposits can be diagnosed relatively easily. The following conditions are present when mineral deposits and scale have coated the interior surface of the water side of the heat exchanger:

- Low temperature differential occurs across the heat exchanger in both heating and cooling modes.
- Low suction pressure occurs in the heating mode.
- High discharge pressure occurs in the cooling mode.
- The pressure drop across the heat exchanger is normal.
- The water flow rate, in gallons per minute, through the heat exchanger is normal.

If mineral deposits are found to be present in the heat exchanger, the heat exchanger must be cleaned chemically, just as in the case of the water-cooled condensers that were introduced in an earlier chapter. Before adding chemicals to the heat exchanger, the technician should check the manufacturer's recommendations for cleaning. The heat exchanger should ideally be flushed and back flushed to remove all scale accumulation and mineral deposits within the coil.

SERVICE NOTE: *If mineral deposits form regularly within a heat exchanger, a water sample test may indicate that a water treatment plan may be in order. Water treatment is more cost-effective than having to remove and clean a heat exchanger every couple of years.*

Improper Water Flow

Improper water flow can be a result of some of the conditions already discussed in this chapter. The symptoms of improper water flow are:

- Small pressure drop across the heat exchanger
- Large temperature differential across the heat exchanger in both the heating and cooling modes

- Discharge pressure that is slightly higher in the cooling mode
- Suction pressure that is slightly lower in the heating mode
- Water flow rate through the heat exchanger is low

A small pressure drop across the heat exchanger is an indication that not enough water flow is occurring within the water circuit. This, teamed with a large temperature differential across the exchanger, indicates that the water is remaining in contact with the refrigerant for a longer period of time than normal. The water is picking up more heat from the refrigerant in the summer and giving up more heat in the winter.

Because the water is remaining in contact with the refrigerant longer, the rate of heat exchange is actually going to decrease. Consider that situation when the system is operating in the cooling mode. Cool water enters the heat exchanger and begins to absorb heat from the refrigerant. Because the water flow rate is low, the water remains in the coil for a longer period of time and its average temperature increases. A higher water temperature results in a reduction in the amount of heat transfer between the water and the refrigerant, because these temperatures are now closer together. This will result in a higher-than-normal head pressure in the cooling season. The same argument, only in reverse, can be made to explain the lower-than-normal suction pressure during the winter season.

AIRFLOW PROBLEMS

Airflow problems are the same regardless of what type of air-conditioning system is being evaluated. Insufficient airflow through a duct system can be a result of:

- A defective blower motor
- Dirty or blocked air filters
- A dirty or blocked indoor coil
- Closed supply registers
- A blocked return-air grill
- Loose duct lining
- Broken or loose belts on the blower
- A dirty squirrel-cage blower

These items are very important as far as system performance goes. A problem with the airflow through the indoor coil can result in system problems, or apparent system problems, with the other heat exchanger as well. Therefore, the technician must check all aspects of system operation before committing to a major system repair that could possibly have been avoided. To establish that flow exists through the air side of a heat-pump system, the preceding items must be checked. Now consider two scenarios in which airflow problems can lead a technician's troubleshooting efforts astray.

Cooling Example

Insufficient airflow through the indoor coil during the cooling mode will result in low suction pressure. This is a result of air coming in contact with the cooling coil for a longer period of time, decreasing the average temperature of the air passing over the coil. As the air temperature over the coil decreases, the suction pressure of the system will decrease as well. How will this affect the rest of the system?

Consider a ground loop system that operates normally with the following conditions in the cooling mode:

- Water supplied to the heat exchanger: 68°F
- Return water from the heat exchanger: 80°F
- System refrigerant: R-22
- Suction pressure: 65 psig
- Head pressure: 210 psig
- Water flow through the heat exchanger: 9 GPM
- Pressure drop across the heat exchanger: 2 psig

If the air filter becomes clogged with dirt, consider that the suction pressure now drops to 55 psig. This lower suction pressure will result in a lower head pressure of 195 psig. The condenser saturation temperature of the system has dropped from 105°F to 100°F. The temperature of the water entering the heat exchanger remains at 68°F. The temperature differential between the refrigerant and the water has become smaller, resulting in a lower heat-transfer rate. As a result of this, the temperature of the water leaving the heat exchanger drops from 80°F to 76°F. The normal temperature differential across the water circuit was 12 degrees (80°F – 68°F) but is now only 8 degrees (76°F – 68°F). The pressure across the heat exchanger and the flow rate through the heat exchanger remain unchanged. This may lead a technician to conclude that the reduction in the temperature differential across the heat exchanger is due to mineral deposits on the interior surfaces of the exchanger when, in fact, the only problem is a dirty air filter.

Service technicians must evaluate all symptoms before making a final diagnosis. In the system just examined, the head pressure of the system dropped and the temperature differential across the heat exchanger dropped as well. If, indeed, mineral deposits were present in the heat exchanger, the head pressure would have risen, not dropped!

Heating Example

Consider a ground loop system that operates normally with the following conditions in the heating mode:

- Water supplied to the heat exchanger: 68°F
- Return water from the heat exchanger: 58°F
- System refrigerant: R-22
- Suction pressure: 65 psig
- Head pressure: 210 psig
- Water flow through the heat exchanger: 9 GPM
- Pressure drop across the heat exchanger: 2 psig

Once again, the air filter has become clogged with dirt. The head pressure, which was normally 210 psig, now rises to 225 psig. This higher head pressure will result in a higher suction pressure of 75 psig. The evaporator saturation temperature of the system has increased from 38°F to 44°F. The temperature of the water entering the heat exchanger remains at 68°F. The temperature differential between the refrigerant and the water has become smaller, resulting in a lower heat-transfer rate. As a result of this, the temperature of the water leaving the heat exchanger increases from 58°F to 62°F. The normal temperature differential across the water circuit was 10 degrees (68°F – 58°F) but is now only 6 degrees (68°F – 62°F). The pressure across the heat exchanger and the flow rate through the heat exchanger remain unchanged. This may lead a technician to conclude that the reduction in the temperature differential across the heat exchanger is due to mineral deposits on the interior surfaces of the exchanger when, once again, the only problem is in fact a dirty air filter.

SUMMARY

- Technicians should be aware that most of their previously acquired troubleshooting skills will come in handy when working on geothermal systems.
- Water circuit problems can come in the form of water leaks, air restrictions, defective water pumps and motors, or mineral deposits.
- Water leaks affect ground loop systems more drastically than open-loop systems.
- One of the most obvious signs of a water leak on a ground loop system is an audible crackling sound at the pump.
- Ground loop water leaks often result in low pressure drops across the heat exchanger and a low flow rate through the heat exchanger.
- Pressure and temperature readings at a heat exchanger can be obtained quickly with a Pete's port.
- Pump motors and linkages should be evaluated separately to avoid errors in diagnosing a system.
- Mineral deposits in a water circuit can result in higher head pressures in the summer and lower suction pressures in the winterseason.
- Airflow problems can result in system conditions that resemble water flow problems.

KEY TERMS

Brazing
Impeller
Pete's port
Pressure drop
Standing-pressure test

FOR DISCUSSION

1. What are some similarities that exist between servicing an air-to-air heat pump and a geothermal heat-pump system? What are some differences?
2. Why must a diagnosis be absolutely correct when dealing with ground loop systems?

REVIEW QUESTIONS

1. When discussing the water circuit on a ground loop system, a decrease in the pressure drop across a water/refrigerant heat exchanger is an indication that:
   a. The flow rate of the water has decreased.
   b. The high-side pressure of the system has increased.
   c. The low-side pressure of the system has increased.
   d. The airflow through the system is not sufficient.
2. If the flow rate of water through a heat exchanger is reduced:
   a. The temperature differential of the water will increase.
   b. The head pressure of the system while operating in the cooling mode will decrease.
   c. The suction pressure of the system while operating in the heating mode will increase.
   d. All of the above are possible.
3. A restricted loop in a parallel ground loop system will:
   a. Reduce the amount of heat transfer between the earth and the water.
   b. Cause an increase in the supply water temperature in the cooling mode.
   c. Cause a decrease in the supply water temperature in the heating mode.
   d. All of the above are correct.
4. A defective circulator pump can result in which of the following?
   a. A frosted evaporator coil in the heating mode
   b. An open high-pressure switch in the heating mode
   c. Both *a* and *b*
   d. Neither *a* nor *b*
5. Mineral deposits on the interior of the water pipes on the water/refrigerant heat exchanger:
   a. Result in a higher temperature differential of the water across the heat exchanger.
   b. Result in a reduction of pressure drop across the heat exchanger.
   c. Result in a more efficient heat-pump system.
   d. Both *a* and *b* are correct.
6. Explain the operation and purpose of a Pete's port.

7. Which of the following is *not* a symptom of a system with a defective water pump?
    a. Reduced pressure drop across the heat exchanger
    b. Increased head pressure in the cooling mode
    c. Decreased suction pressure in the heating mode
    d. Increased temperature differential across the heat exchanger
8. A water leak on an open-loop system is most difficult to fix when it is located:
    a. At the outlet of the heat exchanger.
    b. At the pressure tank pipe connections.
    c. Inside the heat exchanger.
    d. In the dry well.
9. Air accumulation in a ground loop is an indication of:
    a. A water leak.
    b. A damaged pressure tank.
    c. A cracked supply well.
    d. A defective circulator pump.
10. A ground loop system that has no water flow through the heat exchanger may have:
    a. A defective pump motor.
    b. A defective pump linkage.
    c. A defective impeller.
    d. All of the above are correct.
11. Explain why it is important to evaluate the pump and impeller separately.
12. List the symptoms of a ground loop system that is experiencing improper water flow through the heat exchanger.
13. A system that has a large pressure drop across the heat exchanger and a flow meter reading of 0 most likely has a:
    a. Defective circulator pump.
    b. Defective flow meter.
    c. Defective pump linkage.
    d. Defective pressure tank.
14. A defective submersible pump in an open-loop system can be the result of a:
    a. Defective flow meter.
    b. Defective pressure tank.
    c. Defective compressor.
    d. None of the above are possible causes for a defective submersible pump.
15. A standing-pressure test on a heat exchanger can help:
    a. Diagnose a cracked or leaking heat exchanger.
    b. Increase the efficiency of the heat exchanger.
    c. Increase the flow rate through the heat exchanger.
    d. All of the above are correct.

CHAPTER 16

Servicing Geothermal Heat Pumps

OBJECTIVES After studying this chapter, the reader should be able to:

- Explain how a technician can identify and resolve the issue of mineral deposits in the water circuit of a geothermal heat-pump system.
- Explain how a technician can identify and resolve the issue of water leaks within a geothermal system.
- Explain how a technician can identify and resolve the issue of a defective water pump in a geothermal system.
- Explain how a technician can identify and resolve the issue of improper water flow through a geothermal system.
- Explain how a technician can identify and resolve the issue of improper airflow in a geothermal system.

INTRODUCTION

Servicing geothermal heat-pump systems requires the same skills that have been discussed throughout this text. A reader, having reached this portion of the text, now has a solid heat-pump foundation upon which to base examination of geothermal heat-pump-system problems. Presented in this chapter are a number of sample service calls that will help the reader to hone the skills of troubleshooting and system evaluation.

OVERVIEW

When performing service on a geothermal heat pump, the technician must always be aware of the seven steps to completing a successful service call. These steps have been the focus of this text, and successful geothermal heat-pump servicing relies on these steps as well. Of the seven steps, verifying the complaint and gathering information are among the most important to ensure that the technician is steered in the right direction. Geothermal systems involve three separate—yet interdependent—circuits that are easily affected by one another:

- The airflow circuit
- The refrigerant circuit
- The water circuit

The airflow circuit and the water circuit can operate completely independently of the other two, but the refrigerant circuit, in order to operate properly, must have the support of the other two. A failure in either the water or the air circuit will cause the refrigeration circuit to malfunction. In addition, a malfunction in the air circuit may lead a technician to diagnose a nonexistent problem in the water circuit. Technicians must be meticulous in their troubleshooting processes and must be completely comfortable with the following skills:

- Taking and interpreting temperature and pressure readings
- Making superheat and subcooling calculations
- Interpreting pressure drops across a heat exchanger
- Interpreting flow rates through a heat exchanger
- Evaluating the airflow through an indoor coil
- Evaluating the water flow through an open-loop or ground loop system
- Evaluating the operation of refrigeration system components, including compressors, solenoids, reversing valves, check valves, and metering devices

SERVICE CALL 1: INADEQUATE COOLING

Customer Complaint

The owner of a geothermal heat-pump system calls his service company and informs the service manager that his system is not providing adequate cooling to the house. Prior to his calling the company, the home owner has noticed that the outdoor unit is not operating. The system is equipped with a single-layer, horizontal ground loop.

Service Technician Evaluation

Upon arrival at the home, the service technician, Greg, inquires about the system and is told that the unit is not cooling at all. He is also informed that the indoor fan seems to be operating but that the air that is coming from the supply registers is warm. Greg checks the thermostat settings, and he can hear the indoor fan operating. He goes over to a supply register and can feel that a large quantity of air is being discharged into the room. He checks the air-return grill in the hallway and confirms that the filter is clean. Having determined that the airflow is adequate, he then makes his way to the heat-pump unit in the basement.

Upon initial inspection, Greg can see that the outdoor unit is not operating. He opens the service disconnect and checks for power. He reads 218 volts. He then opens the service panel and checks the voltage at the line side of the compressor contactor. Again he reads 218 volts. A check of the voltage on the load side of the contactor yields a reading of 0 volts. Touching the compressor, Greg establishes that the compressor is not overheated, as it is cool to the touch. Since the blower motor is operating, Greg knows that the control transformer is operational. He then decides to check the voltage at the holding coil of the compressor contactor. He obtains a reading of 0 volts.

Close examination of the control circuit shows that both a low-pressure and a high-pressure switch are wired in series with the holding coil. Greg checks voltage across the low-pressure switch and, by a reading of 0 volts, determines that the switch is in the closed position. Taking a reading across the high-pressure switch, a reading of 24 volts is obtained. The system has cut off on high head pressure. Greg now turns his attention to the water circuit.

Checking the circulator pump, he can hear the sound of a motor attempting to turn. With a flashlight, he inspects the shaft of the pump and the linkage connecting the impeller to the shaft. The motor is turning very slowly. After turning off the service disconnect at the unit, Greg removes the motor mounts and disconnects the linkage from the shaft. He then sets the motor on the floor, still connected electrically, and then re-energizes the system. The motor starts immediately. He then turns the system power off again. He next turns to the linkage assembly.

Greg attempts to turn the linkage by hand but is unable to do so. The linkage is jammed, so the pump needs to be replaced. Greg leaves the customer's home to obtain a replacement pump. Upon returning, Greg valves off the ground loop and drains the heat exchanger into a bucket. He then removes the existing pump body. With the pump out of the system, Greg makes certain that the flange connections are clean and free from any gasket material. He then mounts the new pump body, moistening the gaskets with a small amount of oil. This helps ensure a tight seal between the flanges. Once the pump body is mounted, Greg fills the heat exchanger with water and makes certain that all air is removed. He examines the flange connections for leaks and then mounts the motor back in place.

Once the motor is properly mounted, Greg opens the valves to the ground loop and restarts the system. The pump begins to operate. Knowing that the pump is now fully operational, Greg resets the high-pressure control in the low-voltage compressor control circuit. The compressor starts immediately. Using the Pete's port, Greg checks the pressure drop through the heat exchanger and the temperature drop across the exchanger. Finding everything working well, he fills out his paperwork and proceeds to his next job.

Service Call Discussion

Before Greg even set eyes on the system, he had already established that proper airflow existed through the system. He was then able to narrow his search down to the refrigerant and water circuits. Checking the control circuit led him to the high-pressure switch. Since the head pressure is controlled by the amount of heat transfer that takes place between the water and the refrigerant, he immediately zeroed in on the water circuit—no wasted time here, just good solid troubleshooting.

SERVICE CALL 2: INADEQUATE HEATING

Customer Complaint

A customer calls and informs her service company that her geothermal heat-pump system is not providing adequate heating. She can hear the compressor operating, but it does not seem to be doing the job. The customer has an open-loop system connected to a dedicated, single well.

Service Technician Evaluation

Later that afternoon, Hal, the service technician, arrives at the customer's home. The home owner informs the technician that her system is operating but that it does not seem to be heating the house effectively. Hal first goes over to the thermostat and checks the settings. He sets the system to operate in the heating mode. He then goes downstairs to the heat-pump system.

While at the unit, he can hear the blower motor operating. He then checks the air filter and sees that it has been recently replaced. He checks the temperatures of the supply and return air and finds that the temperature differential across the indoor coil is only 8 degrees. Hal, under normal conditions, would expect a temperature differential of approximately 14 degrees. He installs his gauge manifold on the system and obtains a low-side pressure of 50 psig and a high-side pressure of 145 psig. For an R-22 system, these pressures seem very low. Initially, he suspects that the system might be short of refrigerant. Before adding refrigerant to the system, he decides to check the water circuit to confirm that the system is in need of refrigerant. He expects that, since the suction pressure is lower than normal, the heat-exchange rate between the water and the refrigerant will be higher than normal.

Using the Pete's ports on the unit, Hal takes pressure and temperature readings at both the inlet and the outlet of the heat exchanger. The pressure drop through the heat exchanger is about 2 psig, which is normal for that particular system. The temperature differential across the coil is only 4 degrees, far below the normal 10-degree split that was expected. Hal's initial suspicion is incorrect. The system is not short of refrigerant.

Hal soon discovers that a normal pressure drop through the heat exchanger, a low temperature differential across the heat exchanger, and low operating pressures indicate that the heat exchanger has mineral and scaly buildup on its interior surfaces. An appointment has to be made for the coil cleaner to come to the home and clean the coil.

Once the coil has been properly cleaned with the appropriate chemicals, Hal returns to check out the system. The operating pressures are well within the proper ranges, the temperature differential across the heat exchanger is back in the 10-degree range, and the system has been left in good working order.

Service Call Discussion

Hal could have found himself in a real bind had he added refrigerant to the system. However, he had the insight to exhaust all possibilities before altering the refrigerant charge of the system. He had been expecting to find a large temperature differential across the heat exchanger but instead found a low temperature differential. By utilizing his knowledge of the water and refrigerant circuits, he was able to narrow his search down to the heat exchanger itself.

SERVICE CALL 3: INADEQUATE HEATING

Customer Complaint

The owner of a small specialty shop calls her landlord and informs him that her heating system is not providing adequate heat to her store. The landlord, who also performs air-conditioning service, comes to the store to check out the system. The system is a geothermal system with a ground loop system.

Service Technician Evaluation

When the landlord arrives at the store, he goes right to the heat-pump unit. He notices immediately that the compressor runs for a short time and then cuts off. After 3 minutes, the compressor starts and runs for about a minute and then cuts off. Knowing that the low-pressure switch is a control that automatically resets, he suspects that the system is cutting off on low pressure. Before resorting to adding refrigerant, he decides to check the water circuit.

Installing a pressure gauge at the inlet and the outlet of the heat exchanger, he notices that the pressure drop across the heat exchanger is zero, indicating that the pump is not moving any water through the heat exchanger. He visually inspects the motor and linkage assembly and sees that the linkage is not turning. He disconnects the power to the unit and removes the cover of the electrical connection box on the motor. He then loosens the wire nuts on the electrical connections to the motor and, after making certain the wires are not touching each other, re-energizes the unit. Using a voltmeter, he checks for power being supplied to the motor and finds that 115 volts are being supplied. Since the motor is rated at 115 volts, the reading is correct. He again disconnects the power and then completely disconnects the motor wires from the power wires.

Using his ohmmeter, he checks for continuity through the windings and finds that one winding is open. The pump motor is defective and needs to be replaced. He obtains a new pump motor and installs it on the existing pump. He makes all necessary electrical connections and then replaces the cover on the electrical connection box. He then starts up the system and once again takes pressure readings across the heat exchanger. This time, the pressure drop across the heat exchanger is 2 psig, which is perfect for that particular system. On his way out of the store, he feels the air being discharged into the store and finds that it is sufficiently warm to heat the store.

Service Call Discussion

Even though the landlord was not a service technician, per se, he did employ a logical troubleshooting plan of attack. Since he was the landlord, he obviously knew a lot about his own systems. Before adding refrigerant, he wisely decided to check the water system first. This practice should be followed by all technicians since system problems that appear to be refrigerant related may, in fact, be related to either the water or air circuits.

SERVICE CALL 4: INADEQUATE COOLING

Customer Complaint

A residential customer calls and reports that her geothermal heat-pump system is not cooling the space properly. The outside temperature is not overly hot, so she is concerned that the system will not operate well when the outside temperature rises. She feels that the system runs continuously and does not shut down. She is thinking ahead and is concerned about the possibility of increased electric bills in the near future. Her system is tied into a vertical, parallel ground loop water system.

Service Technician Evaluation

A short time later, the service technician, Brandon, arrives at the customer's house and can feel that the air in the house is very warm and that the humidity seems to be high as well. The customer is very quick to tell him that the system has been recently serviced and that a new circulating pump was installed. Needless to say, she is not very happy. Brandon first calls the office to get the details about the prior repair, and he is told that a new technician, James, was on the job and replaced the pump and motor assembly about a week ago. He immediately suspects a water loop problem but does not want to jump to any conclusions.

Brandon checks the air system and finds that the filters are clean and adequate airflow exists through the system. He then goes to the heat-pump system and finds that the compressor is running. Much to Brandon's dismay, he hears a loud crackling sound coming from the new circulating pump. He installs his gauges on the system and finds that:

- The head pressure is too high.
- A low pressure drop is occurring across the heat exchanger.
- The flow rate of water through the heat exchanger is low.
- The temperature differential across the heat exchanger is low.

All of these factors, along with the fact that the loud crackling sound is coming from the pump, indicate to Brandon that the water system has a leak and air has entered the loop.

He visually inspects the pump installation that had just been completed and notices that water is seeping from the return flange on the pump. Upon closer inspection, he sees that no gasket material is apparent between the pump and flange surfaces. He shuts off the power to the system and valves off the ground loop. He then proceeds to drain the heat exchanger, and he removes the pump assembly. He retrieves some gasket material from his truck and proceeds to cut new gaskets for the pump, as he cannot locate the ones originally supplied with the new pump. Wetting the surfaces of the new gaskets with oil, Brandon mounts and secures the pump assembly.

Going to the ground loop, he is fortunate enough to find that hand valves and hose bibs are at each of the parallel loops. He proceeds to isolate each loop and pushes water through each individual loop. Once this is accomplished, he then fills the ground loop. After the loop is completely filled, he then opens the valves that connect the heat-pump system to the ground loop. He restores power to the system and then starts it up. The crackling sound is gone. He then rechecks the pressures and temperatures of the system and proceeds to his next job.

Service Call Discussion

Although Brandon was correct in his initial assumption about the water circuit, he was wise to check the system out as he normally would have. One problem that many seasoned technicians often experience is the fact that they try to troubleshoot a system without even checking it properly first. This poses the problem of overlooking small, important details that could ultimately help resolve the situation. Brandon went through his normal system check and diagnosed the system properly.

SERVICE CALL 5: POSSIBLE BLOWER PROBLEM

Customer Complaint

A residential customer calls his service company and tells the dispatcher that his system is not cooling properly. He does not feel much air coming from the supply registers, but the air that he does feel is relatively cold. He tells the dispatcher that he thinks there might be a problem with the blower. The customer has an open-loop, geothermal heat-pump system with a dedicated, two-well water supply.

Service Technician Evaluation

Two hours later, the service technician, Herb, arrives at the residence. The customer explains his thoughts about the potential problem; Herb listens to the complaint but assumes that the customer does not have a clue as to the actual problem. He goes immediately to the heat-pump unit and checks the pressures and temperatures at both ends of the heat exchanger. The pressure drop through the coil is correct, as is the flow rate through the coil. He then checks the temperature differential across the coil and finds that it is very low. He quickly concludes that the coil has mineral deposits on the interior of the heat exchanger and immediately calls for the coil cleaner to come and flush out the coil. The next day, after the coil has been cleaned, Herb returns to the job to check the system operation. Nothing has changed. The system is doing exactly the same thing it was doing the day before. The customer questions Herb about the repair, and Herb just ignores him.

Herb goes on to check the system further and determines that the crew that came to clean the coil must not have done a good job. The customer calls the service company to express his dissatisfaction. The service manager assures the customer that the problem will be resolved. The service manager sends Herb on another service call and sends another technician to the customer's house.

An hour later, Peter arrives. He listens to the customer's complaint and then goes to the return grill and inspects the filter. It is completely clogged with dirt and dust. As soon as Peter removes the filter, he can hear the rush of air through the duct system. He installs a new filter and checks the operation of the system. All pressures and temperatures are within the desired ranges and the supply air is 19 degrees cooler than the return air. Peter apologizes to the customer and fills out the work ticket.

Service Call Discussion

Herb apparently did not know his crucial seven steps for completing a successful service call. Verifying the complaint is a very important step and should not be skipped. Although Herb assumed that the home owner did not know air-conditioning systems, he should have listened attentively to the customer's complaint. The home owner deals with the system day in, day out and is very familiar with the sounds, vibrations, and air volumes that the system produces. Even though customers may not be able to fix their systems, they can usually provide valuable information that can lead the technician in the right direction.

SUMMARY

- The presence of mineral deposits in a heat exchanger is characterized by a low temperature differential across the heat exchanger, normal pressure drops and flow rates through the exchanger, and low suction pressure in the heating mode.
- One tell tale sign of a system water leak is a crackling sound coming from the pump.
- Defective water pumps can be identified by a lack of water flow through the heat exchanger.
- Improper airflow through a system can give the impression of other, major system problems.
- Always listen carefully to the customer complaint.
- Do not jump to any conclusions about the cause for system failure.
- Always proceed in a logical manner when troubleshooting any air-conditioning system.
- Exhaust all possibilities before opting for a costly, time-consuming solution that may or may not be the correct one.

FOR DISCUSSION

1. Why can the customer be a valuable source of information for the technician?
2. Why should a strong working knowledge of all system aspects be a requirement for all field technicians?

REVIEW QUESTIONS

1. When installing a new circulator pump, how can the technician reduce the possibility of water leaks?
   a. Install the proper gaskets between the pump and flange surfaces.
   b. Make certain that all mounting bolts are secure before introducing water to the circuit.
   c. Make certain that all old gasket material is removed from the flanges before installing the new pump.
   d. All of the above are correct.
2. What should a technician check before recommending a major system repair?
3. The first step in completing a successful service call is __________.
4. Explain how a technician would go about determining whether or not a system has mineral deposits on the interior of the heat exchanger.
5. How would a technician go about checking to see if the circulator pump is operating?
6. A large pressure drop through the heat exchanger is an indication that:
   a. Proper water flow exists through the heat exchanger.
   b. The circulator pump is defective.
   c. The system probably has a water leak.
   d. All of the above are correct.

7. How would a technician properly diagnose a defective pump linkage or impeller?
8. What should a technician do between the time a repair is completed and the time he leaves the job?
9. What is the best thing a technician can do if he finds himself over his head on any particular job?
10. A reduced water flow rate through a heat exchanger is an indication that:
    a. The pump may be defective.
    b. A water leak may be present.
    c. Both *a* and *b* are correct.
    d. Neither *a* nor *b* is correct.

CHAPTER 17

Air-to-Air Heat-Pump-System Installation

OBJECTIVES After studying this chapter, the reader should be able to:

- Explain the requirements for selecting the location of the outdoor unit.
- Explain the requirements for selecting the location of the indoor unit.
- Describe different styles of duct systems.
- List various types of materials used in duct fabrication.
- Explain how various types of duct systems are connected and supported.
- Explain the importance of condensate lines and drain pans.
- Describe the process of running refrigerant lines between the indoor and outdoor units.
- Describe the interconnecting low-voltage wiring between the thermostat, indoor unit, and outdoor unit.
- Explain how to properly leak-check a system before system start-up.
- List important factors that must be considered during the evacuation process.
- List the items that should be checked before a system is initially started up.
- Explain how to perform an initial system start-up.

INTRODUCTION

As with any air-conditioning system, the satisfactory operation of a heat-pump system can be prolonged by a high-quality installation. Skilled installation crews, using high-quality materials, will help ensure that the integrity of the refrigerant circuit, the wiring as well as the actual mechanical equipment is maintained. Systems that are properly installed tend to have fewer malfunctions and, therefore, require fewer service calls. Compressor and other system component failures are also less frequent, ultimately leading to a satisfied customer. Consequently, a good long-term relationship with a customer starts with the proper installation of a new system. This chapter focuses on the basics of system installation, ranging from unit location to system start-up. Because this chapter provides only general installation techniques and procedures, the installation crew must follow the installation guidelines provided by the manufacturer.

OVERVIEW

Even though a new heat-pump system may be of the highest quality, both in construction and **energy efficiency ratio (EER);** the system is only as good as the installation. The energy efficiency ratio is a comparison between the heat-transfer capability of the system and the amount of electrical energy used to transfer that heat. The proper installation of a heat-pump system involves many individual tasks that must all be performed well. A well-trained installation technician has a mastery of general skills, including:

- Carpentry
- Plumbing
- Electrical

Although each member of an installation crew may have a specific area of expertise, each member of the crew must have at least a working knowledge of the entire installation process.

Carpentry skills are useful because the heat-pump system, after installation, becomes a part of the structure in which it is installed. Return grills and supply registers must be cut into the existing walls, floors, or ceiling of the conditioned space; knowledge of the structure of these walls and/or ceilings is therefore extremely important to prevent damage to the structure. These devices must be installed securely to ensure proper operation, as well as to ensure that they remain in place. In addition, the equipment—namely, the indoor and outdoor units—must be mounted or placed in position, which could require the fabrication of brackets or cradles on which the equipment will rest. On split-type heat-pump systems, the shell of the structure will also need to be penetrated to facilitate the running of the refrigerant and condensate drain lines, as well as low-voltage control wiring. Once again, knowledge of the structure is needed.

In addition to carpentry skills, plumbing and piping skills are also needed. For the integrity of the refrigerant circuit to remain intact, the solder joints used to join the sections of tubing should be leak-free and able to stand the test of time. The condensate also must be properly removed from the system via condensate lines, pans, and pumps. The installation of the pans, pumps, and lines helps ensure that the moisture removed from the air in the form of condensation is removed from the system and structure without causing damage to the structure itself. Being able to work with different piping materials, including PVC and copper, is a requirement. Different methods used for joining these piping materials must also be known.

Electrical knowledge is also a must. Field wiring must be installed in every installation. Depending on local codes, the heat-pump installation crew is responsible for making electrical connections between the disconnect boxes and the individual piece of equipment. The installation crew is also responsible for running and connecting the low-voltage control wiring that connects the thermostat to the indoor and outdoor units. In addition to running and connecting the interconnecting wiring, local codes must be followed regarding the size of the individual conductors as well as the size of the conduit carrying these conductors. The methods by which these wires are connected, as well as the means by which the conduits are connected to the disconnect boxes, must also be in accordance with local electrical codes.

Finally, the installation crew must have a working knowledge of the refrigeration system. Ensuring that no excessive peaks and valleys occur in the refrigerant piping and making certain that the piping is properly trapped as needed help ensure that the system will operate well, reducing the possibility of liquid floodback to the compressor. The installing technician must also be familiar with the concepts of airflow. This helps ensure that sufficient airflow exists through the duct system and that no tight bends and twists are present that can ultimately lead to a reduction in airflow. Insufficient airflow can lead to a multitude of system problems, including improper airflow to the occupied space, coil freeze-ups, and, more important, customer discomfort.

As can be seen, the air-conditioning installation crew must be able to perform a wide range of tasks. This crew of skilled individuals can make or break any installation project. Note that getting any air-conditioning or heat-pump system to run upon initial start-up is possible, but a great installation will help ensure that the system continues to run properly for years to come.

SELECTING THE PROPER LOCATION FOR THE OUTDOOR UNIT

Before the actual job of connecting the components of the heat pump begins, the outdoor unit and indoor unit locations must be determined. First, a number of factors must be addressed in order to establish the proper location for the outdoor unit. Some of these factors are:

- Sound transmission
- Wind factors
- Location of electrical power
- Airflow restrictions
- Proximity to the indoor unit
- Ground slope

Sound Transmission

Remember that the outdoor unit of the heat-pump system contains both a compressor and a fan motor that generate noise that could be distracting, to say the least. For this reason, the outdoor unit should be positioned far enough away from any bedrooms or other living areas to reduce the inconvenience created by system noise. Ideally, the outdoor unit should be located close to the structure. It should, however, be located along the side of the house, in a location that is not near a window. If applicable, it should also be placed in a location that will cause the least amount of inconvenience for any other individuals residing nearby.

Wind Factors

To prevent prevailing wind from affecting the operation of the heat-pump system, the outdoor unit should be positioned in a well-shielded location. Shrubs provide a very good shield for the unit, preventing wind from blowing through it. The shrubs, however, must be trimmed regularly to prevent them from blocking the coil surfaces on the unit. The leaves from the bushes can also cause problems if they are pulled into the unit, creating an airflow problem through the coil itself. Fences can also be used to block the wind. The location of an outdoor unit within a fenced-in area should be carefully considered before the piping process is started. Ample room should be available for a service technician to access the electrical panel, as well as the service panel that provides access to the compressor and other system components. In either case, shrubs and fences provide protection from the wind and help to alleviate the eyesore nature of the unit itself.

Location of Electrical Power

Both the outdoor and indoor units require separate electrical power supplies that are protected by either fuses or circuit breakers. On a new installation, these power supplies must be provided from the existing fuse or circuit-breaker panel. For this reason, a new line must be run for each section. Although a relatively minor consideration, the shorter the run, the less expensive the installation will be. Therefore, having the outdoor unit located as close to the electrical service panel as possible is desirable. This line must be installed in accordance with local electrical codes, and a licensed electrician should perform this work.

Airflow Restrictions

Regardless of where the outdoor unit is located, air must be able to flow freely through the coil. The air flowing through the coil must be able to mix freely with the outside air in order to prevent the hot discharge air from the outdoor unit, in the cooling mode, from recirculating through the unit. For this reason, the unit should not be located under any overhangs or porches. Overhangs trap the discharge air from the outdoor unit, causing the head pressure of the system to rise while operating in the cooling mode. In the heating mode, the overhangs will trap the cooled air, thereby reducing the system's heating efficiency. Typical clearance minimums are shown in Figure 17-1. The manufacturer's specifications should always be followed when evaluating the clearances for a specific piece of equipment. Under no circumstances should the outdoor unit be positioned against a structure; it should be at least 30 inches away from any wall.

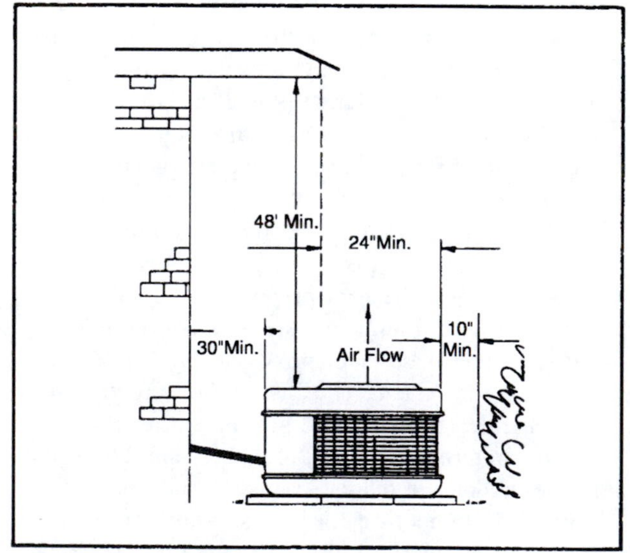

FIGURE 17-1 Outdoor unit location. *(Courtesy Addison Products Co.)*

Proximity to the Indoor Unit

Probably one of the most important factors that will ultimately affect the success of the installation is the outdoor unit's location with respect to the indoor unit. The distance between the indoor and the outdoor unit determines the length of the refrigerant lines that must be installed. The shorter these lines, the better. Longer refrigerant lines can lead to a decrease in efficiency in both the heating and the cooling modes of operation. In the heating mode, the hot discharge gas from the compressor flows to the indoor coil, where the heat is transferred to the occupied space. If the distance that the refrigerant must travel is long, some of the heat can dissipate from the line, reducing the heating efficiency. In the cooling mode, the temperature of the refrigerant flowing back to the compressor from the indoor coil will increase, also reducing the operating efficiency of the system. In both cases, the cost of operating the system will rise. In an effort to reduce the effects of unwanted heat transfer, the vapor line should always be insulated.

For aesthetic purposes, some home owners opt to locate the outdoor unit far away from the structure. The problems with this are threefold. First, the cost of running the electrical power to the unit is higher because of the added distance. Second, the added length of the refrigerant lines can have a negative effect on system performance and operating efficiency. Finally, the refrigerant lines would need to be buried under the ground. This makes the servicing of the equipment much more difficult, especially in the event of a refrigerant leak.

Ground Slope

Wherever the unit is located, it should be perfectly level. This will prevent unnecessary strain on the outdoor fan motor and will help prevent any undesired noise created by excessive vibrations. The unit should never be set directly on the ground. Doing this can lead to the sinking of the unit, as well as the blocking of the coil with ice, snow, leaves, or other debris. The unit should be located above ground level and set on a slab made of one of a number of materials, including concrete and plastic. Prefabricated outdoor unit pads can be purchased in a wide range of sizes, depending on the physical size of the unit. The chosen pad should be large enough to completely support the unit with enough room left on all sides to prevent the unit from vibrating off the pad. If need be, a poured concrete pad can be constructed on site by fabricating a wooden form and then pouring the concrete into it. Obviously, this should be done long before the installation is started to give the concrete ample time to set up. Another good idea is to have the outdoor unit completely isolated from the structure to prevent the transmission of noise into the structure. For example, placing the unit on a patio or deck that is connected to the structure itself would not be wise. In areas that experience heavy snowfall, the outdoor unit should be elevated well above ground level.

SELECTING THE PROPER LOCATION FOR THE INDOOR UNIT

Just as with the outdoor unit, a number of factors must be considered when choosing the location for the indoor unit of the heat-pump system. These factors include:

- The type of air-distribution system
- Location of the electric power supply
- Location of the outdoor unit
- Serviceability
- Indoor unit configuration
- Ease of condensate removal
- Noise level
- Return air
- Location of space to be conditioned

Types of Air-Distribution Systems

A number of different air-distribution systems can be utilized on any given system installation. A few of these configurations are shown in Figure 17-2. In the radial or plenum-type duct system, shown in Figure 17-2(a), the indoor unit is located in a position that is central to the supply register locations. On this type of duct configuration, the supply registers are normally located on the interior walls of the structure. On heat-pump systems, however, this type of configuration is not the most desirable because the temperature of the heated air being supplied to the conditioned space is not as high as with

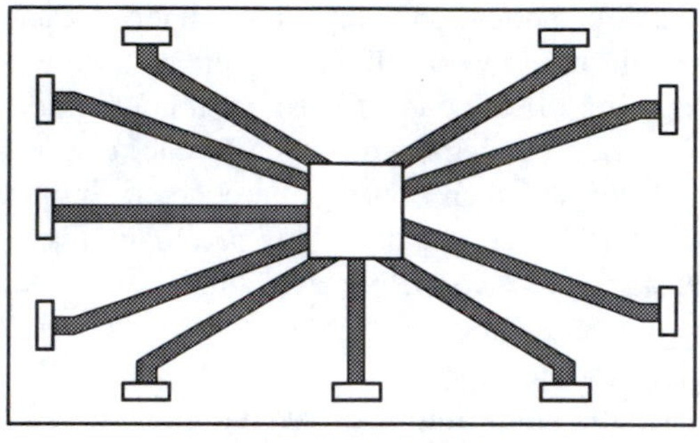

(a) Plenum or Radial Duct System

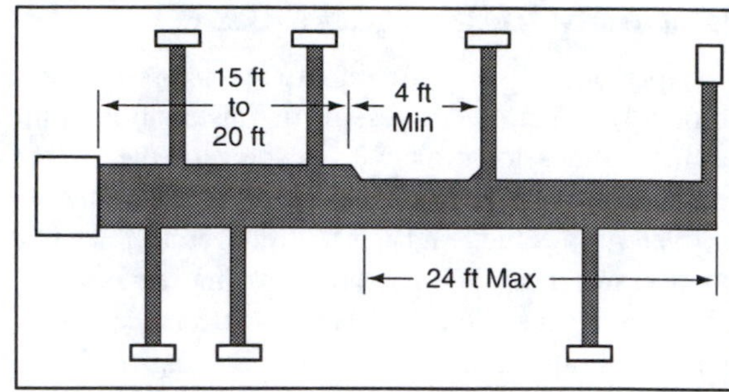

(c) Reducing-Extended-Plenum System

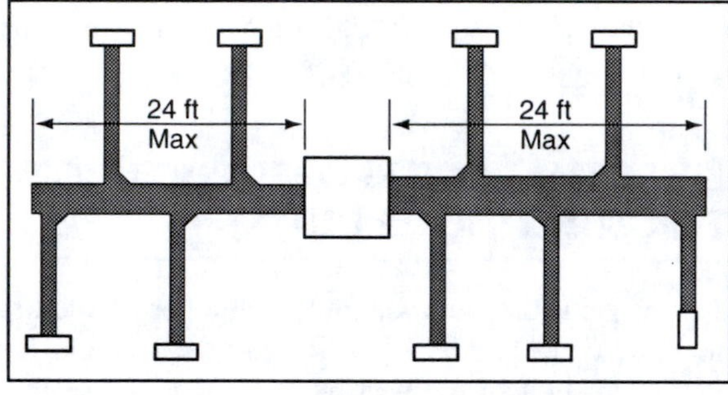

(b) Extended-Plenum System

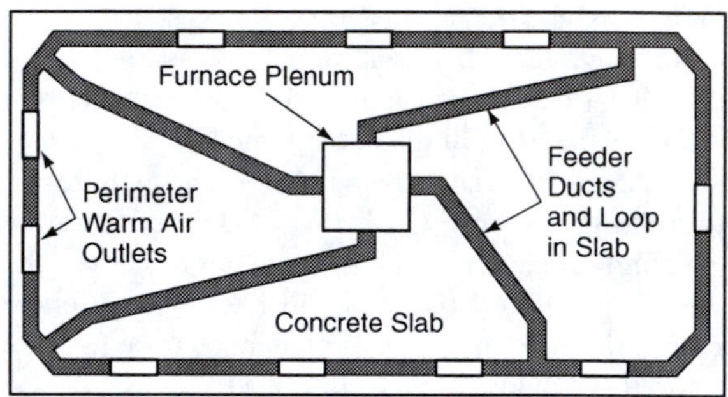

(d) Perimeter Loop System with Feeder and Loop Ducts in Concrete Slab

FIGURE 17-2 (a) Plenum system. (b) Extended-plenum system. (c) Reducing-extended-plenum system. (d) Perimeter loop system.

gas heat or other fossil fuels. The extended-plenum system, shown in Figure 17-2(b), is used primarily when the duct system and the indoor unit are not located on the same level (e.g., if the indoor unit is located on the first floor of the space and the duct system is located in the attic overhead). A large supply plenum is used to bring the supply air up to the level of the remainder of the duct system. The reducing-extended-plenum duct system, shown in Figure 17-2(c), is used on systems that have the indoor unit located at one end of the structure to be conditioned. In this type of duct system, the size of the trunk line gets progressively smaller as the end of the run is reached in an effort to maintain the velocity of the air moving through the duct. In the perimeter duct system, shown in Figure 17-2(d), the indoor unit is typically located below the conditioned space and the ductwork runs around the perimeter of the space. The supply registers, which are floor mounted, are located around the perimeter of the space as well.

Location of the Electrical Power Supply

Another factor that is to be considered when choosing the location of the indoor unit is its location with respect to the electrical panel. Once again, as in the case of the outdoor unit location, power must be brought to the unit from the circuit breaker or fuse panel. The longer the run, the more costly the installation. The running of this power line must be done in accordance with all local electrical codes and should be done by a licensed electrician. The electrical panel should also be inspected to make certain that the power requirements of the system can be satisfied.

The supplementary electric-strip heaters on some systems draw large amounts of current—some as much as 100 amperes.

Length of the Refrigerant Lines

As already mentioned, the length of the refrigerant lines should be as short as possible. For this reason, the indoor unit should be located as close to the outdoor unit as possible. Excessively long refrigerant lines will have a negative effect on both the operation and the efficiency of the entire system. Long refrigerant lines will increase the system superheat in the cooling mode of operation and will reduce the heating efficiency in the heating mode of operation. The vapor line—the suction line in the cooling mode and the discharge line in the heating mode—should always be well insulated to reduce these effects. In addition, the insulating of this line reduces the amount of sweating on the line during cooling operation.

Serviceability

Regardless of where the indoor unit is located, it must be serviced periodically—both during preventive maintenance and during system repair. The location that is ultimately chosen for the indoor unit must be such that all of the service panels are unobstructed (Figure 17-3). In addition to this, ample clearance must be available for the technician to gain access to the unit. Although crawl spaces under the conditioned space may be ideal locations for the unit, it should be positioned as close to the access of the crawl space as possible to make the servicing of the equipment easier. When located in an attic, the path leading to and under the unit should be clear and unobstructed. If possible, plywood sheets should be laid down on the path to the unit to reduce the possibility of damage to the ceiling of the space, not to mention to help ensure the safety of the technician. Ideally, a permanent light fixture should be mounted in the attic as well to help facilitate servicing. Upon installation, members of the installation crew should consider the fact that any component within the unit may need to be replaced in the future and they should therefore plan accordingly.

SERVICE NOTE: *Always refer to and obey local building codes when planning a system installation. Some municipalities require that a solid, permanent path be installed to the indoor unit when installed in an attic or crawl space. Some codes require a platform at least 30 inches wide at the unit location. Most, if not all, municipalities require that a service disconnect be located on or next to the indoor unit.*

Indoor Unit Configuration

The location of the indoor unit also relies, to a lesser degree, on the configuration of the indoor unit or air handler. Common configurations of the indoor unit include:

- Vertical upflow
- Vertical downflow
- Horizontal

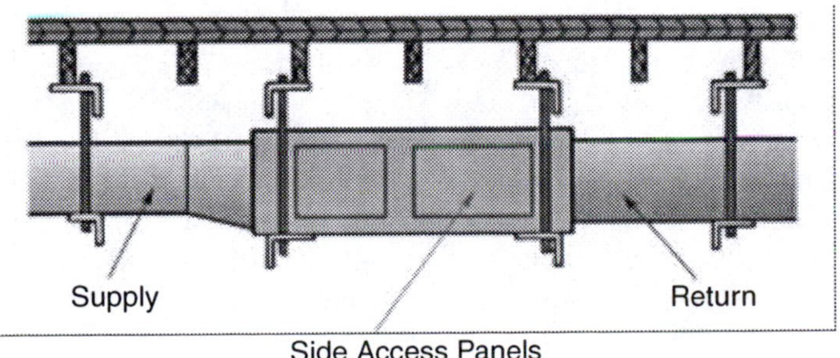

FIGURE 17-3 Access panels must be unobstructed.

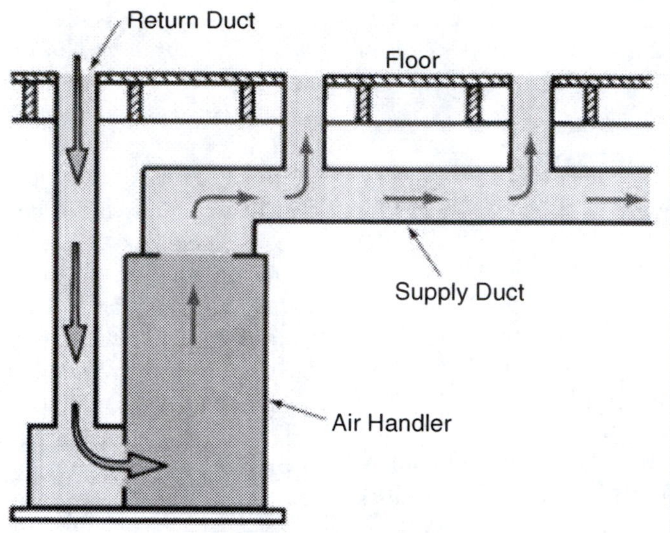

FIGURE 17-4 Vertical upflow configuration

VERTICAL UPFLOW. The vertical upflow unit is designed for applications in which the unit is located below the conditioned space—in a basement, for example—or on the same level as the conditioned space—such as in a utility closet (Figure 17-4). In this configuration, the blower is located downstream of the refrigerant coil and the supplementary electric-strip heaters are located downstream of the blower. When installed in a utility closet, the supply plenum often extends straight up into the attic or enclosed ceiling where the duct system is located. In the case in which the unit is located in the basement, the duct system is normally located in the basement as well, with the supply registers located in the floor. On two-story dwellings, a system is often located in the basement serving the first floor and another unit is in the attic that serves the second floor.

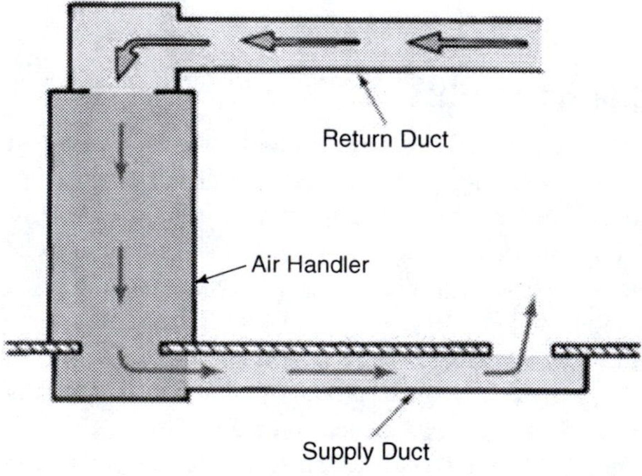

FIGURE 17-5 Vertical downflow configuration

VERTICAL DOWNFLOW. The vertical downflow unit is designed primarily for use with floor registers. This type of indoor unit discharges conditioned air to the basement-mounted duct system or into a floor plenum, which is commonly found in computer rooms (Figure 17-5). The return air enters the unit from the top and is then discharged downward. One benefit of this type of unit is its serviceability. The unit can be serviced from its location in a closet or utility room without having to access the basement or crawl space. This also makes the replacement of any unit components easier.

HORIZONTAL. The horizontal configuration is probably the most versatile as far as installation location is concerned. These units can be configured for right-hand or left-hand flow with only minor alterations made to the position of the drain pan within the unit. Horizontal units can be used in attics, dropped ceilings, utility closets, and basements. When installed in attics, they are normally sus-

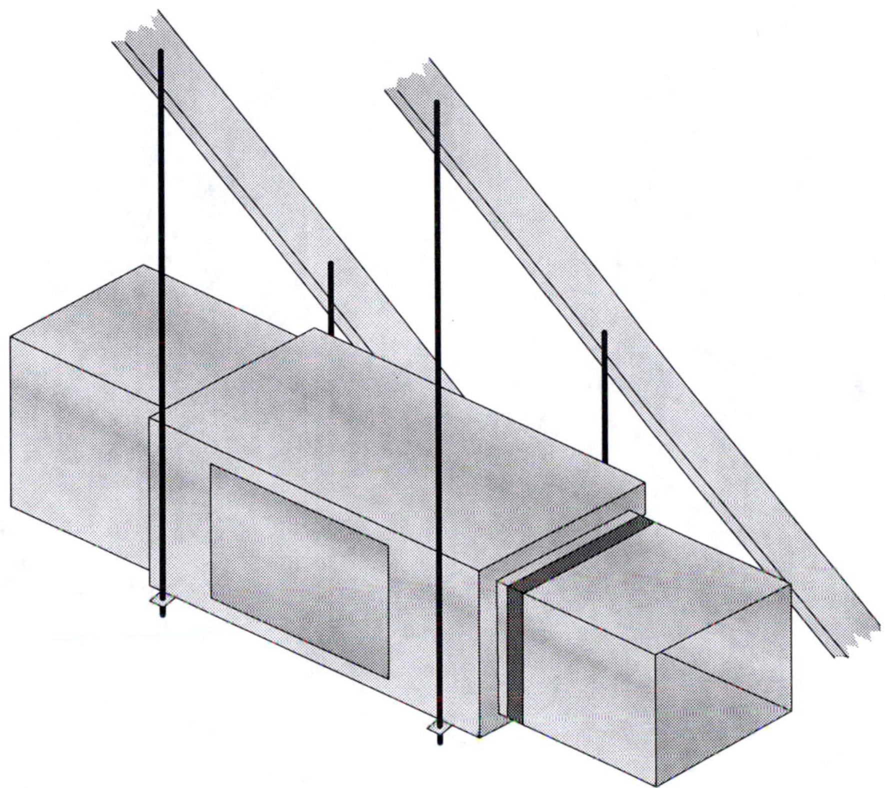

FIGURE 17-6 Indoor unit supported from the rafters in an attic

pended by cradles that are attached to the rafters of the structure itself (Figure 17-6). On these installations, the duct system is normally located in the attic as well. When used in a basement, they are also supported by cradles that are suspended from the floor joists (Figure 17-3). Using horizontal units in basements is beneficial, especially when the potential for flooding exists. By keeping the unit off the floor, the possibility of water damage is greatly reduced. In addition, by suspending the unit, more floor space is available for storage. These units can also be positioned in utility closets, again making more storage space available.

Ease of Condensate Removal

Regardless of where the indoor unit is located, it is designed to perform the same functions. It moves heated air through the duct system in the cooler months and moves cooled air through the duct system in the warmer months. In addition to removing heat from the air in the warmer months, the system also dehumidifies the air, thereby creating condensation. This condensate must be effectively removed from the structure. If not properly removed, water damage will result. To reduce the possibility of water-related damage, the length of the condensate drain line should be as short as possible.

When the indoor unit is located in an attic or overhead area, the condensate drain line should be routed to the outside of the structure if at all possible. If an outside wall is not accessible, the condensate should be directed to a nearby waste line or drain (Figure 17-7), as long as the piping work complies with all local plumbing codes. The same holds true for units installed in utility closets.

When the indoor unit is located in the basement, routing the condensate outdoors is not as easy. This is

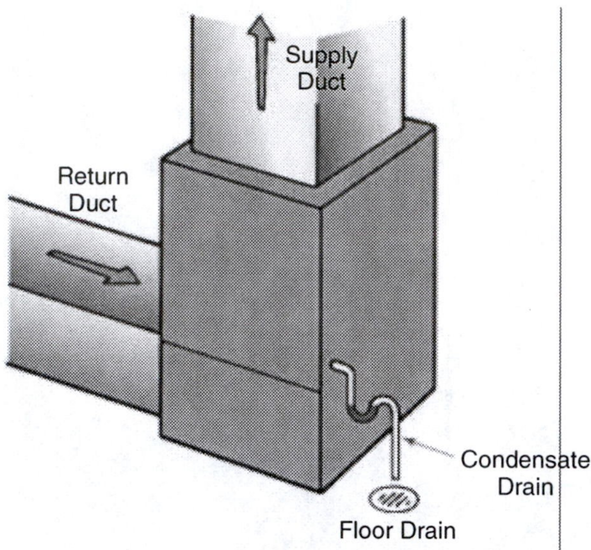
Supply
Duct
Return
Duct
Condensate
Drain
Floor Drain

FIGURE 17-7 Condensate piped to a nearby drain

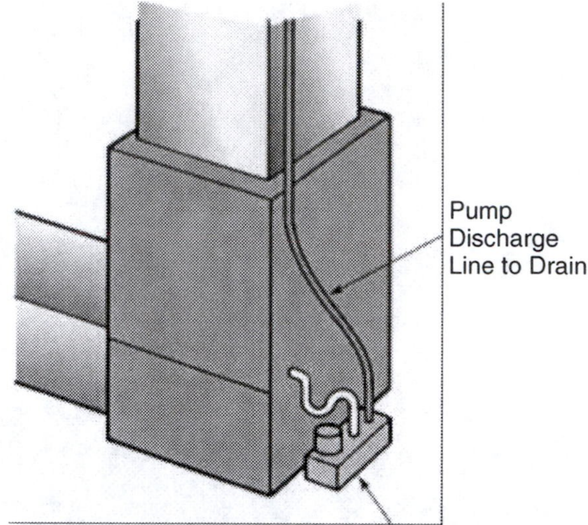
Pump
Discharge
Line to Drain
Condensate pump has a float to turn the pump on.
Some pumps have a second float and switch to stop
the unit if the first float fails.

FIGURE 17-8 Typical condensate pump location

mainly because the basement is below ground level. In this case, a condensate pump (Figure 17-8) is needed to pump the condensate up to an overhead line, which, in turn, is routed to a waste line or utility sink. If the basement is not finished and a floor drain is available, it can be used to accept the condensate from the system. Piping needs to be run from the unit to the floor drain to prevent water from accumulating on the floor. In any event, the location of the unit with respect to the ultimate termination point of the condensate should be evaluated carefully before the unit is set in place.

SERVICE NOTE: *Always check the installation literature that is supplied with a unit before piping in the condensate line. Many indoor units require that the condensate line be trapped at the condensate outlet.*

SERVICE NOTE: *Good field practice requires insulating the condensate line for at least the first 10 feet of the run from the indoor unit. This helps prevent sweating on the surface of the drain line.*

Noise Level

Even though the indoor unit is typically much quieter than the outdoor unit, noise level is another factor that must be considered when choosing the indoor unit's location. When located in an attic, the unit should be located above a common area, such as a bedroom hallway. If possible, the unit should never be located directly above a bedroom. The cycling of the fan, although not loud, is magnified at night when the house is otherwise very quiet. If the unit must be placed over a bedroom, sufficient insulation should be placed in the attic to reduce the effects of the unit noise. In a commercial office setting, the unit should be located away from private offices and conference rooms if at all possible. Stock areas and basements make perfect locations for the indoor unit. If the unit is located in a basement, it should be positioned so that the inconvenience caused by the noise will be minimized. When the unit is located in a utility closet, the walls of the closet should be covered with an acoustical material to reduce the amount of noise transmission to the adjacent areas.

Return Air

A method of returning air from the occupied space to the indoor unit must be provided. The source of the **return air** should be a common location, such as a hallway. When the unit is installed in a utility closet, the easiest way to ensure return air is to place a **return grill** in the door of the closet. This type of **natural return,** which does not require any physical ductwork, per se, can result in more noise transmission to the occupied space. When the system is located in the attic, the return grill is often located in the ceiling of the hallway. In this situation, the return grill is connected to the indoor unit via a **return duct,** which physically connects the return grill to the unit. The unit should not be located directly above the return grill, as the noise levels will be higher. If, however, the unit must be located very close to the return grill, good field practice would involve creating a loop, or indirect path, in the return duct. By doing this, the unit noise will be dampened and will have less effect on the occupied space.

SERVICE NOTE: *The return grill should be sized to provide a minimum of 144 unrestricted square inches per ton of capacity. If restrictive filters are to be used, more unrestricted area is needed.*

Location of the Space to Be Conditioned

Probably one of the most important factors that should be considered when choosing the location of the indoor unit is proximity to the space that is to be serviced. Ideally, the unit should be directly above, below, or next to the space. Excessively long duct runs will have a negative effect on the operation of the system. The interior surfaces of the ductwork offer resistance to airflow, reducing the velocity, or speed, of the air moving through it. An exchange of heat takes place between the air inside the ductwork and the air surrounding it. Even when insulated, a certain amount of air leakage takes place within the duct system, causing a reduction in the system's operating efficiency. If placement in a nearby location is not possible, the duct system should be constructed of low-resistance sections in an effort to maintain proper airflow rates. For example, flexible or spiral duct offers a great deal of resistance to flow and should be avoided if at all possible. The use of an excessive amount of transition sections and turning duct sections should be avoided as well because they add additional resistance to airflow.

INSTALLING DUCT SYSTEMS

Once the indoor unit location has been determined and the unit is actually set in place, the duct system can be installed. This can be a very tricky aspect of the installation and should be performed by qualified individuals. Two scenarios that are addressed when dealing with duct systems relating to the time at which the system is installed are:

- New construction installations
- Existing conditions installations

New Construction Installations

When systems are being installed as part of a new construction project, the problems that are encountered as far as ductwork installations are relatively simple to overcome. During new construction, the walls and ceiling are open, giving the installation crew free access to them. Duct sections can be placed in the walls and ceilings before any plasterboard or other covering material is set in place. All duct sections should be properly joined and insulated before they are enclosed, as access to them after construction is complete will be next to impossible. On new construction installations, wall registers are very popular due to the ease of positioning the duct sections in between the wall studs and then extending these sections either upward to the attic or down to the basement, depending on the indoor unit location.

The duct system layout for the new construction installation is relatively straightforward because the duct system has been designed along with the structure. For this reason, there are typically fewer bends and twists in the ductwork, making the installation much easier. Since the majority of heat-pump-system installations take place in existing structures, discussion follows for those types of installations.

Existing Conditions Installations

When a heat-pump system is installed in an existing structure, many more problems must be addressed than in the instance of the new construction installation. The duct system must be designed and installed around the limitations that are already present in the structure. The walls are already sealed, and access to them is limited. For this reason, floor- and ceiling-mounted supply registers and grills are the most popular for installations in existing structures.

One factor that must be considered is the fact that heat-pump systems require almost 2 ½ times as much airflow as heating systems that operate on fossil fuels such as oil or gas. If a fossil fuel system is being replaced with a heat-pump system, chances are that the duct system is undersized. Although the design engineer should have already addressed this issue, the installation crew should be aware of it as well. An improperly sized duct system can have a number of effects on the system, including noise and comfort issues.

Air that is moving through a duct system too fast will cause excessive noise within the duct system and at the register locations, as well as possibly creating unnecessary drafts in the space. On the other hand, if the speed of the air is too low, comfort issues arise, including cold spots in the colder months and excessive humidity in the warmer months. The ductwork on a heat-pump system should be sized based on the heating mode of operation to ensure that the system is not undersized.

Duct System Configurations

As mentioned earlier in this chapter, four common duct system configurations are the:

- Plenum system
- Extended-plenum system
- Reducing-extended-plenum system
- Perimeter loop system

Regardless of which type of duct system is ultimately chosen for installation, its ultimate purpose remains the same: to provide proper airflow to the occupied space. This poses a potential problem since different spaces have different comfort level requirements and, therefore, different airflow requirements. Generally speaking, an air-conditioning system delivers 400 cubic feet of air per minute per ton of system capacity, which must be divided among all of the supply register locations. A 3-ton system, for example, will circulate approximately 1200 cubic feet of air per minute to the occupied space.

PLENUM SYSTEM. Probably the easiest of all duct systems to install, the **plenum,** or radial, air-distribution system can be installed quickly and efficiently by individuals with a minimum of ductwork experience. This system consists of a closed-ended plenum coming off the supply end of the unit, into which are cut individual takeoffs for the individual supply locations (Figure 17-9). The connections between the plenum and the supply register itself are often made with insulated flexible ductwork (Figure 17-10). This type of system has both benefits and drawbacks, and they should be weighed against each other before a final decision is made as to which type of duct system to use. The benefits of a plenum-type system include:

- They are very economical from a first-cost standpoint.
- They are extremely easy to install.
- They do not require extensive duct measuring and layout.
- They are a very good choice when numerous physical obstacles exist in the area of the unit.

The main disadvantage of this type of system is that substantial friction occurs within the flexible duct itself. If the individual duct runs are very long, the operational effectiveness of the system will be somewhat compromised. In an effort to reduce the frictional losses in the duct system, takeoffs made from rigid sheet metal can be used. Of course, this will greatly increase the cost of the installation. If this turns out to be the case, the extended-plenum system may be a better alternative route to take.

SERVICE NOTE: *When connecting takeoffs to the plenum, the end of the plenum must be left intact. Cutting a takeoff into the end of the plenum will result in an excessive amount of air being discharged through that opening. Needless to say, the other takeoffs will be starved for air, resulting in uneven conditioning of the occupied space.*

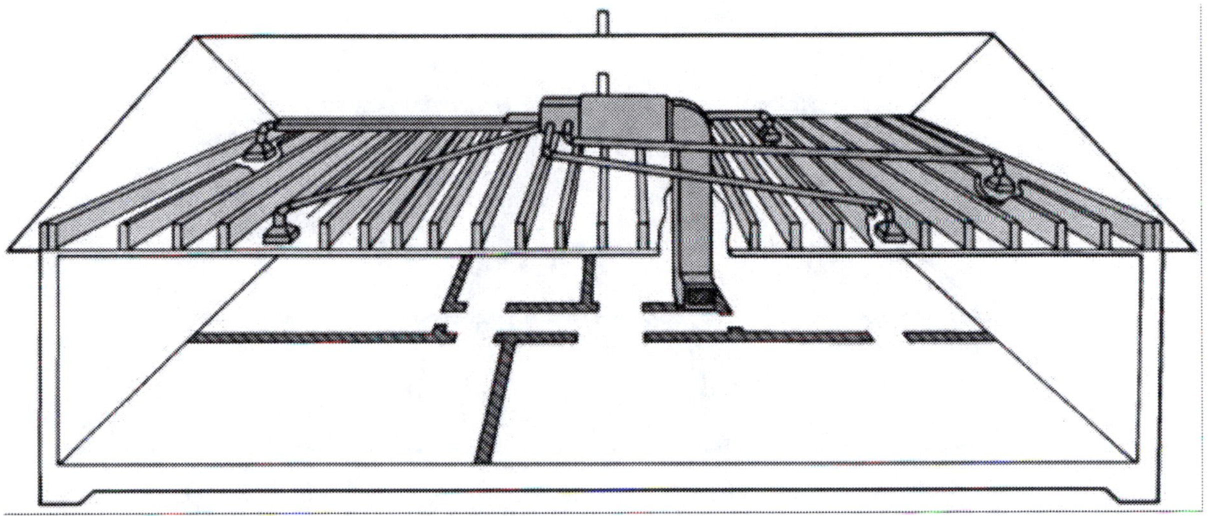

FIGURE 17-9 A plenum duct system

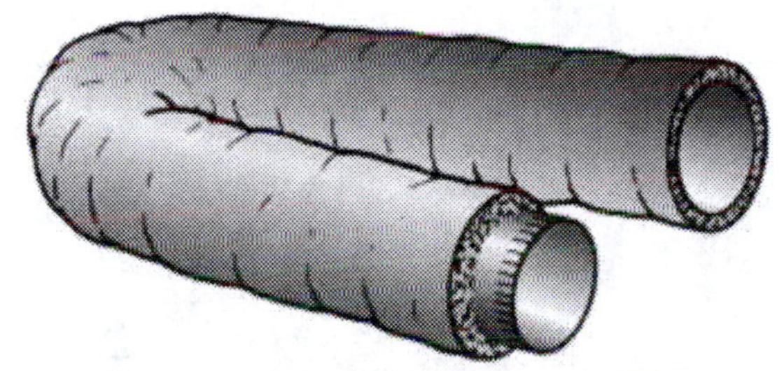

FIGURE 17-10 Flexible spiral duct

EXTENDED-PLENUM SYSTEM. In the event that the individual duct runs are found to be too long, the extended-plenum system is a good alternative to the standard plenum system. This type of system involves the installation of rigid duct sections that are connected to the main plenum box, in effect extending the plenum closer to the individual supply register locations (Figure 17-11). This extension is commonly referred to as a **trunk line.** It is from this plenum extension that the now shorter flexible duct runs are connected. The size of this extended plenum remains the same for the entire length of the run. Although this type of system utilizes more prefabricated duct sections, it is still relatively easy to install but more time-consuming than the plenum system described earlier. This type of air-distribution system is somewhat more costly than the plenum system, but the frictional losses in the system are greatly reduced.

SERVICE NOTE: *As with the plenum system, takeoffs should not be located on the end of the plenum.*

REDUCING-EXTENDED-PLENUM SYSTEMS. Very similar in construction to the extended-plenum system is the reducing-extended-plenum system (Figure 17-12). This type of system is also designed to bring the supply plenum closer to the supply registers. The one main difference between the reducing-extended-plenum system and the extended-plenum system is that the main trunk reduces in size as the end of the plenum is reached. The main benefit of utilizing this type of system is that the velocity and the pressure of the air flowing through the duct are relatively constant throughout the system. Although the cost of producing smaller duct sections is lower, an added cost is involved. This added cost comes from the fact that it costs more to fabricate **transition duct sections** than straight sections. Transition pieces are used to connect two sections of different-sized duct to each other (Figure 17-13).

SERVICE NOTE: *In each of the duct systems discussed, each individual takeoff should be equipped with a damper at the main trunk line by which the volume of air through each branch duct can be controlled. By closing off a branch duct at the register, excessive noise can result.*

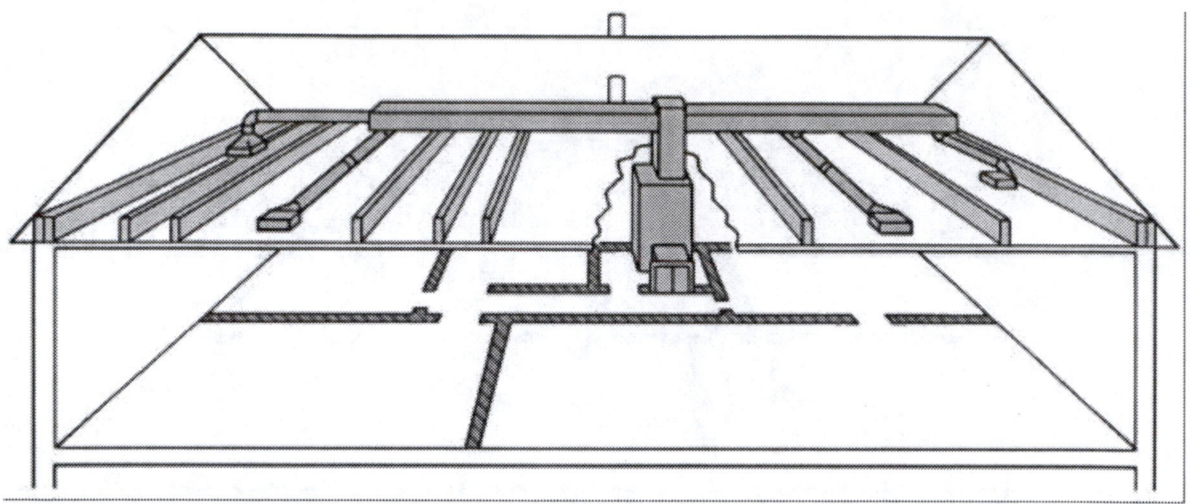

FIGURE 17-11 An extended-plenum duct system

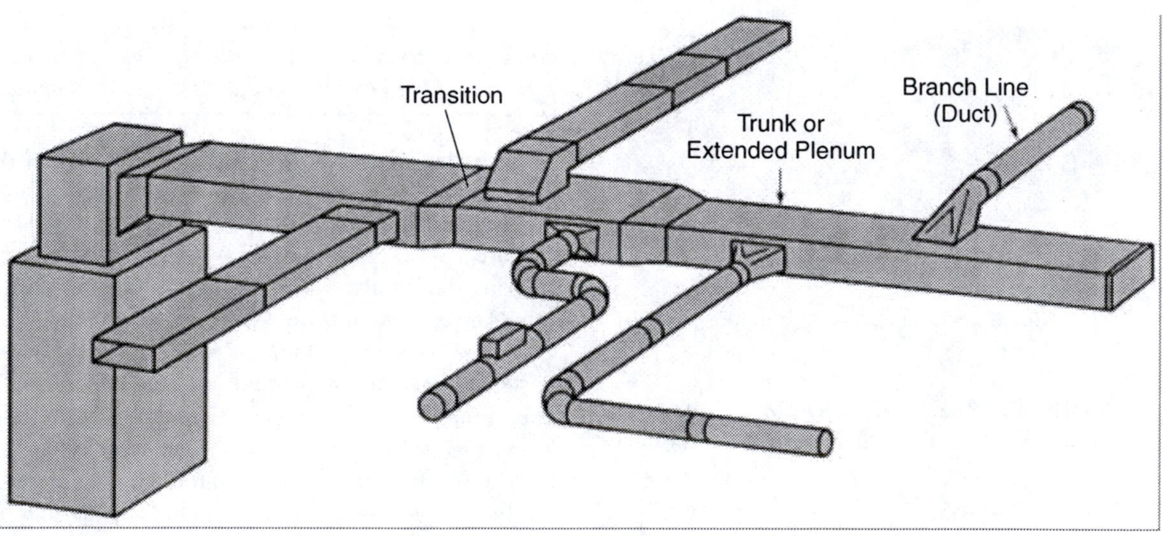

FIGURE 17-12 A reducing-extended-plenum duct system. *(Courtesy Climate Control)*

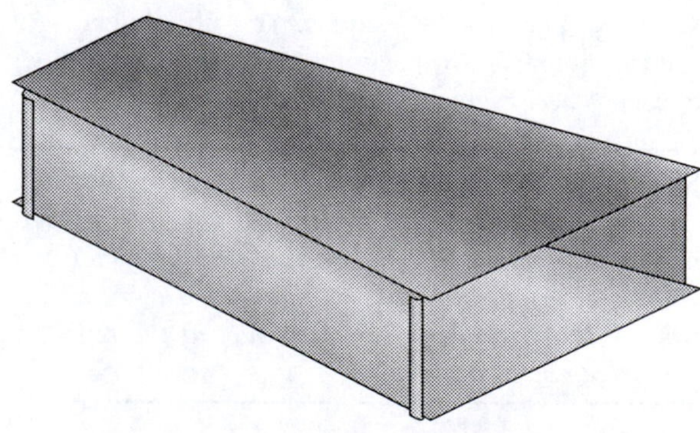

FIGURE 17-13 Transition duct section

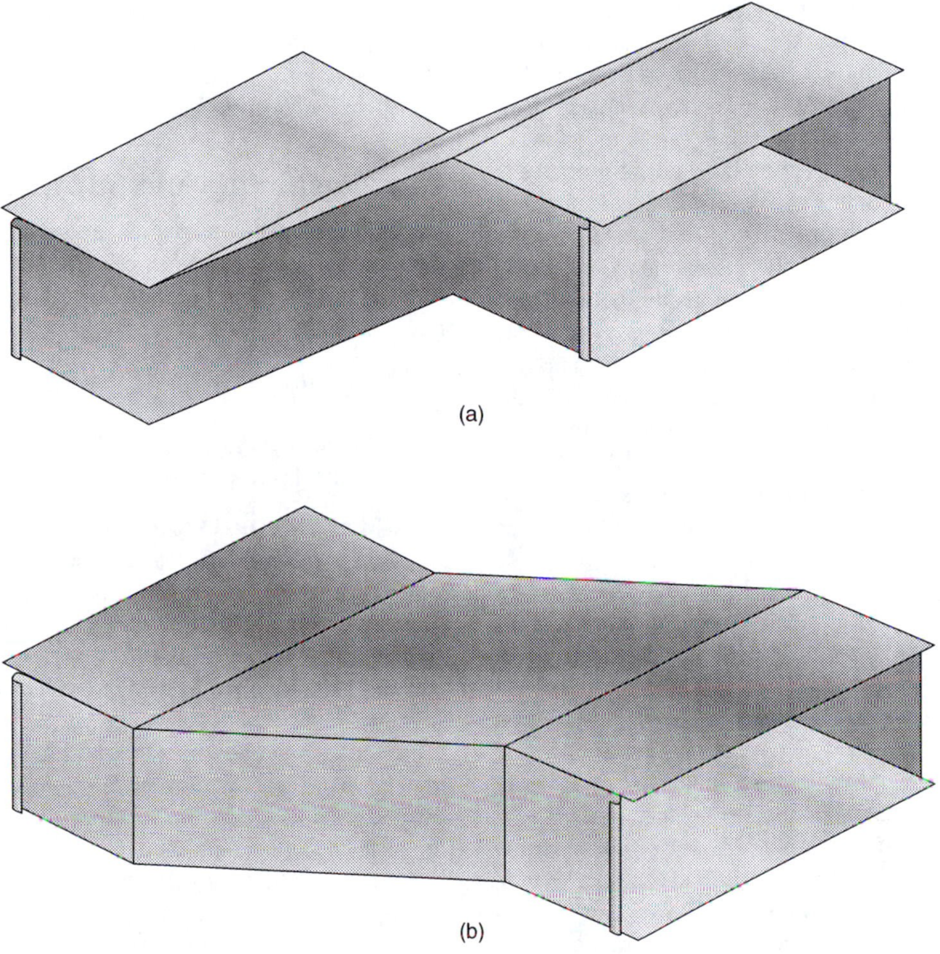

FIGURE 17-14 (a) Vertical offset duct section. (b) Horizontal offset duct section.

In addition to requiring more transition duct sections, extended-plenum and reducing-extended-plenum systems that are installed in existing structures also require the use of more **offset duct sections** (Figure 17-14). Offsets are needed when the duct system must be made to fit around and within the existing conditions of the structure. Figure 17-14(a) shows a vertical offset; Figure 17-14(b) shows a horizontal offset.

PERIMETER LOOP SYSTEMS. In the perimeter duct system, a continuous duct loop runs around the perimeter of the conditioned space, as shown earlier in Figure 17-2(d). This perimeter loop is fed by a number of feeder ducts that extend from the supply plenum on the indoor unit. Although not commonly found on residential applications, the perimeter loop system is a good choice for commercial structures that are built on concrete slabs. In this application, the loop is installed before the slab is poured. This type of system is ideal for conditioning large open areas, as the pressure in the entire loop is the same, delivering a constant flow of air to all of the registers.

Duct Materials

Ductwork can be made of a number of different materials. The choice of materials is often made by considering the location, the cost, and the intricacy of the installation. Some of the materials that can be found on air-distribution systems are:

- Galvanized metal
- Fiberglass sheeting
- Fiberglass-wrapped, spiral duct (flexible duct)

Galvanized sheet metal duct sections are the most costly and must be prefabricated in a duct shop and then installed on the job. Before the duct is installed, the space must be carefully measured and each piece must be fabricated exactly, according to the plan. If, for any reason, a duct section does not fit properly or if an obstruction is in the way of the duct run, a new piece will have to be made up. This could possibly delay the completion of the installation. The use of galvanized ductwork is the most costly route to take, but the life expectancy of the system is also the longest. This type of ductwork is ideal for applications in which ductwork is to be sealed behind walls and in enclosed ceilings. When ductwork is to be run outdoors, this is also the route to take. Under extremely damp conditions, stainless steel or aluminum ductwork is sometimes used, although the cost for these materials is higher.

Fiberglass sheeting, also known as ductboard, is another alternative for fabricating ductwork. This fiberglass is in the form of rigid sheets, and the duct sections are fabricated in the field. A system made of fiberglass boards gives members of the installation crew the flexibility to fabricate as they go. Simply put, the sections are measured and fabricated as they are installed, minimizing the chances of making a mistake. The on-site installation time of this type of system is much greater than that for a prefabricated duct system, but the cost of the material is less. Two major benefits of using fiberglass ductwork are that practically no noise is transmitted through the duct system and that the ductwork will never sweat. Fiberglass ducts are not as durable as galvanized systems and are, therefore, a good choice when the system is located in a low-traffic area. Ductwork that is to be buried behind walls or in ceilings should not be made of fiberglass.

Another material commonly found in duct systems is spiral flexible duct. This duct material is constructed of spring-like shaped metal that is covered with a plastic film. The duct is then sometimes wrapped in fiberglass and then covered with another casing of either foil or plastic (Figure 17-10). Flexible duct is very popular because of its ease of installation. It is cut to length in the field and provides the connection between the plenum takeoff and the supply register. Flexible duct comes in a wide range of diameters, from 4 inches to over 24 inches. As mentioned earlier, the biggest drawback on using this type of duct material is that the friction within the duct is very great. It should, therefore, be used primarily for short runs.

Duct Fastening and Supporting

Depending on the construction of the duct, as well as the material used, a number of different methods can be used to fasten duct sections together. Methods for duct fastening include:

- Slips and drives
- Bar slips
- Reinforced duct tape
- Sheet metal screws

SLIPS AND DRIVES. When installing a duct system that is made of galvanized sheet metal duct sections, the most common method for joining the sections is with **slips and drives.** Slips are simply strips of metal formed into an S shape, while drives are strips of metal that are formed into a flattened C shape (Figure 17-15). The longer edges of the duct slip into the openings in the slip (Figure 17-16), hence the name. Once the duct sections are pushed together, a drive is used to secure the sections (Figure 17-17). To use slips and drives, the shorter edges of the duct must be bent over to form ears (Figure 17-18).

Slips and drives are not needed when the duct system is of the plenum type, because only one piece of ductwork—the plenum—is used. When installing the extended-plenum or reducing-extended-plenum-system, several pieces of ductwork must be joined together. Because the metal ductwork can easily transmit noise and vibrations, a canvas collar (Figure 17-19) should be used. This collar is made of fireproof canvas or other flexible material and prevents unit vibrations from traveling through the duct sections. The ends of the collar are made of metal to facilitate the connection of the collar into the sheet metal duct system. This collar can be installed right on the discharge of the indoor unit or on the duct sections that connect to the supply plenum. If installed on the indoor unit itself, proper support for the plenum must be provided. The flexible canvas collar should not be used to support the weight of the duct.

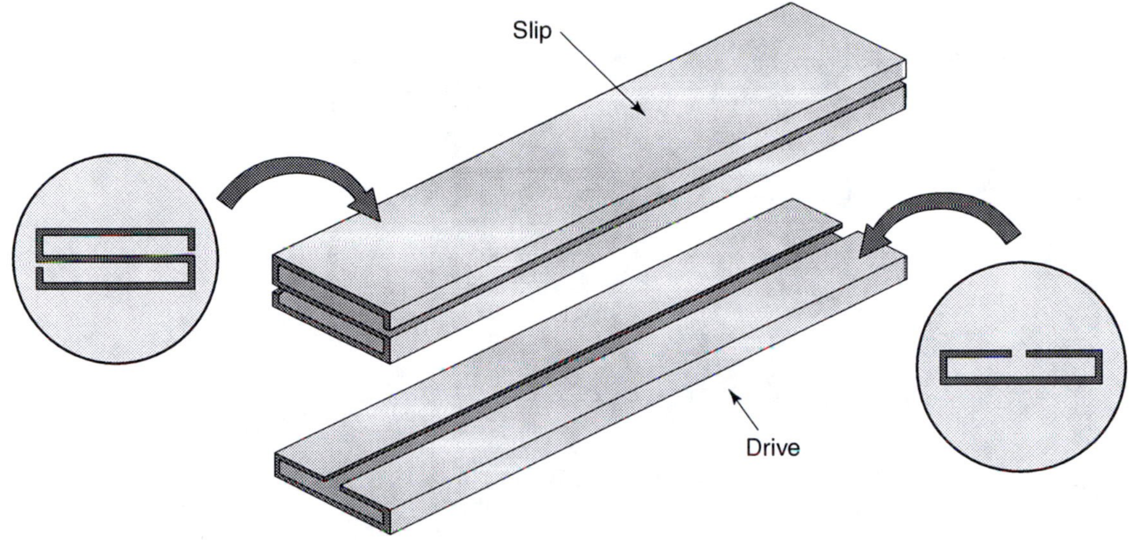

FIGURE 17-15 Slips and drives

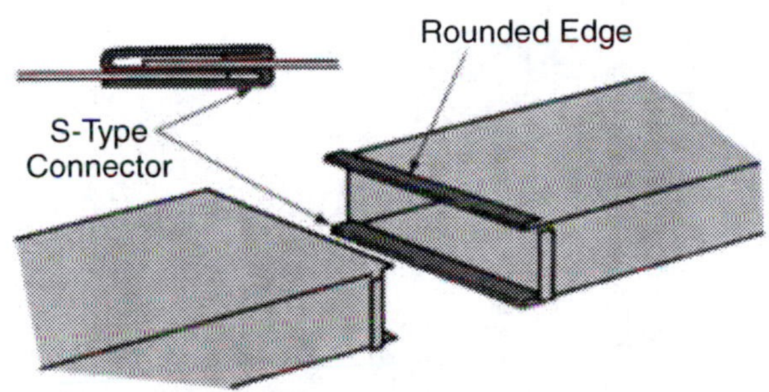

Ducts Prior to Fitting Together and Cross-Section Detail of S-Type Connector

FIGURE 17-16 Long ends of the duct sections slide into the slip S connector.

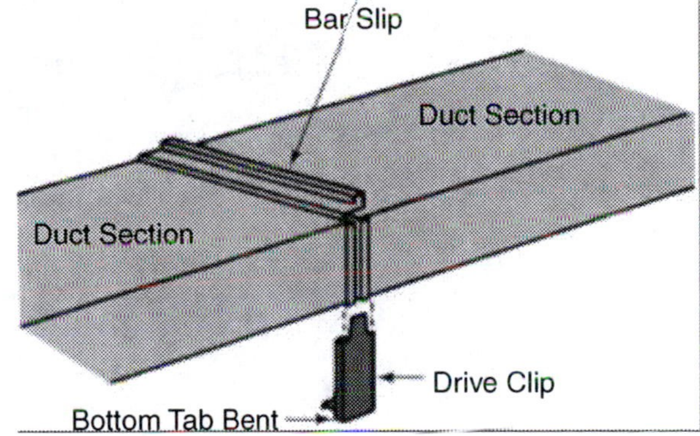

Ducts in Place Ready for Securing with Drive Clips

FIGURE 17-17 Drives are used to hold the sections together.

BAR SLIPS. When duct sections are very wide, they tend to flex and bend when the indoor fan cycles on and off. In an effort to reduce the noise generated by this, a modified slip is often used. This slip is often called a **bar slip** (Figure 17-20). Similar in construction to the slip described earlier, the bar slip has an extra perpendicular tab that prevents the slip, as well as the duct that it is connecting, from bending and flexing during blower start-up and shutdown.

SHEET METAL SCREWS. Even though slips and drives are designed to hold duct sections together, good field practice also requires using sheet metal screws as well. Depending on the size of the duct, a few well-placed screws will ensure that the duct system remains in tip-top shape for

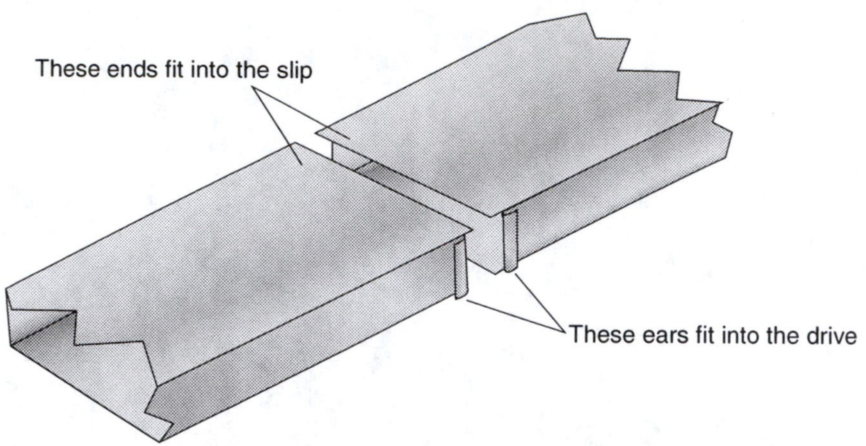

FIGURE 17-18 Short ends of the duct sections are folded back to form ears.

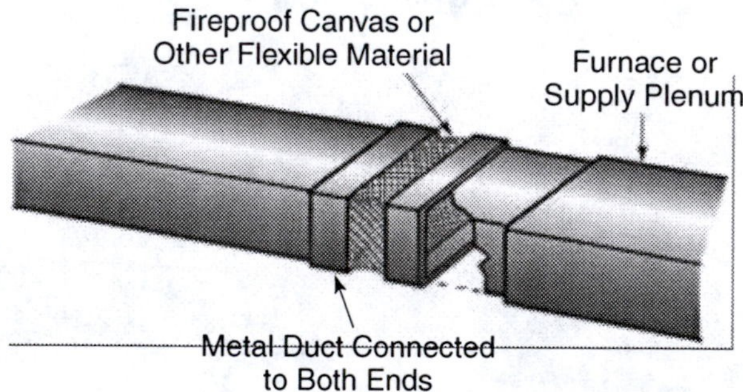

FIGURE 17-19 Canvas collar

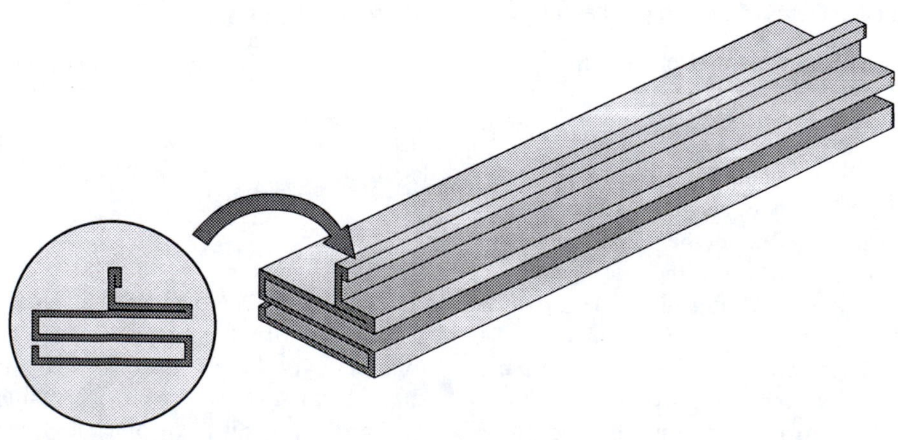

FIGURE 17-20 Bar slips provide more duct rigidity.

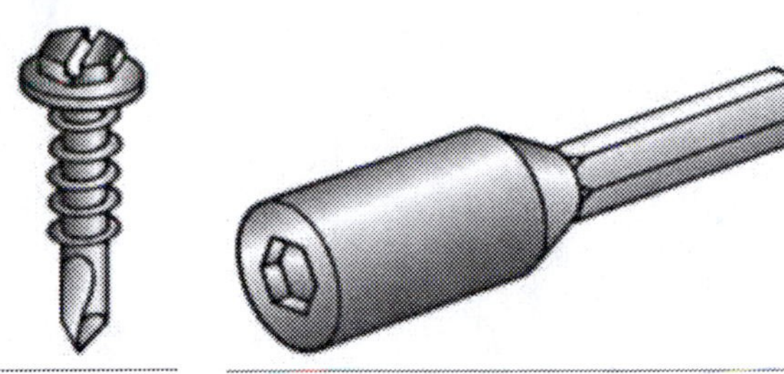

FIGURE 17-21 Self-tapping sheet metal screws with a magnetic screw holder

years to come. The location of the screws should be on the longer edges of the duct where both sections meet at the slip connector. This ensures that the screw penetrates both the slip connector and both sections that are being joined together. When supporting the ductwork, sheet metal screws are also used to secure the strapping material to the ductwork itself.

The most commonly used sheet metal screws are self-tapping. These screws have drill-bit-type tips that are able to make the hole in the metal and install the screw. These screws typically have a hex head, which reduces the chances of slippage. Special magnetic holders fit directly into the drill that will hold the screws in place while they are being positioned (Figure 17-21). The typical head sizes for these screws are ¼ and 5/16 of an inch. When screws are being installed in the ductwork, the drill should be running at a low speed; then, once the hole has been started, the drill speed can be increased. For this reason, a variable-speed drill is desirable for duct installation. Because duct installers are constantly moving from one location to another, a cordless drill should be used.

SAFETY NOTE: *When installing self-tapping sheet metal screws, remember that small metal filings will become airborne as a result of the drilling action. These filings can be very hot! Be sure to wear proper eye protection.*

SUPPORTING SHEET METAL DUCTWORK. As the ductwork is installed, it must be properly supported. The most common method used to support the duct is by utilizing strapping material, which is simply strips of heavier-gauge sheet metal. These straps are secured to the floor joist or rafters and are then screwed into the sides and bottom of the duct sections (Figure 17-22). On smaller-sized ducts, there should be one set of duct straps per duct section; on larger ducts, there should be at least two sets per section.

If the ductwork is located in an attic, for example, and the rafters are too high to use the method just described, the duct can be supported from below. Angle iron or aluminum angle can be used to form legs, which are secured to the sides of the duct sections (Figure 17-23). In this case, the flooring under the duct must be sturdy enough to support the weight of the ductwork. For example, the support legs must not rest directly on plaster board that is, in effect, the ceiling of the occupied space.

STAPLES AND REINFORCED DUCT TAPE. When fiberglass duct sections are fabricated, a flap of the foil material is left on the edges of the section so that one section can be stapled to the next. Once stapled together using outward clinching staples, reinforced duct tape is used to complete the connection (Figure 17-24). This tape must be UL181 type, which is often installed with the aid of a heat iron to activate the adhesive. This tape also has fibers within it, making it very durable. These fibers match the foil backing on the fiberglass boards themselves, creating an airtight seal. One of the major benefits of using fiberglass duct sections is that air leakage is almost completely eliminated. As with any other duct system, the ductwork must be properly supported. Since these duct sections are not made of metal, screwing strapping material to them will not work very well. In this case, a cradle that the duct will rest on can be easily fabricated from strapping material (Figure 17-25). This cradle can be secured to the floor joists or rafters, depending on the installation. Because this duct system is not rigid, the supporting cradles should be located close together to prevent the duct system from sagging.

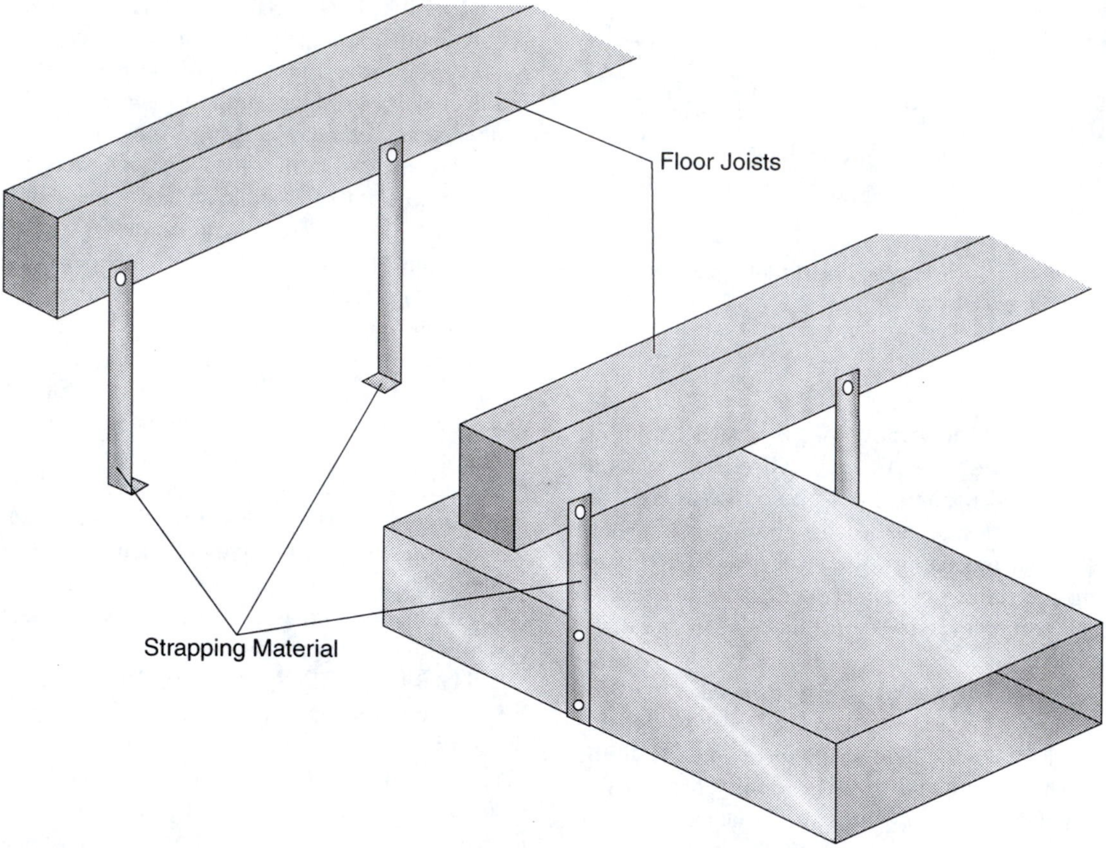

Floor Joists

Strapping Material

FIGURE 17-22 Sheet metal ducts supported with metal strapping material

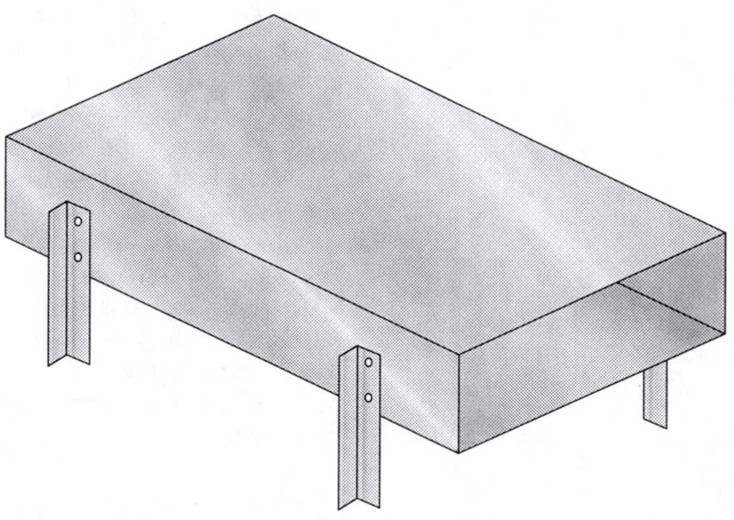

FIGURE 17-23 Angle iron used to form duct supports

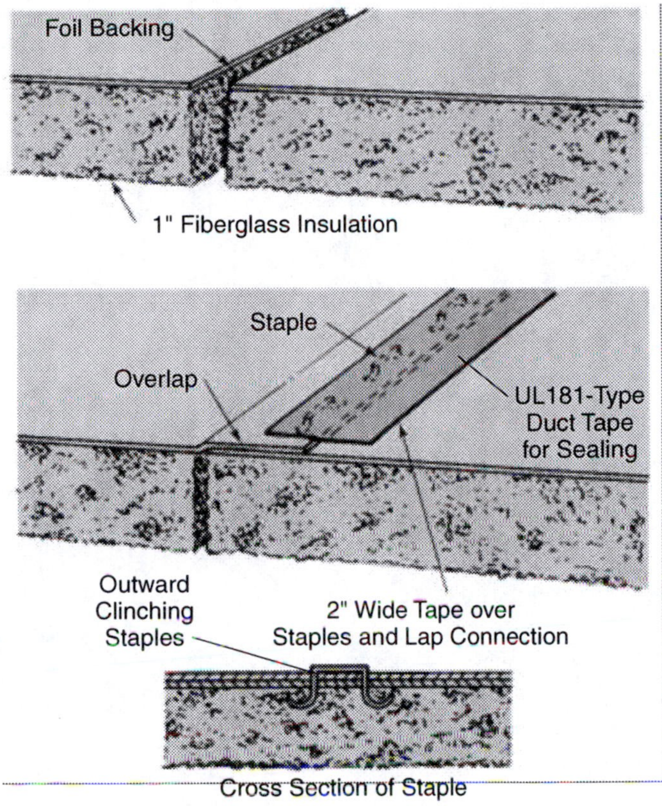

FIGURE 17-24 The joining of fiberboard duct sections

Duct Insulating

When using a sheet metal duct system, heat transfer will take place between the air in the duct and the air surrounding the duct. In the cooling mode, heat from the surrounding air will be transferred to the cooled air inside the duct. In the heating mode, the heat transfer will take place from the heated air to the surrounding air. In either case, a loss of system effectiveness and efficiency will result. In addition, metal ductwork will sweat during the cooling mode of operation. This can result in water damage to the structure. To alleviate all three of these problems, metal air-distribution systems should be insulated. A few options are available to the installation crew with regard to insulating the ductwork. They are:

- Wrapping the duct with foil-covered fiberglass insulation
- Lining the duct with acoustical lining
- Wrapping and lining the ductwork

Once the duct system has been installed, the duct can be wrapped with foil-covered fiberglass insulation. This insulation comes in rolls that are typically 4 feet wide; it is secured around the duct with reinforced duct tape that is similar to that used to secure the fiberglass ductwork sections described earlier. The entire duct run must be insulated to prevent any condensation from forming on the duct surface. Also, the duct system must be wrapped tightly to prevent the formation of large air pockets between the duct and the insulation, which can result in excessive condensate accumulation.

Another option is to line the duct sections with acoustical lining. Duct lining is installed on the sheet metal before the duct section is fabricated. The lining is held in place with pins, as well as a contact adhesive designed specifically for this purpose. The lining must be secure on the inside surface of the duct and must be able to withstand the effects of the high-velocity air moving across it. If the lining should

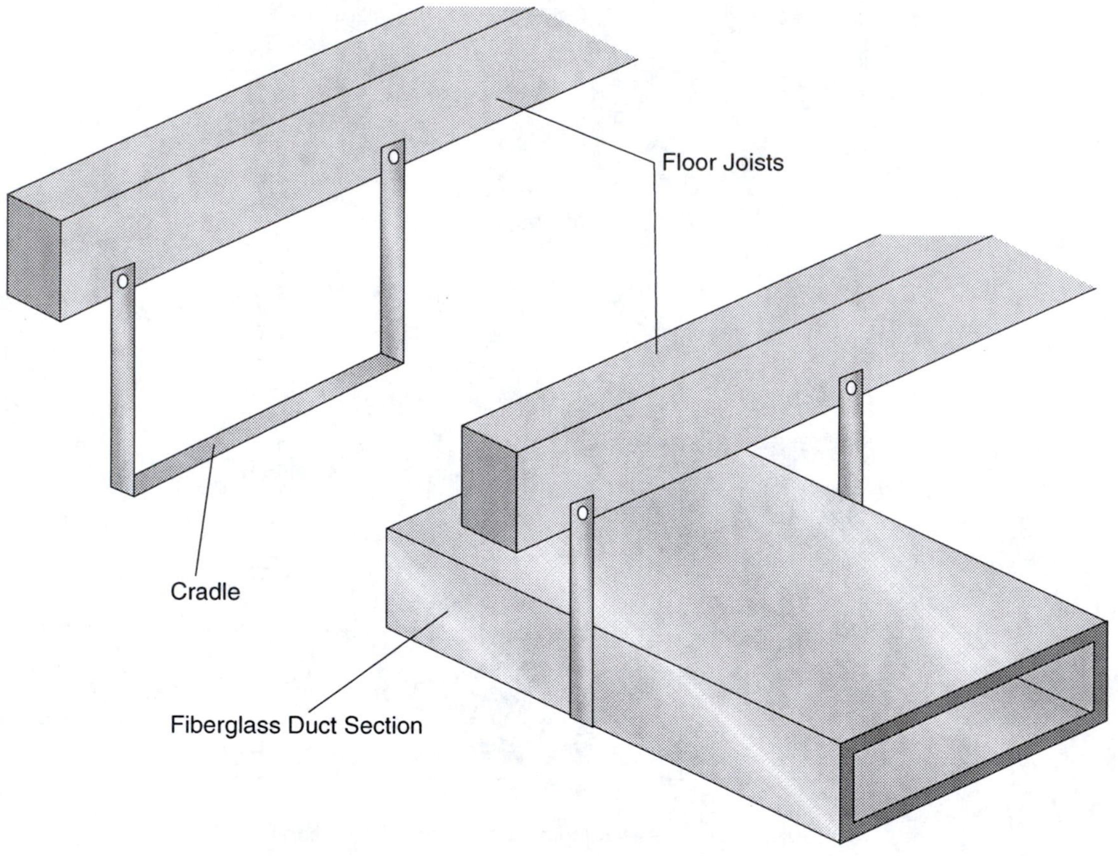

FIGURE 17-25 Fiberboard duct sections rest in cradles made of strapping material.

come loose during system operation, the ducts can become blocked, causing system malfunction. The decision to line the duct system must be made when the system is initially designed. The duct size will have to be made larger to accommodate the lining. Acoustical lining, depending on the application, is approximately 1 inch thick and will reduce the effective cross-sectional area of the duct run. For example, a 10 inch by 10 inch, uninsulated duct will have a cross-sectional area of 100 square inches, while a 10 inch by 10 inch duct insulated with 1-inch thick insulation will have a cross-sectional area of only 64 square inches (Figure 17-26).

Fiberglass duct sections do not need to be insulated after installation since the duct material itself acts as an insulator. This is another benefit of the fiberglass duct system. Spiral flexible ductwork, on the other hand, can be purchased as either insulated or noninsulated. If the determination is made that the duct run should be insulated, then it should be purchased and installed as such.

CONDENSATE REMOVAL

In the cooling mode of operation, the heat-pump system will remove moisture from the conditioned air, as well as lowering its temperature. The moisture from the air must be safely removed from the structure this moisture is called **condensate.** For this reason, a **condensate drain line** must be run in order to carry this condensation away. A number of factors must be considered when laying out the condensate piping for a system, including:

- Drain line size
- Drain line material

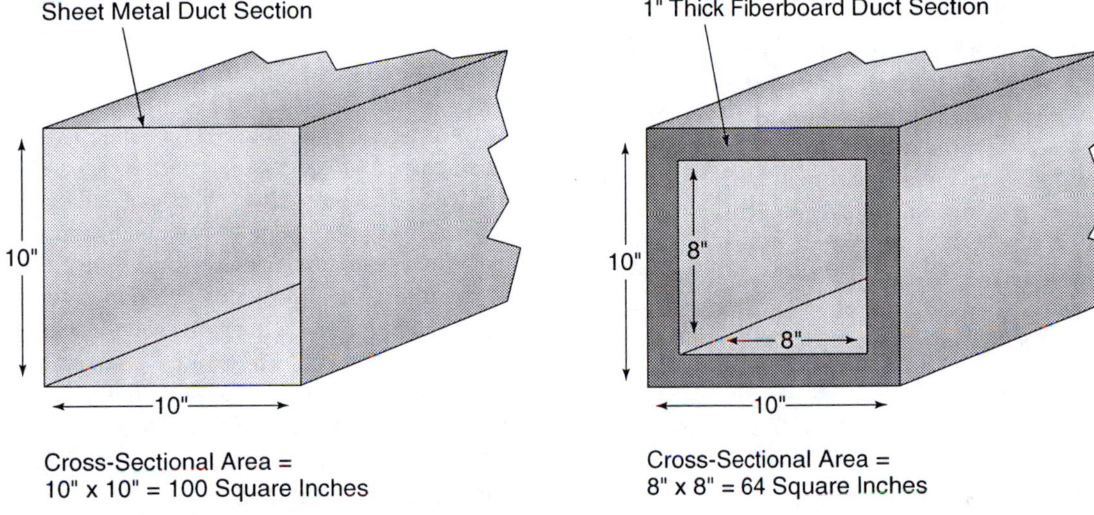

FIGURE 17-26 Fiberboard duct sections reduce the cross-sectional area of the duct.

- Pitching of the line
- Traps
- Auxiliary drain pans
- Condensate pumps
- Safety float switches

Drain Line Size

The drain line of the system must be able to handle the water flow that is generated by the removal of moisture from the air. All air handlers are equipped with internal drain pans and female pipe connections to which the field-installed drain line is to be connected. The size of the drain line that is run should be the same size, or bigger, as the fitting provided with the system. For example, if the unit is equipped with a ¾-inch female pipe thread connection at the unit, under no circumstances should the size of that line be reduced. This line size was determined by the system manufacturer and should not be altered. In areas of extremely high humidity, however, the line size can be made larger than the system design warrants. Common sense dictates that larger units will require larger condensate lines. For example, a 5-ton air-conditioning system will produce, on average, approximately 2 gallons of condensate per hour, while a 10-ton system will produce nearly 4 gallons per hour. Under no circumstances should a drain line be less than ¾ of an inch in diameter.

Service note: To prevent drain line condensation, the primary drain line should be insulated at least 10 feet from the indoor coil's drain pan outlet.

Drain Line Material

The drain line should be constructed of materials that are in accordance with local plumbing codes. Normally, PVC piping is acceptable for most residential applications and some commercial applications. Other larger installations require the use of copper piping. When the condensate line is to be located within a concealed wall or ceiling, it is recommended, and sometimes required, that copper be used as well. When installing a PVC condensate drain line, all fittings and pipe ends should be cleaned first with a PVC primer and then joined using PVC cement. Using cement that is not designed for use with PVC pipe can result in a weak joint that may leak in the future. When installing a copper drain line, all solder joints must be checked for leaks before the system is put into operation. Other lightweight rigid plastic materials should be avoided but can be used if the length of the line is very short.

Pitching of the Line

Regardless which material is chosen for use, the line should always be pitched toward the termination point of the line. The minimum pitch on the line should be ¼ of an inch per 1 foot of horizontal run. For example, a 10-foot horizontal drain line should be pitched a minimum of 2½ inches. Drain lines should be adequately supported to ensure that the proper pitch is maintained. To facilitate proper draining of the condensate, the pitch should be made as large as possible. Improperly pitched drain lines can result in water damage to both the system and the structure.

Traps

Traps in condensate lines are used to help prevent air from flowing up through the drain line. Air flowing up the line will prevent the condensate from draining properly. The **condensate drain trap** functions in much the same manner as the trap under a bathroom sink. The under-the-sink trap prevents waste fumes from entering the home. A trap, however, is not required on all air handlers. If the refrigerant coil is located downstream of the blower, meaning that air is being pushed through the coil, a trap is not needed. In this case, the pressure at the inlet of the drain line is at a higher pressure than at the outlet and the condensate will drain properly. If, however, the blower is downstream of the coil, a trap must be used. In this case, the air is being pulled through the coil and air is also being pulled through the drain line. This causes the pressure at the inlet of the drain line to be lower than the pressure at the outlet, thereby pulling air through the line in the direction opposite to the direction of condensate flow. If the drain line is under a negative pressure and no trap is installed in the line, the condensate will accumulate in the pan and eventually cause it to overflow. This can lead to water damage. This applies to both horizontal and vertical units. When in doubt, install a trap. The following chart (Figure 17-27) provides guidelines for determining whether or not a trap is needed. For example, the system in Figure 17-28 requires a trap on the condensate line because it is an upflow unit with the coil located downstream of the blower.

Drain Line Termination

The condensate that is removed from the system must ultimately be deposited in a location that is not objectionable to the occupant of the occupied space. This location is commonly either outside the structure or down a waste line. If the unit is located in an attic or in a utility closet, the drain line commonly leaves the structure through the same building penetration that the refrigerant lines and low-voltage wiring will ultimately pass through. The drain line should be as short as possible to reduce the possibility of a blockage that would cause the line to back up and cause damage. If running the line to the outside of the structure is not possible, the line should be run to a nearby waste line or utility sink. Once again, all local codes regarding the installation of drain lines should be followed.

	Blower above Coil	**Blower below Coil**
Vertical Upflow	Trap needed	Trap not needed
Vertical Downflow	Trap not needed	Trap needed
	Blower to the Right of the Coil	**Blower to the Left of the Coil**
Horizontal Left-to-Right Flow	Trap needed	Trap not needed
Horizontal Right-to-Left Flow	Trap not needed	Trap needed

FIGURE 17-27 Condensate trap usage

If the indoor unit is located in a basement, running the condensate line to the outside may not be as easy as in the case of the attic or utility closet installation. When in a basement, the unit is below ground level and a gravity-type drain is not possible. In this case, two other options are available to the installation crew:

- Utilize a floor drain
- Pump out the condensate

If the unit is installed in an unfinished basement and a floor drain is nearby, the condensate should be directed toward it (Figure 17-28). The line should be run from the unit, along the floor of the basement, making certain that the line is secured to ensure that the water is directed to the drain. Whether or not a trap is required depends on the configuration of the unit. Before the system is started up, the floor drain should be tested to be certain that it is still operational.

If a floor drain is not available or if the basement is finished, the condensate may need to be pumped from the location. **Condensate pumps** (Figure 17-8) come in a wide range of styles based on voltage, holding capacity, and body style. Low-profile pumps are used when not much clearance space is available for the pump. Low-profile pumps are popular on units installed in ceilings where no termination point is nearby for the condensate. In operation, the pump is normally in the off position and is controlled by a float switch. Once the water in the pump's reservoir reaches a predetermined point, the pump switches on, removing the water from the pump. Once the water level is lowered, the pump switches off.

The pump is located in close proximity to the unit, and the water is then pumped either outside or to a utility sink for disposal. The piping run from the unit to the pump is therefore very short. Once again, the need for a trap is determined by the configuration of the indoor unit.

Some condensate pans are equipped with an overflow control. This will prevent water damage from occurring if the pump should fail. This overflow control is a normally closed switch that will open if the water level rises above the maximum level permitted by the pump. If, for example, the pump is designed to turn on if the water reaches a level of 4 inches, the water level should never be higher than that if the

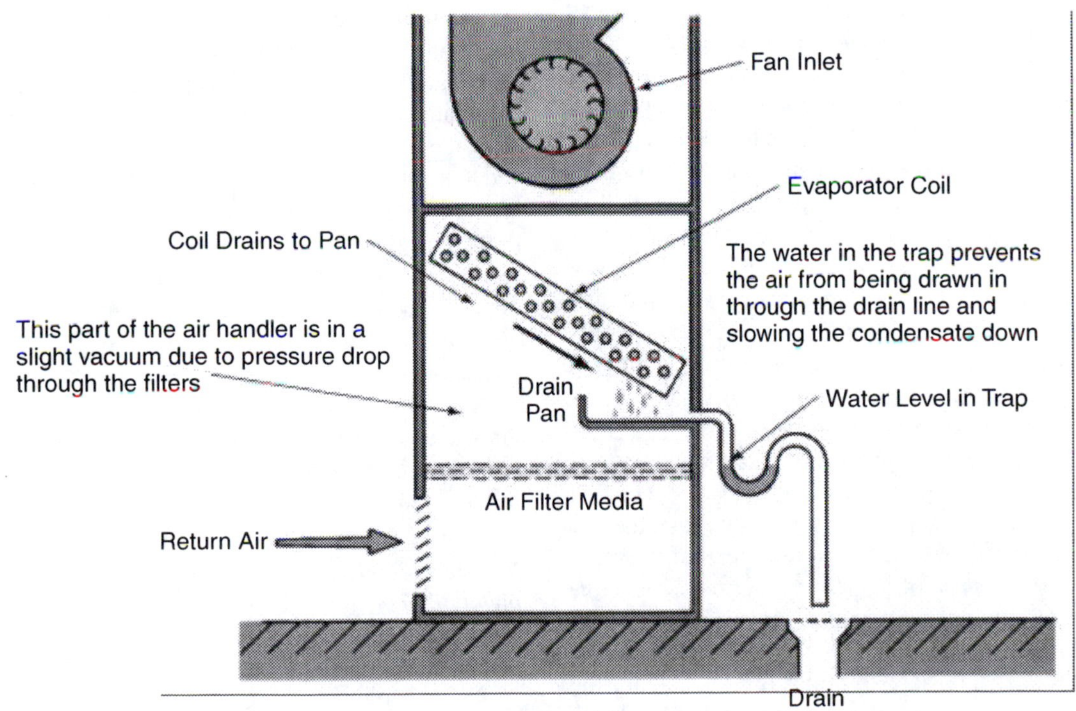
Fan Inlet
Evaporator Coil
Coil Drains to Pan
The water in the trap prevents the air from being drawn in through the drain line and slowing the condensate down
This part of the air handler is in a slight vacuum due to pressure drop through the filters
Drain Pan
Water Level in Trap
Air Filter Media
Return Air
Drain

FIGURE 17-28 This air handler needs a condensate trap.

pump is operating correctly. The overflow control will open its contacts if the water reaches, for example, 4½ inches. This overflow control is normally wired in series with the R wire coming off the control transformer. When the overflow control opens its contacts, the system shuts down, preventing any further moisture removal from the conditioned space. The customer will then place a service call indicating that the system is not operating at all. The shutting down of the system prevents the water from overflowing the pump and causing damage.

Auxiliary Drain Pans

If a problem arises with a drain line—such as a blockage, for example—the condensate cannot be carried away from the unit. This would lead to an overflow condition and could result in water damage to property located under the unit. Preventive maintenance could help prevent the drain line from getting clogged, but this occurrence cannot be predicted. One way to effectively eliminate the possibility of water damage due to an overflow conditions is to install an **auxiliary drain pan** under the unit (Figure 17-29). The purpose of this auxiliary pan is to catch any overflow condensate that may result from a blocked or clogged condensate line and carry it away.

The pan should be sized in a manner that would allow the pan to catch water leaking from anywhere in the unit. In other words, the pan must be larger than the indoor unit. Prefabricated plastic drain pans can be purchased in a wide range of sizes that will be suitable for many systems. Larger pans, however, may need to be custom-made. Pans can be made of sheet metal as long as the corners are sealed to ensure that the pan will hold water. The auxiliary pan should be mounted securely under the unit in the event that it actually has to hold water. In an ideal situation, this pan should remain perfectly dry at all times.

The auxiliary drain pan and the pan manufactured in the air handler have two main differences. The first difference is that the primary condensate line, the one installed on the unit at the time of manufacture, is piped to the outside of the structure or to a waste line in a manner that disposes of the water discretely. The home owner or occupant of the space will not be aware that condensate is being removed from the structure. The auxiliary drain pan, on the other hand,

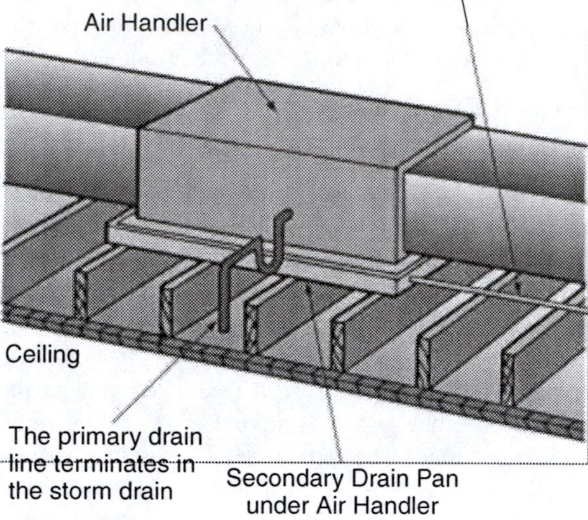
The secondary drain line terminates in a conspicuous place. The owner is warned that if water is seen at this location a service technician should be called.
Air Handler
Ceiling
The primary drain line terminates in the storm drain
Secondary Drain Pan under Air Handler

FIGURE 17-29 Auxiliary drain pan under unit

is piped to a location that is very conspicuous. This is done for one very good reason. If water is seen coming from that line, there is water in the auxiliary pan and there is a problem with the primary drain on the unit. The equipment owner is instructed upon system installation and start-up that, if water is seen coming from the line, a service call must be placed in order to prevent damage to the structure. This drain line is typically piped to a location over a doorway or window, so as to alert the occupant of a potential problem as quickly as possible. The second difference between the two drains is that the drain line from the auxiliary pan does not need a trap. No pressure difference exists between the two ends of the drain line, so water will flow freely from the pan. Of course, the drain line from the pan must also be pitched downward toward its termination point in order to ensure proper drainage.

Service note: The auxiliary drain line must be completely independent of the primary drain line. Never join the auxiliary drain line and the primary drain line into a common line. This will defeat the purpose of the backup line.

Safety Float Switches

Although auxiliary drain pans are designed to help prevent water damage the possibility exists that the auxiliary line can become clogged or blocked as well. Although this is very unlikely, it is possible. In this instance, once the auxiliary pan fills up, the water will overflow and potentially cause damage to the area below. To prevent this, a safety float switch can be installed in the auxiliary pan. Operating in a manner similar to the safety float switch in the condensate pump, this switch is a normally closed device that will open its contacts if the level of the water in the auxiliary drain pan reaches a predetermined level. This switch is wired in series with the low-voltage control circuit and will shut the system down if its contacts open. Once the system shuts down, the customer will call the service company and report that the system will not operate.

RUNNING REFRIGERANT LINES

Probably one of the most important tasks that the installation crew can perform to ensure the long-term satisfactory operation of the heat-pump system is the proper installation of the refrigerant lines. Poorly installed refrigerant lines can result in premature compressor failure, as well as a system that does not operate effectively and efficiently. The individuals installing these lines should keep a number of things in mind, including:

- The run should be as short as possible.
- No excessive piping should be used.
- The number of fittings should be kept to a minimum.
- When needed, long radius elbows should be selected.
- The run should be as straight as possible.
- All solder joints should be as perfect as possible.
- The vapor line should be pitched back toward the compressor.
- Refrigerant traps should be installed when necessary.
- The vapor line should always be insulated.

Length of the Piping Run

Although the preceding list seems to be quite long, the items listed summarize a satisfactory piping job. If any one of the items is neglected, system operation will be affected in a negative way. As was discussed earlier in the chapter, the indoor and outdoor units should be as close to each other as possible. If the location selection is done properly, it follows that the length of the refrigerant lines will be as short as possible. This will increase the system's efficiency in both the heating and cooling modes of operation. To ensure maximum efficiency, though, the piping run should be as direct as possible, with no excessive solder joints or pipe sections. Remember that the shortest distance between two points is a straight line.

Choice of Pipe Fittings

During the piping process, a number of pipe fittings will need to be installed, such as 90-degree elbows, 45-degree elbows, and couplings. When choosing 90-degree elbows, be sure to select those with a wide or long radius, as shown in Figure 17-30(a). Wide radius elbows provide less resistance to refrigerant flow than those with a tighter radius, as shown in Figure 17-30(b). This will help maintain a constant refrigerant velocity. Each fitting that is added to the refrigerant piping circuit adds to the resistance encountered by the refrigerant; therefore, the number of fittings should be kept to a minimum. Reducing the number of fittings in the piping circuit will reduce the number of solder joints required to connect the indoor and outdoor units. The smaller the number of solder joints, the smaller the chance of a refrigerant leak occurring.

Solder and Solder Joints

To ensure a high-quality installation, high-quality materials should be used. This includes high-quality solder and brazing materials. This is one of the easiest areas for the installation crew to cut corners, as it is next to impossible for an equipment owner to tell the difference between brazing material that contains 0 percent silver and one that contains 15 percent silver. Refrigerant lines vibrate to some degree, and a silver-bearing solder will stand up much better to these vibrations than a non-silver-bearing product.

All solder joints should be carefully inspected before a new system is put into operation. Refrigerant lines are often sealed behind walls and in ceilings, so access to them will be very limited after the installation is completed. In addition, the vapor line

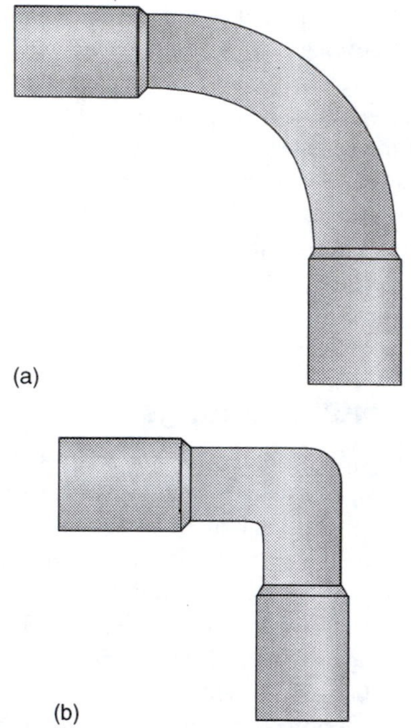

FIGURE 17-30 (a) Long radius elbow. (b) Short radius elbow.

FIGURE 17-31 Proper method to unroll copper tubing. *(Photo by Bill Johnson)*

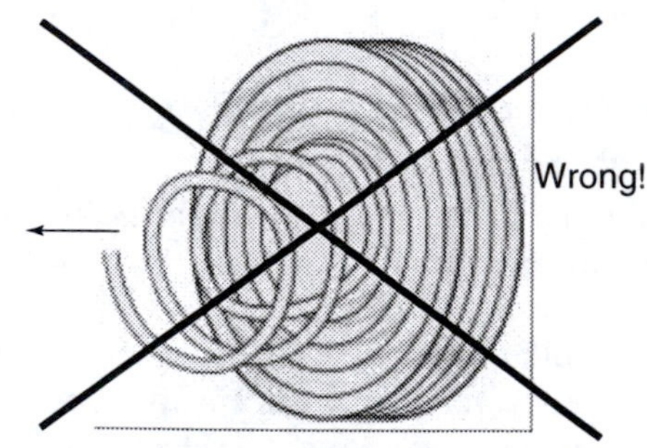

FIGURE 17-32 Wrong way to unroll copper tubing

is insulated, which would make finding a leak later on that much more difficult. Any solder joint that appears to be questionable as to its integrity should be resoldered or brazed.

SERVICE NOTE: *When brazing, 1 to 2 pounds of pressure of nitrogen should be permitted to flow through the lines being joined. This will prevent oxidation from occurring on the inside of the pipe surfaces. Excessive oxidation within the refrigerant circuit can result in the clogging of the metering device.*

Soft-Drawn Copper Tubing

One way to minimize the number of solder joints and fittings required is to use soft-drawn copper tubing whenever possible. Soft-drawn copper tubing comes in rolls, typically 50 feet in length, and should be carefully unrolled to ensure that the tubing does not kink. The best way to unroll the tubing is to place the loose end on the ground and hold it in place with your hand or foot. Then, carefully unroll the tubing, keeping the roll on the ground as it is unrolled (Figure 17-31). Simply pulling the tubing off the coil will result in the twisting of the tubing (Figure 17-32), thereby creating waves in the line. Although this will not have a major effect on system operation if the run is vertical, small refrigerant traps will be formed if the run is horizontal. An excessive number of refrigerant traps can affect system operation and hinder the process of compressor oil return.

Refrigerant Traps

Refrigerant traps are designed to aid in the return of oil to the compressor. Approximately 5 percent of the compressor's oil is traveling through the refrigerant circuit at any given point in time when the system is operating. The trap allows this oil to accumulate in one location, as shown in Figure 17-33(a); then,

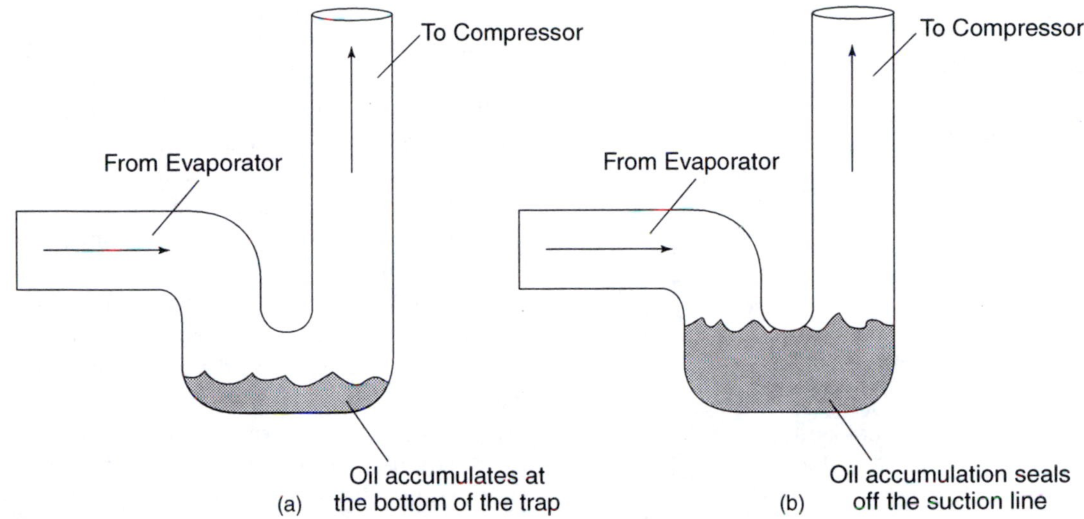

FIGURE 17-33 (a) Oil collects at the bottom of the refrigerant trap. (b) When the trap becomes liquid-locked, the difference in pressure pushes the oil back to the compressor.

when the line becomes blocked or the trap becomes liquid-locked, as shown in Figure 17-33(b), the oil is pulled back to the crankcase by the suction of the compressor. Traps, however, are not required on all system installations. Traps are typically installed on systems in which the indoor unit is located below the outdoor unit. One common example of this is when the air handler is located in the basement of a house and the outdoor unit is located in the backyard. In this instance, a refrigerant trap should be located as close to the outlet of the indoor coil as possible. The outlet of the indoor coil is identified when the system is operating in the cooling mode of operation. The exact measurements of the trap depend on the capacity of the system and the line sizes used. When installing a refrigerant trap, the installation crew should always refer to the manufacturer's installation book for exact trap sizing information.

On larger installations, in which the air handler is located well below the outdoor unit, good field practice involves installing one refrigerant trap every 15 to 20 feet for a vertical run. In these cases, installing an inverted trap at the top of the vertical run is also beneficial (Figure 17-34). This ensures that the oil returns to the compressor once it has reached the top of the run. The line connecting the inverted trap to the compressor should be sloped downward toward the outdoor unit.

INSTALLING LOW-VOLTAGE CONTROL WIRING

During installation, two separate low-voltage lines must be run. One line is run between the thermostat and the indoor unit, and the other is run between the indoor and outdoor units. The easiest way to run the low-voltage wiring between the indoor and outdoor units is to secure the low-voltage cable to the liquid and vapor lines as they are installed. This bundle (Figure 17-35) is made up of the insulated vapor line, the liquid line, and the low-voltage cable. Enough excess wire should be available to ensure that the line connecting the indoor and outdoor units is continuous. Joining sections of wire together to make this run poses potential service problems in the future.

Running the low-voltage wiring between the thermostat and the indoor unit is somewhat more involved. Unlike the running of the indoor/outdoor wiring, where the wire is simply getting a free ride from the piping, the thermostat line must be run by itself. The running of this line involves either running the wire up within a wall from a basement location or down within a wall from an attic location. In either case, the exact wall location must be determined from either the basement or the attic. Once the location has been determined, the exact location of the thermostat must be determined. Then, a hole must be drilled

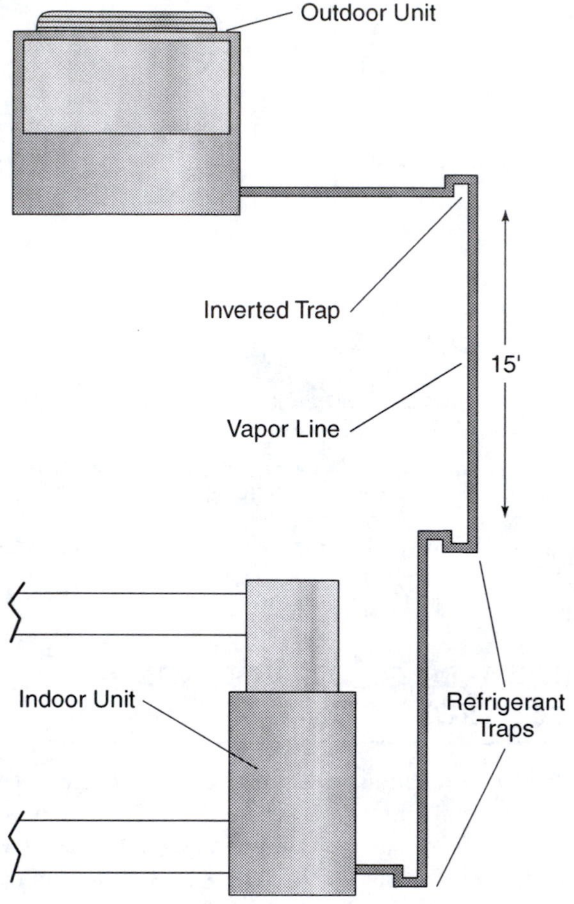

FIGURE 17-34 Refrigerant traps installed every 15 feet on a vertical run. Inverted trap located at the top of the run.

directly above for attic installations or directly below for basement installations in order to gain access to the interior of the wall. Once done, an electrical snake can be used to pull the wire through the wall. Running the wire from the thermostat to the indoor unit is very often a two-man job. When running the line between the indoor unit and the thermostat, the technician must remember that the walls have studs in them. These studs are typically 16 inches apart from center to center. For this reason, it is extremely important to ensure that the hole made in the wall for the thermostat wire penetration is in the same bay as the hole made in the attic or the basement (Figure 17-36). By using a flashlight and a small piece of hooked electrical snake material, one member of the installation crew can catch the snake as it is fed up from the basement by another crewmember (Figure 17-37). Once the electrical snake is brought through the hole in the wall, the low-voltage cable should be attached securely to the snake and pulled down to the basement location. Good field practice requires the use of electrical tape to secure the low-voltage cable to the snake. The same principle can be applied to systems that have their indoor units located in the attic.

LEAK-CHECKING A SYSTEM PRIOR TO START-UP

Leak-checking a system helps to ensure that all solder and flare connections in the system were made properly. A leaking refrigeration system results in

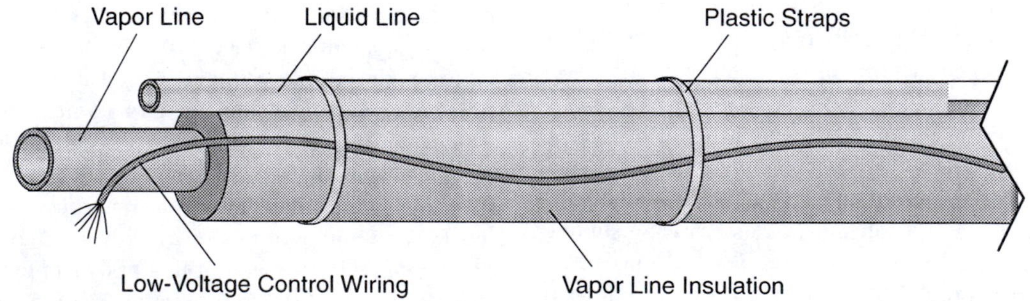

FIGURE 17-35 Tubing bundle consists of liquid line, vapor line, and the low-voltage cable.

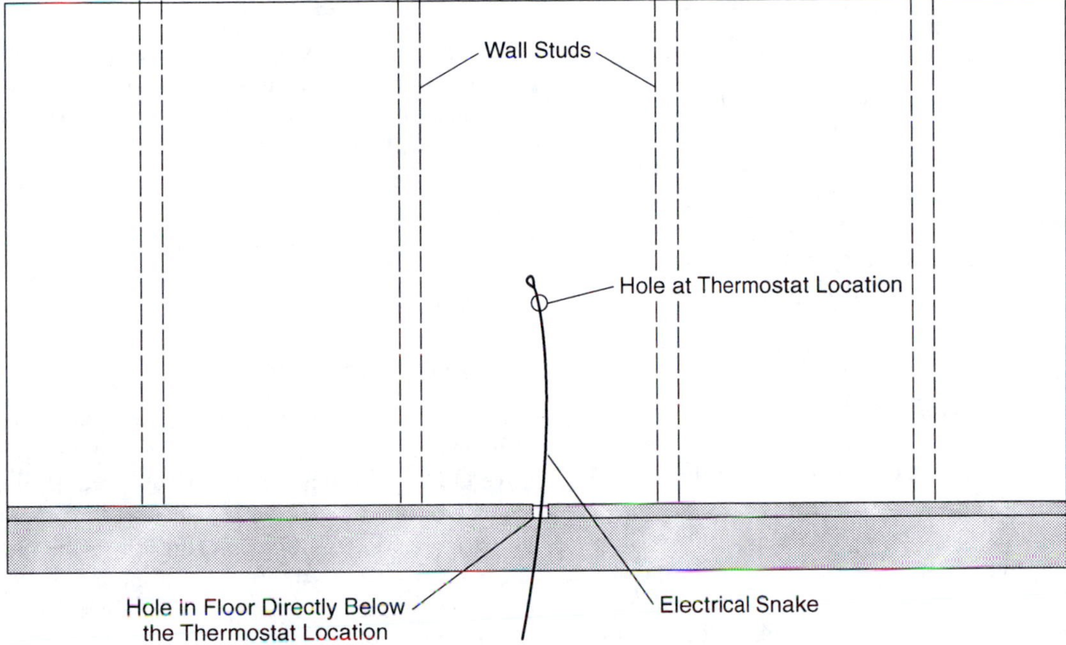

FIGURE 17-36 The hole at the thermostat location must be in the same bay as the hole in the foot of the wall.

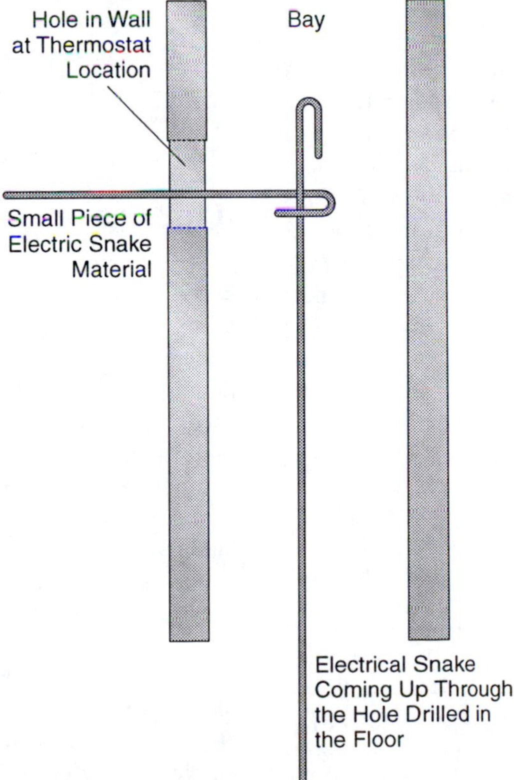

FIGURE 17-37 Fishing the electrical snake out of the wall

improper operation, as well as the introduction of refrigerant to the atmosphere. Therefore, leak-checking is a must prior to system start-up. This is especially true in circumstances in which refrigerant lines will eventually be sealed behind walls or ceilings. Accessing these solder joints will be extremely difficult once the walls and ceilings are finished. The method that is widely accepted in the field for performing a leak-check on a system is the **standing-pressure test.** During a standing-pressure test:

- The pressure within the refrigerant piping circuit is brought up to approximately 150 psig.
- The system is permitted to rest for approximately 5 to 10 minutes.
- The needle on the gauge is marked to indicate its exact position.
- The system is then observed after a period of time to ensure that the position of the needle has not moved.
- If the needle moves, there is a leak. If the position of the needle does not change, the system is tight and there is no leak.

Pressurizing the System

When the pressure is initially introduced to the system, the manufacturer's test pressures should never be exceeded. Exceeding these pressures could result in damage to the equipment. The test pressure on a typical indoor unit is approximately 150 psig, so this pressure is safe under most conditions. When bringing the system pressure up for leak-checking purposes, nitrogen is the gas of choice. Nitrogen is relatively inexpensive, and, after the leak check is complete, it can be safely released into the atmosphere. Small amounts of a trace gas, R-22, can be introduced along with the nitrogen to aid in the location of a system leak if one exists.

SAFETY NOTE: *Never use pressurized air to leak-check a system! Some refrigerants can react with pressurized air and cause an explosion.*

SERVICE NOTE: *Never pressurize a system for leak-checking purposes with 100 percent refrigerant. This is against the laws set forth by the EPA.*

Marking the Gauge

After the pressure has been introduced to the system, good field practice requires letting the system stand for a short period of time. This gives the gas time to settle and will ensure that the pressure in the system will not change due to this settling. Once the gas has settled, the face of the gauge should be marked, indicating the exact position of the needle. By marking the gauge, any small movements in the needle's position will be easy to spot.

System Observation

Once the gauge has been marked, the system should be observed later on to see if the needle has indeed moved. The time that should be allowed to elapse depends on the size of the system and the length of the refrigerant lines. Obviously, the longer the system stays under pressure, the better and more reliable the leak-check will be. If at all possible, the system ideally should be under pressure overnight to ensure that the system is leak-free prior to start-up.

If a large leak is present, the pressure in the system will begin to drop rapidly. Smaller leaks, however, may require much more time to show up on the gauge. When possible, digital gauges should be used for leak-checking a system, as very small changes in the system pressure will be immediately noticeable with that type of instrument.

If the System Leaks

If the pressure in the system drops over time, a leak exists in the system. In this case, the solder and flare joints, if any, should be carefully inspected. If the system was pressurized with 100 percent nitrogen, a liquid-bubble-type, leak-detecting solution should be used. This solution should be applied to all solder joints as well as to all flare connections and Schrader valve stems. The formation of bubbles on the surface of the piping indicates the presence of a leak. Large bubbles indicate a small leak. If the pressure in the system has dropped rapidly and a large bubble is found, chances are that another leak in the system needs to be located as well.

If the system has been pressurized with a mixture of nitrogen and a trace gas, the technician can use an electronic leak detector to help locate the leak. Using an electronic device is much faster and neater than using the bubble method.

After the leak is located, the pressure in the system must be relieved before the leak is repaired. This will prevent the solder from being pushed out of the solder joint as heat is applied. If the leak is found on a flare nut or Schrader valve stem, however, keeping the pressure in the system as the fitting is tightened is a good idea. Once tightened, the area can immediately be rechecked with either the bubble solution or the electronic leak detector. Once the leak has been located and repaired, the system should be repressurized and observed at a later time to ensure that the system is now free from any leaks.

SERVICE NOTE: *The technician must remember that the gauge manifold that is connected to the system may also be a leak source. When checking the system for a leak, the technician should always check the connections on the manifold itself. The system may in itself be leak-free, but the manifold may not be.*

SERVICE NOTE: *Good field practice entails continuing to look for leaks after one has been found. The initial leak should be marked while the rest of the system is leak-tested. This prevents the need to continually pressurize a system over and over again as more leaks are located.*

Leak-Checking a System in a Vacuum

Many technicians feel that by leak-checking a system in a vacuum, they will save time on the installation. They argue that:

- The system must eventually be pulled into a vacuum before start-up anyway.
- If the system holds a vacuum, it must not have a leak.
- Time will be saved by not having to pressurize the system with nitrogen.
- Time will be saved by not having to wait to see if the pressure in the system changes.

Although the argument seems to be a valid one, in reality it is not. Granted, the system must be pulled into a vacuum anyway. But that is as far as that argument goes. Even if the system does hold a vacuum, a

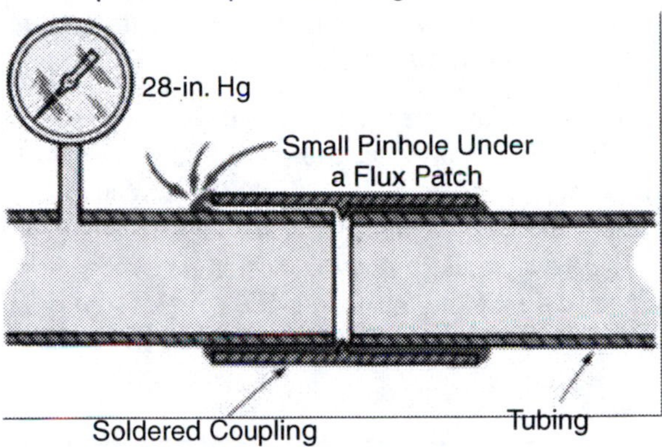

FIGURE 17-38 Flux can be pulled into a leak when pulling a vacuum.

leak still may be present. Consider the situation in which a small leak exists in a soldered piping connection. If the system is pulled into a vacuum, the flux around the leak may be pulled into the joint—in effect, sealing it (Figure 17-38). The gauge now indicates that the system is in a deep vacuum, and the vacuum holds. Once pressure is introduced into the system, the flux is blown out of the joint, reopening the leak.

In addition, if a system is leak-checked in a vacuum, the possibility exists that air can get pulled into the refrigerant piping circuit if a leak is present. The moisture contained in the air will need to be removed during system evacuation. The time required for proper evacuation will be increased because of this added moisture.

One final reason that a system should not be leak-checked in a vacuum is that the pressure differential created by a vacuum pump is very small compared to that created by adding pressurized nitrogen to the system. A leak that may result in substantial refrigerant loss during system operation may not surface during a vacuum-type leak-check.

Eventually, that leak will have to be repaired. Repairing the leak at this stage of the game, after start-up, could involve the following:

- Having to recover the system refrigerant
- Having to open sealed walls and/or ceilings to access the piping circuit
- Having to locate and repair the leak, which would involve pressurizing the system with nitrogen anyway
- Having to perform a standing-pressure test

- Having to evacuate the system for a second time
- Having to go through a second start-up procedure
- Having to supply refrigerant to replace the amount that was lost due to the leak

By trying to save time on the installation, a great deal of additional time, energy, and money can, in fact, be wasted. In addition to those items just mentioned, the inconvenience that the customer is being subjected to is also a major factor. In fact, the competence level of the entire service company may be questioned if the system begins to leak shortly after installation.

Service note: Once a system has been leak-tested by pressurization, a vacuum test can and should be used. A micron vacuum gauge should then be used to determine if the vacuum is holding.

SYSTEM EVACUATION

Before refrigerant is added, the system must be properly evacuated. The process of evacuation removes moisture from the system, which can facilitate the formation of acid within the system. Acid formation can damage the windings of the compressor and lead to the premature failure of the system. Acid can also eat away at the interior surfaces of the copper refrigerant lines. Moisture, if not properly removed from the system, can also freeze within the system. This can lead to a multitude of problems, including:

- Blocked metering devices and strainers
- Blocked filter driers
- Malfunctioning thermostatic expansion valves

When evacuating a system, the technician should:

- Make certain that the system is evacuated from both the high-pressure and low-pressure sides.
- Make certain that the vacuum pump is sized correctly for the system being evacuated.
- Change the vacuum pump oil frequently.
- Apply heat to the system to speed the evacuation process.
- Never operate a compressor while a system is in a deep vacuum.
- Use the largest valve ports possible to speed the evacuation process.

During the evacuation process, both the high- and low-pressure sides of the system should be connected to the vacuum pump. A typical heat-pump system has a number of components, such as check valves, that can be in the normally closed position. Thermostatic expansion valves and other components have small openings and orifices that will slow the evacuation process if the system is evacuated from only the high- or low-pressure side.

To ensure that the system is being evacuated in the most efficient manner possible, the size of the vacuum pump should be adequate to ensure proper evacuation. For example, a portable, compact 1.0-cfm vacuum pump should not be used to evacuate a 10-ton system. Using a pump that is too small will increase the time required to evacuate the system. In addition, an undersized vacuum pump may not be able to remove all of the moisture present in the system.

The moisture that the vacuum pump removes often finds its way into the pump itself. This moisture condenses in the body of the pump and dilutes the oil in the pump. If a system containing a large amount of moisture is being evacuated, the pump can eventually displace the oil. This will leave the pump's crankcase filled with water instead of oil, causing damage to the pump. To help avoid this, good field practice requires replacing the vacuum pump oil frequently.

In an effort to speed the evacuation process, heat can be added to the system. This will help boil off any liquid moisture into vapor, which is much easier for the pump to remove. If this method is used, the entire system should be heated in order to prevent the moisture from vaporizing in one part of the system and condensing back to a liquid in another. This is often the case on a split-type heat-pump system that has both an indoor component and an outdoor component. The best source of heat for use in this situation is a heat lamp. If the technician suspects that moisture is present in the compressor crankcase, heat should be applied directly to the crankcase.

Using larger connectors and hoses can also speed up the evacuation process (Figure 17-39). Larger hoses designed especially for vacuum pump use can be purchased at any refrigeration supply house. These hoses have removable pin depressors and are commonly used in conjunction with a gauge adaptor that serves the purpose of depressing the pin on the Schrader valve. The Schrader valves that are commonly found on system service valves create a

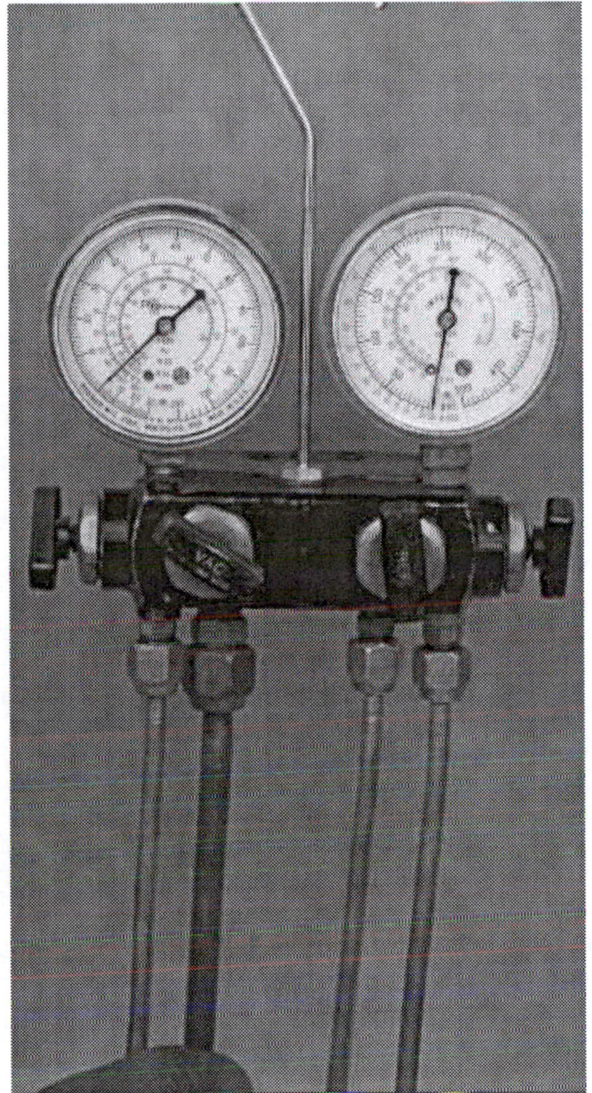

FIGURE 17-39 Gauge manifold with large vacuum hose connector. *(Photo by Bill Johnson)*

restriction even when they are in the open position. Field service valves can be used to remove the Schrader pins for evacuation and to replace the pins after the proper vacuum has been achieved.

SERVICE NOTE: *Once a deep vacuum has been reached, the compressor should not be operated. The compressor may overheat, causing damage to the motor windings.*

SYSTEM HOLDING CHARGES

After the system has been properly leak-tested and evacuated, a refrigerant holding charge can be introduced to the system. In many cases, the outdoor unit comes precharged with enough refrigerant for the entire system, assuming a 25-foot refrigerant line run between the indoor and outdoor units. If the run is longer than 25 feet, additional refrigerant will need to be added. If the run is less than 25 feet, refrigerant will need to be removed. Because this 25-foot run is correct in only a general sense, the specific amount of refrigerant in the outdoor unit, if any, should be confirmed by the installation information provided with the new system.

SERVICE NOTE: *Some outdoor units are pressurized with a holding charge of nitrogen and contain no refrigerant at all.*

PRE-START-UP CHECKLIST

Before the system is started up for the first time, the system should be inspected to make certain that all aspects of the installation have been completed. Following is a checklist that will help to ensure that the system is ready to be energized:

Outdoor unit:

- Is the outdoor unit level?
- Is the unit mounted on a pad that properly supports the unit?
- Is the area around the unit free from debris such as leaves and other potential coil obstructions?
- Are all line-voltage connections between the disconnect box and the unit made correctly?
- Are all field, line-voltage connections within the unit itself made correctly?
- Are all field, low-voltage wiring connections correct?
- Have the service valves been positioned so as to release the holding charge, if any, into the system?
- Are all refrigerant lines leading into the structure properly supported and insulated?
- Is there ample space around the unit for servicing?
- Are all shipping materials (compressor support blocks) removed from within the unit?

- Are the compressor mounting bolts secure?
- Is the compressor resting properly on the support springs? (The compressor mountings are sometimes tightened for shipping purposes and should be loosened before start-up to allow the compressor to vibrate during operation without transmitting noise through the unit.)
- Does the outdoor fan spin freely?

Indoor unit:

- Is the unit correctly mounted and supported?
- Is all supporting hardware tight and secure?
- Are all access and service panels accessible?
- Are all access and service panels in place and secure?
- Is there ample space for system servicing?
- Are all shipping materials removed from inside the unit?
- Are the condensate lines, both primary and auxiliary, run to their proper termination points?
- Do the condensate lines drain properly?
- Are all line-voltage connections between the disconnect box and the unit made correctly?
- Are all field, line-voltage connections within the unit itself made correctly?
- Are all field, low-voltage wiring connections correct?
- Does the indoor blower spin freely?

Duct system:

- Are all duct sections properly connected, supported, and insulated?
- Are all registers and dampers in the open position?
- Are all takeoffs properly connected?
- Are canvas collars, if applicable, properly installed to prevent noise transmission through the duct system?
- Is the duct system properly sized to deliver 450 to 500 cfm per ton?

General:

- Are all refrigerant lines properly supported and insulated?
- If needed, are refrigerant traps installed and properly located?
- Have all electrical lines from the fuse or circuit-breaker panel to the indoor and outdoor units been properly installed?
- Has the thermostat been properly secured and leveled?
- Are air filters installed?
- Has the system been properly leak-checked?
- Has the system been properly evacuated?
- Is there a holding charge of refrigerant in the system?
- Has the crankcase heater been energized prior to system start-up?
- Is the manufacturer's start-up literature for the unit close at hand?

SYSTEM START-UP

After the installation has been completed and all items on the preceding checklist have been accomplished, it is time to start the system for the first time. Even though the heat-pump system is designed to provide both heating and cooling, starting the system up in the cooling mode is easier. In the summer months, this should not pose a problem. In the fall, winter, and spring, however, the problem arises of cooler outdoor ambient temperatures. Starting up a system during times of cooler outdoor temperatures will result in a system that is ultimately grossly overcharged once the warmer months arrive. To compensate for the lower ambient temperatures, the start-up technician can partially block the outdoor coil, thereby reducing the airflow through the coil. This will cause the head pressure of the system to rise. Ideally, the system should be started up in the cooling mode with the following conditions:

- A condenser saturation temperature of approximately 115°F
- An indoor air temperature of approximately 75°F

A number of factors must be considered when a heat-pump system is started up for the first time:

- Airflow through the indoor unit
- Airflow through the outdoor unit
- The high-side pressure of the system
- The low-side pressure of the system
- Temperature differential across the evaporator coil
- Evaporator and system superheat
- Condenser subcooling

Airflow through the Indoor and Outdoor Coils

On heat pumps, as well as on other types of air-conditioning systems, the airflow through both the indoor and outdoor coils must be correct. One way to ensure that the airflow is within design range is to take amperage readings of the motors. On smaller systems, the outdoor unit fan is of the direct-drive type, where the fan blade is connected directly to the shaft of the motor. In these cases, the speed at which the fan blade turns is predetermined at the factory by the speed of the motor itself. At any rate, the amperage of the motor should still be checked and compared to the amperage on the nameplate of the motor. On larger systems where the outdoor fan is belt-driven, the pulley size should be adjusted in order to ensure that the amperage draw of the motor is in line with the amperage rating on the nameplate. If the amperage draw of the motor is too low, the speed of the blower should be increased. This is accomplished by reducing the size of the driven pulley. The *driven* pulley is the pulley connected to the blower, while the *drive* pulley is the pulley connected to the shaft of the motor. Reducing the size of the driven pulley will increase the amperage draw of the motor. Conversely, if the amperage draw is too high, the blower is turning too fast and the driven pulley should be made larger. Most belt-driven blower assemblies are equipped with variable-pitch driven pulleys (Figure 17-40).

These pulleys are designed in a manner that allows the technician to effectively change the size of the pulley without having to replace it. The pulley is constructed of two flanges. One of these flanges is stationary, while the other is movable. By loosening the set screw on the pulley, the movable flange can be twisted either clockwise or counterclockwise to increase or decrease the effective diameter of the pulley. When turned clockwise, the two flanges get closer together, causing the belt to rest closer to the outside of the pulley, increasing the diameter of the wheel, as shown in Figure 17-41(a). Turning the movable flange counterclockwise moves the flanges farther apart, causing the belt to rest closer to the center of the pulley, as shown in Figure 17-41(b).

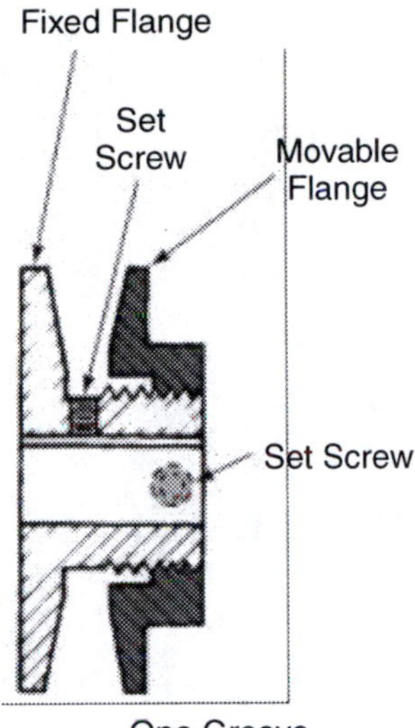

FIGURE 17-40 Variable-pitch pulley

The same holds true for the indoor units. If the blower assembly is direct drive, the amperage reading should be similar to the rating on the nameplate. On larger units, the driven pulley may need to be adjusted in a manner similar to that of the outdoor unit in order to reach the desired amperage draw of the motor.

A motor's amperage draw that is consistent with the nameplate rating is an indication that the correct volume of air is passing through the system. Too much airflow through the indoor coil, for example, will result in a higher-than-normal amperage draw. Insufficient airflow will result in a lower-than-normal amperage draw.

High-side and Low-side Pressures

Once the proper airflow through both the indoor and outdoor units has been established, the operating pressures of the system can be evaluated. When checking the system's operating pressures, thinking in terms of saturation temperatures rather than pressures is more convenient. Operating pressures vary from refrigerant to refrigerant, but the saturation temperatures are more or less constant from system to system. The condenser—outdoor coil for a heat pump operating in the cooling mode—saturation temperature should be in the range

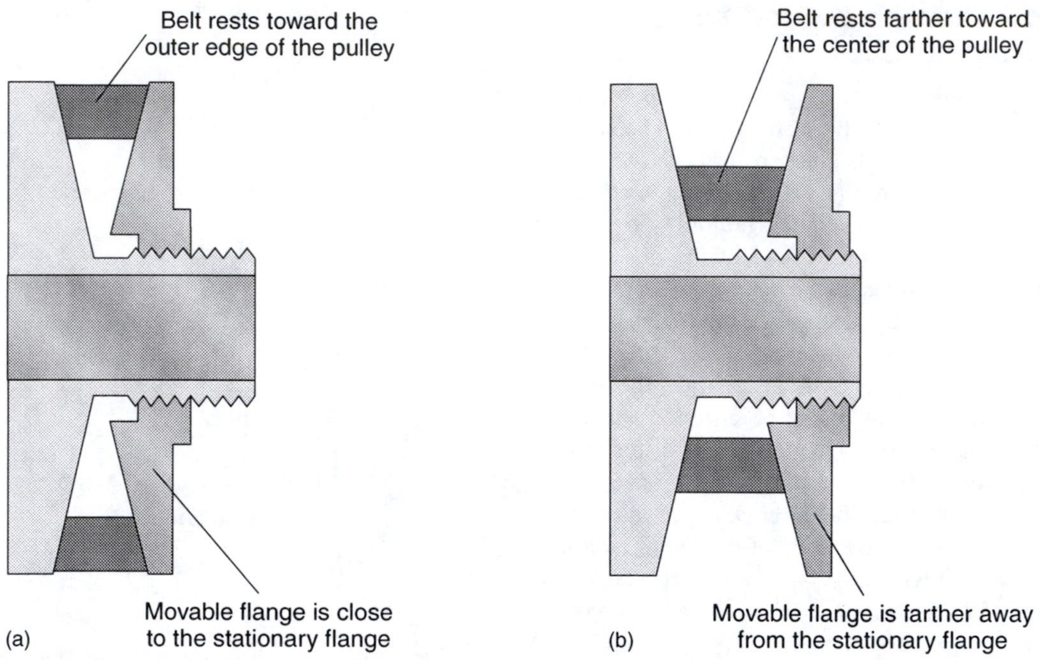
Belt rests toward the outer edge of the pulley
Movable flange is close to the stationary flange
(a)
Belt rests farther toward the center of the pulley
Movable flange is farther away from the stationary flange
(b)

FIGURE 17-41 (a) Variable-pitch pulley adjusted to increase blower speed. (b) Variable-pitch pulley adjusted to decrease blower speed.

of 30 to 35 degrees higher than the outdoor ambient temperature for a standard-efficiency unit and 20 to 25 degrees higher than ambient for higher-efficiency models. On cooler days, as mentioned earlier, a portion of the coil can be blocked to increase the pressure to the desired level. This saturation temperature should be maintained while the start-up process is completed.

The low-side pressure of the system can then be checked. For an indoor air temperature of approximately 70°F, the evaporator saturation temperature should be in the range of from 38 to 42°F. A saturation temperature well below that indicates that more refrigerant needs to be added to the system, while a saturation temperature well above that indicates that refrigerant needs to be removed from the system. Refrigerant, however, should not be added or removed at this time.

Temperature Differential across the Evaporator Coil

Another factor that will help determine the refrigerant charge in a heat-pump system is the temperature differential across the evaporator coil. The difference between the supply and return air temperatures should be in the range of 17 to 20 degrees. A temperature differential lower than 17 degrees is an indication that the system needs to have more refrigerant added to it. A temperature differential of more than 20 degrees indicates that refrigerant may need to be removed from the system.

In addition to the refrigerant charge, humidity also has an effect on the evaporator's temperature differential. If the relative humidity of the indoor air is in the 50 percent range, then the 17- to 20-degree differential is typical. In areas of excessively high humidity, the differential will be lower. In areas of very low humidity, temperature differentials above 20 degrees can be expected.

Evaporator Superheat and Condenser Subcooling

Earlier in the text, methods for determining both evaporator superheat and condenser subcooling were discussed. These two factors, in addition to those just discussed, will help the technician determine if the refrigerant charge needs to be adjusted. An evaporator superheat in the range of 12 to 20 degrees is considered to be well within the desired operating range. An evaporator superheat well below 12 degrees is an indication that the system is overcharged, while an excessively high superheat indicates that the system is undercharged. Condenser subcooling should ideally be in the range of 10 to 20 degrees.

Adding or Removing Refrigerant

After the preceding factors have been evaluated, refrigerant can then be added to or removed from the system. If refrigerant is to be added, it should be done in a slow deliberate manner. The following guidelines should be followed when adding refrigerant:

- Refrigerant should be added to the system in vapor form, whenever possible.
- If a blended refrigerant is used, it should be slowly introduced to the system as a liquid.
- Refrigerant should be added in 4-ounce increments.
- Allow the system to settle for approximately 5 to 10 minutes after each addition.
- Reevaluate the pressures and temperatures before adding more refrigerant.
- Continue adding refrigerant in 4-ounce increments until the desired pressures and temperatures are reached.
- If the system is charged properly in one mode of operation, it will be charged properly in the other mode.

Refrigerant removal from the system must be done in accordance with the laws set forth by the Environmental Protection Agency (EPA). Refrigerant must be removed from the system and stored in a DOT-approved cylinder (Figure 7-11). The following guidelines should be kept in mind when removing refrigerant from a system:

- Wear safety glasses when removing refrigerant from a system.
- Remove refrigerant in 4-ounce increments.
- Remove refrigerant from the high side of the system.
- Make certain that the recovery cylinder or tank that is being used to store the refrigerant has been properly evacuated prior to use.
- Give the system ample time to settle after each 4-ounce removal.
- Reevaluate the system temperatures and pressures after each removal.
- Continue to remove refrigerant until the desired temperatures and pressures are reached.

PUTTING IT TOGETHER

Once the system has been properly charged, all modes of operation should be checked. This includes:

- The cooling mode
- The first-stage heating mode
- The second-stage heating mode
- Safety cutouts
- Thermostat operation

Before leaving the job site, the technician in charge of performing the start-up should make certain that the system operates in all modes. The system should be checked by running the thermostat through its range of cycles. The technician should make certain that:

- The reversing valve switches over from heating to cooling properly, and vice versa.
- The supplementary electric-strip heater energizes during second-stage heating, as well as in defrost.
- The safety controls, such as high-pressure and condensate float switches, are operational.

Once the various modes of operation have been tested, the technician should record technical data for the system. These data include:

- Outside ambient temperature
- Indoor temperature
- Operating pressures in the heating mode
- Operating pressures in the cooling mode
- Amperage draws of the compressor and fan motors
- Temperature differentials across the indoor and outdoor coils
- Evaporator superheat measurements
- Condenser subcooling measurements

This information should then be submitted to the service company's office and filed for future reference in the event that a problem arises.

SUMMARY

- A high-quality installation will prolong the useful life of a heat-pump system.
- Installation crews must have knowledge of refrigeration, carpentry, plumbing, masonry, and electricity.
- The location of the outdoor unit should be chosen after considering a number of factors, including sound transmission, wind factors, airflow restrictions, electrical power supply, and its relation to the indoor unit.
- The location of the indoor unit is determined by a number of factors, including the type of air-distribution system, location of the outdoor unit, serviceability, ease of condensate removal, and noise levels.
- Duct systems can be of various configurations including the plenum, the extended plenum, the reducing-extended plenum, and the perimeter loop.
- Duct systems are commonly constructed from galvanized sheet metal, fiberboard, spiral flexible duct, or a combination thereof.
- Sheet metal duct sections are commonly connected with slips and drives.
- Fiberboard duct sections are connected with special staples and UL181 duct tape.
- Sheet metal duct systems must be insulated to prevent noise transmission and sweating.
- The condensate from a heat-pump system must be carried away via drain lines or pumps.
- Drain lines should be pitched downward and should never be reduced in size.
- Auxiliary drain pans are used to alert the occupant of potential water damage.
- Trapping of condensate lines is necessary if the line is under a negative pressure.
- Refrigerant piping runs should be as short as possible.
- The vapor refrigerant line should always be insulated to reduce the amount of heat transfer between the refrigerant and the surrounding air and to reduce the amount of condensation permitted to drip from it.
- Refrigerant traps are needed when the indoor unit is located below the outdoor unit.
- All systems must be properly leak-checked prior to start-up.
- Leak-checking in a vacuum is not recommended.
- Prior to start-up, the system should be carefully inspected.
- The operating pressures and temperatures should be monitored carefully during start-up.
- Refrigerant should be added or removed in 4-ounce increments.
- The system must be allowed to settle between refrigerant adjustments.

KEY TERMS

Auxiliary drain pans
Condensate
Condensate drain line
Condensate drain trap
Bar slip
Condensate pump
Energy efficiency ratio (EER)
Natural return
Offset duct sections
Plenum
Return air
Return duct
Return grill
Slips and drives
Standing-pressure test
Transition duct sections
Trunk line

FOR DISCUSSION

1. Consider a fossil fuel heating system that has recently been replaced with a heat pump. The occupant of the space notices that certain areas seem to be very drafty during system operation. In addition, during system operation, a constant sound resembling the rushing of air can be heard whenever the indoor fan is operating. What is a likely cause for this situation and how could it have been avoided?
2. Why is it recommended that multiple refrigerant traps be installed on excessively long vertical runs of refrigerant piping?

REVIEW QUESTIONS

1. Which of the following factors will have the biggest effect on the operating efficiency of a heat-pump system?
   a. Location of the electrical power supply with respect to the outdoor unit
   b. Location of the indoor unit with respect to the outdoor unit
   c. Location of the indoor unit with respect to the conditioned space
   d. Location of the outdoor unit with respect to the conditioned space
2. If a heat-pump system is to be installed in a geographic location that experiences a large amount of snowfall, which of the following would be the best choice for mounting the outdoor unit?
   a. Concrete pad
   b. High-impact plastic pad
   c. Metal framework cradle
   d. Plywood sheet cut larger than the size of the outdoor unit
3. Which of the following is the most likely outcome if the outdoor unit is not properly leveled?
   a. The outdoor fan motor may fail prematurely.
   b. The compressor may fail prematurely.
   c. The outdoor coil will become blocked or clogged more quickly.
   d. The reversing valve will not properly switch over from heating to cooling.
4. Which of the following can result in an overflowing condensate drain pan?
   a. An improperly mounted indoor unit
   b. An undersized condensate drain line
   c. A missing trap in the condensate line
   d. All of the above
5. Which of the following would commonly be found on a galvanized sheet metal extended-plenum duct system that has been installed in an existing structure?
   a. Transition duct sections
   b. Offset duct sections
   c. Excessively long spiral flexible duct takeoffs
   d. Both *b* and *c*
6. The most common method of joining sections of galvanized sheet metal duct sections is:
   a. Fiber-reinforced duct tape.
   b. Self-tapping sheet metal screws.
   c. Metal strapping material.
   d. Slips and drives.
7. To reduce the flexibility of wide galvanized sheet metal duct sections, ____________________ are commonly used to provide a more rigid duct system.
8. When installing a reducing-extended-plenum duct system, __________________ are used to prevent noise transmission and vibration in the duct system.
9. One of the biggest drawbacks of using spiral flexible duct is that:
   a. The spiral duct is more costly than prefabricated sheet metal sections.
   b. There is a great deal of friction within the spiral flexible duct.
   c. A highly skilled technician must perform the installation.
   d. The spiral flexible duct is only available in a limited number of sizes.

10. Which of the following is a benefit of fiberboard duct systems?
    a. They are more durable than galvanized sheet metal duct systems.
    b. They do not need to be insulated after installation.
    c. Sections can be fabricated as they are needed right on the job site.
    d. Both *b* and *c* are benefits of fiberboard duct systems.
11. If an indoor unit is equipped with a ¾-inch female pipe connection on the condensate drain pan, which of the following pipe sizes could be used effectively to remove the condensate from the structure?
    a. ½ inch
    b. 1 inch
    c. Either *a* and *b*
    d. Neither *a* nor *b*
12. When the indoor unit is located in a basement, which of the following methods could be used to remove the condensate from the structure?
    a. Utilize an available floor drain.
    b. Utilize a condensate pump.
    c. Pipe the condensate through the foundation of the structure to the outside.
    d. Both *a* and *b* are possible.
13. Explain the purpose of an auxiliary drain pan. Explain how it should be sized and how the drain line from the pan should be terminated.
14. When running refrigerant lines between the indoor and outdoor units, why are long radius elbows preferred over short radius elbows?
    a. Long radius elbows are more economical from a first-cost standpoint.
    b. Long radius elbows offer less resistance to refrigerant flow.
    c. Long radius elbows are easier to braze than short radius elbows.
    d. All of the above are correct.
15. On new construction installations, explain why it is extremely important for all solder joints to be inspected and reinspected.
16. Explain how an uninsulated vapor line can affect system operation in both the heating and cooling modes of operation.
17. If the outdoor section of a heat-pump system is located 30 feet below the indoor unit, which of the following is correct?
    a. A refrigerant trap should be located at the outlet of the indoor coil.
    b. An inverted trap should be located at the top of the vertical refrigerant piping run.
    c. The oil will flow back to the compressor crankcase by gravity.
    d. All of the above are correct.
18. Soft-drawn copper tubing is ideal for use behind sealed walls or ceilings because:
    a. The walls of soft-drawn copper tubing are thicker than rigid pipe and less likely to leak.
    b. The number of solder joints will be greatly reduced.
    c. The resistance to refrigerant flow is much less when soft-drawn copper tubing is used.
    d. All of the above are correct.

19. Leak-checking the refrigerant piping circuit of a new heat-pump system using the vacuum method is not recommended because:
    a. Air and moisture can be pulled into the system.
    b. A small leak may not become evident due to the small pressure differential between the interior of the piping circuit and the surrounding air.
    c. Solder flux may be pulled into the leak, temporarily sealing it.
    d. All of the above are correct.
20. Which of the following are true regarding standing pressure tests?
    a. A decrease in pressure indicates that a leak is present.
    b. High-pressure air is an effective and inexpensive alternative to nitrogen.
    c. The system should ideally be observed after 1 day to ensure that no leaks are present.
    d. Both *a* and *c* are correct.
21. Excessive superheat and a small temperature differential across the evaporator coil are indications that:
    a. The system is overcharged.
    b. No noncondensable gases are present in the system.
    c. The thermostatic expansion valve is overfeeding the evaporator coil.
    d. The system is undercharged.
22. Very low superheat and a large temperature differential across the evaporator coil are indications that:
    a The system is overcharged.
    b. Noncondensable gases are present in the system.
    c. The thermostatic expansion valve is overfeeding the evaporator.
    d. Both *a* and *c* are possible.
23. Upon starting up an R-22 heat-pump system in the cooling mode, it is found that refrigerant needs to be added to the system. The best way to introduce new refrigerant to the system is to:
    a. Introduce vapor refrigerant to the high side of the system.
    b. Introduce liquid refrigerant to the high side of the system.
    c. Introduce liquid refrigerant to the low side of the system.
    d. Introduce vapor refrigerant to the low side of the system.
24. When removing refrigerant from an overcharged system:
    a. The excess refrigerant should be removed in 4-ounce increments.
    b. The excess refrigerant will be stored in the refrigerant receiver and therefore does not need to be removed from the system.
    c. It is quickest to remove vapor from the low side of the system.
    d. The system should not be operating when refrigerant is being removed to prevent compressor damage from occurring.
25. Once the refrigerant charge has been corrected, the technician should:
    a. Check all modes of operation before leaving the job.
    b. Make certain that all debris and rubbish is removed from the site.
    c. Take down all pertinent system data including operating temperatures and pressures.
    d. All of the above are correct.

CHAPTER 18

System Preventive Maintenance

OBJECTIVES After studying this chapter, the reader should be able to:

- Describe the tasks that can be performed to help ensure proper operation of the indoor unit of a heat-pump system.
- Describe the tasks that can be performed to help ensure proper operation of the outdoor unit of a heat-pump system.
- Explain how to properly inspect the electrical circuits of a heat-pump system.
- Explain how to properly inspect the refrigerant circuit of a heat-pump system.

INTRODUCTION

After initial installation and start-up, ongoing preventive maintenance can help ensure that the system continues to operate as initially intended. Routine service can lead to the identification and remediation of small problems before they escalate into larger ones. For example, discovering pitted compressor contactor contacts early can prevent compressor motor damage from occurring. If left unattended, the defective contactor could ultimately lead to compressor failure, resulting in large repair bills and inconvenience for the system owner. This chapter discusses the concept of preventive maintenance and describes many small tasks that can be performed in an effort to keep a system up and running for many years to come. Locating small problems before they become larger ones reduces system downtime, lowers repair costs, and ultimately makes for a satisfied and repeat customer.

OVERVIEW

To keep a heat-pump system operating properly, routine maintenance should be performed twice a year. These preventive maintenance service calls are often referred to as *spring start-ups* and *fall changeovers.* With the first signs of spring comes the sound of humming compressors and ringing phones—good news for the air-conditioning service company. Conscientious customers and equipment owners want to make certain that their air-conditioning systems will make it through another cooling season without problems. As the warm weather fades and is replaced by cool fall breezes, attention is turned to the upcoming heating season. Once again, the phones at the service manager's desk start ringing. Once again, good news for the service company. But what takes place after the phone rings is most important. Equipment owners are entrusting their systems to the technicians that ultimately arrive at the job sites. It is the technician's job to thoroughly check the system and to locate any potential problems and remedy them before they cause system failure. Just as people, ideally, go to the doctor every year and take their pets to the vet, a semi-annual visit from the local air-conditioning technician can provide peace of mind for the upcoming season.

Preventive maintenance should not be treated lightly. Unfortunately, many service technicians feel that a surface inspection will suffice, but a surface inspection fails to do justice to the underlying concept of the service. Customers are, in essence, asking for a complete physical, so to speak, to be performed on their system. This physical involves the inspection of:

- The indoor unit
- The airflow system
- The duct system
- The outdoor unit
- The refrigerant circuit
- The electrical circuits

When performing maintenance on the system the technician should be checking each of the preceding items. The technician should be looking for anything out of the ordinary and anything that could potentially cause problems in the future. Although it may seem that the technician is asked to be something of a fortune teller or clairvoyant, there are telltale signs that a problem is likely to surface. Checking the integrity of wiring connections, the level of refrigerant charge, motor mountings, blowers, fans, and associated components—among other things—helps to ensure that midseason system failure does not occur.

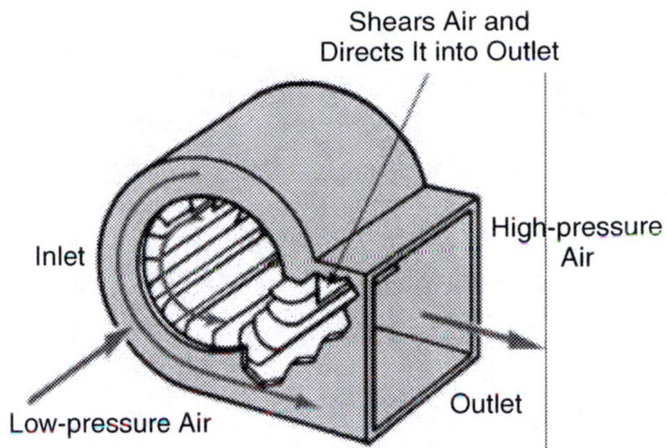

FIGURE 18-1 Forward-curved centrifugal blower and housing assembly

INDOOR UNIT AIRFLOW

Checking the airflow through the indoor unit involves a number of small tasks that can help prevent system trouble. These tasks directly relate to airflow through the indoor unit:

- Inspecting the blower
- Inspecting the belts and pulleys
- Inspecting the motor mounts
- Inspecting the motor shaft
- Inspecting the air-inlet side of the coil
- Inspecting service and access panels
- Checking the air filters
- Lubricating the motor
- Greasing the blower bearings
- Inspecting the duct system
- Inspecting supply registers and return grills

Inspecting the Blower

On heat-pump systems, air is most often moved through the duct system with a forward-curved centrifugal **blower,** or **squirrel-cage blower** (Figure 18-1). These blowers are ideal for moving air through a duct system because they are able to create a large pressure at the blower outlet, making it easy for the air to overcome the friction in the duct system. Upon inspection, the squirrel cage should spin freely without wobbling from side to side. A flashlight will aid the technician in the observing of the spinning blower to ensure that it is well balanced. A wobbling blower puts undue strain on the motor and also results in excessive system noise. Quite often, a whining sound can be noticed on systems that have a blower that is out of balance. If the blower is found to be out of balance, the motor itself should be carefully checked. In addition to the balance of the squirrel cage, the forward-curved blades should also be inspected. These blades (Figure 18-2) provide a perfect place for dirt accumulation. If dirt accumulates on the blades, the ability of the blower to move the proper amount of air is greatly reduced (Figure 18-3), due to the decreased size of the pocket that actually catches and throws the air.

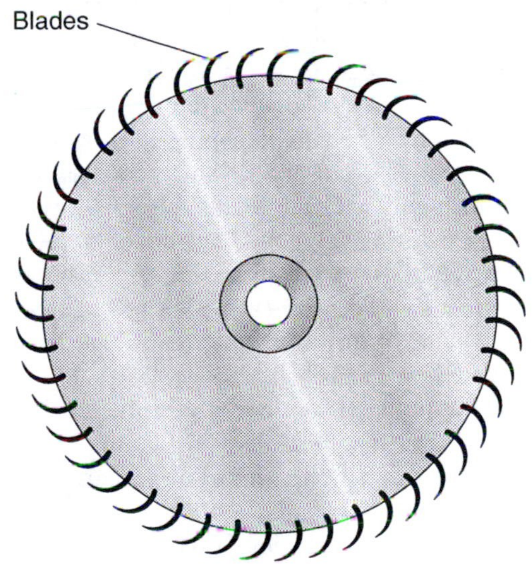

FIGURE 18-2 A cross-sectional view of the blades on a squirrel-cage blower wheel

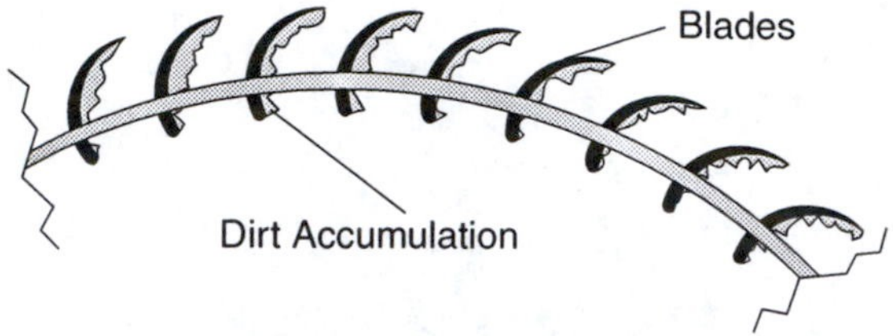

FIGURE 18-3 Dirt accumulation reduces the ability of the blower to move air.

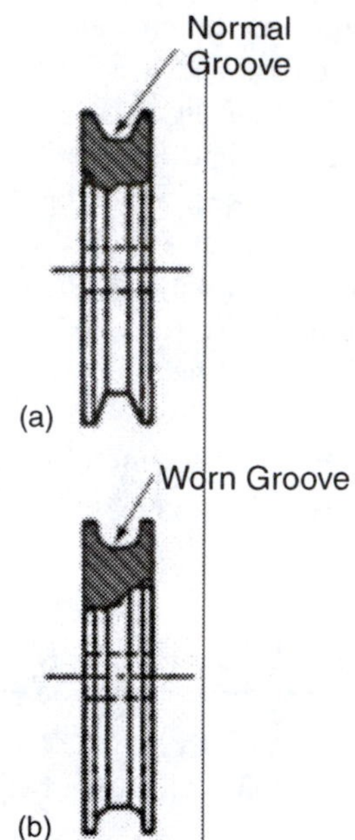

FIGURE 18-4 (a) Normal pulley groove. (b) Worn pulley groove.

Inspecting the Belts and Pulleys

Over time, belts and pulleys require adjustment and/or replacement. Checking the proper belt tension is facilitated by using a belt tension gauge. The proper use of this device was discussed earlier in the text. The belts themselves should also be inspected for cracks and tears that begin to appear as the belt gets older. Any cracks or tears indicate a potential problem, and the belt should be replaced.

In addition to the belt, the pulley itself should also be inspected. A mirrorlike finish on the interior surface of the pulley indicates that the belt has been slipping. In this instance, the pulley should be replaced, as the perfectly smooth surface reduces the amount of friction that should exist between the belt and the pulley. The shine of the pulley may also cause a technician to replace the belt unnecessarily in the future. The groove in the pulley should resemble the letter V with a flat bottom, as shown in Figure 18-4(a). A worn pulley will have a rounded bottom, as shown in Figure 18-4(b). A rounded pulley groove can also lead to the premature tearing of the belt. Therefore, if the belt is seen to have tears along its edges, chances are that the pulley is also worn.

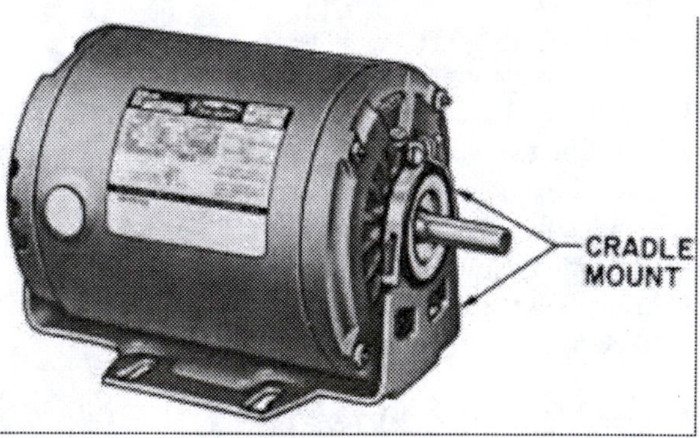

FIGURE 18-5 Cradle-mounted motor. *(Courtesy W.W. Grainger, Inc.)*

Inspecting the Motor Mounts

During normal system operation, vibrations cause screws, bolts, and other mounting hardware to come loose. During a preventive maintenance service call, the motor mounts should be checked and tightened, if necessary, to ensure that the motor remains securely in place. On direct-drive motors, the mounting hardware usually secures the motor directly to the housing of the blower wheel, so a few well-timed screwdriver or nut-driver turns can help prevent costly damage. A loose motor mount can result in damage to both the motor and the blower wheel. In either case, the cost of replacing those components can be substantial. Motor mount styles commonly found on heat-pump systems include the cradle mount (Figure 18-5), the belly-band mount (Figure 18-6), and the end mount with tabs and studs (Figure 18-7).

On belt-driven blowers, the motor base should also be inspected and tightened as needed. A loose motor-mounting bracket will cause the motor to shift, resulting in belt slippage and/or pulley damage. When checking the motor mount, the technician should make certain that the drive and driven pulleys are perfectly aligned. Pulleys that are not properly aligned can cause premature belt wear, primarily tearing, in addition to causing the belt to skip off the pulleys entirely. If the belt, still intact, is found lying next to the motor, chances are that the pulleys are in need of alignment. The technician should also make certain that the motor shaft and the shaft of the blower are parallel to each other. This will help prevent damage to the shaft's drive mechanism. Normally, an adjustment can be made on the motor mount to help align the shafts properly.

Whether the blower is direct drive or belt driven, loose motor mounts can result in excessive noise transmission and vibration throughout the system. System vibration resulting from a loose motor mount can also result in refrigerant leaks that arise from refrigerant pipes rubbing against each other or other surfaces within the unit.

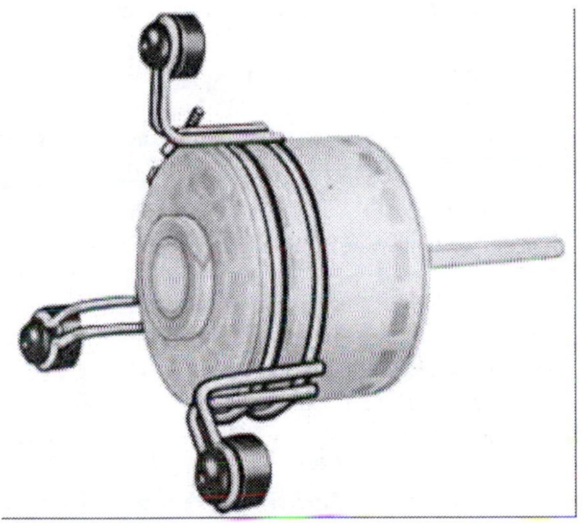

FIGURE 18-6 Belly-band-mount motor. *(Courtesy W.W. Grainger, Inc.)*

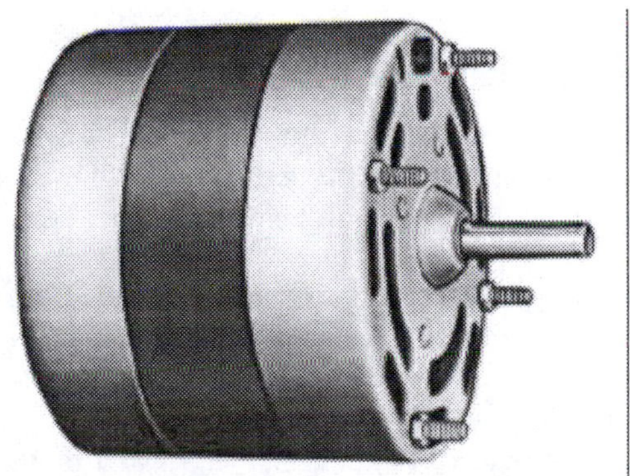

FIGURE 18-7 End-mounted motor. *(Courtesy W.W. Grainger, Inc.)*

Inspecting the Motor Shaft

One way that the servicing technician can determine if the motor is in good working order is to physically inspect the shaft on the motor. Ideally, there should be relatively little play—approximately ⅛ inch at most—in the shaft when it is moved in the direction parallel to the shaft itself (Figure 18-8). Excessive movement indicates that there is likely a problem with the bearings or the internal shaft support. Excessive movement or play in the direction perpendicular to the shaft (Figure 18-9) is an indication that the bearings on the motor are bad.

On smaller, fractional horsepower motors, the best way to remedy the problem is to replace the entire motor. Repairing smaller motors is not economically feasible. Larger motors, on the other hand, often have

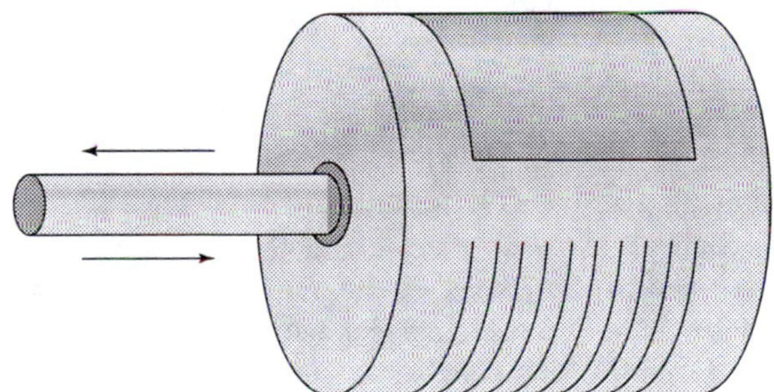

FIGURE 18-8 Motor shafts often have a small amount of parallel play.

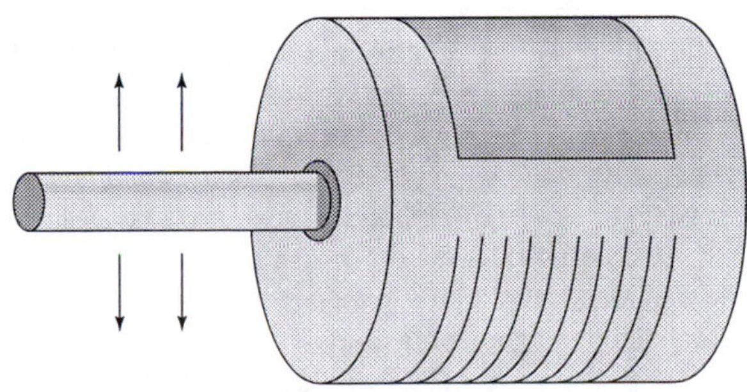

FIGURE 18-9 Motor shaft play in the perpendicular direction is an indication that the bearings are defective.

their bearings replaced upon failure. The cost of replacing the bearings is very small when compared to the cost of replacing the entire motor.

Inspecting the Air-Inlet Side of the Coil

Any blockage in the air stream will reduce the amount of airflow through the duct system and indoor coil. One of the most overlooked locations for a potential air blockage is the inlet side of the indoor coil. Any dirt, dust, or particulate matter that bypasses the filter media will most definitely wind up embedded in the indoor coil. One reason that this location is often overlooked is the fact that it is not often easily accessible. The most useful tool that can be used to inspect the indoor coil is a flashlight. If the flashlight is placed on the discharge side of the coil and light is observed through the coil on the inlet side, the coil is clear. If the coil is found to be dirty, it should be brushed down or chemically cleaned if necessary. Any debris that is removed from the coil surface should be removed from the unit completely to help prevent the dirt from accumulating in the condensate drain.

Inspecting Service and Access Panels

During a preventive maintenance service call, the technician should always make certain that the service and access panels on the air handler are secure. This will help prevent air leakage into and out of the air system. Loose panels can greatly reduce the efficiency of the system. Return-air leaks can result in the introduction of colder air into the duct system in the heating months and warmer air into the system in the cooling months. Supply-air leaks can result in a reduction of heated air being supplied to the occupied space in the heating months and a reduction of cooled air supplied to the space in the cooling months.

Improperly mounted service panels can also increase the noise levels generated by the system. Service panels are often lined with an acoustical material that helps deaden the noise of the motor and blower assembly. If the panels are not secured, the effectiveness of the lining is greatly reduced. Also, the rattling sound of a loose service panel is an unnecessary inconvenience for the customer. If service panels, especially those on older units, continue to rattle even when properly mounted, one or two well-placed sheet metal screws can keep the panels in place.

SERVICE NOTE: *If screws are being used to secure a panel in place, make certain that the screws are long enough to hold the panel in place but short enough to prevent them from coming in contact with any system components, pipes, or tubes.*

Checking Air Filters

One of the most basic of all preventive maintenance service tasks is checking and replacing the **air filters** on the system. At a bare minimum, air filters should be changed during spring and fall service. If the air filters are of the disposable, cardboard-frame type, an ample supply should be available at the location so that the owner or operator of the equipment can change them easily. The customer or system owner should be instructed how to replace the filters, if they are easily accessible. Ideally, the filters should be replaced each month. Frequent filter changes reduce the amount of dirt and particulate matter that ultimately find their way onto the indoor coil's surface.

When checking filters, it is also important to inspect the filter rack configuration to ensure that as little air as possible is permitted to bypass the filter. Improperly sized filters and improperly mounted filter racks permit airflow around the filter, which allows for excessive dirt buildup on the coil, on the blades of the blower, and on the supply registers. Effective air filtration helps to reduce the amount of particulate matter in the air, so a high-quality filter should be used whenever possible.

Lubricating the Motor

The successful operation of the indoor airflow system ultimately relies on the proper operation of the motor. Although the windings of the motor cannot be readily serviced, the mechanical components—namely, the **bearings**—can be. The motor bearings must be well lubricated in order to prevent motor overheating and seizing. Motor bearings, depending on the design, can require periodic lubrication or can be permanently lubricated. Permanently lubricated motors do not require periodic lubrication. These motors have the bearings packed with either grease, on larger heavy-duty motors, or oil, on smaller sleeve-type bearing motors. Under normal operating conditions, these motors are designed to provide years of service without needing lubrication. If certain conditions exist, such as prolonged operation outside of design amperage ranges, the lubricant can break down due to motor overheating. In this instance, additional lubricant must be added in order to save the motor.

Motors that require periodic lubrication are typically smaller in size and are generally equipped with sleeve bearings. These sleeve bearings, identified by their quiet operation, are lubricated with motor-grade oil. Generally, 20-weight, nondetergent, motor oil can be used, but the manufacturer's literature will most likely specify a particular oil type and weight that should be used whenever possible. Sleeve bearings are designed so that the motor shaft actually floats in the lubricant without actually coming in contact with the surface of the bearing. If the oil is too thin, the shaft will rub against the bearing, causing damage. If the oil is too thick, the oil will not flow into the space between the shaft and bearing, causing the motor to overheat.

Greasing the Blower Bearings

Larger motors require the use of grease, as opposed to oil. These motors are equipped with grease fittings (Figure 18-10), and a grease gun is needed to introduce grease to the bearings. A relief plug is located on the bearing, which must be removed before grease is added. The removal of the plug prevents the technician from adding too much grease to the bearing. Excessive grease in the bearing can result in the popping of the bearing, breaking the seal. Broken bearing seals prevent the bearing from being able to contain the grease, resulting in friction-generated motor overheating.

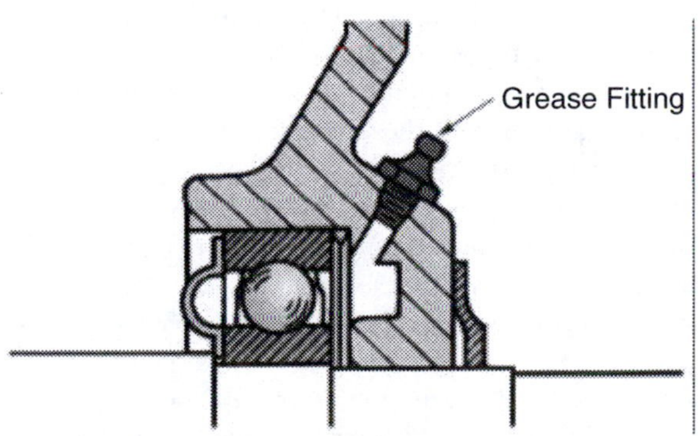

FIGURE 18-10 Typical grease fitting

Inspecting the Duct System

The air-distribution system must be free from air leaks, and the ductwork must be properly insulated to prevent air leakage and condensate formation on the duct surfaces. Air leaks in the supply distribution system can result in the loss of conditioned air to the occupied space. This reduces the efficiency of the system because money is being spent to condition the air surrounding the system. If the duct system runs within the space, the effects of air leakage are greatly reduced. Air leakage on the return side of the air-distribution system will result in the introduction of warmer air to the indoor unit in the cooling mode and cooler air to the unit in the heating mode. In either case, the efficiency of the unit is also reduced. Once again, if the return ductwork is located within the conditioned space, the effects of return-air leaks are reduced as well.

When located in an attic or basement, the duct system is at a temperature lower than the surrounding air when operating in the cooling mode. This leads to condensation formation on the surface of a sheet metal air-distribution system. Condensate formation can eventually lead to water damage, duct deterioration, and a damp musky smell within the occupied space. Properly insulating the duct system will alleviate these potential problems.

When performing a preventive maintenance service call, the technician should inspect the insulation to make certain that the ducts are wrapped securely and that none of the insulation has come loose. Any loose insulation should be resecured. If access to the interior of the duct system is provided, a visual inspection of the duct's acoustical lining is also beneficial. Loose lining can result in the blockage of the main trunk line or individual takeoffs. Loose lining should be resecured using an appropriate adhesive.

Safety Tip:

Duct lining adhesive vapors are dangerous. The adhesive should therefore be used in a well-ventilated area. Since air ducts are often located in confined spaces, it is recommended that a respirator-type mask be worn to reduce the effects of the fumes.

The **branch ducts** on the distribution system must also be checked. In addition to the insulation issues already mentioned, the volume dampers at each takeoff should also be inspected. The dampers should be tight and in the proper position to introduce the desired amount of air to each location. Loose dampers within a takeoff can result in improper airflow and excessive noise levels within the system.

Inspecting Supply Registers and Return Grills

Finally, the supply registers and return grills should be inspected. The supply registers are the locations where the air leaves the air-distribution system, and the return grill is the location where the air initially enters the system. A blockage at either end of the system will reduce the amount of airflow through the system, reducing system efficiency.

SUPPLY REGISTERS. Supply registers should not be blocked. Furniture or other items should not obstruct floor registers. It is very common for occupants to place sofas and other items over the floor registers for aesthetic purposes. The customer should be informed that this prevents the system from operating properly. Proper airflow is a key to proper system operation. The registers should also be inspected for excessive dirt accumulation. Excessive dirt is an indication that the air filters may not be doing their job.

RETURN GRILLS. Return-air grills should be inspected for proper mounting and air leakage around the filter media. The filters should be sized properly to completely cover the daylight opening in the filter rack. Any issues regarding air leakage around the filters should immediately be addressed. The return grill is the location where air, dirt, and particulate matter are introduced to the system. Effectively filtering the air will result in a cleaner system overall. Allowing dirt to enter the system will eventually cause the indoor coil to become blocked, paving the way for more expensive repair tactics.

OTHER INDOOR UNIT FACTORS

Although the indoor unit is primarily responsible for moving the air through the occupied space, a number of other factors must be taken into account when preventive maintenance is being performed on a heat-pump system. These factors include:

- Condensate removal
- Electrical circuits
- Refrigerant circuit

Condensate Removal

Water damage to the structure as a result of condensate accumulation often requires costly repairs to ceilings and walls, not to mention damage to the furnishings within the space. By the time a ceiling starts to leak, it has already become saturated with water. Ceilings have been known to collapse as a result of condensate accumulation. The best thing that a technician can do to help prevent this from occurring is to check the condensate drain and be absolutely certain that the line is able to carry the water out of the structure. Water should be poured into both the primary and auxiliary drain pans to ensure that the lines are clear. Once water has been poured into the drains, the termination points of the lines should then be observed to verify that the water is being removed. Pouring water in the drain pans will also confirm that the pans are properly pitched for adequate removal.

A number of products are on the market that help keep drain lines clear. One of the most popular items comes in tablet form and resembles a large aspirin. The *drain tab* is placed in the drain pan, and, as condensate falls into the pan, the tablet begins to dissolve. The chemicals in the tab help to reduce the accumulation of minerals on the interior surfaces of the drain line. These tablets do dissolve completely over time and should be replaced each year.

Systems that are equipped with overflow safety switches in the auxiliary drain pan should have the operation of these switches tested as well. Blocking off the drain connection on the pan and filling the pan with water best test these switches. Water should be added to the pan slowly, as adding water quickly could lead to splashing and pan overflow. The system should be operating at the time the switch is tested, and it should be verified that the system is disabled once the water level reaches the cutout point. Not only will this test the operation of the switch, it will also check the integrity of the auxiliary pan as well. An improperly mounted pan may tend to tip as it fills with water.

SERVICE NOTE: *If the drain is blocked for testing purposes, make certain that the line is cleared before leaving the job. Leaving the line blocked could result in legal liability.*

As a part of preventive maintenance, the operation of the condensate pump should be tested as well. This is easily done by filling the pump with water and verifying that the pump is able to remove the water. The pump should turn off once the level of the water in the sump has dropped. On condensate pumps equipped with safety cutout switches, the pump should be allowed to fill to above the normal level. This can be accomplished by simply unplugging the pump. With the system operating, the increased water level in the pump should cause the normally closed contacts on the safety to open. This should disable the unit.

SERVICE NOTE: *If the condensate pump is unplugged for testing purposes, make certain that power is restored to the pump before leaving the job.*

Electrical Circuits

The majority of system failures are electrical in nature. Checking the electrical components and the interconnecting wiring is an effective way to reduce future system downtime. Because the unit vibrates during normal operation, electrical connections have a tendency to become loose. Both field- and factory-installed connections should be checked. These connection locations include:

- Terminal board screw connections
- Solderless connectors on relays, capacitors, and other components
- Service disconnect connections
- Molded-plastic connectors
- Wire connections made with wire nuts

Line-voltage connections and low-voltage connections should be checked. Loose electrical connections can result in system malfunction. Wires with damaged insulation should be repaired and or replaced as needed. Missing insulation can result in blown fuses, tripped circuit breakers, and personal injuries resulting from electrical shock. The connections at the service disconnect should not be overlooked. Damaged wiring in the disconnect will prevent the system from operating. All wire connections made with wire nuts (Figure 18-11) should be inspected. Overtightened wire nuts can cause the wires inside to break. Ideally, the wires should be twisted together securely inside the wire nut (Figure 18-12).

Safety Tip:

When checking the integrity of wiring connections, be sure that the power to the unit is turned off. When checking the connections in the service disconnect box, the circuit breaker for the unit should be switched off as well.

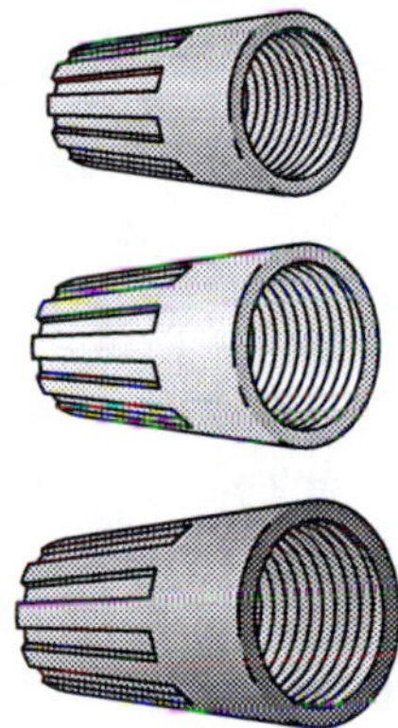

FIGURE 18-11 Screw-on wire connectors or wire nuts. *(Courtesy Klein Tools)*

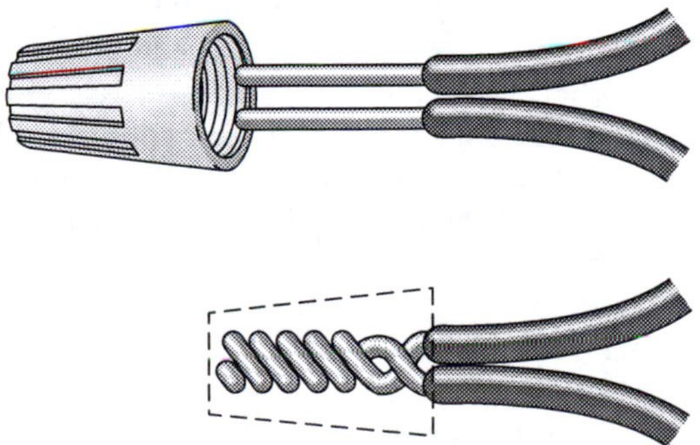

FIGURE 18-12 Installing screw-on wire connectors. *(Courtesy Klein Tools)*

SERVICE NOTE: *Although the electrical connections should be tight, be careful not to overtighten them. Screw connections can break if they are turned too far.*

After the electrical connections are checked, the amperage of the unit and its components should be checked as well. These components include blower motors and electric-strip heaters. The readings obtained from the unit should be recorded and submitted to the service department for future reference.

Refrigerant Circuit

Although the operating pressures are checked from the outdoor unit, the associated piping located within the indoor unit should be visually inspected at the time preventive maintenance is performed. The technician should check for:

- Properly mounted thermal bulbs on thermostatic expansion valves
- Contact between refrigerant pipes and other pipes and/or metallic surfaces
- Oil traces on the piping surface
- Rusting or deteriorating coils
- Tight flare connections

THERMAL BULBS. The thermal bulb on the thermostatic expansion valve should be fastened securely to the outlet of the indoor coil. The bulbs are normally mounted to the piping with copper strapping material that is secured with two machine screws and appropriate nuts. If the thermal bulb can be shifted by hand, the mounting needs to be tightened. An improperly mounted thermal bulb can result in an overfeeding evaporator, which could ultimately lead to liquid floodback to the compressor. The position of the thermal bulb on the line should be in accordance with the manufacturer's specifications. Normally, on lines less than ¾ of an inch in diameter, the bulb should be mounted on top of the line. For more information regarding the position of the thermal bulb, refer to earlier chapters.

CONTACT BETWEEN REFRIGERANT PIPES AND OTHER PIPES AND/OR METALLIC SURFACES. One of the major causes for refrigerant leaks in air-conditioning systems is contact between a refrigerant line and other lines or surfaces. Refrigerant lines should not be in contact with other lines. Normal system vibrations can cause lines to rub together, eventually causing a leak to develop. Lines that are in close proximity to each other should be separated if possible. If this is not an option, foam insulation should be placed between the lines to prevent future contact between them. Long capillary tubes and transmission lines from pressure controls, valves, and sensors should be bundled together and secured with plastic cable ties; this bundle should then be wrapped with an insulating tape in the event that the cable tie breaks. Refrigerant lines should also be far enough away from other surfaces, such as metal edges, screws, flanges, and brackets. Any of these items could also cause leaks in the lines.

OIL TRACES ON THE PIPING SURFACE. Technicians should always be on the lookout for oil traces on refrigerant lines. Oil traces are an indication that a refrigerant leak is present. Oil travels through the system with the refrigerant and will, of course, seep out through a leak that develops. The refrigerant will evaporate rapidly, but the oil will remain behind. If a leak is located, the technician should be sure to wipe up any oil that remains on the lines after the repair so that other technicians do not suspect that a leak is present when there is none.

RUSTING OR DETERIORATING COILS. While performing preventive maintenance, the technician should also be aware of excessive rust buildup and/or deterioration within the unit. This could result in water and refrigerant leaks. Any conditions that could possibly result in future system failure should be immediately reported to the service department. Rust formation in the unit could also be a result of a clogged or blocked condensate drain line. Rust formation or water accumulation in sections of the unit that are not designed or intended for water is also an indication that a problem exists.

FLARE CONNECTIONS. Another common location for refrigerant leaks is at flare connections. Once again, system vibration can cause connections to become loose. The technician should, therefore, check all flare connections on expansion valves, precharged line sets, and pressure sensors. Oil stains on flare nuts and associated connections are also an indication of a refrigerant leak.

THE OUTDOOR UNIT

Just as with the indoor section of a heat-pump system, a number of items need to be checked on the outdoor unit during a preventive maintenance service call. Many of these items take just a few seconds to complete but can help prevent future system malfunction. Other items take longer to complete but once done will provide the customer with peace of mind knowing that the system has been thoroughly inspected and checked, quite literally, from top to bottom. These items include:

- Outdoor airflow
- Refrigerant charge
- Electrical circuits
- Defrost switchover
- Heating/cooling changeovers

Outdoor Airflow

Airflow through the outdoor unit is typically much easier to inspect than that of the indoor unit. If located outdoors, there is no associated duct or air-distribution system to be evaluated. Generally, two areas should be checked when dealing with the airflow through the outdoor unit. They are the motor and the coil.

THE MOTOR. The air system on the outdoor unit is most often of the direct-drive configuration with the fan blade connected directly to the shaft of the motor. When inspecting the motor, the following items should be addressed:

- Motor mounts
- Motor lubrication
- Motor bearings

Motor Mounts. Depending on the configuration of the unit, the motor can be mounted vertically for a top-discharge unit or horizontally for a side-discharge unit. In either case, the motor should be mounted securely to the bracket. The brackets can be either welded to the casing of the unit or screwed in place. If welded, the connections should be checked to ensure that none have broken loose. If the bracket is screwed in place, the screws should be tightened periodically to make certain that the bracket and motor remain in place as well. The individual straps, screws, and brackets that secure the motor to the bracket must also be tight. The motor must not be permitted to move or shift within the bracket.

Motor Lubrication. The guidelines for motor lubrication are the same as for the indoor fan motor. If the motor must be periodically lubricated, 20-weight, nondetergent, motor oil is typically acceptable for this application. When in doubt, refer to the manufacturer's specifications for oiling guidelines. Any caps that are removed from oil ports should be replaced to prevent oil from seeping out of the motor. Properly lubricated motors will operate quietly and will provide many years of uninterrupted service.

Motor Bearings. By listening to the sound the motor makes, as well as physically moving the shaft by hand, the technician can determine if the motor bearings are in good shape. With the power to the unit disconnected, the technician should check the play in the motor shaft. Too much play in either direction, perpendicular to or parallel to the shaft, is an indication that the bearings on the motor are going bad. Too much play can result in a motor that will overheat or may even fail to start. If the bearings on a small, fractional horsepower motor are bad, it is economical to replace the motor. Larger motors will often have the bearings replaced, as it is relatively inexpensive to do so compared to the cost of a new motor.

THE OUTDOOR COIL. A visual inspection of the outdoor coil will conclude the airflow inspection of the unit. The outdoor coil should be free from any debris and should not be blocked by leaves, snow, dirt, or any other obstruction. The coil itself should be clean and can be inspected easily with a flashlight. The light should be easily observable through the coil when it is shone through the fins. A dirty outdoor coil should be hosed down to remove surface dirt but may need to be cleaned chemically if it is extremely dirty or clogged.

SERVICE NOTE: *The outdoor coil ideally should be cleaned annually, even if it appears to be clean. The coil should be flushed with a water hose in the direction opposite to the direction of airflow through the coil.*

Refrigerant Charge

An integral part of preventive maintenance is to check the refrigerant charge in the system. Any refrigerant loss should be dealt with immediately and in accordance with all guidelines and laws set forth by the EPA. A complete loss of refrigerant should be addressed by performing a standing-pressure test, as outlined earlier. Smaller refrigerant leaks that may not have surfaced or cannot be located should be addressed by using either an electronic leak detector or an ultraviolet solution. The system should be allowed to operate in both the heating and cooling modes of operation. The operating pressures and temperatures, along with the indoor and outdoor ambient temperatures, should all become part of the report turned into the office at the completion of the service call.

The operating guidelines discussed in the start-up section of the previous chapter will provide insight as to the correct operating temperatures and pressures that should be expected during a preventive maintenance service call. If a discrepancy arises between the actual and expected readings, the service manger should be contacted so that the original start-up report can be consulted. Discrepancies between the start-up data and those obtained during maintenance should be an indication that the refrigerant charge may not be correct. This, of course, would take into account any variances in indoor and outdoor ambient temperatures.

Superheat and subcooling measurements should also be obtained as part of preventive maintenance and should be a part of the service report. The details of how to obtain these measurements were provided earlier in the text but will be briefly recapped here. To obtain evaporator superheat, two pieces of data are needed: the evaporator saturation temperature and the evaporator outlet temperature. Evaporator superheat is the difference between the two. To measure subcooling, two pieces of data are required as well: the condenser saturation temperature and the condenser outlet temperature. Condenser subcooling is determined by finding the difference between the two.

Electrical Circuits

The electrical components and circuits of a heat-pump system must be properly maintained. The electrical circuits are simply road maps that the electrical current follows. It is therefore important that the circuitry and the associated components, including, relays, contactors, and other switches, remain fully operational. To help keep the electrical end of the system in tip-top shape, the technician should evaluate the following:

- Integrity of electrical connections
- Contacts and switches
- Voltage readings
- Amperage readings

INTEGRITY OF ELECTRICAL CONNECTIONS. Just as with the indoor unit, the integrity of the electrical connections within the outdoor unit must be maintained. Upon inspection, the technician should not overlook the electrical connections made in the factory. Careful attention should be paid to the:

- Terminal board screw connections
- Solderless connectors on compressors, relays, capacitors, and other components
- Service disconnect connections
- Molded-plastic connectors
- Wire connections made with wire nuts

CONTACTS AND SWITCHES. The contacts on the contactors should be visually inspected. Over time, these contacts tend to pit (Figure 18-13) and add additional resistance to the circuit. Pitted contacts should be replaced when economically feasible. Smaller contactors in the 30-ampere range are generally replaced when the contacts are pitted, as the labor cost involved in replacing the contacts far outweighs the cost of a new contactor.

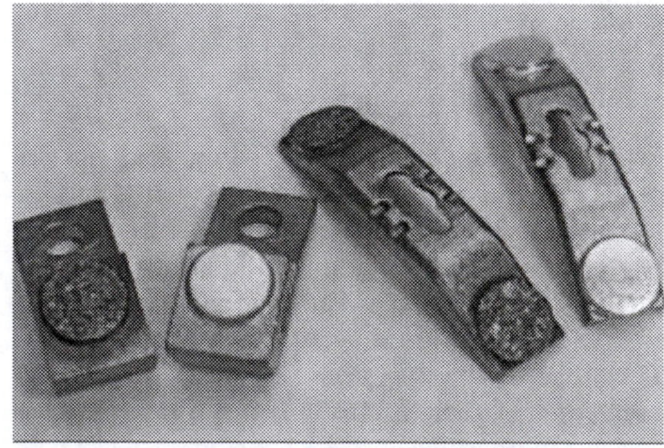

FIGURE 18-13 Clean contacts contrasted with a set of dirty, pitted contacts. *(Courtesy Square D Company)*

The operation of the switches located within the unit should also be tested. Any malfunctioning switch or control should be replaced. The switches and components within the defrost timer and other associated relays should be evaluated as well. All electrical components should also be mounted securely to the cabinet of the unit. Relays, switches, and other components should never be allowed to hang freely. This can lead to short circuits, blown fuses, tripped circuit breakers, and system failure.

VOLTAGE READINGS. A system and its components can operate properly only if the proper voltage is being supplied to them. For this reason, a reading of the voltage being supplied to the unit from the disconnect box is a must. The reading should be within 10 percent of the nameplate voltage. Discrepancies larger than 10 percent must be investigated. Voltage readings should also be taken across relay and contactor contacts to ensure that the contacts are opening and closing properly. There should be a very low, or zero, voltage reading across a set of closed compressor contactor contacts, for example. A large voltage reading indicates that the contacts may be pitted and in need of replacement.

AMPERAGE READINGS. After the integrity of the wiring connections and associated circuits has been checked, amperage readings of the components are to be recorded. With the use of a clamp-on ammeter, taking amperage readings at strategic points throughout the system is a relatively easy task. Amperage readings should be taken and recorded at the following locations:

- Power lines feeding the outdoor unit
- Lines feeding the compressor
- Lines feeding the outdoor fan motor

All amperage readings that are recorded should be within the ranges indicated on the component nameplates.

Defrost Switchover

During the heating mode of operation, the system should be induced to go into the defrost mode. This will confirm that the defrost cycle operates as intended. Initiating a defrost cycle may be difficult, however, if the outside ambient temperature is high. In this situation, a minimum of switches can be jumped out so that the technician can confirm that defrost does indeed initiate.

On pressure-initiated defrost systems, the pressure differential switch can be overridden to initiate a defrost cycle. If the pressure differential switch is a normally closed device, simply removing a wire from the control should put the unit in defrost. If the switch is normally open, a jumper wire placed across the contact terminals will initiate the defrost cycle. During defrost, the outdoor fan motor should cycle off and the reversing valve should switch the system over to cooling.

On time-initiated/time-terminated defrost systems, it is a simple matter to just manually advance the defrost cam on the defrost timer. This will initiate defrost. On time/temperature-initiated defrost systems, the temperature sensor will have to be jumped out and the cam on the defrost timer will have to be advanced into defrost.

Heating/Cooling Changeovers

A major part of a preventive maintenance call is the checking of the space thermostat. The system must be able to switch over freely from heating to cooling, and vice versa. The different modes of operation can be checked from the thermostat, enabling the technician to check not only the system components but the thermostat as well. While at the thermostat, the technician should verify that it is mounted securely and level on the wall. The electrical connections on the subbase must also be tight, as with any other electrical connection in the system. During system changeovers, the operation of the following components is to be monitored:

- The supplementary electric-strip heaters
- The outdoor fan
- The indoor fan motor
- The compressor

The supplementary electric-strip heaters should be energized when the system is operating in either the defrost mode or in second-stage heating. When in heating, the strip heaters should cycle off when the desired space temperature is reached. The heaters should also be energized during the entire defrost cycle. The outdoor fan should be energized during the cooling and heating modes but de-energized during defrost. If the outdoor fan motor is equipped with some type of head pressure control, the fan may not operate all the time in the cooling mode. The indoor blower motor should be operating whenever the fan switch on the thermostat is switched to the ON position and should cycle with the compressor when it is in the AUTO position. The compressor should be operating when the system is calling for either heating or cooling as well as when the system is in defrost mode.

PUTTING IT TOGETHER

Following is a list of tasks that should be part of a complete preventive maintenance service call:

- Check the integrity of all electrical connections at both the indoor and outdoor units.
- Check the airflow through the indoor and outdoor units.
- Inspect the entire duct system.
- Air filters should be inspected and replaced.
- Check the mounting of all motors.
- Lubricate motors as needed.
- Check both the primary and auxiliary drain pans for proper operation.
- Inspect the refrigerant piping circuit for signs of leaks.
- Inspect the insulation on the refrigerant circuit's vapor line.
- Check the system for proper refrigerant charge.
- Check the thermostat operation.
- Check the unit in all modes of operation.
- Take and record voltage, amperage, temperature, and pressure readings.

SUMMARY

- The airflow through the indoor and outdoor units should be unobstructed.
- The motors should be properly mounted and lubricated.
- There should be a minimum of play in the motor shaft.
- Air filters should be checked regularly and replaced as needed.
- All electrical connections should be tight.
- Even factory connections should be inspected.
- Drain lines should be checked by pouring water into the lines.
- All safety float switches should be inspected.
- Refrigerant lines should be properly insulated.
- The refrigerant circuit should be checked for potential leaks.
- The level of refrigerant charge should be checked.
- All modes of operation should be checked.
- Thermostat mounting and location should be checked.

KEY TERMS

Air filter
Bearings
Blower
Branch duct
Squirrel-cage blower

FOR DISCUSSION

1. Why should technicians record system temperature and pressure readings when performing a preventive maintenance service call?
2. When performing routine service, why should the technician inspect electrical connections if the system seems to operate perfectly?
3. Explain the importance of a perfectly operating condensate removal system. Explain the possible results if a condensate drain line goes unchecked during a preventive maintenance service call.

REVIEW QUESTIONS

1. Which of the following could result in the freezing of a heat pump's indoor coil while operating in the cooling mode?
   a. Blocked return-air grill
   b. Defective indoor fan motor
   c. Closed supply registers
   d. All of the above
2. A squirrel-cage blower wheel that is out of balance:
   a. Puts extra stress on the motor.
   b. Will reduce the amount of air that is moved by the blower.
   c. Will decrease the level of noise transmission in the system.
   d. Will not have any negative effect on the system.
3. During a motor inspection, the technician determines that the shaft has a great deal of side-to-side play. The technician correctly concludes that:
   a. The motor is in need of lubrication.
   b. The motor has defective bearings.
   c. Side-to-side play is normal.
   d. The motor mounts are loose.
4. Dirt accumulation on the indoor coil can be reduced by:
   a. Increasing the speed of the indoor blower motor.
   b. Relocating the return-air grill.
   c. Minimizing the amount of air filter bypass.
   d. Removing the air filter from the system.
5. Motor overheating can result from:
   a. Defective bearings.
   b. Improper lubrication.
   c. Excessive airflow.
   d. Both *a* and *b* are correct.

6. A galvanized sheet metal duct system is inspected, and the inspection reveals that the duct is sweating during the cooling mode of operation. Technician A concludes that the duct system should be insulated immediately. Technician B concludes that the sweating of the duct system is an indication that the system is undercharged. Which of the following statements is most likely correct?
   a. Technician A is correct, and Technician B is incorrect.
   b. Technician B is correct, and Technician A is incorrect.
   c. Both Technician A and Technician B are correct.
   d. Both Technician A and Technician B are incorrect.
7. The indoor unit of a heat pump system is being inspected. The unit is located in an attic, and the inspection reveals that the service panel on the return side of the unit is missing. Technician A concludes that the cooling efficiency of the system will be reduced. Technician B concludes that the heating efficiency of the system will be reduced. Which of the following statements is most likely correct?
   a. Technician A is correct, and Technician B is incorrect.
   b. Technician B is correct, and Technician A is incorrect.
   c. Both Technician A and Technician B are correct.
   d. Both Technician A and Technician B are incorrect.
8. The condensate-removal system on the indoor unit of a heat-pump system is being inspected. The unit is located in an attic, and the inspection reveals that the condensate is leaving the structure from a location just above the patio door and is dripping down onto the patio. Technician A concludes that the primary drain line must be clogged. Technician B concludes that the auxiliary drain line must be clogged. Which of the following statements is most likely correct?
   a. Technician A is correct, and Technician B is incorrect.
   b. Technician B is correct, and Technician A is incorrect.
   c. Both Technician A and Technician B are correct.
   d. Both Technician A and Technician B are incorrect.
9. Describe the operation of the safety float switch and explain why these devices should be installed on all systems.
10. Upon inspecting the belt and pulleys on a belt-driven air-distribution system, the technician finds that the belt has skipped completely off the pulley arrangement. The belt is in perfect condition, and the blower motor is operating. Technician A concludes that the belt tension must have been too tight. Technician B concludes that the blower and motor shafts are not perfectly parallel. Which of the following statements is most likely correct?
    a. Technician A is correct, and Technician B is incorrect.
    b. Technician B is correct, and Technician A is incorrect.
    c. Both Technician A and Technician B are correct.
    d. Both Technician A and Technician B are incorrect.
11. A heat-pump system is checked in the cooling mode of operation. The evaporator superheat is measured and found to be 9 degrees. Technician A concludes that the system superheat will most likely be a few degrees higher than the evaporator superheat. Technician B concludes that, if the expansion valve feeds more refrigerant into the evaporator, the evaporator superheat will rise above 9 degrees. Which of the following statements is most likely correct?
    a. Technician A is correct, and Technician B is incorrect.
    b. Technician B is correct and Technician A is incorrect.
    c. Both Technician A and Technician B are correct.
    d. Both Technician A and Technician B are incorrect.

12. Refrigerant leaks can be identified by __________ found on the surface of the refrigerant tubing.

13. During a preventive maintenance service call, a technician finds that the system has lost a substantial amount of refrigerant since the last time that service was performed on the system. Technician A concludes that refrigerant can be safely and legally added to the system only if the system is monitored closely in case of a future leak. Technician B concludes that, after searching for a leak, refrigerant can be added to the system if immediate steps, such as adding a UV leak-detection solution to the system, are taken. Which of the following statements is most likely correct?
    a. Technician A is correct, and Technician B is incorrect.
    b. Technician B is correct, and Technician A is incorrect.
    c. Both Technician A and Technician B are correct.
    d. Both Technician A and Technician B are incorrect.

14. Pitted compressor contactor contacts:
    a. Will result in an increase in the resistance of the circuit.
    b. Should be replaced.
    c. Can ultimately cause compressor failure.
    d. All of the above are correct.

15. Which two pieces of data are needed to calculate the superheat of an evaporator?
    a. Supply-air temperature and return-air temperature
    b. Compressor-discharge temperature and suction-line temperature
    c. Evaporator saturation temperature and evaporator outlet temperature
    d. Evaporator saturation temperature and evaporator inlet temperature

CHAPTER 19

Heat-Pump-System Service Calls

OBJECTIVES After studying this chapter, the reader should be able to:

- Describe sample service calls typical to the industry.
- Explain how technicians should conduct themselves when performing service.
- Explain the importance of a professional relationship with customers.
- Explain the proper procedures for evaluating system components.
- Recall and explain basic refrigeration theory as it is applied to heat-pump technology.
- Review the seven steps for completing a successful service call, as outlined earlier in the text.

INTRODUCTION

Throughout this text, many topics relating to heat-pump-system operation have been discussed. These topics include:

- Basic refrigeration theory
- Basic heat-pump theory
- Servicing heat-pump air-conditioning systems
- Performing preventive maintenance on heat-pump systems
- Troubleshooting heat-pump systems

This chapter presents a number of service calls that are typical in the heat-pump industry. These calls cover a wide range of topics and also reflect the different troubleshooting styles and methods that can be utilized. Each service call is followed by a brief evaluation, calling attention to the actions of each of the technicians. In some cases, the technician did not perform in an ideal manner but ultimately arrived at the correct system diagnosis. In these cases, the discussion includes what the technician could have done to make the service call run more smoothly. Although not explicitly stated in the service calls, the reader should be aware of the seven steps for completing a successful service call as outlined earlier in the text.

At the end of the chapter, you are presented with a number of service calls in which you are asked to be the technician. On these calls, you are to describe what you would do on the call and what you think the system problem is.

OVERVIEW

Throughout this text, the reader has been exposed to a number of sample service calls. These service calls allowed the reader to follow along as the technicians made their way through the troubleshooting process. As was seen in most cases, the technician performed in a logical and concise manner. Successful field technicians must keep their wits about them. In essence, a heat pump is a heat pump is a heat pump. The basic theory behind the system is the same, even though some of the rules have been changed along the way. Field technicians must be aware of the operational differences that exist between air-source heat pumps and water-source units. They must be aware of the differences between open-loop systems and closed-loop systems.

For technicians to keep on top of the industry, continued education is a must. Technologies change, and new ways to accomplish old tasks are uncovered. Every technician *must* keep abreast of these changes. Performing effective service calls requires this knowledge. Continued and ongoing training has become a major concern of service companies worldwide in an effort to build a strong service team. Performing well in the field helps ensure the continued success of the company but also provides a great deal of personal satisfaction.

SERVICE CALL 1: ICE BUILDUP

Customer Complaint

A home owner calls his service company and informs the service manager that his heat-pump system is accumulating a lot of ice on the outdoor unit. The customer tells the dispatcher that the temperature has been roughly 30°F outside during the nighttime hours and approximately 45 to 50°F during the day. Upon inspecting the outdoor unit, the customer noticed that the fan was not operating but a humming or buzzing sound was coming from the unit. The customer turned off the unit to allow the unit to defrost. Once the frost melted, the customer turned the unit on again. The outdoor fan began to operate. Time and time again, though, the unit frosted up. The customer tells the dispatcher that, even though the unit has ice buildup on the coil, the system still provides adequate heat to the occupied space.

Service Technician Evaluation

Approximately 2 hours later, Joe, the technician, arrives at the customer's residence. Joe listens as the customer explains the situation to him. Joe then goes over to the system thermostat and checks the settings. The unit is set in the heating mode, and the space temperature is 68°F. The thermostat is set to maintain the space temperature at 70°F. Joe turns the thermostat setting up to 90°F and then goes outside to the unit.

At the unit, Joe can see that both the compressor and the fan motor are operating. After inspecting the system's defrost system, he establishes that defrost is initiated by a combination of time and temperature. Since frost is already present on the coil, he advances the cam on the defrost timer until the unit clicks into defrost. The outdoor fan motor cycles off, but he does not hear the reversing valve switch over to the cooling mode. Instead, he hears a buzzing sound coming from the unit. He then turns the unit off at the service disconnect switch. Joe goes back into the house and switches the unit over to the cooling mode. Returning to the outdoor unit, he restores power to the unit and the compressor and the outdoor fan motor begin to operate. Once again, he hears the buzzing sound. Going back into the house, Joe feels the air being discharged into the room and it is warm. Once again, he returns to the outdoor unit, disconnects the power, and removes the service panel from the side of the unit.

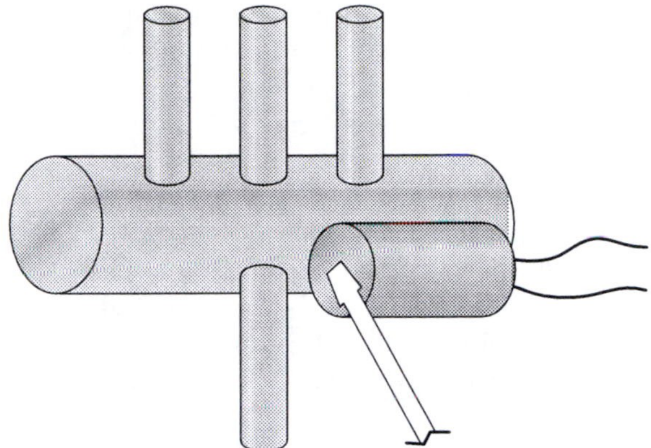

FIGURE 19-1 Screwdriver placed next to solenoid valve

Joe realizes that this system is designed to fail in the heating mode and that the reversing valve is not switching over in either the cooling or the defrost mode. Once again, Joe energizes the unit; and again, he hears the buzzing sound. He is now able to determine that the sound is coming from the reversing valve itself. Taking a screwdriver from his toolbox, be places the blade of the screwdriver next to the coil on the reversing valve and can feel the magnetic field that it is generating (Figure 19-1). Turning the screwdriver around, he lightly taps the body of the valve with the handle. He then hears the swish of the refrigerant as the valve switches over to the cooling mode.

Joe then goes inside again and switches the system over to the heating mode. Upon returning to the unit, he finds that the system is operating properly. He then advances the cam on the defrost timer into defrost again. The fan motor shuts down, and once again the buzzing sound returns. The system does not switch over. Tapping the valve again causes it to switch over.

Joe then concludes that the reversing valve is defective and needs to be replaced. Joe informs the customer of his findings and then calls the office to inform his dispatcher. Joe then goes to pick up a replacement valve from a nearby supply house.

Upon returning to the job, he begins to recover the refrigerant into an evacuated recovery tank. While the refrigerant is being recovered, he disconnects the power to the unit and removes the coil from the reversing valve in preparation for the repair. Once the refrigerant is completely removed from the unit, Joe begins to remove the reversing valve from the system. Using his tubing cutter, he cuts the refrigerant lines feeding into the valve and removes the valve from the unit. Joe then mounts the new valve in place and wraps the valve with a damp rag to protect the valve body from heat-related damage. In addition, he places heat-absorbing paste around the tubes to be brazed to reduce the amount of heat transmission to the valve.

Once the valve is brazed in place, he pressurizes the system with nitrogen and checks the valve for leaks. After the standing-pressure test is complete, he releases the nitrogen from the system and then begins the evacuation process. During the evacuation process, he mounts the coil on the valve. Once the system is evacuated, he reintroduces the refrigerant to the system and restarts the unit.

Joe first starts the unit in the cooling mode and is pleased to hear the swish of the refrigerant as the valve switches over to the cooling mode of operation. He then switches the system over to the heating mode and allows the system to run while he packs up his tools, writes out his work ticket, and informs the customer about the repair. Once completed, he switches the system into defrost to confirm that the valve is switching over properly. After obtaining the customer's signature on the work order, Joe proceeds to his next service call.

Service Call Discussion

In this service call, the technician proceeded in a logical manner in order to establish that the reversing valve was indeed defective. He listened to the customer's complaint and then went about determining the cause of the problem. He was very careful to confirm his conclusion, especially when the proposed repair would have been very time-consuming. He made many trips into the customer's house to switch the system over from heating to cooling, and vice versa, in order to triple-check his initial diagnosis. Joe could have made his diagnosis solely on the fact that the valve did not switch over in the defrost mode until he tapped the valve body with his screwdriver, but, once again, he wanted to be absolutely certain that his ultimate decision was the correct one. It seems from this service call that Joe has learned his trade well and is aware that improper conclusions about a system repair can result in lost time and money for both the customer and the service company.

SERVICE CALL 2: NO HEAT

Customer Complaint

It is the heating season, and a store owner calls his service company and tells the service manager that his heat-pump system is not providing heat to his store. He tells the dispatcher that the outdoor unit seems to be operating. When he went to the indoor unit located in the basement, he heard the fan motor running and saw that the air filter was clean. He can feel no air coming from the supply registers on the ceiling.

Service Technician Evaluation

When Martin arrives at the customer's store, he asks the store owner what the problem seems to be. The equipment owner proceeds to explain exactly what is happening with the system. Martin immediately suspects an airflow problem and goes down to the air handler in the basement. Just as the store owner told him, the indoor fan is operating and the filter is perfectly clean. He then decides to check the volume dampers located on the takeoffs. All of the volume dampers are in the open position, but he cannot hear the sound of air movement in the duct system.

Next, Martin turns the power off on the indoor unit and removes the access panel from the unit to inspect the motor. The motor is connected to a belt-driven blower assembly. The belt is intact and everything looks normal. Opening an access panel in the main trunk line of the duct system, Martin, with the aid of a droplight, checks the interior of the ductwork. He finds that a large piece of the duct lining has come loose and is blocking the supply duct. Luckily, the loose lining section is close to the access

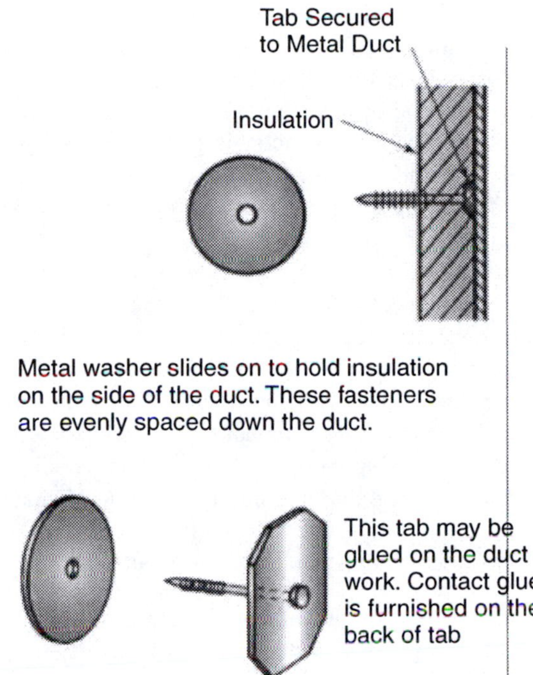
Tab Secured to Metal Duct
Insulation
Metal washer slides on to hold insulation on the side of the duct. These fasteners are evenly spaced down the duct.
This tab may be glued on the duct work. Contact glue is furnished on the back of tab

FIGURE 19-2 Tabs and washers used to hold duct lining material in place

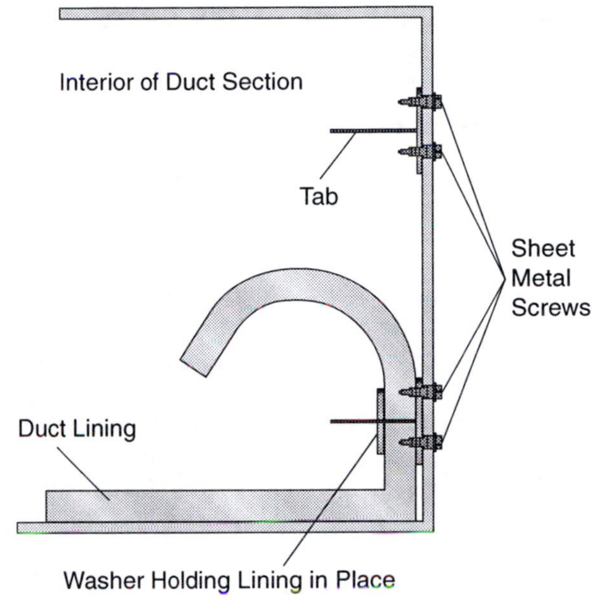
Interior of Duct Section
Tab
Sheet Metal Screws
Duct Lining
Washer Holding Lining in Place

FIGURE 19-3 Sheet metal screws used to mount the tabs on the interior surface of the duct section

panel. Martin tells the customer what he has found and then goes back to his shop to retrieve some tabs and washers (Figure 19-2).

Once he returns with the tabs and washers, Martin proceeds to secure them to the inside of the duct section. Because the store owner needs to use his system immediately, Martin decides not to glue the lining in place, as the fumes from the glue are very strong. Instead, he chooses to use self-tapping sheet metal screw, in addition to the adhesive located on the tabs, to secure them to the interior surface of the duct (Figure 19-3). This ensures that the tabs will not come loose. After mounting the tabs, he then pushes the lining onto the tabs and then pushes the washers onto the tabs, securing the lining to the duct.

Once the lining is secured to the duct, Martin proceeds to close up the access panel and start up the system. He then checks the airflow through the duct system, and everything seems to be working fine. He takes temperature readings across the indoor coil and checks the operation of the outdoor unit to confirm that the system is operating correctly. He fills out his paperwork and proceeds to his next call.

Service Call Discussion

In this service call, Martin immediately determined that there was an airflow problem. The problem was not, however, very common. He first checked for the obvious solutions, such as a broken belt, clogged filter, and a closed volume damper. When he dismissed these as causes, he then proceeded to inspect the duct system. Knowing that the store owner needed to use the system immediately, Martin took an alternative approach to secure the lining to the interior of the duct. This showed respect not only for the equipment owner but for the customers in the store as well. One very important thing that Martin did was that he kept the store owner informed as to what the situation was with the system. This is extremely important on all service calls.

SERVICE CALL 3: HIGHER BILLS AND LESS HEAT

Customer Complaint

The owner of a heat-pump system calls her service company and reports the following information to the dispatcher regarding the heating operation of her unit:

- The compressor always seems to be operating.
- The outdoor fan is never operating.
- The air being supplied to the room is not as warm as usual.
- Her electric bills have been nearly double as compared to last year's bills for the same months.
- The temperature pattern this year is similar to that of last year.

Service Technician Evaluation

About an hour later, Patricia, the technician, arrives at the customer's house. She is given the same information that was given to the dispatcher earlier. Patricia goes over to the thermostat to check the settings of the system. The system is set to the heating mode. The space temperature is 62°F and the thermostat is set to maintain a 70°F space temperature. The system should be operating in the heating mode at this point. Patricia then goes over to a supply register and feels the air being introduced to the room. The air is not warm, but it is not cold either. It seems that the air is just a little warmer than the temperature of the space. The blower seems to be moving enough air through the system, so Patricia decides to check the outdoor unit first.

At the outdoor unit, Patricia notices that the vapor line is frosted back to the compressor, indicating that the system is operating in the cooling mode. She also notices that the outdoor fan motor is not operating. From this, she suspects three possible scenarios:

- The system is not switching over to heating, and the outdoor fan motor is defective.
- The system is stuck in the defrost mode.
- The reversing valve or the solenoid coil is defective, and the outdoor fan is also defective.

Knowing that three possibilities exist for the system malfunction, Patricia decides to eliminate as many scenarios as possible. She decides to check the operation of the outdoor fan motor first. She reasons that, if the outdoor fan motor is good, the system is most likely stuck in defrost. If the fan motor is defective, she can then check the reversing valve and the solenoid coil.

Patricia first disconnects the power to the outdoor unit and connects power directly to the outdoor fan motor, jumping out all controls and switches in series with the motor. She then energizes the unit and observes that the outdoor fan motor is indeed, operational. She is then able to conclude that some control or switch is keeping the fan from operating. By eliminating the first and third possibility, Patricia concludes that the unit is stuck in defrost. To Patricia, this makes perfect sense. A unit stuck in defrost would:

- Have the outdoor fan motor de-energized
- Have the supplementary electric-strip heaters energized
- Be operating in the cooling mode
- Result in abnormally high electric bills due to continuous compressor and strip heater operation.
- Discharge only tempered air to the occupied space

Since all of the pieces seem to fit perfectly, Patricia then decides to check the defrost timer on the unit. Upon inspection, she determines that the system's defrost cycle is initiated by time and temperature and terminated only by temperature. Removing the temperature sensor from its location on the coil, Patricia rubs it vigorously between her hands. The unit remains in defrost. After disconnecting the power to the unit, she removes the electrical connections on the sensor and checks for continuity across the device. She reads infinite resistance. Next, she places a jumper wire across the terminals to which the sensor is connected and energizes the system. The unit begins to operate in the heating mode. The problem is the temperature sensor on the outdoor unit.

Patricia then disconnects the power to the unit and reports her findings to the customer and to her service manager. Upon installation of a newly obtained temperature sensor, the unit resumes normal operation.

Service Call Discussion

Making a list of all possible causes for the system malfunction and then eliminating possibilities one at a time is often beneficial. Patricia noticed that a defective outdoor fan motor appeared in two of the possible causes and decided to check that one first. Because the fan motor was good, she was able to eliminate those two possibilities. This left only the possibility that the system was stuck in the defrost mode. After evaluating the defrost system on the unit, she was able to determine that the temperature sensor was defective. Great job, Patricia!

SERVICE CALL 4: EMERGENCY HEAT

Customer Complaint

A customer places a service call and tells the service manager that her heat-pump system is not heating the space as it used to. She goes on to say that the outdoor unit does not seem to be operating at all and that she has had to put the system into the emergency heat mode in order to heat the space.

Service Technician Evaluation

Later that day, Doug arrives at the customer's house. The customer tells him that she thinks there is a problem with the outdoor unit, as it is not operating. Doug checks the thermostat settings and goes to the outdoor unit. The unit is not operating. Doug first decides to check the power being supplied to the unit so he turns off the disconnect switch and opens the disconnect box. Using his voltmeter, Doug checks the voltage at the disconnect and finds that 220 volts are being supplied to the unit. He then removes the access panel from the unit, restores power, and checks for voltage on the line side of the compressor contactor. Once again, the reading is 220 volts, but the unit is not operating. Checking the low voltage coming from the indoor unit, he finds that 24 volts are being supplied to the outdoor unit's control circuit.

Doug then takes a voltage reading at the holding coil of the compressor contactor and obtains a reading of 0 volts. An inspection of the low-voltage circuit indicates that both a high-pressure and a low-pressure switch are on the unit. Again, with his voltmeter, Doug takes readings across both controls. The readings are 24 volts across the low-pressure control and 0 volts across the high-pressure switch. This indicates that the low-pressure switch is in the open position. He then concludes that there is a refrigerant problem.

Retrieving his gauges from his truck, Doug connects the manifold to the system and finds that the system is completely void of refrigerant. After another trip to his truck, Doug pressurizes the system with 150 psig of dry nitrogen. Armed with his soap bubble solution, he begins to check the solder joints at the outdoor unit. While checking, he hears a hissing sound coming from the bidirectional filter drier on the liquid line. Wrapping his fingers around the flare connections on the drier, he can feel the nitrogen escaping from the connection.

Keeping the pressure in the system, Douglas tightens the flare connections on both ends of the drier. The hissing sound stops. He checks the connections with the soap bubble solution and confirms that the flare connection is no longer leaking. Since the system has pressure, he decides to leak-check the rest of the system to ensure that the system is tight. After leak-testing the system and finding no other leaks, he releases the nitrogen from the system and begins to evacuate the system. While the system is on the vacuum pump, he goes indoors and discusses the situation with the customer. He then checks the indoor unit and the air filters to make certain that everything is in order.

Once the system is properly evacuated, he begins to add refrigerant to the system. First, he introduces a holding charge to the system and then starts the system. Once the system is properly charged, he takes temperature and pressure readings on the system and completes his paperwork.

Service Call Discussion

Doug utilized his time very well on this service call. During the evacuation process, he took the time to discuss the situation with the customer and to check the indoor unit and air filter. He proceeded in a logical manner and was able to locate the system problem quickly. When performing service, time is of the essence. Always try to accomplish the required task in as short a period of time as possible without sacrificing the quality of the repair.

SERVICE CALL 5: INADEQUATE HEAT

Customer Complaint

The owner of a heat-pump system calls his service company and reports that his system is not providing adequate heat to his home. He also states that the outdoor unit does not seem to be cycling off. He also tells the dispatcher that the air that is being discharged from the outdoor unit is warmer than normal and that the air is normally very cool.

Service Technician Evaluation

Later that afternoon, Jason drives up to the residence. He asks the customer to explain the symptoms that he observed and then checks the thermostat settings. Jason then goes out to the outdoor unit. The compressor and the outdoor fan are both operating. Touching the vapor line, he finds that the line is warm but not as hot as it should be. The liquid line is also cooler than normal. Gauging up on the system, Jason establishes that the head pressure is lower than normal and the suction pressure is higher than normal.

Using a strap-on digital thermometer, he then takes a temperature reading at the outlet of the outdoor coil in order to take an evaporator superheat reading. The superheat is 0 degrees, indicating that the evaporator is being overfed. The system is configured with two thermostatic expansion valves and two check valves, as shown in Figure 8-21(b) and (c). Jason then concludes that the problem is with either the check valve or the thermostatic expansion valve at the outdoor coil. He first decides to evaluate the thermostatic expansion valve.

He removes the thermal bulb from its location on the outlet of the outdoor coil and heats it between his hands. No change occurs in the suction pressure. He then places the bulb in a glass of cold water. Once again, no change occurs in the suction pressure. Jason then considers the possibility that the thermal bulb may have lost its charge. He remembers that the bulb pressure in a thermostatic expansion valve is the only pressure that pushes the valve open. If the bulb had lost its charge, the valve would be pushed closed, which contradicts the present situation.

Jason then turns his attention to the check valve. He taps the valve with the handle of a screwdriver, and the suction pressure begins to drop rapidly. Jason Allows the system to operate for a few minutes, the temperature of the vapor line begins to rise, and the superheat in the evaporator reaches 8 degrees. The check valve is defective. After obtaining a replacement valve, he proceeds to pump the system down to perform the repair. After the repair is made, the system is leak-checked, evacuated, and restarted.

Service Call Discussion

During this service call, Jason reached the point where he was able to narrow the cause for the system malfunction to the check valve or the thermostatic expansion valve. If he had not taken the extra time to determine the exact cause, he would have had a fifty-fifty chance of identifying the problem. Remembering his thermostatic expansion valve theory, he was quickly able to eliminate the possibility of a lost bulb charge. After tapping the check valve with the handle of the screwdriver, the system began to operate. Why then did he choose to replace the valve?

SERVICE CALL 6: SYSTEM SHUTDOWNS

Customer Complaint

A repeat customer calls to inform the service manager that the fuses on the outdoor unit of his heat-pump system have blown again. This has been happening every 3 days for the past 2 weeks. Many service technicians have been to the customer's house, and each one has inspected the unit from top to bottom. Everything seemed to be perfectly normal. At the time of each service call, the system's operating temperatures, pressures, amperages, and voltages were checked and recorded. Absolutely nothing seemed out of the ordinary. The customer tells the dispatcher that he is starting to get a little annoyed at the fact that his system keeps shutting down, almost like clockwork. The service manager calls a representative from the manufacturer and requests that one of the factory technicians accompany the technician to the job to help resolve the problem once and for all.

Service Technician Evaluation

The next day, the service technician arrives at the job site along with a service technician from the manufacturer. Together, the two techs inspect and reinspect the unit. Neither one can come up with a solution to the problem. Much time, energy, and money has already been spent to find the cause of this intermittent problem. The manufacturer's representative concludes that everything within the confines of the unit is perfectly operational and that the problem is most likely external to the unit.

The men decide to check the power lines feeding power to the unit. They start at the circuit-breaker panel. Carefully removing the cover from the panel, they check the circuit breaker. The breaker is brand new and in great condition. They next inspect the interconnecting wiring between the circuit-breaker panel and the fused disconnect at the outdoor unit. Again, everything looks normal. They finally arrive at the fused disconnect switch. With the circuit breaker in the OFF position, the technicians open the fused disconnect to continue their inspection of the unit. The fuses are removed from the fuse holders, and the electrical connections in the disconnect box are checked. The screw holding the fuse on the right side of the box is extremely loose. The two techs look at each other and nod in agreement. The fuse clip is tightened and the system is put back together, including the cover on the circuit-breaker panel. Once together, the system is restarted.

The two technicians explain the situation to the customer, and the customer responds with skepticism. He basically tells the technicians that he has been told five times before that the problem was remedied and that only time will tell if indeed the problem is now a thing of the past. The two techs check the operation of the system, report their findings, and leave the site. They also request that the customer call the shop and keep the dispatcher informed as to how the system is operating. About 2 weeks later, the customer calls the service company to report that the system has been working perfectly since the last service call.

Service Call Discussion

A number of technicians had been at this job in the past, and none were able to remedy the problem with the system. When a system is being checked, it is important to check all possible causes for failure. Checking electrical connections is among the most important things that should be done when performing service, especially during preventive maintenance. Technicians must be able to look at the big picture and not limit the search area for system failure to the confines of the unit casing.

SERVICE CALL 7: NONFUNCTIONAL UNIT

Customer Complaint

A heat-pump-system owner calls the service company and reports that the unit is not coming on at all. He has checked all of the circuit breakers in the panel, and they are all in the ON position. The indoor fan does not operate, and the outdoor unit does not operate either.

Service Technician Evaluation

Later that morning, Barry arrives at the customer's home. The customer immediately tells Barry about the problems that he is having with the system. Barry then goes over to check the thermostat settings and verifies that the system is set to operate in the cooling mode. First, Barry switches the fan switch to the ON position but does not hear the blower come on. Immediately, Barry suspects a low-voltage problem because the system is not responding at all. He then makes his way down to the basement, where the indoor unit is located.

With the power to the unit disconnected, he proceeds to remove the service panel from the unit. After restoring power, he checks for voltage being supplied to the unit. At the terminal board, Barry confirmed that 220 volts are being supplied to the air handler. Next, he decides to check the voltage at the secondary side of the control transformer. He obtains a voltage reading of 24 volts, indicating that the control transformer is operational. He next turns to the low-voltage terminal block at the indoor unit.

Checking voltage at the terminal board, he obtains the following readings:

- 0 volts between Terminal R and Terminal G
- 0 volts between Terminal R and Terminal Y
- 0 volts between Terminal R and Terminal W1
- 0 volts between Terminal R and Terminal W2

It seems to Barry that, although there is voltage at the transformer, no voltage is being supplied to the control components. He decides to check the low-voltage line from the transformer. He takes a voltage reading from the power terminal (where the red wire is connected) to ground and obtains a voltage reading. He then takes a voltage reading from the R terminal at the terminal block and reads 0 volts. He determines that the red wire leading from the transformer to the terminal block has a break in it. After turning the power to the unit off, he disconnects both ends of the red wire from the transformer and the terminal block and checks for continuity. He reads infinite resistance, indicating that the wire is indeed broken or damaged.

Barry then goes to his truck to retrieve some low-voltage wire so that he can replace the defective wire. Upon returning from his truck, he notices that the red wire does not run directly from the transformer to the terminal block. After tracing the wire, he realizes that the red wire runs to the condensate pump located at the base of the indoor unit. He inspects the pump and finds that it is filled with water and has been left unplugged. He also notices that a safety float switch has been installed in the pump. He plugs in the pump and it immediately begins to pump the condensate out of the basin. He checks the ends of the red wire for continuity, and this time he obtains a reading of 0 ohms.

With the power off, Barry reconnects the red wire to the transformer and the terminal board and re-energizes the system. The indoor blower motor begins to operate, and Barry can hear the rush of refrigerant through the indoor coil. He takes a temperature reading across the indoor coil and also inspects the air filters. He alerts the customer about what has happened and instructs the home owner to make certain that the condensate pump is never unplugged. Barry obtains the customer's signature on the work order and proceeds to his next call.

Service Call Discussion

During this service call, Barry did not even need to go the outdoor unit. All indications pointed to the control transformer because there was absolutely no system activity whatsoever. Even though the control transformer was operational, Barry was definitely on the right track. Right up until the time he went to his truck for some wire, he was unaware of the fact that the pump was unplugged. This was okay, since he did eventually locate the situation when he began to replace the wire. Note that Barry explained to the customer that the pump was simply unplugged and nothing was wrong with the system. Honesty is a trait that should be inherent in all technicians performing service.

SERVICE CALL 8: OUTDOOR FAN MOTOR NOT OPERATING

Customer Complaint

A heat-pump system is not providing adequate heat to the occupied space. The equipment owner calls his service company and requests that a technician be sent over to check the system out. The customer informs the service manager that it appears that the outdoor fan motor is not operating.

Service Technician Evaluation

The next morning Miguel arrives at the job site. The system is set to operate in the heating mode, but the air that is being discharged into the occupied space is not satisfying the load on the area. Miguel first goes to inspect the outdoor unit. He can hear the compressor operating, but the outdoor fan motor is not operating. He also notices that the coil has a substantial amount of ice buildup on it. He feels the vapor line, and it is only warm to the touch. From this, Miguel is able to establish that the system is operating in the heating mode but the outdoor fan is not operating. After disconnecting the power to the unit, he removes the service panel from the unit.

Miguel notes that a set of normally closed contacts is on the defrost timer in series with the outdoor fan motor. During the heating mode, these contacts are normally closed but open when the system switches into defrost. After restoring power to the unit, Miguel checks the voltage across the normally closed contacts and obtains a reading of 220 volts, indicating that the contacts are in the open position when they should be closed. A jumper placed across the contacts causes the outdoor fan to operate.

After removing the jumper, Miguel advances the timer into defrost. He can hear the reversing valve switch over to the cooling position, and the ice on the coil quickly melts. He then disconnects the power to the unit and obtains a new defrost timer. He replaces the timer and restarts the unit in the heating mode. Both the compressor and the outdoor fan motor begin to operate. A touch test on the vapor line indicates that the line is now much warmer than before, indicating that the system is operating properly. He checks the indoor unit and the air filter before leaving the job.

Service Call Discussion

That Miguel noticed that a set of contacts was in series with the outdoor motor was important. He easily could have concluded that the motor was defective since the system was operating but the motor was not. Properly evaluating the electrical circuits carefully before deciding that a major system component is defective is important. Even though the repair was made to the outdoor unit, Miguel made a point to inspect the indoor unit and the air filter while he was there. A little extra service goes a long way in the eyes of the customer.

SERVICE CALL 9: ICE BUILDUP ON THE COIL

Customer Complaint

A heat-pump system is not providing adequate heat to the occupied space. The equipment owner calls his service company and requests that a technician be sent over to check the system out. The customer informs the service manager that the outdoor unit appears to be operating because the fan is turning, but more ice appears to be on the coil than usual.

Service Technician Evaluation

Later in the day, Charles arrives at the customer's residence. The customer tells him that the system must need some refrigerant because the system is not providing adequate heating. Charles examines the thermostat and checks the control settings. The unit is set to the heating mode, and Charles can hear the indoor fan motor operating. He then goes to the backyard where the outdoor unit is located. He notices an extremely large amount of ice on the outdoor coil. He immediately disconnects the power to the unit and removes the service panel. His immediate assumption is that the system is not going into defrost. He inspects the unit and discovers that the system's defrost is initiated by time and temperature and terminated by temperature alone.

Charles advances the cam on the defrost timer and then re-energizes the system. The system immediately switches over to defrost, and Charles can hear the swish of the refrigerant as the reversing valve changes its position. He allows the unit to completely defrost. Once the unit is defrosted, it automatically switches back over to the heating mode. Charles concludes that the defrost motor must be bad, so he goes to his truck to get a replacement defrost timer. He disconnects the power and replaces the timer. He jumps out the temperature sensor and advances the cam on the timer to a point that will initiate defrost after only a few minutes of operating time. Charles waits and waits and waits. The system does not switch into defrost. He manually advances the cam, and the system switches over to the defrost mode. The new timer is doing exactly what the old one was doing.

He now decides to check the voltage being supplied to the timer motor. He reads 0 volts. He disconnects the power to the unit and takes a closer look at the electrical circuits, where he locates a wire that has come loose from the terminal block. He checks the wiring diagram on the unit and finds that this wire provides power to the timer motor. He replaces the wire on the correct terminal and restarts the unit. He now checks the voltage at the timer motor and obtains a reading of 220 volts. Once again, he disconnects the power and puts the customer's original timer back into the system. He restores power to the unit once again and checks the operation of the timer. This time the defrost timer works fine. He prepares his paperwork and leaves the job.

Service Call Discussion

Although Charles ultimately located and repaired the problem, he made one crucial mistake. The mistake was that he did not confirm that the timer motor was defective before he replaced the timer. He should have established that voltage was being supplied to the component before determining that it was defective. He was lucky in the sense that the component was easily changed out. If the component had been a compressor, for example, a great deal of additional, unnecessary expense would have been incurred. The technician must be absolutely sure that the component is defective before condemning it.

SERVICE CALL 10: BURNING SMELL

Customer Complaint

A customer calls and informs the service manager that her heat-pump system is not operating. She also tells him that she smells something burning in the area directly under the return grill of the unit. She is instructed to immediately turn off the circuit breaker for the indoor unit and await service. The service manager immediately contacts one of his technicians who is working nearby and tells him to go over to the house right away. The service manager also informs the home owner to clear away the area around the attic access hatchway.

Service Technician Evaluation

About 10 minutes later, James appears at the house. He immediately goes up into the attic and inspects the unit. A burning smell is coming from the unit. He shuts off the disconnect switch at the unit and opens the service panel. He notices that the smell is coming from the supplementary electric-strip heaters within the unit. James goes to his truck and retrieves the rest of his tools. He resets the breaker for the unit while he is downstairs. He also sets the thermostat to operate in the heating mode.

Once back in the attic, James turns on the disconnect switch. He can hear the sound of the refrigerant flowing through the indoor coil, but the indoor fan does not begin to operate. He shuts off the power to the unit and inspects the motor. The motor is cool to the touch. He disconnects the motor wires from the terminal block and checks the motor windings. He reads infinite resistance across the windings, indicating that the motor has been burned and is in need of replacement. He takes down all of the data from the motor's nameplate. While he is in the attic, he decides to check the fusible links on the strip heaters, since it is obvious from the smell that they have been energized with no air moving across them. He checks the fusible links and finds that one out of three of the links has been burned. He removes that link to get a replacement as well. Finally, James is curious as to why the heaters could be energized when no air was moving across them. He checks to see if a sail switch is installed in the duct, but he cannot locate one. He decides to install one when he comes back with the other components.

James arrives back at the job about 2 hours later and replaces the defective motor and fusible link. He also installs the sail switch in the duct and wires it in series with the control circuit for the electric-strip heaters. This will prevent the heaters from becoming energized if no air is moving in the duct system. Once the repair is complete, James explains the repair to the customer, informing her about the installation of the sail switch. She thanks him profusely for arriving so quickly, and he returns to his other job.

Service Call Discussion

The almost immediate arrival of the technician on the job was key to this service call. A potentially dangerous situation was present. Although the customer was not in immediate danger, this was not established until the technician arrived. James should be a model for all technicians. Not only did he repair the problem, but he also took preventive steps to help ensure that the heaters would not get energized in the future unless air was flowing in the duct system. The sail switch should have been installed originally, but, for one reason or another, it was not. In addition, James checked the fusible links to be sure that they were still operational.

SERVICE CALL 11: INADEQUATE COOLING

Customer Complaint

The owner of a heat-pump system places a call to his service company, indicating that not much cool air is coming from the supply registers. The outdoor unit seems to be working fine, but the house is uncomfortably warm.

Service Technician Evaluation

A short time later, Anne arrives at the customer's home. The home owner explains the situation, and Anne immediately suspects an airflow problem. Since the air handler is located in the attic and the return-air grill is located in the main floor hallway, she decides to check the air filter first. She makes certain that the fan switch on the thermostat is switched to the ON position and then goes to the filter rack. Upon opening the filter rack, she can see that the filter is completely blocked with dirt. It obviously has not been changed for quite some time. While the filter is removed, Anne goes over to the supply register and finds that, still, very little air is coming through the system. She can see that the supply registers are open all the way, so she next has to inspect the air handler.

Up in the attic, Anne can hear the fan operating but it does not sound like it is running up to speed. She disconnects the power to the unit and removes the access panel. She restores power to the unit, and she observes the fan motor running very slowly. She checks the voltage being supplied to the motor, and it is well within the proper operating range. Before condemning the motor, she decides to replace the run capacitor on the motor to see if that helps the situation. Retrieving one from her truck, she replaces the capacitor and restarts the unit. The motor begins to run at full speed. While in the attic, she inspects the indoor coil because she suspects that it might be dirty due to the condition of the filter. Luckily, the coil is clean. She replaces the access panels and speaks with the customer. She shows the home owner how to check and replace the air filters and also tells him that checking the filters at least once a month is important to help ensure proper system operation. Anne also suggests that preventive maintenance be performed in the near future and explains the importance of maintaining the system.

Service Call Discussion

Anne made certain that she checked the voltage at the motor before condemning it. Additionally, she decided to change the capacitor to be sure that it was the motor and not simply the capacitor. Luckily for the home owner, the technician knew enough to check the capacitor first. This could have resulted in a costly repair had Anne not had the insight to exhaust all other options first. Even though it was not part of the initial repair, Anne also took the time to inform the home owner about filter inspection and replacement. Anne could tell that the customer had not had the system checked out for a long while and recommended that the customer call for an appointment to have preventive maintenance performed on the unit.

Note to the reader: The next three service calls refer to Figure 19-4.

SERVICE CALL 12: YOU BE THE TECHNICIAN

Customer Complaint

It is the middle of the summer, and a customer calls his service company to report that very hot air is coming from the supply registers. The customer also tells the dispatcher that the outdoor unit is running and that his heat-pump system worked well during the winter, although he noticed that more ice than normal built up on the outdoor coil.

Service Technician Evaluation

Later that afternoon, the technician, Marco, arrives at the customer's home. He listens to the customer as he explains what is happening with the system. Marco then goes over to the space thermostat and checks to make certain that the system is set to the cooling mode and is set to maintain a temperature well below the actual space temperature. Everything seems to be set correctly. Marco then goes to the outdoor unit and disconnects the power to the unit. He then removes the service panel and inspects the schematic diagram for the unit, as shown in Figure 19-4.

Marco then restores the power to the unit. He hears the compressor come on and sees the outdoor fan motor begin to operate as well. He feels the vapor line and finds it to be hot. What should Marco's next steps be?

- What system component would be the first he should check?
- According to the wiring diagram, which electrical terminals should Marco check with a voltmeter?

Legend

M1	Compressor Motor
M2	Condenser Fan Motor
M3	Evaporator Fan Motor
DR	Defrost Relay
CR	Cooling Relay
TM	Timer Motor
RV	Reversing Valve
CC	Compressor Contactor
CB	Circuit Breaker
LMC	Limit Control
LOR	Lockout Relay
TDR	Time-Delay Relay
TR	Transformer
FR	Fan Relay
HTR	Heater
FLK	Thermal Fuse
TB	Tab Bushing
DSW	Disconnect Switch (Field Installed)
FU	Fuse (Field Installed)
E15	Defrost Control
HT	Holdback Thermostat
CPR	Capacitor
HPA	High-pressure Control
PRI	Transformer Primary
SEC	Transformer Secondary
——	Power Wire (Factory Installed)
- - - -	Control Wire (Factory Installed)
═══	Power Wire (Field Installed)
= = = =	Control Wire (Field Installed)
D20	Pressure Sensor
R	Red Light
B	Blue Light
CA	Cooling Anticipator
HA	Heating Anticipator
ADR	Auxiliary Defrost Relay

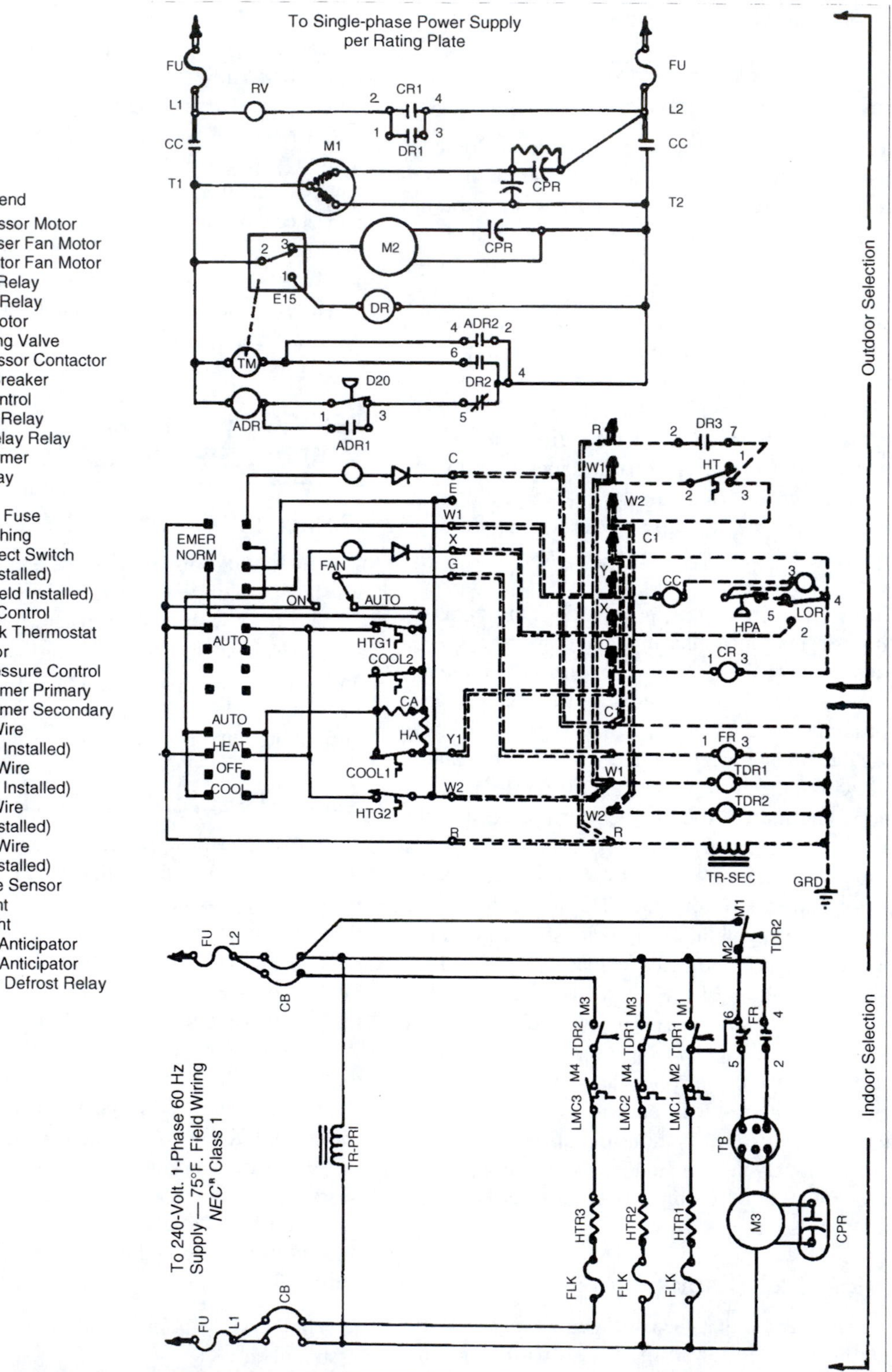

FIGURE 19-4 Schematic diagram of a heat-pump system. *(Courtesy Addison Products Co.)*

- What could be a likely cause for the system malfunction if it is electrical in nature?
- What could be a likely cause for the system malfunction if it is mechanical in nature?

SERVICE CALL 13: YOU BE THE TECHNICIAN

Customer Complaint

The owner of a small shop calls her service company and reports that her heat-pump system periodically blows the circuit breaker. Otherwise, the system works just fine. It is fall, and the weather has been relatively mild, with the exception of the early mornings when it is quite chilly outside when she first opens the store for business in the morning.

Service Technician Evaluation

Later that morning, Bill arrives at the customer's store to check out the system. After listening to the customer's complaint, Bill inspects the circuit-breaker panel and notices that the breaker for the indoor unit has indeed tripped. He turns off the unit at the thermostat and resets the breaker. He then turns the unit back on from the thermostat and can hear the indoor fan begin to operate. He goes around back to the outdoor unit and notices that the compressor and the outdoor fan motor are operating. He then goes to the basement where the indoor unit is located.

Bill disconnects the power to the indoor unit and then removes the service panel, where he sees the wiring diagram shown in Figure 19-4. He restores power to the unit and takes an amperage reading of the indoor fan motor. The amperage is well within design range. During single-stage heating, the only system component that is energized in the unit is the fan motor and it seems to be working fine.

- What system component would be the first Bill should check?
- What could be a likely cause for the system malfunction if it is electrical in nature?

SERVICE CALL 14: YOU BE THE TECHNICIAN

Customer Complaint

A home owner calls and tells the dispatcher that her heat-pump system trips the circuit breaker every so often in the heating mode. When she switches the unit over to the cooling mode, the circuit breaker trips immediately. The following is a list of conditions that are present at the unit:

- The indoor fan operates properly even when the circuit breaker for the outdoor unit trips.
- The heat-pump system is designed to fail in the heating mode.
- The system operates properly in the heating mode but trips the breaker after about an hour of system operation.
- When switched over to the cooling mode, the breaker trips immediately.

Service Technician Evaluation

What would you do?

SERVICE CALL 15: YOU BE THE TECHNICIAN

Customer Complaint

A heat-pump owner calls her service company and informs the service manager that the system is not cooling properly. The following is a list of conditions that are present at the unit:

- The system's operating pressures in the cooling mode are correct.
- The condenser subcooling is approximately 18 degrees.
- The evaporator superheat is approximately 9 degrees.
- The suction line is sweating back to the compressor.
- The liquid-line temperature is approximately 90°F.
- The return-air temperature at the indoor unit is 73°F.

- The supply-air temperature at a supply register is 78°F.
- The unit provides adequate heat when operated in the heating mode.

Service Technician Evaluation

What would you do?

SERVICE CALL 16: YOU BE THE TECHNICIAN

Customer Complaint

A heating customer places a service call and informs the dispatcher that his heat-pump system is not heating properly. The following is a list of conditions that are present at the unit:

- The system is configured with two check valves in parallel with two thermostatic expansion valves.
- The suction pressure at the compressor is in a vacuum.
- The compressor is operating but is very hot to the touch.
- Heating and cooling the thermal bulb on the thermostatic expansion valve does not cause a change in the suction pressure.
- The system operates perfectly in the cooling mode.

Service Technician Evaluation

What would you do?

SUMMARY

- Always strive to do the best possible job on every service call.
- Basic refrigeration theory is useful on all service calls.
- Technicians should be alert and open to the customer's concerns and suggestions.
- Always follow EPA guidelines with respect to refrigerant handling, recovering, and leak-checking a system.
- Remember that time is money. Wasted time costs the customer as well as your service company.
- Rushing through a service call also wastes time and money. Be thorough and complete. Callbacks are the number one enemy of a service company. Try to avoid them!
- Perform all work in a safe manner.
- Always act in a professional, courteous manner.
- Always follow the manufacturer's recommendations when servicing a system and replacing components.
- High electric bills are a sign that a problem exists with an air-conditioning system.
- Always perform a complete system evaluation before committing to any major system repairs or modifications.
- Always be aware of the seven steps for completing a successful service call.

FOR DISCUSSION

1. Review the seven steps for completing a successful service call and describe the importance of each step.
2. Discuss the importance of personal grooming and hygiene as they relate to customer service.
3. Explain the importance of maintaining a professional attitude toward the job, customers, and coworkers.

REVIEW QUESTIONS

1. On a heat-pump system that fails in the heating mode, a defective four-way reversing valve can result in:
   a. Excessive ice buildup on the outdoor coil when operating in the heating mode.
   b. Poor system performance in the heating mode.
   c. Poor system performance in the cooling mode.
   d. All of the above are possible results.
2. A defective outdoor fan motor can result in:
   a. Excessive ice buildup on the outdoor coil when operating in the heating mode.
   b. Low suction pressure in the cooling mode.
   c. High head pressure in the heating mode.
   d. All of the above are correct.
3. A defective indoor fan motor can result in:
   a. High head pressure in the heating mode.
   b. Low suction pressure in the cooling mode.
   c. Excessive ice formation on the outdoor coil during the heating mode.
   d. Both *a* and *b* are correct.
4. The solenoid on a four-way reversing valve can accurately and definitively be checked by:
   a. Checking to see if voltage is being supplied to the coil.
   b. Placing the blade of a screwdriver next to the coil and feeling for the pull of the magnetic field.
   c. Switching the unit from the cooling to the heating mode.
   d. All of the above can definitively determine if the solenoid coil is good.
5. Explain how a sticking reversing valve can affect system operation.
6. Describe a standing-pressure test and explain why this test is recommended for leak-checking a refrigeration system.
7. Explain what tools and equipment are needed to properly recover refrigerant from an air-conditioning system.
8. Loose duct lining can result in:
   a. Improper airflow through the indoor coil.
   b. Insufficient conditioning of the occupied space.
   c. Ice buildup on the indoor coil during cooling operation.
   d. All of the above are correct.
9. List several possible causes of reduced airflow through the indoor coil.
10. Explain why making a list of possible causes for system failure can aid in the troubleshooting process.
11. List the possible symptoms a heat-pump system would exhibit if it was designed to fail in the heating mode and it was stuck in the defrost mode.
12. How can an improperly mounted defrost temperature sensor affect system operation?
13. What precautions should be taken before resetting a heat-pump system's high-pressure switch?
14. What procedures should be followed when an air-conditioning system has completely lost its refrigerant charge?

15. Consider a heat-pump system that is operating with a lower-than-normal head pressure and a higher-than-normal suction pressure. Technician A concludes that one of the metering devices may be stuck open. Technician B concludes that one of the check valves may be defective. Which of the following statements is most likely correct?
    a. Technician A is correct, and Technician B is incorrect.
    b. Technician B is correct, and Technician A is incorrect.
    c. Both Technician A and Technician B are correct.
    d. Neither Technician A nor Technician B is correct.
16. Consider a heat pump that is operating with the suction pressure in a vacuum. Technician A concludes that one of the metering devices may have lost its thermal-bulb charge. Technician B concludes that one of the check valves may be defective. Which of the following statements is most likely correct?
    a. Technician a is correct, and Technician B is incorrect.
    b. Technician B is correct, and Technician A is incorrect.
    c. Both Technician A and Technician B are correct.
    d. Neither Technician A nor Technician B is correct.
17. A defective control transformer gives the impression that the system is in the OFF position. What other system condition can give the impression that the system is off when it is actually energized?
18. Explain the importance of safety float switches on auxiliary drain pans and condensate pumps.
19. Explain how to properly evaluate the operation of a defrost timer.
20. What system component should always be installed on heat-pump systems to prevent the supplementary electric-strip heaters from energizing when no indoor fan motor operation is occurring?
    a. A float switch
    b. A sail switch
    c. A low-pressure switch
    d. A high-pressure switch

GLOSSARY

absolute pressure The sum of the gauge pressure plus atmospheric pressure, which is 14.696 at sea level at 68° F.

accessories System components that are intended to enhance the operation of an air-conditioning or refrigeration system.

accumulator A refrigerant storage tank located in the suction line that allows for liquid refrigerant to boil off into a vapor before returning to the compressor.

air conditioning The process of maintaining acceptable levels of heating, cooling, air movement, humidity, and/or air filtration.

air filter Any device that removes particulate matter from the air.

air handler System component consisting of a heat-transfer surface and a means by which air is moved through the system.

air-cooled condenser A condenser that uses air as the medium to which the system heat is transferred.

air-sensing thermostat A thermostat that opens and closes its contacts depending on the air temperature surrounding the sensing element.

air-source heat pumps Heat pumps that use air as the source of heat while operating in the heating mode and as a heat sink while operating in the cooling mode.

air-to-air heat pump An air-source heat-pump system that is ultimately used to heat and/or cool air.

air-to-liquid heat pump An air-source heat-pump system that is ultimately used to heat/cool a liquid.

ambient temperature The temperature of the air surrounding an object. Often used when referring to indoor or outdoor temperatures.

ammeter A piece of test equipment used to measure the current flowing in an electric circuit. Ammeters can be clamp-on devices or can be installed in series with the circuit being tested.

amperage The amount of current, measured in amperes, flowing in an electric circuit.

ampere The unit of measure of current flowing in an electric circuit.

approach temperature The temperature difference between the refrigerant and the return water leaving a chilled-water evaporator.

aquastat A thermostat that opens and closes its contacts depending on the temperature of a liquid as sensed by the device.

aquifer The water-bearing underground formation that serves to supply water to geothermal heat-pump systems. The water is contained in soil, sand, gravel, or rock; and its depth varies along with its geographical location.

armature The moving or rotating portion of a motor, relay, contactor, or solenoid valve.

automatic changeover A thermostat that has the capability of changing over from the heating mode to the cooling mode, and vice versa, automatically.

automatic control Any device that can open or close its contacts automatically.

automatic defrost A control system that controls the frequency and duration of the defrost cycle that removes ice and/or frost from the evaporator coil.

automatic expansion valve A metering device that feeds refrigerant to the evaporator coil in an effort to maintain a constant evaporator pressure.

automatic reset A control that will return its contacts to their normal position automatically once the system's operating conditions have once again fallen within the desired range. A low-pressure control is an example of a device that automatically resets.

auxiliary drain pan A pan that is located underneath the air handler that will catch and remove condensate if the primary drain pan or line becomes clogged or blocked. This pan is primarily used when the air handler is located above the occupied space.

back pressure Same as suction pressure. This is the pressure of the refrigerant returning to the compressor.

backseated The normal operating position for service valves. The valve stem is turned all the way out of the valve body, sealing off the service port from the system.

balance point The outdoor temperature at which the heating capacity of a heat-pump system, without the use of supplementary electric heaters, is equal to the heating requirements of the structure.

ball bearings Uniformly shaped steel spheres commonly used to provide a low-friction contact surface between a motor and its shaft or between a blower shaft and its support.

ball check valve A device that permits fluid flow in only one direction. A floating ball in the valve body prevents fluid from flowing in the wrong direction.

ball-bearing motor Type of motor that uses ball bearings to provide low-friction contact between the rotating motor shaft and the motor itself. Typically noisier than motors with sleeve bearings.

bar slip Rigid, S-shaped sheet metal strip used to connect sheet metal duct sections.

bearing A low-friction device used to support and align moving parts, as in a motor shaft.

belt tension gauge A tool used to check the belt tension on belt-and-pulley assemblies.

bidirectional liquid-line filter drier A filter drier that is designed for heat-pump applications where refrigerant can flow through the drier in either direction.

bidirectional thermostatic expansion valve A thermostatic expansion valve designed for heat-pump applications where a built-in check valve permits refrigerant to flow through the valve unimpeded while the system is operating in the reverse cycle.

bimetal strip Strip made up of two dissimilar metals that will bend as the temperature it is exposed to is changed. Primarily used in thermostats and other controls.

bimetal thermostat A thermostat that operates on the principle that different metals expand and contract at different temperatures. A *bimetal strip* in the device is used to open and close the thermostat's contacts.

blank ports The ports on a gauge manifold that are used to secure the free ends of the refrigerant hoses. This prevents dirt from entering the hoses.

blowdown The process by which water is removed from a cooling tower to reduce the concentration of minerals in the water system.

blower Device used to move air through or across a heat-exchange surface; often of the centrifugal type, which creates a large pressure difference between the inlet and outlet to overcome static pressure in a duct system.

Boyle's law Law of physics that relates the pressure, temperature, and volume of a gas. At a constant temperature, the pressure will increase as the volume decreases. As the volume increases, the pressure will decrease.

branch duct A smaller duct section or run connected to the main trunk line. Branch ducts normally feed one section of the conditioned space.

brazing The process of joining two metals, such as copper fittings, at a temperature above 800° F with a filler material.

bridge The air space between the earth and the buried ground loops. Bridging reduces the efficiency of the ground loop.

British thermal unit (Btu) The amount of heat required to change the temperature of 1 pound of water 1 degree Fahrenheit.

bulb pressure In a thermostatic expansion valve, the only pressure that pushes to open the valve. As the evaporator outlet temperature increases, the bulb pressure increases, pushing the thermostatic expansion valve open.

bypass A pipe or duct section used to short circuit fluid flow. Usually controlled by a valve or damper.

cable a group of insulated conductors in a common casing. Low-voltage thermostat wire is an example of a cable.

calibrating The process of adjusting test instruments to ensure that the obtained readings are accurate. Refrigerant gauges and multimeters are examples of instruments that should always be calibrated.

capacitor An electrical storage device that is used to increase the starting torque of single-phase electric motors, as well as increasing their operating efficiency.

capacitor-start-capacitor-run (CSCR) motors Single-phase motors that are equipped with both start and run capacitors. The start capacitor is only in the active circuit long enough to start the motor.

capacitor-start-induction-run (CSIR) motors Single-phase motors that are equipped only with a start capacitor. Once the motor is operating, there is no capacitor in the active electric circuit.

capacity The sizing or rating system used in the air-conditioning industry to determine how much heating or cooling can be provided by a given system.

capillary tube A fixed-bore metering device that has no moving parts. Used primarily on systems that experience a relatively constant load. The length of the tube and the size of the bore affect the pressure drop through the device.

centrifugal blower Also called a *squirrel cage.* These blowers are commonly found on air handlers feeding duct systems as they are able to easily overcome the static pressure in the ductwork. They have a low pressure at their inlet but a high pressure at their outlet.

centrifugal switch A starting component that opens and closes its contacts based on the speed of the motor. As the motor reaches its normal operating speed, the contacts open to remove the start capacitor from the active circuit.

centrifugal-type distributor A refrigerant distributor located at the inlet of the evaporator. A swirling action is created within the device to evenly distribute the refrigerant to each of the evaporator circuits.

charge The total amount of refrigerant contained within a vessel. Also refers to the volatile liquid contained in remote thermal bulbs, such as in the case of the thermostatic expansion valve.

charging The process of adding refrigerant to an air-conditioning system.

check valve A piping valve that only permits fluid flow in one direction. Can be of the ball type, which must be installed vertically, or the disc type, which is not position sensitive.

chiller barrels The evaporator section of a chilled-water system. Heat is removed from the water as it flows through the chiller barrel.

circuit The complete path created for flow of a fluid or for electricity. Facilitates the flow from an energy source to a load (point of usage) and back to the energy source.

circuit breaker Device that protects an electric circuit from an overcurrent condition. If the circuit current exceeds the rating of the breaker, the device will open, de-energizing the circuit. Circuit breakers must be manually reset.

circulator Pump used to move liquid at a low velocity through a piping circuit such as a ground loop found in a geothermal heat-pump system.

clearance vapor In reciprocating compressors, the refrigerant trapped between the valve plate and the piston, the clearance volume, when the piston is at top dead center.

clearance volume In reciprocating compressors, the space between the valve plate and the piston when the piston is at top dead center.

closed circuit In electric circuits, a complete path for current to take. All switches in the circuit are in the closed position and the load is energized.

closed system A refrigeration system in which the refrigerant supply does not deplete. As the refrigerant flows through the system, it undergoes a series of changes and, when the cycle is complete, the refrigerant has been returned to its original state.

closed-loop system Geothermal heat-pump system in which the fluid that facilitates the heat exchange between the refrigerant and the earth is contained within a sealed-piping arrangement.

cold anticipator A component part of the space thermostat that causes the system to begin operation in the cooling mode shortly before the thermostat's contacts actually call for cooling. This helps reduce the variation of temperature in the space. The cold anticipator is a fixed resistor that is wired in parallel with the cooling contacts on the thermostat.

comfort zone The combination of dry bulb temperature and relative humidity that satisfy the comfort requirements for most individuals.

compound gauge Device used to measure pressures both above and below atmospheric pressure.

compression The process by which the vapor refrigerant entering a compressor experiences an increase in both temperature and pressure before it is discharged to the condenser.

compression ratio The value obtained by dividing an air-conditioning system's high pressure by its low pressure. The pressures used in calculating the compression ratio must be the absolute pressures. The lower the compression ratio, the more efficiently the compressor is operating.

compressor The air-conditioning or refrigeration system component that performs the task of increasing both the temperature and the pressure of the refrigerant. The increase of temperature facilitates the rejection of system heat, while the increase in pressure facilitates the flow of refrigerant through the system.

compressor cylinder The portion of a reciprocating compressor in which the piston is housed. The valve plate is located at the top of the cylinder.

compressor head The component part of the compressor that houses the compressor valves and the valve plate. The compressor head has both high- and low-pressure compartments, which separate the suction and discharge refrigerant.

concrete slab The pad on which the outdoor unit is mounted. The slab should always be perfectly level.

condensate The moisture that is removed from the air as it passes over or through the evaporator coil.

condensate drain line Piping arrangement used to carry condensate from the drain pan and the structure.

condensate drain pan Used to collect condensate as it falls from the evaporator coil.

condensate drain trap Installed in the condensate drain line when the drain is at a negative pressure to ensure proper drainage of the condensate.

condensate pump Mechanical pump used to carry condensate from the interior of the structure to a waste drain or to a location outside the structure.

condensation Liquid that forms when a vapor is cooled below its condensing temperature or dew point.

condenser The air-conditioning system component that facilitates the state change of the refrigerant from a superheated vapor to a subcooled liquid.

condenser fan motor The motor that is responsible for moving air through the condenser coil. Found only on air-cooled condensers.

condenser saturation temperature The temperature at which a vapor changes to a liquid at a specific pressure. Also referred to as the *condensing temperature.*

condensing medium The fluid, either air or water, that absorbs heat from the discharge vapor refrigerant, permitting it to condense into a liquid.

condensing temperature The temperature at which a vapor changes to a liquid at a specific pressure. Also referred to as the *condenser saturation temperature.*

condensing unit Portion of a split system consisting of the compressor, the condenser (outdoor coil in the cooling mode), and the means by which the condensing medium is passed over or through the condenser.

consolidated formation Materials below the surface of the earth that contain very little water for use in geothermal heat-pump systems. Examples of consolidated formations include granite, sandstone, and limestone.

contactor An electromagnetically controlled relay with contact ratings above 20 amperes. Contacts are controlled by the energizing and de-energizing of a holding coil. A contractor is made up of a number of sets of contacts as well as a holding coil that ultimately controls the movement of the contacts. The term *contactor* is widely used throughout the industry.

continuity Low-resistance connection between two points in an electric circuit.

control Any manual or automatic device used to start, stop, or modulate the flow of a fluid or electricity.

control circuit The electrical path by which the system controls are energized or de-energized.

control fuse The fuse in the control circuit that protects the transformer. A blown control fuse is an indication of a short circuit in the control circuit.

cooling coil The coil that facilitates the absorption of heat into the air-conditioning or refrigeration system.

cooling tower Found on many water-cooled systems, the piece of equipment that cools the water returning from the condenser. Cooling towers can ideally cool return water to a temperature within 7 degrees of the wet-bulb temperature of the air flowing through the tower.

counterflow Used to describe the flow pattern through a tube-in-tube heat exchanger when the water and refrigerant flow in opposite directions.

cracked off the backseat Position of a service valve that enables an electrician to take pressure readings of or add refrigerant to an air-conditioning or refrigeration system.

crankcase heater Electric-resistance heater that heats the crankcase of the compressor to boil refrigerant from the oil. Crankcase heaters can be of the insertion or wrap-around type.

critical charge Term used to describe the amount of refrigerant in air-conditioning or refrigeration systems that are equipped with a capillary-tube metering device. All of the refrigerant is moving through the system, so the amount of refrigerant in the system must be within a very specific range to ensure proper system operation.

cross-charged In thermostatic expansion valves, the term used to identify a thermal bulb assembly that contains a different refrigerant than that contained within the system. Also refers to a mixture of refrigerants that is used to create the desired pressure-temperature relationship.

current The flow of electrical energy in an electric circuit. Current is expressed in amperes, which is the amount of flow when 1 volt is applied to a 1-ohm resistor.

current magnetic relay (CMR) Starting relay that opens and closes its contacts depending on the amount of current flowing through the run winding.

cut-in pressure The pressure at which a pressure control will close its contacts.

cut-in temperature The temperature at which a temperature control will open its contacts.

cut-out pressure The pressure at which a pressure control will open its contacts.

cutout temperature The temperature at which a temperature control will open its contacts.

cycle Used to describe a complete sequence of events before repeating.

cylinder The portion of a reciprocating compressor in which the piston is housed. The valve plate is located at the top of the cylinder.

cylinder unloaders Pressure or electrically controlled devices that enable a compressor cylinder to be effectively removed from the refrigerant circuit. This allows the compressor to operate at different capacities, lowering energy costs. Also called *unloaders.*

dedicated well A single well that is used on geothermal heat-pump systems. The supply water is taken from the top of the well while the return water is reintroduced to the well at the bottom.

defrost The process by which frost and ice are melted from the surface of the evaporator. Defrost is usually achieved by placing the system into the cooling mode for a period of time.

defrost control Device used to sense ice formation on the evaporator and initiate and terminate the defrost cycle.

defrost cycle The period of time that the system is operating in a reverse-cycle mode to melt the ice that has formed on the surface of the evaporator coil.

defrost timer Time-controlled device that has the capability of initiating and/or terminating a defrost cycle at predetermined time intervals.

defrost-termination thermostat Temperature-sensing device that terminates the defrost mode depending on the temperature of the surface of the evaporator coil.

dehumidifying The process by which moisture or humidity is removed from the air. Air-conditioning systems both dehumidify and cool the air in the occupied space.

delta configuration Configuration of three-phase motor wiring. Motors wired in a delta configuration have twelve wires protruding from the motor. These motors can be wired for operation at different voltages, depending on the available voltage.

demand defrost The process by which a heat-pump system will go into defrost mode whenever there is sufficient frost or ice on the coil to warrant defrost. This type of defrost does not employ a predetermined time interval between defrost attempts.

desiccant In an air-conditioning system, a substance such as silica gel or activated alumina used to collect and hold moisture.

desuperheater A heat-exchange surface used to absorb heat from the discharge refrigerant before it enters the condenser in an effort to provide heat reclaim as well as an increase in system efficiency.

desuperheating The process of removing sensible heat from a superheated vapor as found in the compressor discharge line.

device port The port on a service valve that is connected to the device, which is normally the compressor or the receiver.

diaphragm A flexible membrane, usually made of thin metal, rubber, or plastic.

differential The difference between the cut-in and cut-out settings on a pressure or temperature control.

direct-expansion evaporator An evaporator that is designed in a manner so that all of the liquid refrigerant boils off into a vapor before leaving the coil. Also referred to as a *dry-type evaporator.*

discharge The process by which hot vapor refrigerant is expelled from the compressor.

discharge line The refrigerant-piping line that carries discharge gas from the compressor to the condenser or other desuperheating heat-exchange surface.

discharge pressure The pressure of the refrigerant in the discharge line, the condenser, and the rest of the high-pressure side of the system. Also referred to as *high pressure* or *head pressure.*

discharge service valve Service valve located at the discharge port of the compressor used to measure the discharge pressure, to evacuate the system, or to adjust the refrigerant charge.

domestic well A well that is used for heat-pump applications as well as for domestic water usage. The water used in the heat pump is not altered in any way with the exception of temperature.

double pole A switch that opens and closes two sets of contacts when the switch is thrown.

double throw A switch that has two ON positions in addition to the OFF position.

drier System component that removes moisture, acid, and particulate matter from refrigerant as it flows through the system. Driers are filled with a desiccant, which facilitates the moisture and acid removal.

drilled well Small-diameter wells that are drilled into the ground to depths up to 500 feet. These wells are cased and are equipped with electric pumps by which the water is removed and returned to the well.

drip pan Pan used to collect condensate from the evaporator coil. Also called a *condensate drain pan.*

drive pulley The pulley connected directly to the motor shaft. This pulley rotates at the same speed as the motor itself.

driven pulley The pulley connected to the blower shaft in an air-distribution system. This pulley can rotate either faster or slower than the motor, depending on the size relationship between the drive and driven pulleys.

drop-out voltage The voltage at which a potential relay will open its contacts.

dry wells Used in geothermal open-loop heat-pump systems, these wells are used to accept the return water and are constructed as large pits filled with sand and gravel.

dry-bulb temperature The temperature of surrounding or ambient air as measured by a standard, dry thermometer. This temperature is not affected by the moisture content of the air.

dry-type evaporator An evaporator that is designed in a manner so that all of the liquid refrigerant boils off into a vapor before leaving the coil. Also referred to as a *direct-expansion evaporator.*

duct Any structure that is used to carry air either to or from an air-conditioning system. Usually constructed of sheet metal or fiberboard panels.

earth-coupled systems Term often used to refer to closed-loop, geothermal heat-pump systems.

electric heat Heat source generated by resistance-type, electric-strip heaters. Can be used to provide a primary or secondary source of heat.

electrical timer Device used to open or close a set of electrical contacts at predetermined time intervals.

electromagnetic device A device that utilizes the magnetic field created when current flows through a conductor. Coils of wire are employed to strengthen this field to perform work such as opening or closing sets of electrical contacts or mechanical valves. Contactors and solenoid valves are examples of electromagnetic devices.

electromagnetic force The force that is harnessed by an electromagnetic device. This force is used to perform work such as opening or closing mechanical valves or electrical contacts.

electronic leak detector A field instrument used to detect refrigerant leaks in air-conditioning systems by utilizing internal pumps and electronic sensors and circuits.

emergency heat Heat-pump mode of operation that is utilized when the vapor-compression heat-pump system is not operating. Emergency heat is typically provided by electric, resistive-type strip heaters.

endplay The side-to-side movement of a motor shaft within the motor casing. Excessive endplay is an indication that the motor bearings are in need of repair.

energy efficiency ratio (EER) A rating of equipment efficiency that is determined by dividing the system's Btu output by the power input in watts. The start-up and shutdown of the equipment are not taken into account when EER is calculated.

evacuation The process by which moisture is removed from an air-conditioning or refrigeration system prior to the introduction of refrigerant to the system.

evaporation The process by which a fluid changes state from a liquid to a vapor. The temperature at which evaporation occurs changes as the pressure changes. The lower the pressure, the lower the temperature at which evaporation occurs.

evaporator The portion of an air-conditioning system where evaporation takes place. The refrigerant boils from a liquid to a vapor, absorbing heat from the air or liquid passing over it in the process.

evaporator pressure The pressure of the refrigerant at which evaporation takes place. Also referred to as *low-side* or *back pressure.*

expansion device See *metering device.*

expansion valve A modulating device that opens and closes in order to feed the desired amount of refrigerant to the evaporator. One of the dividing points between the high- and low-pressure sides of the refrigeration system.

external equalizer line Portion of a thermostatic expansion valve that enables the pressure at the outlet of the evaporator to determine the valve's position. External equalizer lines are found primarily on evaporators with large pressure drops through them.

externally equalized A thermostatic expansion valve that is equipped with an external equalizer line. Used on evaporators with large pressure drops through them.

factory wiring Any system wiring that is installed on a system at the time of manufacture.

fail position The mode in which a heat-pump system will operate if the reversing valve is in the de-energized position.

fail-safe The programmed or elapsed time after which a heat-pump system will terminate its defrost cycle and return to the normal heating mode.

fan A propeller-type device used to move air across a heat-exchange surface. Usually found on the outdoor portion of a heat-pump system.

fan shroud Protective covering around a fan that reduces the possibility of personal injury resulting from contact with the rotating fan blade.

field wiring System wiring that is performed at the time of system installation. Low-voltage control wiring between the thermostat, indoor unit, and outdoor unit is an example of field wiring.

filter Device used to remove particulate matter from the air flowing through the indoor coil of an air-conditioning system. Filters can be of the disposable or permanent type.

filter drier System component that removes moisture, acid, and particulate matter from refrigerant as it flows through the system. Driers are filled with a *desiccant,* which facilitates the moisture and acid removal.

fins per inch On heat-exchange surfaces, the number used to determine the spacing between the metal fins attached to the refrigerant tubing. The larger the number of fins per inch, the closer the fins are to each other. High-temperature refrigeration systems used for air-conditioning applications typically have twelve to fifteen fins per inch on the indoor coil.

fixed bore Metering device that does not modulate in response to load changes. Capillary tubes are an example of fixed-bore devices.

flash gas The process by which liquid refrigerant boils off immediately upon entering the evaporator coil. Flash gas absorbs a great deal of heat from the vapor refrigerant in the evaporator, increasing the effectiveness of the heat-transfer surface.

flat Description of an air-conditioning or refrigeration system that has completely lost its refrigerant charge.

floodback The process by which liquid refrigerant enters the compressor via the suction line. This can result in major damage to reciprocating compressors.

flooded evaporator Evaporators that are designed to operate full of liquid refrigerant. Flooded evaporators operate with 0 degrees of superheat.

flow switch A switch that is mounted inside a liquid-carrying pipe. The switch will close its contacts when flow is established in the piping circuit. When flow stops, the contacts open.

foaming Term used to describe refrigerant oil in the compressor when liquid refrigerant is boiling from it.

forced draft The fan configuration in which air is pushed through the heat-transfer surface.

forced-draft tower A cooling tower that is configured so that air is pushed through the heat-exchange surface.

four-way reversing valve Electromagnetically controlled valve that facilitates the operation of reverse-cycle refrigeration systems. Four-way valves change the direction of refrigerant flow through the evaporator and condenser coils in heat-pump systems.

freeze-up The accumulation of ice on the surface of the evaporator coil. This can occur during normal operation, as with an outdoor coil freezing during the heating mode, or during a system malfunction, as with the indoor coil freezing during the cooling mode of operation.

frontseated The position of a service valve in which the stem is turned all the way into the valve. This position seals off the line port from the device and service ports. Used primarily during a system pump-down.

frost-back The condition when the suction line has ice accumulation on it right back to the compressor.

fuse A one-time, overcurrent-sensing device. Used to protect electrical circuits from excessive current.

fusible link An electrical safety device located in furnaces and electric duct heaters. The link melts and de-energizes the circuit when an overheating condition exists.

gas A fluid in the vapor state.

gauge A device that is used to measure the level of pressure in a closed vessel.

gauge manifold Field-service tool equipped with more than one gauge and a series of hoses and valves to facilitate the servicing of air-conditioning and refrigeration systems.

gauge port The service port used to attach a gauge for the purpose of servicing the system.

gauge pressure The pressure reading as obtained by a gauge. This pressure does not take atmospheric pressure into account.

geothermal heat pump A heat-pump system that uses the earth as a source of heat during the heating mode of operation and as a heat sink during the cooling mode of operation.

geothermal well An underground well that is used in geothermal heat-pump systems. Water is drawn from the top of the well and returned to the bottom of the same well after flowing through the system's heat exchanger.

grease fittings Located on the bearings of larger motors, they provide a port by which grease is added to motors during lubrication.

grill Located within the occupied space, the decorative cover on supply or return ducts.

ground coil In geothermal heat-pump systems, the buried piping system that facilitates the heat transfer between the earth and the water/glycol solution. Also referred to as a *ground loop*.

ground loop See *ground coil*.

halide leak detector A device used to locate leaks in an air-conditioning system. This device utilizes an open flame that changes color when refrigerant is present. An exploring tube is passed over suspected leak locations, which pulls air and, hopefully, traces of refrigerant over the flame.

halide torch See *halide leak detector.*

head pressure Pressure of the refrigerant being discharged by the compressor. Also referred to as *high-side pressure.*

head-pressure control An electrical pressure-control switch operated by the high pressure of the system. When the head pressure reaches the predetermined set point, its contacts will open and the compressor will be de-energized. This prevents an unsafe pressure condition in the system.

heat anticipator On thermostats, the device that causes the heating system to shut down prior to the space temperature set point so as to prevent the space from getting too warm. The residual heat in the system on shutdown is sufficient to bring the space to the desired temperature.

heat exchanger Any device or component that facilitates the transfer of heat from one substance or fluid to another.

heat pump A vapor-compression refrigeration system that has the capability to provide heating in the cooler months and cooling in the warmer months by reversing the flow of refrigerant through the evaporator and condenser coils.

heat reclaim The process by which heat from the discharge line is used to heat domestic water or the occupied space. Heat reclaim increases the efficiency of the refrigeration system and also reduces heating costs.

heat sink The location where system heat is rejected. In the case of a geothermal heat-pump system operating in the cooling mode, the heat sink is the earth.

heat source The source of the heat that is supplied to the occupied space. In the case of a geothermal heat-pump system operating in the heating mode, the heat source is the earth.

heat transfer The movement of heat energy from a warmer substance to a cooler substance.

heater Any device that has the capability to raise the temperature of a substance. Also used to describe the overload devices found on three-phase motor starters.

heating coil A fluid-carrying device made of tubing or piping designed to facilitate the transfer of heat to a cooler substance.

heating mode The mode of operation in which heat is supplied to the occupied space. In the case of the heat-pump system, discharge refrigerant from the compressor flows first to the indoor coil.

heat-pump pool heater A heat-pump system that is used to heat pool water as it flows through the filter and pump piping loop.

heat-pump spa heater A heat-pump system that is used to heat spa water.

heat-pump water heater A heat-pump system that is used to heat water for domestic use.

hermetically sealed Term used to describe a system, compressor, or device that has no means by which its internal components can be accessed.

hermetically sealed compressor Term used to describe a compressor that cannot be serviced or opened in the field. The shell of the compressor is welded closed.

hermetically sealed system Term used to describe an air-conditioning system that is not equipped with service valves or any other means by which the refrigerant circuit can be accessed.

high-side Portion of an air-conditioning system consisting of the compressor, condenser, discharge line, liquid line, and refrigerant receiver.

high-side pressure The pressure contained in the high side of the system. Also referred to as *head pressure.*

holdback thermostat Outdoor ambient thermostat used to determine the number of supplementary electric-strip heaters that will be energized during defrost mode.

holding charge Small amount of refrigerant or nitrogen that is shipped within the sections of a split system. Ensures that the system is leak-free upon delivery. A loss of the holding charge is an indication that there may be a leak in the system.

holding coil The part of a contactor, relay, or valve that causes the device to switch position. The coil generates an electromagnetic field that causes the contacts or valve mechanisms to shift.

horizontal ground loop On geothermal heat-pump systems, horizontally installed ground loops or coils. Used in locations where there is ample land surface or where digging large trenches is not practical nor possible.

horsepower Measurement of power. One horsepower is equal to 746 watts.

hot gas Refrigerant discharged from the compressor. Also referred to as *discharge gas.* This gas is at a high temperature and a pressure equal to the discharge pressure.

hot gas defrost Defrost method by which the hot gas from the compressor is fed directly into the evaporator, causing the frost to melt. Systems equipped with hot gas defrost systems should also be equipped with suction-line accumulators to prevent liquid floodback.

hot gas line The refrigerant line that carries the discharge refrigerant from the compressor. Also referred to as the *discharge line.*

hunting The fluctuation resulting from a control attempting to establish an equilibrium condition.

ice ring The accumulation of ice at the bottom of the outdoor coil. This is an indication of an incomplete defrost cycle.

impeller The rotating portion of a centrifugal liquid pump.

inches of water column A unit of pressure measurement. One inch of water column is equal to approximately 0.04817 psig.

indoor coil In heat-pump systems, the coil that acts as the cooling coil during summer operation and as the heating coil during winter operation.

induced draft The fan configuration in which air is pulled through the heat-transfer surface.

induced voltage Potential that is produced in one conductor resulting from the magnetic field created by current flow in another conductor.

induced-draft tower A cooling tower that is configured so that air is pulled through the heat-exchange surface.

induction-start-induction-run (ISIR) motor A motor that starts and runs solely on the torque generated by the difference in magnetic field between the start and run windings.

infinite resistance Also referred to as an *open circuit,* term used to describe a circuit with an energized power supply but with no current flow.

in-line fuses Fuses that are connected in series with the loads that they are intended to protect.

intake The process by which suction vapor is introduced to the compressor prior to being compressed. Also referred to as *suction.*

interlock Wiring configuration in which one circuit cannot be energized unless another condition necessary to the system's operation is satisfied. For example, a condenser water-flow switch must be closed in order for the compressor to operate. The compressor circuit and the water flow are said to be interlocked.

internally equalized Type of thermostatic expansion valve that has the pressure at the inlet of the evaporator act on the valve to determine its position. Used on evaporators with very low pressure drops through them.

isolating subbase The mounting portion of a thermostat that has separate power terminals for the heating and cooling circuits. Separate heating and cooling transformers can be used.

king valve A service valve located at the outlet of the refrigerant receiver.

ladder diagram Wiring diagram that is configured with the power lines drawn as vertical lines on the left and right sides of the diagram. Each parallel circuit is represented on a separate, horizontal line connecting to the vertical power lines. When complete, the diagram resembles a ladder. This diagram is very helpful in the troubleshooting process. Also referred to as *line diagrams.*

laminated core Thin plates of laminated steel that are layered together to form the center, or core, of a control transformer.

latent heat Also referred to as *hidden heat,* the heat that causes a change of state without changing the temperature of the substance. For example, it is latent heat that causes water at 212° F to change to steam at 212° F.

leak detector Any device or method used to aid in the location of a refrigerant leak.

learning curve Concept that individuals are able to perform similar tasks faster as they become more experienced and familiar with the tasks.

legend The section of a wiring diagram that explains any abbreviations, color codings, and other symbols that appear in the diagram.

line diagram See *ladder diagram.*

line port The port on a service valve that connects to either the liquid line or the suction line. The other two ports are the device and service ports.

liquid The state of matter that is characterized by a readiness to flow and very low compressibility.

liquid floodback The process by which liquid refrigerant enters the compressor via the suction line. This can result in major damage to reciprocating compressors.

liquid line The refrigerant line that carries the refrigerant from the condenser coil. The refrigerant in this line is ideally in the liquid state—hence, the name.

liquid receiver System component located at the outlet of the condenser. Its purpose is to store high-temperature, high-pressure liquid refrigerant until it is needed by the evaporator. Capillary-tube systems do not have receivers, since all of the refrigerant is flowing through the system whenever it is energized.

liquid-cooled condenser Condensers that use liquid to absorb the heat that the system is rejecting. See also *water-cooled condensers.*

liquid-source heat pumps Heat-pump systems that utilize liquid as the source of heat while operating in the heating mode.

liquid-to-air heat pump Liquid-source heat pumps that are used to ultimately treat air in either the heating or cooling mode.

liquid-to-liquid heat pump Liquid-source heat pumps that are used to ultimately treat another liquid.

load The electrical component that is ultimately being energized by the power supply and associated circuits. Also refers to the amount of heat per hour that a refrigeration system is required to supply at design conditions.

locked-rotor amperage (LRA) The amperage that an electric motor draws on initial start-up before it reaches its design speed. This amperage can be as high as 7 times the normal amperage draw of the motor.

lockout relay A manually reset relay that prevents the system compressor from operating if an unsafe condition, usually pressure, has been established. To reset, the power to the system must be interrupted and then restored.

low ambient controls Devices that are used to allow an air-conditioning system to function within desired parameters when the outdoor temperature is below the desired level. Fan cycling and condenser flooding are commonly used methods to simulate design conditions.

low side Portion of an air-conditioning system that consists of the outlet of the expansion device, evaporator, accumulator, and suction line.

low-pressure control A pressure-operated device located on the low-pressure side of an air-conditioning or refrigeration system that opens its contacts on a drop in low-side pressure; used for temperature control on refrigeration systems and as low-charge protection on air-conditioning systems.

low-side charging The process by which refrigerant is added to a refrigeration system through the low-pressure side of the system. Primarily used for adding small amounts of vapor refrigerant.

low-side pressure The pressure of the refrigerant entering the compressor prior to being compressed by the compressor. Also referred to as *back pressure, suction pressure,* or *evaporator saturation pressure.*

makeup air The air supplied to a structure to replace air that is exhausted. Also referred to as *fresh air.*

makeup water Water that is supplied to a cooling tower or water loop to replace water that has evaporated.

makeup water line The piping arrangement that facilitates the adding of makeup water to a system.

manifold A piping header that has two or more branch lines connected to it, providing a parallel connection for fluid-carrying lines. Commonly found on parallel loops in geothermal heat-pump systems.

manifold-type distributor Device used to distribute refrigerant evenly to parallel circuits in an evaporator.

manual changeover The process by which a system is switched over from the heating to the cooling mode. The switch must be repositioned manually.

manual reset The means by which lockout relays and high-pressure switches are reset. This prevents a system from operating under unsafe conditions.

mechanical-draft tower A cooling tower that employs a fan or series of fans to move large amounts of air over its wetted surfaces. Can be forced or induced draft.

meter Term used to describe a piece of field equipment used to take voltage, amperage, and resistance readings in electric circuits.

metering device System component that feeds refrigerant to the evaporator. Usually used to describe fixed-bore devices, such as a capillary tube. Also referred to as *expansion device.*

micron Unit of pressure measurement. One micron is equal to 1/25,400th of an inch of mercury.

micron gauge Field device used to measure the level of vacuum present in an air-conditioning system. Used to measure pressures close to a perfect vacuum.

midseated The position of a service valve that is midway between the frontseated and the backseated positions. Used primarily for system evacuation.

modulating Term used to describe a valve or switch that has the ability to open or close slowly in response to changing load conditions.

motor Device that converts electrical energy into mechanical energy. Configured with a stationary component, the stator, and a rotating component, the rotor.

motor burnout The condition that results when an electric motor experiences a deterioration of insulation due to overheating.

motor starter Starting component that resembles a contactor but is equipped with overload protection, called heaters, for the motor. Primarily used on three-phase motors, providing protection from single phasing.

multicircuit evaporators Evaporators that are manufactured with two or more parallel paths for refrigerant to flow through. These evaporators are equipped with refrigerant distributors to evenly distribute the refrigerant to each of the paths.

multimeter Piece of troubleshooting equipment used by field technicians to obtain voltage, current, and resistance readings of electric circuits.

multistage thermostat A thermostat that is equipped with multiple heating and/or cooling terminals to control the operation of multiple heating or cooling sources depending on the system requirements as determined by the load conditions.

multitap transformers Transformers that can be used with different supply voltages.

natural draft Term used to describe a heat-exchange surface that does not rely on a fan or blower to move air across or through it. Natural air current moves the air across the heat-exchange surface.

natural return Term used to describe air-conditioning systems that do not employ a physical ductwork arrangement to facilitate the returning of air to the air handler. The space above a dropped ceiling is an example of a natural return.

natural-draft tower A cooling tower that relies on natural air current to move air across the wetted surfaces of the tower. No fan or blower is used.

negative temperature coefficient (NTC) Thermistor that decreases its resistance as the temperature increases.

neoprene Synthetic rubber that is resistant to oils and vapors. Commonly found in refrigerant-grade valves and controls.

nitrogen Inert gas that is commonly used in the air-conditioning industry to leak-test systems. Also introduced in small quantities to refrigerant lines during the brazing process to reduce the amount of oxidation on the interior surfaces of the piping.

noncondensable gas A vapor that does not condense into a liquid within the normal operating pressures of typical air-conditioning systems.

normally closed contacts On relays, contactors, and other electromagnetic devices, a set of electrical contacts that permits current flow through them when the holding coil is de-energized.

normally open contacts On relays, contactors, and other electromagnetic devices, a set of electrical contacts that does not permit current flow through them when the holding coil is de-energized.

offset duct sections Duct sections that have the same cross-sectional measurements at either end but are configured to route the duct around, over, or under an obstruction.

ohm The unit of measurement of electrical resistance. A 1-ohm resistor will permit a current of 1 ampere to flow through a circuit to which 1 volt has been supplied.

ohmmeter Electrical test instrument used to measure electrical resistance of a circuit or electrical component in ohms.

Ohm's law Mathematical relationship between voltage, resistance, and current. Voltage = current × resistance ($E = I \times R$).

oil level The level in a compressor crankcase to which oil must be brought in order to properly lubricate the compressor's moving parts.

oil ports Access ports on electric motors that provide a means of adding oil to the motor bearings.

oil sludge Solid matter mixed with refrigerant oil. Formed by contaminated oils and other impurities in the system.

one-time valve A pressure-relief valve that, after causing the system to release its pressure, must be replaced as it will reset itself.

open circuit An electrical circuit that has been interrupted to stop the flow of electricity.

open winding Term used to describe a motor winding that has an open circuit, preventing the flow of current.

open-loop system Geothermal heat-pump configuration that obtains its supply water from an open source such as a lake, pond, well, or other underground water source.

outdoor ambient thermostat Device that senses the temperature of the outdoor temperature. Used on heat-pump systems to determine the number of supplementary electric-strip heaters that are energized during the defrost mode of operation.

outdoor coil The heat-exchange surface that is located outside the conditioned space. In heat-pump applications, the outdoor coil is the condenser when operating in the cooling mode and the evaporator when operating in the heating mode.

outside air The air that surrounds the outdoor coil.

overfed evaporator An evaporator that is operating with a lower amount of superheat than desired. The expansion device is feeding too much refrigerant to the coil.

overload Term used to describe a load greater than that for which the system, machine, or circuit was designed.

overload protector Electrical protection device that de-energizes a compressor or motor when an overcurrent condition is present.

overload relay Thermal device that opens its contacts when the current through a heater coil exceeds the predetermined limit for a certain period of time.

package unit Self-contained air-conditioning system. Although field wiring may be required, there are no field-installed refrigeration lines.

parallel circuit An electric circuit that provides more than one path for electrical current to flow through.

parallel flow Term used to describe the flow of refrigerant and water through a heat exchanger. Used when the refrigerant and water flow in the same direction.

parallel ground loop In geothermal heat-pump systems, the term used to describe the ground loop configuration that provides multiple paths for the heat-exchange fluid to flow through.

part-wind start Method of motor starting that energizes only a portion of the motor windings during start-up to reduce the amperage draw on the motor, as well as to prolong the life of the contacts on the motor starter. Once the motor is up and running to speed; the rest of the windings are introduced to the active electric circuit.

pass-through device Term used to describe a switch or valve that, when in the open position, adds little if any resistance to flow.

perimeter duct system Duct system configuration in which the main supply duct is run around the perimeter of the occupied space. The individual takeoffs for each supply register location branch off the main perimeter duct.

permanent-split-capacitor (PSC) motor Type of single-phase motor that is equipped with a run winding that is always in the active circuit but with no start capacitor or starting relay.

Pete's port The access port that permits a field technician to take pressure and/or temperature readings in the water loops of geothermal heat-pump systems. The device is self-sealing, which prevents water leakage from the loop.

pick-up voltage Referring to potential relays, the term used to describe the voltage at which the relay's contacts open.

pinch-off tool Service tool used to seal off a process tube after a system repair.

piston The component part of a reciprocating compressor that moves back and forth within the cylinder, facilitating the compression and reexpansion processes.

pitch Used to aid in the drainage process, the term used to describe the slope of a piping run. The greater the pitch, the better the drainage.

pitting The resulting rough surfaces on electrical contacts on contactors, switches, and relays resulting from the heat generated from an overcurrent condition through the contacts. Pitting causes an increase in circuit resistance as well as reduction in the voltage supplied to the circuit load.

plenum The section of duct connected directly to the air handler on either the supply or return side.

positive temperature coefficient (PTC) Thermistor used to aid in the starting of split-phase motors with relatively low starting torque.

potential Term used to describe the amount of voltage or electrical pressure that is available to supply an electric circuit.

potential difference The net difference in voltage between two points in an electric circuit.

potential relay Relay that opens and closes its contacts depending on the induced voltage across the start winding of a split-phase motor. Also referred to as a *potential magnetic relay (PMR).*

pounds per square inch gauge (psig) Pressure existing above atmospheric pressure. Also referred to as *gauge pressure.*

power The rate at which work is done. Electrically speaking, and from Ohm's law, power is defined as the voltage supplied to a circuit times the current that flows through the circuit.

power circuit The electric circuit that feeds the main system loads. Includes the main load, the power supply, and the contacts of any control devices that determine when the load is to be energized.

pressure The energy impact on a unit area. Also defined as the amount of force or thrust against a surface.

pressure differential The difference in pressure between two vessels.

pressure drop The amount by which pressure is reduced as a fluid flows through a vessel or pipe.

pressure switch An electrical pass-through device that opens and closes its contacts depending on the pressure that the device senses.

pressure tanks Used on geothermal heat-pump systems to prevent the submerged well pump from operating continuously. Water is stored in the pressure tank to provide supply water to the heat exchanger.

pressure-drop-type distributor Refrigerant distributor that is configured in a manner that causes a pressure drop through the device, which in turn increases the velocity of the refrigerant as it flows through the device. This ensures that the liquid and vapor entering the evaporator are well-mixed, resulting in even distribution to all of the evaporator circuits.

pressure-relief valve Safety device that facilitates the release of pressure from a vessel in the event that an unsafe pressure level is reached. Relief valves can be spring type, which reset automatically, or one-time, which must be replaced upon release.

pressure/temperature relationship The relationship that exists between the pressure and temperature of a saturated refrigerant. A change in temperature will result in a predictable change in temperature, and vice versa.

preventive maintenance Tasks performed by a technician that will ideally reduce the breakdown rate of a system and also help to reduce the probability of prolonged system downtime.

primary voltage The voltage that is supplied to a control transformer's primary winding. This voltage is then converted to the desired voltage, referred to as the *secondary voltage.*

primary winding The winding of the transformer to which the primary voltage is supplied. The primary winding has fewer turns than the secondary winding on step-up transformers and more turns than the secondary winding on step-down transformers.

process tubes Point in a hermetically sealed system where the refrigerant circuit can be accessed with the aid of an adaptor or line-tap device.

propeller fan Air-moving device used when there is a low-pressure drop across a heat-exchange surface such as the outdoor coil of a heat-pump system.

psig Pressure that exists above atmospheric pressure. Also referred to as *gauge pressure.*

pump-down Process by which system refrigerant is stored in the high side of an air-conditioning system while a repair is performed on the low side. Can also be used to store refrigerant in the high side of a system at the end of the cycle, creating a reduction in suction pressure. These systems are cycled on and off by the electrical contacts on the low-pressure control.

range The limits of the settings on a pressure or temperature control.

receiver A cylindrical, steel tank that stores refrigerant until needed by the evaporator An internal dip tube ensures that only 100 percent liquid refrigerant leaves the device.

reciprocating compressor System component that facilitates the process of compression by the back and forth motion of a piston within a cylinder.

recirculated air Term used to describe the cycle of airflow through the heat-exchange surface to the conditioned space and back to the heat-exchange surface.

recirculating system Term used to describe a water-cooled system that utilizes a cooling tower to cool the heat-laden liquid before returning it to the heat exchanger.

reclaiming The process by which refrigerant is restored to new product specifications.

recovery The process by which refrigerant is removed from an air-conditioning system and stored in an EPA-approved container. The refrigerant is not processed during the recovery process.

recycling Process by which refrigerant is cleaned for reuse. This process may include oil separation as well as filtering to reduce the content of acid, moisture, or particulate matter.

reduced-voltage start Motor starting method by which a lower voltage is supplied to a motor during start-up to reduce the amperage draw of the motor, as well as to prolong the useful life of the starter's contacts.

reexpansion In reciprocating compressors, the process by which the pressure of the clearance vapor is reduced to the level of that in the suction line.

refrigerant Any fluid that has the ability to absorb heat during the process of vaporization and to release heat during the condensing process. Water is an excellent refrigerant.

refrigerant charge The amount of refrigerant that is contained within an air-conditioning system.

refrigerant distributor Located immediately downstream of the expansion device, the system component that divides the refrigerant equally to all of the circuits of the evaporator coil.

refrigerant recovery The process by which refrigerant is removed from an air-conditioning system and stored in an EPA-approved container. The refrigerant is not processed during the recovery process.

refrigeration The process by which heat is transferred from one location to another, in an effort to maintain the desired temperature condition in an enclosed space.

refrigeration service wrench Ratcheting service tool used to open and close service valves.

register Decorative cover or panel located at the point in a duct system where air is supplied to the occupied space.

relay An electric control that consists of a holding coil and at least one set of normally open or normally closed contacts. The energizing and de-energizing of the holding coil determines the position of the contacts.

remote bulb The sensing element of a control that facilitates the locating of the operating portion of the control far away from the fluid whose temperature is being sensed.

remote-bulb thermostat A temperature-sensing device that is equipped with a remote bulb. The actual switching component of the thermostat is located a distance away from the sensing element.

repeating cycle A cycle that can continue indefinitely without loss of fuel or refrigerant. In air conditioning, the refrigerant experiences a number of physical changes as it flows through the system but finishes the cycle in the same state as when it started. During this process, the amount of refrigerant in the system will not change.

return air The air from the occupied space that enters the air handler. Refers to the air before it is either heated or cooled by the air-conditioning system.

return duct Run of ductwork that carries air from the occupied space back to the air handler. Return ducts are typically oversized to reduce the amount of noise at the return grill.

return grill Located in the occupied space, the decorative cover for the return duct. The air filter is typically located just behind this grill.

return plenum The section of the return duct that is connected directly to the air handler.

return water The water leaving the heat-exchange surface. This water typically returns to the cooling tower or well from which it originally came.

reverse-cycle defrost The process by which a heat-pump system switches over to the cooling mode for a period of time in order to allow the ice formation on the evaporator coil to melt.

reverse-cycle refrigeration An air-conditioning or refrigeration system that is capable of reversing its operation and direction of heat transfer. Also called a *heat-pump system.*

reversing valve The heat-pump-system component that makes reverse-cycle refrigeration possible. Mechanically alters the direction in which the refrigerant flows through the indoor and outdoor coils. Reversing valves are controlled by solenoid coils, which are energized and de-energized in order to change the position of the valve.

rotary compressor System component that uses rotary motion to pump fluids through a system. Refrigerant enters as a low-pressure, low-temperature vapor and is discharged as a high-pressure, high-temperature vapor.

rotor The rotating portion of a motor. Includes the motor shaft.

run capacitor Electrical storage component that provides starting torque to split-phase motors. Also helps increase the running efficiency of electric motors.

run winding The motor winding that is energized whenever power is supplied to the motor.

safety control Any electrical, electronic, mechanical, or electromagnetic device that will disable a system if unsafe conditions are present.

sail switch Set of electrical contacts that open and close depending on air movement through a duct section.

sand well In geothermal heat-pump systems, a large, sand-filled pit that is used to receive the water after it has passed through the heat exchanger. The water eventually seeps back to the water table.

saturated The physical state when a refrigerant is a mixture of liquid and vapor.

saturated vapor State at which liquid refrigerant has just changed to a vapor.

scale Mineral deposits that accumulate on the interior surfaces of water-type, heat-exchange surfaces.

schematic diagrams Diagrams that show the electrical components of a system as well as all of the interconnecting conductors.

schrader pin The self-sealing, component part of a Schrader valve that prevents the loss of refrigerant through the valve.

schrader valve Service valve that gives a field technician access to the refrigerant circuit for troubleshooting or servicing purposes. Equipped with a self-sealing pin, the valve automatically seals itself when the gauge is removed.

seasonal energy efficiency ratio (SEER) A rating of equipment determined by dividing the system's Btu output by power input in watts, taking into account the energy used on system startup and shutdown.

secondary voltage The voltage that is measured across the secondary winding of a control transformer. On step-down transformers, the secondary voltage will be lower than the primary voltage.

secondary winding Coil of wire in a control transformer that has an induced voltage generated by the primary winding across it. On step-down transformers, the secondary winding will have fewer turns than the primary winding.

semi-hermetic compressor Hermetic compressor that is bolted together to facilitate the performing of minor service operations.

sensible heat Type of heat energy that causes the temperature of a substance to change.

sensing bulb Portion of a remote temperature control that contains a pressurized fluid. The pressure in the bulb changes as the temperature it is exposed to changes.

sensor Component that relays information regarding temperature, pressure, or other factors to a control device.

series circuit An electrical circuit in which there is only one possible path for current to flow.

series ground loop In geothermal heat-pump systems, the buried piping arrangement that carries the water/glycol mixture between the unit and the earth. The loop provides only one path for the fluid to flow through.

service panels Sections of a unit casing that can be removed to gain access to the system components.

service port Point at which the refrigerant circuit can be accessed. Usually refers to Schrader valves.

service valve Point at which the refrigerant circuit can be accessed. Service wrenches are usually required to change the position of the valve. Also used to pump a system down prior to repair.

set point Temperature or pressure at which a control will change the position of its contacts.

shaded-pole motor Inexpensive, low-starting-torque AC motor used for light-duty applications. These motors are rated in watts as opposed to horsepower because of their small size.

shell-and-coil heat exchanger Heat-transfer surface that has one fluid, usually a liquid, flowing through a coil of tubing or piping that is located within a surrounding shell. Another fluid, usually a vapor, flows within the shell, facilitating the heat transfer between the two fluids. This type of heat exchanger must be cleaned chemically.

shell-and-tube heat exchanger Similar to a shell-and-coil heat exchanger with the exception that a series of straight tubes takes the place of the coil. This type of heat exchanger can be cleaned by mechanical means.

short circuit A complete electric circuit with no resistance. This will cause the current draw of the circuit to increase beyond the limit set forth by the wire and circuit breaker size.

short circuit to ground (short to ground) Term used to describe a no-resistance electrical path between a hot leg or terminal to a ground terminal.

short cycling Term used to describe the constant energizing and de-energizing of an electrical device such as a pump, motor, or compressor.

shroud Protective casing around a blower or fan.

silica gel Material commonly used in filter driers to aid in the removal of moisture and acid from an air-conditioning system.

silver soldering The process by which a silver-bearing solder is used as a filler material to connect tubing, piping, or other metallic surfaces.

single pole Term used to describe a switch that opens and closes one set of contacts.

single pole, double throw Single-pole switch that has two ON positions.

single pole, single throw Term used to describe a switch that controls one set of contacts and has an ON and OFF position.

single throw Switch with only an ON and OFF position.

slab Concrete platform on which the outdoor unit rests.

sleeve bearings Mechanism by which a motor shaft is supported within the motor. Also used to support a driven pulley, blower, and shaft assembly.

slide valve Term used to describe a valve that has an internal slide mechanism that shifts from side to side in order to change the position of the device. A reversing valve is an example of this type of device.

slips and drives Components of a duct system that are used to fasten sheet metal duct sections together.

slow-closing solenoid valve Solenoid valve that moves from the open to closed position slowly to prevent the condition known as *water hammering*.

slugging Term used to describe liquid refrigerant entering a compressor. Also called *liquid floodback*.

snap action Term used to describe a control, valve, or switch that is either fully open or fully closed. Used to describe valves that do not modulate.

solder Filler material with a low melting temperature used to join two other metals together.

soldering The process of joining two metals by using a third metal or alloy, known as *solder*.

solenoid A coil of wire that creates a magnetic field capable of doing work when electric current flows through it.

solenoid valve Mechanical-flow control valve that is controlled by a solenoid coil.

solid state Term used to describe electronic circuitry.

splash lubrication The process of compressor lubrication that relies on the movement of compressor components within the crankcase below the oil level to supply lubricant to the moving parts.

split system An air-conditioning system that has separate indoor and outdoor units. These systems have field-installed refrigerant piping.

split-phase motors Electric motors that are manufactured with a start and a run winding.

spring pressure On thermostatic expansion valves, the pressure that determines the amount of superheat that will be maintained in the evaporator. The higher the spring pressure, the higher the superheat. The spring pressure, along with the evaporator pressure, pushes to close the thermostatic expansion valve. Also referred to as the *superheat spring pressure.*

spring-loaded valve Type of relief valve that will automatically reset itself once the pressure in the vessel has reached a safe level.

squirrel-cage blower Blower typically found on air-handling systems connected to duct runs. These blowers have a low pressure at their inlet but a high pressure at the outlet, which helps the system overcome the resistance in the ductwork. Also called *forward-curved centrifugal blowers.*

standing-pressure test A method for leak detection. Usually accomplished by pressurizing the system with dry nitrogen, sometimes mixed with a trace of R-22 for use with an electronic leak detector.

start capacitor Power-storage device used on split-phase motors that is only in the active electric circuit long enough to start the motor. Once the motor has reached its operating speed, the device is removed from the circuit. Referred to as a *dry* capacitor.

start winding The winding of a split-phase motor that provides starting torque for the motor. In most cases, this winding is removed from the active circuit once the motor has reached its operating speed.

starters Contactors that are equipped with overload protection. Used primarily on three-phase pumps, motors, and compressors.

starting relay Any relay that aids the starting of a motor. Often used to remove the start winding and start capacitor from the circuit after the motor reaches its operating speed.

starting torque The difference in magnetic field that causes a motor to begin turning. Three-phase motors have a very high starting torque, while PSC motors have very low starting torque.

starved evaporator An evaporator that is being supplied less refrigerant than required. These evaporators operate with a higher superheat than desired.

static pressure The pressure exerted on the interior surfaces of a duct system.

stator The stationary portion of a motor.

step-down transformer A transformer that has a secondary voltage that is lower than the primary voltage.

step-up transformer A transformer that has a secondary voltage that is higher than the primary voltage.

strainer System component that removes particulate matter from a fluid as it flows through the device.

subbase Mounting plate to which a wall thermostat is secured. Location where electrical connections are made for the thermostat.

subcooling The process by which a liquid is cooled to a temperature below its condensing temperature.

suction The process by which refrigerant enters a compressor.

suction line The portion of the refrigerant piping arrangement that carries suction gas back to the compressor.

suction pressure The pressure of the refrigerant flowing back to the compressor. Also referred to as *low-side* or *back pressure.*

suction service valve Valve located on the low side of the system, usually on the body of the compressor, that enables a field technician to obtain pressure readings of, add refrigerant to, or evacuate an air-conditioning or refrigeration system.

suction-line accumulator System component located in the suction line near the compressor that facilitates the vaporization of liquid refrigerant prior to entering the compressor.

superheat Additional sensible heat added to a vapor that raises its temperature above its boiling temperature.

supplementary heat Electric-strip heaters that are used to increase the heating capacity of a heat-pump system or temper the air that is being supplied to the occupied space during defrost mode.

supply branch A section of ductwork that connects the main trunk line to the wall, floor, or ceiling register. Also referred to as a *supply duct.*

supply duct See *supply branch.*

supply plenum The duct section that connects directly to the air-handler portion of an air-conditioning system.

supply water The water that is being supplied to a heat-exchange surface, such as the condenser. The water has not yet absorbed or rejected any system heat.

switches Found on three-phase starters, term used to describe the normally closed sets of contacts in the control circuit that open when a higher-than-normal temperature is generated in the power lines. Intended to de-energize the starter's holding coil when a single-phasing condition exists.

system charge The amount of refrigerant that is contained in an air-conditioning or refrigeration system.

system evacuation Using a vacuum pump, the process by which moisture is removed from an air-conditioning or refrigeration system prior to the introduction of refrigerant.

system pump-down The process by which the system refrigerant is stored in the compressor, receiver, and condenser coil in preparation for a repair on the low-pressure side of the system.

temperature differential The difference in temperature between two fluids.

terminal Location on a relay, contactor, circuit board, or other device where electrical connections are made.

thermal cutout Temperature at which a sensing device will open its contacts. Often used to protect motors, pumps, and compressors.

thermistor Used mainly in electronic circuits, a resistor whose resistance changes with changes in temperature.

thermometer Piece of test equipment that measures changes in sensible heat.

thermometer well Port, pocket, or recess in a piping arrangement designed to receive a thermometer for the purpose of obtaining accurate temperature readings of the fluid within the pipe.

thermostat Control device that opens and closes its electrical contacts depending on the temperature it senses.

thermostatic expansion valve A metering device that feeds refrigerant to the evaporator to maintain a constant superheat in the evaporator. Low superheat will cause the valve to close, while a high superheat will cause the valve to open.

three-phase motor An electric motor that employs three windings, each of which is connected to a power source that is 120 electrical degrees out of phase with the next. These motors have the highest starting torque and the highest operating efficiency of all common motor configurations.

time-delay fuse A fuse that will blow only after it has sensed an overcurrent condition for more than the predetermined period of time.

time-delay relays Relays that open and/or close their contacts at a predetermined time after the holding coil has been energized.

timer A system component, either mechanical or electronic, that opens and closes a set of electrical contacts at a predetermined time interval. For heat-pump applications, see *defrost timers.*

total-system charge The amount of refrigerant that is contained in an air-conditioning or refrigeration system. On package units, the total-system charge is indicated on the nameplate of the unit. On split systems, the total-system charge is determined by weighing the refrigerant as it is introduced into the system.

touch test Used primarily as a troubleshooting technique; the technician can determine the general operating conditions of a system by touching the refrigerant lines and sensing their temperatures. A solid understanding of the basic refrigeration system is needed for this method to be effective.

transformer A device used in electrical circuits to convert one voltage to another. Transformers can be of the step-up or step-down configuration.

transition duct section Duct section that is used to join two duct sections with different cross-sectional measurements.

trunk line The main run of a duct system. This run of ductwork is connected to the air handler; the takeoffs, in turn, are connected to the main trunk line.

tube-in-tube A heat-exchange surface that is constructed as two pipes, one located within the other. Refrigerant typically flows in the outer pipe, while the water or other fluid flows through the inner pipe.

two-well system A geothermal heat-pump configuration in which the supply and return water piping for the system are connected to separate, underground wells.

ultraviolet solution A liquid added to air-conditioning or refrigeration systems that aids in the leak-detection process. The liquid glows brightly when seen under UV light.

unconsolidated formation A mixture of granulated material such as soil, clay, or gravel that contains a large amount of easily obtainable water.

underfed evaporator An evaporator that is operating at a low efficiency due to a reduced amount of refrigerant being supplied to it. Also referred to as a *starved evaporator.*

unloaders Pressure or electrically controlled devices that enable a compressor cylinder to be effectively removed from the refrigerant circuit. This allows the compressor to operate at different capacities, lowering energy costs. Also called *cylinder unloaders.*

vacuum Any pressure recorded as being below atmospheric. A vacuum can be measured in either inches of mercury or microns.

vacuum gauge A field tool used to measure the vacuum in a closed vessel. Vacuum gauges can read the level of vacuum in either inches of mercury or microns.

vacuum pump The piece of service equipment that is used to help remove moisture from an air-conditioning or refrigeration system before refrigerant is added to it.

valve Any device that controls, either automatically or manually, the flow of a fluid.

vapor A fluid that exists in the gaseous state. An evaporated liquid is said to be in the vapor state.

vapor compression The process on which the basic refrigeration cycle is based. Refrigerant vapor is compressed to increase its temperature and pressure, after which the heat absorbed into the system can be rejected to a medium at a lower temperature.

venturi effect Effect created as a fluid flows through a reducing orifice, creating an increase in velocity of the fluid as well as a reduction in pressure at the inlet of the device.

venturi-type distributor A distributor that relies on the venturi effect to thoroughly mix the liquid and vapor refrigerant before distributing the refrigerant to the evaporator circuits. These distributors have the advantages of causing little turbulence in the refrigerant and causing a minimum of overall pressure loss as refrigerant flows through the device.

vertical ground loop Used in geothermal heat-pump systems, these buried loops are installed vertically where ground space is limited.

volatile liquid A liquid that exhibits a pressure/temperature relationship, as a refrigerant does.

volt The unit measure of potential to do work in an electrical circuit. One volt is the amount of potential that allows one ampere of current to flow through a device with one ohm of resistance.

voltmeter A field tool used to measure the voltage between two points in an electrical circuit. Voltmeters can be either analog or digital devices.

volumetric efficiency The efficiency rate describing how well a compressor displaces the refrigerant that enters the compression chamber. The smaller the clearance volume between a piston and a valve plate, the higher the volumetric efficiency.

wastewater system A system in which water flows through a condenser and is then discharged to a drain. Wastewater systems do not recycle or reuse the water in the system.

water hammer The hammerlike sound that is produced when water flow is suddenly stopped by a rapidly closing valve.

water loops The underground piping arrangements through which water or water/antifreeze solutions flow in geothermal heat-pump systems.

water table Source of underground water used for geothermal heat-pump systems. Wells must be drilled to reach the water table, which can be up to 200 feet below the surface of the earth.

water treatment The process by which chemicals are added to system water to reduce the rate at which scale and other undesirable effects will occur.

water-cooled condenser A heat-transfer surface in which system heat is rejected to the water that passes through the coils of the device.

watt The unit of electrical power. One horsepower is equal to 746 watts.

wet-bulb temperature An air-temperature reading that takes into account the moisture content of the air. The higher the moisture content in the air, the lower the wet-bulb-temperature reading. The wet-bulb temperature can be lower than or equal to the dry-bulb temperature but never higher than it.

wye configuration Wiring arrangement for three-phase motors. Motors with windings configured in a *wye* typically have nine wires protruding from the motor casing. The windings can be arranged in series or parallel, depending on the voltage at which the motor is to be used.

INDEX

Note: Italicized page numbers indicate illustrations.